Problem Solving in Engineering

Problem Solving in Engineering

Analytical Mathematics and Numerical Analysis

Larry A. Glasgow
Department of Chemical Engineering,
Kansas State University,
Manhattan, KS, USA

WILEY

Published by John Wiley & Sons, Inc., Hoboken, New Jersey.
Published simultaneously in Canada.

Library of Congress Cataloging-in-Publication Data Applied for:

Hardback: 9781394179985

Cover Design: Wiley
Cover Images: Courtesy of Larry A. Glasgow

Set in 9.5/12.5pt STIXTwoText by Straive, Chennai, India

Contents

Preface

This book, *Problem Solving in Engineering: Analytical Mathematics and Numerical Analysis*, is intended to help students acquire key competencies crucial to their ability to solve important practical problems. The text can be used in a formal classroom setting (as I have employed it at Kansas State University) or for self-study, and both students and practicing professionals in STEM fields may use it effectively.

The Programme for International Student Assessment (PISA) has determined that "…students in the United States have particular weaknesses in performing mathematic tasks with higher cognitive demands, such as taking real-world situations, translating them into mathematical terms, and interpreting mathematical aspects in real-world problems." It has been suggested that a factor contributing to this crisis is the lack of integration between science and mathematics in secondary education. It seems that the focus in the U.S. system is often based on content coverage (possibly as a consequence of our emphasis upon standardized testing) and not on the contextual connection between math and real, physical phenomena. For students in the applied sciences—who are often visual and tactile learners—the latter is absolutely critical.

It should also be noted that the life experiences of contemporary adolescents are vastly different from those of previous generations. Although it is easy to attribute problems in U.S. secondary education to PC gaming, social media, and smartphones, this is too conveniently simplistic. Indeed, smartphones *can be used to great advantage in STEM education*, as we demonstrate in Chapter 10. That said, many aspects of contemporary life have conspired to deprive students of opportunities for somatic learning. Rites of passage for teenagers 60 years ago (e.g., adjusting the carburetor or ignition timing on an automobile) simply no longer exist. It is up to educators to step into this void with creative, contextual learning experiences. This is important because of a characteristic noted among many of the most successful engineering professionals: *They typically have an unbridled curiosity about how and why machines and processes work the way they do*. In applied mathematics, we must do our very best to nurture this mindset by demonstrating the connections between theory and important physical phenomena.

With these observations in mind, I have revised this book to emphasize how applied mathematics can be used by engineers to design processes, analyze their dynamic behavior, and interpret their performance in industrial settings. We note that the success of our efforts in this area has profound implications for economics, safety, productivity, and environmental stewardship. Accordingly, many new detailed examples and narrative-type problems have been included in this revision; these additions emphasize the essential connections between math and the physical world around us. I have drawn from diverse fields, including navigation, aerospace, signal processing, fluid flow, heat and mass transfer, chemical reactor operations, the response engineered structures to their environment, process optimization, machinery monitoring, and others. The intent, of course, is to find ways to appeal to the interests and talents of a broad range of engineering students and practicing professionals and to demonstrate to all users of this book why the topics we study are worthy of their attention.

Larry A. Glasgow
Department of Chemical Engineering,
Kansas State University,
Manhattan, KS, USA

About the Companion Website

This book is accompanied by a companion website:

www.wiley.com/go/glasgow/problemsolvingengineering

This website include:

1. Solutions Manual
2. Ancillary Materials and Problems

1

Problem Formulation, Models, and Solution Strategies

1.1 Introduction

Many engineering textbooks describe the steps one should follow when solving engineering problems, and the list given by Eide et al. (2008) is appropriate for our purposes:

1) Recognize the problem and develop a physical understanding of it
2) Accumulate pertinent data if they are available
3) Select an appropriate principle or theory as the basis for the model
4) Apply the simplifications necessary to make the model tractable
5) Solve the problem with the best available tools
6) Verify the results using experience and physical insight.

Most of our work here will be focused upon 4), 5), and 6) since the models we will be working with—for the most part—have already been formulated for us and the data needed are provided in the problem statements. However, we should remember a physical understanding of a particular problem, 1), is a key element of a successful process; we cannot evaluate the results of our solution or our computational procedure without it. We should also observe that there is one aspect of the philosophy of engineering problem solving that does not appear in this list: *Always begin with the simplest possible model and the most direct solution procedure.* You are less likely to make a mistake, and additional complexity can always be added should you find it necessary. An inelegant solution procedure that provides the required result is always better than an elegant one that fails.

In this context, we note that many of the numerical methods described in this text were developed with *PBCC*™ (the Power BASIC Console Compiler). *This has not been done with the expectation that students of this material will learn to write their own programs.* A few may be so inclined, but for most it is more practical to utilize readily available commercial software. Nevertheless, there are important reasons for providing these examples: i) BASIC codes are pretty transparent—the logic is clear even to students with no background in high-level languages, ii) the author has found that *PBCC* generates very fast executables; although this is of little consequence for, say, root finding with Newton–Raphson, it may be critically important if one is trying to solve partial differential equations, and iii) both the structure and the intent of these example programs are obvious, so the solution strategies can be adapted to many different hardware/software combinations.

This touches on an old debate in engineering education: Should engineering students be taught computer programming? Proponents of this idea argue that programming in a high-level language fosters a systematic approach to problem solving. In the era of centralized computing prior to 1980 (in centralized computing users submitted their programs on punched cards to a mainframe computer and then waited for the printed output to be returned) this meant learning FORTRAN, a high-level language designed specifically for scientific computations. Software available for technical calculations included IBM's Scientific Subroutine Package (SSP) and, after 1971, SAS (the Statistical Analysis System). The former had capabilities for common operations like solving algebraic equations, performing numerical quadrature, and solving sets of ordinary differential equations. The SSP routines had to be "called" by a main program written by the user; furthermore, an engineer with an atypical computational problem would need to write their own code from scratch. This situation persisted into the beginning of the personal computing revolution when IBM introduced the PC based upon Intel's 8088 processor. Technical professionals immediately discovered the benefits of accessible computing power; unfortunately, software available for personal computers was limited to the disk operating system (DOS), a BASIC interpreter, and rudimentary spreadsheets. By necessity, engineers developed BASIC programs for many technical computations despite

Problem Solving in Engineering: Analytical Mathematics and Numerical Analysis, First Edition. Larry A. Glasgow.

Companion website: www.wiley.com/go/glasgow/problemsolvingengineering

the speed limitations of the line-by-line interpreter. This changed rapidly, though, with the availability of the 8087 numeric coprocessor and Microsoft's publication of high-level language compilers for both FORTRAN and BASIC. By 1990 the computing power of PCs had evolved to the point where engineers could solve important problems without ever leaving their desk. It is no exaggeration to say that the decentralization of computing power greatly facilitated progress in the technical professions.

It is possible—even likely—that at some point in the near future an AI (artificial intelligence) system for engineering problem solving will exist. We can assume that it will make calculations for routine engineering problems autonomously with a near-zero probability of error. However, this does not mean that skilled engineering problem solvers will disappear. What the incorporation of automation in industrial settings has demonstrated is that when some job functions are eliminated, others are created. We should anticipate a future where highly capable engineering professionals will work in concert with AI to produce solutions that are practical, efficient, and economically optimal.

Exactly what are we trying to accomplish in this course?

1) We want to model a system, process, or phenomenon so that future behavior can be predicted.
2) We want to select an appropriate solution procedure so the model can be used as a tool for process design, improvement, or control. In the case of the latter, we are particularly concerned with problems that may result from transient operations or process excursions. In these instances the fidelity of the model and the accuracy of the solution procedure may have significant implications for safety.
3) We want to interpret data that have already been collected; i.e., we need to develop a framework that allows us to understand what has transpired and confidently apply that to analogous situations.

To accomplish these objectives we will review some mathematical techniques that can be used to solve important problems in engineering and the applied sciences. We will focus upon problem types that are crucial to the analysis and simulation of real, physical phenomena. As noted previously, sometimes our objective will be to predict future behavior of a system and sometimes it will be to interpret behavior that has already occurred. We want to stress that *the author and the readers are collaborators in this effort*, and whether this text is being used in a formal setting, or for self-study, the ultimate goal is the same: We want to be able deal with problems that arise in the applied sciences and do so efficiently. And—this is important—we do not want to rely upon calculation software unless we know something about the method(s) being employed. Too often real problems can have alternative solutions, so it is essential that the analyst be able to exercise some judgment based upon understanding of the problem and of the algorithm that has been selected.

We will redirect our discussion momentarily to underscore a common difficulty faced by engineers engaged in problem solving—and one not described above. Before you commit the time required to solve a particular problem you should ask the following question: Do I have enough information to formulate a model and to find an appropriate solution? Though the answer is usually obvious in instructional examples, it is less likely to be so in real-world problems. All engineers (and managers) want to make efficient use of the time spent solving problems, so allocating effort to determine whether a particular problem as posed is solvable prior to commencing the actual work is always justified. Let us illustrate this with two examples beginning with an elementary DC circuit illustrated in Figure 1.1.

Assume a voltage is applied and the resulting current flow in the circuit is measured. The resistors are parallel–series–parallel, so we write the appropriate equations beginning with Ohm's law:

$$V = IR\frac{1}{R_{AB}} = \frac{1}{R_1} + \frac{1}{R_2}\frac{1}{R_{CD}} = \frac{1}{R_4} + \frac{1}{R_5}R_{eq} = R_{AB} + R_3 + R_{CD} \tag{1.1}$$

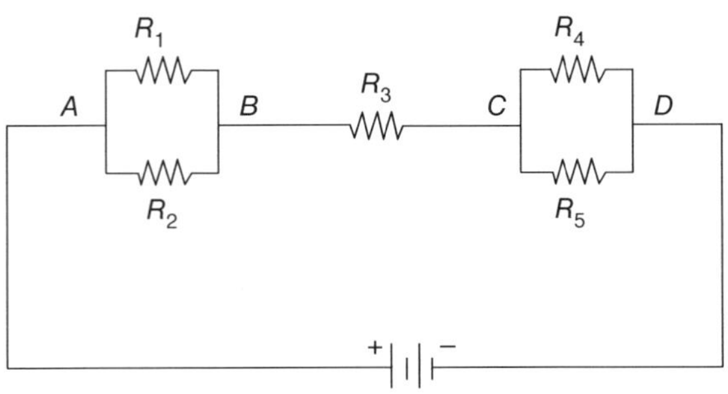

Figure 1.1 A DC circuit with five resistors.

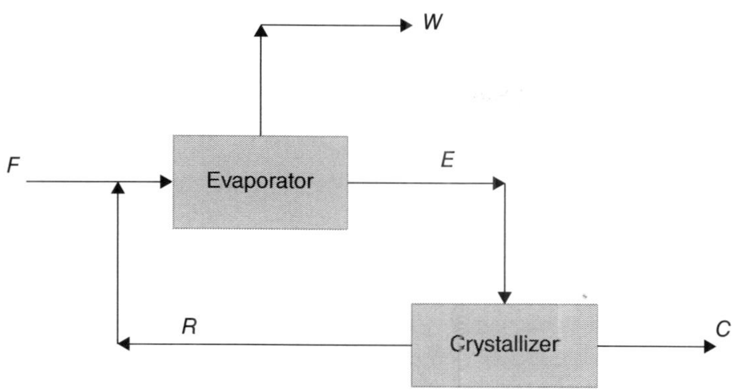

Figure 1.2 A desalination process consisting of an evaporator and a crystallizer.

Now, suppose the following information is available: $V = 75$ V, $I = 2.19$ amps, $R_1 = 8\,\Omega$, $R_2 = 15\,\Omega$, and $R_3 = 10\,\Omega$. Can we determine R_4 and R_5? We cannot since we have two unknowns:

$$75 = 2.19\left[\left(\frac{1}{8} + \frac{1}{15}\right)^{-1} + 10 + \left(\frac{1}{R_4} + \frac{1}{R_5}\right)^{-1}\right],$$

or

$$0.0525 = \frac{1}{R_4} + \frac{1}{R_5}.$$

So what we can calculate is the sum of the inverses—we must know something about one of these resistors if we are to go farther. For example, if we know $R_4 = R_5$, one can then show that $R_4 = 38\,\Omega$. In this simple problem the lack of information necessary for solution was apparent. But that will not always be the case, more complex problems may be missing subtle elements that render solution impossible; e.g., many engineering students have experienced this in the analysis of forces in trusses or solving material balances for chemical processes with recycle streams. Let's look at an example of the latter.

Material balance problems for chemical processes produce sets of algebraic equations, and it is often unclear whether or not a solution can be obtained with the information provided. For example, consider a desalination process in which a salt solution is fed (F) to an evaporator to produce salt crystals (C) and water (W) shown in Figure 1.2.

The process operates at steady state, so we first picture an envelope that encloses both the evaporator and the crystallizer (the entire system); since the mass coming in must exit, the overall balance is:

$$F = W + C. \tag{1.2}$$

Next, we write a balance on the evaporator:

$$F + R = W + E, \tag{1.3}$$

and a balance on the crystallizer,

$$E = C + R. \tag{1.4}$$

You can see that if we add eqs. (1.3) and (1.4) together, we obtain eq. (1.2), so of course only two of these equations are independent. The brine produced by the crystallizer is recycled (R) to the evaporator. There are two components in this process, salt and water, so we can also develop component balances: Let x_F and y_F be the mass fractions of salt and water, respectively, in the feed stream, F; now suppose $x_F = 40\%$ and $F = 2{,}500$ kg/hr. The only place salt exits the process is in the stream from the crystallizer, C. If C contains *only* salt crystals (and more on this momentarily) then:

$$x_F F = C. \tag{1.5}$$

We can also write salt balances for both the evaporator and the crystallizer:

$$x_F F + x_R R = x_E E \quad \text{and} \quad x_E E = C - x_R R. \tag{1.6}$$

It is reasonable to assume that both the mass flow rate of the feed, F, and the salt concentration will be specified, i.e., F and x_F are known. This means that both C and W can be immediately determined. Which quantities are unknown, and do we have sufficient information to calculate them? The four unknowns are E, R, x_E, and x_R; but we have only two remaining equations, the salt balances on the evaporator and the crystallizer—there is insufficient information to complete the solution for this problem. *Now*, assume $x_E = 50\%$ and $x_R = 45\%$ and demonstrate that $R = 10{,}000$ kg/hr: F is provided, so $0.4F = C = 1{,}000$ kg/hr. Since $F = W + C$, $W = 1{,}500$ kg/hr. Next we perform a salt balance on the evaporator: $0.45R + 0.4F = 0.5E$ and as noted previously, $F + R = W + E$ and we eliminate E using the evaporator salt balance: $1{,}000 + R = 0.9R + 2{,}000$ or $0.1R = 1{,}000$, and $R = 10{,}000$ kg/hr. Finally, we observe that $E = 2{,}500 + 10{,}000 - 1{,}500 = 11{,}000$ kg/hr. There are two concluding observations we must make: **i)** There are just two components, salt and water. We should exercise care when using *both* component balances since the mass fractions in a given stream must sum to 1, e.g., for the recycle stream $x_R + y_R = 1$. Thus, if one added the two component balances on the crystallizer, you would simply obtain eq. (1.4), $E = C + R$. **ii)** As a practical matter it is virtually impossible to eliminate all of the water from C (the crystallizer product stream). Could you work this problem if C contained both salt (say 85%) and water (15%)? See if you can verify for the revised problem $R = 8235.3$ kg/hr and $E = 9411.8$ kg/hr.

Many of the problems we will be solving come from transient or equilibrium balances where something (force, momentum, mass, energy, charge, etc.) may be subject to a conservation principle. In their simplest forms such balances might be written as [*Rate in*] – [*Rate out*] = [*Accumulation*] or $\sum F = 0$. Often they will involve forces, fluxes, and the couplings between driving force–flux pairs. Examples of the latter are Newton's, Fourier's, Fick's, and Ohm's laws:

$$\tau_{yx} = -\mu \frac{\partial v_x}{\partial y}, \quad q_y = -k\frac{\partial T}{\partial y}, \quad N_{Az} = -D_{AB}\frac{\partial C_A}{\partial z}, \quad \text{and} \quad J_z = -\sigma\frac{\partial V}{\partial z}. \tag{1.7}$$

where τ_{yx} is the shear stress (acting upon a y-plane due to fluid motion in the x-direction), q_y is the flux of thermal energy in the y-direction, N_{Az} is the molar flux of species "A" in the z-direction, and J_z is the current density. Note that these four fluxes are *linearly related* to the velocity, the temperature, the concentration, and the potential, respectively. Each driving force–flux pair has, under ideal conditions, a constant of proportionality (the viscosity, μ, the thermal conductivity, k, the diffusivity, D_{AB}, and the electrical conductivity, σ); these constants are *molecular properties of the medium* that can be determined from first principles if the right conditions are met. Unfortunately, it is also possible for viscosity to depend upon velocity, for thermal conductivity to depend upon temperature, for diffusivity to depend upon concentration, etc. In such cases the driving force–flux relationships are no longer linear as indicated by the above set of eq. (1.7).

Example ***A Microscopic Thermal Energy Balance***

Let us illustrate how one of the driving force–flux couplings from eq. (1.7) is incorporated into a *balance*; we will formulate a *microscopic* balance in which Fourier's law will be used to describe the thermal energy flux. Consider heat transfer occurring in an electrical conductor, perhaps a copper wire. The conductor is carrying an electric current, so thermal energy will be produced in the interior by dissipation (I^2R heating) and thermal energy will be lost to the surroundings at the wire's surface. We will construct a thermal energy balance upon a volume element shown in Figure 1.3, an annular region extracted from the interior of the wire of length L that extends from r to $r + \Delta r$.

We express the balance verbally in the form:

$$\{\text{Rate of thermal energy in at } r\} - \{\text{Rate of thermal energy out at } r + \Delta r\} + \{\text{Production of thermal energy by dissipation}\} = \{\text{Accumulation}\} \tag{1.8}$$

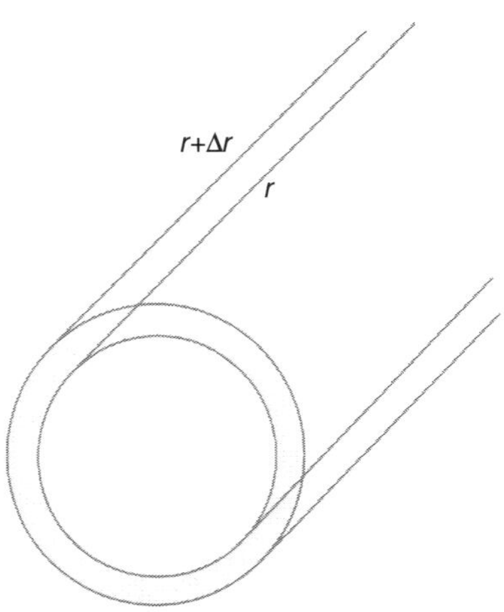

Figure 1.3 Annular volume element extracted from conductor for the thermal energy balance. The thickness of the annular shell is Δr and the length is L.

Since the temperature in the conductor may vary continuously with *both* position and time, the result of this balance will be a partial differential equation. We can rewrite the balance eq. (1.8) introducing the appropriate symbols:

$$+(2\pi rLq_r)_r - (2\pi rLq_r)_{r+\Delta r} + 2\pi rL\Delta rP_e = 2\pi rL\Delta r\rho C_p\frac{\partial T}{\partial t}. \tag{1.9}$$

Now we divide by $2\pi L\Delta r$, take the limit as $\Delta r \to 0$, apply the definition of the first derivative, and substitute Fourier's law for q_r (we also divide by r):

$$k\left[\frac{1}{r}\frac{\partial}{\partial r}\left(r\frac{\partial T}{\partial r}\right)\right] + P_e = \rho C_p\frac{\partial T}{\partial t}. \tag{1.10}$$

Note that we have assumed that the volumetric rate of thermal energy production, P_e, is a constant; this cannot be strictly correct since the resistance of copper wire (e.g., AWG 12) is 1.65 ohms/1,000 ft at 25 °C, but increases to 3.08 ohms/1,000 ft at 250 °C. In our model we neglected the temperature dependence of the conductor's resistance; this would probably be acceptable if the temperature change in the wire is modest. For *steady-state* conditions, the solution for eq. (1.10) is simply

$$T = -\frac{P_e}{4k}r^2 + C_1\ln r + C_2. \tag{1.11}$$

If T is finite at the center then clearly $C_1 = 0$. One question that arises in such problems concerns the speed of approach to steady state: When might we use eq. (1.11) and when must we proceed with the solution for eq. (1.10)? We can illustrate this concern using 2 AWG bare copper wire ($d = 0.6544$ cm) with a constant surface temperature of 30 °C (this is an example of a *Dirichlet* boundary condition). We set $\frac{P_e}{\rho C_p} = 5950$ °C/s and let the wire have a uniform initial temperature of 30 °C. Because copper has a very large thermal diffusivity, $\alpha = \frac{k}{\rho C_p} = 1.14\ \mathrm{cm^2/s}$, the approach to steady state should be quick:

t (s)	0.005	0.01	0.02	0.05	0.075	0.100	0.175
T center (°C)	59.67	86.87	124.51	162.43	168.01	169.20	169.52

As we anticipated, the steady-state condition is attained very rapidly suggesting that for many similar situations eq. (1.11) could be used to find $T(r)$.

The example above is a microscopic balance, i.e., we are modeling a *distributed parameter* system. We will also have occasion to use macroscopic balances for *lumped-parameter* systems in which the field (or dependent) variable does not vary with position. For the electrical wire with dissipation discussed above, this would mean that the temperature would *not vary* in the r-direction. This is clearly not valid for the case we just examined where $T(r = R)$ was forcibly maintained at 30 °C by removing heat at the surface. We will discuss the circumstances under which the temperature might be (nearly) independent of position a little later.

1.1.1 Rationale for Modeling and Some Unwanted Complications

In modern industrial production applied scientists and engineers constantly struggle to meet product specifications, satisfy regulatory constraints, increase output, and maximize the return for investors and stakeholders. A reality of modern industrial operations is that economic survival is often predicated upon continuous process improvement. And because of the scale of industrial processes, even incremental improvements can be very significant to the bottom line. In the early twentieth-century process tweaking was carried out empirically by trial and error; this usually worked since margins were wide, there was less global competition, and product specifications were often loose. Since there was little automatic process control, skilled operators quickly learned through experience how to make adjustments to improve production. That era has passed, and now operational decisions and control strategies are often based upon models or process simulations. As Hanna and Sandall (1995) point out, contemporary economic reality dictates that modeling and simulation be favored over labor-intensive experimental investigations. In this introduction we will examine a few of the possible ways models can be formulated, and we will look at some examples illustrating the underlying principles that are key to modeling and simulation.

Before we do that, however, we need to recognize that a model—however complex—is merely a representation of reality. Though we may understand the governing physical principles thoroughly our mathematical formulation will never be in perfect fidelity with the "real" world. This is exactly what Himmelblau and Bischoff (1968) referred to when they noted, "*.....the conceptual representation of a real process cannot completely encompass all of the details of the process....*"

Nearly always in real processes there are random events, stochastic elements, or nonlinear couplings that simply cannot be anticipated. Nowhere does this become more apparent than in the examination of engineering or industrial catastrophes; the actual cause is almost always due to a chain or cascade of events many of which are quite improbable taken individually. In cases of this kind, the number of state variables may be *very* large such that no mathematical model—at least none that can be realistically solved—will account for every contingency. And even in relatively simple systems quite unexpected behavior can occur, such as a sudden jump to a new state, or the appearance of an aperiodic oscillation. Examples of real systems where such behaviors are observed include the driven pendulum, the Belousov–Zhabotinsky chemical reaction, and Rayleigh–Bénard buoyancy-driven instability. Real systems are always dissipative, i.e., they include "frictional" processes that lead to decay. Where we get into trouble is in situations that include both dissipation and at least one mechanism that acts to sustain the dynamic behavior. In such cases the behavior of the system may evolve into something much more complicated, unexpected, and possibly even dangerous.

There is an area of mathematics that emerged in the twentieth century (the foundation was established by Henri Poincaré) that can provide some *qualitative* indications of system behavior in some of these cases; though what has become popularly known as *catastrophe theory* is beyond the scope of our discussions, it may be worthwhile to describe a few of its features. In catastrophe theory, we concern ourselves with systems whose normal behavior is smooth—i.e., that possess a stable equilibrium, but that may exhibit abrupt discontinuities (become unstable) at instants in time. Saunders (1980) points out that catastrophe theory applies to systems governed by sets (even *very large* sets) of differential equations, to systems for which a variational principle exists, and to many situations described by partial differential equations. In typical applications, the number of state variables may indeed be very large, but the number of *control* variables may be quite small. Let us explain what we mean by *control variable* with an example: Suppose we wished to study the flow of water through a cylindrical tube. We impose a particular pressure gradient (or head, Δp) and then measure the resulting flow rate. The head is the control variable and the flow rate through the tube is established in response to Δp. If the number of control variables is less than or equal to four, then there are only *seven elementary types of catastrophes*. The beauty of catastrophe theory is that it makes it possible to predict the *qualitative* behavior of a system, even for cases in which underlying differential equations are unknown or hopelessly complicated. An excellent review of this field with numerous familiar examples (including biochemical reactions, population dynamics, orbital stability, neural activity in the brain, the buckling of structures, and hydrodynamic instability) has been provided by Thompson (1982). His book is a wonderful starting point for students interested in system instabilities. If you would like to explore catastrophe theory with the objective of gaining greater understanding, please consider the team exercise, problem 1.21, in which a *catastrophe machine* is constructed and tested.

The principal fact we wish to emphasize as we conclude this introduction is that *every model is an idealization* and when we rely upon a mathematical analysis (or a process simulation) it is prudent to keep its limitations in mind. We would do well to remember the statistician George E. P. Box's admonition, "…..*essentially, all models are wrong but some are useful.*" In the modern practice of applied science we must add a corollary: Not only can models be useful, sometimes they are absolutely necessary even when they are wrong in some limited sense.

Let us now look at just a few examples of how problems of the types we wish to solve are actually developed. We will begin with two very different situations—force resolution in statics, and equilibrium for four chemical species partitioned between gas and liquid phases; both problems require solution of sets of algebraic equations.

1.2 Algebraic Equations: Force Resolution and Vapor–Liquid Equilibria (VLE)

Imagine two steel rods of unequal length attached to the underneath side of a horizontal beam; the longer rod ($L = 16$ m) is attached at point "A" and the shorter rod ($L = 9.24$ m) at point "B." The free ends of the two rods are pinned together at point "C" to form a 30-60-90 triangle, and a 1,000 lb. weight is suspended from "C." Our interest is in the tension created in each of the two rods. The system is at equilibrium, so the 1,000 lb_f weight must be balanced by the vertical components of the tensions in the rods; therefore

$$1{,}000 = T_{AC}\sin(30^\circ) + T_{BC}\sin(60^\circ). \tag{1.12}$$

It is also necessary that the horizontal components of the tensions cancel, that is, there can be no net force acting in the horizontally ($\sum F_{horiz} = 0$):

$$-T_{AC}\cos(30^\circ) + T_{BC}\cos(60^\circ) = 0. \tag{1.13}$$

Consequently, $T_{AC} = 0.577 T_{BC}$ and by eliminating T_{AC} from eq. (1.12) we obtain

$$1{,}000 = 0.289 T_{BC} + 0.866 T_{BC}, \tag{1.14}$$

so that $T_{BC} = 866.1\ \mathrm{lb_f}$ and $T_{AC} = 499.7\ \mathrm{lb_f}$. In this elementary example from statics the simultaneous algebraic equations were formulated by noting that the resultant forces must be zero at equilibrium.

Problems in vapor–liquid equilibria require solution of mass balances but in cases where the temperature (T) is unknown (as in this instance), a trial-and-error process can be employed. We will assume that we have a vapor consisting of an equimolar mixture of light hydrocarbons, ethane (1), propylene (2), propane (3), and isobutane (4). The vapor-phase mole fractions are all ¼, i.e., $y_1 = y_2 = y_3 = y_4 = 0.25$. The constant total pressure is 14.7 psia (1.013 bars), and the vapor phase is cooled slightly until the first drop of liquid is formed (this temperature is the *dew point*). Our objective is to find the temperature, T, at which this occurs, and the composition of the liquid that forms (in equilibrium with the vapor). We will solve this problem two different ways and then compare the results.

First, we will use the Antoine equation to get the vapor pressures of all four species as functions of temperature:

$$\log_{10} p^* = A - \frac{B}{C + T}. \tag{1.15}$$

The necessary constants will be obtained from Lange's *Handbook of Chemistry* (Revised Tenth Edition, 1961).

	A	B	C
Ethane	6.80266	656.4	256
Propylene	6.8196	785	247
Propane	6.82973	813.2	248
Isobutane	6.74808	882.8	240

Keep in mind that T must be °C and p^* is mm Hg. We will assume that Raoult's law is applicable such that

$$y_1 P = x_1 p_1^*(T). \tag{1.16}$$

P is the total pressure and it is 760 mmHg. Therefore, the liquid-phase mole fractions are determined from $x_i = \frac{y_i P}{p_i{*}(T)}$, and a solution will be found when $\sum x_i = x_1 + x_2 + x_3 + x_4 = 1$. Such problems are amenable to machine computation, and a simple strategy suggests itself: Estimate T, compute the vapor pressures with the Antoine equation, then calculate the liquid-phase mole fractions and check their sum. If the $\sum x_i \neq 1$, adjust T and repeat. A very short program was written for this purpose, and it shows $T = -34.2\,°\mathrm{C}$ (−29.6 °F), with

$$x_1 = 0.0273 \quad x_2 = 0.1406 \quad x_3 = 0.1789 \quad x_4 = 0.6611 \quad (\Sigma x_i = 1.0079).$$

Of course, vapor–liquid equilibria for light hydrocarbons are enormously important and have been intensively studied. For an alternative procedure we can use data published by C. L. DePriester (1953) which are in the form of nomograms relating the distribution coefficients, *K's* ($K = y/x$), for light hydrocarbons to temperature and pressure. Again, we estimate T, then find the distribution coefficients and use them to calculate the liquid-phase mole fractions. By trial and error we find for $T = -30\,°\mathrm{F}$ (with $P = 14.7$ psia):

	K (y/x)	x
Ethane	7.95	0.031
Propylene	1.92	0.130
Propane	1.43	0.175
Isobutane	0.37	0.672

In this case the summation of the liquid-phase mole fractions is 1.008, and the agreement with our first solution is reasonable (the worst case is ethane, with a difference between values of x_i of about 13%).

This VLE example illustrates a possible outcome when two different solution procedures are available; the results, particularly for the volatile constituents, are slightly different. It is important to note, however that the results obtained from the two solution procedures for the *major components of the liquid* (propane and isobutane) are very close, differing only by about 2%.

1.3 Macroscopic Balances—Lumped-Parameter Models

We observed previously in this chapter that the temperature of a copper conductor carrying electrical current *might not* vary with radial position. Such a situation could arise if there was large resistance to heat transfer at the surface; i.e., if the heat generated by dissipation could not escape to the surrounding fluid phase. In the following example we look at a case that might meet this stipulation, a hot metal casting that is being quenched in an oil bath. It is always good practice for the analyst to begin with a verbal statement of the balance:

$$\{\text{Rate of thermal energy in}\} - \{\text{Rate of thermal energy out}\} = \{\text{Accumulation}\}$$

This is a *macroscopic* balance in which we *assume* that the temperature *throughout the casting is the same* (as we pointed out, this cannot be strictly correct, but under the right circumstances it may be adequate). There is no {*rate in*} since the casting simply loses thermal energy to its surroundings, and we will assume that this loss is approximately described by Newton's law of cooling. Therefore,

$$-hA(T - T_\infty) = MC_p\frac{dT}{dt}. \tag{1.17}$$

h is the heat transfer coefficient, A is the appropriate surface area, M is the mass of the casting, and C_p its heat capacity. This first-order, *lumped-parameter* model is readily separated to yield:

$$-\frac{hA}{MC_p}dt = \frac{dT}{(T - T_\infty)}. \tag{1.18}$$

We integrate and find

$$T - T_\infty = C_1 \exp\left(-\frac{hA}{MC_p}t\right). \tag{1.19}$$

The casting has an initial temperature, T_i, at $t = 0$, and thus:

$$\frac{T - T_\infty}{T_i - T_\infty} = \exp\left(-\frac{hA}{MC_p}t\right). \tag{1.20}$$

According to this first-order model the temperature of the casting will follow an exponential decay, ultimately approaching the temperature of the oil bath. It is worth noting that the quotient, $hA/(MC_p)$ is an inverse time constant for this system, $1/\tau$. When $t = 1\tau$, about 63% of the ultimate change will have been accomplished. When $t = 2\tau$, about 86%, and at $t = 3\tau$, about 95%. Consequently, a common rule of thumb for simple first-order systems is that the dynamic behavior is nearly complete in 3τ.

We should wonder what to look for in more general applications that would indicate that a lumped-parameter model is acceptable. What are the conditions that might allow us to neglect the variation of the field (dependent) variable in the interior of the medium? We have pointed out that in a case like the metal casting considered above, we can answer this question by identifying where the main resistance to heat transfer is located. If the main resistance is in the fluid phase, then the temperature in the interior of the solid may be nearly uniform. But if the main resistance is in the solid material, then T may vary significantly with position. We can assess the location of the resistance to heat transfer through use of the Biot modulus, $Bi = hR/k$, where h is the heat transfer coefficient on the *fluid side* of the interface and k is the thermal conductivity of the *solid* medium. If Bi is very small, then the fluid side of the interface offers the main resistance to heat transfer.

Let us look at a second macroscopic balance example that is slightly more complicated; we will model a continuous, perfectly mixed, stirred tank reactor (CSTR) in which the reactant species, "A" is consumed by a first-order chemical reaction. We will have flow into the tank, flow out of the tank, and depletion of "A" by chemical reaction. In this case, a verbal statement of the mass balance on "A" will appear: {Rate in} – {Rate out} – {Depletion by reaction} = {Accumulation}. Symbolically we write:

$$\dot{v}C_{Ain} - \dot{v}C_A - k_1VC_A = V\frac{dC_A}{dt}. \tag{1.21}$$

We will divide by the volumetric flow rate, $\dot{v}$, obtaining

$$C_A(1 + k_1\tau) - C_{Ain} = -\tau\frac{dC_A}{dt}. \tag{1.22}$$

In this instance the time constant for this system (τ) is merely the volume of the reactor, V, divided by volumetric flow rate ($\tau = V/\dot{v}$); this is called the *mean residence time* of the reactor. For the initial condition we take the concentration of "A" in the tank to be zero ($C_A = 0$ for $t = 0$), and once again the solution for this first-order model is elementary:

$$C_A = \frac{C_{Ain}}{1 + k_1\tau}\left[1 - \exp\left(-(1 + k_1\tau)\frac{t}{\tau}\right)\right]. \tag{1.23}$$

Note that under steady-state conditions the concentration, C_A, is attenuated from the inlet (feed) value by the factor, $\frac{1}{1+k_1\tau}$. If the residence time, τ, in the reactor is large, and if the reaction rate constant, k_1, is large, then the exiting concentration will be *much smaller* than the feed, C_{Ain}.

Both of the examples immediately above led to exponential solutions. This is not the only possible outcome for these elementary balances; consider a *batch* kinetic study of a second-order chemical reaction in which the reactant species is consumed according to

$$r_A = -k_2C_A^2. \tag{1.24}$$

Since this is a batch process there is no flow either into or out of the reactor volume, so the macroscopic mass balance yields:

$$\frac{dC_A}{dt} = -k_2C_A^2. \tag{1.25}$$

This differential eq. (1.25) is nonlinear, but it is easily integrated:

$$-\frac{1}{C_A} + C_1 = -k_2t. \tag{1.26}$$

At $t = 0$, $C_A = C_{A0}$, so the dynamic behavior in the batch reactor is

$$\frac{1}{C_{A0}} - \frac{1}{C_A} = -k_2t. \tag{1.27}$$

If the initial concentration in the reactor is 1 (one), then

$$C_A = \frac{1}{1 + k_2t}; \tag{1.28}$$

we have an inverse relationship between concentration and time. If this batch process is carried out for a *long* period of time, the reactant species "A" will nearly disappear. Of course, the decay in an exponential process is more rapid; for example, taking $kt = 3$, we have $\exp(-3) = 0.0498$, whereas $1/(1 + 3) = 0.25$.

We continue this discussion by noting the study of chemical kinetics in constant volume systems is fertile ground for the exploration of macroscopic balances with less familiar outcomes (i.e., solutions that do not yield a simple exponential result). For example, we could have *parallel* first- and second-order reactions yielding

$$\frac{dC_A}{dt} = -k_1C_A - k_2C_A^2. \tag{1.29}$$

This equation can also be integrated, and the reader may want to use West (1975) to show that

$$\frac{1}{k_1}\ln\left[\frac{k_2C_A}{k_2C_A + k_1}\right] = -t + C_1. \tag{1.30}$$

Other possibilities include rate expressions with fractional orders and Michaelis–Menten kinetics for certain catalyzed systems, where

$$r_A = \frac{dC_A}{dt} = \frac{-K_1C_A}{K_2 + K_3C_A}. \tag{1.31}$$

If eq. (1.31) is integrated we find that the sum of both $K_2\ln(C_A)$ and K_3C_A terms is proportional to time, $-K_1t$ (plus the constant of integration).

1.3.1 Recognizing When a Model Suggests Potential Hazards

We noted previously that circumstances can arise in engineering practice where an unexpected—perhaps even dangerous—event may occur. This is particularly true during transient operations, and it is vital that engineers use models to anticipate when serious problems may arise. In some cases the difficulties will be obvious, but in others the critical features may be very subtle. We conclude this section with an illustration related to our previous discussion of a CSTR (continuous stirred tank reactor) operated isothermally. A mass balance upon the reactant species can be written:

$$V\frac{dC}{dt} = \dot{v}(C_{in} - C) - VR. \tag{1.32}$$

V is the volume of the reactor and R is the rate at which reactant is consumed (per unit volume). Now suppose the kinetic description of the reaction is:

$$R = \frac{kC}{(1+\kappa C)^2}. \tag{1.33}$$

We define a time constant (τ) for the reactor as the quotient of the reactor volume and the volumetric flow rate, $V/\dot{v}$. Then, under steady-state conditions we have the relation:

$$\frac{1}{\tau}(C_{in} - C) = \frac{kC}{(1+\kappa C)^2}, \tag{1.34}$$

a third-degree algebraic equation with the possibility of *multiple solutions* of course. Let us illustrate this using the parameters selected by Matsuura and Kato (1967):

$$C_{in} = 1.32 \quad \kappa = 10 \quad \text{and} \quad \dot{v}/(Vk) = 0.019.$$

These choices result in three possible steady-state values for C which can easily be identified using Newton–Raphson (or any scientific calculator):

$$C(t \to \infty):\ 0.2549 \quad 0.8004 \quad \text{and} \quad 0.0647.$$

A situation like this immediately raises two critical questions: 1) Are these conditions equally likely? 2) How large must a transient (or process upset) be to drive the reactor to a different—and possibly dangerous—equilibrium? The answer to the first question is *no*, and the reader is invited to explore the likelihood of arriving at $C = 0.2549$ during unsteady operation of the reactor. It is the second question that is crucial both for process design and operational safety. To address it we can solve the differential equation numerically using different initial values for C; we quickly discover that if $C(t=0) = 0.26$ we will arrive at 0.80, but if we start with $C(t=0) = 0.24$ we get to 0.0647. If you have just a few moments, see if you can identify the threshold value for C; you will discover that it is 0.25495. It is essential that the analyst tasked with modeling a dynamic process recognize the possibility of steady-state multiplicity and take the necessary steps to ensure safe operation. A worthwhile thought exercise for engineers is: *In cases like this CSTR example how do you protect the process from an undesirable excursion?*

1.4 Force Balances—Newton's Second Law of Motion

Many problems in mechanics require us to set

$$Ma = \sum F. \tag{1.35}$$

For example, consider a mass, M, suspended vertically from the ceiling, through a spring (that exhibits Hookean behavior). We will take the *positive z*-direction to be down. The apparatus also has some kind of viscous damping provided by a dashpot or shock absorber:

$$M\frac{d^2z}{dt^2} = Mg - Kz - A\frac{dz}{dt}. \tag{1.36}$$

In this case we obtain a second-order, ordinary differential equation. We will divide by M and rewrite the equation

$$\frac{d^2z}{dt^2} + \frac{A}{M}\frac{dz}{dt} + \frac{K}{M}z - g = 0. \tag{1.37}$$

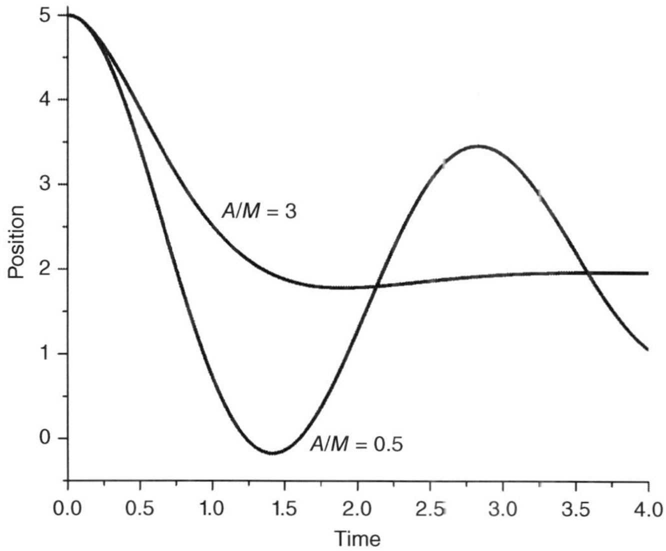

Figure 1.4 Displacement of a suspended weight attached to the ceiling with a spring. Viscous damping is reflected by the magnitude of A/M. The mass is moved to an initial position of $z = 5$, and then released.

The response of this system to an initial perturbation can assume different forms depending upon the values selected for the constants. Suppose for example that we let $A/M = 3$ and $K/M = 5$ ($g = 9.8\ \text{m/s}^2$, of course). We will initiate the dynamic behavior of the system by pulling the mass downward to a new, extended, position. The equilibrium position is $z = 9.8/5 = 1.96$ m, so we will start by extending the assembly to $z = 5$. We will then repeat the solution, but *decrease* the damping coefficient, A/M to 0.5. For this second case, the response will be much more oscillatory; see Figure 1.4.

We will explore options for the solution of differential equations similar to eq. (1.37) in Chapters 5 and 6.

It is worth pointing out that the form of the differential eq. (1.37) can arise in other phenomena, quite unrelated to $F = Ma$. For example, if we place a resistance (R), an inductance (L), and a capacitance (C) in series to form an RLC circuit, and if we supply sufficient energy to offset the dissipative losses, we can see sustained oscillation. The governing equation in this case is:

$$\frac{d^2I}{dt^2} + \frac{R}{L}\frac{dI}{dt} + \frac{1}{LC}I = a$$

which looks exactly like eq. (1.37). The driving force (forcing function), a, appears on the right-hand side. In fact, if we were to choose a large inductance and a small resistance, the circuit would be severely underdamped and oscillation in response to a forcing function would be guaranteed.

1.5 Distributed Parameter Models—Microscopic Balances

We looked at some cases previously where lumped models were used to make difficult problems tractable; these included heat lost by a metal casting and the operation of stirred chemical reactors. We pointed out that in heat transfer we might be able to use the Biot modulus to assess the suitability of the lumped approach. Let us provide a little elaboration by way of additional examples. Consider a *tray* (*plate* or *stage*) in a distillation column; we often treat this type of separation as a cascade of equilibrium stages where the *conditions over the entire tray are the same*. This means that the temperature and the mole fractions in the liquid and vapor *do not vary* over the plate. This is incorrect as local fluid mechanics as well as spatially-variable heat and mass transfer affect the approach to equilibrium for each stage. Sometimes we can account for such variations with a *stage efficiency*, which is the fractional approach to equilibrium for that tray and often this approach works very well.

Similarly, in the case of a continuous stirred tank reactor (CSTR), we normally *assume* that the temperature and concentration are *completely uniform* throughout the reactor volume. This too cannot be correct as we know with certainty that there will be dead zones (in corners and perhaps near baffles) where little mixing takes place. In some instances the idealized treatment of a chemical reactor will result in a suitable model. But what happens if fine-scale variations seriously impact process performance? In such cases we must turn to distributed parameter models, and let us illustrate the process with an example from heat transfer.

Consider a *long* cylindrical rod, initially at some uniform temperature, T. At $t = 0$, the surface temperature of the rod is instantaneously changed to some new, elevated value. Our interest is the flow of heat into the interior of the rod. We will make a shell energy balance on an annular element of length, L, and thickness, Δr.

$$\{\text{Rate in at } r\} - \{\text{Rate out at } r + \Delta r\} = \{\text{Accumulation of thermal energy}\}$$

Thus,

$$+(2\pi r L q_r)_r - (2\pi r L q_r)_{r+\Delta r} = 2\pi r L \Delta r \rho C_p \frac{\partial T}{\partial t}. \tag{1.38}$$

We divide by $2\pi L \Delta r$, take the limit as $\Delta r \to 0$, and apply the definition of the first derivative:

$$-\frac{1}{r}\frac{\partial}{\partial r}(r q_r) = \rho C_p \frac{\partial T}{\partial t}. \tag{1.39}$$

We use Fourier's law, $q_r = -k\frac{\partial T}{\partial r}$, to rewrite this partial differential equation in a more useful form:

$$\frac{\partial T}{\partial t} = \alpha\left[\frac{\partial^2 T}{\partial r^2} + \frac{1}{r}\frac{\partial T}{\partial r}\right], \tag{1.40}$$

where α is the thermal diffusivity of the medium, $\alpha = \frac{k}{\rho C_p}$. Eq. (1.40) is a parabolic partial differential equation, and it is a candidate for solution by a technique known as *separation of variables*. We will now illustrate some of the initial steps in this process, deferring the intricacies for Chapter 7. We take $T = f(r)g(t)$ and introduce this product into eq. (1.40):

$$fg' = \alpha\left[f''g + \frac{1}{r}f'g\right]. \tag{1.41}$$

Dividing by the product, fg, we note that the left-hand side is a function of time and the right-hand side is a function only of r. The only way the two can be equal is if they are both equal to a constant:

$$\frac{g'}{\alpha g} = \frac{f'' + \frac{1}{r}f'}{f} = -\lambda^2. \tag{1.42}$$

Notice that we have purposefully chosen a negative constant—the necessity for this will become apparent in a moment. We have identified two ordinary differential equations (the second of the pair is a form of Bessel's differential equation which we discuss in detail in Chapter 5):

$$\frac{dg}{dt} = -\alpha\lambda^2 g \quad \text{and} \quad \frac{d^2 f}{dr^2} + \frac{1}{r}\frac{df}{dr} + \lambda^2 f = 0, \tag{1.43}$$

and the solutions are

$$g = C_1 \exp(-\alpha\lambda^2 t) \quad \text{and} \quad f = AJ_0(\lambda r) + BY_0(\lambda r). \tag{1.44}$$

Therefore, according to our initial assumption regarding the form of the solution:

$$T = C_1 \exp(-\alpha\lambda^2 t)[AJ_0(\lambda r) + BY_0(\lambda r)]. \tag{1.45}$$

However, $Y_0(0) = -\infty$, and since our temperature must be finite at the center of the cylindrical rod,

$$T = A \exp(-\alpha\lambda^2 t) J_0(\lambda r). \tag{1.46}$$

Let us emphasize that at this point we merely have *a solution* for the partial differential eq. (1.40)—in order to get *the solution* of interest we must apply a boundary condition at $r = R$ and an initial condition for $t = 0$. As we shall see later in the course, these final two steps often cause the analyst the most difficulty. It is entirely appropriate for us to wonder how well this model for transient heat transfer in a cylindrical rod represents physical reality. A comparison is provided below in which an acrylic plastic rod was heated by immersion in a constant temperature bath; part of the discrepancy between model and experiment is caused by the fact that the experimental rod was of finite length, whereas the model is for an *infinitely long* cylinder. The rod was chilled to 3 °C, then immersed in a heated water bath maintained at 69 °C.

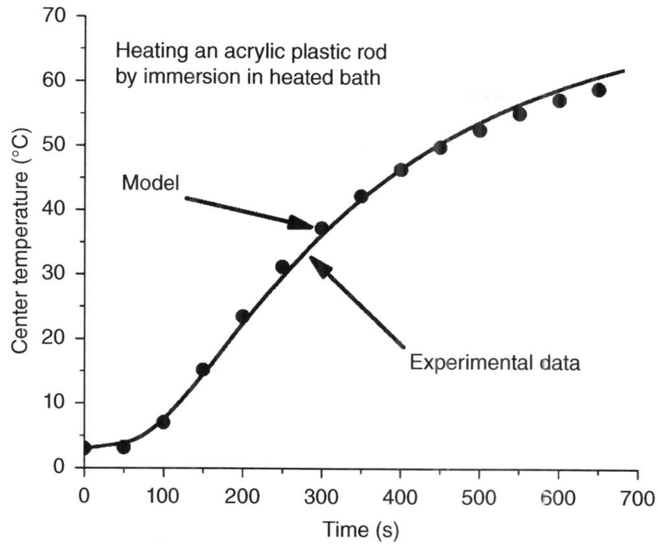

Figure 1.5 Comparison of a distributed parameter model (filled circles) with experimental data (solid curve) for transient heating of a cylindrical rod. The discrepancy between two may be due to the following: the experimental cylinder was not infinitely long (in fact, $L/d = 6$), the heat transfer coefficient is not constant, the thermal contact in the TC well may not be perfect, and the TC has its own thermal inertia.

A thermocouple (TC), embedded at the center of the rod, was used to measure the cylinder's (centerline) temperature as a function of time (Figure 1.5).

Of course, if the cylinder length-to-diameter ratio is not large, then axial (z-direction) transport would have to be included in the model; the term, $\frac{\partial^2 T}{\partial z^2}$, would be added to the right-hand side of eq. (1.40). This illustrates a case where the experience of the analyst comes into play. How might we determine if the axial transport term is required? Of course, we could solve the complete problem for a sequence of decreasing ratios (of L/d) until the solutions begin to show significant differences. We will actually try this below, but we also recognize that such an approach is time consuming and computationally expensive, so we will seek another course of action as well. One possibility is to construct dimensionally correct representations for the second derivatives using appropriate temperature differences and suitable characteristic lengths:

$$\frac{\partial^2 T}{\partial r^2} \approx \frac{\Delta T}{\delta^2} \quad \text{and} \quad \frac{\partial^2 T}{\partial z^2} \approx \frac{\Delta T}{L^2}.$$

If $\delta^2 \ll L^2$, then axial transport is almost certainly unimportant. Let us explore the effect of L/d in a concrete way by revisiting the above example (transient heating of an acrylic plastic rod); we will solve eq. (1.40) numerically, but with the addition of the axial conduction term, $\frac{\partial^2 T}{\partial z^2}$.

Temperature at Center of Acrylic Plastic Rod of Finite Length.

t (s)	*L*/*d* = 10	*L*/*d* = 6	*L*/*d* = 4	*L*/*d* = 2
50	3.164	3.164	3.164	3.164
100	7.365	7.365	7.365	7.365
150	15.952	15.952	15.952	15.955
200	25.107	25.108	25.108	25.131
250	33.242	33.244	33.245	33.322
300	40.040	40.041	40.043	40.208
400	50.112	50.114	50.117	50.482
500	56.702	56.704	56.708	57.214

It is clear from these data that for cases in which molecular transport is dominant, the "tipping" point for the assumption of infinite-length behavior occurs at about $L/d = 4$. In cylindrical geometries, if $L/d \le 4$, we may have to account for axial (z-direction) transport.

1.5.1 Fluid Flow and Heat Transfer Combined

Of course we understand that we can have distributed parameter problems where more than one dependent variable changes with position. Consider the case of an energy balance drawn upon a fluid contained within parallel, planar walls; the space between the walls extends from $y = 0$ to $y = B$. The lower wall is heated, the upper wall is insulated, and the fluid moves in the z-direction (fully developed, laminar flow). The appropriate energy balance is:

$$\rho C_p \left(\frac{\partial T}{\partial t} + V_z \frac{\partial T}{\partial z} \right) = k \left[\frac{\partial^2 T}{\partial y^2} + \frac{\partial^2 T}{\partial z^2} \right]. \tag{1.47}$$

In this case the momentum balance can be solved separately to obtain V_z, and the velocity distribution for the fully developed, laminar flow is:

$$V_z = \frac{1}{2\mu} \frac{dp}{dz} (y^2 - By). \tag{1.48}$$

Therefore, the temperature of the fluid between the parallel walls is governed by:

$$\frac{\partial T}{\partial t} = -\frac{1}{2\mu} \frac{dp}{dz} (y^2 - By) \frac{\partial T}{\partial z} + \alpha \left[\frac{\partial^2 T}{\partial y^2} + \frac{\partial^2 T}{\partial z^2} \right], \tag{1.49}$$

assuming that the fluid properties are constant. Finding a solution for eq. (1.49) will be a more difficult proposition than in our previous examples. However, if we restrict our attention to steady state and neglect axial conduction (i.e., omit the $\frac{\partial^2 T}{\partial z^2}$ term), we can isolate $\frac{\partial T}{\partial z}$ and solve the problem easily by forward-marching in the z-direction using only values from the previous z-row; some computed results are provided in the table immediately below. We will take water as the fluid (it enters at a uniform temperature of 4.44 °C), assume the fluid properties are constant, and let the lower (heated) wall be maintained at 37.78 °C. The maximum (centerline) fluid velocity will be 15 cm/s and we let $B = 2$ cm. There is an important observation to make with respect to these computed results: Note how slowly thermal energy is transported in the transverse (y) direction. This is characteristic of processes in which we rely upon molecular transport. We will see additional examples of this explicit computational process in Chapter 8.

y/B	*z/B* = 10	*z/B* = 100	*z/B* = 1000	*z/B* = 10,000
0.1	8.31	21.91	30.75	36.34
0.2	4.45	10.03	23.96	34.96
0.4	4.44	4.49	12.94	32.53
0.8	4.44	4.44	5.34	29.95

1.6 Using the Equations of Change Directly

In many of the previous examples we formulated balances upon, say, mass or thermal energy, and then developed a governing equation from the balance statement. However, some models can be formulated directly from the *equations of change*; i.e., from the complete balances for momentum, thermal energy, or concentration (of species "A"). To illustrate this process, we will consider mass transfer occurring in laminar flow in a cylindrical tube (of radius, R). We assume species "A" is transported from the tube wall to the fluid which is in motion in the positive z-direction. We have a highly ordered (laminar) flow and we can obtain the needed equations from Bird et al. (2007), or from Glasgow (2010). We will need the continuity (mass balance) equation for species "A" in cylindrical coordinates:

$$\frac{\partial C_A}{\partial t} + V_r \frac{\partial C_A}{\partial r} + \frac{V_\theta}{r} \frac{\partial C_A}{\partial \theta} + V_z \frac{\partial C_A}{\partial z} = D_{AB} \left[\frac{1}{r} \frac{\partial}{\partial r} \left(r \frac{\partial C_A}{\partial r} \right) + \frac{1}{r^2} \frac{\partial^2 C_A}{\partial \theta^2} + \frac{\partial^2 C_A}{\partial z^2} \right] + R_A \tag{1.50}$$

and we also need the z-component of the Navier–Stokes equation (momentum balance):

$$\rho\left(\frac{\partial V_z}{\partial t}+V_r\frac{\partial V_z}{\partial r}+\frac{V_\theta}{r}\frac{\partial V_z}{\partial \theta}+V_z\frac{\partial V_z}{\partial z}\right)=-\frac{\partial p}{\partial z}+\mu\left[\frac{1}{r}\frac{\partial}{\partial r}\left(r\frac{\partial V_z}{\partial r}\right)+\frac{1}{r^2}\frac{\partial^2 V_z}{\partial \theta^2}+\frac{\partial^2 V_z}{\partial z^2}\right]. \tag{1.51}$$

The tube has constant cross-section, the flow is steady, and the fluid is incompressible. Therefore, $V_r = V_\theta = 0$, and $\nabla \cdot V = 0$ so that $\frac{\partial V_z}{\partial z} = 0$. Furthermore, $\frac{\partial p}{\partial z}$ is constant, and for eq. (1.51) we are left with:

$$0=-\frac{\partial p}{\partial z}+\mu\left[\frac{1}{r}\frac{\partial}{\partial r}\left(r\frac{\partial V_z}{\partial r}\right)\right]. \tag{1.52}$$

Integrating the first time:

$$r\frac{dV_z}{dr}=\frac{1}{2\mu}\frac{dp}{dz}r^2+C_1. \tag{1.53}$$

Of course, $\frac{dV_z}{dr} = 0$ at $r = 0$, so $C_1 = 0$. Integrating a second time, we find

$$V_z=\frac{1}{4\mu}\frac{dp}{dz}r^2+C_2. \tag{1.54}$$

We apply the no-slip condition at the tube wall: $V_z = 0$ at $r = R$, so $C_2 = -\frac{1}{4\mu}\frac{dp}{dz}R^2$, and

$$V_z=\frac{1}{4\mu}\frac{dp}{dz}(r^2-R^2). \tag{1.55}$$

In addition to the simplifications already made, we assume that the molecular transport (diffusion) of species "A" in the flow direction is negligibly small (this is often quite reasonable), and we make a rather severe assumption that the flux of "A" from the wall into the fluid is *constant* (we could provide experimental conditions to make this approximately true over a finite tube length). Therefore, $\frac{\partial C_A}{\partial z} = \beta$ and for eq. (1.50) we are left with:

$$V_z\frac{\partial C_A}{\partial z}=V_z\beta=D_{AB}\left[\frac{1}{r}\frac{d}{dr}\left(r\frac{dC_A}{dr}\right)\right]. \tag{1.56}$$

Since we have the parabolic velocity distribution given by eq. (1.55) for V_z, we can multiply by rdr and integrate the first time:

$$r\frac{dC_A}{dr}=\frac{\beta}{4\mu D_{AB}}\frac{dp}{dz}\left(\frac{r^4}{4}-\frac{R^2r^2}{2}\right)+C_1. \tag{1.57}$$

C_A must be finite at the center of the tube, so $C_1 = 0$. We integrate a second time, noting that $C_A = C_{A0}$ at $r = R$. It is also helpful to introduce a new concentration variable by letting $C = C_A - C_{A0}$; the result is

$$C=\frac{\beta}{4\mu D_{AB}}\frac{dp}{dz}\left(\frac{r^4}{16}-\frac{R^2r^2}{4}+\frac{3}{16}R^4\right). \tag{1.58}$$

We can explore the general behavior of this result by making a few calculations:

r/R	$\left(\frac{r^4}{16}-\frac{R^2r^2}{4}+\frac{3}{16}R^4\right)$
0	$0.1875R^4$
0.1	$0.1850R^4$
0.3	$0.1655R^4$
0.5	$0.1289R^4$
0.7	$0.0800R^4$
0.9	$0.0260R^4$
1	0

The reader may want to plot these data to get a better sense of the shape of the profile. Remember that we still have to multiply by $\frac{\beta}{4\mu D_{AB}}\left(\frac{dp}{dz}\right)$ (which is negative) in order to get $C = C_A - C_{A0}$.

1.7 Discretizing a Problem that is Continuous in Time or Space

Circumstances arise where it is desirable to take a continuous model and make it discrete. We will do this routinely in Chapter 8 where partial differential equations will be converted to algebraic equations to facilitate solution. Of course the result in such cases is approximate, but if appropriate care is taken such simplifications may be extremely useful.

A note for the student: The following example will make sense if you have had an introductory course in automatic process control. If not, skip it for now and come back to it when you are ready.

Automatic process control is essential to modern industrial production, whether it be the manufacture of automobiles or baby shampoo. Let us suppose that an engineer wants to control a second-order process using PI (proportional-integral) control. We note that this is certainly not the best choice for a number of reasons, foremost among them is the fact that integral control instantly contributes 90° to the phase lag making instability a very real possibility for anything more complicated than a first-order system. Nevertheless, our intent is to subject the closed feedback control loop to a step change in set point and determine the response by simulation. In the time domain, and in standard second-order form, the differential equation for the process is:

$$\tau^2\frac{d^2Y}{dt^2} + 2\tau\zeta\frac{dY}{dt} + Y = X(t) \tag{1.59}$$

where τ is the time constant for the system and ζ is the damping coefficient (the system is underdamped if $\zeta < 1$). We will discretize this model using backward difference approximations for the derivatives:

$$\frac{d^2Y}{dt^2} \cong \frac{1}{(\Delta t)^2}(Y_i - 2Y_{i-1} + Y_{i-2}) \text{ and } \frac{dY}{dt} \cong \frac{1}{\Delta t}(Y_i - Y_{i-1}).$$

Consequently, in order to compute Y at time i, we need the two immediately preceding values, i–1 and i–2 (that is, $t-\Delta t$ and $t-2\Delta t$); we obtain:

$$Y_i\left[\frac{\tau^2}{(\Delta t)^2} + \frac{2\tau\zeta}{\Delta t} + 1\right] \cong X_i + \frac{2\tau^2}{(\Delta t)^2}Y_{i-1} - \frac{\tau^2}{(\Delta t)^2}Y_{i-2} + \frac{2\tau\zeta}{\Delta t}Y_{i-1}. \tag{1.60}$$

The output from the PI controller (which will be the driving force for the second-order system) can be written:

$$P = p - p_S = K_C\left(\varepsilon + \frac{1}{\tau_I}\int_0^t \varepsilon dt\right). \tag{1.61}$$

Here, K_C is the controller gain, ε is the error provided to the controller by the comparator, and τ_I is the integral time ($1/\tau_I$ is often referred to as the reset rate). Our discretization for the controller is

$$P_i \cong K_C\left(\varepsilon_i + \frac{1}{\tau_I}\sum_{n=1}^{i}\varepsilon_n\Delta t\right) = X_i. \tag{1.62}$$

Historically linear control problems of this type were analyzed in the s-plane using the Laplace transform. However, now such feedback control loops can be easily simulated in the time domain; the results are transparent and the impact of changing controller settings can be quickly evaluated. We will assume that the measurement device (in the feedback loop) is instantaneous and has unit gain. The same assumption will be applied for the final control element, for example, a control valve. Our simulation logic will consist of the following steps:

1) The set point will be changed at $t = 0$ from zero to one (1), i.e., this is a *servo* problem.
2) The resulting error is fed into the discretized PI controller.
3) The controller output then drives the dynamic change in the discretized underdamped second-order system.
4) Output from the second-order system is measured instantaneously and the resulting signal returns to the comparator which produces a new value for the error, ε. We then return to 2) and repeat.

Let the time constant for the second-order system, $\tau = 6$, and the damping coefficient, $\zeta = 0.5$. The critical value for ζ is one (1), so the system is significantly underdamped, making oscillatory behavior a certainty. Consequently we expect this process will prove difficult to control and instability too easily obtained. We set the integral time, τ_I, to 1 (a modest, nonaggressive value) and then simulate the loop with different values for the controller gain, K_C. The necessary steps will look something like this (**R** is the controller set point and **e** is the error):

```
e=R-y(i-1)
 sum=sum+e*dt
 p=kc*(e+1/taui*sum)
   denom=(tau^2/dt^2+2*tau*zeta/dt+1)
   rhs=p+2*tau^2/dt^2*y(i-1)-tau^2/dt^2*y(i-2)+2*tau*zeta/dt*y(i-1)
     y(i)=rhs/denom
        tt=tt+dt
```

We will employ three different controller gains, 0.11, 0.22, and 0.33, and plot the simulation results in Figure 1.6.

We see that $K_C = 0.11$ is stable and 0.22 is just beyond the stability threshold. As an aside we observe that stability for this system can be assessed algebraically in the s-plane using the *Routh test*, which reveals that the *critical* controller gain is 0.20.

Recall that we stated at the beginning of this example that PI control was probably not a great choice for this second-order process, and the results in Figure 1.6 reinforce this observation. Consider the simulation results for $K_C = 0.11$: Although the response is stable and the loop is conforming to the change in set point, the rise time is slow, the overshoot is significant, and the oscillatory behavior is persistent. In fact, at $t = 45$ the output is about 30% below the set point. At $t = 65$ output is still more than 10% above the set point. By no stretch would we deem this satisfactory. Thus, one should ask what might we do to improve loop output? One possibility is to add derivative control (often referred to as anticipatory control, or "pre-act"). The algorithm for the three-mode controller in the time domain is:

$$p - p_S = K_C\left(\varepsilon + \tau_D\frac{d\varepsilon}{dt} + \frac{1}{\tau_I}\int_0^t \varepsilon dt\right), \tag{1.63}$$

where τ_D is the derivative time. If you are comfortable with process control problems, try adding derivative control to the simulation and adjust the controller parameters to obtain improved loop response; see if you can produce something like the example shown here in Figure 1.7.

Why the emphasis upon simulation in the time domain? In this case the discretization yields transparency and improves the speed with which controller strategies can be compared. But there are other important reasons as well; time delays (transportation lags) are readily incorporated and nonlinear process elements (e.g., control valves) can be simulated directly. In classical linear control theory we accommodate nonlinearities through use of a truncated Taylor series; however, this

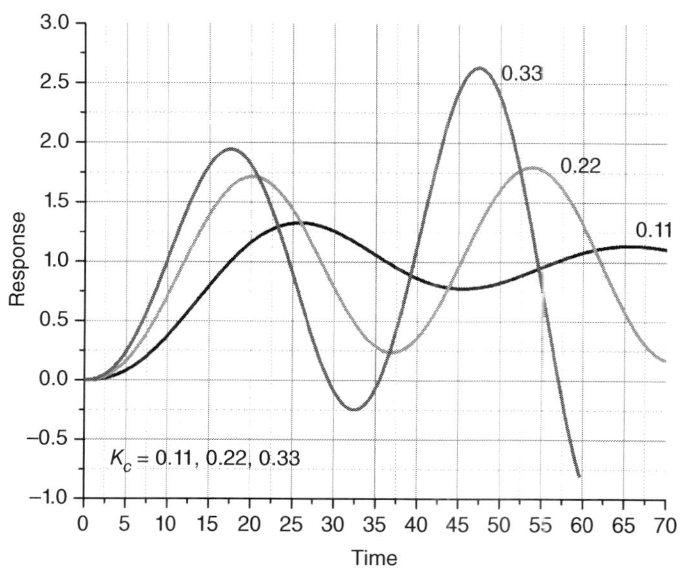

Figure 1.6 Closed-loop response for an underdamped, second-order system with PI control subjected to a unit step change in set point. Three different values of the controller gain were used: 0.11, 0.22, and 0.33.

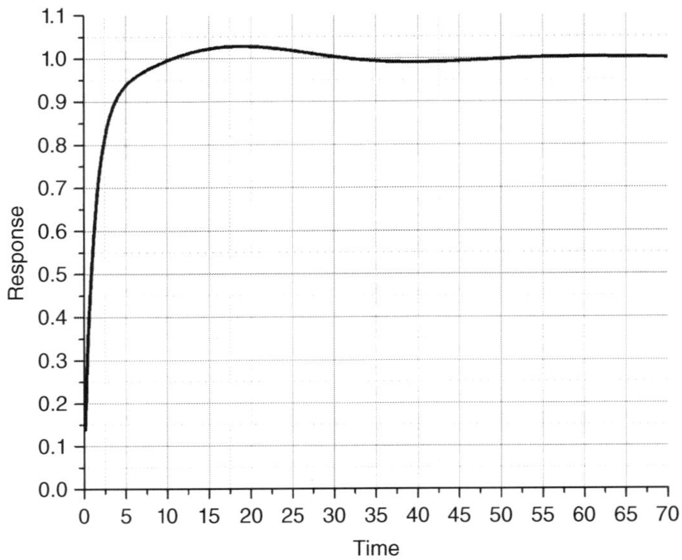

Figure 1.7 System response obtained with PID control. Notice how rapidly the loop response falls within 5% of the changed set point.

approach can be problematic. For example, consider a pneumatic control valve that is quick opening; the relationship between volumetric flow rate, V_{in}, and valve stem position (or lift) is:

$$V_{in} = c_1\sqrt{h}. \tag{1.64}$$

We assume that eq. (1.64) is in dimensionless form for convenience such that both flow rate and lift vary between 0 and 1:

h	0.1	0.2	0.4	0.8	1.0
V_{in}	0.316	0.447	0.632	0.894	1.0

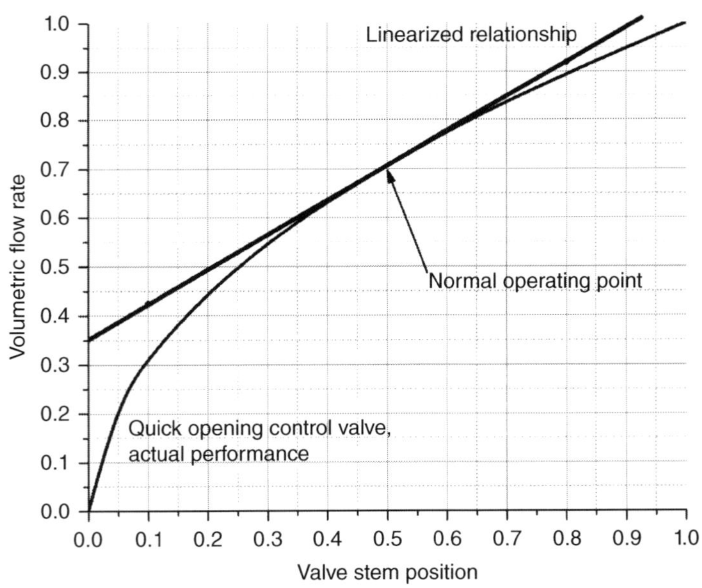

Figure 1.8 Comparison of actual control valve operation with the linearized version. If h remains *sufficiently near* the normal operating point ($h_0 = 0.5$), then the agreement is excellent.

Note that the valve yields 50% flow when h is only 0.25. Now suppose that the normal (steady-state) operating position is h_0; we truncate a Taylor series:

$$V_{in} \cong \sqrt{h_0} + \frac{1}{2}h_0^{-1/2}(h - h_0). \tag{1.65}$$

Let us compare the original function $V_{in}(h)$ with the linearized version assuming h_0 is 0.5; from eq. (1.65) we see that $V_{in} \approx 0.707 + 0.707(h - 0.5)$.

In linear control theory we *assume* that the linearized $V_{in}(h)$ will provide stable system operation as long as h remains in the *vicinity* of the normal operating point and Figure 1.8 suggests this may be the case. But in transient operations h may deviate significantly from 0.5 and as we can see from this figure if $h = 0.1$, the linearized value for $V_{in}(h)$ will be about 0.424 which is about 37% larger than that produced by the control valve. We might take some reassurance from Lyapunov's theorem which we can state in very simple terms: If the linearized model exhibits stable behavior when X is proximate to the normal operating point, then the nonlinear model should also be stable *near* that position. In terms of a differential equation, a system is Lyapunov stable if it stays at an equilibrium point after a *small disturbance*.

But, if we employ discrete time-domain simulation of a dynamical system we may not need to linearize components of the model at all. This means we may be able to assess the transient behavior of a process—even one with nonlinear elements—that is subject to negative feedback control. This is especially significant to transient operations where large deviations from the normal operating (equilibrium) point may occur. And, of course, such excursions always pose the greatest hazards.

1.8 A Contrast: Deterministic Models and Stochastic Processes

In much of the preceding discussion, we formulated (or simply selected) differential equations that expressed the continuous relationship between dependent and independent variables. The appeal of this approach is that given boundary conditions and the initial condition for the system, and the differential equation, the *future behavior of the system is set for all time, t*. This is what we mean when we say that a system is *deterministic*. Often in engineering and the applied sciences, systems of interest do behave exactly this way. But suppose, for example, that we have to concern ourselves with a population of animals. We know that the animals will reproduce, and that some will die, whether it be from old age, disease, or predation. It is also possible that some animals will immigrate, and some may emigrate. Let N be the number (density) of animals, perhaps the number of animals per acre; a simplistic approach to the dynamic problem might be formulated:

$$\frac{dN}{dt} = cN^2 + bN + a, \tag{1.66}$$

with the initial condition, $N = N_0$ at $t = 0$. Therefore, we have

$$\int \frac{dN}{cN^2 + bN + a} = t + C_1. \tag{1.67}$$

The solution for the integral will depend upon our choices for a, b, and c. We will arbitrarily select $c = 1/16$, $b = -1/4$, and $a = 1/2$, and assume we have 9 animals per acre at $t = 0$. We take

$$q = 4ac - b^2 = (4)(1/2)(1/16) - (1/16) = 0.0625.$$

The left-hand side of eq. (1.67) can be found in any standard table of integrals (e.g., in *CRC Handbook of Tables for Mathematics*),; therefore

$$\frac{2}{\sqrt{q}}\tan^{-1}\left(\frac{2cN + b}{\sqrt{q}}\right) = t + C_1, \tag{1.68}$$

such that

$$N = \frac{\sqrt{q}\tan\left[\frac{\sqrt{q}}{2}(t + C_1)\right] - b}{2c}. \tag{1.69}$$

For the initial condition we have selected, $C_1 = 10.34$, and the number of animals per acre, N, behaves as shown in Figure 1.9.

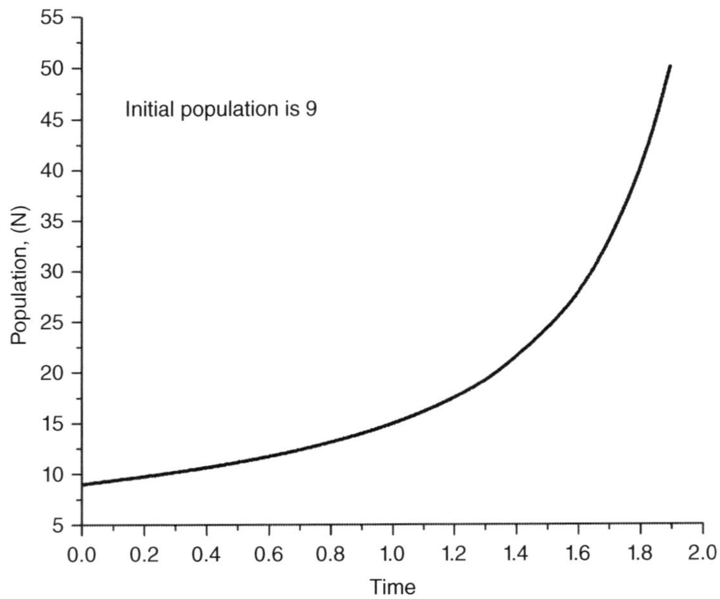

Figure 1.9 The growth of a population of animals (per acre) as described by eq. (1.69). The initial population was 9.

It is clear from these data that $N(t)$ will grow without bound if we employ the deterministic model, eq. (1.69). Obviously this is not correct. In biological systems an explosive growth of numbers leads to collapse—the available food supply simply cannot sustain too many animals. We might even see extinction. This is an important lesson: Certain types of systems exhibit variabilities that should not be represented in familiar deterministic form. Biological systems particularly are affected by behavioral drives—greed, hunger, sex, etc.—that *might* be modeled deterministically if the total numbers are very large, but more generally must be treated stochastically. A stochastic process is one in which the behavior through time is determined by random phenomena. For example, in a population of prey animals, some may have the misfortune to encounter predators. It will be far more appropriate in such cases to base our modeling on the *probability that the population may total N animals at some point in time.* We will discuss this type of formulation in greater detail in Chapter 9.

1.9 Problems with Integer-Valued Variables

In Section 1.7 we considered why one might want to discretize a problem that was continuous in time or space (and often this is done to facilitate computations). Here we observe that certain types of problems are inherently discrete because one or more of the important variables can assume only integer values. For example, in the production of automobiles a manufacturer assembles an SUV from a countable number of parts. Similarly, a herd of cattle will have some integer-valued number of animals, and an electric utility will have a specific number, say n, of substation transformers. In addition, there are problems such as the design of telecommunications networks, scheduling for airlines, and the design of electronic circuits where many alternative arrangements are possible, but only certain ones will have the desired efficiency. A classic example is the machine assignment problem where a production facility has n machines each of which can make m different products. Each product requires a particular fabricating time, t_m, and each has a value, V_m. How does the plant maximize profit or minimize cost? Let's look at an example where we are seeking an optimal solution for a case in which a crucial variable can take on only a limited number of integer values.

The EMD (Electro-Motive Division of GM) SD40-2 was a robust and dependable diesel electric locomotive manufactured from 1972 to 1989, and about 4,000 were produced. Some are still in service and many others have been rebuilt to more modern standards. The prime mover (engine) was a V16 diesel, EMD 16-645E3, capable of 3,000 hp. As one might guess with such a popular locomotive, its efficiency has been studied extensively and reported figures for fuel consumption are provided here (these data were originally posted by *BNSF* engineer Al Krug and are now available on the excelsiorstatistics .com website):

Notch	Gallons/hr	Horsepower	Efficiency (hp·hr/gal)
8	167.7	3,000	17.9
7	145.8	2,375	16.3
6	108.5	1,830	16.9
5	79	1,420	18
4	57.2	1,085	19
3	41.4	710	17.1
2	24.9	390	15.7
1	7.4	200	27
Idle	5.5	0	0

Unlike an automobile, the throttle on a diesel electric locomotive has nine *discrete* positions, referred to as "notches." As you can see in the table above, the most efficient throttle positions are 1, 4, 5, and 8. These efficiency data indicate that *how* a locomotive is operated will significantly affect its fuel consumption; furthermore, the pattern of notch selection during acceleration matters.

Railroads have horsepower requirements for trains based upon the weight of the drag and the route selected (grades are obviously very important). Suppose four SD40-2s are to be lashed together and the train tonnage indicates that 7,300 hp will be required. Assume each locomotive's throttle position can be set independently; what is the combination that yields the least fuel consumption for 7,300 hp? Any combination yielding $7{,}300 \pm 50$ hp is deemed satisfactory. If all four engines are operated at notch 6 (6-6-6-6), we get 7,320 hp but the fuel requirement will be $(4)(108.5) = 434$ gal/hr. If we choose (7-7-7-1) we get 7,325 hp but use 444.8 gal/hr. Contrast these results with the settings: (8-5-5-5) = 7,260 hp with fuel burn of 404.7 gal/hr. Alternatively, with *five* engines in operation we could use (8-4-4-4-4) = 7,340 hp with fuel burn of 396.5 gal/hr.

You might wonder why a Class 1 railroad (examples include BNSF and Union Pacific) would care. Currently, #2 diesel fuel is about $4.60 per gallon retail; using 37 gallons less fuel per hour is an approximate savings of $170 per hour and over an eight-hour drag this amounts to $1,360. If a large railroad operated 100 similar trains per day, the savings amount to $136,000 per day (and over $4,000,000 in one month). Suppose we want to identify solutions for this discrete optimization problem using five SD40-2 locomotives. Of course, there are many different combinations of power settings that will yield the necessary 7,300 hp; we will use sequences of random numbers to identify them. The statement in the code shown below, **RANDOMIZE TIMER**, guarantees that we get a different sequence with each pass (this is necessary because the random number algorithm generates *pseudo-random* numbers). The five whole numbers ranging from 1 to 8 corresponding to notch position for each locomotive are selected in the section of code in boldface (*note that the function*, **INT**, *returns an integer value less than or equal to the argument*). The criterion we use to identify suitable solutions is that the variable, **sum** (the total horsepower of the five engines), must lie in the range: $7{,}250 \leq$ **sum** $\leq 7{,}350$.

```
#COMPILE EXE
#DIM ALL
    REM *** Power settings for five SD40-2 locomotives
    GLOBAL n,j,zran,ntrials,sum,traineff,jj,nsuccess,zz AS SINGLE FUNCTION PBMAIN
        DIM gph(8) AS SINGLE:DIM hp(8) AS SINGLE:DIM eff(8) AS SINGLE:DIM nn(8) AS
INTEGER
            gph(8)=167.7:hp(8)=3000:eff(8)=17.9
            gph(7)=145.8:hp(7)=2375:eff(7)=16.3
            gph(6)=108.5:hp(6)=1830:eff(6)=16.9
            gph(5)=79:hp(5)=1420:eff(5)=18
            gph(4)=57.2:hp(4)=1085:eff(4)=19
            gph(3)=41.4:hp(3)=710:eff(3)=17.1
            gph(2)=24.9:hp(2)=390:eff(2)=15.7
            gph(1)=7.4:hp(1)=200:eff(1)=27
        OPEN "c:trainSD40.dat" FOR OUTPUT AS #1
        RANDOMIZE TIMER
```

```
    100 REM *** continue
                sum=0
                FOR n=1 TO 5
                    zran=RND
                        j=INT(zran*8)+1
                            sum=sum+hp(j)
                                nn(n)=j
                                    NEXT n
                IF sum>7249 AND sum<7351 THEN 200 ELSE 100
    200 REM *** continue
                traineff=0
                FOR n=1 TO 5
                    jj=nn(n)
                        traineff=traineff+gph(jj)
                            PRINT jj
                                NEXT n
                                    PRINT sum,traineff

WRITE#1,nn(1),nn(2),nn(3),nn(4),nn(5),sum,traineff
                                        nsuccess=nsuccess+1
                                        IF nsuccess>100 THEN 300 ELSE 100
    300 REM *** continue
            CLOSE
          INPUT "end program now?";zz
          IF zz>0 THEN END

END FUNCTION
```

We assume that the order of power settings (notches) among the five locomotives is irrelevant. After running this program long enough to achieve 100 successes (meaning we have met the horsepower requirement 100 times) we find two extremes:

(7-7-5-3-2) = 7270 hp burning 436.9 gph, and
(8-4-4-4-4) = 7340 hp burning 396.5 gph.

The difference is 40.4 gal/hr, and the case with greater efficiency provides 70 more horsepower as well. A few lines of typical program output are given here; the first five numbers are notch positions, followed by total horsepower, and finally fuel burned per hour:

```
8, 5, 2, 5, 4, 7315, 407.8
8, 3, 3, 5, 5, 7260, 408.5
3, 3, 6, 4, 8, 7335, 416.2
4, 4, 2, 7, 7, 7310, 430.9
4, 4, 4, 8, 4, 7340, 396.5
7, 7, 5, 2, 3, 7270, 436.9
5, 4, 2, 5, 8, 7315, 407.8
7, 3, 2, 7, 5, 7270, 436.9
7, 6, 6, 4, 1, 7320, 427.4
5, 2, 6, 6, 6, 7300, 429.4
4, 7, 3, 3, 7, 7255, 431.6
4, 1, 7, 6, 6, 7320, 427.4
6, 3, 8, 4, 3, 7335, 416.2
4, 7, 2, 4, 7, 7310, 430.9
5, 8, 3, 4, 4, 7300, 402.5
```

Since the five locomotives' notch settings are being chosen randomly, we can expect to obtain the same notch values occurring in different orders. For example, see the two lines yielding 416.2 gph where the selected notches are 3,3,6,4,8 and 6,3,8,4,3. If we make a scatter plot of these results with horsepower on the x-axis and fuel burn (gph) on the ordinate, we find a discrete number of solutions (16 in total) with the more efficient choices located at the bottom right-hand side of the figure.

Let us make some observations regarding the approach chosen to solve this discrete optimization problem: The technique we employed is an example of *enumeration* where we identify all possible solutions explicitly. The use of sequences of random numbers to identify notch selection for each of the five locomotives—while functional—is not very efficient. It is a bit like the humorous scenario with the literary chimpanzees: If we put 50,000 chimps in a room and give each one a typewriter, we *might* end up with a literary masterpiece, but there will be a lot of chaff to be discarded. We note that more efficient techniques have been devised for problems of this type (see, for example, *Discrete Optimization Algorithms* by Syslo, Deo, and Kowalik, Dover Publications, 2006) and the importance of this area is driving contemporary research; in fact, Elsevier publishes the journal, *Discrete Optimization*, which focuses upon mathematical, computational, and applied aspects of integer programming and combinatorial optimization.

One approach to the problem of discrete optimization that has been the subject of much investigation is based upon the behavior of ant colonies. Ants seeking food will initially wander around randomly, laying down a pheromone trail to mark their paths. Because the pheromones evaporate the level of the attractant diminishes with time such that the longer, inefficient pathways are abandoned. On the shorter, more direct routes, the pheromone trail gets consistently reinforced leading to members of the colony finding the most efficient path to the food source. One of the seminal papers describing ant colony optimization was authored by Dorigo, Caro, and Gambardella (Ant Algorithms for Discrete Optimization, *Artificial Life*, 5:137, 1999).

Our reference to ant colony behavior above brings up the topic of the random walk, sometimes referred to as the drunkard's walk. In two dimensions we can imagine an inebriated person possibly moving to the left, to the right, forward, backward, or remaining stationary. As a modeling technique the random walk has been applied to financial economics, cell migration, genetic drift, Brownian motion, the movement of animals and micro-organisms, electric networks, gambling, movement through porous media, etc. We will see a related process in Chapter 4 in our discussion of numerical integration by the Monte Carlo method. Let us think about the random walk in the context of the motion of ants as described above; in fact, we will construct a simple model on a two-dimensional grid with 40,000 nodes. We assume that for each step in the process, an individual ant can move left, right, up, down, or remain at its previous location. For simplicity, we start the ant from the index (51, 51), corresponding to the location of the colony. We also assume that a "food source" is located at the upper right-hand corner of the domain (201, 201), meaning that the most direct route between the colony and the food is a diagonal path with a minimum distance of $\sqrt{45{,}000}$ or 212.1 units. Of course, any ant arriving at the food source will have traveled a significantly greater distance, and if there is no directional preference (i.e., no learning or pheromone trail to follow), the individual ant may never find the food source at all. Let us try this with the initial assumption that for each stage, every possible step has an equal probability. A simplistic approach might look like this:

```
#COMPILE EXE
#DIM ALL
    REM *** ant migration by random walk in two dimensions
        GLOBAL xpos,ypos,xmove,ymove,ntrial,yellip AS INTEGER
        GLOBAL zx,zy,zz,upperlim,lowerlim,upperlim0 AS SINGLE
FUNCTION PBMAIN
      DIM maprec(201,201) AS SINGLE
          xpos=51:ypos=51:ntrial=0
          upperlim=0.66:lowerlim=0.34:upperlim0=upperlim
              OPEN "c:antwalkR.dat" FOR OUTPUT AS #1
                  LOCAL antwalk AS DWORD
              GRAPHIC WINDOW "testcase",30,30,300,300 TO antwalk
                  GRAPHIC ATTACH antwalk,0
              RANDOMIZE TIMER
  100 REM *** continue
              zx=RND
```

```
17                    IF zx>upperlim THEN xmove=1
18                        IF zx<lowerlim THEN xmove=-1
19            IF zx>lowerlim-0.01 AND zx<upperlim+0.01 THEN xmove=0
20                                xpos=xpos+xmove
21                                IF xpos<1 THEN xpos=1
22                                IF xpos>201 THEN xpos=201
23                zy=RND
24                    IF zy>upperlim THEN ymove=1
25                        IF zy<lowerlim THEN ymove=-1
26            IF zy>lowerlim-0.01 AND zy<upperlim+0.01 THEN ymove=0
27                                ypos=ypos+ymove
28                                    IF ypos<1 THEN ypos=1
29                                    IF ypos>201 THEN ypos=201
30                                        maprec(xpos,ypos)=1
31                  WRITE#1,xpos,ypos,maprec(xpos,ypos)
32                        yellip=201-ypos
33         GRAPHIC ELLIPSE (xpos,yellip)-(xpos+2,yellip+2),%BLUE,-1
34                            ntrial=ntrial+1
35 upperlim=upperlim0-0.0*ntrial/15000
36                            IF ntrial>15000 THEN 200 ELSE 100
37   200 REM *** continue
38                CLOSE
39                    GRAPHIC SAVE "c:antwalkP0"
40                    SLEEP 7000
41                    END
42   END FUNCTION
```

Please note that this code graphically presents the progress of the ant toward the food source; the actual drawing is done by line *33*, **GRAPHIC ELLIPSE**, which draws a small blue ellipse at each node occupied by the ant. We will look at two cases obtained with this logic—in the first no "learning" occurs, i.e., the probabilities of a movement in each direction are equal. This is the situation in the code as it appears above. For the second case an enhanced pheromone trail is simulated by increasing the probabilities of motion to the right, and up, promoting something approaching diagonal movement. We are, in a very crude manner, allowing the ants to "learn" as the number of discrete steps increases. There is an important distinction here: In a normal random walk each step is completely independent of previous outcomes (this is called a Markov process). But in the logic presented here prior experience is allowed to affect future trials, as the probability of positive moves up and to the right can increase during the course of the simulation as illustrated in Figure 1.10.

As you can see in the left-hand side of this figure, in 15,000 random steps employing equal probabilities in each direction, the ant may never find the food source at all. But, by increasing the probability of making discrete steps both up and to the

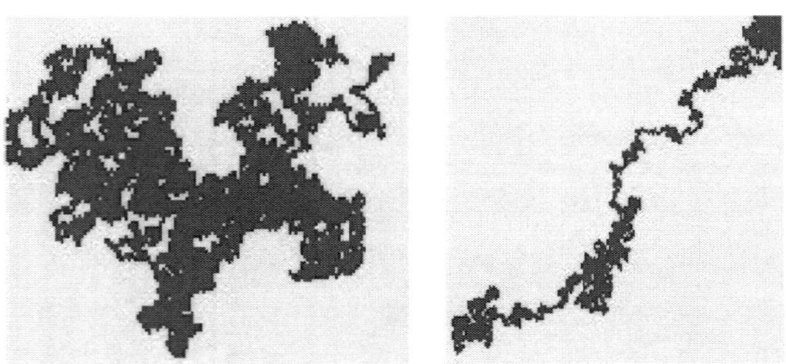

Figure 1.10 Ant travel simulated as a random walk, with no "learning" on the left. On the right, the pheromone trail is reinforced by increasing the likelihood of positive steps both up and to the right. The colony is located at index (51,51), and the objective (the food source) is located at the upper right-hand corner (201,201).

right as the *total number of trials increases*, the ant is more likely to move in the desired diagonal direction. We are simulating "learning" (more precisely, an enhanced pheromone trail) by allowing the probability of a desired step to increase with each trial; this is achieved by adjusting the parameter, **upperlim** as shown in boldface in the code. For example, by setting line *35* to `upperlim=upperlim0-0.5*ntrial/15000`, we can change this limit from 0.66 (the initial value) to a much lower value. This significantly increases the likelihood of positive steps both up and to the right, producing a tendency toward diagonal motion. Of course, we can obtain an endless set of intermediate results by simply changing how quickly the ant "learns" the preferred direction; for example, we could employ `upperlim=upperlim0-0.02*ntrial/15000` to simulate an ant that is a very slow learner.

The student interested in random walks should know that the literature in this field is vast and the important applications nearly limitless. A detailed treatment of random walks is beyond the scope of this course, but should you like to know more about this topic a readable introduction is Jens Feder's book, *Fractals* (Plenum Press, New York, 1988). Feder discusses cluster growth by diffusion limited aggregation, viscous fingering in porous media, and percolation, among other topics. Chapters 3, 4, 7, and 9 provide useful introductions to several aspects of the random walk. We note that since the 1980s diffusion limited aggregation (DLA) has been used as a basis for modeling secondary and tertiary oil recovery from depleted reservoirs. For an insightful historical perspective I recommend that interested students explore the contributions of Stan Ulam and John von Neumann to the development of nuclear weapons.

1.10 Empiricisms and Data Interpretation

A system of interest to us can be so complex that even though we might be able to write down the underlying physical laws, there will be absolutely no thought of actually obtaining a solution. Alternatively, we might be given data with the requirement that a framework for interpretation be developed—what do the data mean, and can they be used to formulate a coherent picture of what has taken place? In either case, the conventional approach of writing out some model equation and then seeking a solution will not be applicable. We are forced to adopt different viewpoints, and we will spend just a little time here considering some of the alternatives.

1.10.1 Energy Output of Candle Flame

We first contemplate a scenario in which experimental data have been supplied to us, and we are required to develop a model that accurately describes those data. In particular, we use a candle flame to heat 300 g or water (in an uncovered beaker) from an initial temperature of about 8 °C to a little over 40 °C. The heating of the water takes place over a period of about 20 min. We will go through the reasoning process that an experienced problem solver might employ, in detail. We examine the experimental data:

t (s)	0	100	200	400	600	800	1,000	1,200
T (°C)	8.2	10	13.6	21	27.1	32.7	37.7	41.9

We note immediately that a single-point temperature is reported, which presumably is the bulk or average water temperature. We can infer that a microscopic balance—a distributed parameter model—is quite inappropriate. There would be no mechanism by which we could critically evaluate the model. Moreover, once the water temperature exceeds that of the surroundings, thermal energy will be lost by several processes including natural convection with induced turbulent flow. Those processes will certainly be three dimensional and time dependent, and we have no practical capability for describing them successfully. It is clear that a macroscopic balance is our only realistic chance to accurately mimic the experimental data. Verbally, our balance will take familiar form: [*Rate of thermal energy supplied*] – [*Rate of loss to surroundings*] = [*Accumulation*], and symbolically:

$$Q_{in} - hA\Delta T = MC_p \frac{dT}{dt}. \tag{1.70}$$

Q_{in} is the energy supplied by the candle, h the heat transfer coefficient (between the heated water and the surroundings), A the appropriate surface area, M and C_p are the mass and heat capacity of the water, respectively. It seems likely that the energy output of the candle is constant, or very nearly so, and it has been reported in the literature that a candle flame produces about 40 W (or 9.55 cal/s). In the early portion of the experiment no heat is lost to the surroundings—in fact, some thermal energy is gained since the air temperature was about 20 °C. However, this is negligible over the first few hundred seconds and since we know both M and C_p, we can obtain a quick estimate for Q_{in}; this turns out to be about 9.84 cal/s (the student should verify this).

Next, we need to obtain a value for the heat transfer coefficient, h. There are no *a priori* means for us to do this; there are only a limited number of cases for which the heat transfer coefficient can be determined from first principles. Therefore, we will use the experimental data at large t to arrive at an estimate for h. We know that the surface area at the water–air interface was 78 cm^2 and that most of the thermal energy lost will come from that surface. The experimental data show that at $t = 1{,}000$ s, the water temperature was about 37.7 °C, and at $t = 400$ s, $T = 21$ °C. We can either integrate the macroscopic balance to obtain $T(t)$ or we can estimate the slope, dT/dt, at specific times. Either way, the heat transfer coefficient can be shown to be about 0.004 cal/(cm^2 s °C)—the analyst should immediately question whether or not this is a reasonable value. For heating or cooling with air, the heat transfer coefficient for low air velocities has been found to be *on the order of* 10 Btu/(hr ft^2 °F), about 56.8 W/(m^2 K), or about 0.0014 cal/(cm^2 s °C), which is *very roughly* in accord with our estimate from the experimental data. Let us now put the pieces of our macroscopic model together to see if it can describe results illustrated in the figure above.

t (s)	0	100	200	400	600	800	1,000	1,200
T (°C)	8	12.3	16.22	22.86	28.21	32.52	36.0	38.81

Compare these data with those shown in the previous table; notice that the projected temperature is too high at small t's, and too small at large t's. What does this suggest to you? Perhaps the estimate for the heat transfer coefficient is too large? Suppose we adjust h to 0.0032 cal/(cm s °C):

t (s)	0	100	200	400	600	800	1,000	1,200
T (°C)	8	12.1	15.88	22.56	28.21	33.0	37.05	40.48

We see that these results are more nearly proximate to the experimental data and certainly improved at larger t's. Moreover, the macroscopic model turns out to represent the data fairly well. There is a final consideration the analyst should take into account: Does the model correspond to our physical expectations as time becomes large? Obviously the temperature of the water cannot exceed 100 °C; with the revised value for h, we see $T(t \to \infty) = 59.4$ °C.

1.10.2 Correlations Developed from Experimental Data

Next, let us think about heat transfer in a tube with turbulent flow. Since turbulence is always three dimensional and time dependent, we would need all three components of the Navier–Stokes equation, the continuity equation (conservation of mass), and the energy equation to completely describe the system. A solution—*without any approximation*—would therefore require that we handle all five of the listed partial differential equations simultaneously, including accurate boundary and initial conditions. This is an impossibility and it is likely to remain so for many decades to come, perhaps forever. Nevertheless, processes must be designed and their anticipated performance estimated. An approach that has often been taken historically in such situations is the development of a correlation based upon experimental data. In the case of heat transfer in tubes, the Dittus and Boelter (1930) correlation provides the Nusselt number in terms of the Reynolds and Prandtl numbers:

$$Nu = \left(\frac{hd}{k}\right) = 0.023\mathrm{Re}^{0.8}\,\mathrm{Pr}^{0.4}, \tag{1.71}$$

where $\mathrm{Re} = \dfrac{d < V > \rho}{\mu}$ and $\mathrm{Pr} = \dfrac{\mu C_p}{k} = \dfrac{\nu}{\alpha}$. Eq. (1.71) allows us to estimate the heat transfer coefficient, h, and then use that value to find the rate at which thermal energy is transferred to the fluid. For example, if Re = 50,000 and if Pr = 10, then the Nusselt number is about 332. We find a very similar situation for mass transfer occurring in wetted-wall towers; the Gilliland-Sherwood (1934) correlation is

$$Sh = \left(\frac{Kd}{D_{AB}}\right) = 0.023\mathrm{Re}^{0.83}Sc^{0.44}. \tag{1.72}$$

In this case the Sherwood number (*Sh*) is given as a function of the Reynolds and Schmidt numbers, the latter is the ratio of the kinematic viscosity and the diffusivity, $Sc = \frac{\nu}{D_{AB}}$. Eq. (1.72) will allow us to obtain the mass transfer coefficient, *K*, and thereby determine the rate of mass transfer. These two correlations illustrate a time-tested approach to the solution of extremely difficult problems in the applied sciences: Identify the pertinent dimensionless groups (either by dimensional analysis of the governing equations of change, or through use of a technique like the *Buckingham Pi method*), use those dimensionless groups to form a correlation, and then use a statistical approach to select the "best" values for the coefficients and exponents using available experimental data. *Please note that it is inappropriate for us to refer to such an exercise as modeling*. What we are doing in such cases is fitting an empirical relationship (often a power law) to measured or observed data. There is nothing inherently wrong with this strategy, and it has been used successfully in engineering practice for more than 100 years. But it is critical that we remember that a correlation thus developed is *valid only for the conditions covered by the experimental work*. We cannot safely extrapolate these results to systems unlike the original experiments.

1.10.3 Frictional Resistance for Transitional Flow

Unfortunately there are cases where even the correlation approach may not be possible. Consider, for example, the frictional resistance experienced by a fluid flowing through a cylindrical tube. And furthermore, assume we are primarily interested in the transition region, i.e., that range of Reynolds numbers in which the flow ceases to be laminar and becomes highly disordered. Although hydrodynamicists have some understanding of the mechanism by which a flow becomes turbulent, no model has been developed that can fully describe what happens in this process. What we can do however is measure the pressure drop for flow through a specified tube at a given flow rate. The friction factor, f, is related to the pressure drop over a length of tube, L:

$$f = \frac{(P_0 - P_L)}{\rho V^2} \frac{R}{L}. \tag{1.73}$$

Figure 1.11 displays a typical set of data obtained for flow through a polycarbonate plastic tube (which is hydraulically smooth) with a diameter of 0.375 in. (9.525 mm). The experimental effort was concentrated upon Reynolds numbers ranging from about 2,000 to 5,000; since the tested fluid was water and since $d = 9.525$ mm, the approximate range of velocities was about 21 to 52 cm/s.

For a nominal Re of 3,000, the friction factor appears to vary from possibly 0.006 to nearly 0.0107, a value which is 70% larger than the low end. Figure 1.11 illustrates a physically elementary situation for which there is no model, nor is

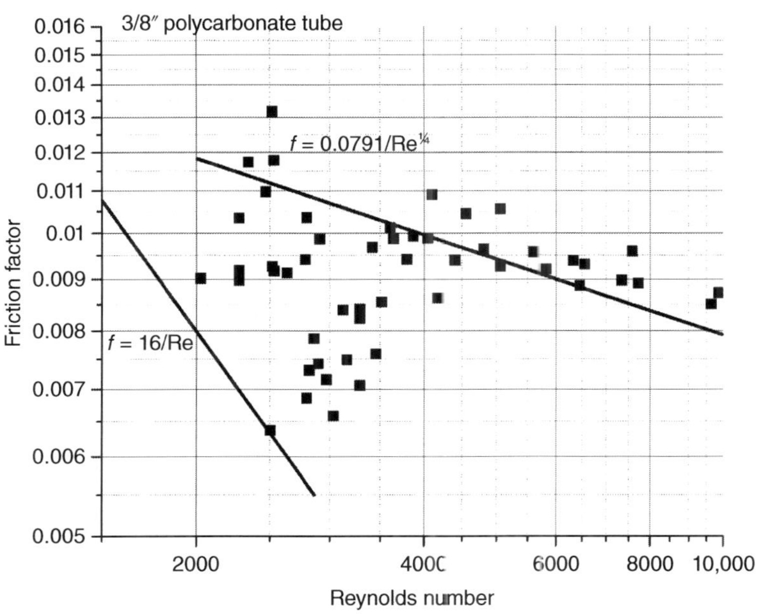

Figure 1.11 Friction factors measured experimentally for flow through a polycarbonate tube with a diameter of 0.375 in. (9.525 mm). Note particularly the scatter of the data in the range 2,000 < Re < 4,000. There is no discernable functional relationship for these data.

there a rational empirical correlation. Naturally, if we had a *sufficiently large data set for a particular apparatus* we might be able to say something about the *probability* that f at Re = 3,000 would be between say, 0.008 and 0.009. This example (merely water flowing through a cylindrical tube) underscores the difference between a deterministic process and one that should be characterized as stochastic. In the transition region the extreme sensitivity to initial conditions (or SIC) yields such large variability in outcome that two trials carried out at the same velocity will rarely (or never) produce identical results. Furthermore, when such experiments are carried out with *viscoelastic* fluids, the results may exhibit *hysteresis*—the results obtained for a sequence of experiments with increasing flow rate may not be the same as those obtained from a sequence with decreasing flow rate. This is a consequence of molecular creep which may occur for certain non-Newtonian fluids.

We conclude that we may be able to say something about the *likely range* for the friction factor, f, but we cannot say with certainty that $f = 0.0085$ at Re = 3,000. Some physical phenomena are not well enough understood (or the governing mathematics are so intractable) that the conventional processes of model development and problem solution are simply inappropriate. Numerous examples—which we pointed out previously—originate from the large variability that is an inherent part of many biological processes.

1.10.4 Speech Recognition

Finally, we will think about a scenario in which some data are received by the analyst along with the requirement that an interpretation of those data be produced *very rapidly*. Speech recognition is a classic example of this problem type, and we will examine the data shown in Figure 1.12, which is a recording of a human voice speaking the word, "*integration*."

There are a few segments in these data where the signal is attractively simple. For example, the first *phone* (a single speech sound) appearing on the left-hand side of the figure is from the "in" part of the word (in'tə-grā'shən) and it exhibits a sharply-defined frequency of about 175 Hz. Note however that much of the balance of the recorded signal is more difficult to characterize in such a direct way. One of the features that does stand out is the high-frequency burst that accompanies the start of the shən-sound of the last syllable. The frequency here is certainly much higher, with components that appear to be around 1,250 Hz. At the very end of the signal the main oscillation returns to about 133 Hz with a small, higher-frequency component superimposed. The question we want to address is: Do these isolated features constitute a "model" or a "pattern" that would permit rapid identification of the word, *integration*? Probably not. We will need a better tool if we are to deal with this case successfully, and toward the end of this book (in Chapter 10) we will consider the treatment and interpretation of time series data and we will learn how useful harmonic analysis can be for this type of situation.

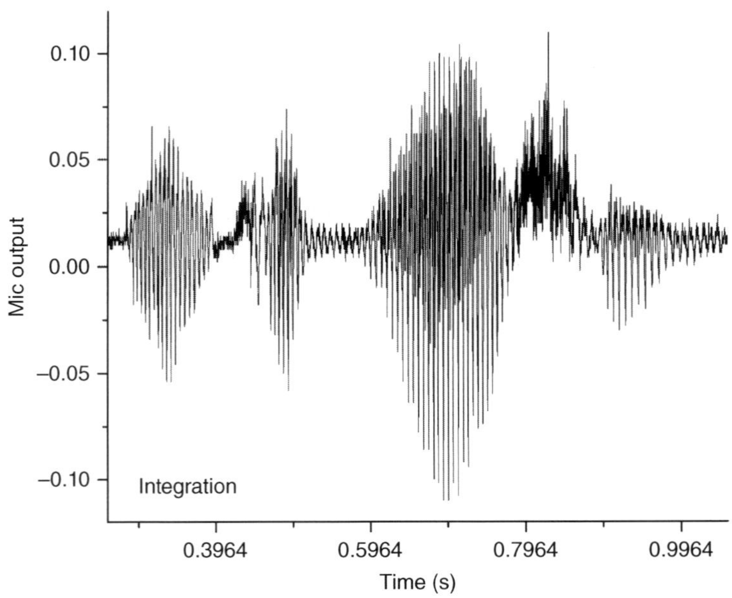

Figure 1.12 A recording of a human voice speaking the word, "integration."

1.11 Conclusion

We have looked at a few examples in which some type of balance (typically from a conservation principle) was used to develop a model for a physical situation of interest. We saw cases in which this technique led to algebraic equations and also to differential equations. This is entirely appropriate since our focus in this course is upon mathematical methods that engineers and scientists routinely use to solve the problems thus formulated. We have also introduced the notion that certain types of problems will not lend themselves to this type of analysis—in some cases, whether it be due to system complexity or the nature of the data that must be interpreted, we will have to employ a different process. Some options for these problem types will be discussed later in this text.

Regardless of the exact nature of the problem, however, there are important aspects to the overall process that we have not mentioned at all and one of those is error.

Consider measurements made by a Bourdon pressure gauge or perhaps a pressure transducer. Such devices can be calibrated with a deadweight test apparatus that may be accurate to within 0.006% of full scale. For a Bourdon gauge with a 500 psi maximum, this would imply an uncertainty of ± 0.03 psi (or ± 0.002 bar) at full scale. Pressure-measuring devices such as these can be easily calibrated and their output corrected, and consequently they are a source of *determinate* error.

In contrast, *indeterminate* errors may arise where neither the magnitude nor the sign of the needed correction will be known, and such errors can originate with both instruments and models. For the former, we can think of a flow meter that has been calibrated at the factory with some standard, perhaps water at 20 °C or air at 273 K and 1 atm pressure. When the process experiences fluctuations in fluid composition, temperature, or pressure (affecting the physical properties of the fluid such as density and viscosity) indeterminate errors will be generated. In the case of models, we may have employed simplifications or boundary conditions that we know are not in accord with reality. For example, we may have assumed that a stirred tank reactor is perfectly mixed, or that a heat transfer surface is at constant temperature. Neither can be strictly correct, and modeling errors thus created can be exceedingly difficult to characterize. In such cases an assessment may need to be based upon prior experiences or engineering judgment. Only in cases where the process is absolutely critical would one consider the extensive experimental investigation necessary to explore and quantify such errors.

We also need to make a final observation by noting that model development, data interpretation, error assessment, and problem solution are merely elements of the larger decision-making process; let's conclude this introduction with an illustration of this broader context using a situation that arises frequently in industrial practice.

Suppose we have a requirement for gas compression. We can assume that the intake and final pressures will be specified, and that the mass flow rate of the compressed gas is dictated by process throughput. What types of questions would we need to address in this situation? One might presume that the critical issues concern the type of compressor and the required power input. However, there are many choices for the compressor (single-stage, multistage, with intercooling or without, reciprocating, centrifugal, etc.). Furthermore, the estimated power requirement will depend upon the assumed thermodynamic pathway (isothermal or isentropic). We will now give some definitive shape to this discussion; suppose we need to compress air from 14.7 psia to 10 atm or 147 psia at a rate of 1,000 ft^3/min (based upon inlet conditions). For a single-stage, *isothermal* compression the power requirement can be estimated:

$$P = (3.03 \times 10^{-5})(14.7)(144)(10^3)\ln\left(\frac{147}{14.7}\right) = 147.7\text{ hp}, \tag{1.74}$$

or 110 kW. For a single-stage *isentropic* compression we find:

$$P = \frac{(3.03 \times 10^{-5})(1.4)}{(1.4 - 1)}(14.7)(144)(10^3)\left[\left(\frac{147}{14.7}\right)^{0.4/1.4} - 1\right] = 208\text{ hp}, \tag{1.75}$$

or 156 kW. And for a *three-stage isentropic* compression with *intercooling*:

$$P = \frac{(3.03 \times 10^{-5})(1.4)(3)}{(1.4 - 1)}(14.7)(144)(1000)\left[\left(\frac{147}{14.7}\right)^{0.4/(1.4)(3)} - 1\right] = 165.1\text{ hp} \tag{1.76}$$

or 123 kW. What compressor power should we specify? We have three solutions for (one aspect of) the problem, but we may be only a little closer to producing an actual answer. Consider that:

- The thermodynamic paths chosen are merely idealizations; the real compression process will be neither isothermal nor isentropic.
- Since the estimates for required power are idealized they do not account for any dissipative effects; the compressor efficiency may be much less than 100%.
- Other factors may weigh upon the decision-making process, including capital cost, operating cost, reliability and maintenance, labor requirements, safety issues, and so on.

This example underscores the fact that while we may be able to select and employ models for a real process, and then find solutions using those models, rarely will that process result in a global answer for the questions being investigated. Most textbook problems and examples are far too "clean" in the sense that there is usually a narrow focus that will yield *one correct solution*. Real situations simply do not work out that way, and since models always entail idealizations, it is prudent to maintain a little skepticism until corroboration is at hand. It is essential that we try to understand the limitations of *both* the model and the solution procedure and at the same time, not lose sight of the complications that distinguish a real-world problem from a textbook construct.

Problems

Investigate the following scenarios and either formulate or simply describe an appropriate modeling approach or data analysis strategy for each, 1.1 through 1.16. *Do not* try to solve these problems unless directed to do so. Problems 1.17 through 1.21 do require specific responses or solutions.

1.1 A colony of prairie dogs numbers 47 individuals at time, t. How many prairie dogs might there be at time $t = t + 30$ days?

1.2 A contaminant that is slightly soluble in water lies at the edge of a flowing stream. How might we predict the concentration distribution at a downstream distance of 1,000 m?

1.3 How could we characterize the number of automobiles expected to arrive at a busy intersection over a two-hour period during mid-day?

1.4 Students drop a tennis ball from the top of a building 375 ft high. What would the velocity of the tennis ball be at impact with the ground?

1.5 A equimolar stream of liquid hydrocarbons containing C_4's, C_5's, C_6's, and C_7's is flashed (partially vaporized) to split off the butanes. If the temperature and pressure are 130 °F and 19 psi, how much vapor will be produced and what will the composition be?

1.6 An intravenous injection of a pharmacologically active agent gets distributed between plasma and tissue (and some is lost by elimination processes). How might we predict the concentrations in both plasma and tissue (as functions of time)?

1.7 Water flowing in a network of pipes reaches a "T" (tee) fitting and is split into two streams. One of the lines leaving the "T" has a diameter of 5 cm and the other line's diameter is 7.5 cm. What system of equations might be used to predict the distribution of flow at the "T" fitting?

1.8 A few data for ¼-mi acceleration times (drag racing) for stock cars and trucks are provided below. If a car weighs 3,000 lbs, what horsepower might be required to provide an ET (elapsed time) of 11.75 s?

Weight	Horsepower	ET(s)/speed(mph)
5,680	300	17.3/81.2
5,440	367	16.2/84.6
3,883	426	13.0/109.8
3,300	460	12.0/119
3,220	330	13.5/107
2,880	173	15.7/91
2,680	170	15.8/90

1.9 Blood flows through a major artery ($d = 0.5$ cm) at a mean velocity of about 40 cm/s. How might we estimate the shear stress at the artery wall? Is blood a Newtonian fluid?

1.10 Air flows past a heated flat plate at 20 m/s. What data and equations should we employ to estimate the rate at which heat is transferred to the moving air stream? How might one determine whether or not the temperature dependence of the fluid properties, that is, $\mu = \mu(T)$ and $\rho = \rho(T)$, should be taken into account?

1.11 You have been directed to study neuron excitability and the generation of electric pulses (spikes) in nerve tissue. What characteristics must a model of this phenomenon have (you may want to start by looking at the FitzHugh–Nagumo model)?

1.12 Atherosclerosis is a disease of the arteries in which lesions form on the artery walls (actually the *arterial, or tunica, intima*). The process is not well understood, but as these lesions develop they begin to obstruct blood flow, lead to plaque formation, and ultimately result in death by heart attack or stroke. Because atherosclerosis is a leading cause of death in much of the developed world, there is great interest in modeling lesion development. It is known that the lesions tend to form in curved arteries or near arterial branches, and it is also known that there is a strong correlation between cholesterol levels and the likelihood of atherosclerosis. What might a model of lesion formation and plaque growth look like?

1.13 It has been suggested oscillations might occur in the operation of a nuclear reactor, perhaps when localized heating of a reactor component resulted in thermal expansion, changing the geometry of the reactor and thereby affecting reactivity. Critics of nuclear power have even suggested that a catastrophic event might be triggered by such a phenomenon. At least one model has been developed purporting to show this mathematically; see Thompson and Thompson, "A Model of Reactor Kinetics" published in *Nuclear Science and Engineering*, Vol. 100, p. 83, September 1988. What are the essential elements of such models, and are they capable of assessing the probability of a catastrophic event? What appears to be the major deficiency of the Thompson–Thompson model?

1.14 Uneven spots in the rails, or eccentricities in the wheels, can lead to side-to-side rocking of cars in normal railroad operations. Suppose you have been assigned the task of figuring out if such motions could ever be amplified, resulting in load shifting, over stressing the car's structure, derailment, etc. What kind of model might be formulated for this purpose, and what would its essential components be?

1.15 Recently researchers from Haverford College (Pennsylvania) have successfully attached "head-cams" to falcons in order to study their hunting behavior. They have obtained video that reveals falcons carrying out attacks on flying crows. The video footage shows that falcons in pursuit of crows fly in such a way that the crow seems nearly stationary (and almost centered in the field of view). Can a mathematical model be formulated for a falcon's flight during pursuit of a flying crow, and what would such a model look like? *Hint: Curve of pursuit.*

1.16 On December 7, 2009, a pressure vessel operated by NDK Crystal Inc. in Illinois ruptured violently resulting in the ejection of large pieces of steel from the property and one off-site fatality. The NDK facility produced synthetic quartz crystals in a batch process by mixing silica with sodium hydroxide in steel pressure vessels 50 ft tall, about 2 ft in diameter, with a wall thickness of 8 in. The mixture was sealed in the vessels for 150 days at temperatures between 570 and 750 °F and pressures of exceeding 20,000 psi. The company felt that the conditions inside the vessels would result in the formation of a protective layer of acmite (aegirine, $NaFe(SiO_3)_2$) on the inside wall, thus protecting the steel from further corrosion. However, no internal inspections were ever carried out, so this hypothesis was unverified. An investigation by the CSB (Chemical Safety Board) following the accident revealed that the conditions in the vessels were favorable for stress-corrosion cracking. Of course such a process would significantly compromise the pressure vessel integrity and increase the likelihood of a catastrophic incident. Note that *Engineersedge.com* has an online calculator that one can use to estimate the minimum thickness of pressure vessel walls: If the internal pressure is 20,000 psi, the internal radius is 12 in., the maximum allowable stress for the steel is 42,000 psi, and the joint efficiency is 100%, then the indicated wall thickness is 8 in. *But*, if stress-corrosion cracking was occurring on the interior walls (as revealed post-accident) then this estimate would no longer be valid.

An engineering disaster like this one provides numerous learning opportunities, as well as a chance to make appropriate design changes to improve the safety of similar processes. Three obvious lessons of the NDK accident are:

1) Pressure vessels should be designed according to ASME code.
2) Interior inspections of high-pressure vessels should be carried out annually.
3) Material selection is critical in high-pressure corrosive environments.

We would like to address the following issues: Were the conditions in the pressure vessel favorable for the formation of an acmite layer on the interior wall? Can stress-corrosion cracking be modeled, and if so, what does such a model look like? Are there steel alloys capable of withstanding high-temperature caustic environments?

1.17 Suppose a tennis ball is released from great height; the ball is subject to gravity, buoyancy, and drag forces. It accelerates and asymptotically approaches its terminal velocity, and its dynamic behavior is governed by the force balance:

$$m\frac{d^2y}{dt^2} = \sum F = mg - F_B - F_D.$$

The mass of the tennis ball is $m = \frac{4}{3}\pi R^3 \rho_b$, and the drag force is written as $F_D = AKf$, where A is the frontal area of the sphere, K is the kinetic energy of the fluid per unit volume, and f is the drag coefficient. Accordingly,

$$AKf = \pi R^2 \frac{1}{2}\rho V^2 f(V)$$

where the velocity of the ball, $V = \frac{dy}{dt}$. Often in problems of this type it is *assumed* that the object or projectile accelerates so quickly that the drag coefficient can be taken to be constant, i.e., independent of velocity. If the Reynolds number is large enough, $f \approx 0.44$. This of course raises the question we might want to address: How much error do we introduce into the computation of the acceleration phase by taking f to be constant?

We are given the following: $R = 3.25$ cm, $\rho_b = 0.3964$ and $\rho = 0.0012$ (both g/cm^3). The acceleration of gravity is 980 cm/s^2.

i) Show that the terminal velocity of the tennis ball is 2,521 cm/s.
ii) If we take f as constant (0.44), how much time is required for the ball to attain 95% of its ultimate velocity (2,395 cm/s)? *Answer: 4.725 s.*
iii) The drag coefficient will be *very large* initially and then decrease as the velocity of the ball increases. Can you think of a way to assess whether or not we must account for $f(V)$—*without* solving the force balance?

1.18 An experiment was carried out in which a horizontal steel plate measuring $10.31 \times 6.31 \times 0.32$ cm was heated from below with an incandescent lamp (50 W) at a distance of 5.4 cm. The mass of the plate was 162 g, so the density of the steel is about 7.78 g/cm^3. The heat capacity of mild steel is about 0.111 cal/(g °C). A small type-T thermocouple was used to record the temperature of the plate as a function of time. These data are shown in Figure P1.18.

Develop a simple macroscopic model for this process and use it to address the following questions:

1) What fraction of the lamp's output (50 W) was utilized in heating the steel plate (show that this is about 22 W, or 44%)?
2) If the laboratory temperature was 25 °C, what is the approximate value of the heat transfer coefficient between the plate and its surroundings?
3) If the experiment is allowed to proceed indefinitely, what will the ultimate temperature of the plate be according to your model?

Can you think of any way to determine what fraction of the total output of the lamp is in the visible portion of the spectrum (from 0.35 to 0.75 μm)?

1.19 *Attempt this problem only if you have prior experience with automatic process control.* In Section 1.7 we described how a feedback control loop could be discretized to facilitate direct, time-domain simulation. Such an approach makes it easy to assess the effects of changes in control strategies and identify the best compromise between response and stability. We will try this in this exercise. Suppose we have a second-order system with $\tau = 7$ and $\zeta = 1$. We want to use PI control, but we have a nonlinear final control element (a control valve), and a transportation delay in the feedback

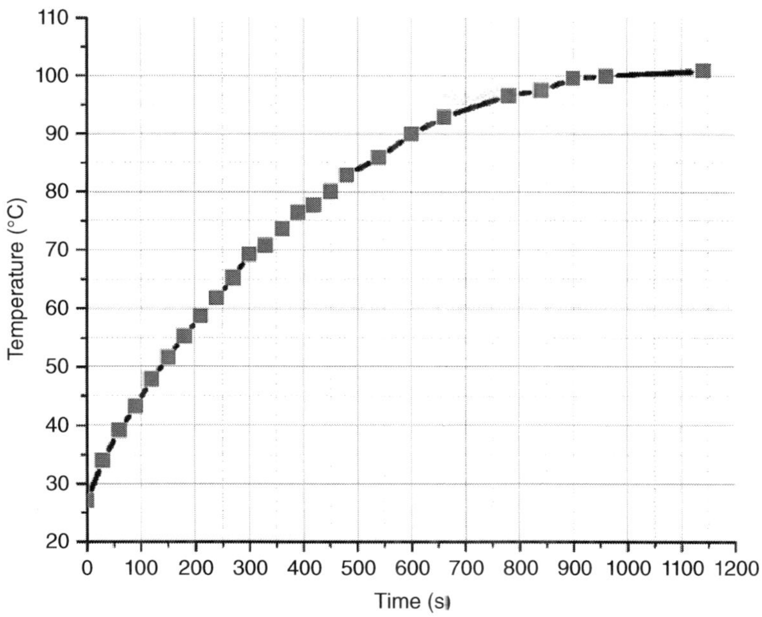

Figure P1.18 Temperature history of the plate during radiant heating.

loop (this is common in industries where sampling requires transport through a tube to an instrumentation room). The second-order system has the form illustrated in Section 1.7:

$$\tau^2 \frac{d^2y}{dt^2} + 2\tau\zeta\frac{dy}{dt} + y = x.$$

Let's assume that the control valve is quick opening such that the output (which is sent to the second-order system) has the form $x = \sqrt{P}$, where P is the signal provided by the controller. We fix the integral time, $\tau_I = 1.5$, and vary K_C. We will fix $\Delta t = 0.2$ and assume the delay in the feedback loop is $6\Delta t$; we should obtain something like the following (Figure P1.19).

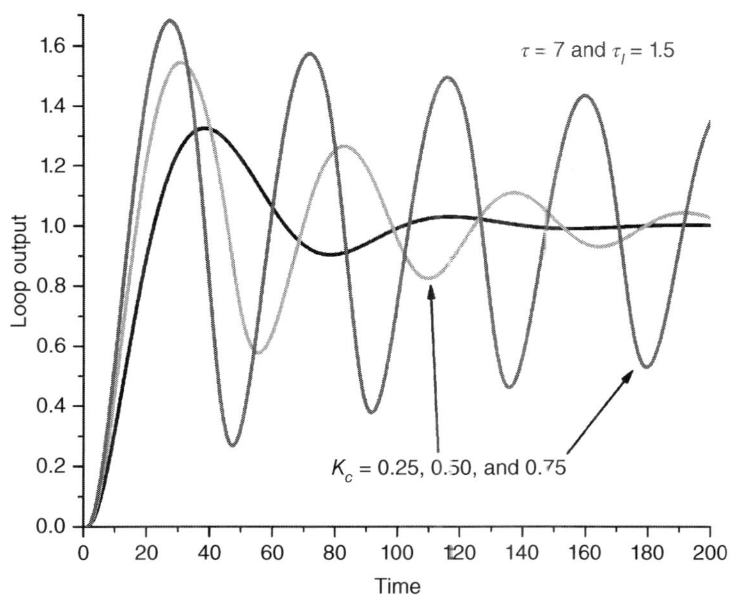

Figure P1.19 Simulated loop output for PI control of a second-order system with a time delay in the feedback loop and a nonlinear final control element (valve). Three different controller gains were employed, 0.25, 0.50, and 0.75; notice the impact of increasing K_C upon rise time, overshoot, and oscillation.

It is clear from this figure that at some point an increased K_C will result in instability. Prepare a discretized simulation and determine the critical controller gain by trial and error. Then, vary the transportation lag and explore the consequences.

Answer: The critical controller gain is $K_C = 0.902$.

1.20 According to CDC estimates, about 35 million American adults have chronic kidney disease, and of that number about 800,000 have end-stage renal disease (ESRD). Most of us recognize that the human kidney is a filtration unit that removes waste products from the blood, and indeed, a healthy kidney will process about 1 l of blood per minute. But in addition to acting as a "simple" filter, the kidney also makes and regulates hormones that affect blood pressure, it regulates the amount of salt in the circulatory system, and it works to maintain fluid levels in the body. The blood enters a kidney through the renal arteries and returns through the renal veins. Each kidney contains about 10^6 *nephrons*, and each has a small filtration unit called a *glomerulus*. One measure of kidney function is the *glomerular filtration rate* (GFR), and a normal GFR is around 100 ml/min per 1.73 m^2 (a GFR of 15 ml/min per 1.73 m^2 is an indicator of kidney failure). The GFR is difficult to measure directly, so physicians often estimate it (eGFR) from the measured level of serum creatinine in the blood (age and gender also figure in the calculation). Should you happen to know your creatinine level you can use *Kidney Health Australia's* website to determine your eGFR. Patients with advanced kidney disease are kept alive with *dialysis* while waiting for transplant—unfortunately, the number of available kidneys each year is only a few percent of the total number needed. Dialysis is a mass transfer process using a semipermeable membrane with blood on one side and dialysate on the other. The unit is operated countercurrently to maintain a more uniform driving force for diffusion. The pores in the membrane allow passage of waste products (from the blood into the dialysate) like urea and creatinine, but they are small enough to block red blood cells and larger proteins. Although hemodialysis may achieve sufficient filtration to keep a patient alive, it cannot replicate all kidney functions. Therefore, much work has been focused upon constructing artificial kidneys, and very recently, researchers have begun to incorporate living kidney cells upon an artificial matrix in efforts to devise kidney units that better mimic real kidney function. Growing mammalian cells in an artificial environment has always been challenging because of their fragile nature; while robust bacteria can be grown nearly anywhere, mammalian cells generally cannot (though certain types of mammalian cells, like Chinese hamster ovary, or CHO cells, used for the production of recombinant therapeutic proteins, have been routinely cultured both in suspension and upon surfaces). Investigate the current status of artificial kidney research and prepare a summary of the principal issues obstructing development of a long-term implantable kidney replacement.

1.21 *This is a group project ideally suited to three-person teams.* In Section 1.1 we mentioned *catastrophe theory* and described in vague terms how it might be used for the analysis of real processes. The descriptions provided did not allow the reader to develop any somatic understanding of catastrophes however, so this exercise is intended to address that deficiency through the construction and exploration of the *Zeeman catastrophe machine* (E. C. Zeeman, A Catastrophe Machine, pages 276 through 282 in *Towards a Theoretical Biology*, C. H. Waddington, Editor, 1972). *You will need*: several square feet of stiff cardboard, a protractor, a ruler with a scale in cm and mm, pushpins, a cutting tool like an Exacto™ knife, and several rubber bands (as physically similar as possible). Begin by cutting a rectangular piece of cardboard roughly 75 cm long and about 20 cm wide. Draw a centerline on the long axis of the rectangle for its entire length. Next, cut a circular disk (be careful) with a diameter of about 16 cm. The circular disk will be fixed to the center of the rectangle using a large pushpin and another large pushpin is installed on the centerline at point "R" as illustrated here:

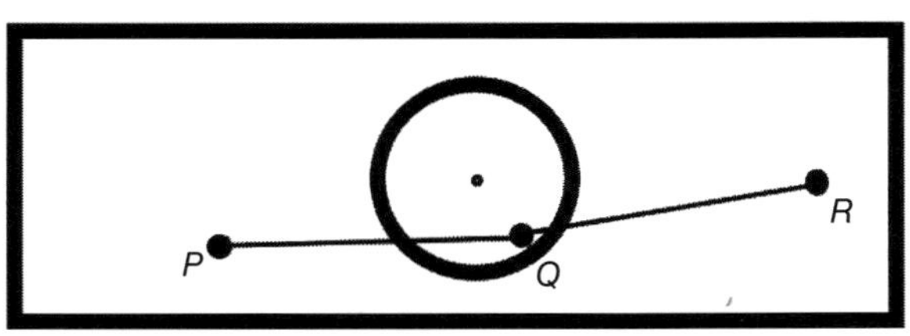

An additional (smaller) pushpin is installed near the edge of the circular disk at point "Q," and one rubber band connects the pins at Q and R. Another rubber band is attached at Q and its free end corresponds to point "P" which will be moved to different positions on the rectangle. Generally, a small movement of P will yield a small rotational movement of Q. However, at certain locations *a small movement of P will produce a jump to a new equilibrium position*. This is the phenomenon we wish to explore.

Each time position P is changed we will make three measurements: The length of rubber band "1" (from Q to R), the length of rubber band "2" from P to Q, and the angle (θ) between position Q and the centerline of the apparatus drawn from R to the center of the rotating disk. Any time a jump transition to a new equilibrium position occurs, we mark the position P on the rectangular base with a Sharpie™.

It's probably clear to you that every time we move position P, we are changing the potential energy of the system which is approximately given by

$$PE = \frac{1}{2}Y\left[(L_1 - L_0)^2 + (L_2 - L_0)^2\right],$$

where Y is Young's modulus, L_1 is the length of rubber band 1, L_2 is the length of rubber band 2, and L_0 is the length of the unstretched rubber bands. Although the modulus of elasticity, Y, is not really constant for rubber bands, we will assume it is. Our objectives in this experiment are: 1) determine the locus of positions, P, for which jump transitions occur, and 2) complete entries in an enlarged version of the table below.

Position (P)	L_1	L_2	θ	PE/Y	Jump?

It is possible to estimate the modulus, Y, from the slope of the linear portion of a stress–strain diagram—this is a common experiment in introductory physics. A rubber band is allowed to hang vertically from a support, weights are attached to the bottom end, and the elongation is measured. Unfortunately, the strain (deformation) will cause the cross-sectional area of the rubber band to change, and so measurements with a micrometer are necessary to obtain numerical values of stress. Although the estimation of Y is interesting, this is not required for our exercise.

As you experiment with your apparatus you will rapidly discover that for some positions, P, only one equilibrium position is possible; however, in a very specific region (which we wish to identify) two equilibrium positions will be possible.

We have two larger goals in mind here:

1) Identification of the shape (locus of points) of the "envelope" inside which two equilibrium positions can occur.
2) Connection between the angle, θ, and the occurrence of a catastrophe.

References

Bird, R. B., Stewart, W. E., and E. N. Lightfoot. *Transport Phenomena,* revised 2nd edition. John Wiley & Sons, New York (2007).

DePriester, C. L. (1953). Chemical Engineering Progress. *Symposium Series* 49: 1.

Dittus, F. W. and Boelter, L. M. K. (1930). University of California Publications. *Engineering (Berkeley)* 2: 443.

Dorigo, M., DiCaro, G., and Gambardella, L. M. (1999). Ant Algorithms for Discrete Optimization. *Artificial Life* 5: 137.

Eide, A. R., Jenison, R. D., Northup, L. L., and Mickelson, S .K. (2008). *Engineering Fundamentals and Problem Solving*, 5the. Boston: McGraw Hill.

Feder, J. (1988). *Fractals*. New York: Plenum Press.

Gilliland, E. R. and Sherwood, T. K. (1934). Diffusion of Vapors Into Air Streams. *Industrial and Engineering Chemistry* 26: 516.

Glasgow, L. A. (2010). *Transport Phenomena: An Introduction to Advanced Topics*. New York: John Wiley & Sons.

Hanna, O. T. and Sandall, O. C. (1995). *Computational Methods in Chemical Engineering*. Upper Saddle River, NJ: Prentice Hall.

Himmelblau, D. M. and Bischoff, K. B. (1968). *Process Analysis and Simulation: Deterministic Systems*. New York: John Wiley & Sons.

Lange, N. A., editor. *Handbook of Chemistry,* revised 10th edition. McGraw-Hill, New York (1961).

Matsuura, T. and Kato, M. (1967). Concentration Stability of the Isothermal Reactor. *Chemical Engineering Science* 22: 171.

Saunders, P. T. (1980). *An Introduction to Catastrophe Theory*. Cambridge: Cambridge University Press.

Syslo, M. M., Deo, N., and Kowalik, J. S. (2006). *Discrete Optimization Algorithms with Pascal Programs*. New York: Dover Publications.

Thompson, J. M. T. (1982). *Instabilities and Catastrophes in Science and Engineering*. Chichester: John Wiley & Sons.

Weast, R. C., Editor-in-Chief. *CRC Handbook of Tables for Mathematics*, revised 4th edition. CRC Press, Cleveland, OH (1975).

2

Algebraic Equations

2.1 Introduction

Algebraic equations are commonly encountered in science and engineering, and a few examples of where they originate include material balances for separation processes, force resolution in structures, flow in pipe networks, application of Kirchoff's rules to electric circuits and networks. radiative exchange in enclosures, solution of discretized differential equations, and balances in chemical equilibria. Though such problems are often thought of as elementary, cases can arise that offer greater challenge than the analyst might expect.

It is impossible to know exactly when an algebraic equation was solved for the first time, but there is evidence indicating that quadratic equations were solved by the Babylonians perhaps 3,700 years ago. Heath (1964) notes that the "father" of algebra, Diophantus of Alexandria, authored the *Arithmetica* in 13 books in the third century CE. Six of the original 13 books still exist, and Heath produced their English translations in 1885 (with the second edition published in 1910). In the *Arithmetica*, Diophantus solves determinate equations of the first and second degree; for quadratic equations he sought only rational, positive solutions in either integral or fractional form. For example, he gives

$$325x^2 = 3x + 18, \tag{2.1}$$

and concludes that $x = 78/325$ or $6/25$. Given an equation of the form $ax^2 - bx + c = 0$, Diophantus would multiply by a to obtain $a^2x^2 - abx + ac = 0$, and then write the solution as

$$ax = \tfrac{1}{2}b \pm \sqrt{\tfrac{1}{4}b^2 - ac}. \tag{2.2}$$

In the twenty-first century nearly any middle-school student confronted by a quadratic equation like

$$x^2 + 2x - 15 = 0, \tag{2.3}$$

will immediately write the quadratic formula,

$$x = \frac{-b \pm \sqrt{b^2 - 4ac}}{2a}. \tag{2.4}$$

Clearly,

$$x = \frac{-2 \pm \sqrt{4 + 60}}{2} = -1 \pm 4;$$

i.e., $x = -5$ and $x = +3$. As we noted above, some determinate equations of first and second degree have been successfully dealt with for several thousand years. However, according to Heath only one *cubic* equation is solved in *Arithmetica*:

$$x^2 + 2x + 3 = x^3 + 3x - 3x^2 - 1, \tag{2.5}$$

which is readily rewritten as

$$x^3 - 4x^2 + x - 4 = 0, \quad \text{or} \quad x(x^2 + 1) = 4(x^2 + 1); \tag{2.6}$$

Diophantus notes that $x = 4$, but it appears that he knew of no general method of solution for cubic equations. Such third-degree equations are much less mysterious today; consider the equation,

$$x^3 - \tfrac{1}{2}x^2 - 6x - \tfrac{9}{2} = 0, \tag{2.7}$$

Problem Solving in Engineering: Analytical Mathematics and Numerical Analysis, First Edition. Larry A. Glasgow.

Companion website: www.wiley.com/go/glasgow/problemsolvingengineering

which we will write as $x^3 + px^2 + qx + r = 0$. If we take $x = y - p/3$, where $p = -1/2$, the second-degree term is eliminated resulting in the form:

$$y^3 + ay + b = 0. \tag{2.8}$$

The solution for the cubic equation can now be written in terms of a and b as described in any standard algebra reference book. For example, noting in eq. (2.7) that p, q, and r are $-1/2$, -6, and $-9/2$, respectively, then $a = \frac{1}{3}(3q - p^2)$ and $b = \frac{1}{27}(2p^3 - 9pq + 27r)$. A and B are now calculated from

$$A = \left[-\frac{b}{2} + \sqrt{\frac{b^2}{4} + \frac{a^3}{27}}\right]^{1/3} \quad \text{and} \quad B = \left[-\frac{b}{2} - \sqrt{\frac{b^2}{4} + \frac{a^3}{27}}\right]^{1/3},$$

and the solutions for eq. (2.8) are then $y = A + B$,

$$y = -\frac{A+B}{2} + \frac{A-B}{2}\sqrt{-3},$$

and

$$y = -\frac{A+B}{2} - \frac{A-B}{2}\sqrt{-3}.$$

The reader may want to show that for our example, eq. (2.7), $a = -6.08333$ and $b = -5.50926$, resulting in $x = -1, +3$, and $-3/2$. This is easily accomplished with a scientific calculator; for example, one might use *cSolve*($x^3 - 1/2 \cdot x^2 - 6 \cdot x - 9/2 = 0, x$) on a TI-*nspire*™ (Texas Instruments).

In the types of applications that are of interest to us, it is not much of a stretch to arrive at more "interesting" cases where we have polynomials of higher degree, or products of functions, such as

$$x^4 - 3x^3 + 2x^2 - 5x + 9.76048 = 0, \tag{2.9}$$

or

$$x^4 \exp(-x) = 4. \tag{2.10}$$

Such problems arise in science and engineering all of the time; the solutions for eq. (2.10), by the way, are $x = -1.07967$, $x = 2.975924$, and $x = 5.23573$. This, too, is easily verified with *cSolve*($x^4 \cdot e^{-x} - 4 = 0, x$). Moreover, in technological fields *sets of simultaneous equations*—both linear and nonlinear—are encountered regularly. Our purpose in this chapter is to review some useful techniques for solving problems of these types. Before we do this however, we will look at one of the simplest applications requiring solution of algebraic equations: interpolation.

2.1.1 An Interpolation Example

A common scenario where algebraic equations must be solved—and one with a long history in the applied sciences—is interpolation. Much of the data needed by engineers for generations appeared in tables in engineering handbooks, e.g., thermodynamic properties of substances like ammonia or carbon dioxide. Only rarely would an entry in such a table correspond exactly to the data point that was required; consequently an engineer would have to interpolate. To illustrate the process, suppose we were given a few discrete data points representing some function $y(t)$:

t	0	0.1	0.2	0.3	0.4
$y(t)$	3	3.959	4.683	4.995	4.819

Now assume we needed reasonably accurate estimates for some intermediate points, perhaps $y(t = 0.235)$ and $y(t = 0.35)$. It is clear from cursory inspection that a straight line connecting adjacent points will not give very good results, so as a next step we consider

$$y = a + bt + ct^2. \tag{2.11}$$

With three unknown parameters, a, b, and c, we can fit three of the given data points exactly by solving three algebraic equations. However, we make a slight modification to simplify the work:

$$y = a + b\left(t - \tfrac{1}{10}\right) + c\left(t - \tfrac{1}{10}\right)^2, \tag{2.12}$$

and therefore $a = 3.959$. This leaves us with two unknown parameters requiring solution of two algebraic equations; if we use the points at $t = 0.3$ and 0.4, we find $b = 9.807$ and $c = -23.133$; you may want to verify this, and the equations that must be solved are

$$4.995 - 3.959 = b\left(\frac{2}{10}\right) + c\left(\frac{4}{100}\right) \quad \text{and} \quad 4.819 - 3.959 = b\left(\frac{3}{10}\right) + c\left(\frac{9}{100}\right).$$

Of course the solution can be found several different ways, but if we multiply the first of the pair by 3/2 and then eliminate b by subtraction, we can immediately determine c. Now we can test eq. (2.12) using $t = 2/10$, and our interpolating parabola produces $y(t = 2/10) = 4.708$, an error of 0.5%. For the two points of interest, $y(t = 0.235) = 4.861$ and $y(t = 0.35) = 4.965$. The quality of our interpolation might be assessed graphically for data of unknown origin, but in this example we know that $y(t) = 3 + 2\sin(5t)$. Therefore, $y(t = 0.235) = 4.845$ and $y(t = 0.35) = 4.968$, and our error at $t = 0.235$ is about 0.33% and at 0.35 it is only 0.06%. For this elementary interpolation, solution of the algebraic equations required very little insight; if you would like to try a similar problem as a refresher, review the enthalpy calculation for liquid ammonia immediately below.

Ammonia has long been used for large-scale refrigeration, and many thermodynamic data are available in the literature; the small table below contains the enthalpy of liquid ammonia at four different temperatures:

T (°F)	−60	−20	0	20
h_f (Btu/lb)	−21.2	+21.4	42.9	54.7

Now, suppose you needed an accurate value for h_f at −12 °F. Select an interpolating parabola, and solve the algebraic equations to determine the needed constants and develop an estimate. If you use the last three entries in the table, you should get 29.96 Btu/lb (the *reported value* in the literature is 30 Btu/lb). What does your equation yield at $T = -30$ °F? You should get about 10.76 Btu/lb; the reported value is 10.7.

2.2 Elementary Methods

2.2.1 Newton–Raphson (Newton's Method of Tangents)

This technique was employed in the seventeenth century by Isaac Newton and Joseph Raphson, both of whom viewed it as an algebraic method to be applied to polynomials. The iterative form that is familiar to us, using derivatives of the function, $f(x)$, appears to have been used in the eighteenth century by Thomas Simpson. Newton–Raphson is a powerful tool that, when it works, can be used to solve a variety of nonlinear algebraic equations.

Let us return to eq. (2.9) in Section 2.1. We wish to find a solution for this equation, and we proceed in the following manner: We select an estimate for x, then construct a line tangent to the curve at that point, and extrapolate it to the x-axis where $f(x) = 0$. We use that point as our new estimate and repeat the process. The algorithm for Newton's method is easily obtained by fitting $f(x) = mx + b$ to two points: x_n, with $f(x_n)$, and x_{n+1} where $f(x_{n+1}) = 0$. Of course the slope, m, is simply $f'(x_n)$ and the intercept b is eliminated by subtraction resulting in

$$x_{n+1} = x_n - \frac{f(x_n)}{f'(x_n)}. \tag{2.13}$$

We illustrate with the example from Section 2.1:

$$f(x) = x^4 - 3x^3 + 2x^2 - 5x + 9.76048$$

such that

$$f'(x) = 4x^3 - 9x^2 + 4x - 5.$$

We choose 3 for an initial estimate for x, and the following sequence of values emerges:

3
2.62469
2.39536

2.27549
2.31760
2.22515
2.225001
2.225001

This seems reasonable, but the polynomial we are trying to solve is fourth degree. Might there be other real solutions as well? An obvious way to explore this is to start the Newton–Raphson method at different values of x. If we try $x = 4$, we end up at the same place, but if we start the sequence at 2 we arrive at 1.875346 in five trials. For confirmation we can use the program *cSolve* on a Texas Instruments™ TI-89 (or a comparable device) and obtain: $-0.550174 \pm 1.42705i$, 1.87535, and 2.225.

Now let us suppose that the function of interest is $f(x) = 2 + \sin(2x)$. Of course we see immediately that $f(x)$ can never be smaller than 1.0; Newton–Raphson does not know this, and if we start with 2 as our initial estimate we get the sequence:

2.000000
2.950975
2.074051
3.153893
2.141288
3.450326
1.868810
2.738003
1.814223
2.681012
1.686051
2.595879
1.389607
2.648622
1.592425
2.571719
1.265582
2.835763
1.965004
2.880478
etc.

Obviously the technique does not always converge! With more complicated algebraic equations such behavior will not be easily anticipated (as it was here). We must exercise care that we do not ask Newton–Raphson to do something it cannot do. The method is best employed in situations where the behavior of the function is well understood (of course this is a trite observation since the statement is valid for just about *every* numerical method).

Let us conclude this section by returning to eq. (2.9) in Section 2.1 for a moment. Recall that we identified two real roots; we also pointed out the existence of a second (complex conjugate) pair of roots. It is occasionally necessary to locate complex zeros for polynomials, and in the case of eq. (2.9), it is possible to accomplish this by polynomial deflation. If we divide eq. (2.9) by $x^2 - 4.10035x + 4.17265$, we obtain

$$(x^4 - 3x^3 + 2x^2 - 5x + 9.76048)/(x^2 - 4.10035x + 4.17265) = x^2 + 1.10035x + 2.33917;$$

we can then use the quadratic equation on the second-degree polynomial that remains (you may want to verify that $x = 0.550175 \pm 1.42705i$). It is also possible to use a search procedure to identify complex zeros of functions, and Hamming (1971) provides a useful discussion of this topic. We will illustrate a direct, brute force approach here using a real function of the complex variable, z (where $z = x + iy$). Let

$$w(z) = e^z - z^2 = e^x[\cos y + i \sin y] - (x^2 + 2ixy - y^2); \tag{2.14}$$

now we identify the real and imaginary parts which are $e^x \cos y - x^2 + y^2$ and $e^x \sin y - 2xy$, respectively. We will use a *preplanned* search (a preplanned search is one in which the experiments or trials are set in advance of the calculations), looking over a range of x and y values: $-\pi \le x \le 2\pi$ and $0 \le y \le 2\pi$, and employing an interval of 0.2 in both directions. We let

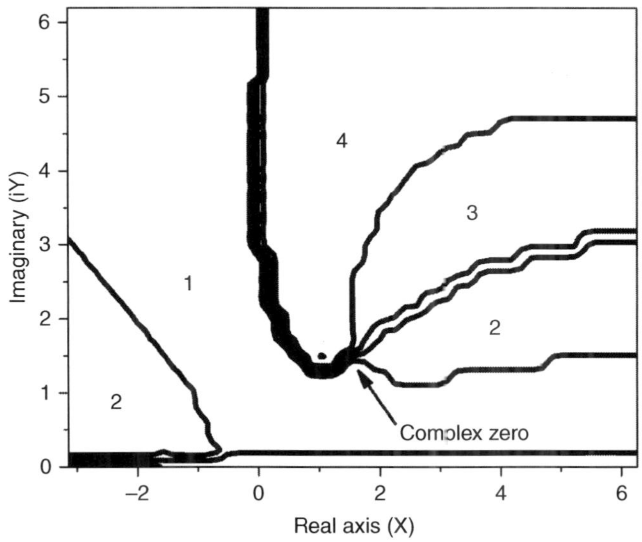

Figure 2.1 Results of the preplanned search in which the appropriate quadrant designations have been determined. The complex zero is located at the intersection where quadrants 1, 2, 3, and 4 meet.

each (x,y) position be represented by discrete values for the indices j and k. Based upon the signs of the computed real and imaginary parts, we assign the appropriate quadrant number for each (j,k) pair:

```
IF realp>0 AND imagp>0 THEN quadr(j,k)=1
IF realp<0 AND imagp>0 THEN quadr(j,k)=2
IF realp<0 AND imagp<0 THEN quadr(j,k)=3
IF realp>0 AND imagp<0 THEN quadr(j,k)=4
```

The result is a two-dimensional array of quadrant values which we can present as a contour plot, and this example is illustrated by the construction of Figure 2.1.

The direct approach that we have employed here is extremely simple, although the actual execution is a bit cumbersome. Naturally if we needed a more accurate estimate for the complex zero, we would simply refine the search array. For example, in Figure 2.1 it is clear that $1 \le x \le 2$ and $1 \le y \le 2$. The reader is encouraged to try this and pin down the location of the complex zero.

Example ***Stability in Automatic Process Control***

In the preceding discussion we observed that it is occasionally necessary to find complex roots of algebraic equations. Let us look at an important application that emphasizes this point; this example is drawn from automatic process control, so if you have had formal exposure to this field it will be familiar. In linear control theory the open-loop transfer function is obtained by forming the product of transfer functions in the negative feedback loop. Suppose we impose PI (proportional-integral) control upon a second-order process, with a first-order lag in the feedback loop (i.e., the measurement device exhibits first-order dynamics). The transfer function for the controller is $G_1 = K_C\left(1 + \frac{1}{\tau_I s}\right)$, and for the second-order process, $G_2 = \frac{1}{\tau^2 s^2 + 2\tau\zeta s + 1}$. The measurement device has the transfer function:

$$G_3 = \frac{K}{\tau_m s + 1}.$$

The *characteristic* equation for this control system is

$$1 + G_1 G_2 G_3 = 0. \tag{2.15}$$

and it is clear that the result will be a *fourth-degree* polynomial in s. The loop output in the time domain will be stable if all complex roots of the characteristic equation have *negative* real parts. This simply means that a disturbance resulting from a change in set point, for example, will decay proportional to $\exp(-at)$ where a is some positive real number. If a complex root has a *positive* real part, then such a disturbance will be amplified as $\exp(+at)$ and the system will be unstable. Since the controller transfer function, G_1, will have adjustable parameters such as the gain, K_C, a direct approach will entail solving eq. (2.15) for various values of K_C. Let's illustrate this process with the characteristic equation:

$$s^4 + 6s^3 + 11s^2 + 6s + K_C + 1 = 0. \tag{2.16}$$

Of course complex roots are a possibility with eq. (2.16), and we have many different options for determining them including commercial software packages and scientific calculators. We will use *cSolve(…)* on a *TI-nspire* to obtain our results.

K_C	Complex Roots	
2	$-0.248 \pm 0.565i$	$-2.752 \pm 0.565i$
4	$-0.157 \pm 0.745i$	$-2.843 \pm 0.745i$
6	$-0.086 \pm 0.866i$	$-2.914 \pm 0.866i$
8	$-0.027 \pm 0.959i$	$-2.973 \pm 0.959i$
9	$0 \pm i$	$-3.000 \pm i$
10	$+0.025 \pm 1.037i$	$-3.025 \pm 1.037i$

Notice what is happening here in the first column of complex roots: As the gain (K_C) is increased, the real part is becoming less negative. This means that the oscillations resulting from a disturbance are becoming more persistent, and at $K_C = 9$, the real part is zero producing sustained oscillation. As a practical matter we note that a gain of 8, although stable, would not be desirable since the exponential damping would be just $\exp(-0.027\,t)$—the decay of a disturbance would be too slow to provide good process control.

2.2.2 Regula Falsi (False Position Method)

Suppose we wish to find a solution for an equation of the type,

$$f(x) = \frac{1}{4}x^3 - x^2 + 3x - 6.50547. \tag{2.17}$$

Furthermore, suppose we know that $f(x=0)$ is negative (actually -6.50547) and that $f(x=5)$ is positive (actually 14.7445). Clearly there must be a sign change in the interval (0,5) that corresponds to a solution for this problem (we know that solution to be $x = 2.93333$). We could take these known end-points, x_1 and x_2, and fit a straight line to them; since

$$f(x_1) = y_1 = mx_1 + b \tag{2.18a}$$

and

$$f(x_2) = y_2 = mx_2 + b. \tag{2.18b}$$

By subtraction we find the slope of this line:

$$m = \frac{y_1 - y_2}{x_1 - x_2},$$

and the intercept can then be written

$$b = y_1 - \left(\frac{y_1 - y_2}{x_1 - x_2}\right)x_1.$$

We are seeking the point where this straight-line approximation crosses the axis, i.e., where $f(x_0) = y = 0$. From our equation for the linear approximation:

$$x_0 = x_1 - \left(\frac{x_1 - x_2}{y_1 - y_2}\right)y_1. \tag{2.19}$$

If the product of the functions $f(x_0)f(x_2) < 0$, then the sign change we are trying to identify still lies to the right which means that we move the left-hand end-point from the original value, x_1 to x_0. If the product is positive, then we move the right-hand end-point inward. Let us apply this method to our example equation using the following logic:

```
#COMPILE EXE
#DIM ALL
   REM *** Application of regula falsi method to an algebraic equation
    GLOBAL x1,x2,y1,y2,x0,f0,fx,x,trial,ZZ AS SINGLE
```

```
FUNCTION PBMAIN
    x1=0:x2=5:trial=0
    OPEN "c:regulaF.dat" FOR OUTPUT AS #1
 100 REM *** continue
    x=x1
    GOSUB 300
    y1=fx
      x=x2
      GOSUB 300
      y2=fx
        x0=x1-(x1-x2)/(y1-y2)*y1
        x=x0
        GOSUB 300
        f0=fx
    IF f0*y2<0 THEN x1=x0 ELSE x2=x0
        trial=trial+1
          PRINT trial,x1,x2
            WRITE#1,trial,x1,x2
            IF trial>20 THEN 200 ELSE 100
 200 REM *** continue
      INPUT "Shall we continue?";ZZ
        IF ZZ>0 THEN CLOSE
        END

 300 REM *** continue
    fx=1/4*x^3-x^2+3*x-6.50547
    RETURN
END FUNCTION
```

This simple code produces the following results:

1,	1.53069877624512,	5
2,	2.17453002929688,	5
3,	2.53261733055115,	5
4,	2.72695684432983,	5
5,	2.82875323295593,	5
6,	2.88081741333008,	5
7,	2.90708756446838,	5
8,	2.92024779319763,	5
9,	2.92681646347046,	5
10,	2.93008899688721,	5
11,	2.93171787261963,	5
12,	2.93252825737,	5
13,	2.93293142318726,	5
14,	2.93313193321228,	5
15,	2.93323159217834,	5
16,	2.9332811832428,	5
17,	2.93330574035644,	5
18,	2.93331789970398,	5
19,	2.93332409858704,	5
20,	2.93332719802856,	5
21,	2.93332862854004,	5

Notice that we have moved the left-hand end-point very close to the actual solution ($x = 2.93333$) in 21 trials. *Regula falsi* is closely related to another linear approximation technique known as the secant method. With the secant method the initial end-points do not have to bracket the zero (the solution). The principal advantage of regula falsi over Newton–Raphson is that the former requires no evaluation of derivatives. A disadvantage of straight-line approximations is their inability to follow curvature; therefore, we might improve this approach by adding one more function evaluation (use an additional point so that the total is three) and fitting a parabola to them. This variation of the technique is known as Müller's method, and it can often provide a better first estimate for the root location. Müller's method will be the focus of one of the exercises at the end of the chapter.

2.2.3 Dichotomous Search

Search procedures can often be used to advantage in the solution of nonlinear algebraic equations, and this is particularly evident for one-dimensional problems. One of the simplest techniques available is the *dichotomous search*, and we will illustrate it by applying it to eq. (2.9) from Section 2.1. We are seeking an optimum, so we take eq. (2.9), square it, and invert it; i.e., the merit function we wish to evaluate will be written as:

$$f(x) = \frac{1}{[x^4 - 3x^3 + 2x^2 - 5x + 9.76048]^2}. \tag{2.20}$$

If problems are created by division by zero, we simply add a small constant to the denominator. The basic idea is that we select an interval to search, find the mid-point (MP) of the interval, and perform two function evaluations located at MP+δ and MP−δ. We reject the "worst" side and move the interval end-point to that location. The logic will look something like this:

```
#COMPILE EXE
#DIM ALL
  REM *** dichotomous search for solution of an algebraic equation
       GLOBAL delta,MP,x1,x2,xl,xh,x,FL,FH,test,trial,FX,ZZ AS SINGLE
FUNCTION PBMAIN
    INPUT "Specify delta:";delta
    INPUT "Select x1:";x1
    INPUT "Select x2:";x2
        OPEN "c:dichoto2.dat" FOR OUTPUT AS #1
  100 REM *** continue
        MP=(x1+x2)/2
         xl=mp-delta
         xh=mp+delta
         x=xl
           GOSUB 300
           FL=FX
         x=xh
           GOSUB 300
           FH=FX
         test=FH-FL
        IF test>0 THEN x1=xl ELSE x2=xh
           PRINT x1,x2
           WRITE#1,x1,x2
            trial=trial+1
               IF trial>25 THEN 200 ELSE 100
  200 REM *** continue
          INPUT "Shall we continue?";ZZ
          IF ZZ>0 THEN CLOSE
          END
```

```
300 REM *** continue for function evaluation
      FX=1/(x^4-3*x^3+2*x^2-5*x+9.76048)^2
      RETURN
```

We will set $\delta = 0.001$ and choose the initial search interval to be $x = 0$ to $x = 22$. The results are:

0,	11.0010004043579
0,	5.50150012969971
0,	2.7517499923706
1.37487494945526,	2.7517499923706
2.06231260299683,	2.7517499923706
2.06231260299683,	2.40803122520447
2.06231260299683,	2.23617172241211
2.14824223518372,	2.23617172241211
2.19120693206787,	2.23617172241211
2.21268939971924,	2.23617172241211
2.22343063354492,	2.23617172241211
2.22343063354492,	2.23080110549927
2.22343063354492,	2.22811579704285
2.22343063354492,	2.22677326202392
2.22343063354492,	2.22610187530518
2.2237663269043,	2.22610187530518
2.22393417358398,	2.22610187530518
2.22393417358398,	2.22601795196533
2.22397613525391,	2.22601795196533
2.22399711608887,	2.22601795196533
2.22399711608887,	2.22600746154785
2.22399711608887,	2.22600221633911
2.22399711608887,	2.22599959373474
2.22399711608887,	2.22599840164184
2.2239978313446,	2.22599840164184
2.2239978313446,	2.22599792480469

We repeat, again with $\delta = 0.001$, but select the initial interval to be $x = 1$ to $x = 21$.

1,	11.0010004043579
1,	6.00150012969971
1,	3.5017499923706
1,	2.25187492370605
1.62493741512298,	2.25187492370605
1.62493741512298,	1.93940627574921
1.78117179870605,	1.93940627574921
1.85928905010223,	1.93940627574921
1.85928905010223,	1.90034770965576
1.85928905010223,	1.88081848621368
1.86905372142792,	1.88081848621368
1.87393605709076,	1.88081848621368
1.87393605709076,	1.87837731838226
1.87393605709076,	1.87715673446655
1.87393605709076,	1.87654650211334
1.87424123287201,	1.87654650211334
1.87424123287201,	1.87639391422272
1.87431752681732,	1.87639391422272
1.87431752681732,	1.87635576725006

1.87433660030365,	1.87635576725006
1.87434613704681,	1.87635576725006
1.87434613704681,	1.87635099887848
1.87434613704681,	1.87634861469269
1.87434732913971,	1.87634861469269
1.87434792518616,	1.87634861469269
1.87434792518616,	1.87634837627411

Clearly more numerical effort was required than needed with Newton–Raphson, but a search technique like this one can often be more revealing with regard to function behavior. In addition, no differentiation is required and for very complicated functions that can be a significant advantage. One might wonder if this approach could be improved with respect to numerical efficiency. For example, what if some of the function evaluations from previous stages could be reused? You will note that in the dichotomous procedure described above, we establish a new mid-point requiring two new accompanying evaluations for every stage. If we use the *golden section* search we will achieve greater efficiency.

2.2.4 Golden Section Search

The *golden section* technique is based upon the *golden ratio*, and its advantages can be made very clear by a simple example. Suppose we wish to locate the maximum value for a function, $f(x)$, over the interval $(0 < x < 10)$. Consider the impact of placing our initial function evaluations at $x = 6.18033989$ and $x = 3.81966011$; assume that the right-most portion of the interval is rejected—that the extremum we seek is located between 0 and 6.18033989. If we multiply 6.18033989 by 0.618033989, we get 3.81966…. This, of course, is the location of one of our initial trials, so we can *reuse that existing function evaluation* in the next stage. Thus, the golden section technique can be used to solve an algebraic equation with far fewer numerical operations than the dichotomous procedure.

For the example we used immediately above, with the initial search interval $(0 < x < 20)$ we find:

7.639319896698,	12.3606796264648
1.80339872837067,	2.22912335395813
2.22912335395813,	2.49223566055298
2.22912335395813,	2.32962322235107
2.22912335395813,	2.26751089096069
2.22912335395813,	2.24378609657288
2.22912335395813,	2.2347240447998
2.22566175460815,	2.22698402404785
2.22484469413757,	2.22515678405762
2.22484469413757,	2.22496390342712
2.22496390342712,	2.22503757476807

Once again we discover that we must change the initial search interval if we wish to locate the other real-valued solution; setting $(1 < x < 21)$ we end up with the final pair 1.875486 and 1.875679. Golden section searches are appealing because they are both efficient and easy to implement. But, for *very complex equations* where execution time is critical, one can go one step farther and use the Fibonacci technique which gives the most rapid reduction in the interval of uncertainty. Fibonacci searches are described in both Shoup and Mistree (1987) and in Beveridge and Schechter (1970); the technique differs from the golden section in that the number of merit function evaluations is selected in advance and the interval of uncertainty can be computed from the Fibonacci numbers (F's), 1, 1, 2, 3, 5, 8, 13, 21, 34, 55, 89, 144, 233, etc. The very last evaluation uses the dichotomous technique since it cuts the final interval to 50% rather than the 61.8 + % possible with golden section. The initial function evaluations are located a distance inside the starting interval end-points corresponding to

$$\frac{F_{n-1}}{F_n} + \frac{(-1)^n}{F_n}\delta.$$

Thus, for $n = 20$ with $\delta = 0.001$, this distance is

$$\left(\frac{6765}{10946}\right) + \frac{1}{10946}(0.001) = 0.618034.$$

Using the same values for n and δ, we can estimate the final interval of uncertainty:

$$I_u = \frac{1}{F_n} + \frac{F_{n-2}}{F_n}\delta = \frac{1}{10946} + \frac{4181}{10946}(0.001) = 0.000473.$$

Example ***Flame Temperature for the Combustion of Methanol***

The combustion of hydrocarbons has been integral to the improvement of living conditions around the world—it has also contributed significantly to global climate change. A key question for engineers concerned with combustion is the expected flame temperature. This type of calculation requires an enthalpy balance in which the heat of combustion is assumed to be absorbed by the product species leaving the combustion apparatus, thereby increasing their enthalpies. If we specify that no thermal energy be lost to the surroundings, then we can determine the *theoretical flame temperature* from this balance. Let us assume that we are interested in burning methanol; the stoichiometric relationship is

$$CH_3OH + \tfrac{3}{2}\,O_2 \rightarrow CO_2 + 2H_2O.$$

We choose as a computational basis 1 mol methanol combusted with 20% excess air; of course, the use of air—as opposed to pure oxygen—means that significant thermal energy will be expended simply heating the accompanying nitrogen. We now construct tables for the inlet and outlet species:

Inlet		Outlet	
CH_3OH	1	CO_2	1
O_2	1.8	O_2	0.3
N_2	6.77	N_2	6.77
		H_2O	2

The heat of combustion for methanol is $\Delta H_C = -726.6\,\text{kJ/mol}$. Because the expected outlet temperature is so large, we cannot assume that the heat capacities of the outlet species are constant—they will vary with T of course. Consequently we will use the following expressions for C_p (all kJ/[mol °C]):

$$CO_2\text{:}\quad 0.0361 + 4.23 \times 1010^{-5}T - 2.887 \times 10^{-8}T^2 + 7.46 \times 10^{-12}T^3$$

$$H_2O\text{:}\quad 0.0335 + 0.688 \times 10^{-5}T + 0.7604 \times 10^{-8}T^2 - 3.593 \times 10^{-12}T^3$$

$$O_2\text{:}\quad 0.0291 + 1.158 \times 10^{-5}T - 0.6076 \times 10^{-8}T^2 + 1.311 \times 10^{-12}T^3$$

$$N_2\text{:}\quad 0.029 + 0.22 \times 10^{-5}T + 0.5723 \times 10^{-8}T^2 - 2.871 \times 10^{-12}T^3$$

Since no thermal energy is lost to the surroundings, $Q = 0$, and our enthalpy balance takes the form:

$$H_{in} + |H_C| = \sum_{i=1}^{n} \int_{25}^{T} n_i C_{Pi} dT. \tag{2.21}$$

Of course the integration process results in a sequence of four fourth-degree polynomials on the right-hand side of the balance. Such equations can be solved with the techniques previously discussed, but we can make the task simpler by observing that the combustion of hydrocarbons with *modest* excess air will usually produce flame temperatures on the order of 2,000 °C. Armed with this information we can employ a pre-planned search or trial and error and quickly determine that $T_{flame} = 1{,}935\,°\text{C}$. If you have not solved this type of problem before, you might repeat this example but with the assumption that 100% excess air is to be employed. Use the dichotomous search and see if you can verify that $T_{flame} = 1{,}329\,°\text{C}$.

2.3 Simultaneous Linear Algebraic Equations

Just about every student in science and engineering has seen a problem of this type:

$$2x_1 + 3x_2 - x_3 = 9 \tag{2.22}$$

$$-5x_1 + 2x_2 + 2x_3 = 4 \tag{2.23}$$

$$x_1 + x_2 + 3x_3 = 17 \tag{2.24}$$

Our goal of course is to find the correct values for the three variables, x_1, x_2, and x_3. We will do this using a technique familiar to many students, *Gaussian elimination*. The objective of this procedure is to obtain a triangular form (with 1's starting in the upper left-hand corner and proceeding diagonally down through the matrix, with zeros in the triangle underneath) where the unknowns are determined by back substitution. We begin by noting that this set of equations can be written equivalently as the *augmented coefficient matrix*,

$$\begin{matrix} 2 & 3 & -1 & 9 \\ -5 & 2 & 2 & 4 \\ 1 & 1 & 3 & 17 \end{matrix}$$

We divide the first row (equation) by 2:

$$\begin{matrix} 1 & 3/2 & -1/2 & 9/2 \\ -5 & 2 & 2 & 4 \\ 1 & 1 & 3 & 17 \end{matrix}$$

Divide the second row by −5 and subtract the first row from the second:

$$\begin{matrix} 1 & 3/2 & -1/2 & 9/2 \\ 0 & -19/2 & 1/10 & -53/10 \\ 1 & 1 & 3 & 17 \end{matrix}$$

Subtract the first row from the third and divide the second row by −19/10:

$$\begin{matrix} 1 & 3/2 & -1/2 & 9/2 \\ 0 & 1 & -1/19 & 53/19 \\ 0 & -1/2 & 7/2 & 25/2 \end{matrix}$$

Divide the third row by −1/2 and then subtract the second row from it:

$$\begin{matrix} 1 & 3/2 & -1/2 & 9/2 \\ 0 & 1 & -1/19 & 53/19 \\ 0 & 0 & -132/19 & -528/19 \end{matrix}$$

Divide the third row by −132/19:

$$\begin{matrix} 1 & 3/2 & -1/2 & 9/2 \\ 0 & 1 & -1/19 & 53/19 \\ 0 & 0 & 1 & 4 \end{matrix}$$

This is the triangular form we were seeking: The coefficient matrix has "1's" on the diagonal and zeros beneath. Therefore, $x_3 = 4$, and by back substitution, $x_2 = 3$, and $x_1 = 2$. This direct elimination scheme is simple to understand and to execute, but there are several potentially serious problems. First, a machine computation will not produce the whole numbers and the fractions we see in the example above. Each number will be represented only with the precision of the machine which means that roundoff error will occur, and for large sets of equations, possibly accumulate. Second, if the system of equations is ill-conditioned, the number of simultaneous equations that can be solved will be quite limited. An ill-conditioned coefficient matrix is one in which small changes to a coefficient have enormous impact upon the solution. Third, if a pivot element is zero, and we do not interchange rows (employ partial pivoting) then division by zero occurs. Fourth, if a pivot element is very small relative to other coefficients in the matrix, the accuracy of the solution will be impaired. For these and other reasons, Gaussian elimination is not used to solve large sets of equations unless the coefficient matrix is sparse (contains many zeros). There are far better elimination methods available, and we will discuss one of those in the next section.

2.3.1 Crout's (or Cholesky's) Method

Crout's method is perhaps the most powerful of the direct elimination schemes—it can be used to solve fairly large sets of simultaneous linear algebraic equations and it can do so efficiently. Let us use the following problem as an example.

$$2x_1 + x_2 - x_3 + 1.26x_4 = 1.3467 \tag{2.25}$$

$$-x_1 + 3x_2 + x_3 + 0.53x_4 = 4.04 \tag{2.26}$$

$$1.082x_1 + 2x_2 + 3x_3 - x_4 = 5.3607 \tag{2.27}$$

$$3x_1 + 2.97x_2 - 1.48x_3 - x_4 = -0.82 \tag{2.28}$$

This set of equations can be written equivalently as:

$$\begin{matrix} a_{11} & a_{12} & a_{13} & a_{14} & x_1 & & c_1 \\ a_{21} & a_{22} & a_{23} & a_{24} & x_2 & = & c_2 \\ a_{31} & a_{32} & a_{33} & a_{34} & x_3 & & c_3 \\ a_{41} & a_{42} & a_{43} & a_{44} & x_4 & & c_4 \end{matrix}$$

or more compactly as **A{X}–{C}** = 0. If we could get the coefficient matrix in the triangular form we discussed previously with "1's" on the diagonal and zeros below, we could find the x's by back substitution. With Crout's method, a nonsingular n by n coefficient matrix is decomposed into lower (**L**) and upper (**U**) triangular matrices such that **A = LU. U**, of course, is *exactly the form we seek*, with 1's (ones) on the diagonal (a_{11}, a_{22}, a_{33}, etc.) and zeros below.

Typical structure for a Crout's (Cholesky's) procedure is shown below for the four-equation example. This routine has been adapted from one given by James et al. (1977).

```
#COMPILE EXE
#DIM ALL
     REM *** Crout's method for solution of simultaneous algebraic equations
         GLOBAL M,NEQ,J,I,II,SUM,JM1,K,IP1,IM1,JJ,NN,L AS SINGLE

         FUNCTION PBMAIN
     DIM A(10,11) AS SINGLE
     DIM X(10) AS SINGLE
          PRINT "********************************************************"
          PRINT "       Crout's method example with four equations:"
          PRINT "       2X1 +  X2 -  X3 + 1.26X4 = 1.3467"
          PRINT "       -X1 + 3X2  + X3 + 0.53X4 = 4.04"
          PRINT "      1.082X1 + 2X2 + 3X3 - X4 = 5.3607"
          PRINT "       3X1 + 2.97X2 - 1.48X3 - X4 = -0.82"
          PRINT "********************************************************"

     REM *** Here's where you must input the coefficient matrix.
         A(1,1)=2:A(1,2)=1:A(1,3)=-1:A(1,4)=1.26:A(1,5)=1.3467
         A(2,1)=-1:A(2,2)=3:A(2,3)=1:A(2,4)=0.53:A(2,5)=4.04
         A(3,1)=1.082:A(3,2)=2:A(3,3)=3:A(3,4)=-1:A(3,5)=5.3607
         A(4,1)=3:A(4,2)=2.97:A(4,3)=-1.48:A(4,4)=-1:A(4,5)=-0.82

     REM *** Crout's method of matrix decomposition
         INPUT "SPECIFY THE NUMBER OF SIMULTANEOUS EQUATIONS:";NEQ
             M=NEQ+1
             FOR J=2 TO M
             A(1,J)=A(1,J)/A(1,1):NEXT J
         FOR I=2 TO NEQ
                 J=I
                 FOR II=J TO NEQ
                 SUM=0!
                 JM1=J-1
                     FOR K=1 TO JM1
                     SUM=SUM+A(II,K)*A(K,J):NEXT K
```

```
                A(II,J)=A(II,J)-SUM:NEXT II
                IP1=I+1
                    FOR JJ=IP1 TO M
                    SUM=0!
                    IM1=I-1
                        FOR K=1 TO IM1
                        SUM=SUM+A(I,K)*A(K,JJ):NEXT K
                A(I,JJ)=(A(I,JJ)-SUM)/A(I,I):NEXT JJ
                NEXT I
            X(NEQ)=A(NEQ,NEQ+1)
            L=NEQ-1
                FOR NN=1 TO L
                SUM=0!
                I=NEQ-NN
                IP1=I+1
                    FOR J=IP1 TO NEQ
                    SUM=SUM+A(I,J)*X(J)
                    NEXT J
                X(I)=A(I,M)-SUM:NEXT NN
                    FOR J=1 TO NEQ
                    PRINT X(J)
                    NEXT J
           PRINT "Shall we continue with program?"
           INPUT "Respond with any positive number";NN
                   IF NN>0 THEN 300
    300 REM *** continue
END FUNCTION
```

For the given example, the program produces the following results for x_1 through x_4: 0.3333475, 0.6666642, 1.66668, and 1.333349.

Crout's method is a particularly effective way to solve sets of simultaneous linear algebraic equations. Moreover, the scheme's logic as laid out above can be easily automated for repetitive calculation. If, on the other hand, an analyst has a one-off application with a limited number of equations there are widely available tools that can be used with less time investment. For example, with Excel™ one can use the *Solver* add-in. For the TI-85™ family, one can use *SIMULT* (for 30 equations or less) and with the TI-*nspire* one could use *linSolve*; we will illustrate the application of the latter on a set of four algebraic equations with the variables x1, x2, x3, and x4:

```
                 2·x1-x2+3·x3+x4=2.625
linSolve         x1+x2+x3+x4=1.375                ,{x1,x2,x3,x4}
                 5·x2-x3-0.5·x4=0.08333
                 3·x1+7·x2-2·x3+3·x4=1.66666

    {0.333333,0.166666,0.624999,0.250001}
```

Clearly, if one has a small set of simultaneous linear algebraic equations that needs to be solved only once, then *linSolve* or similar tool will rapidly produce the desired result.

2.3.2 Matrix Inversion

Consider the following set of equations:

$$X_1 + 2X_2 + 3X_3 + X_4 = 8.98958 \tag{2.29}$$

$$X_1 + X_2 - 9X_3 + X_4 = 0.65625 \tag{2.30}$$

$$X_1 + 6X_3 - X_4 = 4.01042 \tag{2.31}$$

$$5X_1 - 2X_2 - X_3 + 7X_4 = 2.14931, \tag{2.32}$$

with the solution: 1.5, 3.0, 0.444444, and 0.15625 for X_1 through X_4, respectively. As we noted previously, we can write this set of equations alternatively as: $\mathbf{AX} = \mathbf{C}$. If the coefficient matrix, $\mathbf{A}$, is nonsingular, then an inverse matrix exists such that: $\mathbf{A}^{-1}\,\mathbf{A} = \mathbf{I}$. The right-hand side is the identity matrix which consists of 1's on the diagonal and zeros elsewhere. This suggests the following multiplication: $\mathbf{A}^{-1}\mathbf{A}\ \mathbf{X} = \mathbf{A}^{-1}\ \mathbf{C}$. Consequently, $\mathbf{X} = \mathbf{A}^{-1}\ \mathbf{C}$, and the solution is at hand. So, how does one determine the inverse of $\mathbf{A}$?

Since $\mathbf{A}^{-1}\mathbf{A} = \mathbf{I}$, we can obtain the result we seek by applying the Gauss–Jordan method (the forward and backward eliminations are combined into a single procedure) to the *augmented* matrix (note the form of the right-half):

1	2	3	1	1	0	0	0
1	1	−9	1	0	1	0	0
1	0	6	−1	0	0	1	0
5	−2	−1	7	0	0	0	1

The required steps are illustrated by James et al. (1977), but one can also use software tools like *Mathcad*™ for this purpose (matrix toolbar, M^{-1}) The reader should verify that the inverse we are looking for is:

−0.113	0.323	0.548	0 048
0.381	0.084	−0.077	−0 077
0.052	−0.090	0.00645	0 00645
0.197	−0.219	−0.413	0 087

Since $\mathbf{X} = \mathbf{A}^{-1}\mathbf{C}$, we find that $X_1 = -0.113(8.98958) + 0.323(0.65625) + 0.548(4.01042) + 0.048(2.14931) = 1.497+$, $X_2 = 0.381(8.98958) + 0.084(0.65625)–0.077(4.01042)–0.077(2.14931) = 3.0059$, and X_3 and X_4 are, respectively, 0.4481 and 0.1579. Note that the effect of roundoff error is apparent here. Although matrix inversion is seldom used for a single set of equations—it is more commonly thought of in the context of multiple sets of simultaneous equations—there are tools that can be used quickly for small problems. To illustrate this, we turn back to a previous example and we employ Excel. The set of equations is

$$2 \cdot x1 - x2 + 3 \cdot x3 + x4 = 2.625$$

$$x1 + x2 + x3 + x4 = 1.375$$

$$5 \cdot x2 - x3 - 0.5 \cdot x4 = 0.08333$$

$$3 \cdot x1 + 7 \cdot x2 - 2 \cdot x3 + 3 \cdot x4 = 1.66666$$

We put these equations into the spreadsheet as three matrices:

	A			X	C
2	−1	3	1	x1	2.625
1	1	1	1	x2	1.375
0	5	−1	−0.5	x3	0.08333
3	7	−2	3	x4	1.66667

Just as before, we need to invert $\mathbf{A}$ and then multiply by $\mathbf{C}$. From the formular bar in Excel we select *MINVERSE*, which we implement as = MINVERSE(C4:F7); of course, cell C4 is 2 and F7 is 3. The inversion is executed with *ctrl* + *shift* + *enter*. Now suppose we had allocated C9:F12 for $\mathbf{A}^{-1}$:

0.962264	−2.0566	0.075472	0.377358
−0.09434	0.358491	0.188679	−0.0566
−0.07547	0.886792	0.150943	−0.24528
−0.79245	1.811321	−0.41509	−0.07547

Next we multiply $\mathbf{A}^{-1}\mathbf{C}$; again using the formular bar we write = MMULT(C9:F12, K4:K7) followed by *ctrl* + *shift* + *enter*. In the allocated four-cell column we obtain:

0.333335
0.166666
0.624999
0.25

You can see how—with just a little practice—one can execute these steps very rapidly. If you would like to try this yourself, work through the following example.

A					**X**	**C**
1	1	1	1	1	x1	7.121
1	0	−3	2	−2	x2	0.951
3	−1	−1	−1	1	x3	−0.895
6	−2	1	1	−3	x4	0.758
1	1	0	1	−5	x5	2.547

You should get the following for the inverse of **A**, and then the column array for x1 through x5 by the multiplication $\mathbf{A}^{-1}\mathbf{C}$:

0.172973	−0.01622	0.232432	0.010811	0.081081
0.416216	−0.07027	0.340541	−0.28649	0.351351
0.075676	−0.19459	−0.21081	0.12973	−0.02703
0.181081	0.248649	−0.3973	0.167568	−0.24324
0.154054	0.032432	0.035135	−0.02162	−0.16216

1.223
3.27
0.572
1.389
0.667

2.3.3 Iterative Methods of Solution

Iterative methods are often used to solve the types of simultaneous algebraic equations that arise in the numerical solution of elliptic partial differential equations (elliptic PDEs). We will introduce one technique for dealing with such equations here, but will postpone a more extensive discussion until we treat PDEs later in the book in Chapter 8. The *Gauss–Seidel* (GS) iterative method is distinguished from the *Jacobi* method by the fact that the latest iterative values are used immediately in subsequent calculations. GS is easy to understand and easy to apply, and particularly so in cases where the coefficient matrix is sparse. This is exactly what happens when we solve elliptic PDEs numerically. Let us look at an example and then reveal the required computational scheme in detail.

Suppose we have a square domain over which the equation,

$$T_{i,j} = \tfrac{1}{4}[T_{i+1,j} + T_{i-1,j} + T_{i,j+1} + T_{i,j-1}] \tag{2.33}$$

is to be applied at all interior points. This particular form has come about through *discretization* of the governing elliptic PDE and the *isolation* of the term with the largest numerical coefficient (which was 4). Let us assume that we have constant values for T on all four boundaries; in particular, let these edge values be (100,200,200,300) for (L,R,T,B). In other words along the left-hand side of the domain the variable T has the value 100 and across the bottom it is 300; the other two edges are maintained at 200.

This is a *Dirichlet problem*; we apply the computational algorithm eq. (2.33) at each point across the bottom *interior* row successively, then we move up to the next row and repeat, etc. We will assume that our discretization has resulted in a total of 400 mesh points of which 324 are interior points where eq. (2.33) is to be applied—we are solving 324 simultaneous linear algebraic equations. We let these interior points be initialized with the default value of 0 (zero). Obviously we could hasten the convergence of the Gauss–Seidel scheme by making a better initial guess at the distribution, but that is not of much consequence here.

```
#COMPILE EXE
#DIM ALL
  REM *** solution of simultaneous algebraic equations by Gauss-Seidel iterative
method
       GLOBAL Told,eps AS SINGLE
       GLOBAL i,j,iter AS INTEGER
FUNCTION PBMAIN
       DIM T(20,20) AS SINGLE
         iter=0
  10 REM *** continue
     Told=T(10,10)
           FOR j=1 TO 20
            T(1,j)=100:T(20,j)=200
            NEXT j
           FOR i=1 TO 20
            T(i,1)=300:T(i,20)=200
            NEXT i
          FOR j=2 TO 19
            FOR i=2 TO 19
             T(i,j)=1/4*(T(i+1,j)+T(i-1,j)+T(i,j+1)+T(i,j-1))
             NEXT i:NEXT j
          iter=iter+1:eps=ABS(Told-T(10,10))
             PRINT iter,T(10,10),eps
            IF eps<1e-09 THEN 40 ELSE 10
  40 REM *** continue
      OPEN "C:gausseid.dat" FOR OUTPUT AS #1
         FOR j=1 TO 20
           FOR i=1 TO 20
             WRITE#1,i,j,T(i,j)
             NEXT i:NEXT j
              WRITE#1,iter
      CLOSE:END
END FUNCTION
```

Notice that progress toward convergence is being assessed by determining the change in value for a single, central point by comparing old and new iterates. For this example, 249 iterations are required to reach ε (*eps*) = 0.01 and 492 iterations are required to attain $\varepsilon = 1 \times 10^{-8}$. We can better examine the results of the computation by constructing an appropriate contour plot which is provided in Figure 2.2.

The Gauss–Seidel iterative method provides a powerful tool for solving large sets of simultaneous linear algebraic equations, and we will have many opportunities to use the technique in our efforts to solve elliptic PDEs (partial differential equations).

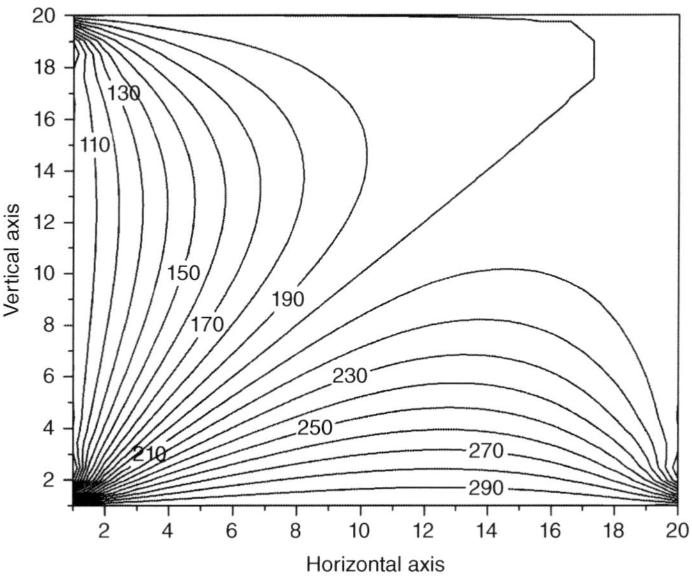

Figure 2.2 Contour plot of the results obtained through application of the Gauss–Seidel iterative method of solution for simultaneous linear algebraic equations. In this example, 324 simultaneous equations (resulting from the discretization of an elliptic PDE) are being solved, and a satisfactory solution is obtained with less than 500 iterations. The bottom is maintained at 300 and the left-hand side at 100; the top and the right-hand side are set to 200.

2.4 Simultaneous Nonlinear Algebraic Equations

2.4.1 Successive Substitution

Unfortunately, we occasionally encounter problems such as the set of equations:

$$3{x_1}^2 x_2 + x_3 = 25.032 \tag{2.34}$$

$$x_1 x_3 / {x_2}^{1.5} = 2.1896 \tag{2.35}$$

$$\sqrt{x_3}\,({x_1}^3 + x_2) = 15.7171 \tag{2.36}$$

In this particular case, we do know that all three unknowns are positive (all greater than 0). The "easiest" way to attempt solution is through *successive substitution*. We first estimate (guess, really) two of the unknowns. We then rearrange the set of equations to solve for the third value. We use that estimate plus one of our guesses to get the second unknown, then use the two revised values to obtain the third; the entire process is repeated successively. For example, we might try to write the equations above as:

$$x_1 = 2.1896 {x_2}^{3/2} / x_3 \tag{2.37}$$

$$x_2 = 15.7171/\sqrt{x_3} - {x_1}^3 \tag{2.38}$$

$$x_3 = 25.032 - 3{x_1}^2 x_2 \tag{2.39}$$

We start by setting x_2 and x_3 equal to 2 and 2, say. The resulting sequence of computed values is

1	3.096561909	−18.578321	559.4556274
2	1.350865364	−1.8006176	34.88750458
3	0.203487575	2.652528763	24.70049858
4	0.382956296	3.106257439	23.66335106
5	0.506576121	3.100982904	22.64268303
6	0.528063238	3.155748367	22.3900528
7	0.548230529	3.156806707	22.18360329
8	0.553610981	3.167327881	22.1177845
9	0.558036625	3.16818738	22.0702343

10	0.559466541	3.170446396	22.05292511
11	0.560504615	3.170782566	22.04154968
12	0.560883105	3.171288967	22.03703499
13	0.561132371	3.171396494	22.03427315
14	0.561231256	3.171513081	22.03310585
15	0.561291933	3.171544313	22.03242874
16	0.561317503	3.171571732	22.03212929
17	0.561332405	3.171580315	22.03196144
18	0.561338961	3.171586752	22.03188705
19	0.561342597	3.171589136	22.03184509
20	0.561344266	3.171590805	22.03182602
21	0.56134522	3.171591282	22.03181458
22	0.561345637	3.171591759	22.03181076
23	0.561345816	3.171591759	22.03180885
24	0.561345875	3.171591997	22.03180695
25	0.561345994	3.171591997	22.03180695

There are obvious problems with this technique, including: 1) the arrangement of the equations is not unique, 2) successive substitution may converge slowly or not at all, 3) the values obtained above may be one solution, but not the one we were actually seeking, and 4) for many nonlinear problems the analyst may have absolutely no prior knowledge of how the equations behave (what the initial estimates should be). We can underscore these points by noting that the "solution" we obtained above by successive substitution is actually not a very good one. The reader may want to try a comparison with 1.473, 2.784, and 6.903 for x_1 through x_3, respectively. Successive substitution is related to *tearing systems of equations*, and in recent years much effort has been expended in the development of tearing algorithms; some of these have been incorporated into commercial simulation software packages. Tearing is particularly useful in the analysis of process flow diagrams where material balances on process units may involve one or more unknown streams. In such cases it is not possible to proceed sequentially through the process units because of the unknown streams (possibly including recycle streams). The reader interested in chemical engineering applications of tearing may want to consult Ramirez (1997); a more general discussion of issues in tearing has been provided by Elmqvist and Otter (1994).

2.4.2 Newton's Method

Fortunately, we have options. For the first of these let us consider applying Newton's method of tangents to problems of this type. We will need the Jacobian (the matrix of partial derivatives) for the system. Let eqs. (2.34) through (2.36) be rewritten as follows:

$$f_1 = 3x_1^{\,2}x_2 + x_3 - 25.032 \tag{2.40}$$

$$f_2 = x_1x_3/x_2^{\,1.5} - 2.1896 \tag{2.41}$$

$$f_3 = \sqrt{x_3}\left(x_1^{\,3} + x_2\right) - 15.7171 \tag{2.42}$$

Therefore,

$$\frac{\partial f_1}{\partial x_1} = 6x_1x_2 \qquad \frac{\partial f_1}{\partial x_2} = 3x_1^{\,2} \qquad \frac{\partial f_1}{\partial x_3} = 1 \tag{2.43}$$

$$\frac{\partial f_2}{\partial x_1} = x_3/x_2^{\,3/2} \qquad \frac{\partial f_2}{\partial x_2} = -\frac{3}{2}\frac{x_1x_3}{x_2^{\,5/2}} \qquad \frac{\partial f_2}{\partial x_3} = \frac{x_1}{x_2^{\,3/2}} \tag{2.44}$$

$$\frac{\partial f_3}{\partial x_1} = 3\sqrt{x_3}x_1^{\,2} \qquad \frac{\partial f_3}{\partial x_2} = \sqrt{x_3} \qquad \frac{\partial f_3}{\partial x_3} = \frac{1}{2}\frac{x_1^{\,3} + x_2}{\sqrt{x_3}} \tag{2.45}$$

We let the set of equations, f_1, etc. be represented as $\mathbf{F(x)}$. The algorithm we want to use is simply

$$\mathbf{x_{new}} = \mathbf{x} - \mathbf{J}^{-1}(\mathbf{x})\mathbf{F}(\mathbf{x}), \tag{2.46}$$

which you will recognize as the equivalent to Newton–Raphson from Section 2.2 which we used previously for the solution of a single equation. As Faucett (2002) points out, it is generally not practical to evaluate the inverse of the Jacobian, so as an alternative we solve the system of linear equations to determine the corrections for x:

$$J(x)y = -F(x). \tag{2.47}$$

The values for y are used to improve x: $x_{new} = x + y$. We choose our initial estimates for x_1 through x_3 to be 2, 2, and 2, and then get the nine initial values for the Jacobian:

24	12	1
0.7071	−1.0607	0.7071
16.9706	1.4142	3.5355

This yields three simultaneous equations for the correction values (the **y**'s):

$$24y_1 + 12y_2 + y_3 = 0.968 \tag{2.48}$$

$$0.7071y_1 - 1.0607y_2 + 0.7071y_3 = -0.7754 \tag{2.49}$$

$$16.971y_1 + 1.4142y_2 + 3.5355y_3 = -1.575 \tag{2.50}$$

The reader may want to verify that the solution for these three equations is 0.6407, −0.9382, and −3.144 for y_1 through y_3, respectively. Therefore our revised estimates for the three unknown variables (the x_i's) are 2.641, 1.0618, and −1.1444. Of course this changes the values in the Jacobian, so it must be recalculated, then used to get improved values for the corrections (the y_i's). The entire process is repeated until the change in the x_i's is sufficiently small to indicate that a solution has been identified. Often we can expect this to require 4 to 6 trials—and possibly more.

There is another approach that is often effective for problems of this type called a pattern search in which we try to minimize an objective function, F. We will use the *sequential simplex* method (we will discuss this technique in greater detail momentarily) on the previous example by combining eqs. (2.40), (2.41), and (2.42):

$$F = \left(3x_1^2x_2 + x_3 - 25.032\right)^2 + \left(\frac{x_1x_3}{x_2^{1.5}} - 2.1896\right)^2 + \left(\sqrt{x_3}\left(x_1^3 + x_2\right) - 15.7171\right)^2.$$

We are trying to find values for x_1, x_2, and x_3 that result in $F = 0$; to use this method we must provide initial estimates for the three variables and the initial length of the simplex side. Our results will often be affected by these choices, so it is prudent to experiment with different combinations. Here are some results obtained with sequential simplex along with the final value for F:

x_1	x_2	x_3	F
1.749	2.264	4.263	3.54×10^{-7}
1.473	2.785	6.909	3.52×10^{-7}
1.474	2.783	6.898	2.25×10^{-7}
1.473	2.784	6.905	2.47×10^{-8}

Of course the last row represents our "best" solution.

2.4.3 An Example Problem with Simultaneous Nonlinear Algebraic Equations

It is worthwhile for us to examine a physically relevant problem that involves nonlinear algebraic equations. Suppose a hydraulically smooth pipe with $d = 3$ in. carries water from a storage tank to a process facility located 985 ft from the inlet. The pipe is horizontal and the inlet pressure (p_1) is 35 psig (pounds force per square inch, gauge); the water is discharged at 0 psig (one atmosphere). There is no significant loss due to fittings, there is no power input (pump), and since the pipe diameter is constant, there is no ΔKE. Under these conditions the macroscopic mechanical energy balance has the form:

$$\frac{p_2 - p_1}{\rho} = -\frac{1}{2}v^2\frac{L}{R_h}f. \tag{2.51}$$

The fluid density, ρ, is 62.4 lb_m/ft^3, v is the velocity of the water in the pipe, L is the pipe length (985 ft), R_h is the hydraulic radius (d/4), and f is the friction factor. For a smooth pipe, the friction factor is approximately, $f \cong 0.0791/Re^{1/4}$. The Reynolds number, Re, is

$$Re = \frac{dv\rho}{\mu},$$

where μ is the viscosity of the water, 1 cp, which must be converted to English units by multiplying by 6.72×10^{-4}. Note that the left-hand side of eq. (2.51) must be multiplied by both 144 (in^2/ft^2) and g_c, 32.17 $ftlb_m/(lb_f\, s^2)$. You can see that once our estimate for f is inserted on the right-hand side of eq. (2.51), the velocity in the pipe will appear as $v^{7/4}$. Solve eq. (2.51) and verify that $v = 9.51$ ft/s for the initial pipeline. *Now*, suppose we discover that two parallel pipes (both 500 ft long) can be connected to the existing 3″ line through a tee located at point "2" (i.e., at 485 ft). One of these pipes has $d = 2$ in. and the other has $d = 1$ in.; the discharge point for the original 3 in. line will be labeled "3," the 2 in. discharge "4," and the 1 in. discharge "5." How much will the original flow rate be increased once the auxiliary lines are placed in service?

We now write eq. (2.51) down four times, neglecting the small losses associated with the fittings at the branch point, "2":

$$\frac{p_2 - p_1}{\rho} = -\frac{1}{2}v_{12}^2 \frac{L_{12}}{R_{h12}} f_{12}$$

$$\frac{p_3 - p_2}{\rho} = -\frac{1}{2}v_{23}^2 \frac{L_{23}}{R_{h23}} f_{23}$$

$$\frac{p_4 - p_2}{\rho} = -\frac{1}{2}v_{24}^2 \frac{L_{24}}{R_{h24}} f_{24}$$

$$\frac{p_5 - p_2}{\rho} = -\frac{1}{2}v_{25}^2 \frac{L_{25}}{R_{h25}} f_{25}.$$

Of course, p_3 p_4 and p_5 are all 0 (zero) psig, and $L_{23} = L_{24} = L_{25} = 500$ ft. Under the new conditions we have four unknown velocities and an unknown intermediate pressure, p_2. The correct solution will satisfy the mass (actually volumetric) balance at the tee where the auxiliary lines are connected:

$$A_{12}\, v_{12} = A_{23}\, v_{23} + A_{24}\, v_{24} + A_{25}\, v_{25}.$$

Four of the five simultaneous equations are nonlinear, and although this problem might appear to be difficult, a simple solution strategy suggests itself: If we estimate p_2 (and of course we know that p_2 must be roughly in the middle of the range 35 to 0 psig) we can calculate all four velocities and then check our result with the mass balance. Therefore, a preplanned search might work well. Try this, and see if you can confirm that $p_2 = 12.9$ psig, and that the flow rate through the system has been increased by 15.3%.

2.4.4 Pattern Search for Solution of Nonlinear Algebraic Equations, Sequential Simplex, and the Rosenbrock Method

We noted previously that search techniques (optimization algorithms) can be used to solve many types of algebraic equations, and this author has found sequential simplex to be particularly useful for the exploration of nonlinear systems. The optimization method we first consider was devised by Spendley et al. (1962) and refined by Nelder and Mead (1964); it is most readily visualized in two dimensions. Suppose we are interested in the system of equations:

$$x_1 x_2^2 + \frac{x_1}{x_2} = 61.4167 \tag{2.52a}$$

and

$$75\frac{x_2}{x_1^3} - \frac{x_1 x_2}{50} = 12.230. \tag{2.52b}$$

We disregard the fact that x_1 can simply be factored out of the first equation, facilitating easy solution. Furthermore, we know that the values we seek for the unknowns lie between 0 and 10. We modify the problem to search for the *minimum* of the function, F:

$$F = \left[x_1 x_2^2 + \frac{x_1}{x_2} - 61.4167\right]^2 + \left[75\frac{x_2}{x_1^3} - \frac{x_1 x_2}{50} - 12.230\right]^2 \tag{2.53}$$

It is convenient—and conceptually useful—to think of F in terms of a contour plot in the x_1–x_2 plane.

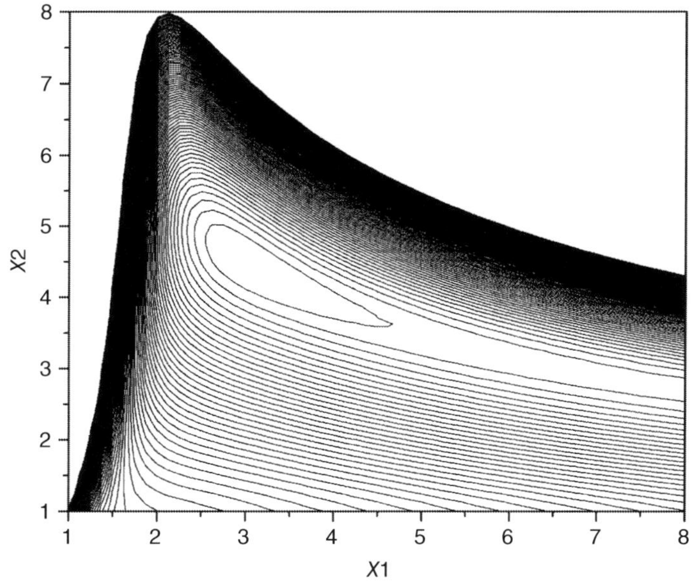

Figure 2.3 Contour plot for the function, F. The actual solution for this problem is $x_1 = 3$ and $x_2 = 4.5$.

Our objective here is to use the simplex pattern search to identify the "valley" that is prominent in the left-center of Figure 2.3; that is where the minimum is located. We will do this in the following way: Place a small equilateral triangle in Figure 2.3 and calculate the function F at each of the three vertices. We reject the "worst-valued" vertex and project the simplex away from it—that is, we form a new triangle using two of the original vertices with the third on the other side. We evaluate F at the new vertex and repeat the process.

There are several observations we should make about this scheme. Obviously, if we use a fixed-size triangle that is not small, we will not get a very good estimate for the solution location, regardless of the number of cycles employed. At the same time, if we make the equilateral triangle very small, it may take a very long time to follow an elongated valley such as the one seen in Figure 2.3. Clearly, we must change the size and shape of the triangle dynamically through the course of the search. This is where the improvements offered by Nelder and Mead (1964) come into play. For example, if we identify a new vertex, and the function F is improved in that direction, then why not accelerate the movement by elongating (stretching) the triangle? In this two-dimensional case there is no reason why we must maintain the equilateral form. If we can both stretch and contract the simplex, as needed, we should be able to follow just about any ridge or valley and do so efficiently.

The Nelder–Mead modifications to sequential simplex produced a powerful tool to find minima of functions. Although Nelder–Mead codes are a bit more complicated than search examples we have used previously in this chapter, a very effective version appears in detail in Chapter 6 in Shoup and Mistree (1987). The critical steps in the logic are as follows:

- The merit function F is constructed and the number of unknowns (dimensions) is specified.
- The initial length of a simplex side is specified and the estimates for the unknowns are provided.
- The vertices of the initial simplex are determined.
- The merit function, F, is evaluated at each vertex.
- The worst and best function values are determined.
- The centroid of the simplex is evaluated and the position of the reflected point is determined (projected away from the worst F of course).
- F is determined at the new point; if it is better, expansion is employed (factor of 2).
- If the reflected F is worse than the other vertices, contraction is implemented (factor of ½).
- A check for convergence (based upon the change in F) is made by comparison to a set value, say 1×10^{-6}.
- Results are directed to output if convergence is attained.

Let us see exactly how this works by applying the technique to the two-dimensional problem we introduced above. Some of the pertinent information along with results and the final value of the function being minimized are presented in the table below. Note that the final value of the merit function F has been *multiplied by* 1×10^7.

Initial Length for Simplex	Starting x_1	Values x_2	Final F ($\times 10^7$)	Final x_1	Values x_2
0.05	0.5	0.5	13.575	0.315976	0.0051447
0.10	1.0	1.0	2.574	0.315987	0.0051449
0.25	2.0	2.0	0.142	2.999995	4.500002
0.50	5.0	5.0	3.4615	2.99996	4.500026
0.75	7.0	1.0	4.7736	2.999953	4.500032
0.05	2.0	5.0	0.590	3.00001	4.499987
0.10	2.0	7.0	1.585	2.999982	4.500025
0.10	8.0	6.0	1.813	3.000005	4.499982
0.10	0.5	9.0	2.971	3.000031	4.499991

There are a couple of important characteristics of this technique that are revealed by this example: We must bear in mind that the starting point matters, and that the initial size of the simplex leg can impact the results. Moreover, the elongated valley apparent in Figure 2.3 is typical of many engineering problems. Although the Nelder–Mead version of sequential simplex can usually adapt to such circumstances, there are specialized search methods *designed specifically* to follow ridges and valleys. One example is the *Rosenbrock* method in which the axis of the search is aligned with the orientation of the ridge, allowing rapid progress toward the location of the optimum. The basic idea is easy to visualize in two dimensions; suppose we have an optimization problem involving two variables, x_1 and x_2. We begin the search with trials in the x_1 direction, identifying successes and failures. Assume we obtain a success. Then we search the x_2 direction, and if a success is achieved, we rotate the coordinate system such that one new axis is the vector sum of the two successes (and the other orthogonal to it, of course). When successes are achieved we increase the step size in that direction in the next cycle (Rosenbrock recommended a factor of 3). For a failure, the step size is contracted (and reversed) using 1/2. The appeal of Rosenbrock's approach is that the analyst obtains acceleration in *both distance and direction*. We will now use the previous example, illustrated by Figure 2.3, to reveal how Rosenbrock works in two dimensions. Our code is one that has been adapted from Shoup and Mistree (1987), and it requires specification of:

ε (test value for minimum step size, 1×10^{-5})
α (scaling factor for step-size increase, 1.25)
β (scaling factor for step-size reduction, 0.85)
ST (initial step size, 0.01)

Please note that we have selected sub-optimal values for both α and β; this has been done to better reveal the Rosenbrock scheme's progress toward finding the valley revealed in Figure 2.3. We are also going to start the search from the initial point, (7,7) and attempt to minimize the function given by eq. (2.53) (Figure 2.4).

A problem that can be encountered with the Rosenbrock method and one that has been mentioned by Beveridge and Schecter (1970) is that it can be slow to start particularly if the initial step size must be reduced repeatedly. It is, however, a very powerful tool that can be used to advantage for certain problems involving nonlinear, algebraic equations and more importantly, examples of programming logic for the Rosenbrock search can be found across the web and throughout the literature.

2.4.5 An Example of a Pattern Search Application: Ship Hull Design

Now let us look at a problem type that is quite different (in both intent and form) from the examples above; we will use the sequential simplex method. We still wish to minimize a merit function developed from a set of simultaneous nonlinear algebraic equations, but our purpose is unlike previous examples in this chapter. Imagine we are seeking an approximate relationship (a correlation) that could be used to predict the likely attainable speed (S) achieved by a conventional ship hull. It is apparent that some of the important factors are displacement, hull length (L), beam (B), available horsepower (P), and hull shape. Of course, we know that certain dimensionless ratios such as Reynolds number, Froude number, and Euler

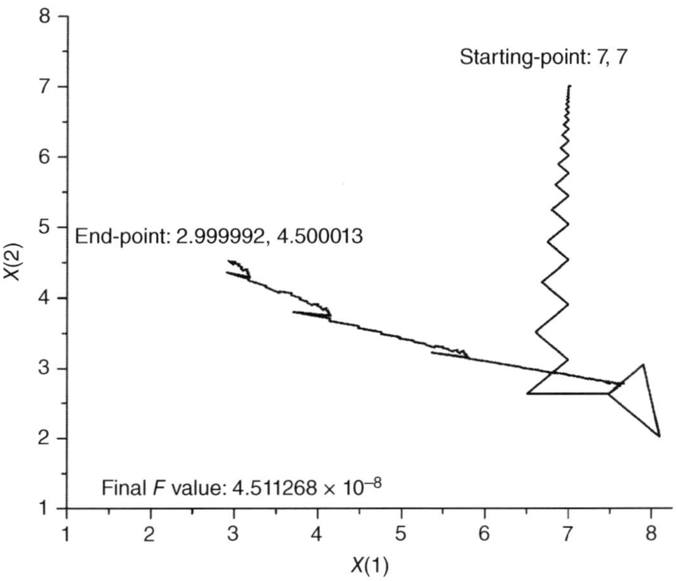

Figure 2.4 Progress of the Rosenbrock search method applied to the problem illustrated previously in Figure 2.3. Because of the very poor choices made for scaling factors and the starting point, 535 steps were required to identify the end-point, 2.999992 and 4.500013. Nevertheless, the final value of the objective function was 4.511268×10^{-8}. Notice how the search direction has changed to follow the valley shown in Figure 2.3.

number are important, but we are going to try a minimalist approach and not worry overtly about hydrodynamic forces. We propose that the ship's likely speed in knots can be determined in the following way:

$$S = aL^b B^c P^d. \tag{2.54}$$

We intend to attempt this by using some available data:

	Length	Beam	P (horsepower)	S (knots)
Titanic	883	92	46,000	21
Bismarck	793	118	150,170	30
Missouri	887	108	212,000	33
Nimitz	1040	134	260,000	31.5
Edmund Fitzgerald	711	75	7500	14
Elco PT boat	80	21	4500	41
Monitor	172	41	320	7–8

Keep in mind that we are using data from ships designed in very different eras for very different purposes, so we expect this may not produce the cleanest result we could imagine. The minimum merit function (F) value was found to be 4.9339 (using all of the data included in the above table), and the correlation that emerged was:

$$S = 34.9084L^{-0.28125}B^{-0.5764}P^{0.3738} \tag{2.55}$$

For the *Titanic* the correlation yields 21.15 knots, for the *Elco* PT boat, 40.84 knots, and for the *Monitor*, 8.34 knots. (An aside: *There are indications that Ericsson's original plans called for the installation of two engines. However, one engine has been recovered from the site of the Monitor's wreck and is now on display at the Mariners' Museum in Newport News, VA. In trials in 1862 the Monitor was able to achieve only 6.25 knots.*) One of the poorer results is the *Edmund Fitzgerald* for which the correlation yields 12.83 knots (instead of 14). For the battleship *Missouri* (BB-63), our equation indicates 34.1 knots and for the *Nimitz* (CVN-68), it yields 31.1 knots. This is encouraging, and the correlation may be good enough to make some rough projections. Suppose for example we wanted a high-speed (50 knot) hull that was 300 ft long; we choose a beam of 40 ft (though such a design will be prone to roll) because the importance of width to speed is clear. Our eq. (2.55) indicates that about 56,445 hp will be required for this hull to make a speed of 50 knots; the fastest ship in the US Navy, the 45+ knot LCS-1 *Freedom*, is 324 ft long with a beam of 43 ft and she has 96,000 hp. The correlation eq. (2.55) suggests that the *Freedom* should be capable of about 57 knots; however, the *Freedom* uses water jet propulsion which is known to be somewhat less efficient than a propeller. The Nelder–Mead version of sequential simplex is extremely useful for solving this and related types of nonlinear problems.

2.5 Algebraic Equations with Constraints

Suppose we have the algebraic equation,

$$f = 2x_1{}^2 + x_2{}^2 - 8x_1 + x_2 + 1 \tag{2.56}$$

accompanied by the constraint, $2x_1 - x_2 = 0$. We wish to find the stationary point (values of the independent variables that cause the derivatives to be zero), and we will do so using two very well-known methods. First, we will make use of the *Lagrange multiplier*, λ. We write a combination of the equation and the constraint as follows:

$$F = 2x_1{}^2 + x_2{}^2 - 8x_1 + x_2 + 1 + \lambda(2x_1 - x_2). \tag{2.57}$$

We now differentiate F with respect to all *three* variables:

$$\begin{aligned} \frac{\partial F}{\partial x_1} &= 4x_1 - 8 + 2\lambda = 0 \\ \frac{\partial F}{\partial x_2} &= 2x_2 + 1 - \lambda = 0 \\ \frac{\partial F}{\partial \lambda} &= 2x_1 - x_2 = 0. \end{aligned} \tag{2.58}$$

These three equations are solved yielding $x_1 = 1/2, x_2 = 1$, and $\lambda = 3$ (the reader may wish to verify this solution). Next, to provide a contrast we will apply *Courant's penalty method* to the very same problem; in Courant's scheme we seek a minimum for the *augmented functional*:

$$F = 2x_1{}^2 + x_2{}^2 - 8x_1 + x_2 + 1 + \frac{\varepsilon}{2}(2x_1 - x_2)^2 \tag{2.59}$$

Note that *no additional variable* has been introduced, only the penalty parameter, ε. Of course this means that the penalty method is computationally simpler than the Lagrange multiplier used above. This time we have two derivatives to set equal to zero:

$$\frac{\partial F}{\partial x_1} = 4x_1 - 8 + 2\varepsilon(2x_1 - x_2) = 0, \tag{2.60a}$$

and

$$\frac{\partial F}{\partial x_2} = 2x_2 + 1 - \varepsilon(2x_1 - x_2) = 0. \tag{2.60b}$$

With a bit of effort one can show that

$$x_2 = \frac{-4 + 12\varepsilon}{8 + 12\varepsilon} = \frac{-1 + 3\varepsilon}{2 + 3\varepsilon},$$

and this result can be used to find $x_1 = f(\varepsilon)$. As the penalty parameter, ε, becomes large, these expressions should converge to the correct values for the two variables. We will explore this process immediately below.

ε	$x_1(\varepsilon)$	$x_2(\varepsilon)$
1,	1.10000002384186	.400000005960464
10,	.59375	.90625
100,	.509933769702911	.990066230297089
1000,	.500999331474304	.999000668525696
10000,	.500100016593933	.999899983406067
100000,	.500010013580322	.999989986419678

We see that Courant's penalty method will approach the correct values for our two variables if the penalty parameter, ε, is sufficiently large.

2.5.1 Further Practice with the Lagrange Multiplier

The real value of the Lagrange multiplier may not be apparent to you, particularly if this is your first exposure to its use. Therefore, we will elaborate a little to better reveal its power for the identification of critical points. Suppose we have the algebraic equation,

$$f(x,y) = x^2 + 2y^2 - 9y, \tag{2.61}$$

with the constraint, $x^2 + y^2 = 9$. Let us define F as

$$F(x,y,\lambda) = x^2 + 2y^2 - 9y + \lambda(x^2 + y^2 - 9). \tag{2.62}$$

We differentiate F with respect to all three variables:

$$\frac{\partial F}{\partial x} = 2x + 2\lambda x \tag{2.63a}$$

$$\frac{\partial F}{\partial y} = 4y - 9 + 2\lambda y \tag{2.63b}$$

$$\frac{\partial F}{\partial \lambda} = x^2 + y^2 - 9, \tag{2.63c}$$

and set all three equal to zero as before. Notice that eq. (2.63a) reveals two possibilities: either $x = 0$ or $\lambda = -1$. If $\lambda = -1$, eq. (2.63b) indicates that $y = 9/2$—which cannot be allowed as it violates our constraint (we presume x must be real). On the other hand, if $x = 0$ then eq. (2.63c) shows $y = \pm 3$ and choosing -3 we find $f(x,y) = 18 + 27 = 45$. Conversely, $y = +3$ yields $f(x,y) = 18–27 = -9$. If we construct a contour plot of $f(x,y)$, recognizing that both x *and* y must be contained within the interval -3 to $+3$, what will we see? Both critical points lie on the vertical line corresponding to $x = 0$, of course, with the maximum (45) at the bottom and the minimum (-9) at the top where $y = +3$.

Next, we will look at a more physically meaningful situation where we can use the Lagrange multiplier: Suppose you are organizing an expedition to Antarctica and you select hemispherical shells that will be used both for shelter and for the storage of supplies and instrumentation. The floors will be simple circles that connect to the bottom of the hemisphere with a series of latches. There are several attractive features of the hemispherical design, including lessened susceptibility to wind loads, and the fact that the shells can be stacked on shipboard and occupy less deck space. Now suppose the floor is placed in the x–y plane with its center at the origin; the *surface* of the hemisphere is given by the equation

$$x^2 + y^2 + z^2 = 49, \tag{2.64}$$

and therefore the height of the structure at its peak (where $x = y = 0$) will be 7 ft. This raises an interesting question: Since most of the supplies will be shipped in rectangular containers, what is the largest "box" that will fit inside the hemisphere? We are interested in maximizing the box volume:

$$V = 2x2yz. \tag{2.65}$$

Remember, the center of the circular floor is located at $x = 0$, $y = 0$. Thus,

$$F = 2x2yz + \lambda(x^2 + y^2 + z^2 - 49). \tag{2.66}$$

Find the maximum volume of the "box" that will fit inside the hemispherical shelter and verify that its length and width will both be 8.08 ft. What is the correct value for λ?

Answer: Maximum volume about 264 ft^3 *and* $\lambda = -8.0829$.

2.6 Conclusion

One of the interesting aspects of the solution of algebraic equations is the spectrum of options available. In many cases the analyst can choose from several solution techniques, all capable of producing the desired result. At the same time, for some cases no method will yield an appropriate result. For students in the applied sciences, balancing the focus between procedure and result may improve the chance of success and lessen the chance of serious error. Obviously for "real" problems we may have little or no information available in advance, and when these problems involve *nonlinear* algebraic equations we may not have much insight into where the solution we are seeking may be found. Although we will encounter cases where (for example) we will know that the volume cannot be negative, or that the summation of the mole fractions must be one (1), situations arise where we have no idea of the appropriate magnitude of the unknowns. Such instances are disconcerting since it is difficult to know whether or not the solution procedure is working properly. These comments are not meant to alarm the student, they are only intended to point out that circumspection is critical; for nonlinear algebraic equations, the analyst should consider the physical situation very carefully *before* initiating an attempt to find a solution.

We can underscore the observations above with a final example for which we will use a pair of widely available tools. Consider the set of nonlinear algebraic equations:

$$x^2 + \frac{y^3}{x} = 1.84401 \quad \text{and} \quad x + xy + x^2y^2 = 2.80076. \tag{2.67}$$

Assume we know that both x and y are positive for the solution we seek (in fact, $x = 1.275$ and $y = 0.653$). We first use a scientific calculator from the TI-89, TI-92 family (TI is the abbreviation for *Texas Instruments*). From the *MATH/Algebra* menu we select *solve*( ), and enter our two equations as shown below:

$$solve(x\wedge 2 + y\wedge 3/x = 1.84401 \quad \text{and} \quad x + x{\cdot}y + x\wedge 2 \cdot y\wedge 2 = 2.80076, \{x = 2, y = 0.8\})$$

and press *ENTER*. Using the guesses, $x = 2$ and $y = 0.8$, the TI-89 produces $x = 1.275$ and $y = 0.653003$. However, if we start with $x = 1$ and $y = 1$, we obtain $x = 0.985357$ and $y = 0.951078$ (verify that these values are a solution). Naturally, a software package like Mathcad also has capabilities for problems of this type, using the *Solve Block*:

$$x := 2$$

$$y := 2$$

Given

$$x^2 + \frac{y^3}{x} = 1.84401$$

$$x + x \cdot y + x^2 \cdot y^2 = 2.80076$$

$$\begin{pmatrix} x\text{ val} \\ y\text{ val} \end{pmatrix} := \text{Find}\,(x, y)$$

$$x\text{ val} = 1.275$$

$$y\text{ val} = 0.653$$

Note that the initial estimates used at the top were $x = 2$ and $y = 2$. Once again, however, if we select 1 and 1 we obtain $x = 0.985$ and $y = 0.951$; when we insert these values into the pair of eq. (2.67), we find 1.84341 (instead of 1.84401) and 2.79921 (instead of 2.80076). Even a very simple nonlinear algebraic problem can produce undesired or incorrect results if a solution tool is used incautiously.

Problems

2.1 Use Newton–Raphson to find a solution for the equation

$$x^5 - 2x^3 + 6x - 0.851372 = 0$$

It is known that the solution we are seeking is between 0 and 1.
Answer: 0.14286. *Of course, there are four other solutions as well, two pair of complex conjugates.*

2.2 A spacecraft with a mass of 2,500 kg is making the trip from the earth to the moon. We want to determine the point at which the gravitational force between the earth and the ship, and the gravitational force between the ship and the moon, are equal in magnitude. The expression for gravitational force is

$$F = -\frac{GM_1M_2}{r^2},$$

and the gravitational constant, G, is 6.674×10^{-11} N m^2/kg^2. The masses of the earth and moon are 5.972×10^{24} kg and 7.348×10^{22} kg, respectively. Assume that the center-to-center distance between the earth and moon is 382,500 km (3.825×10^8 m).

Answer: They are equal when the earth-to-spacecraft distance is 3.443×10^8 *m, about* 90% *of the total distance.*

2.3 Solve the set of simultaneous, linear algebraic equations represented by the augmented coefficient matrix:

5	1	1	−1	−1	0.899171
1	2	1	−4	1	0.223846
2	−1	−1	−1	6	0.49982
3	3	−2	−3	−3	0.072511
−1	−1	−1	7	7	0.979619

2.4 Consider a steel pipe with a nominal diameter of 3″ (actually 3.068″ ID). Water flows through the pipe (which is 980 ft long) at an average velocity of 8.75 ft/s. The Reynolds number associated with this flow is

$$\text{Re} = \frac{d\langle V\rangle\rho}{\mu} = \frac{(3.068/12)(8.75)(62.4)}{(1)(6.72 \times 10^{-4})} = 207,729$$

The friction factor for this flow is approximately given by: $f = 0.0791/(\text{Re})^{1/4}$. Therefore, the value of f for the original flow is about 0.0037. In an effort to increase the flow rate, a parallel (also nominal 3″) pipe is installed *over the last* 600 ft. Find the (increased) flow rate through the new arrangement. The design equation for flow through a horizontal pipe is:

$$\Delta\frac{1}{2}\langle V\rangle^2 + \frac{p_2 - p_1}{\rho} + \sum\frac{1}{2}\langle V\rangle^2\frac{L}{R_h}f = 0.$$

This allows us to calculate the pressure drop through the original 3″ pipe; initially, there is no change in kinetic energy, so

$$\frac{p_1 - p_2}{\rho} = (1/2)(8.75)^2\left(\frac{980}{0.0639}\right)(0.0037) = 2172.3\ \text{ft}^2/\text{s}^2.$$

We now multiply by the fluid's density and take care of the mass–force conversion problem using g_c: The result is 29.26 psi ($p_1 - p_2$). In the revised installation, the *overall pressure drop* is exactly the same, but we have an additional 3″ pipe over the last 600 ft. Find the new flow rate by writing the design equation three times, for the initial 380 ft of 3″ pipe and then once for each of the two parallel legs extending from 380 to 980 ft.

2.5 A piping problem with simultaneous nonlinear equations: Water flows for 800 ft through a straight (horizontal) run of 4 in. ID pipe. The pressure at the entrance (plane 1) is 35 psig. At plane 2 the 4 in. pipe splits into two parallel legs ("A" and "B") of straight pipe for an additional 900 ft. The discharge (plane 3) occurs at 1 atm pressure (0 psig). Pipe "A" has an inside diameter of 1.5″ and pipe "B" has an ID of 3″. Assume that the density of water (ρ) is 62.4 lb$_m$/ft^3 and the viscosity (μ) is 1 cp. Determine the expected flow rate of water through this piping network in gpm (gallons per minute). The mechanical energy balance for the 800 ft of 4″ pipe is:

$$\frac{p_2 - p_1}{\rho} + \frac{1}{2}V_{12}^2\frac{L_{12}}{R_{H12}}f_{12} \cong 0.$$

R_H is the hydraulic radius, $R_H = d/4$, and the friction factor for hydraulically smooth pipe can be approximated by $f \approx 0.0791/\text{Re}^{1/4}$. The Reynolds number for the first 800 ft of 4″ pipe is

$$\text{Re}_{12} = \frac{d_{12}V_{12}\rho}{\mu}.$$

The usual force–mass problem we encounter with English units appears in the mechanical energy balance; note that the frictional loss term has dimensions ft^2/s^2, whereas the pressure term is lb$_f$/ft^2 divided by lb$_m$/ft^3. Thus, the pressure term must be multiplied by g_c, 32.17 ft lb$_m$/(lb$_f$ s^2). You will immediately note that there are four unknowns in this problem, three velocities (V_{12}, V_{23A}, and V_{23B}), and the intermediate pressure, p_2. Therefore, solution will require that we write three mechanical energy balances (one for each pipe leg) and a volumetric flow rate balance at plane 2. The three flow (cross-sectional) areas are 0.08727 ft^2, 0.01227 ft^2, and 0.04909 ft^2 for the 4″, 1.5″, and 3″ pipes, respectively. To convert cp (centipoise) to lb$_m$/(ft s), multiply by 6.72×10^{-4}.

Answer: 219 *gpm*

2.6 We will consider a problem in chemical equilibria in which a compound is put into water and dissociation occurs. Let us preface this problem with a little review.

Compounds like silver chloride (AgCl) are sparingly soluble in water; at 10 °C about 0.00089 g AgCl will dissolve in one liter of water. An equilibrium will be established among the three species:

$$K = \frac{[Ag^+][Cl^-]}{[AgCl]},$$

where K is the equilibrium constant. Water itself is a weak electrolyte for which

$$H_2O \Leftrightarrow H^+ + OH^-.$$

At 25 °C, $K_w = [H^+][OH^-] = 1 \times 10^{-14}$. In contrast, sodium hydroxide (NaOH) is a strong electrolyte that will almost completely dissociate in water. Therefore, if we add 0.005 mol NaOH (about 0.2 g) to a liter of water:

$$1 \times 10^{-14} = [H^+][OH^-],$$

but with complete dissociation, $[OH^-] = 5 \times 10^{-3}$, so

$$[H^+] = \frac{1 \times 10^{-14}}{5 \times 10^{-3}} = 2 \times 10^{-12}.$$

Since $pH = -\log_{10}[H^+]$, $pH = 11.7$.

Now, suppose we place 0.0002 mol HCl in a liter of water at 25 °C. We want to find the concentrations of all four species, as well as the *pH* of the resulting solution. Therefore, we need:

$$[H^+][Cl^-][HCl] \quad \text{and} \quad [OH^-].$$

The following relationships are available to us:

$$[H^+][OH^-] = K_w = 1 \times 10^{-14} \frac{[H^+][Cl^-]}{[HCl]} = K = 1 \times 10^3$$

$$[HCl] + [Cl^-] = 2 \times 10^{-4} \text{ (total concentration)}$$

$$[H^+] = [Cl^-] + [OH^-] \text{ (electroneutrality)}$$

2.7 Use the technique of your choice to find a solution for the following set of equations, given that all of the unknowns lie between zero and one:

$$x_1{}^2 \exp(-x_2) + x_3 = 0.385028$$

$$\frac{x_1}{x_2{}^2 + x_3{}^2} = 2.20022$$

$$x_1 + x_2 - x_3{}^3 = 0.710317$$

2.8 Consider the pair of nonlinear algebraic equations:

$$xy + x^2y^3 = 1.05015 \quad \text{and} \quad \frac{x}{y} + 3xy^2 = 2.70508.$$

It is known that both x and y are positive. Suppose we factor x out of the second equation and use the result to remove x from the first equation. Then it should be possible to use Newton–Raphson to solve the problem. Doing this, one can discover that $y = 1.36365$. Using *two different methods*, verify this solution and find a second one as well.

Answer: Using the TI-nspire procedure, solve(…), we obtain $x = 0.859087$ *and* $y = 0.793379$.

2.9 The equation, $0.5 + \exp(-0.085t)\sin(2t)$, has a finite number of zeros for positive values of t between 0 and 11. Devise a method that will locate and count all of them.

2.10 Most airline travelers would prefer to spend less time in the air. Clearly, one way this could be accomplished is to increase the speed of aircraft. Presently, large passenger airplanes cruise at high-subsonic speeds and at altitude this means about 500 to 600 mph. If longer flights could cruise at Mach 2 or 3 (or more), travel times would be reduced significantly. The problem is twofold: We have the rise in drag accompanying supersonic flight (which dramatically increases fuel consumption) and we have aerodynamic heating. For example, if we look at the Concorde (now retired from service) we find fuel consumption at full power of 6,180 gal (A1 jet fuel) per hour; the Concorde was able to carry one passenger about 17 mi on a gallon of fuel. We can contrast this with a Boeing 747; with 500 passengers, a 747 can carry one passenger between 60 and 100 mi on one gallon of fuel. Of course it will take two and a half times as long to reach the destination! So this raises an interesting question: What must the design characteristics of an advanced aircraft resemble in order for the craft to carry a hundred or more passengers at a cruising speed of Mach 3 or 4? Here are some approximate data from open sources:

Aircraft	Span	Gross Weight	Thrust	Cap	Max Speed
LH C-140	54	38,940	6,000	12	512
DH Comet	115	162,000	42,000	101	525
B 737-700	117	150,000	52,600	149	540
AB 380	262	1,260,000	312,000	575	561
H FGA.9	34	18,000	10,050	1	627
NA F-100D	39	34,832	16,950	1	864
MD F-101B	40	52,400	33,800	2	1130
Concorde	84	408,000	152,000	128	1,330
NA A3J-1	53	49,500	32,300	2	1,385
Con B-58A	57	160,000	62,500	3	1,385
LH F-104G	22	27,000	16,150	1	1,450
Con F-106A	38	35,000	24,500	1	1,525
MiG 25	46	80,950	49,400	1	1,900
NA XB-70	105	534,700	168,000	4	2,056
LH SR-71	56	170,000	68,000	2	2,090
X-15	22	33,000	70,400	1	4,519

Some observations: Note the relatively high take-off weights for the Concorde and the SR-71; this is indicative of the fuel load required if one wants to cruise at high speeds. Even so, the SR-71 had to be refueled routinely in flight in order to accomplish its long-range surveillance missions. The very low weight of the X-15 was possible because it was air launched (dropped from underneath a B-52). One factor we referred to previously that is not reflected here is the trouble caused by aerodynamic heating. For many supersonic aircraft this was/is a serious limitation on the performance envelope. The MiG 25, for example, could only be flown at high Mach for a few minutes before the structural components would attain unsafe temperatures. The X-15 suffered serious thermal damage during its October 1967 flight in which it achieved Mach 6.7. The tabulated data make it pretty obvious that to carry 100 passengers at the desired speeds we will need a wing span of 90 to 100 ft and a take-off weight of least 200,000 to 300,000 lbs.

2.11 Find a solution for the following set of nonlinear algebraic equations:

$$x_1^{\,5}\exp(-2x_2) + 10/x_3 = 2.34284$$

$$x_1 x_2^{\,2} + \sqrt[3]{x_3} = 25.5966$$

$$\frac{x_1 + x_3^{\,3}}{\sqrt{x_2}} = 48.1033$$

Answer: Sequential simplex works well for this problem; by subtracting the constants, squaring each equation, and adding them together we minimize the sum to find 2.2000, 3.2999, and 4.4000 for x_1 through x_3, respectively.

2.12 Müller's method, a root-finding technique in which a parabola is fit to three points, can be modified to find complex roots. Consider the fourth-degree polynomial,

$$x^4 - \frac{17}{2}x^3 + \frac{353}{18}x^2 - \frac{17}{2}x - \frac{185}{18} = 0.$$

Search for real roots using the method of your choice and then use polynomial deflation to locate the complex conjugate pair. Confirm your findings using a calculator (like a TI-89) or suitable commercial software (you should get: $-1/2$, $+5$, and $+2 \pm i/3$). Then, repeat the analysis by consulting the literature (e.g., Section 7.4 in Chapra and Canale's book, *Numerical Methods for Engineers, Fourth Edition*, McGraw-Hill, New York, 2002) and developing a routine to use Müller's method for the same purpose. The following discussion should help you get started.

We assume that the function can be approximated with the polynomial:

$$f(x) = a(x - x_0)^2 + b(x - x_0) + c.$$

This choice insures that $f(x_0) = c$ and thereby lessens our workload. We arbitrarily select 4, 6, and 8 for our three trial points; we find $f(4) = -18.5$, $f(6) = 104.714$, and $f(8) = 920.783$. We choose $x_0 = 8$ for our reference point, so $c = 920.783$. Our functional values at $x = 4$ and $x = 6$ give us two simultaneous equations for a and b: $a = 86.607$ and $b = 581.249$. Of course, we want to identify the value for x where $f = 0$. Thus (after dividing by 86.607):

$$0 = (x - 8)^2 + 6.7113(x - 8) + 10.632.$$

Letting $z = (x-8)$, we have the quadratic: $z^2 + 6.7113z + 10.632$, with solutions at -4.14835 and -2.56295. Remembering to add 8, we get our improved estimates and the closest one is: $+5.43705$. One advantage of Müller's technique is the use of the quadratic equation—we will be able to locate complex roots should they arise. We can also minimize our workload by re-using the trial points closest to the root estimate. In this case we would discard $x = 8$, retaining 4, 5.43705, and 6 for our next cycle.

Answer: $2 \pm i/3$, $-1/2$, 5.

2.13 The cargo ship *MSC Fabiola* is one of the largest container transports ever constructed. She is 366 m (1,200 ft) long with a beam of 48 m; the gross tonnage is 140,259. This class of freighter will not fit through the Panama Canal (which is one of the reasons why the canal is being enlarged). Estimate the maximum speed of the *MSC Fabiola* given that her turbines produce 98,000 hp (single screw). It has been reported that she is capable of 19.5 knots. Is that speed realistic? Does it appear that the correlation given by eq. (2.55), Section 2.4, is appropriate for the *Fabiola*?

2.14 Solve the set of simultaneous algebraic equations:

$$(X_1 - X_2)^2 + X_1X_2X_3 = 61.0229$$

$$X_1 + X_2 + X_3 = 15.8225$$

$$\frac{X_2}{X_3^{\,2}} + \frac{X_1}{X_2} = 11.366$$

We know that for all three variables, $0 \le X_n \le 10$. Use the method of your choice.

2.15 The Lagrange multiplier (λ) is often used for constrained optimization problems. Consider the equation,

$$2x_1^{\,2} - 3x_2^{\,2} - 5x_1 + x_2 = -5.1207,$$

accompanied by the constraint, $2x_1 - 3.066202x_2 = 0$. Use the Lagrange multiplier to solve this problem, i.e., find the stationary point subject to the constraint.

2.16 Use your choice of methods to seek a solution for the two simultaneous equations:

$$\frac{x_1 \exp\left(-x_1^2\right)}{x_2} + x_2^3 = 5.976$$

and

$$x_1^2 x_2 + \frac{x_1 \ln(x_1)}{x_2^2} = 3.554.$$

We know that both unknowns are between 0 and 2.

Answer: 1.377 and 1.803 for x_1 and x_2.

2.17 The first seven *Chebyshev* (or Tschebysheff) polynomials of the *first kind* are:

$$T_0(x) = 1 \qquad T_1(x) = x$$
$$T_2(x) = 2x^2 - 1 \qquad T_3(x) = 4x^3 - 3x$$
$$T_4(x) = 8x^4 - 8x^2 + 1 \qquad T_5(x) = 16x^5 - 20x^3 + 5x$$
$$T_6(x) = 32x^6 - 48x^4 + 18x^2 - 1$$

These polynomials are solutions for the Chebyshev differential equation (for Mathcad™, see *Tcheb(n,x)*). Suppose we need to identify all of the zeros for $T_6(x)$ on the interval, $-1 \le x \le +1$. Prepare a plot of this polynomial and then use the method of your choice to find accurate values for the roots.

Partial answer: $\frac{-(\sqrt{3}+1)\sqrt{2}}{4}$ and $\frac{-\sqrt{2}}{2}$

2.18 A liquid-phase reaction was carried out in a batch reactor to study A + B → C + D. The experiment was conducted for four different trials and stopped at different elapsed times. The data obtained consisted of the duration of the experiment and the concentration of species D:

Time (s)	780	2,040	3,540	7,200
[D] (mol/L)	0.0112	0.0257	0.0367	0.0552

Two different rate expressions have been proposed for this reaction:

$$\frac{d[A]}{dt} = -k_1[A], \quad \text{and} \quad \frac{d[A]}{dt} = -k_2[A][B] = -k_2[A]^2,$$

which when integrated yield

$$\ln[A] = -k_1 t + C_1 \quad \text{and} \quad -\frac{1}{[A]} = -k_2 t + C_2,$$

respectively. Which rate expression is more nearly correct? There is no "D" present in the reactor initially, and the initial concentrations for both "A" and "B" are 0.1.

2.19 Solving the *Blake–Kozeny* equation: At low Reynolds numbers, the pressure drop for flow through a packed bed is approximately described by the Blake–Kozeny (or Kozeny–Carman) equation:

$$\frac{\Delta P}{L} = -\frac{150\mu}{\psi^2 d^2}\frac{(1-\varepsilon)^2}{\varepsilon^3} V_S.$$

L is the length of the packed section, μ is the viscosity of the fluid, ψ is the sphericity of the particles, d is their effective diameter, ε is the void fraction, and V_S is the superficial velocity of the fluid through the column. Assume the column is packed with nearly spherical particles such that $\psi \approx 1$. The bed depth, L, is 3 m, the viscosity of the fluid, μ, is 0.025 g/(cm s), and the superficial velocity (V_S) is 3 cm/s. The pressure drop across the packed bed is 17 psi, or 1.172×10^6 dynes/cm^2 and the approximate particle diameter is 0.25 cm. What is the void fraction (ε) of the bed? Is the Blake–Kozeny equation appropriate for this case?

Answer: $\varepsilon = 0.286$.

2.20 In 1928 B. G. van der Hegge-Zijnen published a study of air flow near the surface of a roughened (waffle-like) plate. The nominal air velocity past the plate was about 16 m/s, and the measurements were made at a horizontal (x) location 175 cm past the leading edge. The y-coordinate was perpendicular to the plate, extending out into the flowing fluid. Some of the data are represented in the following table:

y (cm)	0.03	0.05	0.075	0.1	0.15	0.2	0.25	0.3	0.5	0.7	1	2	3	4
V (m/s)	1.25	2.0	2.93	4.04	6.0	7.5	8.0	8.6	9.8	10.24	11	14	16	16.5

van der Hegge-Zijnen proposed an empirical relationship to describe $V(y)$: $V = ay^b$. Our task is to determine appropriate values for both the prefactor a and the exponent b. Since $\ln(V) = \ln(a) + b\ln(y)$, we could merely select a pair of entries from the table and solve the simultaneous equations. For example, using the y-locations 0.075 and 0.7, we obtain

$$V = 12.504y^{0.56}.$$

Therefore, $V(y = 0.075) = 2.93$ m/s and $V(y = 0.7) = 10.24$ m/s. We see that the proposed power law fits the two points selected—as it must, of course. But we should ask ourselves if the quality of fit is uniformly good over the entire range of y, and we prudently suspect that it is *not*; if we test $y = 0.03$ cm we get 1.75 m/s and for $y = 0.2$ cm, 5.08 m/s. One (possibly) better approach would be to minimize the sum of the squares of the deviations between the proposed power law and the experimental data (using all of the data) just as we did previously in this chapter. However, in this case we have a *linear* relationship between $\ln(V)$ and $\ln(y)$, so we can use linear regression—a tool familiar to nearly every student in the applied sciences. Remember, in what follows $y = \ln(V)$ and $x = \ln(y)$; n is the number of data pairs, and B is the slope of the "best" line.

$$B = \frac{n\sum x_i y_i - \sum x_i \sum y_i}{n\sum x_i^2 - \left[\sum x_i\right]^2} = \frac{\frac{1}{n}\sum x_i y_i - \bar{x}\bar{y}}{\sigma_x^{\ 2}},\ \text{and the intercept is}$$

$$A = \frac{\sum y_i \sum x_i^2 - \sum x_i \sum x_i y_i}{n\sum x_i^2 - \left[\sum x_i\right]^2}.$$

The correlation coefficient (R) tells us something about the quality of the fit, although we must keep in mind that even if $R \approx 1$, indiscriminate use of linear regression can yield a nonsensical correlation.

$$\sigma_x^{\ 2} = \frac{1}{n}\sum x_i^2 - \bar{x}^2 \quad \text{and} \quad \sigma_y^{\ 2} = \frac{1}{n}\sum y_i^2 - \bar{y}^2 \quad \text{and} \quad R = \frac{B\sigma_x}{\sigma_y}.$$

If we create a little program to carry out these calculations and use all of the available data, we find $V = 10.855y^{0.458}$ with a correlation coefficient (actually, the coefficient of determination, R^2) of 0.88. Experiment a little with this equation and verify that it is not really useful. What has gone wrong? Because we are trying to fit an equation to $V(y)$ *in the boundary layer*, we must be careful to avoid using the data points at larger y where we see an asymptotic approach to the freestream velocity (≈16 m/s); the power law will not work there. Limit your analysis to the first 11 data pairs and carry out the linear regression again. What form does the power law assume in this case?

Note: There are many tools available that one can employ for linear regression, including Excel, Origin, and Mathcad. In addition, nearly every scientific calculator has this capability; for example, for the TI-89 family one may use LinReg or LnReg. For the TI-Nspire™ one may use LinRegBx or Mx. If we use a TI-89 on the first nine data pairs we get $V = 22.24y^{0.78}$ with a correlation coefficient of 0.979 and $R^2 = 0.959$. If you wish to try this, carry out the following steps: In function graphing mode, create and store two nine-item lists, L1 and L2, enclosed with brackets {…} with the entries comma delimited. Then, use LinReg from the catalog as: LinReg L1,L2 ENTER. To view the results use ShowStat ENTER. An illustration of the fit obtained from this analysis is shown here (Figure P2.20).

2.21 The space shuttle was lifted into earth orbit primarily by two solid rocket boosters (SRBs). The SRBs used a mixture of ammonium perchlorate, aluminum powder, and iron oxide with a polybutadiene binder, and each produced a maximum thrust of approximately 3.3×10^6 lb_f. The thrust achieved by the SRBs depended mainly upon the rate at which propellant was consumed (dM/dt); however, the data indicate that the thrust was also a sensitive function of the burning area (A). Figure P2.21 illustrates the propellant consumption and the burning area during the two minutes of SRB thrust production.

Suppose we need to develop a correlation that will approximate the thrust, T, produced at any point during the SRB burn time, and we assume that it takes the form:

$$T = a\left(\frac{dM}{dt}\right)^b A^c.$$

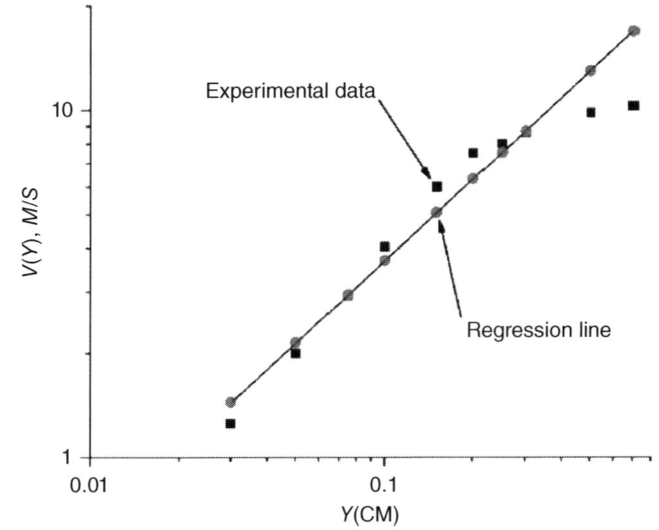

Figure P2.20 Comparison of data with regression line obtained with *LinReg*. Note how linear regression underestimates the slope for the critical data points inside the boundary layer.

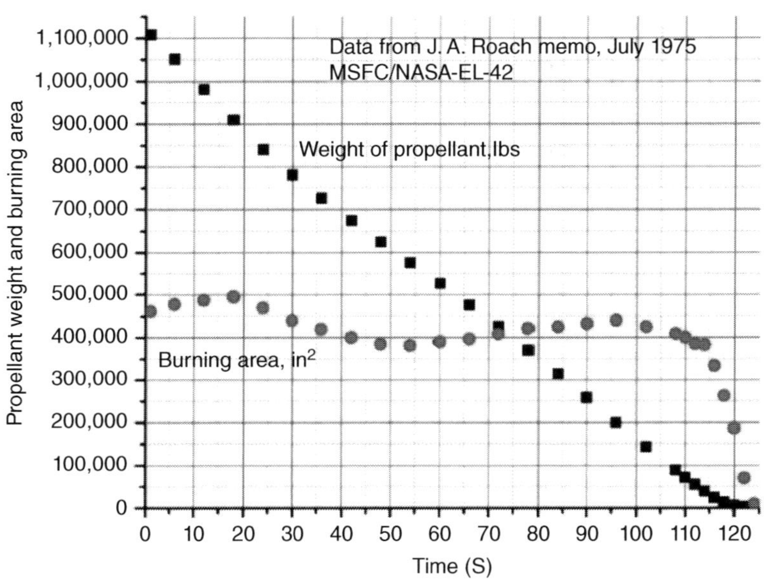

Figure P2.21 SRB performance from J. A. Roach memo (MSFC/NASA-EL-42) of July 1975.

Furthermore, we have the following specific data:

Time (s)	1	20	95
Thrust, lb_f ($\times 10^{-6}$)	2.8	3.3	2.0

Although dM/dt is negative, it will be convenient to treat this rate as an absolute value. Obtain values for the three parameters, a, b, and c, and then use them to estimate the thrust at $t = 112$ s when $dM/dt = (-)7{,}830$ lb/s

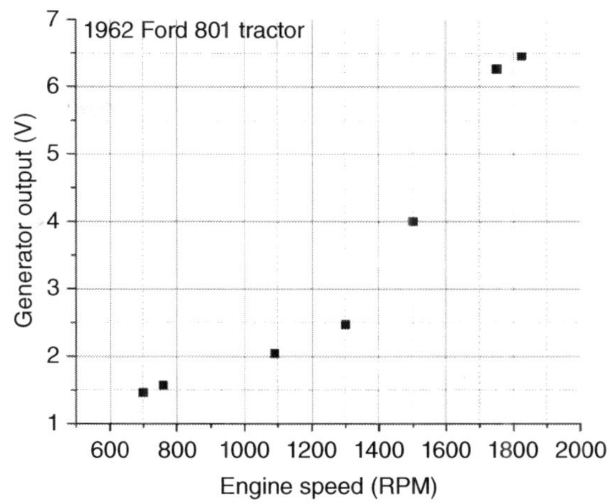

Figure P2.22 Measured generator output as a function of engine speed for an antique Ford tractor.

and $A = 391{,}000$ in^2. At what approximate time does the thrust fall to 175,000 lb$_f$? *Note*: At $t = 0$, dM/dt is about (−)12,200 lb/s and at $t = 50$ s, $dM/dt \approx (-)8{,}700$ lb/s.

One possible answer: $b = 0.765$ *and* $c = 2.568$ *and at 112 s,* $T \approx 1.378 \times 10^6$ lb_f.

2.22 Many antique vehicles were equipped with 6-V electrical systems, and the battery charge was maintained by a generator (as opposed to the alternator found on contemporary vehicles). Typically, generator output was too low at idle speeds to be useful, so prolonged low-speed operation would often result in a run-down battery. The generator output of a 1962 Ford tractor was tested, and the results are provided in Figure P2.22.

Notice that at idle speeds (usually 500–600 rpm) the generator output is only a little more than 1 V. Fit a polynomial (specifically $V = a_1 s + a_2 s^2 + a_3 s^3$) to these data and use it to estimate the generator output at 950, 1,600, and 1,900 rpm. Was the cubic equation well suited to this interpolation problem?

One answer: At 1,900 rpm, 6.67 V.

2.23 A negative feedback control system has the characteristic equation:

$$s^3 + 6s^2 + 11s + 6 + K_C = 0.$$

It is known that the system exhibits stable behavior in response to a bounded input if $K_C = 10$. In an effort to improve loop performance, the controller gain is increased. Solve this cubic equation and determine the value of K_C at the stability threshold (i.e., when the real part of the complex root is zero). What is the value of the *real* root if $K_C = 75$?

Answer: critical gain $K_C = 60$, and -6.296 for $K_C = 75$.

2.24 The previous problem hinted at a technique that was once widely used for the analysis of the behavior of systems equipped with automatic process control, *root locus*. Consider a second-order process provided with straight proportional control and a measurement device in the feedback loop that has first-order dynamics. Suppose the characteristic equation then assumes the form:

$$1.5s^3 + 5s^2 + 4.5s + (1 + K_C) = 0.$$

Obviously, a control engineer would like to know how the dynamic behavior of the loop would be affected by changes in the controller gain, K_C. If the characteristic equation has complex roots, the system behavior will be oscillatory; furthermore, if the real part of those complex roots is positive, any oscillation will grow exponentially in time—the system will be unstable. Therefore, if we determine the roots for various values of K_C we can prepare a plot of the roots in the complex plane that will tell us a lot about the behavior of the closed loop. Let's construct a little table:

K_C	Complex Root	Real Root
1/4	$-0.6229 \pm 0.1055i$	-2.0874
1/2	$-0.5878 \pm 0.3435i$	-2.1578
1	$-0.5313 \pm 0.5522i$	-2.2708
2	$-0.4468 \pm 0.7875i$	-2.4397
4	$-0.3277 \pm 1.0665i$	-2.6779
8	$-0.1667 \pm 1.4044i$	-3
16	$+0.0453 \pm 1.8188i$	-3.4239
32	$+0.3195 \pm 2.3316i$	-3.9723

It is clear that the closed loop is unstable for $K_C = 16$ since the critical part of the output has the form:

$$e^{+0.0453t}\cos(1.8188t) + \ldots.$$

If we plot the locus of the roots (as the controller gain is increased) it is evident that the system will become unstable if K_C is between 8 and 16 (you may want to confirm that the stability threshold occurs at 14). Now suppose that the characteristic equation is fourth degree:

$$2s^4 + 6s^3 + 11s^2 + 5s + (1 + K_C) = 0.$$

Plot the roots of this equation in the complex plane (with increasing values of K_C) and use your figure to estimate the controller gain at the stability threshold.

Answer: $K_C \approx 6.778$.

2.25 *Car and Driver*™ magazine tested a 2022 Corvette (376 cid and 495 hp) and obtained the following acceleration data:

Time (s)	2.8	7.2	13.2	19.4
Speed (mph)	60	100	130	150

These data indicate diminishing acceleration with time which is completely expected as the rising drag force overwhelms the engine's power output (that is, the top speed of the car is drag limited). Use these data to predict the car's top speed as well as the speed attained in 11 s under full-throttle acceleration.

One answer: 185 mph and 117.7.

Note: It is clear that a polynomial like $S = at + bt^2 + ct^3 + dt^4$ will not work. We must get the correct asymptotic behavior if we are to extrapolate successfully; this means that the derivative of speed (dS/dt) should go to zero as t becomes large.

2.26 The vapor pressure of benzene (C_6H_6) varies with temperature as shown in the provided table:

T (°C)	p^* (mmHg)
−36.7	1
−11.5	10
15.4	60
42.2	200
80.1	760
103.8	1,520
178.8	7,600

Estimate the temperatures for which p^* is 100 mmHg and 1,000 mmHg, then use your solution procedure to find p^* at 200 °C. Is this last estimate problematic?

Suggestion: For this interpolation, try an equation of the form: $\ln(p^*) = a + b(T - T_0) + c(T - T_0)^2$.

Answer: $p^* = 1{,}000$ mmHg at 88.7 °C. At 200 °C, $p^* \approx 9{,}723$ mmHg. The use of a polynomial interpolation for extrapolation is likely to produce large errors. In this case 9,723 mmHg is actually close, but only because the 178.8 °C point was used to develop the equation.

2.27 A village must extend its water system to provide 100 gpm to a hilltop development on the edge of town. The distance from the existing storage tank-pump site is 13,500 ft, and there is an increase in elevation of 75 ft. Both the pump inlet and discharge pressures will be 25 psig, so there is no Δp. The PVC pipe is hydraulically smooth, and the entire line will be of uniform diameter (thus, no ΔKE). The village utility has two available pumps, one capable of 145 ft lb_f/lb_m and the other 315 ft lb_f/lb_m. The mechanical energy balance has the form:

$$g(z_2 - z_1) + \frac{1}{2}v^2 \frac{L}{R_h} f = W,$$

where $L = 13{,}500$ ft and R_h is the hydraulic radius, $d/4$. f is the friction factor, given approximately by

$$f = 0.0791/\mathrm{Re}^{1/4}.$$

The Reynolds number, Re, is

$$\mathrm{Re} = \frac{dv\rho}{\mu} = \frac{dv(62.4)}{6.72 \times 10^{-4}}.$$

Our task is to determine the required pipe diameter for each of the two available pumps. Remember, $g_c = 32.17$ ft $lb_m/(lb_f\ s^2)$; therefore, the smaller pump will provide 4,665 ft^2/s^2.

Answer: Smaller pump, d = 3.59″ (so 4″ nominal) and larger pump, d = 2.77″ (so 3″ nominal). The actual choice, of course, must weigh the increased cost of the larger diameter pipe against the reduced energy cost for pumping.

2.28 Consider the function,

$$f = x_1^4 x_2 - 21x_1^3 x_2 + 147x_1^2 x_2 + x_1 x_2^3 - 12x_1 x_2^2 - 307x_1 x_2 - 17.$$

Devise a method to determine values for the variables x_1 and x_2 that will result in $f = 0$. We *know* that the values we seek for both x_1 and x_2 are between 1 and 10, but will it be possible to find a unique pair that yields a definitive best result? It may be useful to prepare a contour plot (f^2 as a function of x_1 and x_2) so that the surface is better revealed.

One answer: $x_1 = 4.9496$ *and* $x_2 = 9.0002$.

2.29 Brown and Associates (*Unit Operations*, John Wiley & Sons, 1950) presented a compilation of experimental data for friction factors (f) for flow through randomly packed beds, where

$$f = \frac{2g_c d_p(-\Delta p)}{F_f L V^2 \rho}$$

(note that F_f is an experimental parameter that is determined by the porosity of the bed and the sphericity of the particles). Here are some selected data points:

Re	10	40	100	400	1,000	4,000	10,000	40,000
f	6.5	1.7	0.7	0.165	0.087	0.04	0.03	0.021

Suppose we need a representation of these data in the form $f(Re)$ that can be subsequently used in repetitive automated computations. It has been suggested that an equation of the form

$$f = a + \frac{b}{Re} + \frac{c}{Re^2} + \frac{d}{Re^3}$$

might be useful for this purpose. Devise a scheme to determine the four constants, a, b, c, and d, and then verify that your approximation is reasonable.

2.30 John Damuth published a study of the relation between population density of mammals and their body size in *Nature* in 1981. The data he used included both herbivores and carnivores, and as one might expect, the carnivores tended toward the lower bound of the data (presumably due to competition for prey species). The herbivores typically had larger population densities. The *approximate center points* of the data presented have been estimated and put into this table:

Body Mass (kg)	***P* (number/km²)**
0.01	9,000
0.1	600
1	35
10	2
100	0.19
1,000	0.012
10,000	0.0018

Paleontologists have long been intrigued by estimations of how many dinosaurs of specific type might have been alive at any particular time. For example, the *Tyrannosaurus rex* that populated western North America about 66 million years ago had an estimated body mass of 5,000–7,000 kg. The *Apatosaurus*, a herbivore living in western North America about 150 million years ago, had an estimated body mass of about 70,000 lbs, or 32,000 kg. Use the rough *mammalian* data in the provided table to construct a correlation, then estimate the population densities for both *Tyrannosaurus rex* and *Apatosaurus*. Based upon the relationship you determine, about how many *T-rex* dinosaurs might have been alive at a given time in North America?

2.31 Adsorption is a physicochemical process in which a constituent (the adsorbate) in a fluid phase is collected upon a surface (the adsorbent). A familiar example is the use of activated carbon adsorbents to remove organic compounds from water supplies to improve taste, control odor, and so on. The Freundlich isotherm is an empirical relationship that relates the amount adsorbed to the concentration of adsorbate in the fluid phase:

$$\frac{x}{M} = kC^{1/n}.$$

x/M represents the amount adsorbed per gram (typically) of adsorbent, and it is dimensionless. k and n are empirical constants that must be determined from experimental data, and C is the concentration (usually molar or molal) of the adsorbate. By molal concentration we mean the number of moles of solute per kilogram of solvent. Temperature will affect the adsorption process, so experimental data must be acquired for the expected adsorption conditions. A study was conducted in which acetic acid was removed from water by adsorption upon activated carbon, and the following eight data points were obtained:

***C* (molal)**	***x/M* (gram per gram)**
0.00625	1.0
0.014	1.52
0.025	2.0
0.05	2.63
0.075	2.99
0.10	3.28
0.125	3.53
0.15	3.76

Determine the "best" values for k and n by regression analysis; of course, we note that

$$\ln\left(\frac{x}{M}\right) = \ln(k) + \frac{1}{n}\ln(C).$$

Does the Freundlich isotherm adequately represent the data? Compare your isotherm with the experimental data graphically, and if it is not acceptable move on to a polynomial regression. You may use any software that is convenient, and there are dozens of options.

2.32 John Laufer carried out an extensive investigation of turbulent flow in a pipe (The Structure of Turbulence in Fully Developed Pipe Flow, *NACA Report 1174*, 1954) at Reynolds numbers of 5×10^4 and 5×10^5. The fluid used was air and the lower Re provided a centerline velocity of 10 ft/s, and the larger Re 100 ft/s. The nominal pipe diameter was 10 in. with an actual radius, R, of 4.86″. One of the objectives of this work was to obtain accurate velocity measurements very near the pipe wall; some of the (approximated) results obtained for time-averaged velocity at Re = 500,000 are presented here. The distance from the pipe wall is s:

s/R	0.002	0.004	0.006	0.008	0.012	0.016	0.020	0.024	0.100	0.200
V/V_{max}	0.401	0.492	0.535	0.577	0.610	0.638	0.660	0.670	0.772	0.842

If the flow was highly ordered (*laminar*) then

$$\frac{V}{V_{max}} = \left(1 - \frac{r^2}{R^2}\right) = \left(1 - \frac{(R-s)^2}{R^2}\right) \tag{2.68}$$

but we see immediately that for $s/R = 0.100$, $V/V_{max} \approx 0.19$ which is only 24% of the measured value. Clearly, eq. (2.68) is *not* valid for turbulent flow.

In the early part of the twentieth century, much effort was devoted to finding an expression for the velocity distribution in turbulent flow that was "universal." However, none of these efforts was completely successful because the mechanism by which momentum is transferred toward the wall changes dramatically with position—very near the wall momentum transfer occurs by viscous friction (a molecular process), but far from the wall the bulk of momentum transport occurs due to the turbulence itself. Consequently, fluid dynamicists were forced to find empiricisms that could be used for practical calculations. One such expression is the power law:

$$\frac{V}{V_{max}} = \left(1 - \frac{r}{R}\right)^{\frac{1}{n}} = \left(\frac{s}{R}\right)^{1/n}.$$

Note that this expression gives the correct velocities at both the pipe wall ($s = 0$) and the pipe center ($s = R$). Use the data provided above and find the "best" value for n.

2.33 Cosmic rays were discovered by the physicist Victor Hess in 1912; he used a balloon to measure radiation levels as a function of altitude. His discovery that radiation intensity increased with height suggested that the main source of the radiation was *not* terrestrial. Most cosmic rays are protons (hydrogen nuclei), some are alpha particles, and a few are nuclei of heavier elements. We measure the energy (E) of these particles in electron volts, eV, and in the case of protons emanating from the sun, a typical energy is about 10^9 eV. Velocities (v) of cosmic rays are very large: For a proton with $E = 10^8$ eV, $v \approx 43\%$ of the speed of light, c. If E = 10 GeV, $v \approx 99.6\%$ of c. Occasionally a particle of extremely high energy is observed; in October 1991, the "OMG" particle was detected in Utah at the Dugway Proving Ground (the Army's center for studies of biological and chemical warfare) with $E \approx 3 \times 10^{20}$ eV, corresponding to a velocity *extremely* close to the speed of light. Rarely do these cosmic rays actually strike the earth's surface because of our atmosphere and the earth's magnetic field. Usually they produce a cascade of secondary particles, and these are what we typically observe at ground level. The fact that we are mostly shielded from cosmic rays is fortunate since the interaction of high-energy charged particles with biological matter is known to be problematic. There are, of course, concerns about human exposure during space flight, as well as the effect of such particles upon microelectronics and telecommunications. Here are some very approximate data for the flux of cosmic rays measured at the upper edge of the earth's atmosphere (as a function of E):

E (eV)	3.3×10^{10}	9×10^{12}	10^{13}	10^{14}	10^{15}	10^{17}
Flux	10^{0}	10^{-6}	2.5×10^{-7}	10^{-9}	1.19×10^{-12}	5.6×10^{-18}

Use these data to estimate the flux of cosmic rays at the top of the earth's atmosphere for two extreme cases: $E = 10^8$ and 10^{20} eV. Note that at very high energies we can expect the cosmic ray flux to be very low—at 3×10^{15} eV we expect about 1 particle *per* m^2 *per year*.

References

Beveridge, G.S.G. and Schechter, R.S. (1970). *Optimization: Theory and Practice*. New York: McGraw-Hill.

Elmqvist, H. and Otter, M. (1994). *Methods for Tearing Systems of Equations in Object-Oriented Modeling*. Barcelona: Proceedings of the European Simulation Multiconference.

Faucett, L. (2002). *Numerical Methods Using Mathcad*. Upper Saddle River, NJ: Prentice Hall.

Hamming, R.W. (1971). *Introduction to Applied Numerical Analysis*. New York: McGraw-Hill.

Heath, S.T.L. (1964). *Diophantus of Alexandria: A Study in the History of Greek Algebra*, 2nde. Dover Publications reprinted from the Cambridge University Press 1910 Edition.

James, M.L., Smith, G.M., and Wolford, J.C. (1977). *Applied Numerical Methods for Digital Computation with FORTRAN and CSMP*, 2nde. New York: Harper & Row.

Nelder, J.A. and Mead, R. (1964). A Simplex Method for Function Minimization. *Computer Journal* 7: 308.

Ramirez, W.F. (1997). *Computational Methods for Process Simulation*, 2nde. Oxford: Butterworth-Heinemann.

Shoup, T.E. and Mistree, F. (1987). *Optimization Methods with Applications for Personal Computers*. Englewood Cliffs, NJ: Prentice-Hall.

Spendley, W.G., Hext, G.R., and Himsworth, F.R. (1962). Sequential Applications of Simplex Designs in Optimization and Evolutionary Operation. *Technometrics* 4: 441.

3

Vectors and Tensors

3.1 Introduction

A significant area of interest in continuum mechanics is the application of forces to material objects and the response of those objects to the applied forces. Most of us tend to think of a force simply as a directed line segment (possessing magnitude and direction), but a little caution is in order here: We must remember that the *effect* of a force applied to a body depends upon both its line of action and the point at which it is applied. For example, if one were using a lever to move a boulder, the distance between the point of application and the fulcrum would have enormous impact upon effectiveness of the action. We also need to emphasize that our discussions here are concerned with continuum mechanics in *three-dimensional Euclidean space*. Thus when we speak of tensors, for example, we mean *Cartesian tensors*. Tensors do figure prominently in non-Euclidean spaces, but those applications are not relevant to our principal objectives.

We will begin by reviewing what we mean when we refer to scalars, vectors, and tensors. A scalar is a quantity that has magnitude only; for example, we might find that an enclosure has a volume of 1.2 m^3 or that the fluid contained within has a temperature of 215 °F (101.67 °C). We also observe that a scalar is a zero-order tensor (we will use *order and rank* synonymously). In contrast, a vector has both magnitude and direction, and we can think of force and velocity as examples. A vector, with three components, is a first-order tensor. By tensor we merely mean an ordered set of numbers; second-order Cartesian tensors (with nine components) are very important in the mechanics of materials and in hydrodynamics. The third-order alternating tensor, ε_{ijk}, has 27 components and is equal to +1 if the subscripts are in cyclic order (e.g., 1,2,3 or 2,3,1), −1 for anticyclic order, and zero otherwise. The alternating tensor is particularly useful for the cross-product of certain vectors. You probably have detected a pattern:

Tensor Order	Number of Components
0	1
1	3
2	9
3	27

To illustrate how second-order tensors come about, let us think about a force acting upon a surface such that the dimensions are F/L^2 or $M/(Lt^2)$. Of course, pressure and shear stress are prime examples, and we recognize that two directions are important for such quantities. Consider a fluid flowing in the x-direction past a fixed solid surface located at $y = 0$; because of the Newtonian no-slip condition (Newton's law of viscosity) a shear stress will be created by the variation of velocity near the surface illustrated in Figure 3.1.

Problem Solving in Engineering: Analytical Mathematics and Numerical Analysis, First Edition. Larry A. Glasgow.

Companion website: www.wiley.com/go/glasgow/problemsolvingengineering

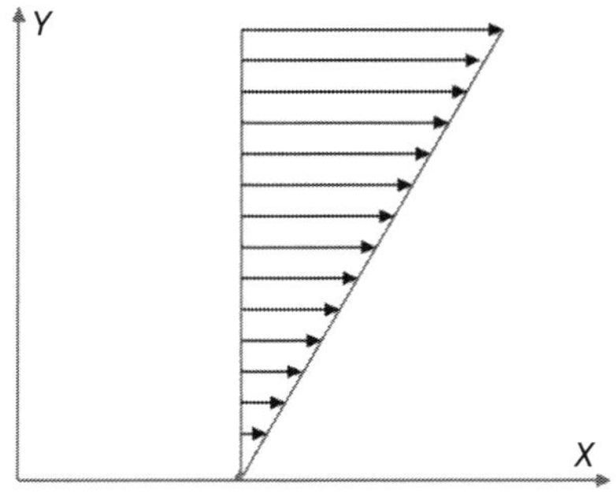

Figure 3.1 Shear stress created by fluid motion past a solid surface. For this illustration, $v_x = my$ and $\tau_{yx} = -\mu \frac{dv_x}{dy} = -\mu m$, where μ is the viscosity of the fluid. Momentum is transferred to the stationary surface (in the negative y-direction).

We will interpret this shear stress, τ_{yx}, as a force acting upon the y-plane (the fixed surface is located at $y = 0$) due to fluid motion in the x-direction. Since we have three principal directions, it is clear that each of the two indices on τ can assume one of three values, corresponding to x, y, and z in rectangular coordinates. Thus, the stress tensor consists of the set of components:

$$\begin{matrix} \tau_{xx} & \tau_{xy} & \tau_{xz} \\ \tau_{yx} & \tau_{yy} & \tau_{yz} \\ \tau_{zx} & \tau_{zy} & \tau_{zz} \end{matrix} \tag{3.1}$$

You will see immediately that the subscripts are repeated on the diagonal; these are called *normal* stresses and their sum is called the *trace* of the tensor. The off-diagonal components are shear stresses, and we should recognize that the corresponding off-diagonal stresses must be equal, i.e., $\tau_{yx} = \tau_{xy}$. If this were not the case, then an infinitely small element of fluid could experience very large (infinite) angular acceleration. This requirement means that the stress tensor is *symmetric* and that it contains just six independent quantities.

It is common practice to write second-order tensors using the Cartesian summation convention. For example, we might write S_{ij} where the indices i and j can each assume the values 1, 2, and 3. Consequently, if we write S_{ii} (with repeated subscripts), we mean

$$S_{ii} = S_{11} + S_{22} + S_{33}, \tag{3.2}$$

which of course is the trace of the tensor as we observed previously.

3.2 Elementary Operations

Let us review some vector algebra, noting that **A**, **B**, and **C** are vectors, and that a and b, are scalars.

$$\mathbf{A} + \mathbf{B} = \mathbf{B} + \mathbf{A} \quad \text{commutative} \tag{3.3}$$

$$\mathbf{A} + (\mathbf{B} + \mathbf{C}) = (\mathbf{A} + \mathbf{B}) + \mathbf{C} \quad \text{associative (addition)} \tag{3.4}$$

$$a(b\mathbf{A}) = (ab)\mathbf{A} = b(a\mathbf{A}) \quad \text{associative (multiplication)} \tag{3.5}$$

$$(a + b)\mathbf{A} = a\mathbf{A} + b\mathbf{A} \quad \text{distributive} \tag{3.6}$$

$$a(\mathbf{A} + \mathbf{B}) = a\mathbf{A} + a\mathbf{B} \quad \text{distributive} \tag{3.7}$$

When we refer to *unit vectors*, we mean unit length. Therefore, for vector **A** with length a, unit length is obtained simply by $\mathbf{A}/a$. In rectangular coordinates it is common practice to write the unit vectors as i, j, k, corresponding to the x, y, z directions.

We will review these basic operations by looking at a few examples. Let the vectors **A** and **B** be given by:

$$A = 4i - 3j + 2k \tag{3.8}$$

$$B = 5i + 5j + 3k. \tag{3.9}$$

First we look at the sum:

$$A + B = (4 + 5)i + (-3 + 5)j + (2 + 3)k = 9i + 2j + 5k. \tag{3.10}$$

The magnitude of A is

$$|A| = [(4)^2 + (-3)^2 + (2)^2]^{1/2} = \sqrt{29} = 5.38516. \tag{3.11}$$

If we multiply B by a scalar, b, letting $b = 3$,

$$bB = 3(5i + 5j + 3k) = 15i + 15j + 9k. \tag{3.12}$$

Now suppose a body, M, is being acted upon by three different forces, $\boldsymbol{F}_1$, $\boldsymbol{F}_2$, and $\boldsymbol{F}_3$: Let

$$F_1 = 5i + 5j - 1k \tag{3.13}$$

$$F_2 = 6i - \frac{1}{2}j + 1k \tag{3.14}$$

$$F_3 = 3i - 3j + 2k. \tag{3.15}$$

The resultant force is $F_1 + F_2 + F_3$,

$$R = F_1 + F_2 + F_3 = 14i + \frac{3}{2}j + 2k. \tag{3.16}$$

If we wished to prevent movement of the body, M, we could apply a force corresponding to $-\boldsymbol{R}$: $-R = -14i - \frac{3}{2}j - 2k$. This force would produce equilibrium, and the body, M, would naturally remain at rest if it were initially stationary. Let us now assume that we would like to identify the unit vector that is parallel to the resultant of two forces given by:

$$F_1 = 3i + 2j + 2k \tag{3.17}$$

and

$$F_2 = 2i - 1j + 1k. \tag{3.18}$$

The resultant is

$$R = F_1 + F_2 = (3i + 2j + 2k) + (2i - 1j + 1k) = 5i + 1j + 3k. \tag{3.19}$$

The magnitude of the resultant is

$$|R| = [(5)^2 + (1)^2 + (3)^2]^{1/2} = \sqrt{35} = 5.91608. \tag{3.20}$$

So the unit vector we are seeking is:

$$\frac{R}{|R|} = \frac{5i + 1j + 3k}{5.91608} = 0.84515i + 0.16903j + 0.50709k. \tag{3.21}$$

Often we are interested in the angle between a given vector and the coordinate axes. For transparency, we will start with a vector in the x–y plane and determine the angle (θ) between this vector and the x-axis: Let

$$F_1 = 6i + 3j. \tag{3.22}$$

It is apparent that $\tan\theta = \frac{3}{6}$ such that $\theta = 0.4636$ rad or 26.57°. More generally, if a vector in three space is given by

$$F_1 = 6i + 3j + 2k, \tag{3.23}$$

then the angles between this vector and the coordinate axes are:

$$\theta_1 = \cos^{-1}\frac{x}{|F_1|} \tag{3.24}$$

$$\theta_2 = \cos^{-1}\frac{y}{|F_1|} \tag{3.25}$$

and

$$\theta_3 = \cos^{-1}\frac{z}{|F_1|}. \tag{3.26}$$

Since $|F_1| = \sqrt{36 + 9 + 4} = 7$, the three angles we are seeking are 0.5411 rad, 1.1279 rad, and 1.2810 rad, respectively. Please note that

$$cos^2\theta_1 + cos^2\theta_2 + cos^2\theta_3 = 1. \tag{3.27}$$

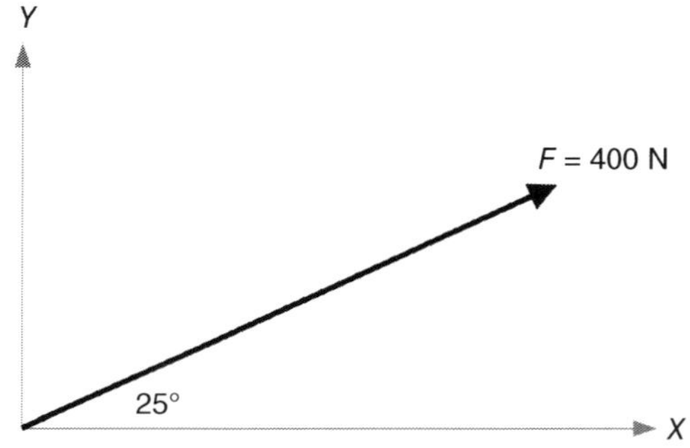

Figure 3.2 A force of 400 N applied in the x–y plane at an angle of 25°.

Now we will consider a force, $\boldsymbol{F}$, applied at the origin in two space (the x–y plane); we will illustrate this in Figure 3.2.

We can resolve $\boldsymbol{F}$ into x- and y-components if we know the angle, θ. Suppose for example that $\boldsymbol{F} = 400\,\mathrm{N}$ and $\theta = 25°$; then

$$F_x = F\cos(\theta) = (400)(0.9063) = 362.5\,\mathrm{N} \tag{3.28}$$

and

$$F_y = F\sin(\theta) = (400)(0.4226) = 169\,\mathrm{N}. \tag{3.29}$$

Of course the resultant of F_x and F_y can be found by rectangular resolution, i.e.,

$$F = \sqrt{F_x^2 + F_y^2} = 399.96\,\mathrm{N}. \tag{3.30}$$

The discrepancy is merely the result of round-off error.

3.2.1 An Illustration

Suppose a light aircraft flies due east with a ground speed of 120 mph; it is subjected to a constant wind from the south blowing at 14 mph; see Figure 3.3. What is the effective ground speed of the airplane, and where will it be after one hour (if the pilot makes no corrections) ?

In this case $\tan(\theta) = 14/120$, so $\theta = 6.654°$. After 1 hr the aircraft will have traveled 120.81 mi, and it will of course be 14 mi north of its intended destination. *This elementary example underscores one of the difficulties of navigation prior to GPS. On a long flight, a small crosswind could result in a catastrophic error if not caught by the navigator. It is thus easy to understand how Amelia Earhart and Fred Noonan missed Howland Island in 1937 on their 2500 mi flight from Lae (to Howland) when their RDF (radio direction finder) did not function properly.*

An important variation of such problems arises when the second vector is not perpendicular to the first. Now assume that the same aircraft, flying east at 120 mph, experiences a quartering wind blowing 22 mph at an angle 18° north of east as shown in Figure 3.4.

In this case the plane's speed to the east will be

$$120 + 22\cos(18°) = 140.92\,\mathrm{mph}. \tag{3.31}$$

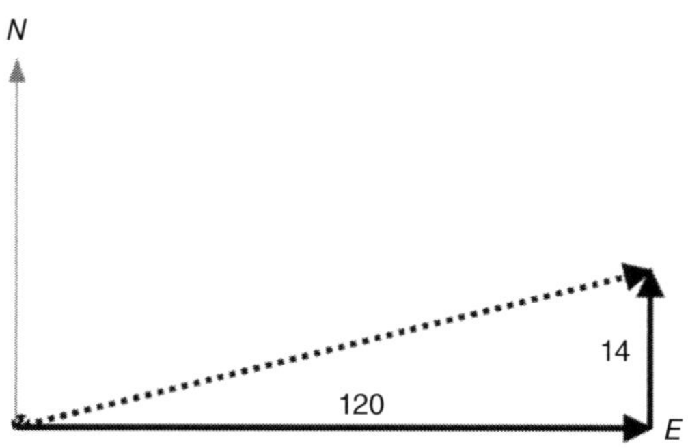

Figure 3.3 Effect of a crosswind upon a light aircraft flying due east at 120 mph.

Figure 3.4 Effect of a quartering wind upon a light aircraft flying east at 120 mph. The wind is blowing to the ENE at 22 mph.

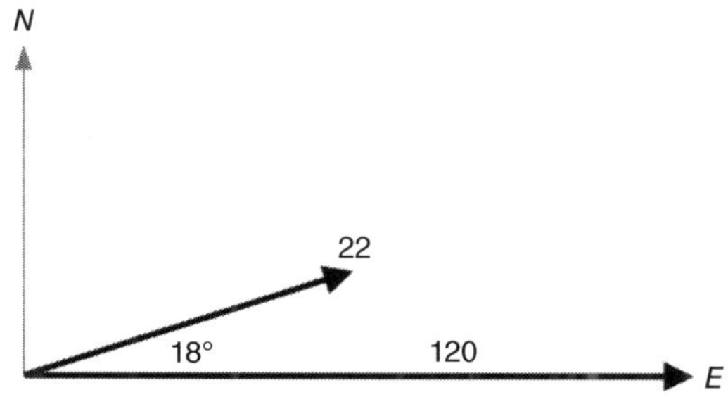

However, its velocity to the north will be

$$22\sin(18°) = 6.80 \text{ mph.} \tag{3.32}$$

After one hour the craft will have flown a distance of about 141.08 mi and it will be 6.8 mi north of the intended point.

3.3 Review of Some Basic Mechanics

3.3.1 Force Equilibrium

We begin with a familiar type of problem in which we utilize equilibrium at a point. Imagine for example that during the erection of a building, a structural member (a beam 30 ft long) weighing 800 lb is temporarily leaned against a vertical wall as illustrated in Figure 3.5 (the angle between the ground and the beam is 50°). At the upper end the wall exerts a horizontal force on the beam indicated by F_1. The ground exerts a force upon the bottom end of the beam indicated by F_2. We recognize that under the equilibrium condition, the summation of forces in the horizontal (x-) direction must be zero. Similarly, the summation of forces acting in the vertical (y-) direction must also be zero; therefore,

$$F_2\cos(\theta) - F_1 = 0 \quad \text{and} \quad F_2\sin(\theta) - W = 0 \tag{3.33}$$

In order for us to solve these equations, we must determine the angle, θ: We do this by using two right triangles, and the first of these is formed using the lower half of the beam such that the hypotenuse is 15 ft; the vertical leg is the distance from the mid-point of the beam to the ground. Therefore, $x = 15\cos(50°) = 9.642$ ft. Now we can determine θ (the angle between the line of action of F_2 and the ground): $\tan(\theta) = 22.98/9.642$, so $\theta = 67.24°$. Our horizontal and vertical force summations are now used to compute $F_2 = 867.56$ lb and $F_1 = 335.64$ lb. Of course, the very same problem can be worked more easily by noting that at equilibrium, the summation of *moments* about the z-axis must be zero.

Figure 3.5 A structural member (or beam) lying against a vertical wall. The beam is 30 ft long and weighs 800 lb_f; the angle of inclination with respect to the ground is 50°.

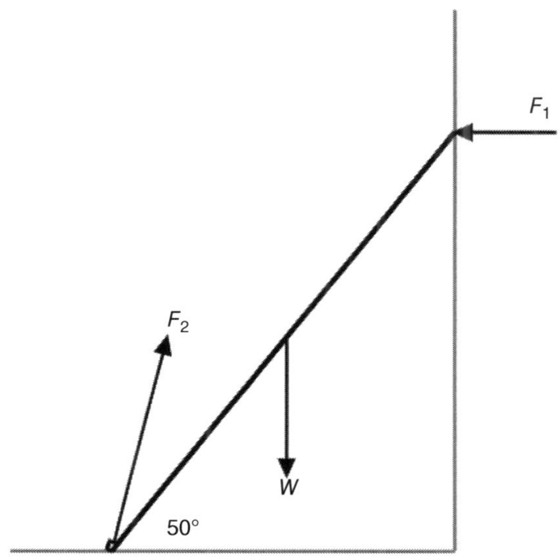

3.3.2 Frictional Force

Figure 3.5 illustrates a beam with its upper end resting against a vertical wall and its bottom lying upon the ground, forming an angle between the inclined beam and the ground surface of 50°. A practical concern in a situation like this is the possibility that the bottom end of the beam might begin to slide away from the wall. As the beam weighs 800 lb_f, it is clear that safety could become a serious issue. Of course there is a frictional force resisting this sliding action, and it is proportional to the normal force at the contact surface. When Coulomb investigated sliding friction in the eighteenth century he discovered that the force was virtually independent of the area of contact and that it was also nearly independent of velocity at low speeds. Thus, for an object with weight, F_w, resting upon a flat surface, the frictional force resisting motion is:

$$F = \mu_f F_w, \tag{3.34}$$

where μ_f is the coefficient of friction. Note that it has a larger value for static conditions; once the motion is initiated, it will decrease significantly. For example, the static μ_f for steel upon steel is about 0.45, and when motion is initiated this value will decrease by ~25%.

Nowhere is this more important than in railroading, where locomotives depend upon steel-upon-steel contact to exert tractive effort. For example, an EMD (Electro-Motive Diesel) SD40 diesel electric locomotive weighs about 360,000 lb_f, and its prime mover produces 3000 hp. An SD40 can generate a starting tractive effort of 92,000 lb_f (with *no* wheel slip); however, we note that

$$\mu_f F_w = (0.45)(360{,}000) = 162{,}000 \text{ lbf}.$$

Why the discrepancy? If the locomotive's full power was applied to wheels in a static condition, dramatic wheel slippage would be immediate. This of course is one reason why locomotives carry sand boxes (for application of sand to the rails) because sand can improve both tractive effort and braking efficiency.

We will now consider an elementary example: A large box (weighing 185 lb_f) lies upon a flat, horizontal surface. We would like to push the box so that it slides along the floor, and we will apply force on a line of action parallel to the floor. Assume that both the box and the surface are wooden and that $\mu_f = 0.55$; the force that must be applied to get to the condition of impending motion will be

$$F = (0.55)(185) = 101.75 \text{ lbf}.$$

This raises an important question: If we apply this force near the top of the box, might it tip over rather than slide? To answer this question we must *balance the moments* about the tipping point (the front bottom edge), and we will consider this process shortly. However, if you are already familiar with this process from your prior exposure to statics, assume the box is 3 ft by 3 ft, and 6 ft high—what is the maximum height (*h*) at which the force (101.75 lb_f) can be applied *without* tipping? You might verify that *h* must be less than 2.73 ft if the weight is uniformly distributed in the box. What if the weight of the box's contents was concentrated in the back half (nearer the vertical surface where we apply the pushing force)? In that case the moment opposing tipping would be increased, and *h* would need to be less than 4.1 ft.

Finally, we observe that coefficients of static friction vary enormously depending upon the materials in contact and the condition of the interface. A few values obtained from *engineeringtoolbox.com* for *clean and dry* surfaces are provided here:

Materials	μ_f
Aluminum–aluminum	1.05–1.35
Plexiglas–Plexiglas	0.8
Wood–concrete	0.62
Brass–steel	0.51
Graphite–graphite	0.1
Ice–steel	0.03
Brass–ice (0 °C)	0.02

3.3.3 Equating Moments

Recall that the moment of a force is the measure of the torque introduced about a chosen point; the moment is the product of the magnitude of the force and the perpendicular distance between the selected origin and the line of action of the force. Furthermore, according to Varignon's theorem, the moment of the resultant of two forces equals the sum of the two moments of the two forces. Therefore, if an object is acted upon by multiple forces—yet remains at equilibrium—then the sum of the moments must be zero. Let us apply this result to an example.

We imagine a cylindrical object with radius, R, that has rolled up against a curb (or step) of height, h. We wish to apply a horizontal force, F, to get the cylinder to roll up over the curb. The force will be applied at the center of the cylinder, although it is obvious that this is not optimal. The cylindrical object has weight, W, and the arrangement is illustrated in Figure 3.6.

The moment arm for the applied force, F, is R–h. The moment arm for the weight (due to gravity) is x, the horizontal leg of the triangle. Therefore,

$$F(R - h) = Wx, \tag{3.35}$$

with x obtained as follows: $\theta = \sin^{-1}\frac{R-h}{R}$, and $x = R\cos\theta$. We can examine this result quantitatively by setting $R = 10$ and $h = 2$ (a fairly small step). For this case $\theta = 0.9273$ rad (53.1°) and $x = 6$. Therefore, $F = 0.75\,W$. This will, of course, change dramatically if the step height is greater; we can repeat the problem but with $h = 7$. The reader may wish to verify that $F = 3.18\,W$ in this case. We observe that the point of the step exerts a force upon the cylinder; at equilibrium the x- (horizontal) component of that force must counter F and the y-component must balance W. For the case in which $h = 7$, this force will total 3.33 W and the angle between the line of action and the upper horizontal surface will be 17.46°

A different situation can arise when coplanar forces act upon a body at different locations. To illustrate, let us take the case where forces $\mathbf{F_1}$ and $\mathbf{F_2}$ are acting upon a rigid body as depicted in Figure 3.7.

We begin by constructing auxiliary forces P_1 and P_2 that are equal in magnitude but opposite in direction. Then the lines of action for Q_1 (the resultant of F_1 and P_1) and Q_2 (the resultant of F_2 and P_2) are constructed and extended back to the point of intersection, B. The force transmitted to the solid body is R, which is the resultant of the pair, Q_1 and Q_2. Naturally,

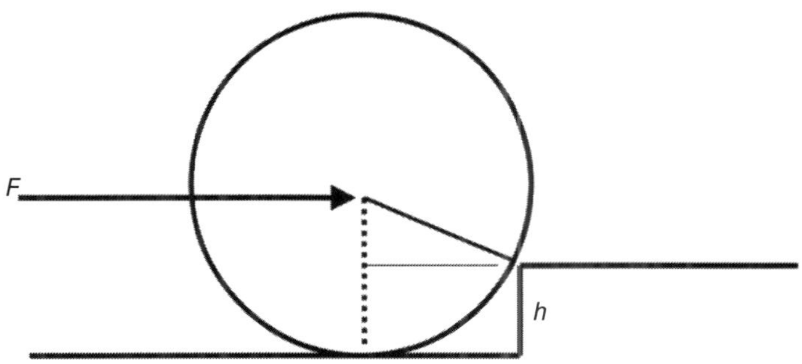

Figure 3.6 A cylindrical object of radius, R, and weight, W, has rolled up against a step of height, h. A force is to be applied at the center with the intent of moving the cylinder to the top of the step.

Figure 3.7 Coplanar forces acting upon a rigid body at different locations.

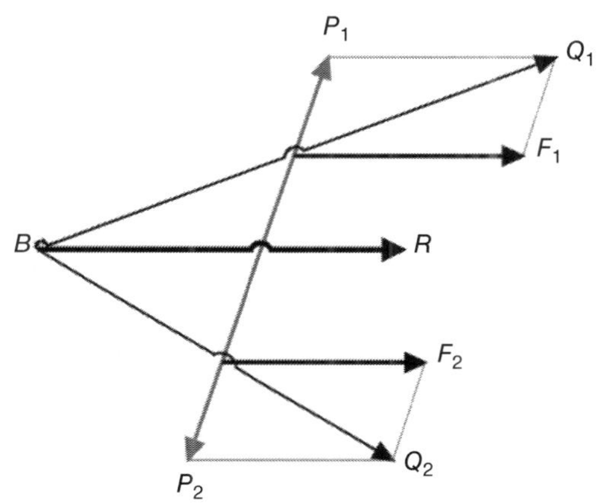

the magnitude of R is just the sum of F_1 and F_2; the line of action of the resultant is obtained by noting that we have similar triangles (created by extending Q_1 and Q_2 back to the point of intersection, B). The case of coplanar, nonparallel forces (applied to a rigid body) is simpler—it is only necessary to extend the lines of action to the point of intersection. Of course, it is entirely possible that a pair of parallel, coplanar forces act in opposite directions. If they have the same magnitude the result will be a couple, and a couple acting upon a rigid body will produce rotation.

3.3.4 Calculation of Centroids

For rigid bodies in the x–y plane the calculation of first moments about the axes may be crucial to the structural analysis, and if *parallel* forces acting upon this two-dimensional shape are distributed but acting *normal* to it, then it is possible to replace them with a single force acting at a particular point (the centroid). In essence we are concentrating the area of some surface at a position (x_c,y_c) and thus greatly simplifying the analysis of forces. The position of the centroid is given by

$$x_c \iint dA = M_y = \iint xdxdy \quad \text{and} \quad y_c \iint dA = M_x = \iint ydydx. \tag{3.36}$$

Suppose an "object" in the x–y plane is described by $y = 5\sqrt{x}$ with $0 \le x \le 10$. The area, which you may want to verify, is 105.409. Let us illustrate by determining y_c:

$$y_c = M_x/A \quad \text{and} \quad M_x = \int_0^{10} \int_0^{5\sqrt{x}} ydydx = \frac{2500}{4} = 625. \tag{3.37}$$

Therefore, $y_c = 625/105.409 = 5.929$. The reader may want to show that $x_c = 6$. Of course this analysis can be extended to three dimensions to find the *center of mass* which is equal to the center of volume *if* the density of the object is spatially uniform.

Now suppose the edge of a 2-D shape lying upon the x-axis is given by $y = 10x - \frac{1}{3}x^3$ with $0 \le x \le \sqrt{30}$. We will work through this example in detail. First, we determine the area, A, by summing up the areas of rectangles of height, $y(x)$, and width, dx:

$$A = \int_0^{\sqrt{30}} \left(10x - \frac{1}{3}x^3\right) dx = 75. \tag{3.38}$$

Since $x_c = M_y/A$, we need to calculate the first moment about the y-axis:

$$M_y = \int_0^{\sqrt{30}} \int_0^{10x-\frac{1}{3}x^3} xdydx = 219.09, \tag{3.39}$$

and therefore $x_c = \frac{219.09}{75} = 2.921$. Now we find the first moment about the x-axis:

$$M_x = \int_0^{\sqrt{30}} \int_0^{10x-\frac{1}{3}x^3} ydydx = 625.969. \tag{3.40}$$

Since the area of the shape is 75 (as shown above), we have $y_c = 625.969/75 = 8.346$. The location of the centroid is (2.921,8.346).

Next, assume the shape of interest in the x–y plane was the area *beneath* the curve, $y = 6x - x^2$, but *above* the straight line, $y = x$. The line intersects the parabola at the point (5,5) so $0 \le x \le 5$. We find the area by summing rectangles of height, $y(x)$, and width, dx for the parabola and then subtracting the area of the triangle at the bottom:

$$A = \int_0^5 (6x - x^2)dx - \int_0^5 xdx = 33.333 - 12.5 = 20.833. \tag{3.41}$$

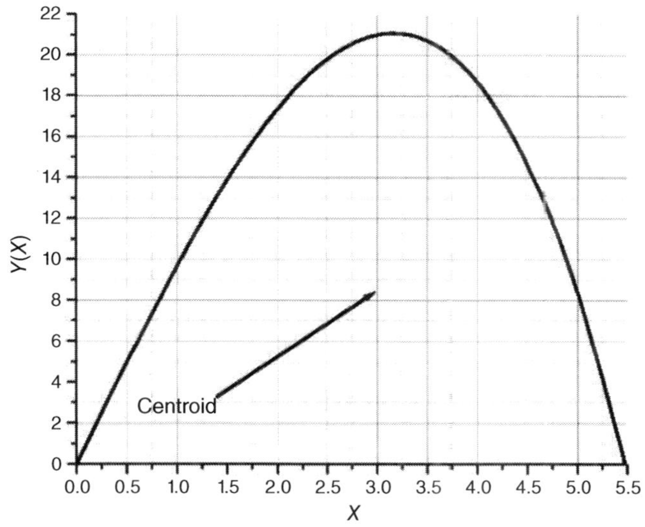

Figure 3.8 Surface in the x–y plane with $y(x) = 10x - \frac{1}{3}x^3$ for $0 \leq x \leq \sqrt{30}$.

Next we determine the first moment about the x-axis:

$$M_x = \int_0^5 \int_x^{6x-x^2} y\,dy\,dx = 104.167. \tag{3.42}$$

Therefore, $y_c = \frac{104.167}{20.833} = 5$. You should verify that $x_c = 2.5$, so the centroid of this shape is located at (2.5,5).

3.3.5 Projectile Motion

Imagine a situation in which a projectile is launched in the x–y plane at an angle (relative to level ground) of θ with an initial velocity of V_0. If we neglect drag (a potentially serious omission) then the initial horizontal and vertical velocity vector components are

$$V_{x0} = V_0 \cos\theta \quad \text{and} \quad V_{y0} = V_0 \sin\theta. \tag{3.43}$$

Since gravity acts in the negative y-direction, we know

$$V_y = V_0 \sin\theta - gt; \tag{3.44}$$

the projectile will reach its apogee when $V_y = 0$, or $V_0 \sin\theta = gt$. During flight, the angle between the instantaneous trajectory and the ground is $\theta = \tan^{-1}(V_y/V_x)$. We can look at a simple example for illustration: Let $V_0 = 300$ ft/s and $\theta = 35°$. The initial horizontal velocity is 245.7 ft/s, and the initial vertical velocity is 172.2 ft/s. The maximum vertical height above ground will occur at

$$t = \frac{(300)(0.574)}{(32.17)} = 5.35\text{ s}, \tag{3.45}$$

and that maximum height will be

$$y_{\max} = \left[(V_0 \sin\theta)t - \frac{1}{2}gt^2\right]_{t=5.35} = 460.9\text{ ft}, \tag{3.46}$$

a number that is actually quite unrealistic. After 8 s, the angle between the ground and the projectile's trajectory would be

$$\theta = \tan^{-1}\left(\frac{(300)(0.574) - (32.17)(8)}{(245.7)}\right) = -19°, \tag{3.47}$$

that is, 19° below horizontal. As we implied above, the effect of drag upon the motion of projectiles is profound. For example, returning to the previous illustration, if the projectile were a sphere launched in air (with $R = 1$ in. and $m = 1\ \text{lb}_\text{m}$) the maximum height would be about 394 ft and that would be attained in about 4.83 s. A larger object of the same mass would experience greater drag, of course. If we let $R = 2$ in., then $y_{\max} = 288$ ft and that height is achieved at $t = 3.93$ s.

3.4 Other Important Vector Operations

3.4.1 Dot and Cross-Products

In physics we often think of the dot product in the context of work: Work is performed when a force is applied to a body, the body moves, and a component of the force vector is acting in the direction of the motion. Thus, $W = \int_1^2 F \cdot ds$. The dot product of vectors **A** and **B** is written **A** · **B**, and it is a scalar. For example, given the situation depicted in Figure 3.9: then $\mathbf{A} \cdot \mathbf{B} = (100)(50)\cos(\theta)$. If the angle $\theta = 30°$, then $\mathbf{A} \cdot \mathbf{B} = 4992.97$, but if θ were changed from 30 to 85°, then $\mathbf{A} \cdot \mathbf{B} = 435.78$.

More generally, given two vectors **A** and **B**:

$$A = A_1 i + A_2 j + A_3 k \quad \text{and} \quad B = B_1 i + B_2 j + B_3 k, \text{ then} \tag{3.48}$$

$$A \cdot B = A_1 B_1 + A_2 B_2 + A_3 B_3. \tag{3.49}$$

Similarly, $A \cdot A = {A_1}^2 + {A_2}^2 + {A_3}^2$. If $A \cdot B = 0$, and neither is null, then A and B *must be perpendicular*. For example, suppose $A = 2i + 3j$, and further, that A is perpendicular to vector B (and both in the x–y plane). We have $A \cdot B = |A||B| \cos\theta$ and since $\theta = 90°$, we must have $(2i + 3j) \cdot (b_1 i + b_2 j) = 0$, or $2b_1 + 3b_2 = 0$. Thus, $\frac{b_1}{b_2} = -\frac{3}{2}$. We also make note of the following properties of dot products:

$$A \cdot B = B \cdot A \qquad \text{commutative} \tag{3.50}$$

$$A \cdot (B + C) = A \cdot B + A \cdot C \qquad \text{distributive} \tag{3.51}$$

$$a(A \cdot B) = (aA) \cdot B = A \cdot (aB) \tag{3.52}$$

The cross-product of two vectors, indicated by $\boldsymbol{A} \times \boldsymbol{B}$, is also a vector, **C**. Thus

$$A \times B = C, \tag{3.53}$$

where **C** is perpendicular to the plane of **A** and **B**. Furthermore, **C** follows the right-hand screw rule, so for the case in which **A** and **B** correspond to the x- and y-axes in the plane of the page force **C** is directed out of the page toward the reader. We also have the following relations for the cross (or vector)-product:

$$A \times B = -B \times A \qquad \text{not commutative} \tag{3.54}$$

$$A \times (B + C) = A \times B + A \times C \qquad \text{distributive} \tag{3.55}$$

$$a(A \times B) = (aA) \times B = A \times (aB) \tag{3.56}$$

We will illustrate the cross-product, **A** × **B**, with an example: Suppose we have two vectors in the x–y plane, $\mathbf{A} = 6i + 2j$ and $\mathbf{B} = 2i + 3j$. The angle between the x-axis and **A** is 0.3217 rad and the angle between the x-axis and **B** is 0.9828 rad (you should verify). Thus, the smaller angle between the two vectors is 0.6611 rad or about 37.9°. The magnitudes of **A** and **B** are $\sqrt{40}$ and $\sqrt{13}$, respectively, so

$$|C| = \sqrt{40}\sqrt{13} \sin(0.6611) = 14.$$

Alternatively, we write

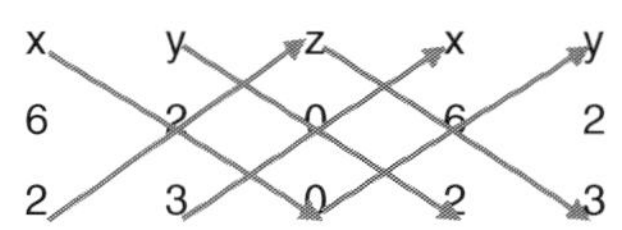

which of course also yields 18–4 = 14. Note that if the vectors are in the x–y plane, only the z-component of **A** × **B** will be non-zero. If you would like an additional review of this process, let $\mathbf{A} = 9i + 3j$ and $\mathbf{B} = 5i + 11j$ (two vectors in the x–y plane) and show that the angle between **A** and **B** is 0.8225 rad and that $|C| = 84$.

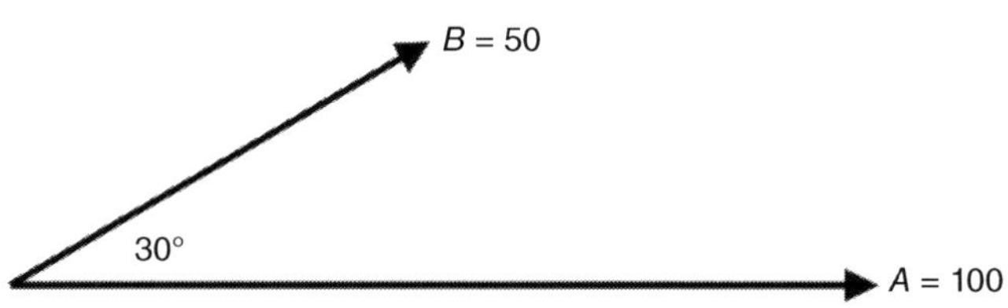

Figure 3.9 Illustration of coplanar vectors **A** and **B** forming an angle of 30°.

3.4.2 Coriolis Effect

An interesting application of the cross-product that is extremely important in meteorology is the Coriolis effect, named after the French mathematician, Gaspard-Gustave de Coriolis. A body moving in a rotating frame of reference will experience a "force" perpendicular to the direction of the motion and to the axis of rotation. It appears that the Coriolis effect was first noted by artillerists in the northern hemisphere when they observed that a precisely aimed shot typically missed the target to the right, although Galileo dismissed this idea in 1632 when he stated that in shooting a cannon the deflection produced by the earth's rotation was too small to be observed. This may have been true in the seventeenth century, but by 1918 the Germans had to correct the aim of the *Kaiser Wilhelm gun*—which could throw a 100 kg shell an astounding 81 mi—by nearly 800 m. See *Notes on the History of the Coriolis Effect* by O. V. Fuentes in *Physics Today*, 68:10 (2015).

The Coriolis acceleration is given by $2\omega \times V$ and the magnitude is $2V\omega \sin \phi$, where V is the velocity of the object, ω is the angular velocity of the earth (about 7.27×10^{-5} rad/s), and ϕ is the latitude. Of course, this means that the Coriolis acceleration will be significantly affected by position on the earth's surface: At Key West (24.555°), $\sin \phi = 0.416$ whereas at Point Barrow, Alaska (71.388°), $\sin \phi = 0.948$. Clearly, $\mathbf{A} \times \mathbf{B}$ will be zero if ϕ is zero and a maximum if $\phi = \pi/2$.

The Coriolis acceleration is relatively small and anyone interested in exploring it may enjoy experimenting with the Coriolis calculator devised by Dominic Czernia and available on the *OMNI Calculator* site. For example, if the mass of a moving object is 20 kg, its velocity is 300 m/s, and the latitude is 60°, the Coriolis acceleration is about 0.0378 m/s^2. We can contrast this with the acceleration of gravity, 9.8 m/s^2, which is more than 250 times larger. Nevertheless, the Coriolis effect plays a significant role in atmospheric dynamics; pressure gradients cause air to move from regions of higher pressure toward regions of lower pressure, but in the northern hemisphere the Coriolis acceleration deflects wind to the right, producing flow parallel to the isobars. Students interested in this topic should read the article by Anders Persson in the *Bulletin of the American Meteorological Society*, 79:1373 (1998). Also, a very early attempt to model the weather numerically and include the Coriolis effect was published in 1950 by Charney et al. (Numerical Integration of the Barotropic Vorticity Equation, *Tellus*, 2:237, 1950). This work is remarkable for several reasons, including the collaboration with John von Neumann and the use of an early electronic computing device, the *Eniac* (*the Electronic Numerical Integrator and Computer* located at the Ballistic Research Laboratories in Aberdeen, Maryland).

There are applications requiring triple products of the vectors, A, B, and C; the *scalar triple product* is:

$$A \cdot (B \times C) = B \cdot (C \times A) = C \cdot (A \times B). \tag{3.57}$$

This relation gives us the volume of a parallelpiped defined by the three vectors. Given $A = 2i + 4j + k$, $B = 6i - j + k$, and $C = 0 + j + 3k$, for example, we have:

$$\begin{matrix} 2 & 4 & 1 \\ 6 & -1 & 1 \\ 0 & 1 & 3 \end{matrix} = -6 + 0 + 6 - 0 - 2 - 72 = -74.$$

The *vector triple product* is:

$$A \times (B \times C) = B(A \cdot C) - C(A \cdot B). \tag{3.58}$$

Eq. (3.58) is often referred to as the "*BAC-CAB*" rule—a useful mnemonic device as long as one remembers the minus (−) sign.

3.4.3 Differentiation of Vectors

Let us suppose that a vector is a function of a scalar like time, t. In particular let $A = A_1 i + A_2 j + A_3 k$, where A_1, A_2, and A_3 depend upon t. Then

$$\frac{dA}{dt} = \frac{dA_1}{dt} i + \frac{dA_2}{dt} j + \frac{dA_3}{dt} k. \tag{3.59}$$

Naturally, $\frac{dA}{dt} = 0$ if A is a *constant* vector. We also want to consider the case in which $B = B_1(x, y, z)i + B_2(x, y, z)j + B_3(x, y, z)k$, and spatial derivatives are required. Let $B = 3xy^2 i - 4xyzj + 7y^2z^2k$, then

$$\frac{\partial B}{\partial x} = 3y^2 i - 4yzj \tag{3.60a}$$

$$\frac{\partial B}{\partial y} = 6xyi - 4xzj + 14yz^2k \tag{3.60b}$$

and

$$\frac{\partial B}{\partial z} = -4xyj + 14y^2zk. \tag{3.60c}$$

3.4.4 Gradient, Divergence, and Curl

We define the *del* operator (∇) in rectangular coordinates in the following way:

$$\nabla = i\frac{\partial}{\partial x} + j\frac{\partial}{\partial y} + k\frac{\partial}{\partial z}. \tag{3.61}$$

Now let us suppose that T is a scalar (perhaps temperature), then

$$\nabla T = i\frac{\partial T}{\partial x} + j\frac{\partial T}{\partial y} + k\frac{\partial T}{\partial z}. \tag{3.62}$$

This is called the *gradient* of T and it is a vector. When we speak of the *divergence* of a vector (say, **A**), we mean

$$\nabla \cdot A = \frac{\partial A_1}{\partial x} + \frac{\partial A_2}{\partial y} + \frac{\partial A_3}{\partial z}, \tag{3.63}$$

and this quantity is a scalar. Therefore in the case of the velocity vector, V, the divergence would be written

$$\nabla \cdot V = \frac{\partial v_x}{\partial x} + \frac{\partial v_y}{\partial y} + \frac{\partial v_z}{\partial z}. \tag{3.64}$$

For an incompressible fluid, the divergence of the velocity vector is a statement of conservation of mass, and since the fluid density (ρ) is constant, $\nabla \cdot V = 0$. A vector field for which the divergence is zero is said to be *solenoidal*. In cylindrical and spherical coordinates:

$$\nabla \cdot V = \frac{1}{r}\frac{\partial}{\partial r}(rv_r) + \frac{1}{r}\frac{\partial v_\theta}{\partial \theta} + \frac{\partial v_z}{\partial z}, \quad \text{and} \tag{3.65a}$$

$$\nabla \cdot V = \frac{1}{r^2}\frac{\partial}{\partial r}(r^2 v_r) + \frac{1}{r\sin\theta}\frac{\partial}{\partial \theta}(v_\theta \sin\theta) + \frac{1}{r\sin\theta}\frac{\partial v_\phi}{\partial \phi}. \tag{3.65b}$$

The *curl* of a vector, A, is written as $\nabla \times A$, and it is a vector. In much of the older (particularly German) literature, the curl of A is written *rot*(A). It is convenient for us to construct a determinant to help with our interpretation:

$$\begin{vmatrix} i & j & k \\ \frac{\partial}{\partial x} & \frac{\partial}{\partial y} & \frac{\partial}{\partial z} \\ A_1 & A_2 & A_3 \end{vmatrix} = \left(\frac{\partial A_3}{\partial y} - \frac{\partial A_2}{\partial z}\right)i + \left(\frac{\partial A_1}{\partial z} - \frac{\partial A_3}{\partial x}\right)j + \left(\frac{\partial A_2}{\partial x} - \frac{\partial A_1}{\partial y}\right)k \tag{3.66}$$

We also often need the curl of a vector in cylindrical and spherical coordinates as well, so it is worthwhile to provide them here:

$$\left(\frac{1}{r}\frac{\partial A_3}{\partial \theta} - \frac{\partial A_2}{\partial z}\right)i + \left(\frac{\partial A_1}{\partial z} - \frac{\partial A_3}{\partial r}\right)j + \left(\frac{1}{r}\frac{\partial}{\partial r}(rA_2) - \frac{1}{r}\frac{\partial A_1}{\partial \theta}\right)k \tag{3.67a}$$

$$\left(\frac{1}{r\sin\theta}\frac{\partial(A_3\sin\theta)}{\partial \theta} - \frac{1}{r\sin\theta}\frac{\partial A_2}{\partial \phi}\right)i + \left(\frac{1}{r\sin\theta}\frac{\partial A_1}{\partial \phi} - \frac{1}{r}\frac{\partial(rA_3)}{\partial r}\right)j + \left(\frac{1}{r}\frac{\partial}{\partial r}(rA_2) - \frac{1}{r}\frac{\partial A_1}{\partial \theta}\right)k. \tag{3.67b}$$

In fluid mechanics, $\nabla \times V$ is particularly significant—it is the *vorticity* vector, ω_i.

$$\left(\frac{\partial v_z}{\partial y} - \frac{\partial v_y}{\partial z}\right)i = \omega_x \tag{3.68a}$$

$$\left(\frac{\partial v_x}{\partial z} - \frac{\partial v_z}{\partial x}\right)j = \omega_y \tag{3.68b}$$

$$\left(\frac{\partial v_y}{\partial x} - \frac{\partial v_x}{\partial y}\right)k = \omega_z \tag{3.68c}$$

The three components of the vorticity vector are directly related to the angular velocity of the fluid (rotation) about the three axes (in fact, each is exactly twice the rate of rotation about that particular axis). A vector field is said to be *irrotational* if $\nabla \times V = 0$. Let us suppose for example that an incompressible fluid undergoes two-dimensional motion which occurs in the x–y plane such that

$$v_x = 3xy^2 + 2y \quad \text{and} \quad v_y = -y^3. \tag{3.69}$$

We observe immediately that $\nabla \cdot V = 3y^2 - 3y^2 = 0$, and that

$$\frac{\partial v_y}{\partial x} - \frac{\partial v_x}{\partial y} = 0 - 6xy - 2 = \omega_z. \tag{3.70}$$

Thus, there will be rotation about the z-axis unless the product $xy = -1/3$. Now, suppose that an arbitrary vector field is described by

$$A = x^2yi + xz^2j + xyzk. \tag{3.71}$$

We begin by finding the divergence, $\nabla \cdot A$:

$$\frac{\partial}{\partial x}(x^2y) + \frac{\partial}{\partial y}(xz^2) + \frac{\partial}{\partial z}(xyz) = 2xy + xy. \tag{3.72}$$

Therefore, at the point (1,1,1), we find $\nabla \cdot A = 2 + 1 = 3$. We can also evaluate the curl of this vector field:

$$\left(\frac{\partial}{\partial y}(xyz) - \frac{\partial}{\partial z}(xz^2)\right) i + \left(\frac{\partial}{\partial z}(x^2y) - \frac{\partial}{\partial x}(xyz)\right) j + \left(\frac{\partial}{\partial x}(xz^2) - \frac{\partial}{\partial y}(x^2y)\right) k = (xz - 2xz)i + (0 - yz)j + (z^2 - x^2)k. \tag{3.73}$$

At the point (1,1,1) this yields $-1i-1j$.

Let us summarize some of the common operations involving the del-operator here:

$$\nabla(a + b) = \nabla a + \nabla b \tag{3.74}$$

$$\nabla^2 a = \nabla \cdot \nabla a \tag{3.75}$$

$$\nabla^2 A = (\nabla \cdot \nabla)A \tag{3.76}$$

$$\nabla \cdot (A + B) = \nabla \cdot A + \nabla \cdot B \tag{3.77}$$

$$\nabla \times \nabla a = 0 \tag{3.78}$$

$$\nabla \cdot \nabla \times A = 0 \tag{3.79}$$

$$\nabla \times (A + B) = \nabla \times A + \nabla \times B \tag{3.80}$$

$$\nabla \cdot (aA) = \nabla a \cdot A + a\nabla \cdot A \tag{3.81}$$

$$\nabla \times (aA) = \nabla a \times A + a(\nabla \times A) \tag{3.82}$$

$$\nabla \cdot (A \times B) = B \cdot (\nabla \times A) - A \cdot (\nabla \times B) \tag{3.83}$$

$$\nabla(A \cdot B) = (B \cdot \nabla)A + (A \cdot \nabla)B + Bx(\nabla \times A) - A \times (\nabla \times B) \tag{3.84}$$

$$\nabla \times (A \times B) = (B \cdot \nabla)A - B(\nabla \cdot A) - (A \cdot \nabla)B + A(\nabla \cdot B) \tag{3.85}$$

$$\nabla \cdot (\nabla a) = \nabla^2 a = \frac{\partial^2 a}{\partial x^2} + \frac{\partial^2 a}{\partial y^2} + \frac{\partial^2 a}{\partial z^2} \quad \text{(the Laplacian operator)} \tag{3.86}$$

$$\nabla \times (\nabla \times A) = \nabla(\nabla \cdot A) - \nabla^2 A \tag{3.87}$$

Recall that we previously identified two types of vector fields, *solenoidal* for which $\nabla \cdot A = 0$, and *irrotational* for which $\nabla \times A = 0$. In the case of the latter, if a vector field is irrotational, then a scalar function (ϕ) exists such that $A = \nabla\phi$. We refer to ϕ as a *potential* function; if we consider the x-component of the velocity vector in a potential flow problem, $v_x = \frac{\partial \phi}{\partial x}$. In an electric field, the component of (negative) electric intensity in a given direction is equal to the derivative of the potential in that direction. We also know by Stokes' theorem that if $\nabla \times A = 0$, then

$$\oint_C A \cdot dr = \iint_S (\nabla \times A) \cdot ndS = 0. \tag{3.88}$$

Let us spend a little time contemplating this statement prior to discussing Green's and Stokes' theorems. In physics when we consider work (W), we mean that a force has been applied to a moving body in such a way that there is a non-zero component of F in the direction of the motion. The work done in moving an object along the x-axis from position, x_1, to x_2 would be written:

$$W = \int_{x_1}^{x_2} F_x dx. \tag{3.89}$$

More generally, we can write a *line integral* of the vector, F, from point 1 to point 2 (along a curve, C) as

$$\int_{P_1}^{P_2} F \cdot dr = \int_C F \cdot dr = \int_C F_1 dx + F_2 dy + F_3 dz. \tag{3.90}$$

To illustrate, suppose we have a vector (force) in the x–y plane: $F = -3x^2 i + 5xyj$. This force is to be applied to an object moving along a path described by $y = 2x^2$. Therefore,

$$\int_C (-3x^2 dx + 5xydy). \tag{3.91}$$

Since $y = 2x^2$, $dy = 4xdx$; consequently, the integral of interest is

$$\int_{x=0}^{x=1} [-3x^2 dx + 5x(2x^2)4xdx] = \int_0^1 (-3x^2 + 40x^4)dx = -1 + 8 - 0 = 7. \tag{3.92}$$

A line integral that is particularly important in fluid mechanics is the integral of the velocity vector around a closed path; it is called *circulation*, $\Gamma = \oint V \cdot dr$. Of course, it is this circulation that is necessary for a wing (airfoil) to generate lift.

Let us take a moment to explore *circulation* a bit further. Consider *potential* flow about a circular cylinder where the velocity of approach is V; the velocity potential for this case is

$$\phi(r, \theta) = Vr\left(1 + \frac{R^2}{r^2}\right)\cos\theta. \tag{3.93}$$

Keep in mind that the *vector field for a potential flow is irrotational*. Of course, at the surface of the cylinder ($r = R$) there can be no radial component of velocity, only tangential:

$$v_\theta = \frac{1}{r}\frac{\partial\phi}{\partial\theta} = -2V\sin\theta \quad \text{and} \quad v_\theta^2 = 4V^2 sin^2\theta. \tag{3.94}$$

The pressure distribution at the surface of the cylinder is then

$$p - p_\infty = \frac{1}{2}\rho\left(V^2 - v_\theta^2\right) = \frac{1}{2}\rho V^2(1 - 4sin^2\theta). \tag{3.95}$$

Consider this result for $0 \le \theta \le \pi$; if $\theta = 0$ (the stagnation point) then $p - p_\infty = \frac{1}{2}\rho V^2$, at $\theta = \pi/2$, $p - p_\infty = -\frac{3}{2}\rho V^2$, and for $\theta = \pi$, $p - p_\infty = \frac{1}{2}\rho V^2$. This means that the pressure distribution at the cylinder's surface is symmetric fore and aft—we have complete recovery of the stagnation pressure at $\theta = \pi$, and therefore if we integrate the pressure over the cylinder surface we find there is no form drag on the object. This is contrary to physical experience and is known as *d'Alembert's paradox*; of course since this is a potential flow, $\mu = 0$, which means there can be no frictional drag either. But suppose we add an ideal flow called a vortex to our flow around a circular cylinder: For a vortex every fluid particle travels a circular path about the origin and

$$v_\theta = -\frac{A}{r} \tag{3.96}$$

where A is the *strength* of the vortex; note the singularity at the origin. Using polar coordinates the circulation for this case is given by:

$$\Gamma = \int_0^{2\pi} v_\theta r d\theta = -2\pi A \tag{3.97}$$

(since for a circular path, neither dr nor dz contribute to the path length). Note that positive Γ implies counter-clockwise rotation. You might be concerned that although we specified a potential flow where the vector field is irrotational, we find we have a *nonzero* circulation. Here we must recognize that although the *fluid as a whole is rotating*, individual fluid particles are *not*. We can think of it this way: For an irrotational field $\nabla \times V = 0$; thus, vorticity is zero on a microscopic level, whereas circulation is a macroscopic property of the flow.

Now suppose we combine the vortex *with* flow around a cylinder; we have two velocity vector components,

$$v_r = V\left(1 - \frac{1}{r^2}\right)\cos\theta \quad \text{and} \quad v_\theta = -V\left(1 + \frac{1}{r^2}\right)\sin\theta + \Gamma/(2\pi r). \tag{3.98}$$

We focus our attention on the surface of the cylinder where $r = 1$ and $v_r = 0$, resulting in

$$v_\theta = -2V\sin\theta + \Gamma/(2\pi). \tag{3.99}$$

Of course, this time $v_\theta \neq 0$ for $\theta = 0$ because of the cylinder's rotation, and we can calculate the circulation by integration using eq. (3.99) and obviously the result is nonzero. Once again there can be no frictional drag since $\mu = 0$, but the pressure distribution reveals a net force acting in the direction of higher velocity. If the flow is left to right, and the cylinder is rotating in the clockwise direction, the fast-moving fluid is at the top of the cylinder and the slow-moving fluid at the bottom. Consequently a lift force is generated; this phenomenon is known as the *Magnus effect*, and it figures prominently in many sports (tennis, golf, soccer, baseball, etc.) in which a spinning object experiences translational motion.

Finally, when we say that a force field is *conservative*, we mean that the work required to move an object, say, from point 1 to point 2, is *independent of path*. If the path is closed, i.e., if we start from point 1 and return to 1, then the work is zero. This is where our result above, eq. (3.88), comes into play. Obviously, based upon this discussion the curl of a conservative force field must be zero—a conservative force field is *always irrotational*. We note that gravity is an example of a conservative force, but friction is not. Consider a force field given by:

$$F = (2xz^3 + 6y)i + (8x - 2yz)j + (3x^2z^2 - y^2)\kappa. \tag{3.100}$$

We would like to know if it is conservative:

$$\nabla \times F = \begin{array}{ccc} i & j & k \\ \dfrac{\partial}{\partial x} & \dfrac{\partial}{\partial y} & \dfrac{\partial}{\partial z} \\ 2xz^3 + 6y & 8x - 2yz & 3x^2z^2 - y^2 \end{array} \tag{3.101}$$

Our result is $(-2y + 2y)i + (6xz^2 - 6xz^2)j + (8 - 6)k \neq 0$; this force field is *not* conservative (but nearly so—it would be if $(8x - 2yz)j$ were changed to $(6x - 2yz)j$).

3.5 Green's Theorem

Green's theorem plays an important role in the solution of many problems in mathematical physics, and examples are found in areas such as electrostatics and hydrodynamics. Green's theorem provides a relationship between line integrals on closed paths (denoted by C) and double integrals over the region enclosed by C (which we denote by D). We stipulate that the path C has positive orientation, which means that the enclosed region, D, is always on the left-hand side as we traverse the path. We also require that the functions P and Q appearing in the theorem have continuous first-order partial derivatives over D.

$$\oint_C Pdx + Qdy = \iint_D \left(\frac{\partial Q}{\partial x} - \frac{\partial P}{\partial y}\right) dA. \tag{3.102}$$

An example that is frequently used to introduce students to Green's theorem eq. (3.102) is the determination of the area of a conic section—the line integral is replaced by a double integral which is often easier to evaluate. There is a mechanical analog that provides some physical insight, a device called the planimeter; in the nineteenth century Jakob Amsler-Laffon constructed the first planimeter, an instrument used by draftsmen to compute the area enclosed by some curve automatically as the outline of the shape is traced. At one time just about every draftsman had a planimeter, and many used the device routinely. There are numerous pictures of vintage *Keuffel and Esser* planimeters online if you have never seen one; of course, such devices have been pretty much replaced by digital instruments today.

For emphasis, we point out that Green's theorem has several important uses including:

- calculating areas enclosed by a simple curve
- determining the circulation of F around C

- translating difficult line integrals into easier double integrals
- calculating the work done on a particle by a force field.

Some elaboration on the application of eq. (3.102) would be useful; suppose for example that we have a vector field in the x–y plane given by $F(x,y) = (F_1, F_2)$; the integrand on the right-hand side of eq. (3.102) would be

$$\frac{\partial F_2}{\partial x} - \frac{\partial F_1}{\partial y}. \tag{3.103}$$

Now, revisit eq. (3.21) in Section 3.4—you will see that eq. (3.49) is the z-component of the *curl* which is the circulation about the z-axis. So in this case Green's theorem provides a means by which we can calculate the total circulation inside a closed path denoted by C. In hydrodynamics, eq. (3.49) would be the z-component of the vorticity vector; but if the fluid in question was *inviscid* (frictionless) then the vector field would be conservative and the curl would be zero. Of course in real fluids, Newton's law of friction requires that the fluid velocity at stationary boundaries be zero, creating velocity gradients and a non-zero curl.

Now assume we had a vector field $F(x,y) = (y^2, 3xy)$ such that $\frac{\partial F_2}{\partial x} - \frac{\partial F_1}{\partial y} = 3y - 2y = y$, and further, assume this field is enclosed in a half-circle region, D, where $-1 \le x \le +1$. Since the half circle is located in quadrants 1 and 4, and since $x^2 + y^2 = R^2 = 1$, $0 \le y \le \sqrt{1-x^2}$, Green's theorem provides

$$\oint_C y^2 dx + 3xy dy = \int_{-1}^{+1} \int_0^{\sqrt{1-x^2}} y dy dx. \tag{3.104}$$

Accordingly,

$$\int_{-1}^{+1} \frac{1}{2} [y^2]_0^{\sqrt{1-x^2}} dx = \frac{1}{2}\left[x - \frac{x^3}{3}\right]_{-1}^{+1} = \frac{2}{3}.$$

If you would like more review and practice with Green's theorem, repeat this example with the same vector field but change the radius of the half circle to 2 so that y ranges from 0 to $\sqrt{4-x^2}$. Show that the result in this case is 16/3. Then, *change* the vector field to $F(x,y) = y^3, -x^3$, where D is a *complete circle* about the origin with $R = 2$. You should get -75.4 (or -24π).

We will also illustrate the use of Green's theorem on a different shape; suppose the (enclosed) positively oriented region under consideration is a triangle, and we wish to evaluate

$$\int_C xy dx + x^2 y^3 dy. \tag{3.105}$$

Let the triangular region be bounded by the x-axis from 0 to 1, a vertical line extending from (1,0) to (1,2), and the hypotenuse which returns from (1,2) to the origin (0,0). In this case the hypotenuse is represented very simply as $y = 2x$, and the x,y region of interest is enclosed by $0 \le x \le 1$ and $0 \le y \le 2x$. We can see by inspection of (3.102) and (3.105) that $P = xy$ and $Q = x^2 y^3$. Therefore,

$$\frac{\partial Q}{\partial x} = 2xy^3 \quad \text{and} \quad \frac{\partial P}{\partial y} = x. \tag{3.106}$$

By virtue of Green's theorem we write:

$$\iint_D (2xy^3 - x) dA = \int_0^1 \int_0^{2x} (2xy^3 - x) dy dx. \tag{3.107}$$

It is left to the reader to show that eq. (3.107) is equal to 2/3.

We previously discussed *conservative* vector fields, and we observed that in such cases there is no circulation. In the x–y plane this means that there would be no tendency for fluid particle (or entity) rotation about the z-axis. For example, if a vector field was uniform, say, $F = 3j$, there clearly could not be any circulation. Similarly, if

$$F_2 = xy^2 \quad \text{and} \quad F_1 = \frac{1}{3}y^3, \quad \text{then} \quad \frac{\partial F_2}{\partial x} - \frac{\partial F_1}{\partial y} = 0,$$

and again there is no circulation for a positively oriented enclosure. Take a moment to think about this force field; if an object was centered on the origin and subjected to F, how would it react? Might there be translation but no circulation? Next, consider the effect of the force field $F(y^2, 0)$, where we have vertical symmetry with respect to the origin. In this case, $\frac{\partial F_2}{\partial x} - \frac{\partial F_1}{\partial y} = -2y$ and for a circle centered on the origin with $R = 1$,

$$\int_0^{2\pi}\int_0^{1} -2r\sin\theta\, r dr d\theta = 0; \tag{3.108}$$

there cannot be any circulation about the origin with this choice for F. As a straightforward exercise, change this force field to $F(y^2 - y, 0)$—what is the result for this modification? You may be able to see by inspection that the answer is π (the circulation is counter-clockwise).

Now we will look at a case where the tendency toward clockwise rotation about the z-axis is obvious: Suppose we have a vector field given by $F = (y^3,0)$ on a positively oriented circle with radius of 1 as illustrated in rectangular coordinates in Figure 3.10:

We see immediately that this vector field will produce circulation about the origin in the *clockwise* (CW) direction. Since $\frac{\partial F_2}{\partial x} = 0$ and $\frac{\partial F_1}{\partial y} = 3y^2$, we have

$$\oint F ds = \iint -3y^2 dA. \tag{3.109}$$

We could work with either rectangular or polar coordinates. For the former, $x^2 + y^2 = R^2 = 1$ so $y = \sqrt{1 - x^2}$, and therefore

$$\int_{-1}^{+1}\int_{-\sqrt{1-x^2}}^{+\sqrt{1-x^2}} (-3y^2) dy dx = -2.3562. \tag{3.110}$$

But since $y = r\sin(\theta)$ we could write equivalently

$$\int_0^{2\pi}\int_0^{1} -3r^2\sin^2\theta\, r dr d\theta = \int_0^{2\pi} -\frac{3}{4}\sin^2\theta\, d\theta = -2.3562. \tag{3.111}$$

Finally, suppose we have an annular region formed by two concentric circles located at the origin. We let the radius of the inner circle be 1 and the radius of the outer circle be 2. Since Green's theorem *does not apply to regions with holes*, we

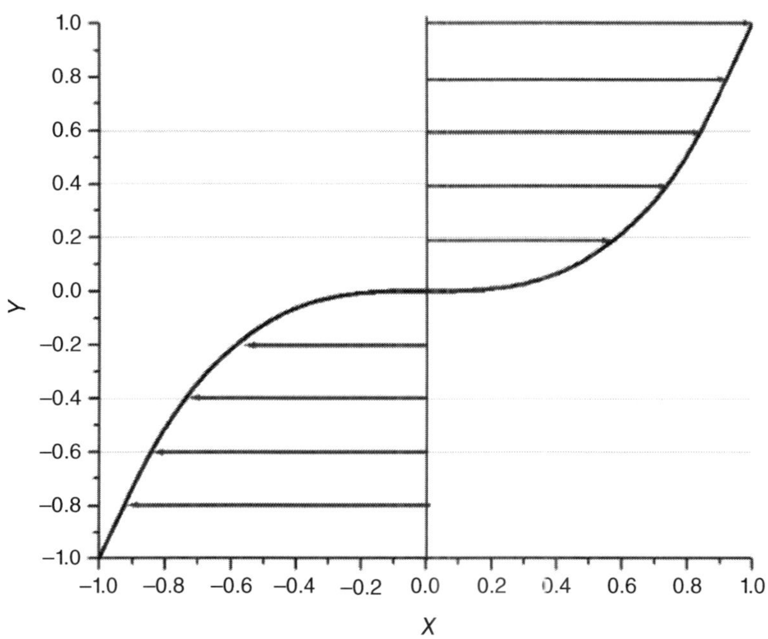

Figure 3.10 Vector field for $F = (y^3,0)$.

must take a different approach and there are two possibilities: We can take the annular region and break it into two halves by drawing a horizontal line just through the middle. We choose the orientation such that as we travel on the path(s) the two regions always lie to the left. But we could also look at the difference between the outer and inner circles; we will take this approach for illustration. By inspection we see that

$$P = y^3 \quad \text{and} \quad Q = -x^3, \tag{3.112}$$

therefore,

$$\frac{\partial P}{\partial y} = 3y^2 \quad \text{and} \quad \frac{\partial Q}{\partial x} = -3x^2. \tag{3.113}$$

Thus we have

$$\iint_D (-3x^2 - 3y^2)dA \quad \text{with} \quad dA = rdrd\theta. \tag{3.114}$$

Since $x^2 + y^2 = r^2$, we have

$$-3\int_0^{2\pi}\int_0^{2} r^3 drd\theta = -12\ \theta|_0^{2\pi} = -24\pi. \tag{3.115}$$

Now we apply the same technique to the inner circle (but with a radius of 1 rather than 2):

$$-3\int_0^{2\pi}\int_0^{1} r^3 drd\theta = -\frac{3}{2}\pi. \tag{3.116}$$

So for the integral of interest to us: $-24\pi - (-1.5\pi) = -22.5\pi$.

3.5.1 The Divergence Theorem of Gauss

The divergence theorem is an extension of Green's theorem:

$$\iint A \cdot n\, dS = \iiint \nabla \cdot A\, dV, \tag{3.117}$$

where V is a volume bounded by a closed surface, S. A is a vector function of position with continuous derivatives and n is the *positive normal* to S. We interpret this equation as the flux of a vector field out through a closed surface, S, and we note that the volume, V, has no holes. Of course, the utility of eq. (3.117) lies in the fact that an often difficult surface integral is converted into an easier volume (triple) integral. The volume element, dV, in rectangular, cylindrical, and spherical coordinates is $dxdydz$, $rdrd\theta dz$, and $r^2\sin\phi d\phi d\theta dr$, respectively.

Let us suppose for example that

$$\vec{A} = \frac{x}{z}, \frac{y}{z}, 0 \tag{3.118}$$

and assume that the volume in question is a cube with:

$$1 \le x \le 4, 2 \le y \le 5, \quad \text{and} \quad 1 \le z \le 4.$$

We note that $\frac{\partial A_1}{\partial x} = \frac{1}{z}$, $\frac{\partial A_2}{\partial y} = \frac{1}{z}$, and $\frac{\partial A_3}{\partial z} = 0$. Therefore, we have

$$\int_1^4\int_2^5\int_1^4 \frac{2}{z} dz\, dy\, dx = \int_1^4\int_2^5 (2ln4)dydx = 2ln4\int_1^4 3dx = 24.953. \tag{3.119}$$

We next apply the divergence theorem to a *tetrahedron* with one vertex located at the origin, and the other three vertices placed at $x = 1$, $y = 1$, and $z = 1$. Assume the vector field is

$$\vec{V} = x, y, z \quad \text{such that} \quad \nabla \cdot V = 1 + 1 + 1 = 3. \tag{3.120}$$

Since $x+y+z=1$ and the volume element is $dV = dxdydz$, we have

$$\int_0^1 \int_0^{1-x} \int_0^{1-x-y} 3dzdydx. \tag{3.121}$$

After integrating with respect to z and then y, we find

$$3\int_0^1 \left(\frac{1}{2} - x + x^2/2\right) dx = \frac{3}{6} = \frac{1}{2}. \tag{3.122}$$

For problems in cylindrical coordinates, the volume element is $dV = rdrd\theta dz$. Suppose we have an annular solid (a cylindrical shell) with a height of 3 ($0 \le z \le 3$) and a radial thickness of 1 with $1 \le r \le 2$. The vector field is $\vec{F} = 9, x, z^2$ so that $\nabla \cdot F = 2z$:

$$\int_0^{2\pi} \int_1^2 \int_0^3 2zdzrdrd\theta = \int_0^{2\pi} \int_1^2 9rdrd\theta = 27\pi. \tag{3.123}$$

Let's try another example employing a different surface; take $\vec{V} = xy^2, yz^2, x^2z$ and suppose the object in question is a sphere centered at the origin with $R=3$. In this case, $\frac{\partial V_1}{\partial x} + \frac{\partial V_2}{\partial y} + \frac{\partial V_3}{\partial z} = x^2 + y^2 + z^2$, and of course this is just r^2, suggesting we change to spherical coordinates where

$$0 \le r \le 3, 0 \le \theta \le 2\pi, \quad \text{and} \quad 0 \le \phi \le \pi.$$

Since $dV = r^2 \sin \phi \, d\phi \, d\theta \, dr$, we get

$$\int_0^3 \int_0^{2\pi} \int_0^{\pi} r^4 \sin \phi \, d\phi \, d\theta \, dr. \tag{3.124}$$

The reader may want to show that the result is 610.7.

The above illustration using a spherical surface brings up an important point: We must distinguish between a hemispherical *solid* and a hemispherical *shell*. Suppose the surface was a *hemispherical shell* (we will take the *upper half* of the sphere's surface) with $\vec{V} = xy, yz^2, 1$; since $\nabla \cdot \vec{V} = y + z^2$, we are interested in the triple integral

$$\iiint (y + z^2) dV. \tag{3.125}$$

However, in this case S is *not* a closed surface; i.e., the "bottom" of the hemisphere is an open circle of radius R. We first *assume* the open bottom circle is a surface and apply the divergence theorem to the *hemispherical solid*; then we correct our result by subtracting the flux through the bottom from the result. Keep in mind that the outward-directed normal will be pointed *down*, so it will be *negative*. In spherical coordinates,

$$x = r \sin \theta \cos \phi, y = r \sin \theta \sin \phi, z = r \cos \theta, \quad \text{and} \quad dV = r^2 dr \sin \theta \, d\theta \, d\phi.$$

Therefore,

$$\iiint (y + z^2) dV = \int_0^4 r^2 dr \int_0^{\pi/2} \sin \theta \, d\theta \int_0^{2\pi} (r \sin \theta \, \sin \phi + r^2 \cos^2 \theta) \, d\phi. \tag{3.126}$$

See if you can verify that this triple integral yields 428.9. Now, consider the flat, circular bottom of this hemisphere—any flux across that surface *can only be due* to the z-component of the vector field. Since this is the constant, 1, $\frac{\partial V_3}{\partial z} = 0$. What does this tell us about the nature of the "correction" that must be made to the first part of our estimate?

Of course, Gauss's divergence theorem can be applied to two-dimensional problems as well:

$$\oint_C F \cdot n \, ds = \iint_D \nabla \cdot F \, dA. \tag{3.127}$$

Note that on the right-hand side we have the dot product instead of the curl, and on the left we have "flow" of the vector field *normal* to the closed curve C instead of tangent to it. So we should think of the difference this way: with the divergence theorem we are determining flow *across* the boundary of a simple curve, whereas with Green's theorem, we are determining flow *along* the boundary. Let's illustrate how Gauss's theorem might be used for problems in the plane. Suppose we have a *point source* from which a fluid is issuing (for example, we might think of a spring or an artesian well), and we assume it is surrounded by a simple closed curve, C. We are interested in determining the total flux across the closed curve in the x–y plane. In this case our "force field" is the velocity vector consisting of components V_x and V_y, and $\nabla \cdot V = \frac{\partial V_x}{\partial x} + \frac{\partial V_y}{\partial y}$. Now suppose

$$(V_x, V_y) = \langle 5x + y, x + 3y \rangle \tag{3.128}$$

such that $\nabla \cdot V = 5 + 3 = 8$. This vector field is enclosed by a rectangle with $-1 \le x \le +1$ and $-2 \le y \le +2$, and therefore

$$\int_{-1}^{+1}\int_{-2}^{+2} 8\, dydx = \int_{-1}^{+1} 32\, dx = 64. \tag{3.129}$$

Note that if the units of the velocity field were ft/s, then this result would be 64 ft^2/s—remember we are working in a plane so we obtain a flux, not a volumetric flow rate (i.e., *not* ft^3/s). Next, let us suppose the same vector field is enclosed by a circle centered at the origin with radius, R. Since $x^2 + y^2 = R^2$, we have $y = \sqrt{R^2 - x^2}$ and

$$\int_{-R}^{+R}\int_{-\sqrt{R^2-x^2}}^{\sqrt{R^2-x^2}} 8\, dydx. \tag{3.130}$$

If $R = 2$, the result is $32\pi = 100.53$. Finally suppose we wanted to identify the size of the circle that would result in the *same flux* (64 ft^2/s) that we determined for the original rectangle. You might spend a few minutes verifying that the correct R is 1.596.

3.6 Stokes' Theorem

Stokes' theorem applies to a closed curve, C, in three-dimensional space that bounds some surface, S. It reduces to Green's theorem when the enclosing curve lies in a plane (is two-dimensional). It has been pointed out repeatedly that Stokes' theorem was not actually developed by Sir G. G. Stokes; Lord Kelvin (William Thomson) is credited with having done so after he discovered George Green's 1828 essay. Neeley (2008) states that it became known as Stokes' theorem due to the frequency with which Stokes placed it on the Cambridge prize examinations.

Given a vector field, F, where S is any surface bounded in three-dimensional space by closed curve, C,

$$\oint F \cdot dr = \iint_S (\nabla \times F) \cdot nds. \tag{3.131}$$

This is a powerful relationship that finds application in the connection between electric (E) and magnetic (B) fields. For example, you may have seen the Maxwell–Faraday equation previously:

$$\oint_C E \cdot dl = -\int_S \frac{\partial B}{\partial t} \cdot dS. \tag{3.132}$$

A verbal statement will help our comprehension here, and we cite the one provided by Sears and Zemansky (1964), "….. *the line integral of an induced electric field around a closed path, or the electromotive force (ε) in the path, equals the time rate of change of magnetic flux across the area bounded by the path.*"

Let us look at some examples to better reveal the utility of Stokes' theorem. To begin, we assume we have a vector field, $F = -yi + xj + x^2k$, with a surface corresponding to a right circular cylinder of height, h, and radius, R (with its bottom placed at the origin). For the right-hand side of eq. (3.131) we need the curl of F:

$$\begin{matrix} i & j & k \\ \frac{\partial}{\partial x} & \frac{\partial}{\partial y} & \frac{\partial}{\partial z} \\ -y & x & x^2 \end{matrix} \tag{3.133}$$

Therefore, $\nabla \times F = -2xj + 2k$. The surface integral will take into account the top (upper end) and the side of course. For example for the top we have

$$\iint_S (\nabla \times F) \cdot nds = \int_0^{2\pi} \int_0^{R} (-2xj + 2k) \cdot krdrd\theta = 2\pi R^2. \tag{3.134}$$

For the side of the cylinder, $ds = Rd\theta dz$, so in this case we note $x = R\cos\theta$ and $y = R\sin\theta$, to obtain

$$\iint_S (\nabla \times F) \cdot nds = -2R \int_0^{h} \int_0^{2\pi} \sin\theta \cos\theta R d\theta dz = 0 \tag{3.135}$$

The two parts combined yield $2\pi R^2$.

Now we will work through an example adapted from Spiegel (1971); in this case the vector field, F, is $3yi - xzj + yz^2k$. The closed curve is a circle located at $z = 2$ and the surface is a paraboloid given by $2z = x^2 + y^2$. We can determine the curl:

$$\nabla \times F = \begin{vmatrix} i & j & k \\ \dfrac{\partial}{\partial x} & \dfrac{\partial}{\partial y} & \dfrac{\partial}{\partial z} \\ 3y & -xz & yz^2 \end{vmatrix} = (z^2 + x)i - (z + 3)k \tag{3.136}$$

We could set about computing $\iint_S (\nabla \times F) \cdot nds$, but in this case we note that the enclosing curve is a circle in the plane corresponding to $z = 2$; accordingly, $4 = x^2 + y^2$. Since $x = R\cos\theta$ and $y = R\sin\theta$, where $R = 2$, we write

$$\oint_C 3ydx - xzdy + yz^2dz. \tag{3.137}$$

We substitute for x and y and note that $dz = 0$:

$$\oint_C - 3R^2\sin^2\theta d\theta - 2R^2\cos^2\theta d\theta. \tag{3.138}$$

The reader may wish to verify:

$$-R^2 \int_{2\pi}^{0} (3\sin^2\theta + 2\cos^2\theta)d\theta = -4(-5\pi) = 20\pi. \tag{3.139}$$

It will be left as a student exercise to verify that the integral of the curl of F over the surface yields the same result (20π).

3.7 Conclusion

We have assumed that everyone who comes to this study of applied mathematics has exposure to basic physics and therefore has some familiarity with vectors and the resolution of forces. Accordingly, our purpose with this chapter is to provide a brief review a few concepts that the student has seen previously. For the reader who needs more, excellent treatments of vectors, tensors, and vector manipulation has been provided by Spiegel et al. (2009) and Danielson (1992). These books are especially useful for self-study because they contain numerous examples that are completely solved.

For the reader interested in the connection between vectors, tensors, and fluid mechanics, it would be difficult to find a better starting point that Rutherford Aris's (1962) book, *Vectors, Tensors, and the Basic Equations of Fluid Motion*. This book is particularly important because it consistently tries to reveal the connection between mathematics and the physical reality of fluid flow. For applied scientists and engineers—many of whom are mainly visual and tactile learners—this is critical. G. K. Batchelor pointed out about 50 years ago that educators should always underscore the relation between "*.... analysis and the behavior of real fluids; fluid dynamics is much less interesting if it is treated largely as an exercise in mathematics.*" This observation applies to effective learning for students in all areas of applied science and engineering.

Problems

3.1 On July 2, 1937 Amelia Earhart (AE) and Fred Noonan left Lae (New Guinea) in a Lockheed Electra 10E Special and headed for Howland Island 2556 mi across the Pacific. The Electra was capable of making about 150 mph airspeed, but it was overloaded with 1000 gal of aviation gasoline, so its average speed initially was considerably less. Keep in mind that all angles given in the following description are compass bearings, e.g., due east is 90° and straight south is 180°.

After 20 hr and 14 min of flight the *USCGC Itasca* (standing by at Howland Island) received a final message from AE: "we are on line of position 157-337, we are running north and south." This would have been a line determined by the navigator, Fred Noonan, and clearly he thought it should intersect Howland; however, they failed to see either the island or the black smoke generated by the *Itasca* as a visual aid. No trace of the crew or the Electra has ever been found, so the plane almost certainly went into the water and sank when its fuel was expended.

We want to try to develop an explanation of what may have happened using the following (approximate) information. Take the Electra's average *airspeed* as 145 mph and assume a compass bearing of 72° was flown. The wind was out of the ESE, coming from 107° at an estimated speed of 14 mph. Of course, this quartering headwind would slow the Electra and push it north of its intended target. The maximum radio signal strength, as recorded by the *Itasca*, was received from AE after 19 hr and 12 min of flight. Fred Noonan may have estimated the wind strength in flight, but this cannot be known with certainty. What is known, is that Noonan liked to miss his intended target on one side or the other intentionally (a navigational theory of deliberate error). Then, when you arrived at what you believed to be the correct line of position, you would know the destination was north (or south if you decided to err to the north) of your position—therefore you would know which way to turn. If no in-flight correction was made (meaning that Fred Noonan was deliberately trying to miss Howland to the north), how far north of the island would the Electra have been after 19 hr and 12 min of flight? And, would the calculated line of position appear to intersect Howland?

Answer: The Electra would have been about 78.6 mi north (and west) of the island. AE reported that they were flying at just 1000 ft above sea level—probably due to clouds—far too low to have any chance of seeing Howland or the Itasca.

Please note: This problem was formulated as a "flat earther" exercise, and was not intended as an accurate portrayal of long distance navigation. Because the earth is nearly spherical, the closest distance between two points is determined with great circle navigation using spherical trigonometry. Thus, the actual heading one employs changes continuously enroute. Lae is located at about 6.72° S, 147° E, and Howland Island at 0.81° N, 176.62° W. The initial heading is given by:

$$\tan(\alpha_1) = \frac{\cos\theta_2 \sin\lambda_{12}}{\cos\theta_1 \sin\theta_2 - \sin\theta_1 \cos\theta_2 \cos\lambda_{12}}$$

where $\lambda_{12} = 176.62 - 147 = 29.62°$. *Therefore, the heading leaving Lae by the great circle route would have been 76.84°. At Howland Island the course would have been 75.28°. The modest change is due to proximity to the equator.* $\theta_1 = -6.7°$ *and* $\theta_2 = 0.81°$.

3.2 A commercial aircraft flying straight east with a ground speed of 495 mph encounters a jet stream flowing at 118 mph at 75° (15° north of east). If no corrective action is taken, what is the speed of the aircraft (relative to the ground) and what is the aircraft's modified course?

3.3 The great pyramid at Giza was built during the Fourth Dynasty for Khufu (in Greek: Cheops) and construction was begun about 2550 BC. According to Herodotus about 100,000 laborers (*not* slaves) built the pyramid over a period of 20 years. However, twentieth century archaeologists have discovered that the workforce was significantly smaller, perhaps only 10,000 to 20,000 workers. Khufu's pyramid is about 230 m on each side at the base, and it was originally about 147 m high. It was declared a UNESCO World Heritage site in 1979.

The materials of construction were granite and limestone blocks and slabs, and the pyramid was originally faced with smooth white limestone—which was stripped off the monument in modern times for buildings in Cairo. The main construction of the great pyramid was accomplished with about 2.3 million limestone blocks, each weighing about 2.5 tons over a period of about 20 years. This means that an average of about 300 blocks had to be moved *each day*. The construction was remarkably precise, with the position of each face corresponding to a cardinal compass direction and the angle of the faces with respect to horizontal is 51°52′.

It is clear that the limestone blocks were floated to the site using a canal linked to the Nile. However, exactly how the blocks were moved into position on the pyramid is less certain. There are two principal theories: the blocks were dragged into position on sleds, or they were floated into position using buoyancy achieved by immersion in water. It seems likely that the former is correct as the use of sleds and brute manpower is depicted in ancient wall decorations. We want to determine whether or not this is really feasible.

A 5000 lb block of limestone would have a volume of about 31 ft^3, and if it were an actual cube it would measure about 1 m on each side. So an immediate question concerns the number of men who could realistically get into a position to impose force upon the block for movement. Let us presume that a system of ropes would allow ~100 laborers to drag the block—making each man responsible for about 50 lb if they were to try to move it *vertically*. We will assume the coefficient of static friction on sand is about 0.7, and when in motion, perhaps 0.60 (it is possible that water was used to wet the soil in front of the block, decreasing friction, but this is not certain). It is likely that earthen ramps were built along the side of the pyramid to raise the block to the desired height. Suppose that early in the construction the ramp was inclined at 7°. The approximate height and weight of the Egyptian laborers was 5 ft 6 in. (168 cm) and 135 lbs (61 kg). Estimate the minimum number of men required to drag the block up the ramp. If the ramp angle was later increased to 15°, how many men would be required?

An aside: The scale of what was achieved at Giza is staggering. If the workday was 12 hr, and 300 blocks were to be moved into position daily, a single ramp would require the transit time for each block to be roughly 2.4 min which strains credibility. Of course, there may have been multiple ramps which would allow more time and alleviate congestion.

3.4 A ladder (24 ft long) leans against a house, forming an angle with the ground of 68°. A man stands midway on the ladder, and the ladder with the man weighs 219 lb_f. Determine the forces transmitted to the ladder by the vertical wall and the ground. At what angle with respect to the ground does the bottom force act?

3.5 Two vectors, A and B, are given by:

$$A = 6i + 3j + 7k \quad \text{and} \quad B = -2i + 5j - 1k.$$

Are these two vectors perpendicular?

3.6 Two vector fields are described by

$$A = xy^2 i + xz^3 j - xyzk \quad \text{and} \quad B = x^2y^3 i - y^2z^2 j + y^3 zk.$$

Is either vector field solenoidal? Irrotational? What simple change could be made to B to make it irrotational?

3.7 A structural component from an aircraft lies in the x–y plane; its upper edge is given by the equation $y = x + \frac{1}{4}x^3$ and the bottom edge by $y = \frac{2}{3}x$. This shape extends from $x = 0$ to $x = 4$, and it is illustrated in Figure P3.7:

Find the area, A, the first moments about the x- and y-axes (M_x and M_y), and the centroid.

3.8 Consider the three vectors

$$A = 25i + 3j\ B = 5i + 12j \quad \text{and} \quad C = 5\,k.$$

The volume of the parallelpiped formed by these three vectors is 1636.66. Confirm that this is equivalent to

$$|A \cdot (B \times C)|.$$

3.9 An elementary—but quite interesting—demonstration that is frequently used in physics classrooms results in the collision of two spheres (when the experiment is successful). Sphere 1 is dropped vertically from some height, h, above a flat, planar surface. Sphere 2 is launched, at the exact instant that sphere 1 is released, from the flat surface some horizontal distance away. Sphere 2 is aimed as precisely as possible at the initial vertical position of sphere 1. If gravity is the only significant force acting upon the spheres, then they should collide, irrespective of the initial velocity of sphere 2. Now, suppose we wanted to attempt this experiment on a larger scale; in particular, assume that

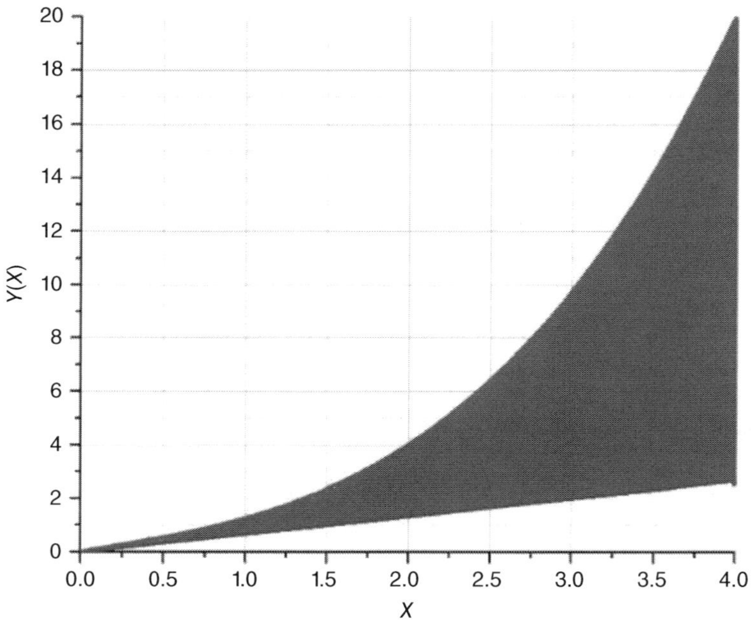

Figure P3.7 Two-dimensional object in the x–y plane.

sphere 1 (S1) is raised to an initial position 100 ft above the ground surface. Sphere 2 (S2) is to be launched at S1 from a (horizontal) distance 275 ft away. We want the collision between the spheres to occur *exactly* 16 ft above the ground surface. What are the launch parameters for S2 that *may* achieve the desired impact? In the classroom drag and buoyancy can usually be neglected. Will drag affect the two spheres differently (the dropped sphere's velocity is low relative to the launched projectile)? Explain the likely impact of drag upon the large-scale experiment.

3.10 An archer releases an arrow at an angle (with respect to level ground) of 70°. If the arrow's initial velocity is 260 ft/s, how far will the arrow have traveled *horizontally* when it reaches its apogee? At $t = 10.2$ s, what is the angle between the arrow's trajectory and the ground? Neglecting drag, what is the *total distance* traveled by the arrow after its return to level ground?

3.11 As part of a construction project, an HVAC unit in a wooden crate must be transported to the peak of a 6/12 pitch roof by sliding. The roof is unfinished, i.e., it is wooden sheathing secured to the rafters. The coefficient of static friction for wood-on-wood is 0.55, and the crated component weighs 325 lbs. Assume the crate is $5 \times 5 \times 5$ ft, and that the workers intend to push it up the slope. How much force will be required to reach the condition of impending motion? And, if the force is applied *parallel* to the roof but 4 ft *above* it, will the crate slide, or tip? For a 6/12 roof, $\tan(\theta) = 1/2$, so $\theta = 26.6°$.

3.12 A two-dimensional flow field has velocity vector components (in the x–y plane) described by

$$v_x = ay\exp(-by) \quad \text{and} \quad v_y = c\sqrt{x}.$$

Evaluate $\nabla \times V$ at the points (x,y): (1,1), (2,2), (3,3), and (4,4), if $a = b = c = ½$.

3.13 Elementary physics texts often cite 45° as the initial angle (of a projectile relative to the earth's surface) that produces maximum range. This is incorrect for spheroidal or cylindrical objects that rotate during their translational motion. Obvious examples include golf balls, tennis balls, baseballs, etc. In golf, impact between the clubhead and the ball imparts reverse spin on the ball, generating a lift force due to the *Magnus effect*. This has the tendency to steepen the trajectory of the ball during ascent. If a golf ball leaves the clubhead at an initial angle of 16°, but with reverse spin of 900 rpm, what will the steepest angle achieved during ascent be? Assume the initial ball velocity is 145 mph, neglect drag, and assume the rotational motion does not decay. The lift force acts perpendicular to the ball's trajectory, and we will estimate its magnitude with the following expression (actually valid for a *right-circular cylinder* on a per unit length basis):

$$\text{Lift} = 2\pi R\rho_{\text{air}} V_\theta V_{\text{trans}}.$$

We will take the density of the air to be 0.0012 g/cm^3, the tangential velocity (ball's surface) to be 188.4 cm/s, and the translational velocity to be 6482 cm/s. Assume the effective radius of the ball is 2 cm. Since the lift force is expressed on a per-unit-length basis, take the "length" of the golf ball to be 3 cm.

3.14 Given

$$A = x^2yi - 3yzj + xyzk$$

and

$$B = 2xzi + y^3zj + x^2y^2zk,$$

show that

$$\nabla \times (A \times B) = (B \cdot \nabla)A - B(\nabla \cdot A) - (A \cdot \nabla)B + A(\nabla \cdot B).$$

3.15 Show that $\iint_S (\nabla \times F) \cdot n ds = 20\pi$ where $F = 3yi - xzj + yz^2k$. See Section 3.5 for elaboration.

3.16 Use Green's theorem to evaluate $\oint xydx + x^2y^3dy$ for a region enclosed by a positively -oriented triangle with vertices located at (0,0), (1,0), and (1,2). Thus, $0 \le x \le 1$ and $0 \le y \le 2x$. Verify that the correct result is 2/3. Then, using the *same* vector field, evaluate the line integral for a rectangular region, D, with corners located at (1,–1), (4,–1), (4,2), and (1,2).

3.17 A vector field in the x–y plane is enclosed by two curves; the lower one extends from (0,0) to (1,1) and is described by $y = x^2$, and the upper curve is $y = \sqrt{x}$ and it extends from (1,1) back to the origin. Evaluate $\oint (2xy - x^2)dx + (x + y^2)dy$ on this positively oriented region by writing

$$\iint \frac{\partial}{\partial x}(x + y^2) - \frac{\partial}{\partial y}(2xy - x^2)dxdy.$$

Show that the result is 0.0333 (1/30).

3.18 The components of a force field are given by $F = y - \cos x,\ e^y - 2x$. Determine the total circulation on a positively oriented circle of radius 4, centered upon the origin.

3.19 Determine the total circulation for a vector field given by $F_1 = x^2y$ and $F_2 = xy + 3x^2$ on a closed path described by the positively oriented ellipse, $\frac{x^2}{a^2} + \frac{y^2}{b^2} = 1$, with $a = 5$ and $b = 2$.

3.20 On May 22, 1963 Mickey Mantle hit a fastball pitched by Bill Fischer of the Kansas City A's that bounced off of the façade at the very top of old Yankee Stadium. In an interview after his retirement Mantle stated that that homerun was the hardest he had ever struck a baseball. The straight-line distance from home plate to the façade was 370 ft, and the point of impact was 118 ft *above* right field. One website has estimated that this ball would have traveled 734 ft, making it the longest homerun ever hit in MLB! This estimate is flawed for a number of reasons, although some witnesses did report that the ball was *still rising* when it struck the façade. Given that a baseball has a circumference of 9 in. and weighs 5 oz, estimate the distance this ball might have traveled taking drag into account:

$$F_{\text{drag}} = AKf = (\pi R^2)\left(\frac{1}{2}\rho V^2\right)(0.44).$$

For point of reference, the initial velocity of a very well-struck baseball is on the order of 45 to 50 m/s. At 47 m/s, the initial Reynolds number is:

$$\text{Re} = \frac{d\rho V}{\mu} = \frac{(7.28)(4700)}{(0.151)} = 226{,}489.$$

In this region the drag coefficient is approximately constant, and therefore we will use 0.44 for f. Prove that this baseball *could not have* traveled 734 ft. The kinematic viscosity of air (μ/ρ) is 0.151 cm^2/s and the density (ρ) is 0.00121 g/cm^3.

3.21 Consider a vector with initial and final points (in terms of x,y,z values) of (5,5,5) and (3,2,2), respectively. Find the magnitude of this vector. Then, given the vector, $A = 2i - 3j + k$, demonstrate that $\nabla \cdot (\nabla \times A) = 0$.

3.22 The distribution of a two-dimensional scalar field in a rectangle with $0 \leq x \leq 10$ and $0 \leq y \leq 5$ is given by

$$\frac{50}{y^2 - 8y + x^2 - 14x + x^2y^2 - 36xy + 389}.$$

Find the maximum gradient at the point $x = 7$ and $y = 3$, and find the corresponding direction of this maximum gradient. Do your answers correspond with estimates made using Figure P3.22?

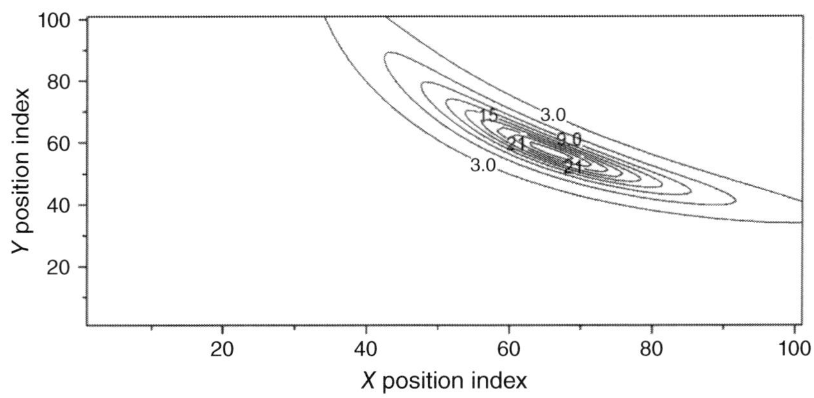

Figure P3.22 Scalar field for problem 3.22.

3.23 A *cardioid* is a figure consisting of the locus of points given by: $x^2 + y^2 + Bx = B\sqrt{x^2 + y^2}$. This shape bears some resemblance to a human heart, hence the name. Suppose $B = 6$, and assume we need to determine the area enclosed by the curve. We should be able to find the area of a plane region through use of a line integral over the boundary according to Green's theorem; specifically

$$A = \iint_R dxdy = \frac{1}{2}\int_C (xdy - ydx).$$

Prove that the area of the cardioid is $\frac{108\pi}{2}$. *Hint: Try putting the problem in polar form.*

3.24 A force field acting in the x–y plane has components F_1 and F_2: $y^2 - 3y, 0$. *i*) We want to find the total circulation for a circle of radius, $R = 3$, centered on the origin and, of course, we note $\frac{\partial F_2}{\partial x} = 0$. Before you make any calculations, determine whether the circulation is positive or negative. *ii*) Now assume the force field is enclosed by an *ellipse* where $\frac{x^2}{a^2} + \frac{y^2}{b^2} = 1$ with $a = 6$ and $b = 1$. *Prior* to making any calculations for part *ii*) *estimate* which case will produce the largest total circulation.

3.25 Suppose we have a velocity vector in the x–y plane: $(V_x, V_y) = (x^2 + 2y, x^2 + 3y^3)$, and furthermore, that this field is enclosed by a circle centered at the origin with $R = 4$. Determine the total flow across this simple curve. Can you tell by inspection whether or not the computed flux will be negative?

3.26 A vector field is given by: $\vec{A} = \langle 4x, -2y^2, z^2 \rangle$. Apply the divergence theorem to a specified volume: $x^2 + y^2 = 4$ and $0 \leq z \leq 3$. Show that the result is 263.9.

3.27 A cube, with one vertex located at the origin, has unit dimensions in each direction:

$$0 \leq x \leq 1, 0 \leq y \leq 1, 0 \leq z \leq 1.$$

It is subjected to a vector field, $\vec{F} = x^2y - x^3, y^2 - xy^2, 3x^2z$. Apply the divergence theorem to this object, assuming that the bottom is missing; do this by treating the cube as an enclosed solid, then correcting for the flux through the bottom surface. Next, suppose the open-bottomed box was *rectangular with a height of 2*: $-1 \leq z \leq 1$ (x and y dimensions are unchanged). What is the result in this case?

References

Aris, R. (1962). *Vectors, Tensors, and the Basic Equations of Fluid Mechanics*. Englewood Cliffs, NJ: Prentice-Hall.
Danielson, D.A. (1992). *Vectors and Tensors in Engineering and Physics*. Reading, MA: Addison-Wesley.
Neeley, M. (2008). *Exploring Stokes' Theorem*. Knoxville: University of Tennessee.
Sears, F.W. and Zemansky, M.W. (1964). *University Physics*, 3rd e. Reading, MA: Addison-Wesley.
Spiegel, M.R. (1971). *Advanced Mathematics for Engineers and Scientists*, Schaum's Outline Series. New York: McGraw-Hill.
Spiegel, M.R., Lipschutz, S., and Spellman, D. (2009). *Vector Analysis*, 2nd e, Schaum's Outline Series. New York: McGraw-Hill.

4

Numerical Quadrature

4.1 Introduction

Definite integrals must be evaluated routinely and for cases in which an antiderivative cannot be found, or for cases in which the analytic process is simply too difficult, a numerical scheme (numerical integration, or quadrature) may be our only recourse. Consider the integral,

$$\int_0^\infty x^2 \exp(-x^2)dx, \tag{4.1}$$

which we know to have the value $\frac{\sqrt{\pi}}{4}$. Let us consider a plot of the integrand as a function of x, shown in Figure 4.1.

We will conduct an elementary experiment with the graph of this integrand: First, we will count the smaller rectangular regions under the curve and add their areas together to obtain (70)(0.25)(0.025) = 0.4375. Next, we will cut out the region under the curve and weigh it and compare that weight to the average weight of the graph paper per rectangle. For the paper under the curve we find W = 0.5754 g, and for each box, 0.008003 g. Therefore, the area under the curve is approximately 71.89 rectangular boxes or (71.89)(0.25)(0.025) = 0.4493. Our first estimate is in error by about 1.3% (under) and our second estimate by about 1.4% (over). For many applications a rough evaluation like this would be acceptable, but suppose it was not. We might have a critically important problem where accuracy was paramount. Therefore it will be productive for us to review some techniques that may be capable of producing significantly improved estimates.

4.2 Trapezoid Rule

One of the simpler schemes for making this evaluation is the trapezoid rule; we use a straight-line approximation for $f(x)$ over a finite interval, Δx, and then multiply the average value for $f(x)$ by the width. For example, let $f_1 = f(x)$ and $f_2 = f(x+\Delta x)$. Our approximation, therefore, is

$$\int_x^{x+\Delta x} f(x)dx = \left(\frac{f_1+f_2}{2}\right)\Delta x.$$

If Δx is sufficiently small, this may work well. We will apply it first to the example immediately above; notice in the code provided below that the upper limit for the integral has been set to 8. We conclude that this *may be reasonable* since $(8)^2 \exp(-8^2) \cong 1.03 \times 10^{-26}$. This is an example of *limit truncation*; we have changed the upper limit of the integral to 8 from ∞. One might want to assess this by using different upper limits to better gauge the contribution of the right-hand tail. Of course limit truncation may introduce error, but in this particular case the behavior of the integrand is easily anticipated.

```
#COMPILE EXE
#DIM ALL
  REM *** Numerical integration by trapezoid rule
```

Problem Solving in Engineering: Analytical Mathematics and Numerical Analysis, First Edition. Larry A. Glasgow.

Companion website: www.wiley.com/go/glasgow/problemsolvingengineering

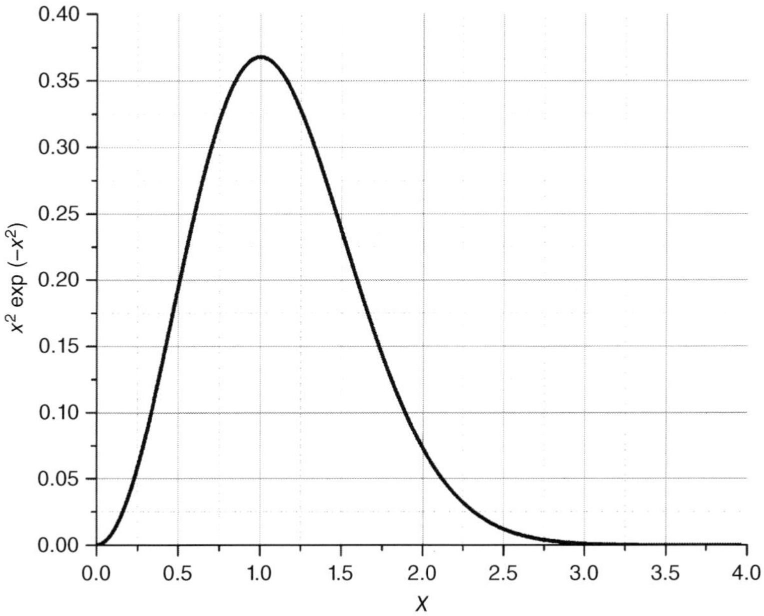

Figure 4.1 Behavior of $x^2 \exp(-x^2)$. The value of the integral (4.1) is known to be $\frac{\sqrt{\pi}}{4} = 0.443113$.

```
      GLOBAL x,dx,fx,f1,f2,a,b,fbar,sum,zz AS DOUBLE
FUNCTION PBMAIN
      a=0:b=8.0:x=a
      dx=(b-a)/1000
        GOSUB 300
          f1=fx
  50 REM *** continue
         x=x+dx
           IF x>b THEN 200
         GOSUB 300
         f2=fx
     fbar=(f1+f2)/2
        sum=sum+fbar*dx
          f1=f2
          GOTO 50
  200 REM *** continue
          PRINT "Value of definite integral is: ";sum
        INPUT "Shall we continue?";ZZ
        IF ZZ>0 THEN END
   300 REM *** function evaluation for integral
           fx=x^2*EXP(-x^2)
           RETURN

END FUNCTION
```

The code listed immediately above produces a result of 0.4431135 using double precision and1,000 intervals between 0 (zero) and 8. Despite the very small error produced in this case, one might wonder if a more demanding application could require a more accurate approximation. After all, there is nothing to prevent us from using a *polynomial* to approximate $f(x)$, as opposed to a straight line. Before we attempt that, however, let us examine one other case where the value of the definite integral is known so we can explore the impact of increasing the number of intervals using both single- and double-precision arithmetic. Again we will use the trapezoid rule to consider

$$\int_{0}^{\pi} \sin^2(3x)dx = \frac{\pi}{2} = 1.570796327. \tag{4.2}$$

n	Single Precision, *I*(*x*)	Double Precision, *I*(*x*)
125	1.570797	1.5707963268
250	1.570800	1.5707874013
500	1.570789	1.5707952107
1,000	1.570781	1.5707961870
2,000	1.570817	1.5707963270
4,000	1.570853	1.5707963250
8,000	1.570871	1.5707963264

Notice that once we exceed $n \approx 2{,}000$ with the single-precision calculations, the result deviates increasingly from the correct value given in eq. (4.2). We will consider this unwelcome development—cumulative roundoff error—more fully later in this chapter.

4.3 Simpson's Rule

We saw that we could employ a linear approximation with the trapezoid rule and get satisfactory results for integrals that were not particularly "difficult." An obvious extension that might yield better results could be obtained by merely increasing the degree of the polynomial. For example, we might let $f(x)$ be approximated by

$$y = f(x) = ax^2 + bx + c \tag{4.3}$$

If we place the three necessary function evaluations (for y_1, y_2, and y_3) at $x = -\Delta x$, $x = 0$, and $x = +\Delta x$, then we find

$$y_1 = a(\Delta x)^2 - b\Delta x + c \tag{4.4}$$

$$y_2 = c \tag{4.5}$$

$$y_3 = a(\Delta x)^2 + b\Delta x + c \tag{4.6}$$

Now we will actually integrate the polynomial (4.3) from $-\Delta x$ to $+\Delta x$:

$$\int_{-\Delta x}^{+\Delta x} (ax^2 + bx + c)dx = \left(\frac{a}{3}x^3 + \frac{b}{2}x^2 + cx\right)\Bigg|_{-\Delta x}^{+\Delta x} = \frac{2a}{3}(\Delta x)^3 + 2c\Delta x. \tag{4.7}$$

We use the three function evaluations above to obtain a, b, and c. The resulting algorithm for *Simpson's rule* is

$$\frac{\Delta x}{3}(f_1 + 4f_2 + f_3) \quad \text{or} \quad \frac{\Delta x}{3}(y_1 + 4y_2 + y_3). \tag{4.8}$$

For an easy comparison, let us apply this method to the integral (4.2) examined in the previous section, $\int_0^{\pi} \sin^2(3x)dx = \frac{\pi}{2}$. Using 500 intervals and single precision we obtain 1.570791, and with 500 intervals and double precision, 1.57079632679; note how these values compare to the appropriate entries in the table provided previously.

An illustration of logic required for Simpson's rule is shown immediately below for the slightly more difficult definite integral:

$$\int_{4.75}^{5} \frac{\exp(2x)}{x} dx = 887.20057$$

(numerical result from the code as it appears here). Mathcad™ produces 887.201 for this integral using its *adaptive* scheme and you may want to contemplate this integrand to see if you can determine why we might need an adaptive algorithm. For transparency we will make no effort to be computationally efficient.

```
#COMPILE EXE
#DIM ALL
  REM *** Integration by Simpson's rule
      GLOBAL xi,xf,dx,x,y1,y2,y3,fx,sum,ZZ AS DOUBLE
FUNCTION PBMAIN
      xi=4.750:xf=5.0:sum=0
        dx=(xf-xi)/64
           x=xi
   50 REM *** continue
          GOSUB 300
            y1=fx
          x=x+dx
          GOSUB 300
            y2=fx
          x=x+dx
          GOSUB 300
            y3=fx
              sum=sum+dx/3*(y1+4*y2+y3)
            IF x<xf THEN 50 ELSE 100
   100 REM *** continue
           PRINT sum
           INPUT "Shall we continue?";ZZ
           IF ZZ>0 THEN END
   300 REM *** function evaluation
               fx=1/x*EXP(2*x)
               RETURN
END FUNCTION
```

The simple code shown above uses double-precision floating-point numbers with about 15 or 16 digits of precision; the actual program output is 887.20056962…

So far we have examined definite integrals that pose no particularly great challenge, but in order to see the benefits offered by increasing the order of the polynomial used for the approximation we need something else. Consider the integral

$$\int_{0}^{2.99} \frac{dx}{(3-x)^2} = 99.6667. \tag{4.9}$$

We will apply Simpson's rule to this problem, using single precision and starting with a very modest number of intervals, successively doubling n.

n	100	200	400	800	1,600	3,200	6,400	12,800
Definite Integral	140.82	109.82	1,779.02	429.62	158.65	122.10	101.83	99.65

It is a worthwhile exercise for the reader to test Simpson's rule on the integral in eq. (4.9) using other values for n, perhaps 600. The results we see here suggest that we might encounter cases for which Simpson's rule will not really be adequate. What would happen if we increased the order of the polynomial approximation again?

4.4 Newton–Cotes Formulae

We posed a question regarding the impact of increasing the order of the polynomial approximation used for numerical quadrature. After all, we used a straight line with the trapezoid rule, and a quadratic approximation with Simpson's rule. There is nothing to prevent us from adding more function evaluations in an effort to make our curve more accurately reflect the behavior of the integrand. This succession of polynomials is often referred to collectively as the Newton–Cotes formulae. Let us tabulate these expressions for orders 1 through 5:

$$\frac{\Delta x}{2}(y_1 + y_2) \text{ trapezoid rule} \tag{4.10}$$

$$\frac{\Delta x}{3}(y_1 + 4y_2 + y_3) \text{ Simpson's rule} \tag{4.11}$$

$$\frac{3\Delta x}{8}(y_1 + 3y_2 + 3y_3 + y_4) \text{ Simpson's } 3/8 \text{ rule} \tag{4.12}$$

$$\frac{2\Delta x}{45}(7y_1 + 32y_2 + 12y_3 + 32y_4 + 7y_5) \tag{4.13}$$

$$\frac{5\Delta x}{288}(19y_1 + 75y_2 + 50y_3 + 50y_4 + 75y_5 + 19y_6) \tag{4.14}$$

There is an obvious cost associated with this process: The number of function evaluations required per step for an n-th order approximation will be $n+1$. Will the extra effort required be worthwhile? We can illustrate the effectiveness of this approach by using the fourth-order Newton–Cotes algorithm on the integral,

$$\int_0^{\pi/2} \frac{dx}{(4\sin^2 x + 9\cos^2 x)} = 0.26179939. \tag{4.15}$$

If we specify the number of intervals as 50, 100, and 200, successively, we obtain the results 0.2774816, 0.2617994, and 0.2617994. We have obtained a high-quality result using a limited number of intervals *with single precision*! Of course, this was not a very demanding example, so we may want to ask if the increased complexity of the algorithm will help us deal with difficulty we encountered in Section 4.3. The unfortunate answer is a qualified no. There are cases where merely increasing the order of the polynomial will not materially improve our estimate for the definite integral. We may occasionally need better tools, and in the following material we will look at several candidates. Before we explore some alternatives, however, we should consider some of the consequences of the types of calculations we have proposed for numerical quadrature.

4.5 Roundoff and Truncation Errors

Let us take the Simpson's algorithm (in single precision) and apply it to an elementary definite integral. We have an intuitive impression at this point that we can always improve our estimate by reducing the size of the interval (increasing the number of steps employed). But we also know that floating-point variables are only represented to the precision of the computing device we use. If the *roundoff* errors associated with that process accumulate, we might get results that are incorrect, or unexpected. We will conduct this experiment with

$$\int_0^{\pi/2} \frac{dx}{4\sin^2 x + 9\cos^2 x} = 0.26179939.$$

Let us increase n by a factor of 10 successively to see what happens:

n	$I(x)$
50	0.2617994
500	0.2617993

n	I(x)
5000	0.2619558
50,000	0.2615938
500,000	0.2633153
5,000,000	0.2490056
50,000,000	2.1934250

Our effort to get an extremely accurate value for the definite integral was misguided as we have discovered that roundoff error can corrupt even a very simple calculation. There are two important lessons here: Use double precision whenever possible, and do not blindly assume that decreasing the interval size (Δx) will produce a better result.

We now shift our focus to *truncation* errors that result from terminating an infinite series expression (this will occur anytime a continuous function is represented discretely). Numerical quadrature is susceptible to this problem too and it would be extremely useful to be able to assess the size of this error. We will begin by using the trapezoid rule to illustrate this phenomenon. Let us contemplate the integral:

$$\int_{3}^{4} (2 + 5x - x^2)dx = 7.16667. \tag{4.16}$$

We set $\Delta x = 1$ and we evaluate the *integrand* at both $x = 3$ and $x = 4$; let those values be represented by $f'(3) = 8$ and $f'(4) = 6$. We also find the *antiderivative* of the integrand and evaluate it at both $x = 3$ and $x = 4$: The values are $f(3) = 19.5$ and $f(4) = 26.6667$, respectively. Of course, if we subtract the former from the latter we get the exact value provided in eq. (4.16), 7.16667. For the trapezoid rule we would take $[f'(3) + f'(4)]/2$ and multiply by Δx (which is 1); the result is 7 of course which means that the absolute error is 0.16667 (or about 2.3%). Now let $f(3)$ be represented by $y(x)$ and $f(4)$ be $y(x + \Delta x)$—we expand the latter in a Taylor series:

$$y(x + \Delta x) = y(x) + \left(\frac{dy}{dx}\right)_x \Delta x + \left(\frac{d^2y}{dx^2}\right)_x \frac{(\Delta x)^2}{2} + \left(\frac{d^3y}{dx^3}\right)_x \frac{(\Delta x)^3}{6} + \cdots. \tag{4.17}$$

We subtract $y(x)$ noting that the remainder on the right-hand side is approximately the area we seek:

$$y(x + \Delta x) - y(x) = f_x'\Delta x + f_x''\frac{(\Delta x)^2}{2} + f_x'''\frac{(\Delta x)^3}{6} + \cdots \tag{4.18}$$

For the trapezoid rule we need both f_x' and $f_{x+\Delta x}'$, so using the Taylor series expansion again,

$$f_{x+\Delta x}' \cong f_x' + f_x'(\Delta x) + f_x'''\frac{(\Delta x)^2}{2} + \cdots \tag{4.19}$$

Therefore,

$$f_x''\frac{(\Delta x)^2}{2} = \frac{\Delta x}{2}\left[f_{x+\Delta x}' - f_x' - f_x'''\frac{(\Delta x)^2}{2} - \cdots\right]. \tag{4.20}$$

Now we take this result back to eq. (4.19):

$$y(x + \Delta x) - y(x) = \left[\frac{f_{x+\Delta x}' + f_x'}{2}\right]\Delta x - f_x'''\frac{(\Delta x)^3}{12} - \cdots \tag{4.21}$$

Notice that the *exact* area is on the left-hand side, and the trapezoid rule approximation is the bracketed term on the right. Therefore, the error associated with this *one interval* example is on the order of

$$e_1 \cong -f_x'''\frac{(\Delta x)^3}{12}. \tag{4.22}$$

Remember, f_x''' is the *second derivative* of the integrand, and it is constant in this case:

$$f_x' = 2 + 5x - x^2 \tag{4.23}$$

$$f_x'' = 5 - 2x, \tag{4.24}$$

and

$$f_x''' = -2. \tag{4.25}$$

Therefore we have

$$f_x'''\frac{(\Delta x)^3}{12} = \frac{(2)(1)^3}{(12)} = \frac{1}{6}. \tag{4.26}$$

In this instance, the estimate of error is borne out exactly by our previous result. Generally we will not be that fortunate, and f_x''' will vary over the range of the independent variable, x. We will rewrite the error so that

$$-f_x'''\frac{(\Delta x)^3}{12} \approx -f_\chi'''\frac{(\Delta x)^3}{12}, \tag{4.27}$$

where $x \leq \chi \leq x + \Delta x$ (think of the mean-value theorem from calculus). If we wish to employ our estimate of error for an arbitrary case with n-intervals,

$$\sum e_n = -\frac{(\Delta x)^3}{12}[f'''(\chi_1) + f'''(\chi_2) + f'''(\chi_3) + \cdots f'''(\chi_n)]. \tag{4.28}$$

How useful this expression is depends entirely upon how difficult it is to evaluate the second derivative of the integrand, f''. If one were integrating tabular data, for example, both f' and f'' would be unknown. There is a great deal of information available related to error estimation in numerical quadrature, and the interested reader might begin by reviewing Chapter 5 in James et al. (1977). We will also make error analysis the focus of one of the exercises at the end of this chapter.

4.6 Romberg Integration

The Romberg scheme utilizes an extremely powerful technique known as Richardson's extrapolation. Suppose for example that we were able to estimate the error associated with one of our numerical quadrature procedures. Furthermore, suppose we then decreased the step size (increasing the number of intervals) and found the error associated with *that* estimate. If we extrapolated the diminishing error to zero, we might be able to obtain a high-quality estimate with far less numerical work than would need to be carried out otherwise. We will illustrate this concept with a simple example, using only the trapezoid rule to prepare our successive estimates. Let us consider the integral,

$$\int_1^5 x \ln x dx = 14.118 \text{ (from the analytic solution)}. \tag{4.29}$$

We begin by using the trapezoid rule, and we will start with just one interval (such that $\Delta x = 4$). We then cut Δx in half and evaluate the integral again. We repeat this process several times, resulting in a series of estimates that appear in the table below:

n	Estimate for Integral, $I(x)$
1	16.094379
2	14.638863
4	14.250903
8	14.151423
16	14.126351
32	14.120069

In this procedure we are cutting the interval in half, successively. An improved estimate for the definite integral is obtained using pairs of values from the table according to:

$$I_{\text{improved}} = \frac{2^2 I_{\Delta x/2} - I_{\Delta x}}{2^2 - 1} = \frac{1}{3}(4I_{\Delta x/2} - I_{\Delta x}). \tag{4.30}$$

We will apply this procedure to the definite integral above, using the subsequent pairs of values from our table.

Pair ID	$I_{improved}$
2–1	14.1537
4–2	13.4549
8–4	14.1183
16–8	14.1180
32–16	14.1180

Notice that by the time we get to the 8–4 pair, we have reduced the error of our estimate down to 0.002%. The attractive nature of the Romberg scheme is now apparent: We have dramatically improved the results of our numerical integration without much additional computational effort! You should be aware of the fact that if we simply apply the trapezoidal rule with 32 intervals, our error is still about 0.015%. This example demonstrates the superiority of Romberg integration—at least for this case—very conclusively.

We would like to know if the outstanding results obtained in the preceding example can be expected more broadly (universally would be nice, but we will settle for less than that). Recall our exploration of the integral,

$$\int_{0}^{2.99} \frac{dx}{(3-x)^2} = 99.6667;$$

let us apply the Romberg scheme here to find out if it is comparably effective for this more difficult case. The integrations will be carried out with Simpson's rule this time.

n	$I(x)$	Extrapolated Value
1	39,866.9658	—
2	4,984.2782	−6,643.28
4	2,493.8525	1,663.71
8	1,250.4297	835.95
16	632.0535	425.93
32	328.6733	227.55
64	185.9160	138.33
128	125.6151	105.52
256	105.2509	98.46
512	100.4610	98.86
1,024	99.7431	99.50

By using the 1,024–512 interval pair, we were able to use extrapolation to reduce the error of our estimate to just 0.17%. It should be apparent to you that for certain types of problems where the computational burden is really significant, Romberg integration can be used to great advantage.

4.7 Adaptive Integration Schemes

The "difficult" integral we have examined previously,

$$\int_{0}^{2.99} \frac{dx}{(3-x)^2} = 99.6667,$$

is one for which the integrand varies by four orders of magnitude (1/3 to 10,000) as x varies from its lower limit (0) to the upper one (2.99). This kind of change in magnitude indicates that a different approach may be needed, namely, an *adaptive algorithm*. The underlying logic is straightforward: When we encounter a region in which the integrand begins to change dramatically, we decrease the interval size. There are many adaptive implementations available in commercial software packages. Mathcad's™ adaptive algorithm, for example, produces 99.667 for the integral in question.

4.7.1 Simpson's Rule

It is reasonable for us to wonder if we could make a few simples changes to one of the elementary quadrature schemes and produce something effective. We will try this using Simpson's rule where we are using the variable, *sum*, to accumulate increments of area. We will look at the per-step change in *sum*; if it exceeds a set value, we will reduce Δx. In particular, we will set the threshold for change in sum at 10% initially and if that size change occurs, we reduce the size of the interval by cutting it in half.

If Change Exceeds:	Divide *dx* by:	Result, *I*(*x*), Using Double Precision
10%	2	109.9738
5%	2	104.5676
2.5%	2	99.66676
1.25%	2	99.66676
0.625%	2	99.66676

We quickly discover that our very crude modification to Simpson's rule will work for this case, as long as we select the *proper values*. The reader is invited to explore other combinations! If we do not know a lot about our integral in advance, it may be difficult to get this to work as we anticipate. After all, we have used two arbitrary choices: The threshold change for the accumulating sum and the size of the reduction of Δx. We must have something more reliable and much more broadly applicable.

Fortunately, a number of options are available to us designed specifically to deal with this type of problem. Kuncir (1962) suggested a method using Simpson's rule in which an interval would be subdivided if the estimated error exceeds some threshold value. Consider the integral,

$$\int_a^b f(x)dx = I(a,b). \tag{4.31}$$

The interval (a,b) has a mid-point, $(a+b)/2$; we rewrite integral eq. (4.31), splitting it into two pieces:

$$\int_a^b f(x)dx = \int_a^{\frac{a+b}{2}} f(x)dx + \int_{\frac{a+b}{2}}^b f(x)dx. \tag{4.32}$$

A strategy is now apparent: We compute the left-hand side of eq. (4.32) and compare it to the right-hand side which was obtained by subdividing (a,b). If the estimates do not agree within a specified tolerance, further subdivision is required. We stop the successive subdividing process when the criterion suggested by Lyness (1969 and 1983) is met:

$$\frac{\left|I\left(a,\frac{a+b}{2}\right)+I\left(\frac{a+b}{2},b\right)-I(a,b)\right|}{15} < \varepsilon. \tag{4.33}$$

We can easily anticipate how this will work for elementary functions, but we should apply it to an integral that we know to be problematic. Once again we consider

$$\int_0^{2.99} \frac{dx}{(3-x)^2} = 99.667; \tag{4.34}$$

we apply Simpson's rule with just three function evaluations (the minimum). Then we will bisect the interval (mid-point 1.495) and apply Simpson's rule to each half. We will put the results in a "tree" structure to illustrate the progress of the calculations:

$$I_0^{2.99} = 4984.278 \tag{4.35}$$

$$I_0^{1.495} = 0.33413 \quad I_{1.495}^{2.99} = 2493.518 \tag{4.36}$$

Of course the sum in (4.36) is very different from level 1 (2,493.85 as opposed to 4,984.28); subdivision is necessary.

$$I_0^{0.7475} = 0.11065 \quad I_{0.7475}^{1.495} = 0.22074 \quad I_{1.495}^{2.2425}0.66153 \quad I_{2.2425}^{2.99} = 1249.437 \tag{4.37}$$

Note that the sum of the first two is 0.33139—this is a change of 0.8% from level 2, line (4.36), and we will assume for this illustration that this is adequate. We now subdivide the two quarters in the right half:

$$I_{1.495}^{1.86875} = 0.21959 \quad I_{1.86875}^{2.2425} = 0.43662 \quad I_{2.2425}^{2.61625} = 1.29678 \quad I_{2.61625}^{2.99} = 629.769 \tag{4.38}$$

The sum of the first pair is 0.6562 which is a bit of a change from 0.66153 (once more about 0.8%). In order to fit the space available we will again assume this is sufficient and subdivide the two intervals of the right-half:

$$I_{2.2425}^{2.42938} = 0.43246 \quad I_{2.42938}^{2.61625} = 0.85425 \quad I_{2.61625}^{2.80313} = 2.49337 \quad I_{2.80313}^{2.99} = 323.899 \tag{4.39}$$

The sum of the first pair is 1.28671 as opposed to 1.29678 from (4.38) above. Once again we assume this meets our (very relaxed) criterion and subdivide the right-half intervals:

$$I_{2.61625}^{2.70969} = 0.83894 \quad I_{2.70969}^{2.80313} = 1.63643 \quad I_{2.80313}^{2.89657} = 4.62063 \quad I_{2.89657}^{2.99} = 176.537 \tag{4.40}$$

Compare the sum of the first pair (2.47537) with 2.49337 from above (the difference is once again about 0.8%); we will subdivide the right-half intervals again:

$$I_{2.80313}^{2.84985} = 1.58086 \quad I_{2.84985}^{2.89657} = 3.01076 \quad I_{2.89657}^{2.94329} = 8.00725 \quad I_{2.94329}^{2.99} = 108.261 \tag{4.41}$$

We will repeat the process one more time, assuming that 4.59162 is sufficiently "close" to 4.62063:

$$I_{2.89657}^{2.91993} = 2.82120 \quad I_{2.91993}^{2.94329} = 5.14751 \quad I_{2.94328}^{2.96665} = 12.3914 \quad I_{2.96665}^{2.99} = 75.550 \tag{4.42}$$

The sum of these four integrals is 95.9101, and we are rapidly closing in on a much better approximation for the definite integral (4.34). This detailed example reveals the power of the adaptive version of Simpson's rule. Since we stop making calculations in intervals where our accuracy is sufficient, we can significantly reduce the total number of operations relative to cases where we merely continually reduce Δx over the entire interval (a,b).

4.8 Gaussian Quadrature and the Gauss–Kronrod Procedure

It is important that the reader be aware of other recent developments in adaptive numerical quadrature. The Gauss–Kronrod procedure is a powerful method that can be used in this context; in fact, Kahaner et al. (1989) state that the Gauss–Kronrod algorithm "*…is currently one of the most effective methods for calculating general integrals.*" To set the framework for a useful description we must begin with an n-point Gaussian quadrature, where a definite integral is approximated by the formula

$$\int_a^b f(x)dx \cong \sum_{j=1}^{n} \omega_j f(x_j). \tag{4.43}$$

The ω_j's are weight factors that are applied to specific x-positions. We can make Gaussian quadrature more transparent with an illustrative example. Consider the definite integral,

$$\int_a^b f(x)dx = \int_0^5 (5 + 4x - x^2)dx = 33.3333. \tag{4.44}$$

Our first task is to transform this integral to $\int_{-1}^{+1} F(t)dt$; we do this by setting $x = mt + c$. Therefore, we have the two equations, $0 = -m + c$ and $5 = m + c$. By adding the equations together we find c = 5/2 and m = 5/2; thus

$$x = \frac{5}{2}t + \frac{5}{2} \quad \text{and} \quad dx = \frac{5}{2}dt. \tag{4.45}$$

Our definite integral is now written

$$\frac{5}{2}\int_{-1}^{+1} \left(\frac{35}{4} - \frac{5}{2}t - \frac{25}{4}t^2\right) dt, \tag{4.46}$$

and the reader can easily verify that this is also equal to 33.3333. Now we return to eq. (4.43) and fix $n = 2$. The specific t-positions are $-1/\sqrt{(3)}$ and $+1/\sqrt{(3)}$, and these locations correspond to zeros of the Legendre polynomial, P_2 (a limited table of Legendre polynomials is provided here for the reader's reference). Remember, Legendre polynomials are orthogonal on the interval (−1,+1).

$n = 0$	Legendre Polynomials (P_n)
0	1
1	x
2	$\frac{1}{2}(3x^2 - 1)$
3	$\frac{1}{2}(5x^3 - 3x)$
4	$\frac{1}{8}(35x^4 - 30x^2 + 3)$
5	$\frac{1}{8}(63x^5 - 70x^3 + 15x)$
6	$\frac{1}{16}(231x^6 - 315x^4 + 105x^2 - 5)$
7	$\frac{1}{16}(429x^7 - 693x^5 + 315x^3 - 35x)$

The weight factors we are to employ are both equal to 1, so our approximation for the definite integral is now written

$$\int_a^b f(x)dx = \frac{5}{2}\sum_{j=1}^{2}\omega_j F(t_j) = \frac{5}{2}\left[(1)F\left(-\frac{1}{\sqrt{3}}\right) + (1)F\left(+\frac{1}{\sqrt{3}}\right)\right]. \tag{4.47}$$

The right-hand side is equal to 33.3333—one might ask why does the method yield the exactly correct result? The answer is that an n-th order Newton–Cotes formula will integrate a polynomial of order n, or less, *exactly*. With Gaussian quadrature the exactness of the result is extended to polynomials of degree $2n + 1$.

For more general cases we can improve the quality of the approximation by simply increasing the order, and the reader interested in using Gaussian quadrature should consult the extensive table (Table 25.4) in Abramowitz and Stegun, 1964, pages 916 through 919. For example, if we choose $n = 6$, then t-positions are found from the roots of the appropriate Legendre polynomial above; these positions (nodes) and the corresponding weight factors are

t-Positions	Weight Factors
±0.238619186083	0.467913934573
±0.661209386466	0.360761573048
±0.932469514203	0.171324492379

In order to apply this to

$$\int_0^{2.99} \frac{dx}{(3-x)^2},$$

we would rewrite the integral as

$$\int_{-1}^{+1} \frac{1.495dt}{2.26503 - 4.49995t + 2.23503t^2}$$

and use the values in the table immediately above. This results in an estimate for the definite integral of 23.6594, which we know to be much too low (not at all surprising since n is only 6). Somewhat better results are obtained for $\int_0^1 x\ln(1-x)dx = -3/4$, which we transform to

$$\frac{1}{2}\int_{-1}^{+1} \left(\frac{1}{2}t + \frac{1}{2}\right) \ln\left(-\frac{1}{2}t + \frac{1}{2}\right) dt.$$

In this case Gaussian quadrature with $n = 6$ yields -0.734846 which is only about 2% smaller (in magnitude) than the correct value. And remember, this is achieved with only 6 function evaluations!

As we indicated above, we can always take Gaussian integration one step further by increasing the order of the procedure; let's raise n to 24 and begin by evaluating the simple definite integral (a fifth-degree polynomial):

$$\int_0^3 \left(x - \frac{1}{3}x^3 + \frac{1}{5}x^5\right) dx. \tag{4.48}$$

For confirmation we can perform the integration directly or use the *TI-nspire* to get the answer: 22.05.

```
#COMPILE EXE
#DIM ALL
  REM *** Gaussian quadrature for n=24
  REM *** evaluating integral of (x-1/3*x^3+1/5*x^5)dx
        GLOBAL fx,sum,zz AS DOUBLE
        GLOBAL n AS INTEGER
FUNCTION PBMAIN
        DIM t(30) AS DOUBLE
        DIM w(30) AS DOUBLE
   t(1)=-0.9951872199:t(2)=-0.974728556:t(3)=-0.938274552:t(4)=-0.886415527:t(5)=
-0.820001986:t(6)=-0.740124192
   t(7)=-0.648093652:t(8)=-0.545421471:t(9)=-0.433793508:t(10)=-0.31504268:t(11)=
-0.191118867:t(12)=-0.064056893
   t(13)=0.9951872199:t(14)=0.974728556:t(15)=0.938274552:t(16)=0.886415527:t(17)=
0.820001986:t(18)=0.740124192
   t(19)=0.648093652:t(20)=0.545421471:t(21)=0.433793508:t(22)=0.31504268:t(23)=
0.191118867:t(24)=0.064056893
   w(1)=0.01234123:w(2)=0.028531389:w(3)=0.04427744:w(4)=0.059298585:w(5)=
0.0733464814:w(6)=0.0861901615
   w(7)=0.097618652:w(8)=0.10744427:w(9)=0.115505668:w(10)=0.1216704729:w(11)=
0.125837456:w(12)=0.1279381953
   w(13)=0.01234123:w(14)=0.028531389:w(15)=0.04427744:w(16)=0.059298585:w(17)=
0.0733464814:w(18)=0.0861901615
   w(19)=0.097618652:w(20)=0.10744427:w(21)=0.115505668:w(22)=0.1216704729:w(23)=
0.125837456:w(24)=0.1279381953


      sum=0
        FOR n=1 TO 24
         GOSUB 300
          sum=sum+fx*w(n)
          NEXT n
```

```
        PRINT sum
        INPUT "end program now?";zz
         IF zz>0 THEN END
  300 REM *** subroutine for function evaluation

          fx=1.5*((1.5*t(n)+1.5)-1/3*(1.5*t(n)+1.5)^3+1/5*(1.5*t(n)+1.5)^5)

          RETURN
END FUNCTION
```

The Gaussian integration with $n = 24$ yields 22.05000001 as expected for the fifth-degree polynomial. The reader is encouraged to repeat this example for verification and then use the procedure to perform the following integration:

$$\int_0^4 \left(\frac{1}{4}e^{-x^2} + \frac{3}{x^2+9}\right) dx. \tag{4.49}$$

You should obtain 1.1488519 (the *TI-nspire* produces 1.14885 for this integral). Keeping in mind that this example is not an elementary polynomial—for which we would expect to get the exact answer—how close is your estimate?

Now that we have some familiarity with Gaussian quadrature, we can proceed to our real objective which is to modify the process in such a way that it can be effective for "difficult" integrals. One clever way to accomplish this goal is with an *embedded algorithm* for which error estimation is automatically available (embedded schemes are also sometimes referred to as *nested* or *progressive*). Such an approach will be more efficient if we do not have to start over with respect to node placement; i.e., we will make far fewer calculations if we can reuse some of the existing function evaluations. This is exactly what the Gauss–Kronrod procedure was designed to do. Each step requires $(2n + 1)$ function values, so for this example we will consider the (7–15) Gauss–Kronrod (GK) nodal placements. We begin with Gaussian quadrature for $n = 7$ (see the table of Legendre polynomials). The Gaussian placements and weights are:

±0.9491079123	0.1294849662
±0.7415311856	0.2797053915
±0.4058451514	0.3818300505
0.0000000000	0.4179591837

Since we have 7 nodes here, the Kronrod modification will have $(2n + 1) = 15$, but 7 of them are identically the Gaussian values. Note that the weight factors are different (they are about *half* of the Gaussian weights); however:

±0.9914553711	0.0229353220
±0.9491079123	0.0630920926
±0.8648644234	0.1047900103
±0.7415311856	0.1406532597
±0.5860872354	0.1690047266
±0.4058451514	0.1903505781
±0.2077849550	0.2044329401
0.0000000000	0.2094821411

We will explore the now familiar integral,

$$\int_0^{2.99} \frac{dx}{(3-x)^2},$$

one more time. First, we apply the (7–15) GK to the entire integral remembering that because of the change in limits, we must make a variable change as $x = 1.495\,t + 1.495$. The results are

G	K
30.15765	84.13157

The discrepancy is enormous, so we cut the interval in half, applying the (7–15) GK to each (0–1.495) and (1.495–2.99). The results are:

Left Half		Right Half	
G	K	G	K
0.331620	0.331371	39.730679	27.053353

As you can see, the significant error is in the right half, so we repeat the process for (1.495–2.2425) and (2.2425–2.99):

Left Half		Right Half	
G	K	G	K
0.656664	0.656175	71.875407	98.943049

Of course we already know that the difficulty with this integral is a consequence of the upper limit, so the right half is divided into two once again (2.2425–2.61625) and (2.61625–2.99):

Left Half		Right Half	
G	K	G	K
1.2876488	1.2867079	88.4937846	98.0094297

You will observe that we are very rapidly acquiring a much-improved estimate for this definite integral; in fact, if we take these last two Kronrod values and add to them the left halves that have been dropped, we will see that we are not that far from the correct value (99.6667). The Gauss–Kronrod imbedded procedure is powerful indeed, and it automatically provides us with a means to estimate our error. Consequently, it lends itself very nicely to adaptive quadrature for "difficult" integrals.

4.9 Integrating Discrete Data

Suppose we have the following set of discrete data:

x	$y(x)$
11	0.338
19	0.340
27	0.335
35	0.333
43	0.326

Our task is to produce an accurate integration of these data for $11 \le x \le 43$. We note immediately that the data are highly nonlinear as is evident in Figure 4.2.

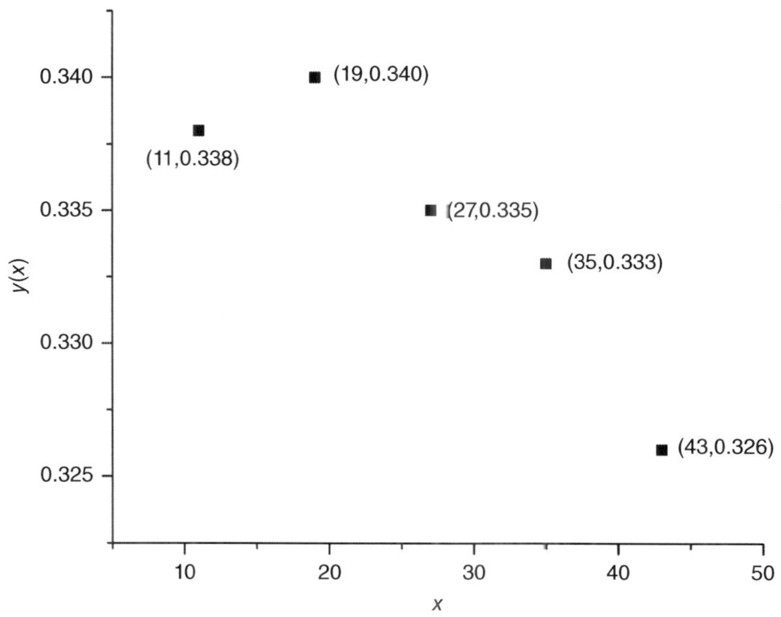

Figure 4.2 Nonlinear discrete data for numerical quadrature. Note that the ordinate, $y(x)$, has been greatly expanded, and only a small segment of the axis has been plotted.

We have several options for performing the necessary integration, including some crude methods discussed at the very beginning of the chapter. But because we have stipulated that we need a very accurate value, we should look for a better technique.

One powerful method for dealing with such situations is through integration of a *cubic spline interpolation*. *Spline* refers to a draftsman's tool used to generate smooth curves connecting multiple points; for a picture of such a device, see Grandine (2005), who describes how extensively splines are being used at Boeing (to represent complicated geometries or geometric boundary conditions). The idea with spline interpolation is a very simple one: Polynomial *pieces* are fit together in such a way that the function is continuous and is continuously differentiable. Moreover, these pieces connect smoothly, and the derivatives are continuous at the interval boundaries. The process employed is distinctly different from merely fitting a polynomial of higher degree to all of the available data—and if you have ever tried this you already know that such curve fits are often unphysical. They usually show too much oscillation (curvature) for intermediate values and unbounded (non-asymptotic) behavior at the extremes.

The data set of interest to us has five points (so four interior intervals) with each separated by a Δx, or h, of 8. We will use a cubic polynomial to represent the data in each interval so for $11 \leq x \leq 19$ we write:

$$y = a_1(x-11)^3 + b_1(x-11)^2 + c_1(x-11) + d_1. \tag{4.50}$$

Naturally, when $x = 11$, we see $d_1 = 0.338$. You will notice immediately that we have four sets of constants, one set for each interior interval. In every case, d will be given by the left-hand end-point value of y, so for the second interval,

$$y = a_2(x-19)^3 + b_2(x-19)^2 + c_2(x-19) + 0.340, \tag{4.51}$$

etc. A key feature of spline interpolation is that the first and second derivatives will be continuous from one interval to the next. Accordingly, we will write the cubic polynomial in generic form:

$$y = a_i(x-x_i)^3 + b_i(x-x_i)^2 + c_i(x-x_i) + d_i. \tag{4.52}$$

Therefore,

$$y' = 3a_i(x-x_i)^2 + 2b_i(x-x_i) + c_i \tag{4.53}$$

and

$$y'' = 6a_i(x-x_i) + 2b_i. \tag{4.54}$$

It is apparent for our problem that we will have (3) × (4) = 12 unknown coefficients. Of course, the relation (4.52) provides four equations. We will also require that the first and second derivatives match at the interval boundaries; that provides six more equations through application of (4.53) and (4.54), but we are still *two short*! One way we can obtain the final two relations is to specify the second derivative at the ends of the interval (i.e., for $x = 11$ and 43, in our case). If we just choose to set $y'' = 0$, we get what is referred to as a *linear spline*. Though this seems incredibly arbitrary, Hanna and Sandall (1995) note that it does not seem to lead to significant errors in actual application. We will now look at the complete set of equations for our cubic spline curve fit:

$$\begin{aligned} 512a_1 + 8c_1 &= 0.002 \\ 512a_2 + 64b_2 + 8c_2 &= -0.005 \\ 512a_3 + 64b_3 + 8c_3 &= -0.002 \\ 512a_4 + 64b_4 + 8c_4 &= -0.007 \end{aligned} \tag{4.55a,b,c,d}$$

$$\begin{aligned} 192a_1 + c_1 - c_2 &= 0 \\ 192a_2 + 16b_2 + c_2 - c_3 &= 0 \\ 192a_3 + 16b_3 + c_3 - c_4 &= 0 \end{aligned} \tag{4.56a,b,c}$$

$$\begin{aligned} 48a_1 - 2b_2 &= 0 \\ 48a_2 + 2b_2 - 2b_3 &= 0 \\ 48a_3 + 2b_3 - 2b_4 &= 0 \end{aligned} \tag{4.57a,b,c}$$

And finally, by setting the second derivative equal to zero at the end-points,

$$b_1 = 0 \quad 48a_4 + 2b_4 = 0. \tag{4.58a,b}$$

Since b_1 was eliminated, we have 11 equations and the same number of unknowns. The solution for this set of equations is:

		Each Multiplied by 10^6
1	a_1	−4.255
2	c_1	522.32
3	a_2	7.60323
4	b_2	−102.12
5	c_2	−294.643
6	a_3	−6.62667
7	b_3	80.3571
8	c_3	−468.75
9	a_4	3.27846
10	b_4	−78.683
11	c_4	−455.357

We can assess the quality of the representation in Figure 4.3.

Now let us see exactly how well this cubic spline curve fit performs. We can use eq. (4.50) as our example, integrating over interval 1:

$$\int_{11}^{19} y_1(x)dx = \frac{a_1}{4}(8)^4 + \frac{c_1}{2}(8)^2 + 0.338(8) = 2.71636. \tag{4.59}$$

Remember that b_1 is zero! The values for the remaining three integrals are 2.70093, 2.66514, and 2.63936, respectively, and the sum of all four is 10.7218. We have, through use of cubic spline interpolation, obtained a very accurate value for the integral of the original discrete data set.

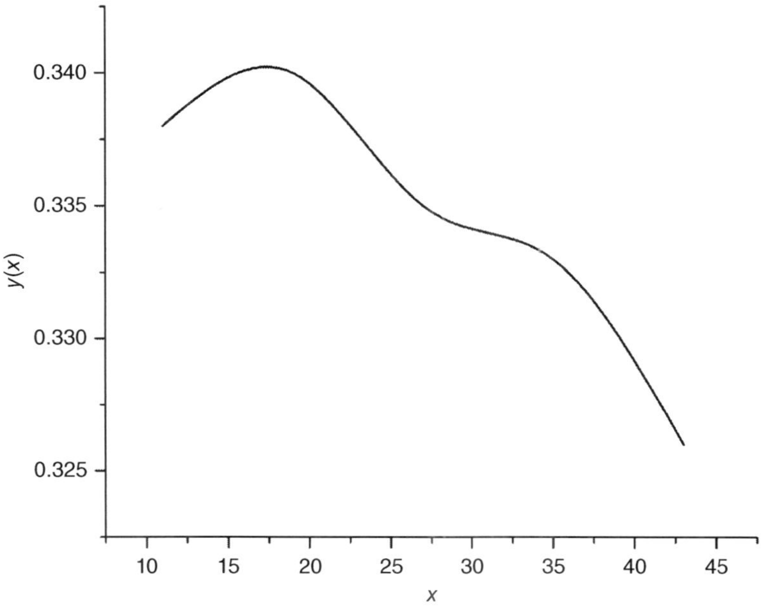

Figure 4.3 Cubic spline curve fit for the nonlinear data set. We now have a very nice spline-function representation of the original data, facilitating integration.

4.10 Multiple Integrals (Cubature)

Recall that the definite integral, $\int_a^b f(x)dx$, can be approximated by a Riemann sum, i.e., the sum of areas of a number of rectangles. For the case in which we use n rectangles (but allow n to become very large) and $x = a \leq \phi \leq x = b$,

$$\int_a^b f(x)dx = \lim_{n\to\infty}\left[\sum_{i=1}^{n} f(\phi_i)\Delta x\right]. \tag{4.60}$$

This process can also be extended to multiple integrals:

$$\iint_R f(x,y)dxdy = \lim_{\delta\to 0}\left[\sum_{i=1}^{n} f(x_i,y_i)\Delta A_i\right]. \tag{4.61}$$

Note that δ is the norm (diagonal) of the partition or rectangle; by allowing it to approach zero we are employing an infinite number of rectangles. If $a \leq x \leq b$ and if the limits for the variable y are set by two functions of x, say, $g_1(x)$ and $g_2(x)$, then for this *type I* region, we have the *iterated integral*:

$$\int_a^b \int_{g_1(x)}^{g_2(x)} f(x,y)dydx. \tag{4.62}$$

For a *type II* region the terminal values of y are constants, $c \leq y \leq d$, and the limits for the variable x are two functions of y. We can review the evaluation process for an iterated integral with a straightforward example:

$$\int_3^5 \int_{-x}^{x^2} (4x + 10y)dydx = \int_3^5 (4xy + 5y^2)\Big|_{-x}^{x^2} dx = \int_3^5 (4x^3 + 5x^4 - x^2)dx$$

$$= 5^4 + 5^5 - \frac{1}{3}5^3 - 3^4 - 3^5 + \frac{1}{3}3^3 = 3393.33 \tag{4.63}$$

This was an extremely simple double integral that could be easily evaluated analytically. But our real interest here is to acquire some tools that might allow us to evaluate multiple integrals that defy elementary solution.

Multiple integrals can be handled, although somewhat inefficiently, using the one-dimensional numerical quadrature techniques that we discussed previously in this chapter. Let us illustrate with the generic double integral:

$$\int_a^b \int_c^d f(x, y)dxdy. \tag{4.64}$$

We will proceed in the following way:

1) Set increment value for y, e.g., $(b - a)/100$
2) Set the increment value for x, e.g., $(d - c)/100$
3) Let y assume its first value, y_1
4) Accumulate the incremental areas for $f(x,y_1)$ as x varies from c to d
5) Set this sum equal to $F(y_1)$
6) Increment y and repeat quadrature on x from c to d
7) Continue process until $y = b$, i.e., until we find $F(y_n)$
8) Perform quadrature on the complete set of $F(y_i$'s).

The process sketched here is referred to as a *product rule*. To illustrate its application we will consider the elementary example,

$$\int_0^2 \int_1^4 (x + 2y)dxdy = 27. \tag{4.65}$$

We will start with 50 intervals for both x and y, and use the trapezoid rule for the quadratures. Using single precision, this direct approach yields 27.72359 (an error of about 2.7%). If we merely increase the number of intervals in *both* the x- and y-directions to 100, the very same process yields 27.36090; with 200 intervals the result is 27.00002. Although the procedure we have described above works (very well for this particular example), it requires *many* function evaluations. In fact, Smyth (1998) notes that the number of evaluations required for product rules grows with the number of dimensions exponentially. And you will also observe that in our example we used the same number of intervals in both directions, which is generally a mistake; Kahaner et al. (1989) suggest that a rule of thumb for application of one-dimensional algorithms to double integrals is that the inner integral be computed such that its accuracy is about 10 times better than the outer. We can demonstrate this using the trapezoid rule (with double precision) for an elementary double integral,

$$\int_0^2 \int_1^4 \frac{x^2}{4y} dydx = 0.924196.$$

We will increase the number of intervals for the inner integral while maintaining a constant number of evaluations for the outer integral. The results are summarized in the table below.

Number of Intervals (*n*)		
Inner	**Outer**	$\int_0^2 \int_1^4 \frac{x^2}{4y} dydx = 0.924196$
10	10	1.3058895
20	10	1.2695594
40	10	1.2355796
80	10	1.2352861
160	10	1.2393796
100	100	0.92927096
200	100	0.92425417
400	100	0.92549427
800	100	0.92424318

Number of Intervals (n)		
Inner	**Outer**	$\int_0^2 \int_1^4 \frac{x^2}{4y} dydx = 0.924196$
1,600	100	0.92424263
1,000	1,000	0.92419717
2,000	1,000	0.92419682
4,000	1,000	0.92419673
8,000	1,000	0.92419671
16,000	1,000	0.92419670

These data show that the accuracy of the estimate of the double integral improves (generally) as the number of intervals for the inner integral is increased. But these calculations also reveal that application of a one-dimensional scheme using the trapezoid rule is not very efficient. An extremely large number of calculations is required in order to obtain the correct sixth decimal place.

In recent years the search for modern *nonproduct methods* for multi-dimensional integrals has been intensive. A non-product method is a quadrature (or more appropriately, cubature) rule that *does not* require us to apply a one-dimensional method successively to the different directions. As we stated above it is computationally inefficient to apply a product rule to the evaluation of multiple integrals. You may recall our discussion of Gaussian quadrature; with proper selection of the nodes (the t-positions) and weight factors we can exactly integrate a polynomial of degree $2n+1$. An obvious question is whether such a process could be extended to polynomials of two (or three or more) dimensions? If this were possible, multiple integration could be made much more efficient. Stroud (1972) notes that the principal problem is identification of formulae of the form

$$\int \cdots \int w(x_1, x_2, \cdots x_n) f(x_1, x_2, \cdots x_n) dx_1 \cdots dx_n \cong \sum_{i=1}^{N} B_i f(\phi_1, \phi_2, \cdots \phi_n). \tag{4.66}$$

The fifth-degree formula of Radon (1948) was one of the first results of this type, and Stroud (1972) developed Gauss–Legendre formulae for iterated integrals of two and three dimensions. His book includes FORTRAN programs for both in Chapter 10. Also, in one important success in efforts to find efficient nonproduct methods, Laurie (1977) was able to extend the Gauss–Kronrod method to double integrals.

4.11 Monte Carlo Methods

Monte Carlo methods were first used extensively by John von Neumann and Stanislaw Ulam to estimate neutron diffusion in connection with the Manhattan Project of World War 2. This technique has been used by many investigators since to evaluate multiple integrals. It is attractive because of its stark simplicity; all one really needs is a mechanism for the generation of random numbers. Originally random numbers were obtained from tables, and for an example one may consult Table 26.11 in Abramowitz and Stegun (1964) which was reprinted from numbers compiled by the Rand Corporation. These days high-level language compilers have built-in capability for generating "random" numbers, but the reader should recognize that these numbers are being generated typically by a pseudo-random algorithm. This means that the same seed value will produce the *exact same sequence* of "random" numbers. Commercial software packages like Mathcad function similarly—the seed value must be changed if a different sequence of random numbers is desired.

We begin this discussion with an elementary example that will illustrate just how straightforward this technique really is; consider the one-dimensional integral,

$$I = \int_a^b f(x)dx. \tag{4.67}$$

Now envision a rectangle that extends from a to b on the x-axis, and from 0 to some value c on the y-axis. That is, we think of a box (an area) that corresponds to

$$a \leq x \leq b \quad \text{and} \quad 0 \leq f(x) \leq c.$$

The area of this box is of course merely $(b-a)c$. Suppose we begin selecting random locations inside the box—the probability that we pick a point *under* the curve, $f(x)$, will be just

$$p = \frac{I}{(b-a)c}. \tag{4.68}$$

We obtain an estimate for p from the ratio of the number of successes divided by the number of trials, n_{hits}/n. For obvious reasons this approach is referred to as the "hit-or-miss" method:

$$I \cong (b-a)c\frac{n_{\text{hits}}}{n}. \tag{4.69}$$

Suppose we were interested in obtaining a value for

$$\int_1^4 \left(2 + x + \frac{1}{3}x^2\right) dx$$

(we know this integral has the value, 20.5). We will conduct our trials on the rectangular area that is bounded by $1 \le x \le 4$ and $0 \le y = f(x) \le 20$. We use random numbers to "shoot" at the box and just keep track of the number of "hits" that fall under $f(x)$; our estimate then comes from eq. (4.69). Here is a typical sequence of results from such an experiment.

Number of Trials	Estimate for I
100	16.2
1,000	20.7
10,000	20.676
100,000	20.4888
1,000,000	20.53326
10,000,000	20.51026

It is extremely easy to obtain a *rough* estimate in this fashion, but it is apparent that an accurate result will require *many* trials. The reader can obtain greater detail regarding the use of the Monte Carlo method in this context from Rubinstein (1981), for example.

Now we focus upon use of the Monte Carlo technique for the evaluation of multiple integrals; we should emphasize the important difference between the "hit or miss" method used above and the approach employed here (we will be using the Monte Carlo method to *place* the function evaluations). To illustrate, we use the simple double integral,

$$\int_0^2 \int_0^3 (x^2 + 2xy^2)dxdy = 42. \tag{4.70}$$

We proceed in the following way: We select the number of trials (n), then use random numbers to select the (x,y) location for integrand evaluation. This function evaluation, $f(x,y)$, is multiplied by the area determined from the limits of the integration and the number of trials chosen, and the results are accumulated. For the double integral that we are using as our example, this accumulation will be performed n^2 times. A simple code written for eq. (4.70) is provided below.

```
#COMPILE EXE
#DIM ALL
  REM *** Monte Carlo integration of double integral
  GLOBAL dx,dy,rnx,rny,x,y,sum,fxy,xlimit,ylimit,ZZ AS SINGLE
  GLOBAL n,ntrial AS INTEGER
FUNCTION PBMAIN
        n=0:xlimit=3:ylimit=2
        ntrial=100
        dx=xlimit/ntrial:dy=ylimit/ntrial
        RANDOMIZE TIMER
  50 REM *** continue
```

```
        rnx=RND
        rny=RND
          x=rnx*xlimit
          y=rny*ylimit
            GOSUB 200
          sum=sum+fxy*dx*dy
            n=n+1
            PRINT n,sum
            IF n>ntrial^2 THEN 100 ELSE 50
100 REM *** continue
      INPUT "Shall we continue?";ZZ
          IF ZZ>0 THEN END
200 REM *** subroutine for function evaluation
          fxy=x^2+2*x*y^2
          RETURN

END FUNCTION
```

Please make note of the ninth line down from the top, *RANDOMIZE TIMER*. This insures that the random number generator obtains a different seed each time it is invoked. You will also observe that the number of trials, *ntrial*, is set to 100; this is a ridiculously small number for a Monte Carlo integration (just 10^4 function evaluations). To illustrate, we ran this program five times, just as it appears here, and we obtained the set of estimates: 41.95124, 41.96887, 42.31614, 41.66427, and 42.33305. The average of the set is 42.0467 which reveals the unpleasant truth that a Monte Carlo integration—if accuracy is required—is computationally expensive. In fact, we can monitor the results from a sequence of realizations in which we increase the number of trials each time. Such an experiment results in a graph similar to Figure 4.4.

It is known that the error for this method diminishes as $\approx \frac{1}{\sqrt{ntrial}}$; if we need a high level of decimal precision, then *many* trials will be necessary. To illustrate, we can see from Figure 4.4 that $2{,}000^2 = 4 \times 10^6$ function evaluations produce just 41.91697 for the double integral—an error of about 0.2%. Just as we observed previously, a rough result is often very easy to achieve with the Monte Carlo technique, but the computational burden may be prohibitive if we need an extremely accurate answer.

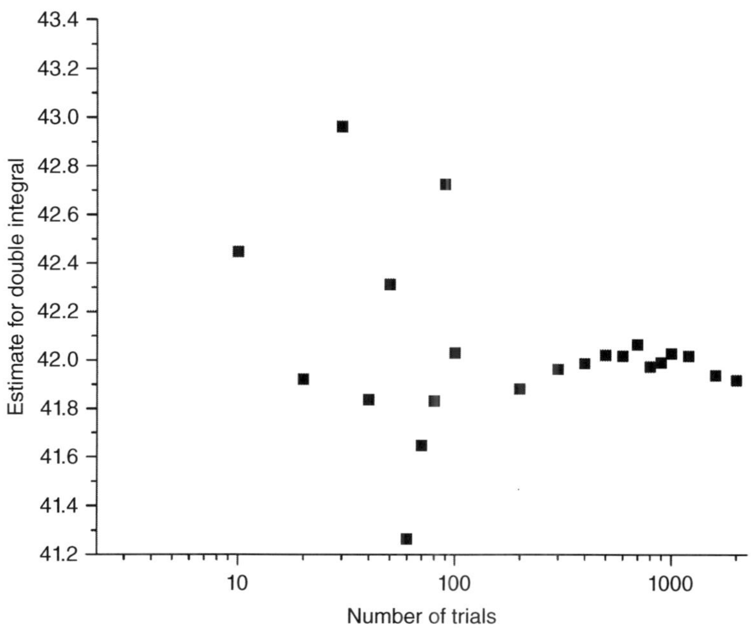

Figure 4.4 Progress of the Monte Carlo integration of the double integral, eq. (4.70) as a function of the number of trials. The result should be 42, of course.

Lattice methods (sometimes referred to as number theoretic, or quasi-random) have been applied to multiple integration, and a useful reference is Sloan and Joe (1994).

With a lattice method the integration region is transformed to a unit cube, and a multiple sum yields an unweighted mean of the integrand evaluated over a regular lattice. Smyth (1998) notes that lattice methods function best when the integrand is transformed so that it is periodic over the cube. There is some evidence (including Sloan and Joe) indicating that lattice techniques may outperform other methods when the number of dimensions is large.

Should one have a critical need to evaluate multidimensional integrals accurately (and particularly if this must be done often), the best starting point is the *Cuba Library* which has been developed at the Max Planck Institut fur Physik (Munchen). *Cuba* has four routines, Vegas, Suave, Divonne, and Cuhre, which provide the analyst with the choice between quasi- and pseudo-Monte Carlo methods as well as lattice and deterministic approaches too. Several interfaces are available (in FORTRAN, C/C++, and Mathematica) and the open-source package is available at http://www.feynarts.de/cuba. Because the algorithms are invoked similarly, the package can be used to very quickly compare methods and thereby assess likely error. The *Cuba* website is maintained by Thomas Hahn, and it is updated frequently.

4.12 Conclusion

Our focus in this chapter was the numerical evaluation of definite integrals. We have not discussed cases in which the integrand is undefined at either one of the limits (a singular end-point), nor have we given very much consideration to cases in which one of the limits is infinite.

A singular end-point may sometimes be handled by using an *open* Newton–Cotes formula such as Milne's rule. Recall that most of the simple quadrature techniques we discussed previously made use of the interval end-points for function evaluation; such quadrature procedures are said to be *closed*. An *open* algorithm such as the fourth-degree Milne's rule does not require function evaluations at the interval end-points, thereby avoiding the problem of an end-point singularity. With Milne's rule the function evaluations required for the integration of $\int_a^b f(x)dx$ are placed at locations denoted by

$$x_i = a + \frac{i(b-a)}{n}, \tag{4.71}$$

where $n = 4$ and the index i assumes the values 1, 2, and 3. Therefore, if we wished to evaluate $\int_1^5 (x^2 - 2x)dx$ using Milne's rule just once, we would place the integrand evaluations at positions $x = 2$, 3, and 4, and "estimate" the definite integral with

$$\frac{b-a}{3}(2f_1 - f_2 + 2f_3) = \frac{5-1}{3}[2(0) - 3 + 2(8)] = \frac{52}{3} = 17.3333\cdots \tag{4.72}$$

Of course, this is precisely the value of the definite integral. Let us apply the technique to an integral we examined much earlier that does *not* have a singular end-point, $\int_1^5 x\ln(x)dx = 14.118$. This will provide us with a performance comparison relative to conventional, closed methods. Since the interval is the same as above (1,5), the function evaluations are again placed at 2, 3, and 4, resulting in

$$\frac{5-1}{3}[2(1.38629) - (3.29584) + 2(5.54518)] = 14.0895, \tag{4.73}$$

a surprisingly good result (0.2% low) obtained with only a single application of the rule. Obviously we could subdivide the interval, applying Milne's rule to each piece, then add the results together (this is called a *composite* rule) to improve our estimate. We will try this using 8 pieces:

```
#COMPILE EXE
#DIM ALL
  REM *** Open Newton-Cotes for quadrature
  REM *** This is the 4th degree Milne's rule for the interval (a,b)
        GLOBAL a,b,fx,del,area,zz,bf,sum AS DOUBLE
        GLOBAL i,n AS INTEGER
FUNCTION PBMAIN
        DIM x(3) AS DOUBLE
```

```
        DIM f(3) AS DOUBLE
          a=1:bf=5
          b=bf:sum=0
        REM *** refine estimate by splitting interval into n-pieces
                n=8
                  del=(b-a)/n
                  b=a+del
    50 REM *** continue
               FOR i=1 TO 3
                 x(i)=a+i*(b-a)/4
                 NEXT i
               FOR i=1 TO 3
                 GOSUB 200
                 f(i)=fx
                 NEXT i
               area=(b-a)/3*(2*f(1)-f(2)+2*f(3))
                  sum=sum+area
               PRINT area,sum
               a=b
               b=a+del
                 IF b>bf THEN 100 ELSE 50
  100 REM *** continue
              INPUT "Shall we continue?";zz
                IF zz>0 THEN END
  200 REM *** subroutine for function evaluation
        fx=x(i)*LOG(x(i))
        RETURN
END FUNCTION
```

This code produces a value for our example integral of 14.117957. Now we can try Milne's algorithm on an integral with a singular end-point; consider

$$\int_0^1 \frac{\ln(x)}{1+x}dx = -\frac{\pi^2}{12} = -0.822467. \tag{4.74}$$

We will apply Milne's method in composite form to this integral by subdividing the interval (0,1). We will begin with 8 intervals and successively double n to gauge the approach to the correct value.

8	−0.8078678
16	−0.8152081
32	−0.8188470
64	−0.8206596
128	−0.8215639
256	−0.8220156
512	−0.8222414
1,024	−0.8223542
2,048	−0.8224106
4,096	−0.8224388
8,192	−0.8224529
16,384	−0.8224600
32,768	−0.8224635

Note that we obtain the first four correct decimal places with $n = 2{,}048$, and we get five decimal places with $n = 16{,}384$; the open Milne's rule is able to cope with this much more difficult case. We should make one final observation in the context of this example: This integration would appear to be an ideal application for Richardson's extrapolation, where we would obtain improved accuracy with far fewer computations.

In the case of a definite integral with an infinite limit, you may recall that we used truncation (of the upper limit) for the very first example in this chapter:

$$\int_0^{\infty} x^2 \exp(-x^2)dx = \frac{\sqrt{\pi}}{4}.$$

We will assess the error involved in a similar situation in one of the exercises at the end of the chapter. Occasionally difficulties like this can be handled through transformation. Given $\int_0^{\infty} f(x)dx$, one might let $x = \frac{t}{1-t}$, or $x = -\ln(t)$. For the former we would integrate from 0 to 1, and for the latter from 1 to 0. Kahaner et al. (1989) point out that this kind of transformation may succeed in producing finite limits, but result in a significantly more difficult integrand.

Problems

4.1 The volume of a solid of revolution (V) generated by rotation of a curve about the y-axis is given by

$$V = 2\pi \int_a^b xf(x)dx.$$

Let $f(x)$ be the continuous function of x, $f(x) = 10 - \frac{1}{4}x^2$, defined for $1 \le x \le 5$. Find V by numerical quadrature.

4.2 Consider a horizontal cylindrical tank with a capacity of 600 gal. The tank is 6 ft long with a radius of 2.065 ft (ID = 4.13 ft). Let h be the depth of liquid in the tank and R be the inside radius. We know that the (area of a) segment of a circle is

$$A_S = R^2\cos^{-1}\left(\frac{R-h}{R}\right) - (R-h)\sqrt{2Rh - h^2}.$$

Therefore, when $h = R$, the occupied portion of the circle is $\frac{\pi R^2}{2}$. If fluid enters the initially empty tank at a rate of $\dot{V} = (30 - 9\sqrt{h})$ gpm, through a small orifice at the bottom of one end, how long will it take for the tank to fill (within 1 in. of the top)?

4.3 The moment of inertia about the y-axis for a continuous function, $f(x)$, defined for $a \le x \le b$, is

$$MI_y = \int_a^b x^2 f(x)dx.$$

Let $f(x) = 1 + x^4\exp(-2x)$; find MI_y for $0 \le x \le 5$ by numerical quadrature.

4.4 We want to find the work required to compress nitrogen (N_2) adiabatically from atmospheric conditions to a pressure of 8 atm, at a rate of 50 gmols/min. Assume that nitrogen behaves ideally such that $PV = nRT$. We know that $W = \int PdV$, and for an adiabatic process, $PV^{\gamma} = C_1$. Since the initial and terminal pressures are specified, we will rearrange the latter equation to yield:

$$dV = -\frac{1}{\gamma}\left(\frac{C_1}{P}\right)^{\frac{1-\gamma}{\gamma}}\frac{dP}{P^2}.$$

Find the work of compression by numerical quadrature, using a normal ambient temperature.

4.5 Evaluate the definite integral,

$$\int_0^{5.95} \frac{1+x}{(6-x)^3} dx.$$

Mathcad's adaptive algorithm yields 1,380 and the *TI*-89™ produces 1,380.07. These results are certainly very close, but which is more nearly correct?

4.6 An initially empty 2,000 gal tank is being filled by gravity; as the available head in the supply diminishes, the flow rate into the tank decreases. The flow rates were observed periodically resulting in the following table of data:

Time (min)	Flow Rate (gpm)
1	348
3	195
5	117
7	72
9	49
11	36
13	26

If the filling process started at $t = 0$, when will the tank contain precisely 1,615 gal?

4.7 The definition for the gamma function is:

$$\Gamma(n) = \int_0^{\infty} x^{n-1} \exp(-x) dx.$$

Find $\Gamma(1.525)$ and $\Gamma(0.525)$ by numerical quadrature. The values you obtain should be related by the recurrence formula,

$$\Gamma(n+1) = n\Gamma(n).$$

Are they?

4.8 Use Gaussian quadrature with $n = 6$ to find the approximate value for the definite integral,

$$\int_0^4 x^{2.25} \exp(-2x) dx.$$

4.9 Use the Gauss–Kronrod (7–15) procedure to evaluate the definite integral,

$$\int_0^{10} \frac{100dx}{(x-2.5)^2+1}.$$

Will it be necessary in this case to use an adaptive modification of GK (7–15)? You can see by inspection that the integrand will have a fairly strong peak centered about $x = 2.5$.

4.10 Consider the definite integral,

$$\int_0^{\infty} \exp\left(-\frac{1}{4}x\right) \cos(3x) dx.$$

This integral has the value 0.027586 and the integrand behaves as illustrated by Figure P4.10.

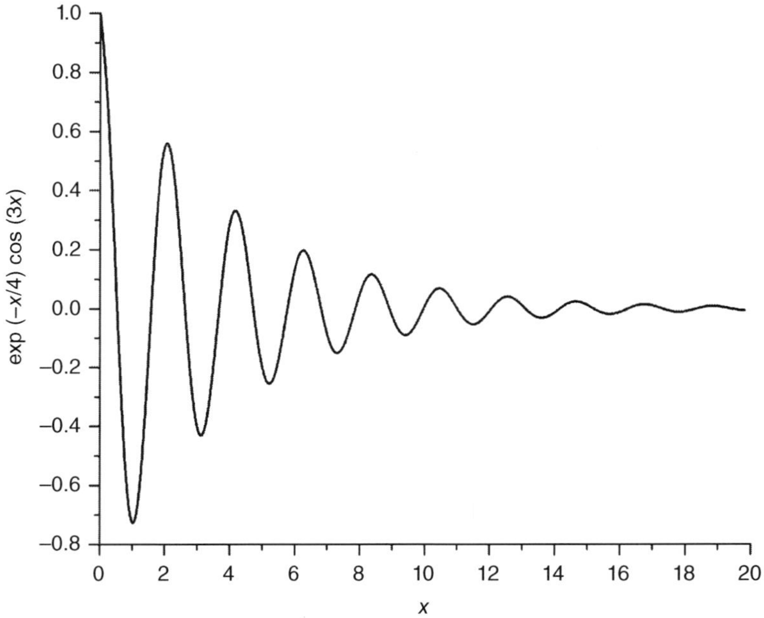

Figure P4.10 Plot of the function, $\exp\left(-\frac{1}{4}x\right)\cos(3x)$.

We want to determine the error associated with truncating the upper limit of the integral; use the quadrature routine of your choice and prepare a plot that shows the absolute value of the error as a function of the finite upper limit.

4.11 The definite integral,

$$\int_0^\infty \frac{dx}{(1+x)\sqrt{x}} = \pi\csc(\pi) = 3.1415926\cdots,$$

illustrates both of the difficulties we described at the very end of the chapter: We have a singular end-point at $x = 0$ *and* an infinite upper limit. Use the open fourth-degree Newton–Cotes formula (composite Milne's rule) to evaluate the integral with an appropriately truncated upper limit. Can you actually obtain as answer close to π? If you cannot, what quadrature method would you recommend?

4.12 Consider the definite integral,

$$\int_0^1 \sqrt{\ln\left(\frac{1}{x}\right)}dx = \frac{\sqrt{\pi}}{2} = 0.8862269\cdots$$

Use an open composite Newton–Cotes formula to evaluate this integral. Employing the example at the end of the chapter as your guide, determine how many subdivisions will be necessary (using the fourth-degree Milne's rule) to obtain the correct value for the seventh decimal place (9).

4.13 We want to evaluate the double integral,

$$\int_0^\pi \int_{-1}^{+1} x^2y^2 dxdy = 6.89028$$

using Simpson's rule (one-dimensional quadrature applied sequentially in the two directions). Explore the accuracy of your technique by adjusting the number of intervals employed for both the inner and outer evaluations. Does Kahaner et al.'s (1989) observation that the accuracy of the inner integral is critical to the success of this approach seem to be upheld?

4.14 Use the Monte Carlo "hit-or-miss" method to evaluate the integrals,

$$\int_0^{2\pi} \frac{dx}{1+\frac{1}{4}\cos x} = 6.48925$$

and

$$\int_0^{\pi} \cos^2\left(\frac{x}{3}\right) dx = 2.22032.$$

4.15 Use the Monte Carlo method to confirm the given value for the double integral,

$$\int_1^2\int_0^1 \left(yx^{1/2} + \frac{x^2}{y}\right) dxdy = 1.23105.$$

How many function evaluations are necessary to verify the second and third decimal places (3 and 1)? If you model your procedure on the sample code provided in Section 4.11, you must set *ylimit* = 2, *dy* = (*ylimit* − 1)/*ntrial*, and *y* = *rny**(*ylimit* − 1) + 1. A typical result obtained with these changes for *ntrial* = 400 is 1.230775.

4.16 Consider the definite integral,

$$\int_{0.01}^{9} \frac{dx}{x\sqrt{1+2x}} = 4.8411.$$

Our objective is to learn about the error produced with the estimation of this integral using different quadrature techniques. First, solve the definite integral analytically, verifying that 4.8411 is the correct value. Then, run through the list of Newton–Cotes formulae given in Section 4.4 generating five different approximations for the integral. Use the minimum number of function evaluations for each; i.e., two for the trapezoid rule, three for Simpson's rule, etc. Assess the error for each approximation using eq. (4.28) in Section 4.5. Are these estimates in accord with your numerical experiments?

4.17 We have two points in the $x-y$ plane located at (1,1) and (6,3), and these two points are to be connected by a function, $y(x)$. If we take this function to be a straight line, then $y = \frac{2}{5}x + \frac{3}{5}$. From elementary calculus, the distance between these two points is given by

$$L = \int_1^6 \left(1 + \left(\frac{dy}{dx}\right)^2\right)^{1/2} dx,$$

and we can easily verify that for this straight line, $L = 5.38516$. But now suppose we want to create a volume by rotating $y(x)$ about the x-axis:

$$V = \pi \int_{x_1}^{x_2} [y(x)]^2 dx,$$

and furthermore, suppose we require that the function—while still connecting the points (1,1) and (6,3)—be of the form, $y(x) = ax^2 + bx + c$. We would like to determine what function of this class yields the smallest volume, V, with the constraint that it pass through (x_1,y_1) and (x_2,y_2) while touching the x-axis *at only one point* (such that $y(x) \geq 0$ of course). First, determine if there is more than one possibility for $f(x)$; for example, if $y(x)$ is zero at $x = 3$, then $y(x) = 0.3x^2 - 1.7x + 2.4$, but we see immediately that $\frac{dy}{dx} = 0.6x - 1.7 \neq 0$ at $x = 3$ (as it should). Then find the swept volume, V, and the length, L, by numerical quadrature. Repeat, but using the function type, $y(x) = ax^3 + bx^2 + cx + d$. With this problem type we are anticipating one of the uses for the calculus of variations (COV) which will be introduced in Chapter 11.

4.18 An elliptic integral of the first kind is:

$$F(x,k) = \int_0^x \frac{dx}{\sqrt{(1-x^2)(1-k^2x^2)}},$$

where k is the modulus (of the elliptic integral). Let $\alpha = 80°$ and $\phi = 89°$, noting that $k = \sin(\alpha)$ and $x = \sin(\phi)$; Davis (1962) gives the corresponding value of $F(x,k)$ as 3.053. Evaluate F using Simpson's rule with $\Delta x = (x_f - x_i)/500$. What happens? Repeat, but this time decrease Δx by a factor of 10.

Answer: With $\Delta x = (x_f - x_i)/5{,}000$, $F = 3.0599$; *note that the TI-nspire CX produces* 3.05358.

4.19 Evaluate the following definite integral (letting $a = 1$) using Simpson's rule:

$$\int_0^9 \frac{xe^{ax}}{(1+ax)^2}dx.$$

Are you able to obtain the correct value (809.308)? Which quadrature technique is better suited to this problem?

4.20 Calculate the value of the definite integral,

$$\int_0^1 \left(ln\frac{1}{x}\right)^n dx = n!$$

for $n = 6$, using three different quadrature techniques to obtain your estimates. Which one yields the best result? Of the methods described in this chapter, which one is likely to produce an accurate result?

Answer: Trapezoid rule with $\Delta x = (x_f - x_i)/30{,}000$, 579.28
Simpson's rule with $\Delta x = (x_f - x_i)/50{,}000$, 681.87
Adaptive Simpson's rule: 712.66

4.21 Evaluate the following double integral using the method of your choice:

$$\int_0^5 \int_0^x \ln(x+y)dydx.$$

Answer: 18.6967

4.22 The velocity attained by a rifle bullet in the barrel depends upon the mass of the bullet, the gas pressure developed by the burning powder, and the barrel length. And of course gas pressure is a function of the amount of powder used and the speed with which it burns. Hornady® provided muzzle velocity data for the 308 Winchester using a variety of commercial powders and an M1A rifle with a 22 in. barrel; using 42.3 gr of powder and 155 gr bullets ($m = 10.044$ g), powder "A" produced 2,700 ft/s and powder "B" 2,450 ft/s. If we neglect frictional losses, the work expended on pushing the bullet down the barrel is reflected by the kinetic energy of the bullet:

$$\int_0^x F(x)dx = \frac{1}{2}m\left(\frac{dx}{dt}\right)^2.$$

The force on the bullet, $F(x)$, is the result of the dynamic gas pressure behind it acting upon the bullet's cross-sectional area ($A = 0.456\ \text{cm}^2$). Experimental data for the gas pressure ($P(x)$ in GPa) as a function of bullet position (x, in meters) can be represented by:

$$P(x) = \frac{0.0115}{0.01 + x^{2.5}}(1 - e^{-5x}).$$

Thus, for example, $P(x = 0.38\text{ m}) = 0.099$ GPa. It is clear that the bullet velocity can be computed for different barrel lengths by integration:

$$\frac{dx}{dt} = \sqrt{\frac{2}{m}\int_0^x F(x)dx}.$$

Determine dx/dt for barrel lengths ranging from 18.5″ to 24″. Remember to convert GPa to N/m², 1 GPa = 1×10^9 N/m². Bill Marr used a chronograph to measure muzzle velocities obtained with commercial ammunition and a 308 rifle with a progressively shortened barrel:

x (in.)	28	26	24	22	20	18	16.5
dx/dt (ft/s)	2,823	2,801	2,766	2,718	2,679	2,595	2,561

Are these data in general accord with the model for $P(x)$ that we employed? We recognize that there will not be quantitative agreement since a different powder was used.

Answer: For $x = 0.6096$ *m,* $dx/dt = 922$ *m/s or* 3,024.9 *ft/s.*

4.23 The fugacity, f, can be thought of as an effective partial pressure—in essence, a correction to ideal gas behavior. So for an ideal gas, $f = p$, and for a perfect gas mixture, $f_i = p_i$. The quotient, f/p, is commonly referred to as the fugacity coefficient, and it can be calculated (relative to a reference pressure, usually 1 atm) with the integral:

$$ln\left(\frac{f}{p}\right) = \frac{1}{RT}\int_{P_0}^{P}\left(V - \frac{RT}{pM}\right)dp.$$

R is the gas constant, 0.729, T is the absolute temperature, °F + 460, and M is the formula weight for ethylene, 28.05. Therefore, if one has access to volume–pressure data for a species of interest, the fugacity can be computed. Ethylene is an alkene used to make polyethylene, and it is one of the most important organic compounds in the chemical process industries (CPI), and as such, its properties have been studied extensively:

Determine the fugacity for ethylene at 180 °F with a pressure of 50 atm, and also for 100 °F and a pressure of 80 atm using the data provided in Figure P4.23.

Answer: For 180 °*F and* 50 *atm,* $f = 49.48$ *atm: for* 100 °*F and* 80 *atm,* $f/p = 0.9765$.

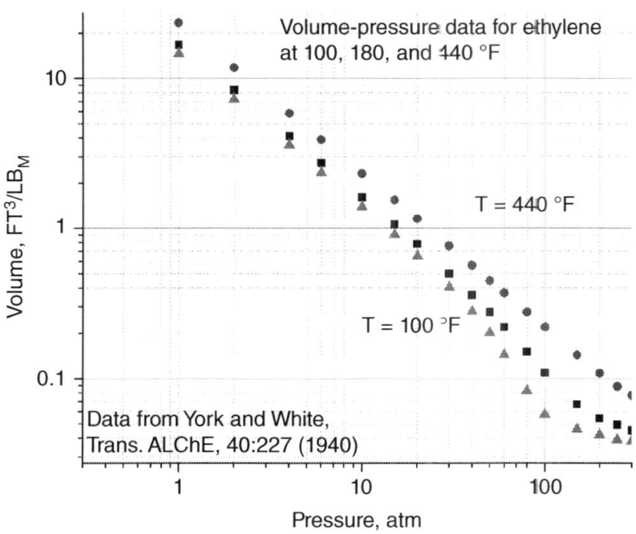

Figure P4.23 Volume–pressure data for ethylene from York and White (1940).

4.24 A special divergent nozzle assembly was designed and a prototype prepared using a 3D printer; the nozzle was to be used to produce thin shear layers with very large velocity gradients. Two trials were conducted in which point velocities were measured at volumetric flow rates of air at 200 and 300 L/min (as indicated by a rotameter). A hot-wire anemometer was used for the velocity measurements, on a vertical traverse through the middle of the flow. The data obtained are shown in Figure P4.24.

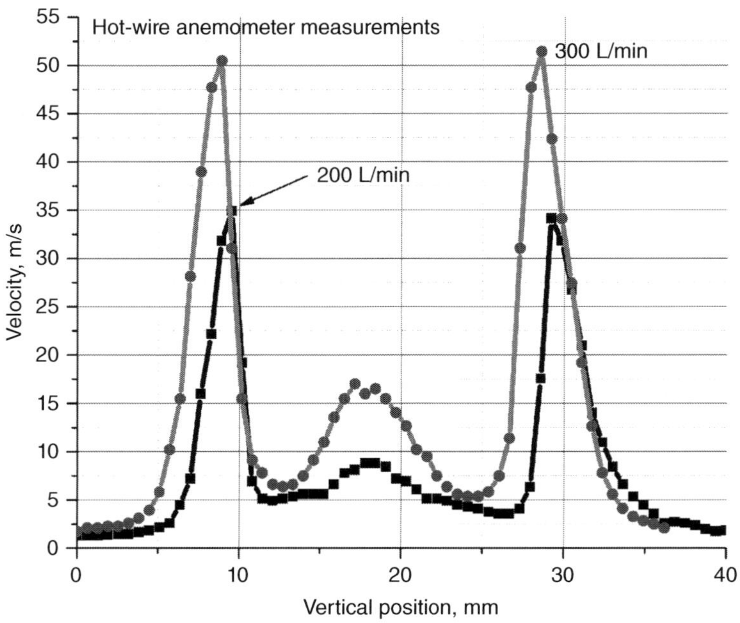

Figure P4.24 Velocity measurements for air discharged through an axisymmetric divergent nozzle at indicated flow rates of 200 and 300 L/min. These data were obtained on a vertical traverse of a hot-wire anemometer through the center of the annular-shaped jet flow.

We want to integrate these two velocity distributions to obtain the volumetric flow rates for comparison with the rotameter indications. The centerline of the experimental nozzle is located at a vertical position corresponding to 18.5 mm in the figure; the spacing between measurement points was 0.63 mm. If the data were perfectly symmetric, then the volumetric flow rate, Q, could be determined from:

$$Q = \int_0^R 2\pi r V_z(r) dr.$$

However, the above graph reveals some minor discrepancies side to side, and these may need to be taken into account. For example, at 300 L/min, the maximum measured velocity was about 50 m/s on the left half, but about 53 m/s on the right. At 200 L/min, the corresponding maximum velocities were about 35 and 34 m/s. One way to account for these differences would be to obtain two summations (one for the left half of the data and one for the right) of the form

$$\sum_{i=1}^{n} \pi r_i \Delta r V_z(r_i).$$

However, this is not the only option you have for solving this problem.

Note: The results you obtain may not be in close agreement with the reported values of 200 *and* 300 *L/min. The hot-wire anemometer was calibrated at significantly lower velocities than measured here; furthermore, the shoulder velocities—vertical positions of* 0 *and* 40*—should have been zero. Finally, the rotameter flow rates were at the low end of the scale, so they are suspect as well.*

4.25 The *PFTR* (plug flow tubular reactor) is an idealized model of a chemical reactor in which it is *assumed* that the velocity distribution is flat, i.e., there is no variation of velocity with respect to radial position. Thus, every fluid parcel has exactly the same residence time—of course this cannot be correct but under certain conditions it may be useful for reactor design. Suppose we are interested in the gas-phase decomposition of acetaldehyde:

$$CH_3CHO \rightarrow CH_4 + CO.$$

Normally we think of working with moles in the context of chemical reactions, but it is convenient in this case to base our balance upon mass flow rate (G) since the number of moles is changing as we move down the reactor. The design equation that allows us to determine the required volume of the reactor is:

$$V = G\int_0^{y_{exit}} \frac{dy}{r},$$

where y is moles of *product* per unit mass and r is the rate expression for the reaction:

$$r = k\left[\frac{1-y}{1+y}\left(\frac{P}{RT}\right)\right]^2.$$

The rate constant, k, has the value 0.43 $m^3/(kgmol\ s)$. P is 1 atm and $T = 793$ K. We assume the reactor operates isothermally with constant pressure; therefore:

$$V = \frac{G}{k}\left(\frac{RT}{P}\right)^2\int_0^{y_{exit}}\left[\frac{1+y}{1-y}\right]^2 dy.$$

Complete conversion of the reactant, acetaldehyde, would result in $y \rightarrow 1$, so we expect the integrand to become very large as the reaction proceeds. Perform the integration using the method of your choice, and prepare a figure that shows the required reactor volume for y_{exit} ranging from 0.2 to 0.95. Remember that to achieve consistent units we must divide the mass flow rate by the formula weight of acetaldehyde: $G = 0.1/44$.

4.26 The *log-normal* density function is used for right-skewed distributions and thus is often employed for particle size distributions since no particles can have a diameter less than zero. The density function has two parameters, μ and σ, which should *not* be confused with mean and standard deviation:

$$f(x) = \frac{1}{\sigma x\sqrt{2\pi}}\exp\left[-\frac{1}{2}\left(\frac{\ln(x)-\mu}{\sigma}\right)^2\right].$$

Ryu, Wang, and Chang (*Journal of Aerospace Engineering*, 2018) presented data for the size distribution of lunar regolith particles; the approximate data below were interpreted from one of their figures. Note that these data are presented as a *cumulative* distribution.

Grain Size (mm)	Percent Finer Than
0.0006	0
0.016	10
0.029	20
0.04	30
0.056	40
0.077	50
0.15	60
0.27	70
0.42	80
0.83	90
1.95	95
5	100

Use these data to find the two needed parameters for the log-normal distribution by integration. For example, we are seeking values for μ and σ such that

$$\int_0^{0.15} f(x)dx = 0.60 \quad \text{and} \quad \int_0^{0.42} f(x)dx = 0.80\text{etc.}$$

4.27 A Fresnel integral is defined by

$$C(x) = \int_0^x \cos\left(\frac{\pi}{2}t^2\right) dt. \tag{4.75}$$

Values for the function $C(x)$ can also be obtained by series expansion:

$$C(x) = \sum_{n=0}^{\infty} \frac{(-1)^n \left(\frac{\pi}{2}\right)^{2n} x^{4n+1}}{(2n!)(4n+1)}. \tag{4.76}$$

Compute $C(x)$ for $0 < x < 2.2$ by numerical integration of (i) and by series expansion using (ii). Use a method of your choice for the numerical quadrature. Compare the results; is there reason to think that one approach is more accurate? Finally, determine as precisely as possible the first two values of x for which $C(x) = 0.600$.

Partial answer: $C(x) = 0.6000058$ *for* $x = 0.6227$

4.28 Commercial software was used to evaluate the following definite integral, $\int_1^{2.95} x^{1.5}/(3-x)^4 dx$. The result obtained was 13,341. Use at least two different methods to assess the accuracy of this answer.

References

Abramowitz, M. and Stegun, I. (1964). *Handbook of Mathematical Functions*. New York: Dover Publications.

Davis, H.T. (1962). *Introduction to Nonlinear Differential and Integral Equations*. New York: Dover Publications.

Grandine, T.A. (2005). The Extensive Use of Splines at Boeing. *SIAM News* 38 (4).

Hanna, O.T. and Sandall, O.C. (1995). *Computational Meethdos in Chemical Engineering*. Upper Saddle River, NJ: Prentice Hall.

James, M.L., Smith, G.M., and Wolford, J.C. (1977). *Applied Numerical Methods for Digital Computation*, Seconde. New York: Harper & Row.

Kahaner, D., Moler, C., and Nash, S. (1989). *Numerical Methods and Software*. Englewood Cliffs, NJ: Prentice Hall.

Kuncir, G.F. (1962). Algorithm 102: Simpson's Rule Integrator. *Communications of the ACM* 5: 347.

Laurie, D. P. Automatic Numerical Integration over a Triangle. *Special Report WISK 273*, NRC Mathematical Sciences, CSIR (1977).

Lyness, J. N. AUG2-Argonne National Lab Report, *ANL/MCS-TM-13* (1983).

Lyness, J.N. (1969). Notes on the Adaptive Simpson Quadrature Routine. *Journal of the ACM* 16: 483.

Radon, J. (1948). Zur mechanische Kubatur. *Monatsh. Math.* 52: 286.

Rubinstein, R.Y. (1981). *Simulation and the Monte Carlo Method*. New York: John Wiley & Sons.

Sloan, I.H. and Joe, S. (1994). *Lattice Methods for Multiple Integration*. Oxford: Oxford University Press.

Smyth, G.K. (1998). Numerical Integration. In: *Encyclopedia of Biostatistics* (ed. P. Armitage and T. Colton). Chichester: Wiley & Sons.

Stroud, A. (1972). *Approximate Calculation of Multiple Integrals*. Englewood Cliffs, NJ: Prentice-Hall.

5

Analytic Solution of Ordinary Differential Equations

5.1 Some Introductory Examples

Ordinary differential equations (ODEs) figure prominently in every branch of mechanics—as well as in chemistry, biology, electronics, process control, etc. Indeed, progress in the applied sciences would be exceedingly difficult without this area of mathematics. We begin our discussion of ODEs by looking at four simple examples that highlight a few of the characteristics of such problems. Our first example will involve a resistor (R) and a capacitor (C) placed in series; a voltage, V, is applied to the circuit and the charge on the capacitor (q) will increase with time.

5.1.1 The RC Circuit

The ordinary differential equation for this RC circuit is:

$$V = \frac{dq}{dt}R + \frac{q}{C}. \tag{5.1}$$

We rearrange the equation as follows:

$$\frac{dq}{dt} = \frac{V}{R} - \frac{q}{RC} = a - bq, \tag{5.2}$$

where $a = V/R$ and $b = 1/(RC)$. The first-order ordinary differential equation is separable, so

$$\frac{dq}{a - bq} = dt \tag{5.3}$$

which we integrate to yield

$$-\frac{1}{b}\ln(a - bq) = t + c_1. \tag{5.4}$$

The time constant for this elementary circuit is $\tau = RC$, and therefore

$$q = VC - c_1 RC\exp(-t/\tau). \tag{5.5}$$

The capacitor has no charge at $t = 0$, so $c_1 = V/R$ and

$$q = VC(1 - e^{-t/\tau}). \tag{5.6}$$

The maximum charge that will be developed is VC, and the corresponding q_{max} will be approached asymptotically according to this exponential limiting law. Now suppose that $R = 5\,\Omega$ and $C = 2$ F such that $\tau = 10$ s; we then apply 12 VDC to the circuit. Eq. (5.6) indicates the following $q(t)$ behavior:

t (s)	1	2	5	10	20	50	100
$q/(VC)$	0.0951	0.1813	0.3935	0.6321	0.8647	0.9933	0.9999

As we observed at the very beginning of this course, the transient behavior we see here is typical of elementary first-order systems, and we note that at $t = 10$ s (or 1τ), 63% of the ultimate change is achieved and at 20 s (or 2τ), 86 + %. As a rule of thumb, we often say that the transient behavior of a first-order system is virtually complete in 3τ.

Problem Solving in Engineering: Analytical Mathematics and Numerical Analysis, First Edition. Larry A. Glasgow.

Companion website: www.wiley.com/go/glasgow/problemsolvingengineering

5.1.2 Cooling by Natural Convection

Suppose we place about 500 mL of water in a beaker and heat the contents with a candle flame until the water begins to boil. We then remove the heat source and allow the contents of the beaker to cool simply by exposure to the ambient air. An approximate energy balance for this situation (the cooling process) can be written:

$$\{\text{Accumulation}\} = -\{\text{Rate of loss to surroundings}\},$$

or symbolically as

$$mC_p \frac{dT}{dt} = -hA(T - T_\infty). \tag{5.7}$$

In this first-order, ordinary differential equation (ODE), m is the mass of heated water, C_p the heat capacity of the water, h the heat transfer coefficient, and A is the surface area for heat transfer. T_∞, of course, is the temperature of the air surrounding the beaker. Actual experimental data for this process are shown in Figure 5.1.

We want to solve eq. (5.7) and then see if the result is capable of describing the cooling process shown in Figure 5.1. It will be convenient to let $\theta = T - T_\infty$, such that

$$\frac{d\theta}{dt} = -\frac{hA}{mC_p}\theta. \tag{5.8}$$

This equation is separable and easily integrated to yield:

$$\ln \theta = -\frac{hA}{mC_p}t + C_1, \tag{5.9}$$

or more conveniently:

$$\theta = C_1 \exp\left(-\frac{hA}{mC_p}t\right). \tag{5.10}$$

We should make note of the fact that hA/mC_p is the reciprocal time constant $(1/\tau)$ for this first-order system, and once again we observe that after an elapsed time of 1τ about 63% of the change has been accomplished, after 2τ about 86%, and after 3τ about 95%. In our example, the cooling process was initiated at $t = 865$ s when the temperature of the water was 99.8+ °C. Since the ambient temperature was 22 °C, we have

$$T = 77.8 \exp\left(-\frac{hA}{mC_p}t\right) + 22. \tag{5.11}$$

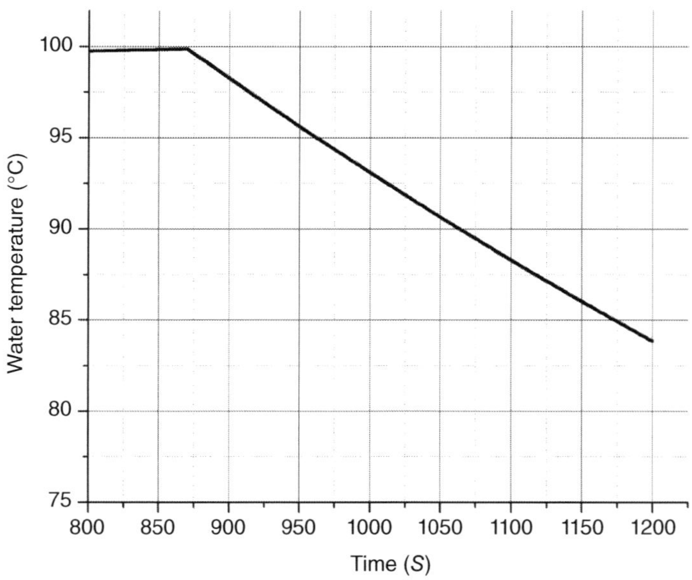

Figure 5.1 Cooling process for 500 mL water, initially heated to incipient boiling. The heat source (flame) was removed at $t = 865$ s.

We can use the experimental data shown in Figure 5.1 to estimate the unknown parameters in the quotient (hA/mC_p); at $t = 1{,}200$ s, the water temperature had fallen to 83.75 °C. So when $t = 335$ s (1,200–865), we find:

$$\ln\left(\frac{83.75 - 22}{77.8}\right) = -\frac{hA}{mC_p}(335), \tag{5.12}$$

which produces a value for hA/mC_p (0.00069 1/s). The time constant, τ, is about 1,449 s, so 3τ is about 1.2 hr. How good is our crude model? At $t = 1{,}050$ s, the data in Figure 5.1 indicate that $T = 90.7$ °C; the model shows that at $t = 185$ s (1,050–865), $T = 90.5$ °C.

Let us make some concluding remarks about this second example. The differential equation we formulated by energy balance, eq. (5.7), was an extremely simple first-order, linear, ordinary differential equation. However, it was *not homogeneous* due to the presence of the constant, T_∞. Notice that we eliminated the inhomogeneity by redefining the dependent variable: $\theta = T - T_\infty$. Of course, this allowed us to follow the sequence: separate the variables, integrate the equation, and evaluate the constant of integration, C_1, through use of the initial condition.

5.1.3 Heat Loss from an Insulated Steam Pipe

A steel pipe carrying saturated steam is covered with a layer of insulation in an effort to minimize heat losses to the surroundings. A steady-state energy balance upon an annular layer (thickness Δr) of the *insulation* shows:

$$-\frac{d}{dr}(rq_r) = 0, \tag{5.13}$$

where the flux of thermal energy, q_r, is given by Fourier's law: $q_r = -k\frac{dT}{dr}$. Therefore,

$$\frac{d}{dr}\left(r\frac{dT}{dr}\right) = 0 \quad \text{and} \quad T = c_1 \ln(r) + c_2. \tag{5.14}$$

Let R_1 be the radius corresponding to the interface between the pipe wall and the insulation, and let R_2 correspond to the outer surface of the insulation. *If* the temperatures are specified at these two radial positions, then

$$T_1 = c_1 \ln(R_1) + c_2 \quad \text{and} \quad T_2 = c_1 \ln(R_2) + c_2$$

such that

$$c_1 = \frac{T_1 - T_2}{\ln(R_1/R_2)}. \tag{5.15}$$

Since $dT/dr = c_1/r$ we can evaluate the thermal energy flux at any r-position (between R_1 and R_2). Now, suppose $T_1 = 200$ °C and $T_2 = 30$ °C with $R_1 = 10$ cm and $R_2 = 16$ cm. If the thermal conductivity, k, of the insulation is 0.0003 cal/(cm s °C), what is the rate of energy loss per unit length of pipe?

$$q_{r=R_1} = -k\frac{T_1 - T_2}{\ln\left(\frac{R_1}{R_2}\right)}\frac{1}{R_1} = 0.01085\ \text{cal}/(\text{s cm}^2). \tag{5.16}$$

The surface area at $r = R_1$ is $A = 2\pi R_1 L$, so the rate of heat loss per meter of length is 68.2 cal/s, or 245,422 cal/hr/m. This corresponds to 296 W lost per meter; if $L = 30$ m and if plant energy costs \$0.05/kwhr, the "cost" of this energy loss would be about \$10,660/day!

We should note that the way this problem was posed was not entirely realistic because the air surrounding the insulated pipe will provide an additional resistance to heat transfer; a "better" boundary condition can be written for $r = R_2$ by equating the thermal energy fluxes at that outer surface:

$$-k\left(\frac{dT}{dr}\right)_{r=R_2} = h(T_2 - T_\infty), \tag{5.17}$$

where h is the heat transfer coefficient and T_∞ is the temperature of the air surrounding the pipe. All of the preliminary steps are exactly the same, but the values of the constants of integration will be changed:

$$c_1 = \frac{-hR_2}{k}(T_2 - T_\infty) \quad \text{and} \quad c_2 = T_2 + \frac{hR_2}{k}(T_2 - T_\infty)\ln R_2. \tag{5.18}$$

If $T_\infty = 25$ °C and $hR_2/k = 10$, then $T_2 = 55.7$ °C. Notice that the temperature gradient across the insulation layer has been reduced resulting in a lower rate of heat loss; in fact, it is about 15% lower at 57.87 cal/s/m due to the insulating effect of the air. Any decrease in air temperature or increase in wind speed would increase the loss rate of course. And finally we observe that the revised boundary condition has increased the temperature of the outer surface (about 132 °F). T_2 may now be large enough to pose a safety concern.

5.1.4 Dissolution of an Inorganic Metal Salt

An inorganic metal salt dissolves in water, and the rate of dissolution is given by the product of the mass undissolved, M, and the difference between the saturation condition, M_S, and the amount present in solution at some time, t. The rate law governing this process is

$$\frac{dM}{dt} = -kM\left(\frac{2}{10} + \frac{M}{100}\right), \tag{5.19}$$

and the rate constant, k, is known to be 0.46 hr^{-1}. Suppose we determine that $M = 20$ g when $t = 2$ hr; what will M be after 5 hr? Of course, this first-order ordinary differential equation is nonlinear, but it is separable:

$$\int \frac{dM}{M\left(\frac{2}{10} + \frac{M}{100}\right)} = -kt + c_1. \tag{5.20}$$

The integral on the left-hand side looks like it might pose a problem, but we simply use a *TI-nspire* in the *calculus/integral* mode:

$$\int \frac{dx}{x(ax+b)} = \frac{1}{b}\ln\left[\frac{x}{ax+b}\right]. \tag{5.21}$$

The *log* of a quotient is equal to the difference between logarithms, so we could rewrite the right-hand side as two *log* terms if so desired. Alternatively,

$$\frac{M}{\frac{M}{100} + \frac{2}{10}} = c_1 \exp\left(-\frac{kt}{5}\right). \tag{5.22}$$

Since $M = 20$ at $t = 2$ hr, $c_1 = 50/0.832 = 60.1$; and after 5 hr, $M = 12.227$ g.

These four introductory examples provide merely an indication of the importance of ordinary differential equations. Although all four equations were separable, we saw a few of the complications that can arise in such cases; these include an equation that was not homogeneous, a less-familiar boundary condition involving fluxes at an interface, and a nonlinearity. But by no means do these examples give us a comprehensive picture of the possible variations—we need more tools to help us deal with less familiar situations.

5.2 First-Order Ordinary Differential Equations

We begin by considering equations of the type

$$\frac{dy}{dx} = f(x, y). \tag{5.23}$$

Let us assume that this equation can be written as: $Mdx + Ndy = 0$. Such an equation is said to be *exact* if

$$\frac{\partial M}{\partial y} = \frac{\partial N}{\partial x}. \tag{5.24}$$

Now suppose

$$\frac{dy}{dx} = f(x, y) = \frac{y+2}{x^2}, \tag{5.25}$$

which we rewrite as $(y + 2)dx - x^2dy = 0$. Therefore,

$$\frac{\partial M}{\partial y} = 1 \quad \text{and} \quad \frac{\partial N}{\partial x} = -2x; \tag{5.26}$$

this equation is not exact. However, it is separable and we can easily show that

$$y = C_1 \exp\left(-\frac{1}{x}\right) - 2. \tag{5.27}$$

We contrast this case with the equation,

$$\frac{dy}{dx} = -\frac{3x^2y^2 + 2xy}{2x^3y + x^2}; \tag{5.28}$$

this time separation is not possible. We rewrite (5.28) as $(2x^3y + x^2)dy + (3x^2y^2 + 2xy)dx = 0$, and observe

$$\frac{\partial M}{\partial y} = 6x^2y + 2x \quad \text{and} \quad \frac{\partial N}{\partial y} = 6x^2y + 2x. \tag{5.29}$$

We see that eq. (5.28) is exact and because we have an exact differential, $Mdx + Ndy = dU$, where

$$\frac{\partial U}{\partial x} = M \quad \text{and} \quad \frac{\partial U}{\partial y} = N. \tag{5.30}$$

Thus, $\frac{\partial U}{\partial x} = 3x^2y^2 + 2xy$, and integrating with respect to x:

$$U = x^3y^2 + x^2y + F(y), \tag{5.31}$$

since F cannot be a function of x. From the second part of eq. (5.30), we differentiate U with respect to y and set the result equal to N:

$$2x^3y + x^2 + F'(y) = 2x^3y + x^2. \tag{5.32}$$

Clearly, $F'(y) = 0$, so $F = C$, and we note:

$$x^3y^2 + x^2y = C. \tag{5.33}$$

When a first-order ODE is *not exact*, we look for an integrating factor. For example, if

$$\frac{\frac{\partial M}{\partial y} - \frac{\partial N}{\partial x}}{N} = f(x), \tag{5.34}$$

where $f(x)$ is a function *only* of x, then an integrating factor (IF) is: $IF = e^{\int f(x)dx}$. Let us illustrate with an example; suppose we have

$$\frac{dy}{dx} + \frac{1}{x}y = \frac{2}{x} + x. \tag{5.35}$$

Therefore,

$$dy = \left(\frac{2}{x} + x - \frac{y}{x}\right) dx,$$

so that $M = \frac{2}{x} + x - \frac{y}{x}$ and $N = -1$; the equation is not exact. By eq. (5.34), $f(x) = 1/x$, and $e^{\int \frac{1}{x}dx} = e^{\ln x} = x$. We multiply eq. (5.35) by the IF:

$$x\left(\frac{dy}{dx} + \frac{y}{x}\right) = (xy)' = \frac{2}{x} + x. \tag{5.36}$$

We integrate and divide by x:

$$y = \frac{2}{x}\ln x + \frac{x}{2} + \frac{C_1}{x}. \tag{5.37}$$

We need to see this process employed for something that is more characteristic of a real problem encountered in applied science. Suppose we have two identical stirred-tank reactors in series, each with a capacity of 100 gal. Both tanks are perfectly mixed, and tank 1 contains a solute at a concentration of 0.8 lb/gal, initially. Pure solvent flows into the first tank at a rate of 4 gal/min (gpm), then the discharge (overflow) from 1 flows into tank 2. Our interest is the concentration of solute in tank 2 as a function of time. The situation we are describing corresponds to two, non-interacting first-order systems in series. By mass balance we formulate two first-order ODEs: For tank 1:

$$\dot{V}C_{in} - \dot{V}C_1 = V_T\frac{dC_1}{dt}, \tag{5.38}$$

and for tank 2,

$$\dot{V}C_1 - \dot{V}C_2 = V_T\frac{dC_2}{dt}. \tag{5.39}$$

Of course, $C_{in} = 0$ since pure solvent is being fed to the first tank. Also, if we divide both equations by the volumetric flow rate, $\dot{V}$, and note that the volume of each tank, V_T, divided by the volumetric flow rate is the time constant, τ $\left(\tau = \frac{V_T}{\dot{V}}\right)$, then we can immediately find the solution for eq. (5.38):

$$C_1 = C_0 \exp\left(-\frac{t}{\tau}\right), \tag{5.40}$$

where C_0 is the initial concentration in tank 1, 0.8 lb/gal. We take this result back to the ODE eq. (5.39) and rearrange the equation so that:

$$\frac{dC_2}{dt} + \frac{1}{\tau}C_2 = \frac{C_0}{\tau}\exp\left(-\frac{t}{\tau}\right). \tag{5.41}$$

Now we recognize that this eq. (5.41) is of the form, $y' + a(x)y = b(x)$, so that the integrating factor (IF) can be determined from $e^{\int a(x)dx}$, which in our case is $e^{\int \frac{1}{\tau}dt} = e^{t/\tau}$. Therefore,

$$\frac{d}{dt}(e^{t/\tau}C_2) = \frac{C_0}{\tau}\exp\left(-\frac{t}{\tau}\right)\exp\left(\frac{t}{\tau}\right). \tag{5.42}$$

The concentration in tank 2 then is simply $C_2 = \frac{C_0}{\tau}te^{-t/\tau}$ (remember, the concentration in tank 2 was zero at $t = 0$). So for this example, after one hour ($t = 60$ min), the concentration of solute in tank 2 will be 0.1742 lb/gal. The maximum concentration in tank 2 occurs at $t = 25$ min, $C_{2max} = 0.2943$.

5.2.1 An Example Exercise: The Streeter–Phelps Model for Dissolved Oxygen Sag

One measure of the level of pollution in surface waters like rivers and streams is the biological oxygen demand, or BOD. BOD is an indicator of dissolved oxygen consumed by micro-organisms in the biochemical oxidation of organic material present in the water. In highly polluted waters, microbial activity may consume oxygen more rapidly than it can be replenished at the water's surface. If the level of dissolved oxygen (DO) is significantly depressed from the normal saturation level of about 8 to 9 ppm, it may signify that the water is carrying high levels of organic pollution. And, should the level of DO fall to around 1 to 2 ppm, the health of aquatic species like fish may be severely impaired.

The Streeter–Phelps model was developed in 1925 to simulate dissolved oxygen consumption and reaeration in the Ohio River. It is important to recognize that this model does not account for oxygen production by photosynthesis nor does it include oxygen demand reduction by sedimentation of organic material. We will let C represent the concentration of organic material in a stream; the decay of C by microbial action is approximately first order:

$$\frac{dC}{dt} = -k_1 C. \tag{5.43}$$

The oxygen *deficit*, D, is roughly described by the ordinary differential equation,

$$\frac{dD}{dt} = k_1 C - k_2 D. \tag{5.44}$$

The reaeration constant, k_2, can range from 0.10 for sluggish streams and lakes to values greater than 0.5 for rapids and waterfalls. Of course, the driving force for reaeration (which occurs by dissolution of atmospheric oxygen at the water's surface) is roughly proportional to the difference between saturation and the actual DO level in the water. There are two important features of this model to note: *i*) Time and distance are essentially equivalent since we are assuming the water in the stream flows at a constant and uniform velocity. *ii*) The rate constants appearing in eq. (5.44) may change if the conditions of the streamflow change; for example, any disturbance at the surface will increase the reaeration rate.

Now suppose some stream is subjected to pollution by an organic material. Microbial activity will be initially elevated as the organic substrate is metabolized and the dissolved oxygen level in the water will decline. As the easily digested materials are consumed, the rate of microbial utilization of oxygen will also diminish allowing reaeration at the stream's surface to replenish the DO. A plot of the dissolved oxygen concentration versus time (or distance) is often referred to as the DO sag.

Let $k_1 = 0.5$ and $k_2 = 0.3$ and assume that the initial oxygen deficit is essentially zero. If $C(t = 0) = 10$, when will the DO level attain its minimum value, and what will that minimum value be? Eq. (5.43) depicts a simple first-order decay of organic material:

$$C = C_0 \exp(-k_1 t),$$

and therefore,

$$\frac{dD}{dt} = k_1 C_0 \exp(-k_1 t) - k_2 D. \tag{5.45}$$

You will immediately recognize this differential eq. (5.45) from our preceding discussion. We want to solve this ODE assuming that the initial dissolved oxygen level is 6 ppm Verify that the minimum DO level occurs at $t = 2.519$ hr, and that the maximum deficit results in a minimum DO concentration of 1.265 ppm Such a low level of dissolved oxygen will put some fish species under severe stress. A few results are given here so you can check your work:

Time (hr)	0	1	4	7	12	15
DO (ppm)	6	2.51	1.80	3.57	5.37	5.73

When and where will the stream return to 99% of its initial DO level (5.94 ppm)?
Answer: 20.06 *hr, if the uniform stream velocity is* 3 *ft/s,* 99% *recovery will occur at a downstream distance of* 41 *miles.*

5.3 Nonlinear First-Order Ordinary Differential Equations

A *nonlinear* first-order ordinary differential equation is one that has the form

$$\frac{dy}{dx} = f(x, y), \tag{5.46}$$

where $f(x,y)$ includes a term in which the dependent variable, y, occurs to some power other than one. Examples could include

$$\frac{dy}{dx} = x^2 + y^2 \quad \text{or} \quad \frac{dy}{dx} = xy\sin(y). \tag{5.47}$$

The second equation of this pair is particularly interesting, and the reader may want to compare what happens when $y(x=0) = 1/4$ with the case where $y(x=0) = 4$. Certain nonlinear, first-order ODEs have seen a great deal of study. For example, Bernoulli's equation which is usually written as

$$\frac{dy}{dx} + p(x)y = q(x)y^n \quad \text{where } n \neq 0, 1 \tag{5.48}$$

has attracted the attention of mathematicians for more than 300 years. One of the reasons this equation is interesting is because of the effect of the transformation, $z = y^{1-n}$; we will illustrate with an example. Let us consider

$$\frac{dy}{dx} + \frac{y}{x} = y^3. \tag{5.49}$$

Applying the transformation ($z = y^{-2}$), we find

$$-\frac{1}{2}z^{-3/2}\frac{dz}{dx} + \frac{z^{-1/2}}{x} = z^{-3/2},$$

which leads directly to

$$\frac{dz}{dx} - 2\frac{z}{x} + 2 = 0. \tag{5.50}$$

This is a familiar form and we know that the integrating factor (*IF*) is just $e^{\int -\frac{2}{x}dx} = e^{-2\ln x} = \frac{1}{x^2}$. It is straightforward to show that

$$z = 2x + C_1 x^2, \tag{5.51}$$

and therefore,

$$y = \sqrt{\frac{1}{2x + C_1 x^2}}. \tag{5.52}$$

5.3.1 Riccati Equations

One of the best-known examples of a nonlinear first-order ODE is the *Riccati equation*, which is generally written as

$$\frac{dy}{dx} + a(x)y + b(x)y^2 + c(x) = 0. \tag{5.53}$$

This equation is reducible to the form

$$\frac{dw}{dx} + c(x)w^2 + d(x) \tag{5.54}$$

by use of the transformation, $w = e^{\int_a^x a(x)dx} y$. Riccati equations arise regularly in applied mathematics, and we will look at two examples for illustration.

Consider a constant-volume reactor in which the reactant species "A" is consumed by a second-order (bimolecular) process. The reactant is added to the process vessel continuously, but in a concentrated form such that the volume (V) is nearly unaffected. A mass balance upon the reactant "A" yields:

$$V\frac{dC_A}{dt} = \dot{n}_A - Vk_2C_A^2. \tag{5.55}$$

Therefore

$$\frac{dC_A}{dt} = \frac{\dot{n}_A}{V} - k_2C_A^2 \quad \text{or equivalently,} \quad \frac{dy}{dx} = a - by^2. \tag{5.56}$$

In this case the solution can be found very easily using a table of integrals (see Selby, page 546 in the *CRC Handbook of Tables for Mathematics*, Revised Fourth Edition, 1975 or Dwight, 1957):

$$y = \frac{a}{\sqrt{ab}} \tanh\left[\sqrt{ab}(x + C_1)\right]. \tag{5.57}$$

The initial concentration of reactant is zero, so $C_1 = 0$. We assign the constants, a and b, the values 3 and 0.475, respectively, resulting in

$$C_A = 2.5131 \tanh(1.19373t). \tag{5.58}$$

Thus, the concentration of "A" in the reactor, which will approach 2.52312 as $t \to \infty$, behaves as shown in the following table:

Time	0	0.1	0.2	0.6	1.0	2.0
Concentration	0.00	0.2986	0.5888	1.5445	2.0902	2.4710

For our second example, suppose a pilot must leave an airplane at very high altitude. The pilot will begin to fall vertically under the influence of gravity and that motion will be opposed by drag. We are interested in his/her approach to terminal velocity, what that velocity will be, and how long it will take the pilot to attain that velocity. An approximate model for this free-fall will be

$$m\frac{dV}{dt} = mg - KV^2. \tag{5.59}$$

m is the mass of the pilot (plus gear) and K is the product of the frontal area of the falling pilot, the fluid density, and the drag coefficient. Generally, the drag coefficient is a function of velocity, but we are taking it to be approximately constant. Dividing by the mass we get

$$\frac{dV}{dt} = g - \frac{K}{m}V^2 \quad \text{or alternatively,} \quad \frac{dy}{dx} = a - by^2, \tag{5.60}$$

which, conveniently, is the same form as the constant-volume chemical reactor. Experiments conducted in the World War 2 era indicated that the terminal velocity of the falling pilot from high altitude would be roughly 250 ft/s (about 170 mph); therefore, $a = 32.2$ and $b = 0.000515$ (1/ft). The solution is exactly the same as given above by eq. (5.57), and the initial (vertical) velocity is zero.

Time (s)	0.5	1	5	10	20	50
Velocity (ft/s)	16.07	32.02	141.88	214.63	247.12	249.99+

Thus, the falling pilot attains terminal velocity in about 50 s.

Both of the equations examined above were of the form: $y' = a - by^2$, and the solution(s) was easily obtained. This is certainly not the norm with nonlinear ODEs, so let us continue the discussion by looking at variations of the Riccati equation:

$$y' + y^2 = -2 \tag{5.61}$$

$$y' + y^2 = -x \tag{5.62}$$

$$y' + y^2 = -x^2 \tag{5.63}$$

The first of these, eq. (5.61), can be solved by proceeding just as we did in the immediately prior examples, and we obtain $\frac{\sqrt{2}}{2}\tan^{-1}\left(\frac{\sqrt{2}y}{2}\right) = c_1 - x$. We can invert the result with *tan* and obtain a solution; if $y(0) = 1$, then $c_1 = 0.4352$. But what about eqs. (5.62) and (5.63)? Eq. (5.63) has a particularly interesting history: Davis (1962) notes that John Bernoulli tried to solve it unsuccessfully around 1694. About nine years later James Bernoulli found a power series solution for it, but the intractable nature of these deceptively simple ODEs was evident. Take a moment to think about these three ODEs qualitatively—if $y(0) = 1$, all three equations will exhibit negative slope at small x. Furthermore, as x becomes large it is clear that all three slopes will be increasingly negative. Their behavior can be explored numerically, and the results of such a study are provided in Figure 5.2.

It is worthwhile for us to think a bit more about the behavior of eq. (5.62). Suppose we are interested in $y(x)$ the fourth quadrant; we set $y(x = -3) = 1$. Since the slope is initially positive, y will increase at first, but as we approach the y-axis it is clear that y' will become negative at some point What will happen when we cross over into positive x-values? We will discuss a means by which such questions can be explored *without* solving the equation numerically in a moment.

In addition to Bernoulli and Riccati equations, many other nonlinear ordinary differential equations have been studied intensively. For example, in the first third of the twentieth century there was great interest in predator–prey problems (populations in conflict). Generally such problems were formulated in terms of two simultaneous ordinary differential equations, one for population "1" and one for population "2." Nonlinear terms come about through interaction between the animals (or species). In some cases the two equations could be combined to produce one nonlinear, second-order ODE. In a typical formulation for populations in conflict we might see:

$$\frac{dN_1}{dt} = aN_1 - bN_1N_2 \quad \text{and} \quad \frac{dN_2}{dt} = -c_1N_2 + c_2N_1N_2. \tag{5.64}$$

Of course it is possible to rewrite the former as

$$N_2 = \frac{\frac{dN_1}{dt} - aN_1}{-bN_1},$$

and use that to eliminate N_2 from the second of the pair of ODEs. We note in passing that the model eq. (5.64) for the predator–prey problem is not very realistic for reasons that will be made clear in Chapter 9.

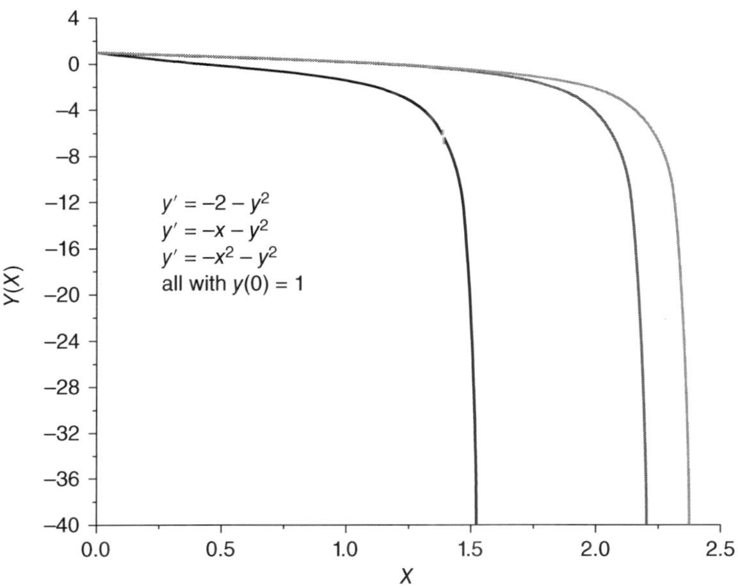

Figure 5.2 Comparison of the solutions for the three ODEs: $y' = -2 - y^2$, $y' = -x - y^2$, and $y' = -x^2 - y^2$, all with $y(0) = 1$.

5.3.2 Graphical Interpretation

Davis (1962) points out that it can often be exceedingly difficult to find an analytic solution for a *nonlinear* ODE. But the analyst may learn a great deal about the behavior of such an equation through simple graphical interpretation as follows. First, we recognize that given a nonlinear first-order ODE:

$$\frac{dy}{dx} = f(x, y), \tag{5.65}$$

we have a means for finding the slope of tangents at any point(s) we wish. Let us examine the first quadrant (on the diagonal) for the equation,

$$\frac{dy}{dx} = xy(y - 2). \tag{5.66}$$

x	0.125	0.250	0.500	1	1.95	2	3	4	5
y	0.125	0.250	0.500	1	1.95	2	3	4	5
Slope	−0.0293	−0.1094	−0.375	−1	−0.190	0	9	32	575

It is a simple matter for us to construct line segments with the indicated slope at the points listed in the table. We extend those segments a short distance, estimate the new position (x,y), calculate a new slope, and repeat. To illustrate, we will take the point (1,1) and try this.

x Position	*y* Position	Slope
1	1	−1
1.1	0.9	−1.089
1.2	0.791	−1.148
1.3	0.676	−1.164
1.4	0.560	−1.129
1.5	0.447	−1.041
1.6	0.343	−0.909
1.7	0.252	−0.749
1.8	0.177	−0.581
1.9	0.119	−0.425
2.0	0.076	−0.294

We repeat this process for various starting positions and plot the results in Figure 5.3.

The reader is encouraged to apply this method in the first quadrant, but *above* $y = 2$; in fact, we will make it one of the exercises at the end of this chapter. We note from our original table of slopes that for large (x,y) pairs, the slopes are very large—the tangents are essentially vertical lines. Furthermore, for very small values of x (but $y > 2$), the slopes are very small (the tangents are nearly horizontal). Thus, the behavior of $y(x)$ *above* $y = 2$ is very different from what we observe in Figure 5.3 for values of $y < 2$.

We should also point out that we can always differentiate our first-order nonlinear ODE and identify the locus of points of inflection by setting the second derivative equal to zero. Thus, a great deal of information regarding the behavior of the first-order, nonlinear ODE may be obtained rather easily even though we may not be able to find a solution for the equation itself.

5.3.3 Solutions with Elliptic Integrals and Elliptic Functions

We want to consider a class of nonlinear ODEs that can be initially represented by the equation,

$$\left(\frac{dy}{dx}\right)^2 = a_0 + a_1 y + a_2 y^2 + a_3 y^3 + a_4 y^4. \tag{5.67}$$

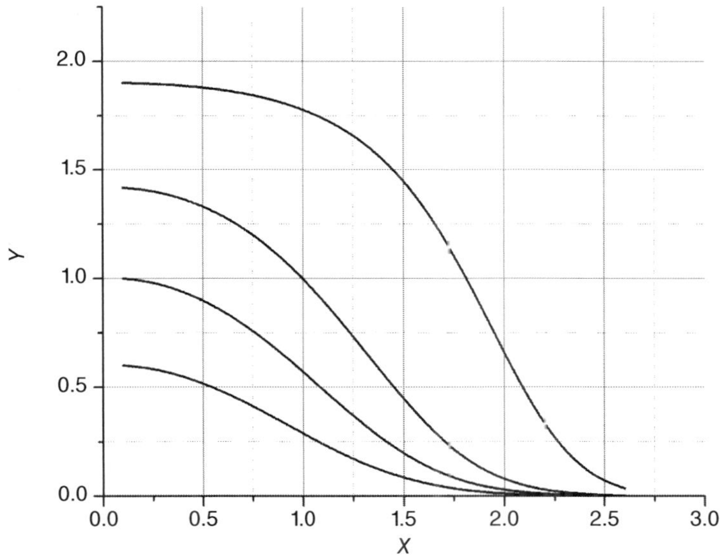

Figure 5.3 Local behavior of $y(x)$ for eq. (5.66) as estimated through the construction of tangents. Please note that slope is always zero for $y=2$; one of the loci is a horizontal line passing through $y=2$. What would we see if we started at $(x=0, y=1.99)$?

We now restrict our attention (for the right-hand side) to quartic polynomials that can be written in the form

$$1-(1+k^2)y^2+k^2y^4 \tag{5.68}$$

such that $a_0=1$, $a_2=-(1+k^2)$, and $a_4=k^2$ by comparison with eq. (5.67). We also limit consideration to values for k less than 1. By taking the square root of the modified ODE,

$$\frac{dy}{dx}=\sqrt{(1-y^2)(1-k^2y^2)}. \tag{5.69}$$

It is a simple matter for us to rearrange this equation so that

$$x=\int_0^{\beta}\frac{dy}{\sqrt{(1-y^2)(1-k^2y^2)}}. \tag{5.70}$$

This is an *elliptic integral of the first kind*, which can be written equivalently as an inverse Jacobi elliptic sine, sn^{-1}. If we now substitute $y=\sin\phi$, we obtain a more compact version of the right-hand side:

$$x=\int_0^{\theta}\frac{d\phi}{\sqrt{1-k^2\sin^2\phi}}. \tag{5.71}$$

k is the modulus of the elliptic integral and θ is the amplitude. Extensive tables of values for elliptic integrals are available; for example, if we set $k=1/4$ and $\theta=\pi/4$, we can find the value for the elliptic integral, 0.8044 (Davis has a small table and *CRC Handbook of Tables for Mathematics* a much larger one). The reader may want to confirm his/her facility with such tables by checking the following value:

$$F(k,\theta)=F\left(\frac{1}{4},\frac{\pi}{6}\right)=\int_0^{\pi/6}\frac{d\phi}{\sqrt{1-\frac{1}{16}\sin^2\phi}}=0.525. \tag{5.72}$$

At this point you might be thinking that the scope of application is strictly limited to equations where the right-hand side is similar to eq. (5.68). However, let us go back to eq. (5.67) and differentiate, obtaining:

$$2\frac{dy}{dx}\frac{d^2y}{dx^2}=a_1\frac{dy}{dx}+2a_2y\frac{dy}{dx}+3a_3y^2\frac{dy}{dx}+4a_4y^3\frac{dy}{dx}. \tag{5.73}$$

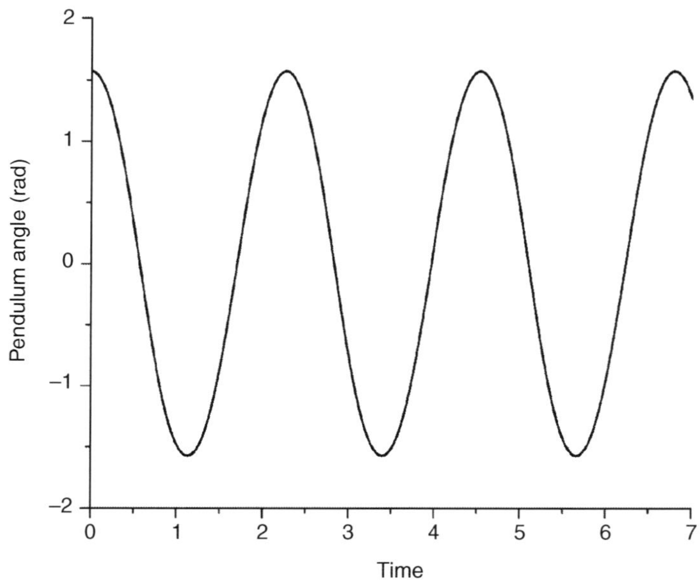

Figure 5.4 Behavior of frictionless pendulum with a starting position of 90° (1.5708 rad) and $L = 3$ ft.

If we divide by $2dy/dx$, we see a nonlinear, second-order ODE of the type that may be solved using elliptic functions; Davis provides a number of examples and Milne-Thomson (1950) points out that application of elliptic functions for solution of nonlinear ODEs includes analysis of pendulum oscillations, capillary phenomena, bending of an elastic rod, viscous flow in a convergent channel (the subject of an exercise at the end of this chapter), and the potential of an electrified ellipsoid, among others. We will look at an elementary example to illustrate this. Consider a simple, *frictionless* pendulum with oscillatory motion governed by

$$\frac{d^2\theta}{dt^2} + \frac{g}{L}\sin\theta = 0. \tag{5.74}$$

L is the length of the pendulum and g is the acceleration of gravity. Let us suppose that $\theta(t=0) = 90°$ and the $d\theta/dt = 0$. We set $g = 32.17$ ft/s, and take $L = 3$ ft. First we will find the solution for this problem numerically, and the resulting behavior is shown in Figure 5.4.

Now we turn our attention back to eq. (5.74), noting that it can be integrated once to yield

$$\frac{1}{2}\left(\frac{d\theta}{dt}\right)^2 - \frac{g}{L}\cos\theta = C_1. \tag{5.75}$$

Since $\left.\frac{d\theta}{dt}\right|_{t=0} = 0$, $C_1 = -\frac{g}{L}\cos\theta_0$, and therefore

$$\frac{d\theta}{dt} = \sqrt{\frac{2g}{L}(\cos\theta - \cos\theta_0)}. \tag{5.76}$$

If we now let $\cos\theta = 1 - 2k^2\sin^2\phi$ and take $k = \sin\left(\frac{\theta_0}{2}\right)$, we find

$$dt = \sqrt{\frac{L}{g}}\frac{d\phi}{\sqrt{1 - k^2\sin^2\phi}}. \tag{5.77}$$

Of course, when integrated, this is exactly the form of eq. (5.71). We can use this elliptic integral to find the period of the pendulum's motion (T_p); we will integrate from the zero position to our maximum angle of $\pi/2$ rad (which is one-quarter of a cycle), and then multiply by 4:

$$T_P = 4\sqrt{\frac{L}{g}}\int_0^{\pi/2}\frac{d\phi}{\sqrt{1 - k^2\sin^2\phi}}. \tag{5.78}$$

Since the initial displacement (angle) is 90°, this means that our modulus is sin(45°) and our amplitude is $\pi/2$. Consulting a table for elliptic integrals, we find a value of 1.8541 and therefore:

$$T_P = (4)(3/32.17)^{1/2}(1.8541) = 2.265.$$

Please compare this value with the results presented in Figure 5.4; you will find that our analytic solution ($T_P = 2.265$) corresponds exactly with the numerical calculation.

5.4 Higher-Order Linear ODEs with Constant Coefficients

We want to examine some higher-order ordinary differential equations, but initially with the stipulations that they be linear, homogeneous, and have constant coefficients. Let us begin with a set of three second-order equations:

$$\frac{d^2y}{dx^2} + 3\frac{dy}{dx} + 2y = 0 \tag{5.79}$$

$$\frac{d^2y}{dx^2} + 4\frac{dy}{dx} + 4y = 0 \tag{5.80}$$

$$\frac{d^2y}{dx^2} + 3\frac{dy}{dx} + 4y = 0 \tag{5.81}$$

We rewrite eq. (5.79) using linear differential operator notation:

$$(D^2 + 3D + 2)y = 0, \tag{5.82}$$

which we can readily factor as

$$(D + 2)(D + 1)y = 0. \tag{5.83}$$

Anytime we can successfully factor the linear operator for an ODE in this manner we can immediately write down the *primitive*:

$$y = C_1 \exp(-2x) + C_2 \exp(-x). \tag{5.84}$$

Values for the two constants of integration can be obtained from the initial (or boundary) conditions. For an equation like eq. (5.79) these might take the form,

$$y(x = 0) = a \quad \text{and} \quad \left.\frac{dy}{dx}\right|_{x=0} = b. \tag{5.85}$$

In the case of ODE eq. (5.80), we have

$$(D^2 + 4D + 4)y = (D + 2)(D + 2)y = 0, \tag{5.86}$$

where we see *repeated* roots. In a situation like this, the primitive is written:

$$y = C_1 \exp(-2x) + C_2 x \exp(-2x). \tag{5.87}$$

For an equation of higher order it is certainly possible to obtain:

$$(D + \alpha)^n y = 0 \tag{5.88}$$

The primitive in this case is written

$$y = C_1 \exp(-\alpha x) + C_2 x \exp(-\alpha x) + C_3 x^2 \exp(-\alpha x) + \cdots C_n x^{n-1} \exp(-\alpha x). \tag{5.89}$$

Eq. (5.81) presents a different challenge; in linear differential operator notation we find

$$(D^2 + 3D + 4)y = 0. \tag{5.90}$$

When we look at $b^2 - 4ac$ for the quadratic formula, we see: $9-16=-7$. Consequently,

$$\left(D + \frac{3}{2} + \frac{\sqrt{7}}{2}i\right)\left(D + \frac{3}{2} - \frac{\sqrt{7}}{2}i\right)y = 0. \tag{5.91}$$

We can, of course, write down the primitive just have we have done for both eqs. (5.79) and (5.80). However, this is a case where Euler's relation will be useful; you may recall

$$e^{(a+ib)x} = e^{ax}(\cos bx + i \sin bx). \tag{5.92}$$

This will allow us to put our result into more appropriate form:

$$y = C_1 e^{-\frac{3}{2}t}\left(\cos\frac{\sqrt{7}}{2}x + i \sin\frac{\sqrt{7}}{2}x\right) + C_2 e^{-\frac{3}{2}x}\left(\cos\left(-\frac{\sqrt{7}}{2}x\right) + i \sin\left(-\frac{\sqrt{7}}{2}x\right)\right). \tag{5.93}$$

Since *cosine* is an even function and *sine* is odd, we can write

$$y = C_1 e^{-\frac{3}{2}x}\left(\cos\frac{\sqrt{7}}{2}x + i \sin\frac{\sqrt{7}}{2}x\right) + C_2 e^{-\frac{3}{2}x}\left(\cos\frac{\sqrt{7}}{2}x - i \sin\frac{\sqrt{7}}{2}x\right). \tag{5.94}$$

For a real (physical) problem such as one arising through a force balance, the imaginary parts will cancel, leaving a real result. We can very conveniently confirm the result, eq. (5.94), by using *deSolve(....)* on the *Texas Instruments* TI-89™; we obtain

$$y = C_1 \exp\left(-\frac{3x}{2}\right)\cos\frac{\sqrt{7x}}{2} + C_2 \exp\left(-\frac{3x}{2}\right)\sin\frac{\sqrt{7x}}{2}.$$

The procedure we have sketched above is valid for comparable equations of any order. For example, consider

$$\frac{d^4y}{dx^4} + 2\frac{d^3y}{dx^3} + 3\frac{d^2y}{dx^2} + 5\frac{dy}{dx} + 8y = 0. \tag{5.95}$$

In this case we have two pair of complex conjugates (roots), which are $D + 0.457547 \pm 1.56529i$ and $D - 1.45755 \pm 0.940024i$. And again we can immediately write down the primitive by inspection. We should note the implication of this result; if eq. (5.95) described a real system, part of the oscillatory response would be multiplied by exp(+1.45755*x*)—we have an unbounded *y*(*x*).

Unfortunately, not every ODE of higher order will be homogeneous; often if the differential equation represents some dynamic process, a forcing function will be applied on the right-hand side to drive the model. For example, we might have

$$\frac{d^2y}{dx^2} + 3\frac{dy}{dx} + 2y = 5 \tag{5.96}$$

In such cases the solution is the sum of the primitive for the *homogeneous* equation—which we refer to as the complementary function (*CF*)—plus a particular integral (*PI*). Thus, for eq. (5.96) above, we have

$$y = C_1 \exp(-2x) + C_2 \exp(-x) + PI. \tag{5.97}$$

Many times the form of the particular integral can be determined by inspection. For example, for eq. (5.96), once the *dynamic* behavior is completed the terms involving derivatives will be zero (for large *x*, the exponentials disappear leaving just the *PI*). Therefore, the *PI* must be 5/2, and the solution can be written

$$y = C_1 \exp(-2x) + C_2 \exp(-x) + \frac{5}{2}. \tag{5.98}$$

We should actually complete this problem by including boundary (or initial) conditions. We assume that the physical process described by this model was drifting along at an equilibrium state until the independent variable, *x*, or very commonly time, *t*, was zero, when the forcing function (the constant 5) was applied. Of course, this means that both *y* and its derivative can be set to zero for $x = 0$—we can always do this for models of this type by using *deviation* variables. For example, if y_{eq} is the equilibrium (or steady) value of the dependent variable, then we simply write:

$$Y = y - y_{eq}. \tag{5.99}$$

This guarantees that the dependent variable was zero at the instant the forcing function was applied. Consequently, we will now assume that the dependent variable was in fact written in this form, such that

$$y(x = 0) = 0 \quad \text{and} \quad \left.\frac{dy}{dx}\right|_{x=0} = 0. \tag{5.100}$$

Therefore,

$$C_1 + C_2 + \frac{5}{2} = 0 \quad \text{and} \quad C_2 = -2C_1. \tag{5.101}$$

For these conditions the solution for our problem becomes:

$$y = \frac{5}{2}\exp(-2x) - \frac{10}{2}\exp(-x) + \frac{5}{2}. \tag{5.102}$$

Of course, the inhomogeneity does not have to be a simple constant as it was in the previous example. Let us examine a method that allows us to deal with cases in which the inhomogeneity is a function of x. For example,

$$\frac{d^2y}{dx^2} + 4\frac{dy}{dx} + 4y = x. \tag{5.103}$$

Again, we find the complementary function from the *homogeneous version* of eq. (5.103) and add the particular integral:

$$y = C_1\exp(-2x) + C_2 x\exp(-2x) + PI. \tag{5.104}$$

The technique we will employ for determination of this particular integral is called *variation of parameters*. We begin by rewriting the complementary function, but we insert unknown functions of x in place of the constants of integration that appear in eq. (5.104):

$$y = L_1(x)\exp(-2x) + L_2(x)x\exp(-2x) \tag{5.105}$$

We now differentiate with respect to x, obtaining

$$Dy = L_1'\exp(-2x) - 2L_1\exp(-2x) + L_2' x\exp(-2x) + L_2\exp(-2x) - 2L_2 x\exp(-2x) \tag{5.106}$$

The terms involving derivatives (of L_n's) are set equal to zero:

$$L_1'\exp(-2x) + L_2' x\exp(-2x) = 0. \tag{5.107}$$

Now we differentiate again, retaining only those terms involving derivatives of the L_n's. But this time, the order of the derivative (2) matches the order of the ODE, so the sum is set equal to the inhomogeneity (which is x); thus

$$-2L_1'\exp(-2x) + L_2'\exp(-2x) - 2L_2' x\exp(-2x) = x. \tag{5.108}$$

By virtue of eq. (5.107) the sum of the first and third terms is zero, leaving us with two equations to solve,

$$L_2' = x\exp(+2x) \quad \text{and} \quad L_1' = -x^2\exp(+2x). \tag{5.109}$$

These equations are integrated and taken back to eq. (5.105). The reader should show that the particular integral turns out to be $PI = \frac{x}{4} - \frac{1}{4}$; therefore, the solution we were seeking is written:

$$y = C_1\exp(-2x) + C_2 x\exp(-2x) + \frac{1}{4}(x-1). \tag{5.110}$$

Since $y(x=0) = 0$, $C_1 = 1/4$. The first derivative is also zero at $x = 0$, so it is easy to show that $C_2 = 1/4$ as well. We can easily confirm our solution using "*deSolve*" on a *TI-nspire* which results in: $(c_1 x + c_2)\exp(-2x) + x/4 - 1/4$.

5.4.1 The LRC Circuit

Let us look at how a second-order ODE with constant coefficients might arise using an elementary DC circuit for the example. An inductor, resistor, and capacitor are connected in series and a constant voltage, V, is applied to the circuit. The dynamic behavior of the circuit can be determined with Kirchhoff's laws where the inductance, resistance, and capacitance are L, R, and C, respectively. The governing second-order ODE is:

$$L\frac{d^2q}{dt^2} + R\frac{dq}{dt} + \frac{1}{C}q = V. \tag{5.111}$$

Since the current in the circuit, I, is $I = dq/dt$, we can differentiate eq. (5.111) to obtain

$$L\frac{d^2I}{dt^2} + R\frac{dI}{dt} + \frac{1}{C}I = \frac{dV}{dt}. \tag{5.112}$$

Now suppose we let $L = 0.05$ H, $R = 15\,\Omega$, and $C = 0.025$ F, and apply 20 V to the circuit; eq. (5.111) becomes

$$\frac{d^2q}{dt^2} + 300\frac{dq}{dt} + 800q = 400. \tag{5.113}$$

It is evident from eq. (5.113) that a particular solution is $40q = 20$, or $q = 1/2$. The left-hand side of eq. (5.113) is rewritten using linear differential operator notation:

$$(D^2 + 300D + 800)q = 0,$$

and we use the quadratic formula to find the roots which are −2.69 and −297.31. Consequently,

$$q = \frac{1}{2} + c_1 \exp(-2.69t) + c_2 \exp(-297.31t). \tag{5.114}$$

The initial conditions to be applied are $q(t = 0) = 0$ and $q'(t = 0) = 0$, and therefore we have two simultaneous equations:

$$-\frac{1}{2} = c_1 + c_2 \quad \text{and} \quad 0 = -2.69c_1 - 297.31c_2.$$

The reader may want to verify that $c_2 = 0.004565$. Let's examine our result eq. (5.114) for a moment; note that there are two characteristic times, 1/2.69 and 1/297.31, and they differ by about two orders of magnitude (0.3717 and 0.00336). What does this tell us about the dynamic behavior of both q and I? It may be clear to you that $q(t)$ will exhibit a point of inflection corresponding to the maximum current in the circuit (I will initially increase, and then diminish as charge accumulates on the capacitor). The reader should confirm that I_{max} occurs at $t = 0.01597$ s, or just under 16 msec.

Problems of a type similar to the example immediately above are excellent candidates for solution through use of the Laplace transform, and it is worthwhile for us to spend a little effort reviewing that technique. But keep in mind that the procedure we are about to explore can only be used to solve a very limited class of ordinary differential equations—we are restricted to *linear* ODEs with *constant coefficients*.

5.4.2 Use of the Laplace Transform for Solution of ODEs

We normally think of the Laplace transform within the context of problems involving change or evolution in time. Accordingly, for decades the Laplace transform was a staple of classical linear process control as it allowed the control engineer to explore the dynamic behavior of a system in the s-plane using only algebraic manipulations. With this perspective in mind, we will use t as our standard independent variable in this section. The formal definition of the Laplace transform of a function, $f(t)$, follows:

$$f(s) = \int_0^\infty f(t)e^{-st}dt. \tag{5.115}$$

This transform is often written more compactly as $f(s) = L\{f(t)\}$. Notice that the transformation takes us from the time domain to the s-plane. Let us actually apply the definition to something—a numerical constant for example: We let $f(t) = 1$, then

$$f(s) = \int_0^\infty (1)e^{-st}dt = -\frac{1}{s}\, e^{-st}\Big|_0^\infty = -\frac{1}{s}(0 - 1) = \frac{1}{s}. \tag{5.116}$$

We envision needing to find the transform for many different types of terms, and we will use the formal definition one more time, applying it to the independent variable, t:

$$f(s) = \int_0^\infty te^{-st}dt = \frac{e^{-st}}{s^2}\,(-st - 1)\Big|_0^\infty = \frac{1}{s^2}. \tag{5.117}$$

If our focus is the solution of ordinary differential equations, then we must also be able to transform derivatives of various orders. For example,

$$L\left\{\frac{df(t)}{dt}\right\} = sf(s) - f(t = 0), \tag{5.118}$$

$$L\left\{\frac{d^2f(t)}{dt^2}\right\} = s^2f(s) - sf(t = 0) - f'(t = 0), \tag{5.119}$$

and more generally,

$$L\left\{\frac{d^nf(t)}{dt^n}\right\} = s^nf(s) - s^{n-1}f(t = 0) - s^{n-2}f'(t = 0) - s^{n-3}f''(t = 0) \ldots$$

The reader will observe that in the transformation of derivatives, differentiation is replaced by multiplication by s. However, we must also subtract off the initial conditions (in the time domain). This makes it clear why it is so important (and convenient) to put problems into deviation form whenever possible. Of course, we do not actually use the formal definition of the Laplace transform, eq. (5.115), in practice since nearly every conceivable form of function has already been transformed. We merely need to look them up in a suitable reference; one very useful source is Abramowitz and Stegun (1972). Here is an abbreviated listing of some very common transforms.

***f*(*t*)**	***L*{*f*(*t*)}**
1	$\frac{1}{s}$
t	$\frac{1}{s^2}$
$\exp(-at)$	$\frac{1}{s+a}$
$t^n \exp(-at)$	$\frac{n!}{(s+a)^{n+1}}$
$\exp(-at)\sin(bt)$	$\frac{b}{(s+a)^2+b^2}$
$\sin(at)$	$\frac{a}{s^2+a^2}$
$\cos(at)$	$\frac{s}{s^2+a^2}$
$t\cos at$	$\frac{s^2-a^2}{(s^2+a^2)^2}$
$\exp(-at)\cos(bt)$	$\frac{s+a}{(s+a)^2+b^2}$
$\sinh(at)$	$\frac{a}{s^2-a^2}$
$\cosh(at)$	$\frac{s}{s^2-a^2}$
$\frac{1}{2a^2}\sin(at)\sinh(at)$	$\frac{s}{s^4+4a^4}$
$\frac{1}{\sqrt{\pi t}}\cosh(\sqrt{4kt})$	$\frac{1}{\sqrt{s}}\exp\left(\frac{k}{s}\right)$
$J_0(\sqrt{4kt})$	$\frac{1}{s}\exp\left(-\frac{k}{s}\right)$
$erfc\left(\frac{k}{\sqrt{4t}}\right)$	$\frac{1}{s}\exp(-k\sqrt{s})$ where $k \geq 0$
$\sqrt{\frac{4t}{\pi}}\exp\left(-\frac{k^2}{4t}\right)-(k)erfc\left(\frac{k}{\sqrt{4t}}\right)$	$\frac{1}{s^{3/2}}\exp(-k\sqrt{s})$ where $k \geq 0$
$\frac{1}{\sqrt{\pi t}}\exp\left(-\frac{k^2}{4t}\right)$	$\frac{1}{\sqrt{s}}\exp(-k\sqrt{s})$ where $k \geq 0$

We will illustrate the application of the Laplace transform to the solution of ordinary differential equations (with constant coefficients) by reconsidering the previous example, eq. (5.103). However, we will change the independent variable to t (from x) in order to be consistent with the table of transforms above. Accordingly,

$$\frac{d^2y}{dt^2}+4\frac{dy}{dt}+4y=t. \tag{5.120}$$

Applying the transform to each term we get:

$$y(s)(s^2 + 4s + 4) = \frac{1}{s^2}, \tag{5.121}$$

and therefore,

$$y(s) = \frac{1}{s^2(s+2)(s+2)}. \tag{5.122}$$

To solve such problems the old-fashioned way (manually) we use partial fraction expansion. There is a slight complication in this case since we have repeated roots that will require us to write:

$$\frac{1}{s^2(s+2)^2} = \frac{A}{s^2} + \frac{B}{s} + \frac{C}{(s+2)^2} + \frac{D}{s+2}. \tag{5.123}$$

We begin by multiplying by s^2 and setting $s = 0$; we find immediately that $A = 1/4$. In order to find B, we return to eq. (5.123) and multiply by s^2 again. But this time we differentiate with respect to s, which isolates B, and we find $B = -1/4$. We proceed analogously for the remaining two terms and find $C = 1/4$ and $D = 1/4$. Therefore, we now know the right-hand side of eq. (5.123):

$$y(s) = \frac{1/4}{s^2} - \frac{1/4}{s} + \frac{1/4}{(s+2)^2} + \frac{1/4}{s+2}. \tag{5.124}$$

Our work is almost complete; we need only to consult the abbreviated list of transforms provided above so that we can write $y(t)$:

$$y(t) = \frac{1}{4}t - \frac{1}{4} + \frac{1}{4}t\exp(-2t) + \frac{1}{4}\exp(-2t). \tag{5.125}$$

Compare this result with eq. (5.110); you will see that they are identical (except for the change in independent variable). Our use of the Laplace transform has significantly reduced the work necessary to find the solution for the ODE eq. (5.103).

We also want to illustrate the procedure with a variation that arises regularly with this type of ordinary differential equation. Consider the second-order ODE:

$$\frac{d^2y}{dt^2} + 3\frac{dy}{dt} + 5y = 2 \tag{5.126}$$

with $y(t = 0) = 0$ and $y'(t = 0) = 0$. Proceeding as before:

$$s^2y(s) + 3sy(s) + 5y(s) = \frac{2}{s}, \tag{5.127}$$

and

$$y(s) = \frac{2}{s(s^2 + 3s + 5)} = \frac{2}{s\left(s + \frac{3}{2} + \frac{\sqrt{11}}{2}i\right)\left(s + \frac{3}{2} - \frac{\sqrt{11}}{2}i\right)}. \tag{5.128}$$

Just as we did previously, we expand the right-hand side,

$$y(s) = \frac{A}{s} + \frac{B}{s + \frac{3}{2} + \frac{\sqrt{11}}{2}i} + \frac{C}{s + \frac{3}{2} - \frac{\sqrt{11}}{2}i}. \tag{5.129}$$

The reader may wish to verify that $A = 2/5$, $B = -0.150943–0.068267i$, and $C = -0.150943 + 0.068267i$. It remains for us to take the following three terms back to the time domain:

$$y(s) = \frac{\frac{2}{5}}{s} + \frac{-0.150943 - 0.068267i}{s + \frac{3}{2} + \frac{\sqrt{11}}{2}i} + \frac{-0.150943 + 0.068267i}{s + \frac{3}{2} - \frac{\sqrt{11}}{2}i}. \tag{5.130}$$

At this point you should be able to identify the essentials of this system's behavior in the time domain: We will obtain a constant (2/5) combined with an exponentially damped $\exp\left(-\frac{3}{2}t\right)$ oscillatory response. Specifically, we find

$$y(t) = \frac{2}{5} - 0.4\exp\left(-\frac{3}{2}t\right)\cos\frac{\sqrt{11}}{2}t - 0.361814\exp\left(-\frac{3}{2}t\right)\sin\frac{\sqrt{11}}{2}t. \tag{5.131}$$

The behavior produced by eq. (5.131) is illustrated in Figure 5.5.

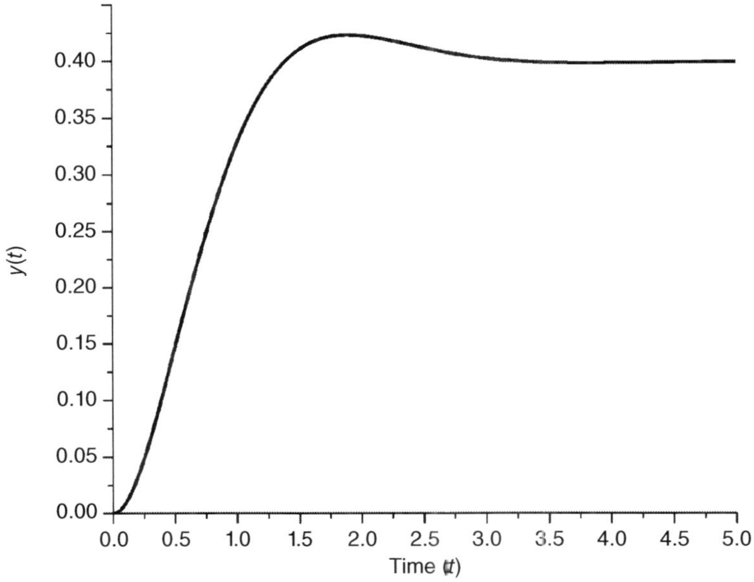

Figure 5.5 Characteristic behavior of a slightly underdamped second-order system as produced by the solution, eq. (5.131). Note the limited overshoot and the rapidly damped oscillation about the ultimate value for $y(t)$ of 0.4.

5.4.3 Finding Time-Domain Response by Inversion of Laplace Transform

We pointed out previously that the Laplace transform has been used extensively in linear control theory as it allows the analyst to work algebraically in the s-plane rather than dealing with a system of differential equations in the time domain. Let's look at a detailed example that will clarify what we mean. Suppose we have a process (described by a differential equation) that is subjected to negative feedback control. At $t = 0$ we impose a unit step change on the controller—we change the set point from zero to 1. The controller will try to adjust the process (for example, by changing the stem position on a valve) such that the new loop output conforms to 1. We proceed by applying the Laplace transform to all of the elements in the loop and by using loop algebra to obtain the obtain the response variable, that is, the loop output, $C(t)$, as a function of s:

$$C(s) = \frac{K}{s^3 + s^2 + Ks}. \tag{5.132}$$

K is the controller parameter. If we assign K a particular value (say, $K = 6$) we can invert $C(s)$ and find the dynamic behavior in the time domain:

$$C(s) = \frac{6}{s^3 + s^2 + 6s}. \tag{5.133}$$

First, we break down the polynomial in the denominator into linear factors using *TI*'s *cSolve*, ignoring the fact that we could factor out s initially:

$$s = -\frac{1}{2} \pm \frac{\sqrt{23}}{2}i \quad \text{and} \quad s = 0.$$

Therefore,

$$\frac{6}{s\left(s + ½ + \frac{\sqrt{23}}{2}i\right)\left(s + ½ - \frac{\sqrt{23}}{2}i\right)} = \frac{A}{s} + \frac{B}{s + ½ + \frac{\sqrt{23}}{2}i} + \frac{C}{s + ½ - \frac{\sqrt{23}}{2}i}. \tag{5.134}$$

We multiply by s and set $s = 0$; this reveals $A = 6/(24/4) = 1$. Next we multiply by $s + ½ + \frac{\sqrt{23}}{2}i$ and set $s = -½ - \frac{\sqrt{23}}{2}i$, this shows that $B = -½ - 0.1043i$. Of course C must be the other half of the complex conjugate pair, so our work here is nearly done. Consulting our table of Laplace transforms we find that if

$$L\{f(t)\} = \frac{1}{s + a} \text{ then } f(t) = \exp(-at). \tag{5.135}$$

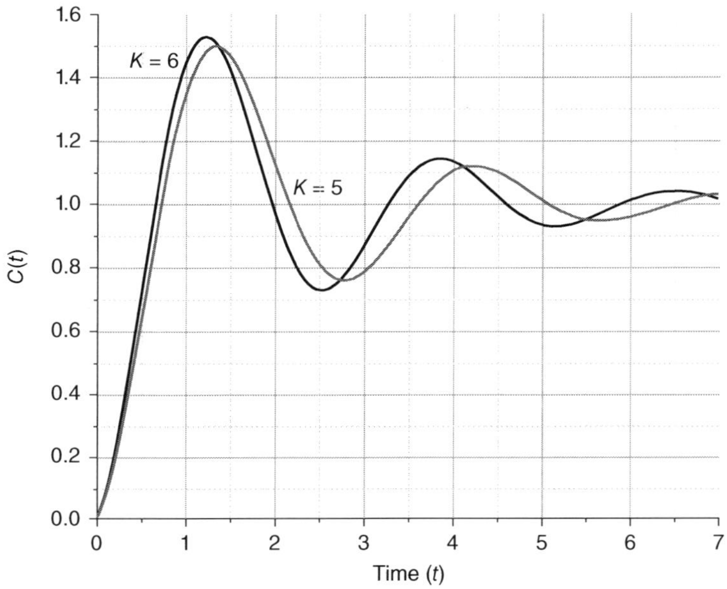

Figure 5.6 Response of negative feedback control loop to a unit step change in set point for $K = 5$ and $K = 6$.

Consequently, for the second term in the expansion:

$$\left(-\frac{1}{2} - 0.1043i\right) \exp\left[\left(-\frac{1}{2} - \frac{\sqrt{23}}{2}i\right)t\right]. \tag{5.136}$$

Using the Euler relation and noting that *cos* is an even function and *sin* is odd:

$$C(t) = 1 - \exp\left(-\frac{1}{2}t\right)\cos\left(\frac{\sqrt{23}}{2}t\right). \tag{5.137}$$

The response of the loop to the set point change is oscillatory but exponentially damped, and ultimately $C(t \to \infty) = 1$. Our control scheme may not be optimal, but it is working as intended. Now suppose we wanted to explore the effects of changing the controller setting upon loop response; for example, let's set $K = 5$ and repeat the process. See if you can verify that

$$C(t) = 1 - \exp\left(-\frac{1}{2}t\right)\cos\left(\frac{\sqrt{19}}{2}t\right). \tag{5.138}$$

We should note that the speed of response (that is, the radian frequency of the oscillation) has been diminished—it is now 2.179 rad per time unit as opposed to 2.398, a 9% decrease. The effect of a larger K is very clear in Figure 5.6.

To conclude this section, it is essential that we reiterate that the Laplace transform is a linear operator, and it can only be used to *solve linear, ordinary differential equations with constant coefficients*. Historically this method was of enormous importance in the development of automatic process control because *idealized* models for the dynamic behavior of processes could be dealt with algebraically (and the computational solution of differential equations could be avoided). But the restriction of the Laplace transform method to linear ODEs with constant coefficients is a significant one since so many of the equations of interest to us in applied mathematics have variable coefficients (or are nonlinear).

5.5 Higher-Order Equations with Variable Coefficients

We want to focus upon second-order ODEs of the form:

$$\frac{d^2y}{dx^2} + P(x)\frac{dy}{dx} + Q(x)y = R(x). \tag{5.139}$$

This differential equation is linear, but the coefficients are not constant. We must preface this part of our discussion with a warning: *There is no general procedure that will permit the analyst to solve every differential equation of this type.* Ayres (1952) laid out a plan of attack for such equations that will sometimes yield success, and we will use the strategy he recommends on the following example. Let us suppose the equation of interest is:

$$\frac{d^2y}{dx^2} - \frac{3}{x}\frac{dy}{dx} + \frac{3}{x^2}y = 2x - 1. \tag{5.140}$$

Our plan is to render the equation homogeneous (removing the $2x-1$ from the right-hand side), and then see if we identify a particular integral, which we will call $u(x)$. Consider trying $PI = x$; the second derivative is zero, the first derivative is 1, and we find $-\frac{3}{x} + \frac{3}{x^2}x = 0$, so the function x is a particular integral for this equation. We now let y be rewritten as the product of the particular integral and a new dependent variable, v: $y = xv$. Therefore,

$$\frac{dy}{dx} = v + x\frac{dv}{dx} \quad \text{and} \quad \frac{d^2y}{dx^2} = 2\frac{dv}{dx} + x\frac{d^2v}{dx^2}. \tag{5.141}$$

Of course this leads to:

$$2\frac{dv}{dx} + x\frac{d^2v}{dx^2} - \frac{3}{x}\left(v + x\frac{dv}{dx}\right) + \frac{3}{x^2}(xv) = 2x - 1, \tag{5.142}$$

which means that the terms in which v is divided by x cancel, leaving us with just

$$x\frac{d^2v}{dx^2} - \frac{dv}{dx} = 2x - 1. \tag{5.143}$$

Anytime we are confronted by an ordinary differential equation that includes *only* derivatives, we should immediately think about reducing the order. In this case we achieve this by letting $\beta = \frac{dv}{dx}$, resulting in

$$\frac{d\beta}{dx} - \frac{1}{x}\beta = 2 - \frac{1}{x}. \tag{5.144}$$

This first-order ordinary differential equation is familiar—see eq. (5.35) in Section 5.2 for example. We know that an integrating factor for this equation is $e^{\int -\frac{1}{x}dx} = e^{-\ln x} = \frac{1}{x}$. Consequently,

$$\frac{d}{dx}\left(\frac{1}{x}\beta\right) = \frac{2}{x} - \frac{1}{x^2}, \tag{5.145}$$

and therefore,

$$\beta = \frac{dv}{dx} = 2x\ln x + C_1x + 1. \tag{5.146}$$

We integrate with respect to x,

$$v = 2\left[\frac{x^2}{2}\ln x - \frac{x^2}{4}\right] + \frac{C_1}{2}x^2 + x + C_2 \tag{5.147}$$

and since $y = xv$,

$$y = x^3[\ln x + C_1] + x^2 + C_2x. \tag{5.148}$$

The procedure we sketched above *will only work if* we can identify a particular integral for the homogeneous equation. Obviously we might fail to find a *PI* and in that case Ayres suggests the following: Return to eqs. (5.139) and (5.140) and compute $Q - \frac{1}{4}P^2 - \frac{1}{2}\frac{dP}{dx}$; if this difference is a constant or alternatively a constant divided by x, then the transformation,

$$y = v\exp\left(-\frac{1}{2}\int P(x)dx\right) \tag{5.149}$$

will reduce the ODE to a linear equation with constant coefficients (or to a Cauchy equation of the form

$$A_nx^n\frac{d^ny}{dx^n} + A_{n-1}x^{n-1}\frac{d^{n-1}y}{dx^{n-1}} + \cdots = F(x)).$$

Let us see how this might work in practice using the following equation for our example:

$$\frac{d^2y}{dx^2} - 4x\frac{dy}{dx} + 4x^2y = x\exp(x^2). \tag{5.150}$$

We see $Q = 4x^2$ and $P = -4x$; therefore,

$$Q - \frac{1}{4}P^2 - \frac{1}{2}\frac{dP}{dx} = 4x^2 - \frac{1}{4}(16x^2) - \frac{1}{2}(-4) = 2.$$

Accordingly, we let

$$y = v\exp\left(-\frac{1}{2}\int Pdx\right) = v\exp(x^2).$$

Proceeding, we differentiate appropriately and substitute into eq. (5.150). All terms involving dv/dx cancel, as well as the terms that involve x^2v, leaving us with:

$$\frac{d^2v}{dx^2} + 2v = x. \tag{5.151}$$

By inspection the *PI* is $x/2$ and from the homogeneous equation we observe

$$(D^2 + 2)v = (D + \sqrt{2}i)(D - \sqrt{2}i)v = 0. \tag{5.152}$$

Thus the solution for v is:

$$v = C_1\exp(-\sqrt{2}ix) + C_2\exp(+\sqrt{2}ix) + \frac{x}{2}. \tag{5.153}$$

Of course we must remember to multiply eq. (5.153) by $\exp(x^2)$ to obtain our original dependent variable, y. We can also use the Euler relation to rewrite the solution in terms of cosine and sine, should we desire to do so; for example, since $e^{(a+ib)t} = e^{at}(\cos bt + i\sin bt)$, then $\exp(-\sqrt{2}ix) = \cos(\sqrt{2}x) - i\sin(\sqrt{2}x)$.

5.6 Bessel's Equation and Bessel Functions

Among the more important second-order ordinary differential equations with variable coefficients appearing in applied mathematics is *Bessel's equation*. The solution for Bessel-type ordinary differential equations can be written in terms of Bessel functions, and *the definitive source* of information regarding Bessel functions is *A Treatise on Bessel Functions and Their Applications to Physics* by Gray and Mathews (and later, Gray et al., 1931). The reader should be aware of the fact that Bessel functions are often referred to—quite appropriately—as "cylinder functions" in the older German literature. The form for Bessel's ODE seen frequently throughout applied mathematics is

$$x^2\frac{d^2y}{dx^2} + x\frac{dy}{dx} + (x^2 - n^2)y = 0, \tag{5.154}$$

where $n \geq 0$. The general solution for this particular form is

$$y = C_1J_n(x) + C_2Y_n(x), \tag{5.155}$$

where J is the Bessel function of the first kind and Y is Neumann's Bessel function of the second kind (both of order n). However, the form of eq. (5.154) is quite restrictive, and we should recognize that many ODEs with forms *similar* to eq. (5.154) have solutions that can be written in terms of Bessel functions. In fact, we need to make the following observation: *Whenever we encounter a radially directed flux in cylindrical coordinates*, the operator:

$$\frac{1}{r}\frac{d}{dr}\left(r\frac{d\phi}{dr}\right)$$

will arise. Depending upon the exact nature of the problem, this can result in some form of Bessel's differential equation, which for the generalized case, can be written as shown by Mickley et al. (1957):

$$r^2\frac{d^2\phi}{dr^2} + r(a + 2br^v)\frac{d\phi}{dr} + [c + dr^{2s} - b(1 - a - v)r^v + b^2r^{2v}]\phi = 0. \tag{5.156}$$

For many real, physical problems in applied mathematics we find that $a = 1$, $b = 0$, and $c = 0$. The nature of the solution is then determined by the sign of d. If the $\sqrt{d}$ is real then the solution is written in terms of J_n or J_n plus Y_n. If $\sqrt{d}$ is imaginary, then the solution will be either I_n or I_n plus K_n. In more general cases the Bessel functions are preceded by the product, $r^{\frac{1-a}{2}} \exp\left(-\frac{br^v}{r}\right)$, but as noted, often $a = 1$ and $b = 0$ so this prefactor is just 1 (one). We will examine a scenario with a different outcome at the end of this section. The order of the Bessel functions, n, is determined by:

$$n = \frac{1}{s}\sqrt{\left(\frac{1-a}{2}\right)^2 - c}.$$

As an illustration, consider steady conduction of thermal energy in an infinitely long cylinder with a production term that is linear with respect to temperature. The governing differential equation has the form:

$$r^2\frac{d^2T}{dr^2} + r\frac{dT}{dr} + r^2\frac{\gamma T}{k} = 0, \tag{5.157}$$

where γ is a positive constant. Note that $a = 1$, $b = 0$, $c = 0$, $s = 1$, and $d = \gamma/k$. In this case the solution is:

$$T = AJ_0\left(\sqrt{\frac{\gamma}{k}}r\right) + BY_0\left(\sqrt{\frac{\gamma}{k}}r\right). \tag{5.158}$$

For a solid cylindrical domain, $T(r = 0)$ would have to be finite and therefore $B = 0$ (since $Y_0(0) = -\infty$). But of course for an *annular region*, no boundary condition could be written for $r = 0$, and both terms (A *and* B) would remain in the solution. Note that if the production term in eq. (5.157) were replaced by a sink (disappearance) term, then γ/k would have been negative and the solution would have been written in terms of the modified Bessel functions, I_0 and K_0. To illustrate this, consider a catalytic reaction in a long, cylindrical pellet; the reactant species, "A" is being consumed by a first-order reaction. A homogeneous model (for heterogeneous catalysis) results in the differential equation:

$$r^2\frac{d^2C_A}{dr^2} + r\frac{dC_A}{dr} - r^2\frac{k_1a}{D_{eff}}C_A = 0 \tag{5.159}$$

with the solution:

$$C_A = AI_0\left(\sqrt{\frac{k_1a}{D_{eff}}}r\right) + BK_0\left(\sqrt{\frac{k_1a}{D_{eff}}}r\right). \tag{5.160}$$

Note that k_1 is the rate constant and a is the specific surface area of the catalyst; D_{eff} is the "effective" diffusivity, an empirical quantity that arises due to our approximate treatment of a catalytic process that is heterogeneous. Let us apply eq. (5.160) to a situation where the concentration of reactant at the surface ($r = R$) is 1 and the concentration at the center of the cylindrical catalyst pellet is finite. We will assume that $\sqrt{\frac{k_1a}{D_{eff}}} = 6$ and that $R = 1$. Before we begin to work on the constants of integration, A and B, we ought to know a little about the behavior of these Bessel functions. Therefore, a short table of numerical values is provided below for zero-order Bessel function of the first and second kinds, as well as the modified Bessel functions I_0 and K_0; more extensive tables are provided by Abramowitz and Stegun (1972). *Note that neither Y_0 nor K_0 can be part of the solution for a problem in cylindrical coordinates if the field variable (V, T, or C_A) is finite at the center ($r = 0$).*

An abbreviated table of zero-order Bessel functions.

R	$J_0(r)$	$Y_0(r)$	$I_0(r)$	$K_0(r)$
0.0	1	$-\infty$	1	∞
0.2	0.99	−1.0811	1.01	1.7527
0.4	0.9604	−0.606	1.0404	1.1145
0.6	0.912	−0.3085	1.092	0.7775
0.8	0.8463	−0.0868	1.1665	0.5653
1.0	0.7652	0.0883	1.2661	0.421
1.2	0.6711	0.2281	1.3937	0.3185
1.4	0.5669	0.3379	1.5534	0.2437
1.6	0.4554	0.4204	1.7500	0.188

(Continued)

R	$J_0(r)$	$Y_0(r)$	$I_0(r)$	$K_0(r)$
1.8	0.34	0.4774	1.9896	0.1459
2.0	0.2239	0.5104	2.2796	0.1139
2.2	0.1104	0.5208	2.6291	0.0893
2.4	0.0025	0.5104	3.0493	0.0702
2.6	−0.0968	0.4813	3.5533	0.0554
2.8	−0.185	0.4359	4.1573	0.0438
3.0	−0.2601	0.3769	4.8808	0.0347
3.2	−0.3202	0.3071	5.7472	0.0276
3.4	−0.3643	0.2296	6.7848	0.022
3.6	−0.3918	0.1477	8.0277	0.0175
3.8	−0.4026	0.0645	9.5169	0.0139
4.0	−0.3971	−0.0169	11.302	0.0112
4.2	−0.3766	−0.0938	13.443	0.0089
4.4	−0.3423	−0.1633	16.010	0.0071
4.6	−0.2961	−0.2235	19.093	0.0057
4.8	−0.2404	−0.2723	22.794	0.0046
5.0	−0.1776	−0.3085	27.239	0.0037
5.2	−0.11029	−0.33125	32.584	0.00297
5.4	−0.04121	−0.34017	39.009	0.002385
5.6	0.02697	−0.33544	46.738	0.00192
5.8	0.0917	−0.317746	56.038	0.00154
6.0	0.15065	−0.28819	67.234	0.00124
6.2	0.20174	−0.24831	80.718	0.001
6.4	0.24331	−0.19995	96.962	0.00081
6.6	0.27404	−0.14523	116.54	0.00065
6.8	0.2931	−0.08643	140.14	0.00053
7.0	0.3001	−0.02595	168.59	0.00042
7.2	0.29507	0.03385	202.92	0.000343
7.4	0.2786	0.09068	244.34	0.000277
7.6	0.2516	0.1424	294.33	0.0002
7.8	0.2154	0.1872	354.69	0.000181
8.0	0.1717	0.2235	427.56	0.000146
8.2	0.1222	0.25012	515.59	0.000118
8.4	0.06916	0.26622	621.94	0.000096
8.6	0.01462	0.27146	750.5	0.000077
8.8	−0.0392	0.26587	905.8	0.000063
9.0	−0.0903	0.2498	1094	0.000051
9.2	−0.13675	0.22449	1321	0.000041
9.4	−0.17677	0.19074	1595	0.000033
9.6	−0.20898	0.15018	1927	0.0000271
9.8	−0.23277	0.10453	2329	0.0000219
10.0	−0.2459	0.05567	2816	0.0000178

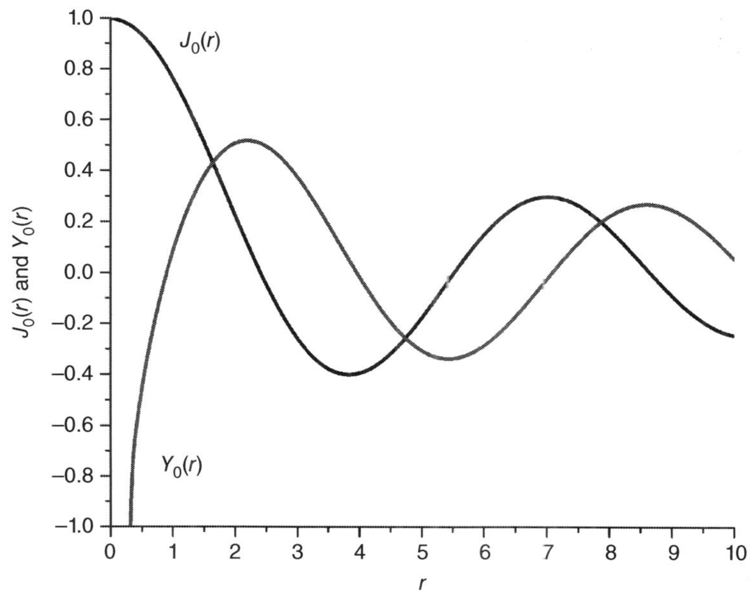

Figure 5.7 Bessel functions $J_0(r)$ and $Y_0(r)$ for r from 0 to 10.

$J_0(r)$ and $Y_0(r)$ are also shown graphically in Figure 5.7.

Now we are ready to proceed with our example. Since $K_0(0) = \infty$, we note $B = 0$. Using the surface concentration, $C_A(r = 1) = 1$, we can compute A:

$$A = \frac{1}{I_0\left(\sqrt{\frac{k_1 a}{D_{eff}}} R\right)}. \tag{5.161}$$

From the table above we find $I_0(6) = 67.234$, and we can then calculate the following values for the concentration distribution in the interior of the catalyst pellet:

r	0.0	0.1	0.2	0.4	0.6	0.8	1.0
C_A	0.01487	0.01624	0.02073	0.04535	0.1194	0.3390	1.000

Naturally it would be more efficient in terms of reactant conversion to use an *annular geometry* for the pellet (providing increased surface area per unit volume, and if you are familiar with separation processes, think of a *Raschig ring*) so we should also look at the case where the reactant can flow both through the center of the annulus and past the outer surface. We let the two radii be denoted by the subscripts, 1 and 2, respectively, such that $R_1 \leq r \leq R_2$, and we set the concentrations at both (inner and outer) surfaces to 1. In this case K_0 must be retained in the solution. Again we let $\sqrt{\frac{k_1 a}{D_{eff}}} = 6$, just as before, and the constants of integration are found to be $A = 801.011$ and $B = 5.3 \times 10^{-5}$.

r	1.0	1.2	1.4	1.5	1.6	1.8	2.0
C_A	1.000	0.2856	0.1098	0.0987	0.1238	0.3254	1.000

Notice that we do not have symmetry about the mid-point of the annular region ($r = 1.5$). This is to be expected because of the greater surface area available to the reactant at the outer surface, $r = 2$.

The need to differentiate Bessel functions arises frequently, particularly when a boundary condition involves a specified flux (Neumann or Robin's type BC). For J, Y, and K, we have

$$\frac{d}{dr}[Z_p(\alpha r)] = -\alpha Z_{p+1}(\alpha r) + \frac{p}{r} Z_p(\alpha r). \tag{5.162}$$

Accordingly, we note that:

$$\frac{d}{dr}[J_0(\beta r)] = -\beta J_1(\beta r) \text{ since } p = 0. \tag{5.163}$$

For I_p we have:

$$\frac{d}{dr}[I_p(\alpha r)] = \alpha I_{p+1}(\alpha r) + \frac{p}{r}I_p(\alpha r). \tag{5.164}$$

To illustrate, suppose we have to differentiate $J_0(\lambda r)$; by eq. (5.162), $\frac{d}{dr}[J_0(\lambda r)] = -\lambda J_1(\lambda r)$. Now assume that $\lambda = 1$ and that $r = 1.5$; if we use Figure 5.7 for a graphical estimate we get about −0.51. We can reassure ourselves by looking up the exact value too; it is −0.558. Next, suppose we wanted to identify the maximum value for $Y_0(\lambda r)$; based upon Figure 5.7, it appears to occur at about 2.1. We set

$$\frac{d}{dr}[Y_0(\lambda r))] = -\lambda Y_1(\lambda r) = 0,$$

and find that $\lambda r = 2.19715$. Let us look at another example: We have an annular catalytic pellet in which $C_A = AI_0(\beta r) + BK_0(\beta r)$, with inner and outer radii (R_1 and R_2) of 2 and 4, respectively. We take $\beta = 0.3$ and assume that the concentration at both exposed surfaces will be 1 (one). Our interest is the driving force for diffusion in the interior of the annular solid: $\frac{dC_A}{dr} = A\beta I_1(\beta r) - B\beta K_1(\beta r)$. The boundary conditions are used to show $A = 0.7175$ and $B = 0.2784$, and we can now evaluate the derivative at any r-location of interest. We are going to illustrate this at the annular mid-point, $r = 3$, keeping in mind that the arithmetic center is *not* a surface of symmetry in this case; the result is $\frac{dC_A}{dr} = 0.0471$ for $r = 3$ (small, but *not* zero). There are many useful sources of information for Bessel's equation and Bessel functions, and a few of them are included in the references provided at the end of this chapter.

5.6.1 Bessel's Equation in Extended Surface Heat Transfer

We observed previously that we might expect to see Bessel's equation in problems in cylindrical coordinates with a radially directed flux. However, this ODE can arise in other situations as well, and we wish to illustrate one possibility with this example. Suppose we have a fin with a triangular cross-section extending from a hot surface; the purpose of this projection is to cast off unwanted thermal energy to the surroundings. The fin extends from the hot wall (located at $x = 0$) into the fluid, and the tip corresponds to $x = L$. The base of this wedge is some fraction of its length, bL. The included angle at the tip of the fin is 2θ, such that

$$\theta = \tan^{-1}\frac{bL}{2L} = \tan^{-1}\frac{b}{2}. \tag{5.165}$$

Therefore, if $b = 1/7$, then $\theta = 0.0713$ rad or just less than 4.1°. We will assume that the temperature in the highly conductive fin varies only in the x-direction; this approximation is reasonable if the main resistance to heat transfer is in the fluid phase and not in the wedge-shaped fin. Of course, in extended surface heat transfer we would normally select a highly conductive material for the fin. Let us construct a thermal energy balance upon a volume element (with trapezoidal shape) with thickness Δx (W is the fin width, and we presume $W >> L$):

$$\underbrace{-kb(L-x)W\frac{\partial T}{\partial x}}_{\text{rate in at } x} \quad \underbrace{+kb(L-x)W\frac{\partial T}{\partial x}}_{\text{rate out at } x \pm \Delta x} \quad \underbrace{-A_{\text{exp}}h(T-T_\infty)}_{\text{loss at} f \text{ in surface}} \tag{5.166}$$

The *rate in* term is negative due to our convention with Fourier's law (heat is conducted in the direction of decreasing temperature), and A_{exp} is the exposed surface area of the volume element (remember that we must account for both the top and bottom). At steady state the sum of these three terms must be zero, so we divide by $kbW\Delta x$ and take the limit as $\Delta x \to 0$. It is left to the reader to show that

$$A_{\text{exp}} = \frac{2W\Delta x}{\cos\left[\tan^{-1}\frac{b}{2}\right]}.$$

Letting $z = L - x$ and $\mathbf{T} = T - T_\infty$ we obtain

$$\frac{\partial^2 T}{\partial z^2} + \frac{1}{z}\frac{\partial T}{\partial Z} - \frac{\beta}{z}T = 0. \tag{5.167}$$

We multiply by z^2 and compare the result with eq. (5.156) in this section; accordingly,

$$\mathbf{T} = AI_0(2\sqrt{\beta z}) + BK_0(2\sqrt{\beta z}). \tag{5.168}$$

We use an approximation for $x = L$ (i.e., $z = 0$) by *assuming* that $T = T_\infty$ such that $\mathbf{T} = 0$. Since $K_0(0) = \infty$, we require that $B = 0$. Please note that this boundary condition is strictly valid only if the fin is infinitely long. For $x = 0$ ($z = L$) we have $\mathbf{T} = \mathbf{T}_0$ and therefore

$$\mathbf{T} = \mathbf{T}_0 \frac{I_0(2\sqrt{\beta z})}{I_0(2\sqrt{\beta L})} \tag{5.169}$$

Now let's give this solution some quantitative form; due to the boundary condition employed for $x = L$, the fin must be relatively long so we set $L = 9$ and $b = 1/6$ such that $\cos[\tan^{-1}(b/2)] = 0.9965$. We will take $h/k = 0.083$ (remember, the main resistance to heat transfer must be in the fluid phase surrounding the fin and not in the wedge material itself). Consequently,

$$\mathbf{T} = \mathbf{T}_0 \frac{I_0(2\sqrt{(L-x)})}{I_0(2\sqrt{L})}. \tag{5.170}$$

We put the numerical results in a table:

x	$I_0(2\sqrt{(L-x)})$	T/T_0
0	67.23	1
1.4375	42.69	0.635
2.75	27.24	0.405
5	11.302	0.168
6.75	4.881	0.073
8	2.280	0.034
9	1	0.015

You will see that at the end of the fin where $x = 9$, $\mathbf{T}/\mathbf{T}_0 = 0.015$. This is at variance with the simplifying BC that we employed—we treated the fin as though it were infinitely long such that $T(x = 9) = T_\infty$. Can you think of an alternative boundary condition that might be more appropriate physically?

5.6.2 An Example from Lubrication Theory

We want to explore a case where a particular form of Bessel's differential equation leads to a solution of different order and a prefactor that is *not* equal to 1. An ordinary differential equation that arises in some planar lubrication problems is:

$$\frac{d}{dx}(x^3\phi') + \lambda^2 x^3 \phi = 0 \tag{5.171}$$

with $x_1 \le x \le x_2$ and $\phi(x_1) = \phi(x_2) = 0$. Performing the differentiation:

$$x^3\phi'' + 3x^2\phi' + \lambda^2 x^3 \phi = 0. \tag{5.172}$$

Now we compare this ODE with eq. (5.156) in Section 5.6; dividing by x we obtain:

$$x^2\phi'' + 3x\phi' + \lambda^2 x^2 \phi = 0. \tag{5.173}$$

In this case $b = 0$, $a = 3$, $d = \lambda^2$ and $s = 1$; therefore, the order is $\sqrt{\left(\frac{1-a}{2}\right)^2} = \sqrt{(-1)^2} = 1$. We note that $\sqrt{d} = \lambda$ (which is real) and the prefactor,

$$x^{\frac{1-a}{2}} \exp\left(-\frac{bx^r}{r}\right) = x^{-1} \tag{5.174}$$

since $a = 3$ and $b = 0$. We can now write out the solution for eq. (5.171):

$$\phi = \frac{1}{x}[AJ_1(\lambda x) + BY_1(\lambda x)]. \tag{5.175}$$

Applying the boundary conditions,

$$0 = AJ_1(\lambda x_1) + BY_1(\lambda x_1) \quad \text{and} \quad 0 = AJ_1(\lambda x_2) + BY_1(\lambda x_2).$$

Therefore,

$$A = -\frac{BY_1(\lambda x_1)}{J_1(\lambda x_1)} = -\frac{BY_1(\lambda x_2)}{J_1(\lambda x_2)},$$

so

$$J_1(\lambda x_1)Y_1(\lambda x_2) - J_1(\lambda x_2)Y_1(\lambda x_1) = 0. \tag{5.176}$$

This transcendental eq. (5.176) is used to find the characteristic values for λ; for example, if $x_1 = 1$ and $x_2 = 3$ then the first value of λ is: 1.6356. Typical numerical values for the solution using this first eigenvalue are shown in this table:

x	1.0	1.2	1.4	1.6	1.8	2.0	2.2	2.4	2.6	2.8
ϕ/ϕ_{max}	0	0.574	0.873	0.990	0.983	0.888	0.736	0.552	0.357	0.168

5.7 Power Series Solutions of Ordinary Differential Equations

We observed previously that for higher-order ODEs with variable coefficients, there is no guaranteed method of solution. In such cases, an approximate analytic solution may sometimes be useful, and we will review one approach here. The technique we describe in this section has been used by mathematicians for a very long time, and it has produced some important results. *However*, before an analyst undertakes a power series solution of an ODE he or she should consider the following points:

1) A highly accurate solution may require *many* terms unless the range of the independent variable is quite limited.
2) The solution *might* be written in terms of familiar functions *if one recognizes* the progression as some kind of Taylor or Maclaurin series.
3) This approach is *probably* justifiable if the analyst needs a quick, rough, analytic approximation, or, if a particular ODE must be solved many times such that a tabulation of results might be used repetitively.

Let us begin with the very familiar equation,

$$\frac{d^2y}{dx^2} + y = 0, \tag{5.177}$$

with $y(x=0)=0$ and $y'(x=0)=1$, for our initial exploration because we know the solution to be

$$y = C_1 \sin x + C_2 \cos x. \tag{5.178}$$

In order to satisfy the two boundary conditions, it is necessary that $y = \sin x$, of course. We are going to propose that the solution be represented by the series

$$a_0 + a_1 x + a_2 x^2 + a_3 x^3 + \cdots \quad \text{or} \quad \sum_{n=0}^{\infty} a_n x^n. \tag{5.179}$$

We note immediately that to conform to the boundary conditions, $a_0 = 0$ and $a_1 = 1$. Let us differentiate the series once, $\sum n a_n x^{n-1}$, and then again to obtain $\sum (n-1)(n) a_n x^{n-2}$. We take this last result back to our ordinary differential equation, resulting in

$$\sum_{n=0}^{\infty} (n-1)(n) a_n x^{n-2} + \sum_{n=0}^{\infty} a_n x^n = 0. \tag{5.180}$$

We need to have the first summation written in terms of x^n for consolidation, so we let $m = n-2$ such that $n = m+2$ and $n-1 = m+1$; therefore,

$$\sum_{n=0}^{\infty} [(n+1)(n+2)a_{n+2} + a_n] x^n = 0. \tag{5.181}$$

In order for this power series to be zero as indicated here, it is necessary for the coefficients to be zero and so,

$$(n+1)(n+2)a_{n+2} + a_n = 0. \tag{5.182}$$

However, the boundary conditions require that $a_0 = 0$ and $a_1 = 1$, as we observed above. The first of these also requires $a_2 = 0$, and thus all even a_n's must be zero. We now use eq. (5.182) to compute (by recursion) coefficients for $n = 3, 5, 7$, etc. This results in:

$$a_3 = -\frac{1}{(3)(2)} a_5 = \frac{1}{(5)(4)(3)(2)} a_7 = -\frac{1}{(7)(6)(5)(4)(3)(2)} \tag{5.183}$$

Therefore, our series approximation for $y(x)$ looks like this:

$$y(x) \approx x - \frac{x^3}{3!} + \frac{x^5}{5!} - \frac{x^7}{7!} + \cdots \tag{5.184}$$

How well can the truncated series represent the actual solution? If we limit n to values less than 27, we obtain the values shown in table immediately below.

x	$y(x)$ Series	$y(x) = \sin(x)$
⅛	0.1246747	0.1246747
¼	0.2474039	0.24740396
½	0.4794255	0.47942554
1	0.841471	0.84147098
2	0.9092975	0.90929743
4	−0.7568024	−0.756802495
8	0.9893362	0.98935825
9	0.4116477	0.41211848
10	−0.5542731	−0.54402111
11	−1.158563	−0.99999021
12	−2.469321	−0.53657292

These results make it clear that the truncated series does an excellent job of representing $y(x)$ for *small* values of x, as expected. Discrepancies do begin to appear as x exceeds about 8.

5.7.1 Power Series Example for a First-Order ODE

Let's carry out a detailed exploration of a first-order ordinary differential equation that offers a slightly different wrinkle. Consider

$$\frac{dy}{dx} = x^2 - 4x + y + 1 \quad \text{with} \quad y(x = 2) = 3. \tag{5.185}$$

We propose:

$$y = \sum_{n=0}^{\infty} a_n (x-2)^2 = a_0 + a_1(x-2) + a_2(x-2)^2 + a_3(x-2)^3 + \cdots \tag{5.186}$$

We note immediately that $a_0 = 3$ and then differentiate to obtain

$$y' = \sum_{n-1}^{\infty} n a_n (x-2)^{n-1}.$$

Substituting into the ODE:

$$\sum_{n=1}^{\infty} n a_n (x-2)^{n-1} - \sum_{n=0}^{\infty} a_n (x-2)^n = x^2 - 4x + 1. \tag{5.187}$$

The second summation on the left-hand side starts at $n=0$, but we already know $a_0=3$ so we add 3 to both sides observing that $x^2-4x+4=(x-2)^2$. We also select a dummy variable, u, letting $u=n-1$ in the first summation:

$$\sum_{u=0}^{\infty}(u+1)a_{u+1}(x-2)^u - \sum_{n=1}^{\infty}a_n(x-2)^n = (x-2)^2. \tag{5.188}$$

Of course, u can assume any value including n, so therefore

$$a_1 + \sum_{n=1}^{\infty}[((n+1)a_{n+1} - a_n)(x-2)^n] = (x-2)^2. \tag{5.189}$$

Let's write out a few terms just to make things very clear:

$$a_1 - (x-2)^2 + [(2a_2 - a_1)(x-2) + (3a_3 - a_2)(x-2)^2 + (4a_4 - a_3)(x-2)^3 + \cdots] = 0 \tag{5.190}$$

Since the coefficients on the successive powers of $(x-2)$ must be zero, we find $a_1=0$ $2a_2-a_1=0$ $3a_3-a_2-1=0$ $4a_4-a_3=0$, and so on, resulting in

n	0	1	2	3	4	5	6	7	8
a_n	3	0	0	1/3	1/12	1/60	1/360	1/2520	1/20160

Obviously, we can use as many terms as necessary for the required accuracy of the problem; if we terminate the series at $n=6$ and compare with both analytic and numerical (using the Euler predictor–corrector scheme from Chapter 6) solutions we find:

x	2	3	4	5	6	7
Series with $n=6$	3	3.436	7.711	24.825	74.111	192.236
Analytic solution	3	3.437	7.778	26.171	86.196	262.826
Euler pred–corr	3	3.435	7.767	26.132	86.076	262.480

These results show that the truncated power series produces acceptable results up to about $x\approx 5$. Obviously solutions valid for larger values of x will require additional terms in the series.

5.7.2 A Complete Power Series Example for a Second-Order ODE with Numerical Confirmation

Consider the second-order ordinary differential equation

$$y'' + xy' + y = 0 \tag{5.191}$$

with $y(0)=0$ and $y'(0)=1$. We note that this ODE is linear, but it does not have constant coefficients. We will try for a solution of the form

$$y = \sum_{n=0}^{\infty}a_n x^n \tag{5.192}$$

and we begin by differentiating the power series:

$$y' = \sum_{n=1}^{\infty}na_n x^{n-1} \tag{5.193}$$

$$y'' = \sum_{n=2}^{\infty}n(n-1)a_n x^{n-2}. \tag{5.194}$$

Substituting into eq. (5.191):

$$\sum_{n=2}^{\infty}n(n-1)a_n x^{n-2} + \sum_{n=1}^{\infty}na_n x^n + \sum_{n=0}^{\infty}a_n x^n = 0.$$

We set the dummy variable $u = n-2$ so that $n = u + 2$ producing:

$$\sum_{u=0}^{\infty}(u+2)(u+1)a_{u+2}x^u + \sum_{n=1}^{\infty} na_n x^n + \sum_{n=0}^{\infty} a_n x^n = 0.$$

Of course, u can assume any integer value, including n:

$$\sum_{n=0}^{\infty}(n+2)(n+1)a_{n+2}x^n + \sum_{n=1}^{\infty} na_n x^n + \sum_{n=0}^{\infty} a_n x^n = 0. \tag{5.195}$$

Obviously we need all of the lower indices to be the same rather than $n = 0,1,0$ as they appear here. If $n = 0$, the second summation disappears and from the first and third we get $2a_2 + a_0$ so consequently:

$$2a_2 + a_0 + \sum_{n=1}^{\infty}[(n+2)(n+1)a_{n+2} + na_n + a_n]x^n = 0. \tag{5.196}$$

The quantity inside the brackets [...] in the summation must be zero for all n; furthermore, we have assumed that y could be represented like this:

$$y = a_0x^0 + a_1x^1 + a_2x^2 + \ldots \text{etc.} \tag{5.197}$$

Since $y(0) = 0$, $a_0 = 0$, and since $y'(0) = 1$, $a_1 = 1$. As we observed above, the quantity inside the brackets in eq. (5.196) must be zero, and therefore we have

$$a_{n+2} = \frac{-a_n}{(n+2)}, \tag{5.198}$$

indicating that all even a_n's are zero and that the odd a_n's follow the pattern:

$$a_1 = 1, a_3 = -1/3, a_5 = 1/15, a_7 = -1/(3 \cdot 5 \cdot 7), \text{etc.}$$

What we need to know now is exactly how well this infinite series represents the solution to the ODE; i.e., is it going to be useful? We can answer this question by solving the ODE numerically and comparing the results to our truncated series (Figure 5.8).

This is where we learn an unpleasant truth: If we terminate the series at $n = 7$, the results are only good up to $x \approx 1.3$. If we take the summation up to $n = 59$, we begin to see significant deviations at about $x \approx 4.5$. What we hope for in a power series solution of an ODE is that the series converges to some known, familiar function. If it does not, the numerical work required to produce something quantitatively useful may not be justified.

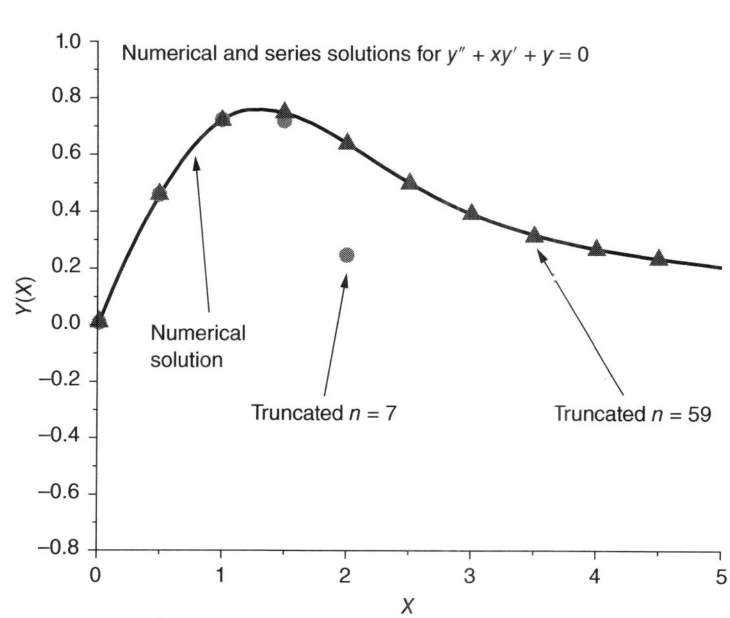

Figure 5.8 Comparison of power series truncated at $n = 7$ and $n = 59$ with the numerical solution for $y'' + xy' + y = 0$.

5.7.3 Guided Exercise: Power Series Solution of a Second-Order ODE

If you would like to experiment with power series solutions, consider the second-order linear ODE (there are additional exercises in the ***Problems*** at the end of the chapter):

$$x^2y'' + (x^2 - x)y' + 2y = 0. \tag{5.199}$$

We will write this equation in the form

$$f_0(x)y'' + f_1(x)y' + f_2(x)y = 0, \tag{5.200}$$

and note that $f_0(x=0)=0$; therefore, $x=0$ is a *singular* point. If we attempt a power series solution about the point $x=0$ (with $y=\sum a_n x^n$) using our previous work as a template—and the reader is encouraged to try this—the effort will fail. Contrast the second-order ODE above with:

$$(1 + x^2)y'' + xy' - y = 0. \tag{5.201}$$

Now, $f_0(x)=1+x^2$ and $f_1(x)=x$; in this case, $f_0=1$ for $x=0$ and therefore $x=0$ is an *ordinary* point. A power series solution about an *ordinary* point is possible; however, for second-order ODEs with polynomials (in x) as coefficients, the solution will be in the form of two power series that are linearly independent. Let us now illustrate with eq. (5.201) by writing:

$$y = a_0 + a_1x + a_2x^2 + a_3x^3 + a_4x^4 + \cdots$$

$$y' = a_1 + 2a_2x + 3a_3x^2 + 4a_4x^3 + \cdots \text{ and}$$

$$y'' = 2a_2 + 6a_3x + 12a_4x^2 + \cdots$$

We substitute into the ODE eq. (5.201):

$$(1 + x^2)[2a_2 + 6a_3x + 12a_4x^2 + \cdots + x[a_1 + 2a_2x + 3a_3x^2 + 4a_4x^3 + \cdots - [a_0 + a_1x + a_2x^2 + a_3x^3 + \cdots = 0, \tag{5.202}$$

and we group like powers of x:

$$x^0: 2a_2 - a_0 = 0$$

$$x^1: 6a_3 + a_1 - a_1 = 0$$

$$x^2: 2a_2 + 12a_4 + 2a_2 - a_2 = 0,$$

and so on. Therefore, $a_2=a_0/2$ and $a_3=0$ and $12a_4=-3a_2$, or $a_4=-\frac{1}{4}a_2=-\frac{1}{8}a_0$. The relationship between a_4 and a_2 suggests the following: $(n+2)(n+1)a_{n+2}=-(n^2-1)a_n$ or $(n+2)a_{n+2}=-(n-1)a_n$. Since $a_3=0$ it follows that $a_5=a_7=0$, and only a_1 remains of the odd n's. All of the even a_n's can be written in terms of a_0, e.g., $6a_6=-3a_4=\frac{3}{8}a_0$ such that $a_6=\frac{1}{16}a_0$ and $8a_8=-5a_6=-\frac{5}{16}a_0$ producing $a_8=-\frac{5}{128}a_0$. Verify that the next two terms are $a_{10}=\frac{35}{1{,}280}a_0$ and $a_{12}=-\frac{21}{1{,}024}a_0$. We are ready to write out the solution of eq. (5.201):

$$y = a_0\left(1 + \frac{1}{2}x^2 - \frac{1}{8}x^4 + \frac{1}{16}x^6 - \frac{5}{128}x^8 + \frac{35}{1{,}280}x^{10} - \frac{21}{1{,}024}x^{12} + \cdots\right) + a_1x. \tag{5.203}$$

Now suppose $y(0)=1$ and $y'(0)=0$; this fixes the values of a_0 and a_1 at 1 and 0, respectively. We can solve eq. (5.201) numerically for comparison and the results appear in the table below.

x	0	1/8	1/4	3/8	1/2	3/4	1
$y_{numeric}$	1	1.008	1.031	1.068	1.118	1.250	1.414
y_{series}	1	1.008	1.031	1.068	1.118	1.250	1.405

The agreement between the numerical solution and the truncated series is excellent for small x, but we must remember that we developed this power series about the *ordinary* point, $x=0$; we are therefore limited to small values of x and, in fact, the series is divergent if $x>1.15$.

In this guided exercise we sought a power series solution of eq. (5.201), a second-order linear ODE that *did not* have constant coefficients. The value of the solution, eq. (5.203), is that we obtained an analytic expression that can accommodate different boundary conditions. Thus, if we needed to solve eq. (5.201) repeatedly we would only need to change the values of the two coefficients. For example, suppose $y(0) = 3$ and $y'(0) = 3$; this means $a_0 = 3$ and $a_1 = 3$, and the truncated power series produces the following results:

x	0	1/8	1/4	3/8	1/2	3/4	1
y_{numeric}	3	3.399	3.842	4.330	4.854	6.000	7.243
y_{series}	3	3.398	3.842	4.329	4.854	6.001	7.257

If one were required to solve the same ODE repeatedly, a power series solution could be used to compile a table of values of the required accuracy—just as mathematicians did many years ago for functions like sin and ln (in fact, the first sine tables were assembled in the ninth century). The reader is encouraged to look at the application of the *method of Frobenius* to Bessel's differential equation, and a useful description of the process is provided by Mickley et al. (1957) in Chapter 5.

5.8 Regular Perturbation

There are occasions when one simply must obtain an approximate analytic solution for a nonlinear, ordinary differential equation. In the previous section we looked at the power-series technique applied to linear ODEs including some with variable coefficients; now we turn our attention to regular perturbation. Perturbation is a powerful method that is especially useful when a *nonlinear* problem contains a parameter that is in some sense small. Let us consider steady-state heat transfer (conduction) in a slab of pure iron. The thermal conductivity of pure iron between −200 °C and +700 °C varies approximately as:

$$k \cong a + bT = 76.333 - 0.0633T \text{ W/(m °C)}. \tag{5.204}$$

Therefore, at −200 °C, $k = 89$ W/(m °C), and at +700 °C, $k = 32$ W/(m °C); from high-to-low temperature then, k increases by a factor of about 2.78. Such a variation will have a profound impact upon the temperature distribution as we shall now see. Assume we have a slab of pure iron that extends from $y = 0$ to $y = 1$ m. The temperature at $y = 0$ is maintained for all time t at −200 °C and at $y = 1$ m, the constant temperature is +700 °C. Under these steady conditions, the temperature distribution is governed by

$$\frac{d}{dy}\left[(a + bT)\frac{dT}{dy}\right] = 0, \tag{5.205}$$

or

$$(a + bT)\frac{d^2T}{dy^2} + b\left(\frac{dT}{dy}\right)^2 = 0. \tag{5.206}$$

One thing that we notice immediately about eq. (5.206) is that if the parameter, b, is *very* small, then the two nonlinearities in the ODE are eliminated! We will assume that T can be represented as a sequence of unknown functions:

$$T = T_0 + bT_1 + b^2T_2 + b^3T_3 + \cdots \tag{5.207}$$

Our goal of course is to determine the unknown functions, T_0, T_1, etc. We begin by writing out the derivatives we need:

$$\frac{dT}{dy} = \frac{dT_0}{dy} + b\frac{dT_1}{dy} + b^2\frac{dT_2}{dy} + \cdots \tag{5.208}$$

and

$$\frac{d^2T}{dy^2} = \frac{d^2T_0}{dy^2} + b\frac{d^2T_1}{dy^2} + b^2\frac{d^2T_2}{dy^2} + \cdots \tag{5.209}$$

We now take eqs. (5.207), (5.208), and (5.209) and substitute into eq. (5.206):

$$[a + b(T_0 + bT_1 + b^2T_2 + \cdots)]\left(\frac{d^2T_0}{dy^2} + b\frac{d^2T_1}{dy^2} + b^2\frac{d^2T_2}{dy^2} + \cdots\right) + b\left[\frac{dT_0}{dy} + b\frac{dT_1}{dy} + b^2\frac{dT_2}{dy} + \cdots\right]^2 \cong 0. \tag{5.210}$$

Now let us assume b can take on any value, including a *very small* one; if b is vanishingly small then we are left with:

$$a\frac{d^2T_0}{dy^2} = 0 \text{ and consequently, } T_0 = C_1y + C_2. \tag{5.211}$$

We treat the boundary conditions exactly the same way—we have −200 at $y = 0$ and + 700 at $y = 1$. Therefore,

$$T_0 = 900y - 200. \tag{5.212}$$

We differentiate eq. (5.212) as needed, and take those results back to eq. (5.213). Of course,

$$\frac{dT_0}{dy} = 900 \quad \text{and} \quad \frac{d^2T_0}{dy^2} = 0. \tag{5.213}$$

If we divide by b and again allow b to become very small, then

$$a\frac{d^2T_1}{dy^2} = -(900)^2. \tag{5.214}$$

Therefore,

$$T_1 = \frac{-(900)^2}{2a}y^2 + C_1y + C_2. \tag{5.215}$$

Now we return to eq. (5.207) to look at the boundary conditions, subtracting the results obtained with eq. (5.212) from T, and then dividing by b: Thus, $C_2 = 0$ and $C_1 = \frac{(900)^2}{2a}$. At this point our approximate solution has the form

$$T \cong 900y - 200 + b\frac{(900)^2}{2a}(y - y^2),$$

and inserting the numerical values for a and b:

$$T \cong 335.851y^2 + 564.149y - 200. \tag{5.216}$$

Our process guarantees that the boundary conditions are satisfied, but our real interest is whether or not this approximation will reproduce the expected curvature for intermediate values of y. Figure 5.9 provides a comparison of the exact solution with our truncated approximation.

Perturbation is a powerful technique that can produce very good approximate solutions under the right circumstances. Experience suggests, however, that if one must go past the third unknown function (past T_2 for the example above) then the time invested for the likely return will probably be excessive. There are very good sources of information available for the applied scientist who must generate an analytic approximation to the solution of a nonlinear ordinary differential equation; for example, this author has found Finlayson (1980) to be quite useful.

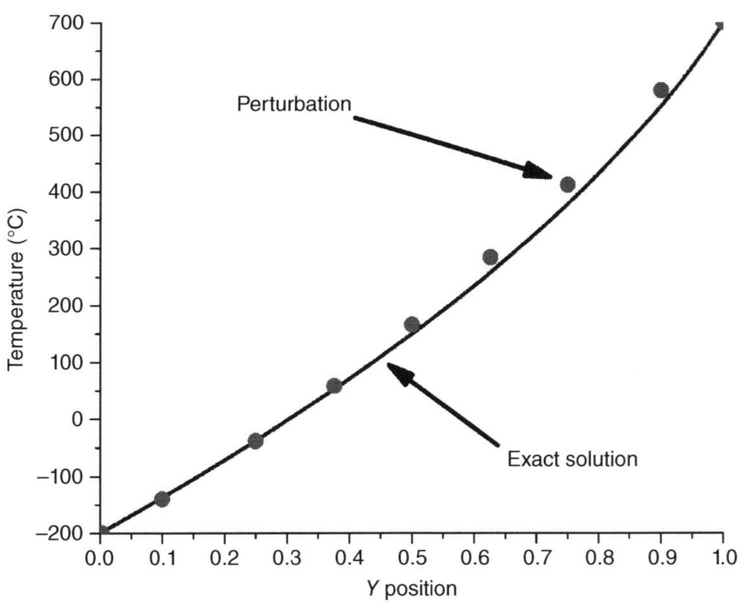

Figure 5.9 Comparison of the approximate solution obtained with regular perturbation with the exact solution for steady conduction in an iron slab with variable conductivity. Even though the perturbative solution was truncated, the agreement between the two is good enough for many practical purposes.

5.9 Linearization

Often the removal of nonlinear terms in ordinary differential equations will yield an equation(s) that can be solved analytically. Of course the analyst must decide if the approximation is adequate for the intended purpose. We will look at a couple of alternatives here, and we will try to give the reader a sense of when such an approach may work and when it may not. Let us begin with the simple first-order ODE:

$$\frac{dh}{dt} = \frac{1}{2}h^{7/4} \tag{5.217}$$

with $h(t=0)=1$ and $0 \leq t \leq 2$. Eq. (5.217) is easy to solve as it stands, so it will provide a very nice means of comparison. The analytic solution, which we will make use of later, is

$$h(t) = \left[1 - \frac{3}{8}t\right]^{-4/3}. \tag{5.218}$$

Let us focus our attention upon the right-hand side of (5.217), and we will use a truncated Taylor series to approximate this function:

$$f(h) = h^{7/4} \cong f(h_0) + f'(h_0)(h - h_0). \tag{5.219}$$

It is evident that our strategy is to place a tangent line on the $h^{7/4}$ curve at a particular value, h_0. It also seems likely that this will work well only if h does not deviate radically from h_0—the further we are away from that point, the worse the approximation will be. Since $f' = \frac{7}{4}h^{3/4}$, we can select an appropriate point and find the slope; we know from the problem statement that $h \geq 1$ so we will begin by trying $h_0 = 2$ such that $f'(h_0) = 2.9431$ and $f(h_0) = 3.3636$. Our approximation is therefore $h^{7/4} \cong 3.3636 + 2.9431(h-2)$. First, we will see how well this linearized version of $f(h)$ represents the original function by constructing a little table.

h	1	1.5	2	2.5	3	4
$f(h)$	1.0000	2.0331	3.3636	4.9704	6.8385	11.3137
$f(h)$ Linearized	0.4205	1.8921	3.3636	4.8352	6.3067	9.2498

The truncated Taylor series corresponds to the nonlinear function nicely *near* the selected point, $h=2$; however, it begins to deviate dramatically for both smaller and larger values. These differences will impact the approximate solution of eq. (5.217) in a very significant way as we shall now see. The linearized differential equation is

$$\frac{dh}{dt} = 1.4716h - 1.2613 \tag{5.220}$$

and the solution is

$$h = \frac{1}{1.4716}[C_1 \exp(1.4716t) + 1.2613]. \tag{5.221}$$

The initial condition is $h(0)=1$, so accordingly $C_1 = 0.2103$. Now we can compare the two sets of results:

t	*h* Actual	*h* Linearized
0	1.00	1.00
0.25	1.1403	1.0636
0.5	1.3190	1.1554
0.75	1.5532	1.2880
1.0	1.8714	1.4796
1.25	2.3242	1.7565
1.5	3.0109	2.1564
1.75	4.1528	2.7342
2.0	6.3496	3.5689

What have we learned from this example? If the transient operation of the nonlinear system takes us far away from the expected value for h (which we decided might be 2) then we will have serious problems. On the other hand, if we are able to stay close to h_0 the linearized ODE may be completely acceptable. In terms of the table shown above, this probably means that we would need to constrain the independent variable, t, to small values if we wanted to get satisfactory results.

5.9.1 Beam Deflection Under Load

In the beginning of this section we used a truncated Taylor series to make a nonlinear ODE linear. Another possibility that *may* yield acceptable results in very particular cases involves simply omitting the nonlinear term; we might do this if we believed that the nonlinearity was *very* small relative to the rest of the differential equation. To illustrate, imagine a simply supported rectangular aluminum beam subjected to a uniformly distributed load. The nonlinear ordinary differential equation describing this situation is:

$$EI\frac{d^2y}{dx^2} = M(x)\left[1 + \left(\frac{dy}{dx}\right)^2\right]^{3/2}, \tag{5.222}$$

where E is the modulus of elasticity, I is the moment of inertia of the cross-section, and M is the bending moment. The x-coordinate runs through the center of the undeformed rectangular beam on the longitudinal axis and y is the transverse position. In the linear theory, we *assume* that dy/dx is small such that $(dy/dx)^2$ is negligible. Let's take a moment to assess this in an approximate way; we will construct a table for the calculation:

Nominal dy/dx	$[1 \pm (dy/dx)^2]^{3/2}$
0.005	1.00004
0.01	1.00015
0.05	1.00375
0.10	1.015
0.20	1.061

Of course if the linear assumption was perfect, the entire second column would be identically 1.0. This calculation merely gives us a rough picture of the deviation from linearity; dy/dx should have its largest magnitude at the beam ends, $x=0$ and $x=L$ and be zero at the center ($x=L/2$) by symmetry. Let us initially presume that linearization is warranted:

$$\frac{d^2y}{dx^2} = \frac{M(x)}{EI}. \tag{5.223}$$

For a simply supported beam with a uniformly distributed load, W, kg/m:

$$M(x) = \frac{Wx}{2}(L-x) = \frac{W}{2}(xL - x^2). \tag{5.224}$$

Therefore,

$$\frac{d}{dx}\left(\frac{dy}{dx}\right) = \frac{W}{2EI}(Lx - x^2) \tag{5.225}$$

and

$$\frac{dy}{dx} = \frac{W}{2EI}\left(\frac{Lx^2}{2} - \frac{x^3}{3}\right) + c_1. \tag{5.226}$$

Since the derivative is zero at $x=L/2$,

$$c_1 = -\frac{W}{2EI}\left(\frac{L^3}{12}\right) \tag{5.227}$$

and

$$y = \frac{W}{2EI}\left(\frac{Lx^3}{6} - \frac{x^4}{12} - \frac{L^3x}{12}\right) + c_2. \tag{5.228}$$

At the supported ends $y(x = 0) = 0$, so $c_2 = 0$. Maximum deflection will occur at the center, $x = L/2$, so

$$y_{max} = \frac{-0.026042WL^4}{2EI} = \frac{-5WL^4}{384EI}. \tag{5.229}$$

For aluminum, $E = 69$ GPa, and we select a beam with a width of 5 cm and a vertical height of 20 cm so that $I = 3.33 \times 10^{-5}$. Letting $W = 100$ kg/m and $L = 15$ m, we calculate the maximum deflection to be (−)2.86 cm. Naturally, the critical question is: Would we get a significantly different result if we solved the nonlinear ODE, eq. (5.222), numerically? This has been done to provide the comparison, and the necessary boundary conditions were $y(0) = 0$ and $y'(0) = -6.12035 \times 10^{-3}$.

x (m)	**Nonlinear y (m)**	**Linearized y (m)**
0	0	0
1	−0.00613	−0.00607
2	−0.01183	−0.01183
3	−0.01704	−0.01704
5	−0.02493	−0.02493
7	−0.02853	−0.02854
7.5	−0.02869	−0.02869

For this aluminum beam deflection problem, *with the specified loading*, the nonlinear and linearized solutions are nearly identical; the simplification was completely justified. An interesting exercise would entail increasing the loading, W, to identify the threshold deformation where solution of the nonlinear ODE would be necessary. The reader may want to verify that for $W = 800$ kg/m the *correct* maximum deflection is −0.229689 m (with $y'(0) = -4.90208 \times 10^{-2}$); this is 0.08% larger than the deflection computed with eq. (5.229). We will examine ways to carry out such calculations in the next chapter.

Of course linearization can be extended to higher-order problems too. Consider the nonlinear second-order system,

$$\frac{dx}{dt} = y - (x^2 + y^2)x \tag{5.230}$$

and

$$\frac{dy}{dt} = -x - (x^2 + y^2)y. \tag{5.231}$$

We will begin by letting $x_i = y_i = \frac{3}{4}$, and solving the pair of eqs. (5.230) and (5.231) numerically to obtain the system trajectory in the phase plane (merely a convenient way to visualize the system's dynamic behavior by cross-plotting pairs of values for $x(t)$ and $y(t)$). The phase-plane construction shows that the system is stable and that both x and y oscillate with decaying amplitude. This results in an inward-directed spiral which is ultimately centered about the equilibrium point, $x = 0$ and $y = 0$, and the trajectory is provided in Figure 5.10.

Now let $u = x - x_0$ and $v = y - y_0$ such that $x = u + x_0$ and $y = v + y_0$. Our plan is to approximate the two nonlinear functions (just as we did for the first-order system in the initial example):

$$f(x, y) \cong f(x_0, y_0) + \left.\frac{\partial f}{\partial x}\right|_{x_0, y_0} u + \left.\frac{\partial f}{\partial y}\right|_{x_0, y_0} v \tag{5.232}$$

and

$$g(x, y) \cong g(x_0, y_0) + \left.\frac{\partial g}{\partial x}\right|_{x_0, y_0} u + \left.\frac{\partial g}{\partial y}\right|_{x_0, y_0} v. \tag{5.233}$$

The necessary partial derivatives correspond to the Jacobian matrix, and the system can be represented in an alternative way:

$$\begin{matrix} \dfrac{du}{dt} \\ \dfrac{dv}{dt} \end{matrix} = \begin{matrix} -3x^2 - y^2 & 1 - 2xy \\ -1 - 2xy & -x^2 - 3y^2 \end{matrix} \begin{matrix} u \\ v \end{matrix} \tag{5.234}$$

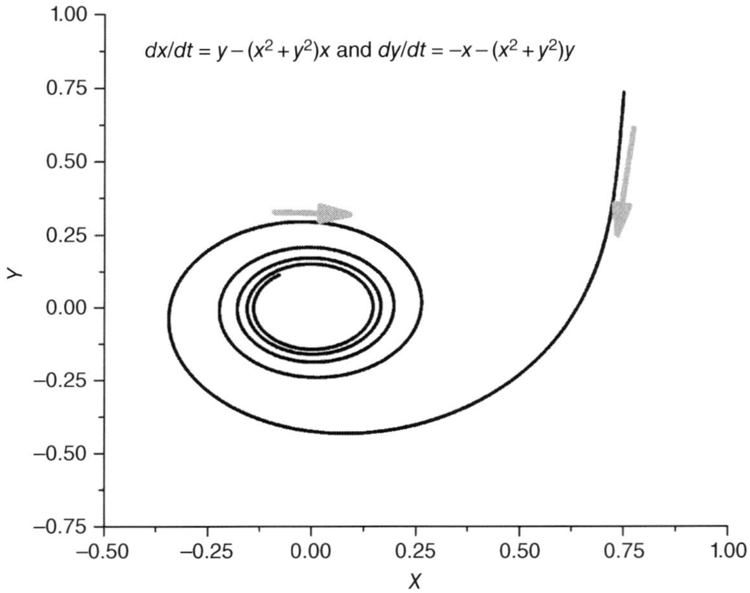

Figure 5.10 Phase-plane portrait of the nonlinear, second-order system revealing stable behavior with an inward-directed spiral.

The equilibrium point is $(x_0, y_0) = (0, 0)$, so the Jacobian matrix is simply

$$\begin{matrix} 0 & 1 \\ -1 & 0 \end{matrix}$$

and consequently,

$$\frac{du}{dt} = v \quad \text{and} \quad \frac{dv}{dt} = -u. \tag{5.235}$$

Equivalently, $\frac{d^2u}{dt^2} + u = 0$ such that $u = C_1 \sin t + C_2 \cos t$ (we obtain the identical result for v). This means that both u and v are purely periodic with *constant amplitude*. In the phase plane, therefore, we will get a closed curve (a limit cycle) such that the system trajectory merely orbits around the equilibrium point. This is distinctly different from the nonlinear case where the oscillations exhibited decreasing amplitude resulting in an inward-directed spiral as shown in Figure 5.10. Although the differences are worrisome in the absolute sense, it is important to note that *qualitatively* the original system and the linearized simplification exhibit similar oscillatory behaviors.

We conclude this discussion by looking at a different approach, one that has been used in fluid mechanics to eliminate nonlinear inertial terms from the Navier–Stokes equation (the interested reader should investigate *Oseen's correction*). Suppose we have the following differential equation,

$$\frac{d^2y}{dx^2} - y\frac{dy}{dx} = 2, \tag{5.236}$$

with $y(x = 0) = \frac{1}{2}$ and $y'(x = 0) = 0$ and let us assume we are interested in $0 \leq x \leq 1.75$. Our plan is too replace the $y\frac{dy}{dx}$ term with $y_0\frac{dy}{dx}$, where y_0 is an appropriately chosen constant; therefore,

$$\frac{d^2y}{dx^2} = 2 + y_0\frac{dy}{dx}. \tag{5.237}$$

Eq. (5.237) can be conveniently solved by reducing the order; e.g., let $\phi = \frac{dy}{dx}$. The solution is (and the reader should verify this result)

$$y = \frac{2}{y_0}\left[\frac{1}{y_0}\exp(y_0x) - x\right] + C_2. \tag{5.238}$$

Since $y(x = 0) = 1/2$, $C_2 = \frac{1}{2} - \frac{2}{y_0^2}$. How well our approximate solution corresponds to the exact numerical results depends upon our choice for y_0 of course. The main difficulty is that with nonlinear differential equations we may not know much about $y(x)$ over the complete range of x—thus making a good choice for y_0 problematic. Here is a comparison of the actual computed and the linearized (approximate) results.

x	y (Exact)	y (Linearized with $y_0 = 2$)
0	0.5000	0.5000
0.3	0.5949	0.6111
0.5	0.7760	0.8591
0.7	1.0754	1.3276
0.9	1.5387	2.1248
1.2	2.8070	4.3116
1.5	6.3878	8.5428
1.75	40.8730	14.8077

Whether or not the examples of linearization applied to ODEs in this section will be satisfactory depends upon the intended application. For cases requiring precision these approaches will probably not be sufficient, but if we are merely interested in learning a little about the behavior of a difficult nonlinear problem (with a quick analytic approximation), then linearization may prove useful.

5.10 Frequency Response for Model Development

Our purpose in this chapter was to review some analytic methods for solving ordinary differential equations, and in most cases the example ODEs were simply presented to us along with appropriate boundary or initial conditions. However, in one important sense the presentation was incomplete—you cannot solve the differential equation if you do not know what it is. In professional practice we might encounter a complex process that is not amenable to a balance statement of the type we have seen so often. It might be a process consisting of a number of units interconnected in such a way as to preclude writing {Rate in} − {Rate out} = {Accumulation}, or $\sum F = ma$, etc.

As we pointed out in Chapter 1, the first step to process improvement and control in industry is usually model development followed by simulation. There are several reasons for this and foremost among them are: 1) We avoid the labor costs associated with an extensive experimental investigation, 2) We do not have to take the process or unit offline and therefore production is not interrupted, 3) We enhance safety for both personnel and the facility. Imagine that you are an engineer at an industrial plant and your responsibility is the efficient operation of some particular process. Furthermore, assume you have been tasked with improving the operation and its response to system upsets. As we noted above, if the system is especially complex there may be no model at hand that can be used to simulate the effects of proposed process changes. In such cases we might need to find a differential equation that can accurately mimic the dynamic response of the process to disturbances. This is where frequency response can be useful.

Suppose we take some complicated "black-box" system and perturb it with a sinusoid of frequency, ω. This is surprisingly easy to do by driving the main input variable (perhaps flow rate) with a sine-wave generator; the required data can often be obtained without removing the unit from production. When we examine the system's output in response to the sinusoidal forcing function we immediately note two features of interest; the amplitude of the response will be attenuated from that of the input, and the response will lag the input, that is, the peaks of the oscillatory response of the system will be delayed relative to the peaks of the forcing function. Let us illustrate with an elementary case, a first-order system for which $\tau \frac{dY}{dt} + Y = X$ with $Y(0) = 0$. Using the Laplace transform we obtain the transfer function relating system response to the forcing function:

$$\frac{Y(s)}{X(s)} = \frac{1}{\tau s + 1}. \tag{5.239}$$

We let $s = i\omega$ and obtain the complex number, $\frac{1 - i\omega\tau}{1 + \omega^2\tau^2}$, or $a + bi$. Our plan is to write this complex number in polar form (that is, a magnitude and an angle) by noting

$$a = \frac{1}{1 + \omega^2\tau^2} \quad \text{and} \quad b = \frac{-\omega\tau}{1 + \omega^2\tau^2}.$$

Since $r = \sqrt{a^2 + b^2}$ and since $\cos\theta = \frac{a}{r}$ and $\sin\theta = \frac{b}{r}$, we find

$$r = \sqrt{\frac{1}{1+\omega^2\tau^2}} \quad \text{and} \quad \tan\theta = \frac{\sin\theta}{\cos\theta} \quad \text{such that} \quad \theta = \tan^{-1}(-\omega\tau). \tag{5.240}$$

The amplitude of the response will be diminished as the frequency of the forcing function is increased; in fact, as $\omega \to \infty$, $r \cong 1/\omega\tau$. If $\tau = 1$ and $\omega = 100$ rad/t, then $r = 0.01$, and the phase lag will be $\theta = -1.56$ rad or $-89.4°$. Note that the phase lag for this first-order system will approach $-90°$ as the frequency of the sine wave is increased.

Of course first-order models like the preceding example are commonplace, and the appropriate differential equation (model) is usually obvious. Our interest is in more abstruse cases where frequency response is likely to be of greater utility. Consider a process represented by

$$\tau^2 \frac{d^2Y}{dt^2} + 2\tau\zeta\frac{dY}{dt} + Y = X(t) = A\sin(\omega t), \tag{5.241}$$

with $Y(0) = Y'(0) = 0$. For ease of interpretation, let $\omega = 1$, $\tau = 1$, and $\zeta = 1$; we prepare Figure 5.11 showing both the sinusoidal forcing function and the response of this second-order system:

Of course, eq. (5.241) lends itself to solution via the Laplace transform, and if you would like to review this technique you might consult Section 5.4 then proceed as indicated here: First, carry out the partial fraction expansion,

$$y(s) = \frac{\omega}{(s+1)^2(s+i\omega)(s-i\omega)} = \frac{A}{(s+1)^2} + \frac{B}{s+1} + \frac{C}{s+i\omega} + \frac{D}{s-i\omega}.$$

Please note our treatment of the repeated roots. Since $\omega = 1$, you can verify that $A = 1/2$, $B = 1/2$, and $C = -1/4$. Therefore, we find

$$y(t) = \frac{1}{2}(t+1)\exp(-t) - \frac{1}{2}\cos(t).$$

This result should mimic the data shown in Figure 5.10, and you may want to confirm the values shown in this small table:

t	3	5	7	9
$y(t)$	0.5956	−0.1216	−0.3733	0.4562

Now, could we *on the basis of* Figure 5.11 *alone* assemble a suitable process model—a differential equation—that might be used for simulation? That is, could we use to information in Figure 5.11 to accurately reproduce ODE (5.241)? The answer

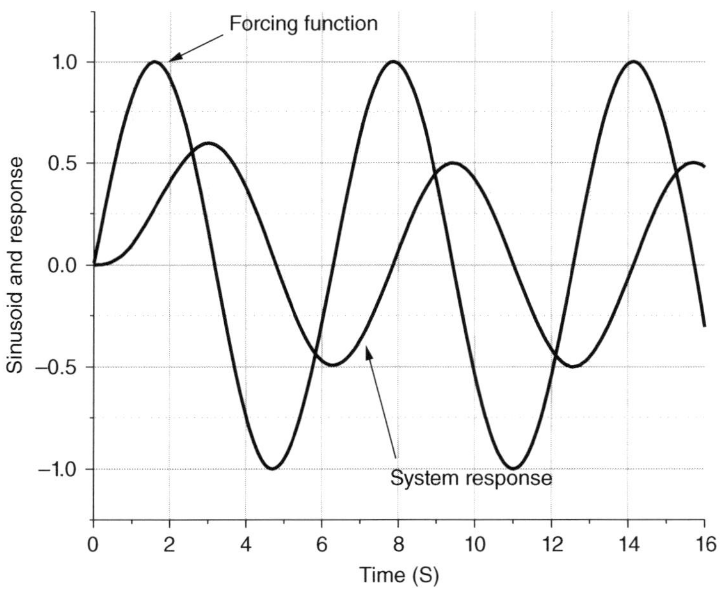

Figure 5.11 Response of a second-order system to a sinusoidal forcing function with $A = 1$, $\omega = 1$, $\tau = 1$, and $\zeta = 1$. The initial transient portion of the response is played out by about $t = 6$ s. Note the attenuation of the response (50%) and the phase lag (about 1.57 s); since the period is about 6.28 s, the phase lag corresponds to about 90°.

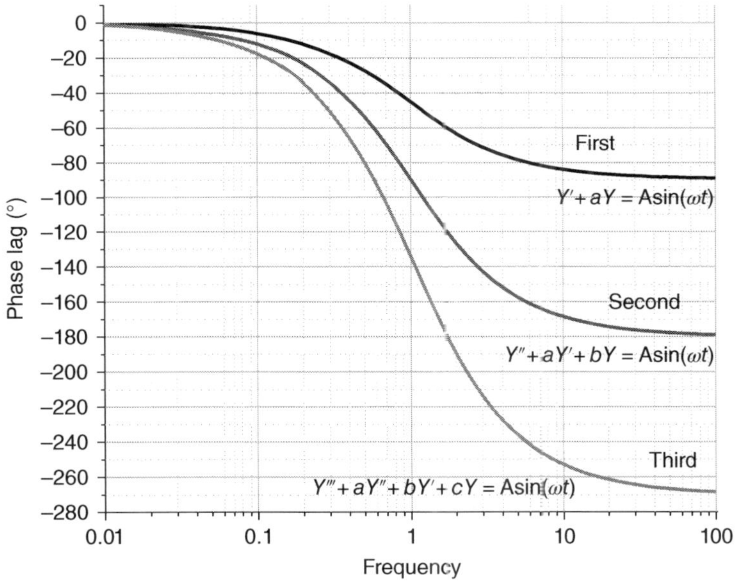

Figure 5.12 Phase lag for first-, second-, and third-order systems, *all* with $\tau = 1$.

is no; we do not have sufficient data to reliably represent this process. *But*, if we also had the phase lag (the angle θ) as a function of the frequency of the forcing function, ω, we might. In Figure 5.12 the phase lags for first-, second-, and third-order systems (all with $\tau = 1$) are shown as functions of frequency.

Much useful information can be extracted from Figure 5.12. We note immediately that as the frequency of the forcing function increases, the phase lag of a first-order system will asymptotically approach $-90°$, θ for a second-order process will approach $-180°$, and for a third-order system, $-270°$. Furthermore, note that in each case, one-half of the ultimate lag is recorded at $\omega = 1/\tau$; thus, for the third-order system with $\tau = 1$, we get $-135°$ at $\omega = 1$ rad/t; for the second-order system with $\tau = 1$ we see $-90°$ at $\omega = 1$ rad/t.

Let us return to the second-order ODE, eq. (5.241), and set down expressions for both the magnitude (or amplitude ratio, AR) and the angle (or phase lag):

$$r = \frac{1}{\sqrt{(1-\omega^2\tau^2)^2 + (2\zeta\omega\tau)^2}} \quad \text{and} \quad \theta = \tan^{-1}\frac{-2\zeta\omega\tau}{1-\omega^2\tau^2}. \tag{5.242}$$

In this case as the frequency becomes very large, $r \sim 1/\omega^2$, and $\theta \to -180°$; for the former this means a *log–log plot* of $r(\omega)$ will have a slope of -2 as ω becomes large. Please note that a strongly underdamped second-order system can complicate interpretation of frequency response data; it will create a spur in the magnitude (*AR* or *r*) plot, and it will significantly steepen $\theta(\omega)$ near the corner frequency. In the case of r, you will observe from (4) that if $\omega^2\tau^2 \approx 1$, then $r \approx 1/(2\zeta\omega\tau)$ such that a really small ζ will produce an elevated value for r.

There is another aspect of frequency response that figures prominently in many processes where fluids must be moved around in pipes, the time delay or transportation lag. This is particularly common in the chemical process industries, but it may also appear in other types of manufacturing. For a time delay,

$$r = 1 \quad \text{and} \quad \theta = -\omega\tau_{delay}. \tag{5.243}$$

This means that the AR is unaffected by a delay, but the phase lag increases without bound as the frequency of the *sine* forcing function is increased. Transportation lags promote instability and make a process much more difficult to control since "old" information must be used to implement corrective action. At the same time, they are easy to detect using frequency response since $\theta \to -\infty$ as the frequency increases. For example, consider a first-order system with a time delay; in particular, let $\tau = 10$ and $\tau_{delay} = 1/8$. Let's compute both the AR and the phase lag and compile the results in this table:

ω	0.01	0.04	0.1	0.4	1	4	10	40	100
AR	0.995	0.928	0.707	0.243	0.099	0.025	0.01	0.0025	0.001
$\theta°$	−5.78	−22.09	−45.72	−78.84	−91.44	−117.2	−161.0	−376.3	−806.2

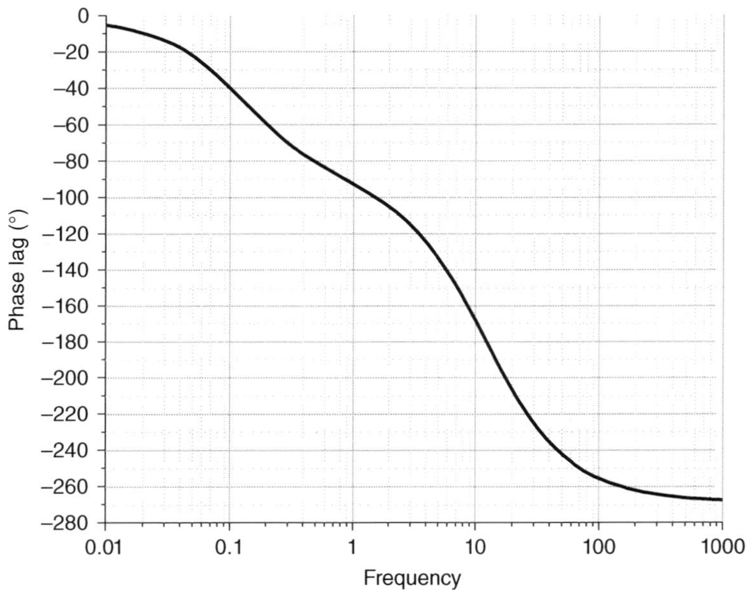

Figure 5.13 Phase lag for a "black-box" system.

Note how the amplitude ratio at very low frequencies is asymptotic to 1; as ω increases beyond about 1 rad/s, AR decreases by a factor of 10 for every 10-fold increase in frequency. Thus, a log–log plot of AR(ω) will ultimately exhibit a slope of −1. This, of course, tells us that we have a simple first-order system with a corner frequency of about 0.1 rad/s—such that $\tau \approx \frac{1}{0.1} = 10$. The phase lag initially appears to be headed toward −90°, but it begins to steepen dramatically at higher frequencies revealing the presence of a time delay. It is apparent that this system lag really begins to take effect at about 4 to 10 rad/s, but we can easily refine our estimate as follows: Consider θ at $\omega = 100$ rad/s, −806.2°. We know that the contribution of the first-order system is complete at 100 rad/s, so we remove 90° from the phase lag: −806.2 + 90 = −716.2°. Therefore,

$$-\frac{716.2}{57.296} = -12.5 \text{ rad} \quad \text{and consequently,} \quad -12.5 = -\omega\tau_{delay} = (-100)\tau_{delay}.$$

The time delay in the process must be about 12.5/100 = 0.125 (remember, we chose 1/8).

Now, let's see if we can use some frequency response data to develop the differential equation(s)—a process model—successfully. Assume we have collected the following data for $\theta(\omega)$ (Figure 5.13):

We see immediately that the process is third order overall since $\theta \to -270°$ as the frequency becomes large. However, there are two distinct parts to these phase lag data; it appears that θ was initially approaching −90° *before* the precipitous decline to −270°, meaning that we almost certainly have first-order and second-order systems working in series. Working backward from the right-hand side of the figure, we add 90° to get to one-half of the contribution of the second-order system: −180° occurs at $\omega = 11$ or 12 rad/s, meaning that the time constant for the second-order system is about 1/11 s. Next we move to the top of the figure and note that −45° occurs at about 0.12 rad/s such that the time constant for the first-order system is about 1/0.12 or 8.3 s. We have a rough model for this process consisting of two ordinary differential equations:

$$8.3\frac{dY}{dt} + Y = X(t) \quad \text{and} \quad \left(\frac{1}{11}\right)^2 \frac{d^2Z}{dt^2} + \frac{2}{11}\frac{dZ}{dt} + Z = Y(t). \tag{5.244}$$

An equivalent system could also be written as three first-order differential equations, where the last two would have the same time constant. Note that the differential equations we have obtained are linear with constant coefficients (as they must be with this approach), and they can be solved with methods described earlier in this chapter.

Let us summarize in a very precise way what our objective was in this section: When an analyst must develop a simulation for a complicated process for which no model exists, and it appears unlikely that the governing differential equation(s) can be determined from physical principles, then frequency response may be of value. In this procedure, the process is driven with a sinusoidal forcing function, and the output is measured as a function of frequency, ω. The data obtained will consist of a magnitude (or amplitude ratio) and a phase lag. These quantities are plotted as functions of ω; the graph of $r(\omega)$ should be a log–log plot, and we use a semi-log plot for the phase lag, $\theta(\omega)$. The two graphs are stacked vertically to produce a

Bode plot which makes extraction of the desired information easy. $\theta(\omega)$ will tell us immediately if there is a time delay or transportation lag in the process, and the ultimate slope of $r(\omega)$ will reveal the overall order of the system. With just a little practice, an analyst can learn to make reasonable estimates for the time constants of the model and, subsequently, by solving the ODEs, predict the behavior of the process in response to adjustments, disturbances, or excursions. We also observe that in many cases, the Bode plot will also give us an indication of system stability: If the *AR* is greater than 1 (one) when the phase lag is $-180°$, the system may be unstable (this is called the *Bode stability criterion*).

Finally, it may have occurred to you that the differential equations we are developing in this manner will be linear with constant coefficients, whereas the real system may be nonlinear. Can we still obtain a useful result in such cases? We must remember that with nonlinear systems *i*) superposition and scaling are not applicable, *ii*) a simple periodic input may result in generation of harmonics, *iii*) the response may be aperiodic, and, *iv*) in some cases we might even see chaotic behavior.

In the chemical process industry many systems—even complex ones—can be adequately modeled as first-order processes with a time delay (FOPTD). But what about other nonlinear systems that simply cannot be represented that way? Consider a process governed by the ODE:

$$\frac{dY}{dt} + aY + bY^2 + cY^3 = X(t) = A\sin(\omega t) \quad \text{with} \quad Y(0) = 0 \quad \text{and} \quad a = 1/2, b = 1/8, \text{and } c = 1/3. \tag{5.245}$$

We can conduct some numerical experiments with this ODE and obtain frequency response data:

Freq (rad/s)	AR	Phase Lag
0.03	0.7720	−0.93°
0.08	0.8725	−2.9°
0.1	0.7712	−3.15°
0.3	0.7634	−9.97°
0.8	0.6875	−31.17°
1	0.6414	−38.37°
3	0.3108	−70.49°
8	0.1236	−77.92°
10	0.0994	−80.22°
14	0.0713	−87.91°
20	0.0498	−86.85°

If we plot these data we will see that an elementary linear model cannot reproduce the tabulated behavior in detail. The reader is encouraged to try this and discover that although the frequency response *appears* to be roughly similar to an elementary first-order system, such a model will not be adequate—it will be incapable of reproducing the observed variations. And, should the forcing function assume some other form—which is a certainty in an actual process—the simple linear first-order model will *not* represent the process dynamics. Indeed, a serious problem with nonlinear systems is that the transfer function relating system output to input *may change* as the input frequency is increased. It may be possible to apply frequency response locally, but the results will not be applicable globally. The interested reader should be aware of the fact that in recent years much effort has been expended on the use of nonlinear frequency response as a tool for model identification.

5.11 Conclusion

Toward the end of this chapter we introduced a couple of techniques by which the analyst could obtain an approximate solution for an ordinary differential equation, solution by power series and regular perturbation. Both can be used to advantage, but there are other more modern, and perhaps more expedient ways to obtain analytic approximations. We will briefly introduce one such technique here, the *variational iteration method* (VIM), but we will defer a more complete discussion for Chapter 9, *Integro-Differential Equations*, where VIM will prove to be extraordinarily powerful.

Consider the linear, first-order ordinary differential equation,

$$\frac{dy}{dx} + 2xy = 4x, \tag{5.246}$$

with the analytic solution, $y = 2 + C_1 \exp(-x^2)$. We will choose $y(x=0) = 4$ such that $C_1 = 2$, and assume we are interested in the behavior of $y(x)$ for $0 \leq x \leq 1$. To use VIM, we start with a simple estimate for y that satisfies the boundary condition, and we will select $y_0 = x + 4$. Though this is not an optimal choice, we will still get to an acceptable approximation as we shall see. We now rearrange the ODE eq. (5.246) and use it—integrated—to improve our initial guess:

$$y_{n+1} = y_n - \int_0^x \left[y_n' + 2sy_n - 4s\right] ds. \tag{5.247}$$

Of course, $y_0' = 1$, and after we carry out the integration we find

$$y_1 = -\frac{2}{3}x^3 - 2x^2 + 4, \tag{5.248}$$

and then successively:

$$y_2 = \frac{4}{15}x^5 + x^4 - 2x^2 + 4 \tag{5.249}$$

and

$$y_3 = -\frac{8}{105}x^7 - \frac{1}{3}x^6 + x^4 - 2x^2 + 4. \tag{5.250}$$

Let us find out how well this sequence of approximations actually describes $y(x)$.

x	y_{analytic}	$y_0 = x + 4$	y_1	y_2	y_3
0	4	4	4	4	4
¼	3.8788	4.25	3.86458	3.87917	3.87882
½	3.5576	4.50	3.41667	3.57083	3.5567
¾	3.13957	4.75	2.59375	3.25469	3.12191
1	2.73576	5	1.33333	3.26667	2.59048

Even though we made a poor choice for our initial trial function, $y_0 = x + 4$, we have rapidly closed in on the values provided by the analytic solution. With just a few more iterations we would have an exceptionally good match to $y(x)$. The student may wish to repeat this example by setting $y_0 = 4 - x$; how well do y_2 and y_3 agree with the analytic solution now? VIM is an incredibly useful tool for finding approximate solutions for difficult equations, and we will gain a much greater appreciation for its value in Chapter 9.

Ordinary differential equations are common in applied mechanics, as many force, mass, and energy balance problems lead to ODEs. The spectrum of application is broad, ranging from chemical kinetics to electronic circuits to mechanical flutter (aeroelasticity). It is obviously important that technical professionals be able to solve such problems with speed and accuracy. Our intent with this chapter is to provide the reader with a practical review of solution methods that have proven to be of value for frequently encountered ordinary differential equations.

Problems

5.1 Use the integrating factor approach to solve the ordinary differential equations,

$$\frac{dy}{dx} + 3xy = 2x$$

and

$$x^2\frac{dy}{dx} - xy = x^3 + 4$$

5.2 Check the following differential equations to see if they are exact. If they are not exact, solve using the integrating factor approach.

$$x\frac{dy}{dx} = y - x^2 + 1$$

$$xy^3\frac{dy}{dx} = x^4 + y^4$$

5.3 A rifle bullet (130 grain WIN 270) leaves the barrel with an initial (muzzle) velocity of 3060 ft/s. Suppose the rifle is fired at an angle of 45° over level ground and that the only force acting upon the bullet in flight is gravity. Find the maximum elevation achieved by the bullet, and its maximum horizontal travel.

It may be obvious to you that the answers you obtained for the first part of this problem are quite unrealistic. Drag is very important! Fortunately, we have ballistic information available for the WIN 270 as follows:

Distance Yards	0	100	200	300	400	500
Velocity (ft/s)	3,060	2,802	2,559	2,329	2,110	1,904

Notice that the rifle bullet loses about 38% of its initial velocity over the first 500 yards of travel. Devise a strategy to use these data to add drag force to your original differential equation and repeat your analysis. What is the maximum height achieved by the bullet under these more realistic conditions?

5.4 The example that accompanied Figure 5.3 was based upon the differential equation,

$$\frac{dy}{dx} = xy(y-2).$$

We want to use the technique described in the text to explore the behavior of this nonlinear equation; connect a sequence of very short tangent lines obtained from the slope(s) at particular points (x_i, y_i) to develop a semi-quantitative picture for this ODE. We will focus upon the first quadrant, but only look at values of *y greater than* 2.

5.5 Consider the wheel-tire assembly on a vehicle equipped with IFS (independent front suspension). The assembly is subjected to a bump (perhaps a pothole), and the motion of the mass is constrained by a spring-shock absorber (or strut) combination. The resulting motion is described approximately by the second-order differential equation,

$$\tau^2\frac{d^2z}{dt^2} + 2\tau\zeta\frac{dz}{dt} + z = F(t).$$

z corresponds to the vertical position of the assembly, τ is the time constant, and ζ is the damping coefficient. The challenge in suspension design, of course, is to keep the tire in contact with the pavement (for consistent traction) yet provide a tolerably smooth ride for the occupants. Suppose the time constant has the value 0.125 s and that the damping coefficient is 0.55; we then find

$$\frac{d^2z}{dt^2} + 8.8\frac{dz}{dt} + 64z = G.$$

If $F=1$ for $t>0$ (thus $G=64$), what is the ensuing motion of the assembly? If the shock absorber is completely worn out, the damping coefficient will be smaller, say, 0.15. What impact will that have upon the motion if the forcing function is exactly the same? Two examples are shown in Figure P5.5 to provide some guidance on what is expected from this ODE.

5.6 We need to analyze the performance of two identical CSTRs in series; this arrangement is being used to carry out the first-order, irreversible decomposition of species "*A*." Both reactors are perfectly mixed and each has a volume of 500 gal. Reactant is introduced into the first reactor (and the initial, steady-state feed concentration is 2 mol/gal) at a volumetric flow rate of 25 gpm. The mass balance for species "*A*" in the first reactor takes the form:

[Rate in] – [Rate out] – [Rate consumed] = [Accumulation]

$$\dot{V}C_{Ain} - \dot{V}C_{A1} - V_T k_1 C_{A1} = V_T\frac{dC_{A1}}{dt}$$

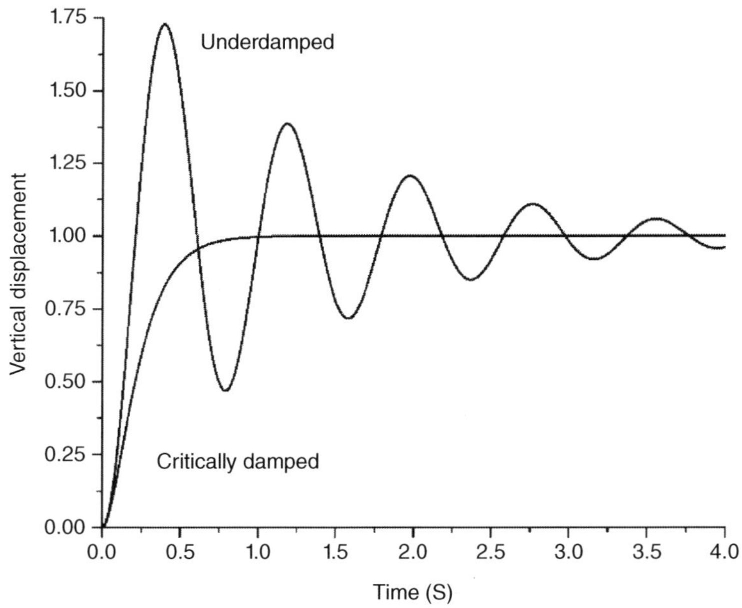

Figure P5.5 Suspension behavior for two cases, underdamped with $\zeta = 0.10$ and critically damped with $\zeta = 1.0$. The compromise inherent in suspension design is obvious: Very rapid response produces overshoot and oscillation.

We divide by the volumetric flow rate, $\dot{V}$, and rewrite the equation,

$$C_{Ain} - (1 + k_1\tau)C_{A1} = \tau\frac{dC_{A1}}{dt}.$$

Let $k_1 = 0.033\ \text{min}^{-1}$ and note that $\tau = 20\ \text{min}\ (V_T/\dot{V})$. At $t = 0$, the feed concentration to CSTR 1 doubles; it is instantaneously increased from 2 mol/gal to 4 mol/gal. Find the analytic solution for $C_{A1}(t)$ and then use that result in the mass balance (for "A") on reactor 2. When will $C_{A2}(t)$ achieve 75% of its ultimate change?

5.7 A pipe carries a very hot fluid, and circular fins (actually annular) are installed on the outer pipe wall to cast off unwanted thermal energy to the surroundings. Each fin extends from the outer pipe wall (at $r = R_1$) to the radial position, R_2. By assuming that the fin is thin and that it is made from a material with high thermal conductivity, the principal variation in temperature will occur in the r-direction (i.e., the temperature is nearly uniform in the transverse, or z-, direction). Assuming that thermal energy is lost from the fin's surface according to Newton's law of cooling, an energy balance reveals:

$$\frac{d^2\theta}{dr^2} + \frac{1}{r}\frac{d\theta}{dr} - \frac{2h}{wk}\theta = 0,$$

where h is the heat transfer coefficient, k is the thermal conductivity of the fin, and w is the width of the fin in the z-direction. $\theta = (T - T_\infty)$, where T_∞ is the temperature of the surrounding air. Solve this ordinary differential equation and find the temperature distribution in the fin given the following parametric values:

$$T(r = R_1) = 500\,^\circ\text{F}\ T_\infty = 60\,^\circ\text{F}$$

$$R_1 = 0.25\ \text{ft}\ R_2 = 0.50\ \text{ft}\ w = 0.0104\ \text{ft}$$

$$h = 7.5\ \text{Btu/(hr ft}^{2\,\circ}\text{F)}\ k = 30\ \text{Btu/(hr ft}\,^\circ\text{F)}$$

Assume that no thermal energy is lost through the end of the fin, that is,

$$-k\left.\frac{dT}{dr}\right|_{r=R_2} = 0.$$

5.8 A cylindrical tank with diameter of 12 ft and a depth of 10 ft is initially filled with water. At $t = 0$ a circular hole with a diameter of 2 in. is opened in the side wall at the bottom. The velocity through this orifice is approximately given by Torricelli's theorem, $V_0 = \sqrt{2gh}$; therefore, the initial velocity will be $V_0 = \sqrt{(2)(32.17)(10)} = 25.37$ ft/s. Since the orifice diameter is 2 in., the initial volumetric flow rate will be about 0.55 ft^3/s or 248 gpm. Find out when the tank will be 90% empty (when $h = 1$ ft). Then, in the second part of this problem, the tank is drained from its initial depth of 10 ft just as before, but this time water is also added continuously at a rate of 55 gpm. When will the depth of water in the tank get within 3% of its ultimate value?

5.9 A particle with a diameter of 1 mm and a density of 1.45 g/cm^3 is released at the surface of a flowing stream of water. The velocity of water in the stream varies from 130 cm/s at the surface ($y = 0$) to 0 cm/s at the bottom ($y = 175$ cm) according to: $V = 130 - 0.004245\,y^2$ (cm/s). Assume the particle is perfectly entrained by the moving water (its velocity in the stream direction always corresponds to the velocity of the water). If the only forces acting upon the particle in the y-direction are gravity and drag, determine how long it will take the particle to reach bottom, and how far it will travel downstream during the settling process. The drag force is given by $F = AKf$, where A is the frontal area of the particle, K is the kinetic energy of the fluid (relative to the particle) per unit volume, and f is the drag coefficient:

$$A = \pi R^2 K = \frac{1}{2}\rho v_y^2 \quad \text{and} \quad f \approx 0.5$$

5.10 Consider the ordinary differential equation.

$$(1 - x)\frac{dy}{dx} = 2x - y \quad \text{with} \quad y(x = 0) = 2.$$

Find a power series solution for this ordinary differential equation and compare your result to the solution you obtain using an integrating factor.

5.11 A thermocouple (TC) well projects into a hot gas stream from a vertical wall located at $z = 0$. The radius of the cylindrical well is R and its length is L. The wall is maintained at a temperature, T_w, which is significantly lower than that of the hot gas stream (T_g). We want to develop a differential equation by making a thermal energy balance on a segment of the well of length, Δz. The temperature of the well is virtually constant in the r-direction. Assume that heat transfer *in the well itself* occurs by conduction (in the negative z-direction); heat transfer *from the gas stream to the well* occurs by several mechanisms approximately described by Newton's law of cooling:

$$Q = 2\pi R \Delta z h(T_g - T).$$

Show that this steady-state balance results in the differential equation,

$$\frac{d^2T}{dz^2} - \frac{2h}{kR}(T_g - T) = 0.$$

The boundary conditions are: at $z = 0$, $T = T_w$, at $z = L$, $dT/dz = 0$ (there is very little heat transfer through the end of the TC well). Because the conductivity of the aluminum well is large, there is concern that the measured temperature will not accurately represent the gas stream (T_g). Given the following parameters:

$$L = 7.5\text{ cm}\; R = 0.5\text{ cm}\; T_w = 100\,^\circ\text{C}\; T_3 = 325\,^\circ\text{C}\; k = 270\text{ W/(mK)}\; h = 55\text{ W/(m}^2\text{K)},$$

what temperature will be recorded at the end of the well where the TC junction is located? If the error in the measurement is significant, what changes would you recommend? *Note: The thermal conductivity of type 304 stainless steel is 16.2 W/(mK).*

5.12 We want to use regular perturbation to find an approximate analytic solution for the example we explored at the end of the chapter where:

$$(a + bT)\frac{d^2T}{dy^2} + b\left(\frac{dT}{dy}\right)^2 = 0.$$

We will use the same terminal temperatures (−200° and +700°); however, in this case we will take $k = 87.7778 - 0.1111T$. This will cause a significant change in the profile, $T(y)$, which is illustrated in Figure P5.12. In this case it will be necessary for us to find T_0, T_1, and T_2. Is the resulting approximation an adequate representation for the profile, $T(y)$, shown in Figure P5.12?

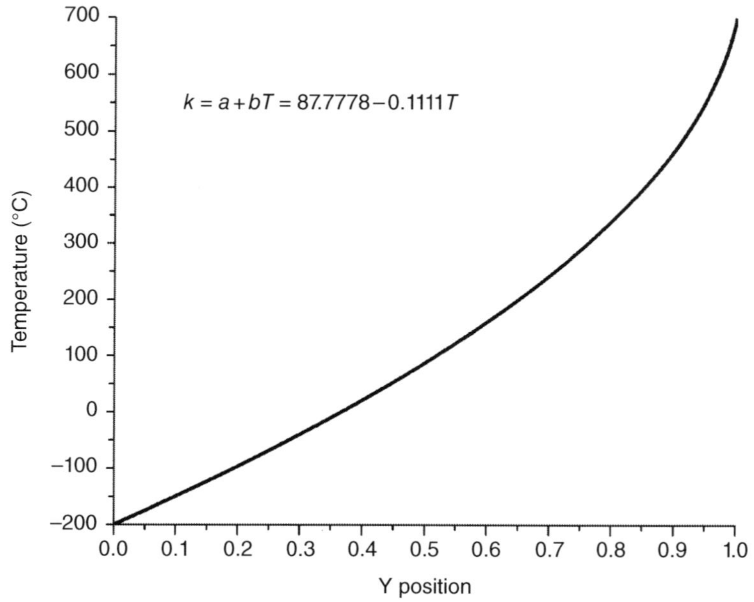

Figure P5.12 Exact solution for steady conduction in a slab of material for which $k = 87.7778 - 0.1111\,T$. The ends of the slab are maintained at −200° and +700° for all t.

5.13 *Please note: This problem is more difficult! You may find Section 2 of Chapter XVI in Gray and Mathews to be helpful.* Suppose we have a length of metal rod or wire positioned vertically. The bottom end is clamped in a vise; we want to determine the length of the wire segment that just results in instability (i.e., when the top of the wire is just ready to fall over). Let β be the rigidity of the wire or rod, and let W be the weight per unit length. The variable x is the vertical position above the clamped end and y is the lateral deviation from vertical (rigid) position. The total length of the wire is L. If the lateral deviation is not too large then,

$$\beta\frac{d^2y}{dx^2} \cong \int_x^L W(y' - y)dx'.$$

Differentiate this equation with respect to x and set $p = \frac{dy}{dx}$:

$$\frac{d^2p}{dx^2} + \frac{W}{\beta}(L - x)p = 0.$$

Now set $L - x = r^{2/3}$ and take $p = r^{1/3}z$. Show that these substitutions result in

$$\frac{d^2z}{dr^2} + \frac{1}{r}\frac{dz}{dr} + \left(\frac{4W}{9\beta} - \frac{1}{9r^2}\right)z = 0,$$

which is a form of Bessel's differential equation. At the top of the wire (the free end) where $x = L$ (or $r = 0$) we must have $\frac{dp}{dx} = 0$ or $r^{1/3}\frac{dp}{dr} = 0$. Find the critical length of the vertical wire.

5.14 The Leveque analysis applies to heat and mass transfer near a surface where the velocity distribution can—at least locally—be described by $v_x = ay$. If we confine our attention to the region very close to the wall, this linear approximation should be valid. For steady-state conditions then, the mass transfer process will be governed by:

$$ay\frac{\partial C}{\partial x} = D\frac{\partial^2 C}{\partial y^2}.$$

By appropriate scaling of the independent variables x and y, namely:

$$\eta = y\left(\frac{a}{9Dx}\right)^{1/3},$$

this partial differential equation can be transformed into a second-order ODE:

$$\frac{d^2C}{d\eta^2} + 3\eta^2 \frac{dC}{d\eta} = 0.$$

Prove that this transformation yields the result shown immediately above, and solve this ordinary differential equation by first reducing the order; that is, let $\phi = \frac{dC}{d\eta}$. Then integrate twice to confirm that:

$$C = \int_0^{\eta} A_1 \exp(-\eta^3) d\eta + A_2.$$

Apply the appropriate boundary conditions: $C = C_0$ at the wall where both y and $\eta = 0$, and $C = C_\infty$ for both y and $\eta = \infty$, to show that $C_2 = C_0$ and $C_1 = \frac{C_\infty - C_0}{\Gamma\left(\frac{4}{3}\right)}$. Please note that Γ is the *gamma function* defined by

$$\Gamma(n) = \int_0^{\infty} x^{n-1} \exp(-x) dx.$$

The recurrence formula is particularly useful, $\Gamma(n+1) = n\Gamma(n)$, and a short table of values for $\Gamma(n)$ is provided here to assist you:

n	$\Gamma(n)$
1.0	1.000
1.1	0.951
1.2	0.918
1.3	0.897
1.4	0.887
1.5	0.886
1.6	0.894
1.7	0.909
1.8	0.931
1.9	0.962
2.0	1.000

5.15 Two identical vented tanks, each 10 ft high and 6 ft in diameter, are interconnected by a straight length of pipe (with a valve in the middle) at the bottom (i.e., at $z = 0$). Tank 1 also has an opening equipped with a valve at the bottom to permit discharge into a floor drain. Tank 1 is completely full of water initially, and tank 2 is completely empty. At $t = 0$, *both* valves are opened (wide) and water begins to flow. Develop a model for this system by mass balance and solve the ordinary differential equations so we can predict both $h_1(t)$ and $h_2(t)$. We will assume that the velocity in the interconnecting pipe can be described by: $V_2 = b(h_1 - h_2)$. The velocity of the water leaving tank 1 destined for the floor drain is given by: $V_0 = ch_1$. Assume that $b = c$ and that both are equal to 0.027 cubic foot per second per foot ($ft^3/s/ft$). What would the maximum value of h_2 be? It would be somewhat more realistic to assume that the velocities were proportional to the square root of the head, e.g., $V_0 = \sqrt{ch_1}$. If these changes are made, can the problem still be solved analytically?

5.16 A horizontal beam is being designed to carry a "skywalk" from one side of a shopping mall to the other, a span of L ft. The intent is to allow shoppers easy passage across a large, open, common area. It is anticipated that the maximum load to be carried will be W lb_f/ft (that is per foot of length, of course). We need to determine the maximum amount of deflection, and for the sake of customer confidence it is not to exceed 0.5 in. The beam is rigidly clamped at both ends (at both $x = 0$ and $x = L$). The beam's deflection—if any occurs—will be in the negative y-direction. Assume that the appropriate model has the form,

$$EI\frac{d^2y}{dx^2} = K + \frac{1}{2}WLx - \frac{1}{2}Wx^2,$$

where E is the modulus of elasticity and I is the moment of inertia. What is the maximum value of the deflection in terms of L?

5.17 The Chebyshev (Tschebysheff) differential equation is:

$$(1-x^2)\frac{d^2y}{dx^2} - x\frac{dy}{dx} + n^2y = 0.$$

This ODE is of the Sturm–Liouville (SL) type, and of course it does not have constant coefficients; problems of the Sturm–Liouville type have the general form:

$$\frac{d}{dx}\left[p(x)\frac{dy}{dx}\right] + [q(x) + \lambda r(x)]y = 0$$

with $a \le x \le b$, $A_1y(a) + A_2y'(a) = 0$, and $B_1y(b) + B_2y'(b) = 0$. Solutions for an SL problem can only be found for certain values of the parameter, λ; hence, they are termed *eigenvalue* problems. Solutions for the Chebyshev ODE are given by Chebyshev polynomials and written as: $y = T_n(x)$. Consequently, if $n = 4$ then $y = T_4(x) = 8x^4 - 8x^2 + 1$. Set $n = 4$ such that $n^2 = 16$ in the Chebyshev ODE and then verify that $T_4(x)$ is a solution. Then let $n^2 = 64$ so that the solution is $T_8(x)$. *Without* looking up the T_8 polynomial, try to determine the form of the solution for this case (it would comprise *even* powers of x added to the constant 1). We note that the Chebyshev polynomials are orthogonal with respect to a weighting function:

$$\int_{-1}^{+1} T_n(x)T_m(x)\frac{1}{\sqrt{1-x^2}}dx = 0$$

if $n \neq m$. Would you need additional information to arrive at a solution for the $n = 8$ case?

5.18 Laguerre's differential equation is:

$$x\frac{d^2y}{dx^2} + (1-x)\frac{dy}{dx} + ny = 0.$$

This ODE arises in the study of harmonic oscillators in quantum mechanics and solutions (for integer values of n) are the Laguerre polynomials: $L_0 = 1$, $L_1 = 1 - x$, $L_2 = 1 - 2x + \frac{1}{2}x^2$, $L_3 = 1 - 3x + \frac{3}{2}x^2 - \frac{1}{6}x^3$, etc. Laguerre polynomials are orthogonal with respect to the weight function, $\exp(-x)$, on the interval, $0 \le x < \infty$, so for example,

$$\int_0^\infty e^{-x}(1-x)\left(1 - 2x + \frac{1}{2}x^2\right)dx = 0.$$

Investigate the *method of Frobenius* (you might find Kreyszig's book, *Advanced Engineering Mathematics*, Wiley & Sons, 1972, very useful), and see if you can use it to verify the solution for $n = 2$.

5.19 A cable is suspended between two supports located at $x = \pm L$, and naturally it will sag due to its own weight. Let W_C be the weight of the cable per unit length. We place the origin at the center of the cable, letting y represent the vertical position at x. The shape assumed by the sagging cable—called a catenary—is governed by

$$\frac{d^2y}{dx^2} = \frac{W_C}{F}\sqrt{1 + \left(\frac{dy}{dx}\right)^2}.$$

F, the horizontal force in the cable, is a constant. We want to solve this ordinary differential equation to find $y(x)$. If W_C is 10 kg/m, and $F = 40$ N, what is the value of $y(x = 2\text{ m})$?

Answer: 0.5105 m

5.20 Marshall and Pigford (*The Application of Differential Equations to Chemical Engineering Problems*, University of Delaware, 1947) presented an approximate analysis of a jet of viscous liquid falling vertically under the influence of gravity. Of course the diameter of the jet will diminish as it stretches (is elongated) in the x-direction, and the result is an increase in velocity with distance from the orifice. The governing nonlinear differential equation is:

$$3\frac{d^2\phi}{dx^2} - \frac{3}{\phi}\left(\frac{d\phi}{dx}\right)^2 + \phi\frac{d\phi}{dx} = 1$$

with $\phi(x=0)=\phi_0$ and $d\phi/dx=0$ at $x=0$. ϕ is the dimensionless velocity, $\phi = V/(gv)^{1/3}$ and x is the dimensionless distance from the orifice: $x = z(g/v^2)^{1/3}$. This is a formidable nonlinear ODE, but a numerical solution can be obtained rather easily; results are shown in the accompanying figure with $\phi_0 = 0.5$ (Figure P5.20).

Marshall and Pigford suggest that an analytic solution can be found for this problem *if* one neglects the second (inertial) term, in effect, limiting the applicability of the model to low velocities. Try this, and compare your approximate results to the computed $\phi(x)$ shown above.

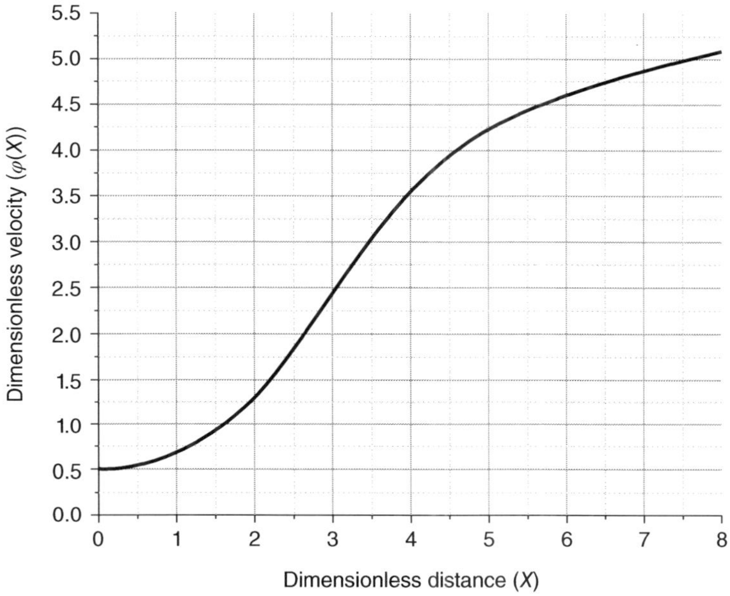

Figure P5.20 Numerical solution for the nonlinear ODE, $3\phi'' - \frac{3}{\phi}\phi'\phi' + \phi\phi' = 1$.

5.21 Let's carry out a power series solution for the first-order ordinary differential equation,

$$(x+1)y' - (x+2)y = 0$$

with $y(0)=1$. We will take

$$y = \sum_{n=0}^{\infty} a_n x^n$$

and note that $a_0 = 1$. This problem is appealing for a couple of reasons: First, we can get a reasonable approximation with a limited number of terms, and second, this ODE can be solved rather easily to given us a nice comparison. Find the proper coefficients a_0 *through* a_4 at a minimum, then compare the truncated series to the analytic solution for $0 \le x \le 2$. How large is the error at $x = 2$?

5.22 Consider the following second-order ordinary differential equation:

$$(1-x^2)y'' + (1+x)y' + 3y = 0,$$

with $y(0)=1$ and $y'(0)=0$ for $0 \le x < 1$. We observe that the equation is linear but that it does not have constant coefficients. Our plan is to attempt a power-series solution, so we let

$$y = a_0 + a_1x + a_2x^2 + a_3x^3 + a_4x^4 + \ldots = \sum_{n=0}^{\infty} a_n x^n.$$

Of course, we immediately note that $a_0 = 1$ and $a_1 = 0$. Identify a reasonable number of coefficients such that the series may adequately represent the solution. Can you—at least partially—assess the accuracy of your series without solving the equation numerically?

Next, use the power series technique to solve the homogeneous ODE:

$$y'' + xy' + \left(1 - \frac{2}{x^2}\right) y = 0,$$

for $1 \leq x \leq 10$ with $y(1) = 0$ and $y'(1) = 1$. Some numerical results are provided here to assist you:

x	1	1.25	2	3	4	5	6	8	9	10
y	0	0.220	0.558	0.532	0.413	0.331	0.276	0.207	0.184	0.166

5.23 A negative feedback control loop is subjected to a unit step change in set point. The loop output variable, $C(t)$, will deviate from its initial value of zero in an effort to accommodate the new setting—ideally, $C(t)$ should approach 1 (one). After the differential equations modeling the components of the loop have been transformed, and block algebra has been applied, the response variable in the s-plane is:

$$C(s) = \frac{K(2s + 6)}{s^4 + 6s^3 + 11s^2 + (6 + 6K)s}.$$

K is the controller parameter (gain), and we are particularly interested in the values $K = 5$, 10, and 15. Find the loop response by partial fraction expansion and determine whether any of the gains result in unstable operation. If not, is the $C(t)$ behavior desirable? To answer this question, consider overshoot (by how much does $C(t)$ exceed 1), speed of response, and permanent offset (by how much does $C(t \to \infty)$ differ from 1).

Helpful hint: Inversion of a Laplace transform by partial fractions can be laborious, and it is all too easy to make an arithmetic mistake, producing an incorrect result. There are some simple things we can do to make this less likely. For example, in a process control problem like this one, $C(t = 0) = 0$; i.e., the system is at equilibrium until the controller set point is changed. The solution for $K = 5$ is given below, and we note that for $t = 0$, $C(t = 0) = 0.833 + 0.141 – 0.974 = 0$. We can also apply the final value theorem to the transform itself:

$$\int_{t\to\infty} [f(t)] = \int_{s\to 0} [sf(s)].$$

Therefore, we multiply $C(s)$ given above by s and set $s = 0$; this results in $6K/(6 + 6K)$, so if $K = 5$, the ultimate value of $C(t)$ should be $30/36 = 0.833$. This is confirmed by the answer given below. Some caution must be exercised in applying this theorem; however,—note that for this problem

$$sf(s) = \frac{K(2s + 6)}{s^3 + 6s^2 + 11s + (6 + 6K)}.$$

If $K = 15$, the cubic polynomial in the denominator is $s^3 + 6s^2 + 11s + 96$ with roots $s = 0.278 \pm 3.817i$ and $s = -6.556$. The complex root has a positive real part, and this means that the behavior in the time domain is *unbounded*, violating the conditions of applicability for the final value theorem.

Answer for $K = 5$: $C(t) = 0.833 + 0.141 \exp(-5.214t) - 0.974 \exp(-0.393t) \cos(2.598t)$ *and* $C(t \to \infty)$ *is 16% under the new set point.*

5.24 We consider viscous flow in a convergent channel where the fluid is moving toward the origin in the x–y plane. Two different approaches have been used for this problem: we consider first the boundary-layer simplification where the fluid velocity is zero at the wall (one wall corresponds to the x-axis, $y = 0$), then asymptotically approaches the potential flow velocity (V) as y increases. *Note: This is one of the very few boundary-layer problems where a closed-form solution is possible.* In this treatment we assume the fluid is moving toward a *sink* located at the origin (this sink is a singular point where we assume the fluid is being removed at *very large* velocity). The usual approach in boundary-layer theory is to use a similarity transformation, that is, we obtain an ordinary differential equation by using a scaling between the independent spatial variables x and y: $\eta = \frac{y}{x}\sqrt{\frac{Q}{2\pi\nu}}$. A new dependent variable (f) is chosen that relates to the stream function such that the derivative of f is the quotient of the local and freestream (potential) velocities: $f' = v/V$. Therefore, $f' \to 1$ as we move across the boundary layer (away from the wall). Schlichting (1968) shows that the resulting ODE is

$$f''' - f'^2 + 1 = 0 \tag{5.251}$$

with the boundary conditions: $f'(0) = 0, f'(\infty) = 1$, and $f''(\infty) = 0$. If we multiply (1) by f "we can immediately integrate to obtain

$$f''^2 - \frac{2}{3}(f' - 1)^2(f' + 2) = 0.$$

Our interest here is the velocity distribution, so we set $\beta = f'$:

$$\left(\frac{d\beta}{d\eta}\right)^2 = \frac{2}{3}(\beta - 1)^2(\beta + 2).$$

See if you can verify that

$$\beta = \frac{v}{V} = 3\tanh^2\left(\frac{\eta}{\sqrt{2}} + 1.1462\right) - 2.$$

5.25 *Please note: This problem requires additional effort and it is intended for students with a particular interest in the solution of more difficult fluid flow problems.* In the previous problem we alluded to the fact that two different solution approaches to the problem of viscous flow between convergent walls have been employed, the boundary-layer simplification used in 5.24. and an analytic solution using the Navier–Stokes (NS) equations in cylindrical coordinates. The latter appears to have been first carried out by G. B. Jeffery (Steady Motion of a Viscous Fluid, *Phil. Mag.*, Series 6, 29:455, 1915) and G. Hamel (Spiralförmige Bewegung zäher Flüssigkeiten, Jahresber d. Dt. *Mathematiker-Vereinigung*, 25:34, 1916). We assume that the radial component of velocity can be written

$$v_r = \frac{\nu}{r}F(\theta);$$

this is introduced into the NS equations and by eliminating pressure from the r- and θ-components we obtain an ordinary differential equation:

$$2FF' + 4F' + F''' = 0.$$

This ODE can be integrated once by inspection:

$$F^2 + 4F + F'' + k = 0. \tag{5.252}$$

The k appearing here can be a function of r, but not of θ; in fact, it is the radial pressure gradient at the wall. The included angle between the convergent walls is 2α such that one wall is located at $-\alpha$ and the other at $+\alpha$; therefore, the boundary conditions are

$F = 0$ for $\theta = \pm\alpha$ and $F' = 0$ for $\theta = 0$ (symmetry at the centerline).

We want to find an analytic solution for eq. (5.252), and some guidance is provided here: Consider the nonlinear ODE,

$$\left(\frac{dy}{dx}\right)^2 = a + by + cy^2 + dy^3 + ey^4, \tag{5.253}$$

where a, b, c, d, and e are constants. Now suppose we differentiate eq. (5.253):

$$2y'y'' = by' + 2cyy' + 3dy^2y' + 4ey^3y'.$$

and divide by y':

$$2y'' = b + 2cy + 3dy^2 + 4ey^3 \text{ or } y'' = \frac{b}{2} + cy + \frac{3}{2}dy^2 + 2ey^3.$$

Compare this equation with eq. (5.252): we find $b/2 = -k$, $c = -4$, $d = -2/3$, and $e = 0$. Putting these values into eq. (5.253):

$$\left(\frac{dy}{dx}\right)^2 = -2ky - 4y^2 - \frac{2}{3}y^3.$$

Of course, the significance here is that eq. (5.253) can be transformed to a standard form so that the solution may be written in terms of the Jacobi elliptic functions.

You will find it helpful to consult Chapter 7, Section 11, pages 209–211, in Davis (*Introduction to Nonlinear Differential and Integral Equations,* 1962). Rosenhead (*Proceedings of the Royal Society of London*, Series A, 175:436–467, 1940) and Landau and Lifshitz (*Fluid Mechanics,* Pergamon Press, 1959) may also be useful.

5.26 *This exercise ties together concepts and elements from Chapters 4, 5, and 7.* You may recall that Bessel's ordinary differential equation was introduced in Section 5.6; like many similar ODEs, solution of Bessel's equation was obtained by expansion in power series. These efforts produced results such as:

$$J_n(x) = x^n \sum_{m=0}^{\infty} \frac{(-1)^m x^{2m}}{2^{2m+n} m!(n+m)!}. \tag{5.254}$$

Therefore,

$$J_0(x) = 1 - \frac{x^2}{2^2(1!)^2} + \frac{x^4}{2^4(2!)^2} - \frac{x^6}{2^6(3!)^2} + \dots \tag{5.255}$$

This series is easily automated to produce desired values for J_0:

```
#COMPILE EXE
#DIM ALL
    REM *** Bessel function calculation for J0 by infinite series
    REM *** Program designed to compute Jo(x) for 0<x<10
            GLOBAL x,J0,sum,zz AS DOUBLE
            GLOBAL n,m,i,mfact,mfact0 AS INTEGER
FUNCTION PBMAIN
            OPEN "c:BesselJ0.dat" FOR OUTPUT AS #1
            x=0.1:mfact0=1
    50 REM *** continue
            n=0:mfact=1:sum=1
                FOR m=1 TO 24
                    IF m<2 THEN 200 ELSE 100
    100 REM *** continue
                    mfact=mfact0
                    FOR i=m TO 1 STEP -1
                 mfact=mfact*(i)
                    NEXT i
                        GOTO 300
    200 REM *** continue
                mfact=1
    300 REM *** continue
                        sum=sum+(-1)^m*x^(2*m)/(2^(m*2)*mfact^2)
                        REM print x,sum
                        NEXT m
                            WRITE#1,x,sum
                            PRINT x,sum
                                x=x+0.1
                                IF x<10 THEN 50 ELSE 400
    400 REM *** continue
                INPUT "end program now?";zz
                IF zz>0 THEN CLOSE
                END
END FUNCTION
```

For example, this series calculation produces values such as:

x	0.5	1	2	4	8
$J_0(x)$	0.93847	0.765198	0.223891	−0.39715	0.17165

You can compare these results with the table provided in Section 5.6. For larger values of the argument x, one can easily get estimates for J_0 using $J_0(x) \cong \sqrt{\frac{2}{\pi x}} \cos\left(x - \frac{\pi}{4}\right)$.

Thus, $J_0\ (10) \approx -0.2468$ (with an error of about 0.4%), $J_0\ (20) = 0.1667$, and $J_0\ (30) = -0.0859$.

Although many computational devices and commercial software packages can compute values for Bessel functions, some cannot. This can be an important liability in certain applications as we shall see clearly in Chapter 7. For example, consider the integral

$$\int_0^2 (100r - 50r^2 + 30r^4) J_0(1.2024r)dr. \tag{5.256}$$

The *TI-nspire CAS* can compute definite integrals, but Bessel functions are not included in the catalog (however, third-party libraries with Bessel function capability can be downloaded). This expression eq. (5.256) has come about in the solution of a *partial differential equation* governing the behavior of the temperature distribution in an infinitely long cylinder of radius 2. In this case there was an initial temperature distribution in the cylinder, which was subjected to an instantaneous change in surface temperature (to 0°) at $t = 0$ (note that a few elementary integrals involving Bessel functions are described in Carslaw and Jaeger, 1959). The initial $T(r)$ in the cylinder was

$$T(r) = 100 - 50r + 30r^3;$$

thus, the center temperature was 100° and at the cylinder surface ($r = 2$), 240°. The minimum was 75.155° located at $r = 0.74536$.

The integral eq. (5.256) above must be evaluated in order to determine the analytic solution for the problem. If one does not have access to a software package with built-in Bessel function capability, then it will be necessary to compute $J_0(x)$ from the infinite series. Let us begin by computing a value for the integral (for the first eigenvalue, $\lambda = 1.2024$) using *Mathcad*™:

$$\int_0^2 (100 \cdot r - 50 \cdot r^2 + 30 \cdot r^4) \cdot J_0(1.2024 \cdot r)dr.$$

This yields 84.958; we can compare the *Mathcad* result with the value produced by computation of J_0 from the series coupled with trapezoid rule integration (using double precision): 84.9577. Use the method of your choice to evaluate the integral eq. (5.256) for the sequence of eigenvalues: 1.2024, 2.7605, 4.3269, 5.8958, 7.4655, 9.0355, 10.6058…

We will make use of these values later in an exercise in Chapter 7.

5.27 Frequency response data were obtained from a "black box" system:

ω (rad/s)	AR	$\theta°$
0.01	5	−2
0.04	4.95	−11
0.1	4.87	−29.5
0.4	2.35	−109
1	0.48	−154
4	0.03	−179
10	0.005	−191
40	0.000297	−250

Find a model—an ordinary differential equation—that yields agreement with the data in this table, then find the response $Y(t)$ of this system to a step function input: $X(t) = 1$.

5.28 Suppose we are concerned with cooling system design for an automobile equipped with a V8 engine of 402 cubic inches displacement (or 6.6 L). Sources of thermal energy production include combustion (although most of this heat is expelled in the exhaust) and friction (viscous dissipation) in the bearings and upon the cylinder walls. It is the latter that is the focus of this exercise. Assume the cylinder bore and the stroke are both 4″ (this is called a *square* design), and take the clearance between the piston (ring) and the cylinder wall as 0.002″ (0.00508 cm). We will estimate the frictional production of thermal energy in the cylinder with

$$P_V \cong \mu(dv_z/dr)^2.$$

μ is the viscosity of the lubricating oil and v_z is the estimated velocity arising from piston motion. For this shear-driven flow the velocity distribution in the oil film is approximated with

$$-\frac{d}{dr}(r\tau_{rz}) = 0 \quad \text{with} \quad \tau_{rz} = -\mu\frac{dv_z}{dr}.$$

For boundary conditions we have: at $r = R_1$, $v_z = V_{\text{piston}}$ and at $r = R_2$ (the cylinder wall), $v_z = 0$. Assume the engine is operating at 3,000 rpm (or 50 revolutions per second); each rotation of the crankshaft produces 8″ of piston travel, so the *nominal* velocity at R_1 is 400 in./s (the maximum velocity is over 600 in./s, and with the piston at TDC—top dead center—$v_z = 0$). The viscosity of the lubricating oil at operating temperature is 11 cP (0.011 N · s/m^2 or 0.11 g/(cm · s)). *Estimate* the total production of thermal energy in the oil film(s) on the cylinder walls in Btu/hr. *An aside: 3,000 rpm is a bit high for a passenger car in normal operation, but if this engine was used for dirt track racing, it might operate at much higher speeds, around 8,000 rpm.*

References

Abramowitz, M. and Stegun, I.A. (1972). *Handbook of Mathematical Functions*. New York: Dover.

Ayres, F. (1952). *Theory and Problems of Differential Equations*. New York: McGraw-Hill.

Carslaw, H.S. and Jaeger, J.C. (1959). *Conduction of Heat in Solids*, 2nde. Oxford: Oxford University Press.

Davis, H.T. (1962). *Introduction to Nonlinear Differential and Integral Equations*. New York: Dover Publications.

Dwight, H.B. (1957). *Tables of Integrals and other Mathematical Data*, 3rde. New York: Macmillan.

Finlayson, B.A. (1980). *Nonlinear Analysis in Chemical Engineering*. New York: McGraw-Hill.

Gray, A., Mathews, G.B., and MacRobert, T.M. (1931). *A Treatise on Bessel Functions and their Applications to Physics*, 2nde. New York: Macmillan and reprinted by Dover (1966).

Kreyszig, E. (1972). *Advanced Engineering Mathematics*, 3rde. New York: Wiley.

Mickley, H.S., Sherwood, T.K., and Reed, C.E. (1957). *Applied Mathematics in Chemical Engineering*, 2nde. New York: McGraw-Hill.

Milne-Thomson, L.M. (1950). *Jacobian Elliptic Function Tables*. New York: Dover Publications.

Selby, S.M. (ed.) (1975). *Handbook of Tables for Mathematics*. Cleveland, OH: CRC Press.

6

Numerical Solution of Ordinary Differential Equations

6.1 An Illustrative Example

In engineering and the applied sciences, transient balances for force, charge, mass, energy, and so on arise frequently, often leading to ordinary differential equations (ODE's). Suppose, for example, we have a jacketed process vessel in which an exothermic chemical reaction may occur. The entering (feed) stream has a temperature, T_{in}, the well-mixed contents have temperature, T, and the steam used to heat the vessel has temperature, T_s. A verbal statement of the appropriate energy balance might appear:

[Rate in] − [Rate out] + [Rate supplied] + [Rate of production] = [Accumulation].

And written out symbolically we would expect something like this:

$$\dot{M}_{\text{in}}C_p(T_{\text{in}} - T_{\text{ref}}) - \dot{M}_{\text{out}}C_p(T - T_{\text{ref}}) + UA(T_s - T) + |\Delta H_{rxn}|(-r_A)V = \rho V C_p \frac{dT}{dt}. \tag{6.1}$$

In the product, UA, U is the overall heat transfer coefficient and A is the heat transfer surface area. The rate at which the reactant species, "A," is consumed is r_A. We will let the mass flow rates in and out be the same, and we set the reference temperature equal to the inlet (feed) temperature and divide by $\dot{M}C_p$:

$$-(T - T_{\text{in}}) + \frac{UA}{\dot{M}C_p}(T_s - T) + \frac{|\Delta H_{rxn}|V}{\dot{M}C_p}(-r_A) = \tau\frac{dT}{dt}. \tag{6.2}$$

Please note that every term in the equation has the dimension of temperature. The characteristic time, τ, that appears on the right-hand side is the total mass in the vessel divided by the mass flow rate, and it is the time constant for this system. This is a first-order, ODE that will generally be coupled to a comparable mass balance upon the reactant species, "A." If the heat of reaction is negligibly small, then this equation may be written:

$$aT - b = -\tau\frac{dT}{dt}. \tag{6.3}$$

The result is separable and easily integrated:

$$-\frac{dt}{\tau} = \frac{dT}{aT - b} \rightarrow -\frac{t}{\tau} + C_1 = \frac{1}{a}\ln(aT - b). \tag{6.4}$$

Consequently we find:

$$T = \frac{C_1}{a}\exp\left(-\frac{at}{\tau}\right) + \frac{b}{a}. \tag{6.5}$$

The constant of integration is determined from the initial condition: at $t = 0$, $T = T_0$; therefore, $C_1 = aT_0 - b$. Let us now suppose that $T_0 = 100$, $\tau = 4$, $a = 1$, and $b = 40$:

$$T = 60\exp(-t/4) + 40. \tag{6.6}$$

Problem Solving in Engineering: Analytical Mathematics and Numerical Analysis, First Edition. Larry A. Glasgow.

Companion website: www.wiley.com/go/glasgow/problemsolvingengineering

We see an exponential decrease from the initial temperature (100) to the ultimate value of 40.

Time, t	Temperature, T
0	100
½	92.9498
1	86.7289
2	76.3918
4	62.0728
8	48.1201
16	41.0989
32	40.0201

While this particular differential equation was extremely simple to solve analytically, that will not always be the case. For example, we might have a set of simultaneous differential equations, or a nonlinear ODE for which no analytic solution is known. Therefore it is entirely appropriate for us to ask if a solution for our elementary example could also be obtained numerically. In the next section, we contemplate the simplest possible numerical approach and revisit the chemical reactor problem.

6.2 The Euler Method

Recall that the definition of the first derivative is given by:

$$\left.\frac{dy}{dx}\right|_x = \lim_{\Delta x \to 0} \frac{y(x+\Delta x) - y(x)}{\Delta x}. \tag{6.7}$$

Suppose we render this definition discrete; that is, we let Δx assume some finite (but hopefully small) value, and rearrange the result:

$$y(x+\Delta x) \cong \Delta x\left(\frac{dy}{dx}\right)_x + y(x). \tag{6.8}$$

Therefore, given an initial (or beginning) value for $y(x)$, and an expression for dy/dx, we can simply forward march in the x-direction, computing new values for y as we go. Let us try this for the preceding energy-balance example. We begin by isolating dT/dt:

$$\frac{dT}{dt} = -\frac{1}{\tau}(aT - b) = -\frac{1}{4}(T - 40). \tag{6.9}$$

Next, let $\Delta t = ½$ and proceed:

Time, t	Temperature, T
0	100
½	92.5000
1	85.9375
2	75.1709
4	60.6165
8	47.0840
16	40.8364
32	40.0117

We should graphically compare these values computed with the Euler method to those obtained previously from the analytic solution; Figure 6.1 will illuminate the principal shortcoming of the technique.

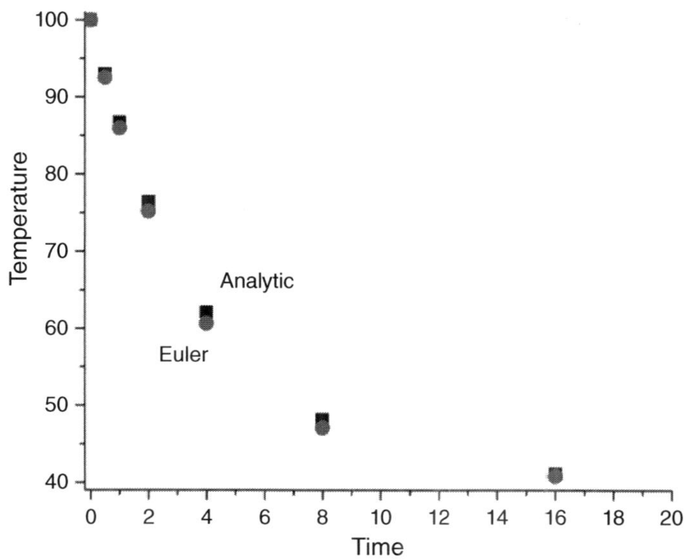

Figure 6.1 Comparison of the analytic and numerical (Euler method) solutions of the energy-balance example. The discrepancies between the two are particularly apparent at $t = 2, 4$, and 8.

Because the Euler method is a straight-line, piecewise approximation, it cannot accurately follow a function with curvature unless we make Δx (or Δt) *very* small. In some extreme cases it might have to be prohibitively small, leading to cumulative roundoff error. Nevertheless, the Euler algorithm is often useful because it is easy to understand and easy to implement. We will illustrate this last point with a slightly more complicated example.

We applied the Euler method to a simple, first-order, ODE, and we obtained reasonable results as evidenced by Figure 6.1. Suppose, however, we are interested in the second-order equation:

$$\frac{d^2y}{dx^2} + \frac{dy}{dx} - 6y = 0 \quad \text{with } y(0) = 1 \text{ and } y'(0) = 0. \tag{6.10}$$

The corresponding analytic solution is: $y = C_1 \exp(2x) + C_2 \exp(-3x)$, where $C_1 = 3/5$ and $C_2 = 2/5$. Now, we will find a numerical solution for this differential equation for values of x ranging from 0 to 0.5 using the Euler method. Let $y_1 = \frac{dy}{dx}$; therefore, $\frac{dy_1}{dx} = \frac{d^2y}{dx^2} = -y_1 + 6y$. Of course, any n-th order ODE can be written as n-first-order equations. The essential part of the algorithm appears simply as:

```
        x=0:dx=0.00390625:y=1:y1=0
20  REM *** find the new value for y
                  y=dx*y1+y
    REM *** find the new value for y1
                  y1=dx*(-y1+6*y)+y1
                       x=x+dx
                  PRINT x,y,y1
```

The initial values for x, y, and y' (which is y_1) allow us to find a new y which in turn is used to find a new y_1. In this case, we will begin by letting Δx be 1/32 (0.03125); we will carry out the calculation from $x = 0$ to $x = 1/4$. The analytic solution indicates that $y(x = 1/4) = 1.17818$.

Δx	$y\,(x = 1/4)$
1/32	1.158303
1/64	1.168356
1/128	1.173297
1/256	1.175746

These results show that by making Δx as small as 1/256 (or 0.00390625), we can get within 0.21% of the correct answer. For many practical calculations this might be adequate; however, we can very easily encounter differential equations where the Euler scheme is not suitable. We will need a better tool, and one possibility is the modified Euler method (a self-starting, predictor–corrector approach).

6.2.1 Modified Euler Method

Let the differential equation of interest be:

$$\frac{dy}{dx} = f(x, y), \tag{6.11}$$

and we take $y(x = 0)$ to be a known value. A *predicted* value for $y(x = \Delta x)$ is found with the Euler method:

$$P = \Delta x f(x, y) + y(x = 0). \tag{6.12}$$

This estimate for $y(x = \Delta x)$ is improved (corrected) by using the average slope:

$$C = \Delta x \left(\frac{y'(x = 0) + P(y')}{2} \right) + y(x = 0). \tag{6.13}$$

The *corrected* value for y (which is C) can be taken back to $f(x,y)$ to get a better estimate of the new slope. This process can be repeated until a desired criterion is satisfied (until the change in $y[x = x + \Delta x]$ is sufficiently small). We can illustrate this process with an example that is easily solved analytically; suppose

$$\frac{dy}{dx} = 2y, \ \text{with} \ y(x = 0) = 1. \tag{6.14}$$

Of course, the solution is just $y = C_1 \exp(2x)$, with $C_1 = 1$. We set $\Delta x = 1/4$ and begin:

$$P = \tfrac{1}{4}(2) + 1 = 1.5$$

and therefore,

$$P(y') = 2(1.5) = 3.$$

Accordingly,

$$C = \frac{1}{4}\left(\frac{2+3}{2}\right) + 1 = 1.625. \tag{6.15}$$

Note that the correct value for $y(x = 1/4)$ is 1.64872. It is clear that our estimate for y has been improved significantly by the addition of the corrector step! Eq. (6.14) was extremely simple, and it is prudent to wonder if the modified Euler method can successfully deal with a more challenging differential equation. Consider

$$\frac{dy}{dx} = ay^2 + by \ \text{with} \ a = -1 \ \text{and} \ b = 2. \tag{6.16}$$

Let $y(0) = -1$ and assume we are interested in $0 \le x \le 0.5489$. We will compare Euler with the Euler predictor–corrector as well as with the analytic solution, using fixed step size ($\Delta x = 0.0002$) and single precision; then we will tabulate the results for $y(x)$:

x	y_{Euler}	$y_{\text{Euler PC}}$	y_{analytic}
0.2027	−1.999	−2	−2
0.3811	−4.99	−5	−5
0.4581	−9.95	−10	−10
0.5394	−92.4	−99.9	−100
0.5488	−599.6	−1,862	−2,179

In this instance neither the Euler method nor the Euler predictor–corrector produces an acceptable value for $y(x = 0.5488)$. If we cut Δx in half, the Euler PC produces a much improved estimate.

6.2.2 A Recommended Exercise

The attractive feature of the Euler predictor–corrector method is its simplicity—it can be implemented easily with many hardware–software combinations. The reader is encouraged try it on the following first-order ODE:

$$\frac{dy}{dx} = \frac{1}{\sqrt{x+\frac{1}{8}}}\left(y^2 - \frac{1}{9}y\right) \text{ with } y(0) = 1 \text{ and } 0 \le x \le 0.653. \tag{6.17}$$

For your initial calculation set $\Delta x = 0.0005$; you should find $y(0.653) = 370.6$. However, pay careful attention to the behavior of $y(x)$ as x approaches the upper end of the interval as you may find reason to question the accuracy of the result. Cut Δx in half (0.00025) and repeat the calculation; what is your revised answer? You should get 348.5 which is still suspect. Halve Δx one more time and try again. We will revisit this recommended exercise as a problem at the end of the chapter.

We have observed that the Euler predictor–corrector can greatly improve our results for *some* ODEs. Even so, some of the deviations we have seen for particular problems are cause for concern. We will almost certainly need something better, and as a first step in that direction we turn our attention to a technique that was originally proposed by Carl Runge.

6.3 Runge–Kutta Methods

Once again let us consider the differential equation

$$\frac{dy}{dx} = f(x, y). \tag{6.18}$$

For the sake of making things a little more compact, we let $\Delta x = h$. A Runge–Kutta method is one in which we employ a series of calculations (to advance from x to $x + \Delta x$) of the form:

$$k_0 = hf(x, y) \tag{6.19}$$

$$k_1 = hf(x + h/2, y + k_0/2) \tag{6.20}$$

$$k_2 = hf(x + h/2, y + k_1/2) \tag{6.21}$$

$$k_3 = hf(x + h, y + k_2), \text{with} \tag{6.22}$$

$$y(x + h) = y(x) + \frac{1}{6}(k_0 + 2k_1 + 2k_2 + k_3). \tag{6.23}$$

The latter (6.23), in fact, is equivalent to a truncated Taylor-series expansion for $y(x)$. Note that four evaluations for the k's are necessary to compute the new value for y; this is an illustration of the *fourth-order* Runge–Kutta algorithm which has long been a standard tool for science and engineering. It is a forward-marching, *self-starting* technique that is easy to implement. All one needs is an initial value for $y(x)$. We can explore this process with the following differential equation:

$$\frac{dy}{dx} = f(x, y) = x^2 + y \text{ with } y(x = 1) = 1. \tag{6.24}$$

We choose $h = 1/2$ and find:

$$k_0 = (0.5)(2) = 1 \tag{6.25}$$

$$k_1 = (0.5)(3.0625) = 1.53125 \tag{6.26}$$

$$k_2 = (0.5)(3.3281) = 1.66406 \tag{6.27}$$

$$k_3 = (0.5)(2.25 + 2.66406) = 2.45703. \tag{6.28}$$

Therefore, we find $y(x = 1.5) = 2.6413$. How does this compare with the analytic solution?

We can use the integrating factor technique to solve this equation, resulting in

$$y = -x^2 - 2(x + 1) + C_1/e^{-x}, \tag{6.29}$$

where $C_1 = 2.20728$. Therefore, $y(x = 1.5) = 2.64234$; the Runge–Kutta approach with $h = 1/2$ (*a single step*) produced an error of about 0.039%.

The procedure we just examined is a standard fourth-order Runge–Kutta (RK) method, and as we indicated, it is used extensively throughout the world for scientific calculations. Of course, one can construct and use RK routines of any order. We will demonstrate this by solving the same ODE using a *third-order* RK procedure adapted from James et al. (1977). The necessary steps are shown immediately below:

```
     #COMPILE EXE
#DIM ALL
   REM *** Example of 3rd order Runge-Kutta solution of an ODE
   GLOBAL H,XI,YI,XF,X,Y,K1,K2,K3,YNEW,XNEW,FXY,XX AS SINGLE
FUNCTION PBMAIN
   H=0.10:XI=1:YI=1:XF=1.5
   X=XI:Y=YI
  80 REM *** CONTINUE
     GOSUB 200
       K1=H*FXY
       X=XI+H/2:Y=YI+K1/2
     GOSUB 200
       K2=H*FXY
       X=XI+H:Y=YI-K1+2*K2
     GOSUB 200
       K3=H*FXY
       YNEW=YI+1/6*(K1+4*K2+K3)
       XNEW=XI+H
         PRINT XNEW,YNEW
         XI=XNEW:YI=YNEW
       IF XI>XF THEN 190 ELSE 80
  190 REM *** CONTINUE
        INPUT "Shall we continue?";XX
          IF XX>0 THEN 195
  195 END
  200 REM *** here is the differential equation
       FXY=X^2+Y
       RETURN
END FUNCTION
```

Once again we will integrate out to $x = 1.5$; but this time we will start with a step size of 0.10, then cut h in half repeatedly to see the effect on the computation.

h	$y\,(x = 1.5)$
0.10	2.644123
0.05	2.642912
0.025	2.642494
0.0125	2.642373
0.00625	2.642340

Note that the last value (using $h = 0.00625$) is essentially identical to the value obtained from the analytic solution. Of course, this third-order RK procedure requires only *three* k-function evaluations per step; if h is appropriately small, we can get quite satisfactory results for many ODEs.

Now we will examine a second-order ODE using an elementary example from mechanics where we utilize a force balance, setting $ma = \Sigma F$. Suppose we have a mass (m) that is suspended from a horizontal surface; it is attached to the surface with a spring and a viscous dashpot (a shock absorber). Position is represented by the variable, y. A simple model for this situation is:

$$m\frac{d^2y}{dt^2} + C\frac{dy}{dt} + Ky = F(t). \tag{6.30}$$

We divide by Hooke's constant (K), resulting in

$$\frac{m}{K}\frac{d^2y}{dt^2} + \frac{C}{K}\frac{dy}{dt} + y = \frac{F(t)}{K}. \tag{6.31}$$

This equation is now in standard form which can be written equivalently as

$$\tau^2\frac{d^2y}{dt^2} + 2\zeta\tau\frac{dy}{dt} + y = X(t). \tag{6.32}$$

τ is the time constant for the system, ζ is the damping coefficient, and $X(t)$ is the forcing function. Let $\tau = 1$ and $\zeta = 0.3$ (underdamped); we will drive the system by giving the mass some initial displacement, and we can expect oscillatory behavior to result. *Please note the effect that the damping coefficient has upon the solution of this differential equation;* that is, *look at $b^2 - 4ac$, or $4\tau^2(\zeta^2 - 1)$ alternatively*. Typical program logic for a fourth-order RK procedure is shown immediately below.

```
#COMPILE EXE
#DIM ALL
   REM ** 4th-order Runge-Kutta scheme for 2nd order ODE.
   GLOBAL xi,yi,y1i,h,xt,xbeg,pk1,x,y,y1,pk2,pk3,pk4,j,f,D,tau,Xforc,cc AS SINGLE
FUNCTION PBMAIN
   XI=0:YI=1.00:Y1I=0.0:H=.1:XT=18:tau=1:D=0.9
   OPEN "c:Dampp9.dat" FOR OUTPUT AS #1
   XBEG=XI:J=0:X=XI:Y=YI:Y1=Y1I
 120 REM *** continue
   GOSUB 570
    PK1=H*F
    X=XI+H/2
    Y=YI+H/2*Y1I
    Y1=Y1I+PK1/2
   GOSUB 570
    PK2=H*F
    Y=Y+H/4*PK1
    Y1=Y1I+PK2/2
   GOSUB 570
     PK3=H*F
     X=XI+H
     Y=YI+H*Y1I+H/2*PK2
     Y1=Y1I+PK3
   GOSUB 570
     PK4=H*F
     J=J+1
     Y=YI+H*Y1I+H/6*(PK1+PK2+PK3)
     Y1=Y1I+(PK1+2*PK2+2*PK3+PK4)/6
       PRINT x,y,y1
        WRITE#1,x,y,y1
     XI=X:YI=Y:Y1I=Y1
       IF X>XT THEN 380 ELSE 120
 380 REM *** continue
            INPUT "Shall we continue?";CC
            IF cc>0 THEN 500
```

> The differential equation appears in line 570 as $\frac{d^2y}{dt^2} = -\frac{2\zeta}{\tau}\frac{dy}{dt} - \frac{y}{\tau^2} + \frac{x(t)}{\tau^2}$

```
500 REM *** continue
       CLOSE:END
570 F=-2*D/tau*y1-y/tau^2+Xforc/tau^2
580 RETURN

END FUNCTION
```

Results obtained from this program are shown in Figure 6.2 for some values of the damping coefficient ranging from 0.3 to 0.9. Note how increasing the value for ζ makes the solution less oscillatory but more sluggish. This has important implications in examples from mechanics (such as an automobile suspension), but it is also significant to many other fields like automatic process control where a decaying oscillatory response to a forcing function (like a step function or an impulse) is quite common.

Let's look at one more example to assess some of the features of the fourth-order Runge–Kutta scheme; we will use a model from automatic process control in which a system at equilibrium is suddenly perturbed by a change in set point. The linear, ODE is:

$$\frac{d^3y}{dt^3} + 2\frac{d^2y}{dt^2} + 3\frac{dy}{dt} + 4y = 4, \tag{6.33}$$

with the initial conditions, $y(0) = 0$, $y'(0) = 0$, and $y''(0) = 0$. It is evident that the ultimate state of the system is $4y = 4$, i.e., $y(t \to \infty) = 1$. An attractive feature of this example is that the third-order ODE has constant coefficients, enabling us to find the analytic solution through use of the Laplace transform:

$$y(s) = \frac{4}{s(s^3 + 2s^2 + 3s + 4)}. \tag{6.34}$$

Using *cSolve* on a *TI-nspire* to factor the cubic polynomial we expand in partial fractions:

$$\frac{4}{s(s + 1.6506)(s + 0.1747 - 1.5469i)(s + 0.1747 + 1.5469i)} = \frac{A}{s} + \frac{B}{s + 1.6506} + \frac{C}{s + 0.1747 - 1.5469i} + \frac{D}{s + 0.1747 + 1.5469i} \tag{6.35}$$

If you would like a little review with this process, you might verify that $A = 1$, $B = -0.5301$, $C = -0.2349 + 0.3094i$, and $D = -0.2349 - 0.3094i$. The analytic solution is therefore:

$$y(t) = 1 - 0.5301e^{-1.6506t} - 0.4698e^{-0.1747t}\cos(1.5469t) - 0.6188e^{-0.1747t}\sin(1.5469t) \tag{6.36}$$

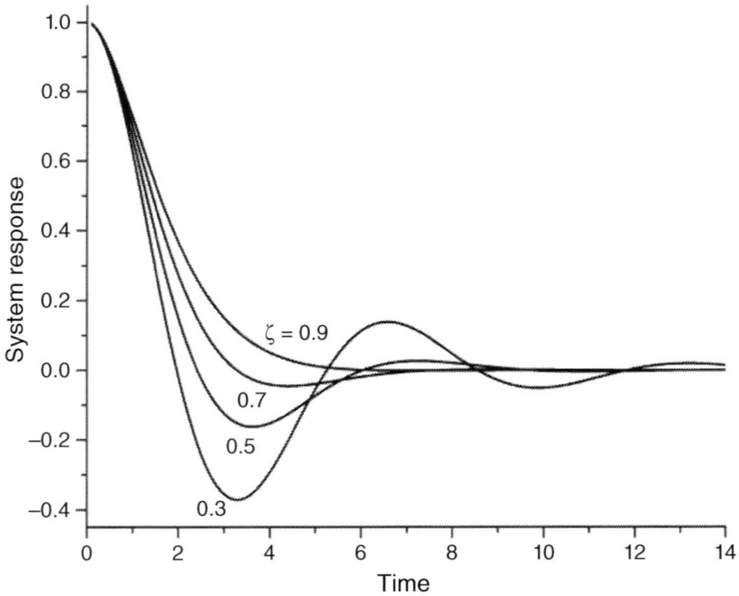

Figure 6.2 Effect of the damping coefficient, ζ, upon the dynamic behavior of the initially displaced mass. Results obtained from the fourth-order Runge–Kutta method with fixed step size. The significantly underdamped case, $\zeta = 0.3$, exhibits large overshoot and very oscillatory response.

Now we will use a fourth-order *RK* scheme in single precision to solve the ODE numerically and explore the impact of changing Δt (h in the notation used above). As we indicated previously, any n-th order ODE can be written as n-first-order ODEs; thus, $y' = dy_1/dt = y_2, y'' = dy_2/dt = y_3$, etc. The results of this exercise are shown in the following table; the analytic solution is shown in the top row, and beneath that are the computations from the fourth-order Runge–Kutta with *increasing* Δt. Although significant discrepancies begin to appear when Δt exceeds 0.2, this table makes the advantages provided by four computations per step very clear; a rather large Δt (or h) can produce good results for ODEs that are not difficult.

t	1	2.7	6	9	12
y_{analytic}	0.36937	1.47549	1.13198	0.85371	0.96613
y, $\Delta t = 0.002$	0.3694	1.4755	1.1320	0.8537	0.9661
y, $\Delta t = 0.01$	0.3694	1.4755	1.1320	0.8537	0.9661
y, $\Delta t = 0.05$	0.3694	1.4755	1.1320	0.8537	0.9661
y, $\Delta t = 0.2$	0.3693	—	1.1318	0.8536	0.9663
y, $\Delta t = 0.333$	0.3687	—	1.1305	0.8528	0.9673

Runge–Kutta methods have been used for a very long time, and it is important that the reader understand that newer, more powerful techniques have been developed for solving ODEs. Indeed, Press et al. (1989) have offered the slightly severe view that "*Runge-Kutta is what you use when you don't know any better.…or you have a trivial problem where computational efficiency is of no concern.*"

They advocate strongly for the Bulirsch–Stoer method which we will discuss later in this chapter.

6.4 Simultaneous Ordinary Differential Equations

Often we must solve sets of simultaneous ODEs. This may come about from reducing an n-th order ODE to n-first-order ODEs as we pointed out previously, or it could result from a problem in which several related phenomena are occurring at the same time. As an example of the latter, consider a series of chemical reactions carried out in a batch reactor:

$$A \rightarrow B \rightarrow C \tag{6.37}$$

A situation can occur in series reactions in which we might wish to maximize production of the intermediate product, B. Naturally, if this reaction sequence is allowed to proceed for a long period of time, we will get mostly C. The mass balances for this system are written as:

$$\frac{d[A]}{dt} = -k_1[A] \tag{6.38}$$

$$\frac{d[B]}{dt} = k_1[A] - k_2[B] \tag{6.39}$$

and

$$\frac{d[C]}{dt} = k_2[B]. \tag{6.40}$$

We assume that we begin the process with only species A present in the reactor. Note that if the rate constants are such that $k_2 \gg k_1$, it will be extremely difficult to obtain much B from the process. Consequently, we will focus our attention upon the case in which the rate constants are comparable. We employ a fourth-order Runge–Kutta scheme designed specifically to handle sets of simultaneous ODEs. We let the initial concentration of A be 1, and set the reaction rate constants $k_1 = 0.85$ and $k_2 = 0.55$ (both with dimensions of reciprocal time). The results are shown in the accompanying Figure 6.3; note that the maximum concentration of the desirable intermediate, B, occurs at approximately $t = 1.45$.

6.4.1 Some Potential Difficulties Illustrated

The focus of this section is simultaneous ODEs, and in the preceding example we saw computational results for a series of chemical reactions carried out in a batch process. The model was elementary, and the dynamic response of the system

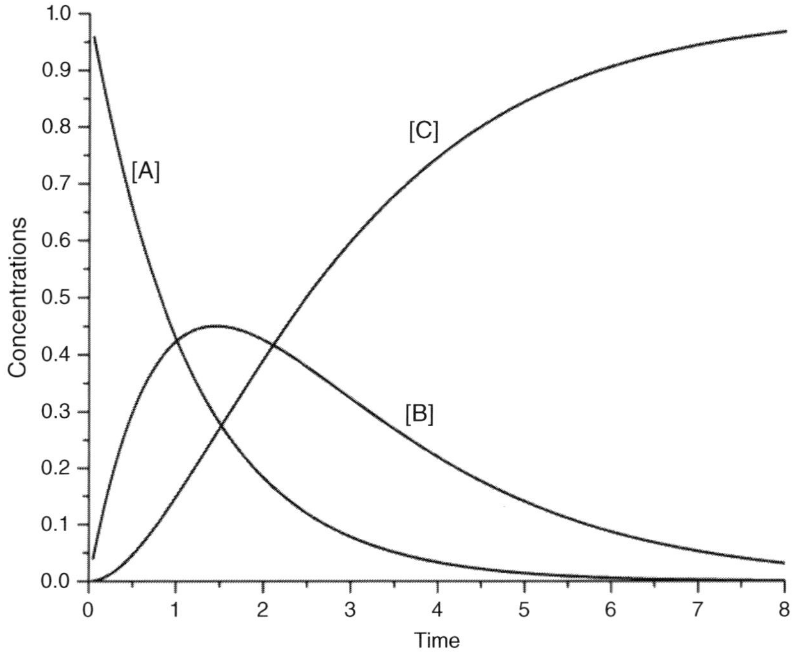

Figure 6.3 Concentrations of reactants for series reactions carried out in a batch reactor. The desirable intermediate, *B*, achieves a maximum concentration of about 0.45 at $t \approx 1.45$.

was easy to anticipate even with no prior exposure to problems in chemical kinetics. However, we may be required to seek solutions for more difficult cases—for example, ones that exhibit sharp fluctuations. In the example we are about to explore we will look at oscillatory behavior associated with cellular cycles.

Recent work published by Ferrell et al. (2011) sought to understand the dynamic behavior of cell cycles driven by a protein circuit in which the activation of cyclin-dependent protein kinase (CDK1, represented by C) drives a cell into mitosis, and the activation of the anaphase-promoting complex (APC, represented by A) leads the cell back out. Ferrell et al. note that proteins such as Polo-like kinase 1 (Plk1, represented by P) may also play a role in the activation of A, and thus they directed their attention to a three-component system governed by the ODEs:

$$\frac{dC}{dt} = \frac{1}{10} - \frac{3CA^8}{\left(\frac{1}{2}\right)^8 + A^8}, \tag{6.41}$$

$$\frac{dP}{dt} = 3(1 - P)\frac{C^8}{\left(\frac{1}{2}\right)^8 + C^8} - P, \tag{6.42}$$

and

$$\frac{dA}{dt} = 3(1 - A)\frac{P^8}{\left(\frac{1}{2}\right)^8 + P^8} - A. \tag{6.43}$$

These three ODEs are to be solved numerically with the initial conditions (for $t = 0$) $C=P=A=0$. The problem is clearly nonlinear and if the analyst is unfamiliar with such systems, he/she may not have any idea what to expect. This is where the potential difficulty lies as we shall see. Suppose we decide to use the Euler method due to the ease with which it can be applied here. We start with a Δt of 0.0125 time units and plot the results in Figure 6.4.

The sharp changes in slope produced by this model are cause for concern. Had we had selected a larger Δt, say $\Delta t = 0.4$ instead of 0.0125, the resulting activity of C (CDK1) would then have a *negative* (unphysical) value at $t = 20$ and the maximum value for A (APC) at $t = 19.6$ would be about 0.76. Neither is correct, of course. This example underscores the fact that we need to be vigilant when dealing with an unfamiliar system of nonlinear differential equations and that we should not be reluctant to employ a better solution technique for such problems even if the algorithm is a little more complicated.

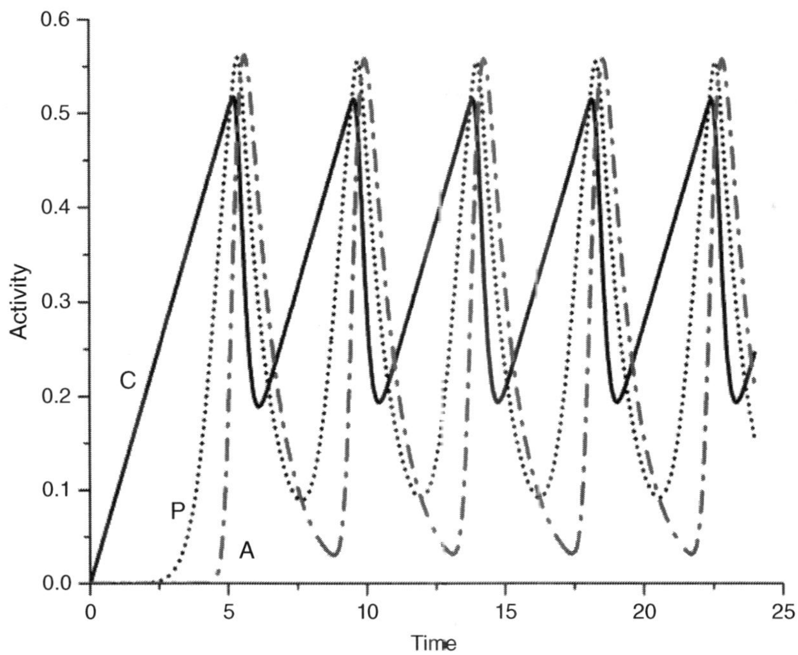

Figure 6.4 Oscillations in the cellular activity of CDK1 (*C*), APC (*A*), and Plk1 (*P*). These results are in *qualitative* agreement with Ferrell et al.

Indeed, if we selected the pedestrian fourth-order Runge–Kutta procedure we could prepare a little comparison of the effect of step size upon the calculations:

h	$C(t = 20)$	$A(t = 19.6)$
0.8	0.4402	0.019
0.4	0.2439	0.3021
0.2	0.2487	0.3004
0.1	0.2413	0.3174
0.01	0.2409	0.3186

Thus it is clear that abrupt changes in activity produced by this model can be problematic for the numerical procedures that we have discussed so far. In the material to follow we will examine some better tools for problems of this type. However, we should also observe that the dynamic behavior of systems similar to (this example of) cellular oscillation can often be very effectively revealed with *phase-space analysis*. In phase space, an oscillatory signal of sustained amplitude produces a limit cycle (a closed-loop trajectory). We will explore this technique later in this chapter and discover its value, particularly in the investigation of nonlinear systems. We will also have an opportunity in a student exercise at the end of the chapter to explore oscillating chemical reactions for the classic two-phase reactor problem of Schmitz and Amundson (1963). This chemical reactor problem bears some (dynamic) similarity to the cellular oscillation example above.

6.5 Limitations of Fixed Step-Size Algorithms

The methods we have discussed so far in this chapter have at least one feature in common: All employ a fixed step-size, i.e., a constant value for h. This can produce an incorrect result for certain equation types; more importantly, the discrepancy can be very difficult to detect unless one has some prior knowledge of that particular problem. To illustrate this point, consider the elementary differential equation,

$$\frac{dy}{dx} = xy^{3/2} \text{ with } y(0) = 2. \tag{6.44}$$

Now suppose we want to compute the value of y for $x = 1.65$. The analytic solution for this problem can be found easily:

$$y = \left\{ -\frac{1}{2}\left(\frac{x^2}{2} - 1.41421\right)\right\}^{-2}. \tag{6.45}$$

Consequently, $y(x = 1.65) = 1{,}426.15$. Let us now use the *third-order* Runge–Kutta scheme described (and used) previously. We begin with $h = 0.05$; the computation produces $y(x = 1.65) = 1{,}211.9$, which corresponds to an error of about 15%. If we were to proceed without prior knowledge of the problem, or without a computed error estimate, this discrepancy would almost certainly remain undetected. And if we were to base a system design, or perhaps a complex system model upon such a result, we might blunder into a disaster. Fortunately, there are very simple means we can use to avoid such errors, and we will first describe here an approach developed by Bailey (1969). The underlying idea is straightforward: If the change in y (for $x = x + h$) is "small" we double the step size, and if the change in y is "large" we cut h in half. Therefore, if we encounter a region in which y is growing very rapidly (which is exactly what happens in our example), h can (and will) become *very* small. Note that there is a certain arbitrariness in play here—the analyst decides what changes are "large" or "small." The logic to be inserted into the RK code is just:

```
REM *** BAILEY METHOD FOR VARIABLE STEP SIZE
  DELY=ABS(YI-YOLD)
   IF DELY<0.0005 THEN 180
   TESTY=ABS(DELY/YI)
    IF TESTY<0.002 THEN H=2*H
    IF TESTY>0.02 THEN H=H/2
REM *** CONTINUE
```

This simple change to the third-order RK procedure yields $y(x = 1.65) = 1{,}434.24$, corresponding to an error of about 0.57%, which would often be tolerable for practical calculations. As an alternative, one might think that simply using a fixed step-size procedure would work if h were suitably small. We can test this hypothesis on eq. (6.44) with the fourth-order RK algorithm. We will start with $h = 0.05$ and use *double precision*:

h	$y(x = 1.65)$
0.05	1,222.466
0.01	1,423.938
0.005	1,425.794
0.001	1,425.961
0.0005	1,425.961+

This is an important lesson: For some types of ODEs, fixed step-size methods such as the Runge–Kutta scheme used previously will be at best computationally expensive and possibly quite ineffective.

In addition to the step-halving approach utilized above, there are other strategies that can be employed when *adaptive* step-size control is required. One very appealing alternative is the Runge–Kutta Fehlberg scheme which is sometimes referred to as an *embedded* RK procedure since both fourth- and fifth-order estimates come from the same sequence of calculations. The essential idea is to compute both fourth- and fifth-order estimates, then compare them to obtain an estimate of the local error. That error is then used to make a step-size adjustment if needed. The particular algorithm we will employ comes from Cash and Karp (1990). Once again we will take

$$\frac{dy}{dx} = xy^{3/2} \text{ with } y(x = 0) = 2, \text{where we are interested in } y(x = 1.65). \tag{6.46}$$

The required sequence of calculations follows:

$$k_1 = f(x_i, y_i) \tag{6.47}$$

$$k_2 = f\left(x_i + \tfrac{1}{5}h, y_i + \tfrac{1}{5}k_1 h\right) \tag{6.48}$$

$$k_3 = f\left(x_i + \frac{3}{10}h, y_i + \frac{3}{40}k_1 h + \frac{9}{40}k_2 h\right) \tag{6.49}$$

$$k_4 = f\left(x_i + \frac{3}{5}h, y_i + \frac{3}{10}k_1 h - \frac{9}{10}k_2 h + \frac{6}{5}k_3 h\right) \tag{6.50}$$

$$k_5 = f\left(x_i + h, y_i - \frac{11}{54}k_1 h + \frac{5}{2}k_2 h - \frac{70}{27}k_3 h + \frac{35}{27}k_4 h\right) \tag{6.51}$$

$$k_6 = f\left(x_i + \frac{7}{8}h, y_i + \frac{1{,}631}{55{,}296}k_1 h + \frac{175}{512}k_2 h + \frac{575}{13{,}824}k_3 h + \frac{44{,}275}{110{,}592}k_4 h + \frac{253}{4{,}096}k_5 h\right). \tag{6.52}$$

The fourth-order estimate is computed from

$$y_{i+1} = y_i + \left(\frac{37}{378}k_1 + \frac{250}{621}k_3 + \frac{125}{594}k_4 + \frac{512}{1{,}771}k_6\right)h, \tag{6.53}$$

and the fifth-order estimate comes from

$$y_{i+1} = y_i + \left(\frac{2{,}825}{27{,}648}k_1 + \frac{18{,}575}{48{,}384}k_3 + \frac{13{,}525}{55{,}296}k_4 + \frac{277}{14{,}336}k_5 + \frac{1}{4}k_6\right)h. \tag{6.54}$$

We will now use this technique for our example equation using constant step-size ($h = 0.05$), determining the local error from the difference between the two estimates at each step. These computations reveal that the error begins to grow objectionably when x exceeds about 1.2; see Figure 6.5.

Now that we understand how the estimated error is behaving in this problem, we can make suitable changes to h. One possibility, suggested by Press et al. and discussed by Chapra and Canale (2002), is:

$$h_{\text{revised}} = h\left|\frac{\varepsilon_{\text{crit}}}{\varepsilon}\right|^{0.25}, \tag{6.55}$$

where $\varepsilon_{\text{crit}}$ is the threshold (desired) accuracy level. Suppose we take $\varepsilon \sim error$, and further, assume "typical" values for both ε's:

$$h_{\text{revised}} = h\left|\frac{10^{-7}}{10^{-6}}\right|^{0.25} = 0.5623\,h. \tag{6.56}$$

Now we are in position to return to our Runge–Kutta Fehlberg routine with the addition of adaptive step-size control. We will track the impact of this modification upon h as we approach $x = 1.65$ and provide the result in Figure 6.6.

The Runge–Kutta Fehlberg (RKF) procedure offers us a very effective way to implement adaptive step-size control, and when applied to nonlinear systems where little is known about the behavior of the dependent variable(s), it can save us a lot

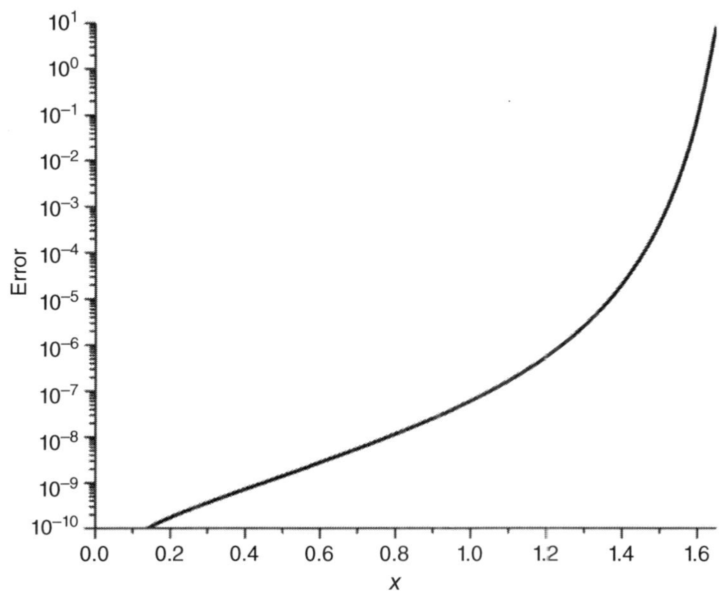

Figure 6.5 A measure of the local error for the equation, $dy/dx = xy^{3/2}$, obtained from the Runge–Kutta Fehlberg computation (fourth- and fifth-order) using fixed step size with $h = 0.05$.

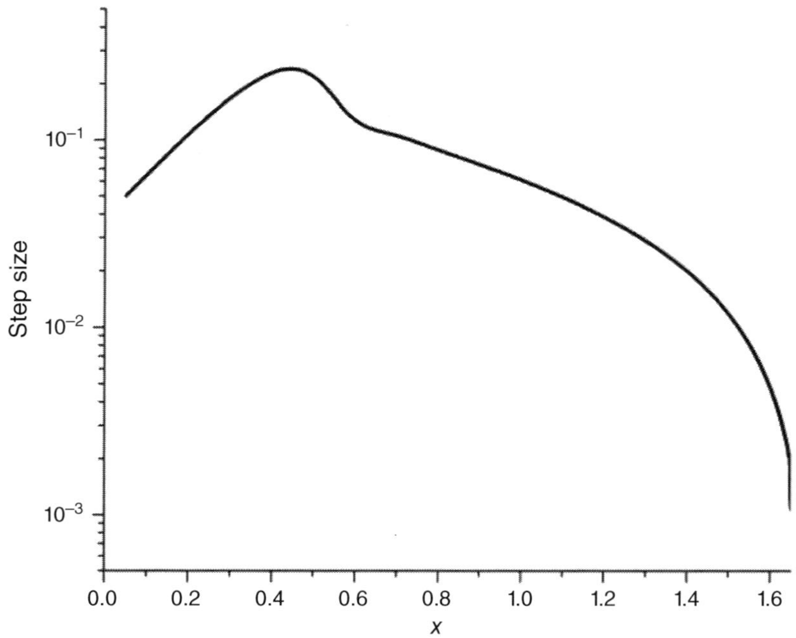

Figure 6.6 Behavior of the step size, h, using the Runge–Kutta Fehlberg scheme with step-size adaptation and an initial value of h of 0.05. With this technique, the computed value for $y\,(x = 1.65)$ is between 1,426 and 1,427, within about 0.03% of the correct value.

of grief. The principal advantage of RKF is that both fourth- and fifth-order estimates are obtained from the same sequence of calculations, providing us with a built-in estimate of error.

6.6 Richardson Extrapolation

Suppose we take the two Taylor series expansions:

$$f(x+h) = f(x) + hf'(x) + \frac{h^2}{2}f''(x) + \frac{h^3}{6}f'''(x) + \cdots \tag{6.57}$$

and

$$f(x-h) = f(x) - hf'(x) + \frac{h^2}{2}f''(x) - \frac{h^3}{6}f'''(x) + \cdots \tag{6.58}$$

We subtract the latter from the former:

$$f(x+h) - f(x-h) = 2hf'(x) + \frac{h^3}{3}f'''(x) + \cdots \tag{6.59}$$

Now we rearrange the equation to solve for the derivative, $f'(x)$:

$$f'(x) = \frac{f(x+h) - f(x-h)}{2h} - \frac{h^2}{6}f'''(x) - \frac{h^4}{120}f'''''(x) - \cdots \tag{6.60}$$

You may recognize that this expression—when truncated—is the second-order, central difference approximation for the first derivative. We rewrite this equation with a small modification, letting the difference approximation be represented by $\phi(h)$:

$$f'(x) = \phi(h) - \frac{h^2}{6}f'''(x) - \frac{h^4}{120}f'''''(x) - \cdots \tag{6.61}$$

Now we cut the interval in half; i.e., we replace h with $h/2$:

$$f'(x) = \phi(h/2) - \frac{h^2}{24}f'''(x) - \frac{h^4}{1920}f'''''(x) - \cdots \tag{6.62}$$

We multiply this equation by 4 and then subtract the previous expression from it, isolating $f'(x)$; the truncated result is

$$f'(x) \cong \frac{4}{3}\phi\left(\frac{h}{2}\right) - \frac{1}{3}\phi(h). \tag{6.63}$$

By adding one more function evaluation, we have significantly improved the quality of our estimate of $f'(x)$. Instead of the neglected term being of the order of h^2, it is now of order h^4! Let us see how well this works with a simple example: Take $f(x) = x^2 \sin x$, such that $f'(x) = 2x \sin x + x^2 \cos x$. We will use the latter to evaluate $f'(x = 1/2)$, which is 0.698821. Now we take $h = 1/8$ and evaluate the derivative using the second-order central difference expression $(f(5/8) - f(3/8))/(1/4)$; the result is 0.708188, which is about 1.3% too large. We repeat but add the additional evaluation at $h/2$ (1/16) using eq. (6.63), finding $f'(1/2) = 0.698828$ which is about 0.001% too large. At the cost of one additional function evaluation, we have dramatically reduced the error in our estimate of the derivative. This is a very powerful procedure: We made two calculations (one using h and one using $h/2$), and then extrapolated to $h = 0$, obtaining an extremely accurate estimate for the derivative at $x = 1/2$. This is an example of the *Richardson extrapolation*, and it has been characterized by Press et al. as "*.... turning straw into gold.*" It is so powerful that it has become a critical element in some newer ODE solvers, and we will see an example later in this chapter.

6.7 Multistep Methods

You should recall that a forward-marching techniques like Euler's method take the slope of the function to be constant over some finite interval. We cannot expect this to work very well unless h is small—perhaps *very* small. But suppose we were able to evaluate the slope at multiple points; if we could fit a polynomial to these values, we could extrapolate this function. By accounting for the change in slope over the interval in this way, we might significantly improve our results. This is the idea behind the Adams' *four-point* formula (also known as the Adams–Bashforth method). The algorithm for this technique is

$$y_{n+1} = y_n + \frac{h}{24}\left(55y'_n - 59y'_{n-1} + 37y'_{n-2} - 9y'_{n-3}\right). \tag{6.64}$$

In the notation being used here, y'_n is simply $y'(x_n)$ or $\left(\frac{dy}{dx}\right)_{x=x_n}$. Note that three points ($n-1$, $n-2$, and $n-3$) *lie to the left* of the interval; the Adams–Bashforth method is not self-starting, so another technique must be used to get underway. The reader interested in the derivation of this formula (which is obtained easily through use of the Gregory–Newton backward interpolation) may consult Southworth and Deleeuw (1965). Now, suppose we have the differential equation,

$$\frac{dy}{dx} = (x - x^2)y \text{ with } y(0) = 2. \tag{6.65}$$

We wish to know the value of $y(x = 1)$; we begin by using the Euler method with $h = 0.2$, which results in $y(1) = 2.33944$ (the analytic result is 2.36272). Now let us apply the Adams–Bashforth method using Euler's technique to get started. But this time we begin the calculation with $h = 0.002$ so that the initial values for the slopes (y'_n) will be more accurate. In this case, the Adams–Bashforth method yields $y(1) = 2.36274$ corresponding to an error of about 0.0008%. The technique can be further improved by addition of a corrector step (which can be applied repeatedly) to the predictor that we have used here. The corrector is:

$$y_{n+1} = y_n + \frac{h}{24}\left(9y'_{n+1} + 19y'_n - 5y'_{n-1} + y'_{n-2}\right). \tag{6.66}$$

Let us apply the Adams–Bashforth predictor–corrector scheme to a slightly more demanding problem, a resistor and capacitor placed in series and driven by a 60 Hz sinusoidal voltage. The linear ODE describing the charge on the capacitor (q) is:

$$\frac{dq}{dt} + 100q = 10\sin(120\pi t), \tag{6.67}$$

with $q(t = 0) = 0$. We observe that the initial transient will die out rather quickly which we confirm by applying the Laplace transform to the first-order ODE:

$$q(s) = \frac{1{,}200\pi}{(s+100)(s+120\pi i)(s-120\pi i)} = \frac{A}{s+100} + \frac{B}{s+120\pi i} + \frac{C}{s-120\pi i}. \tag{6.68}$$

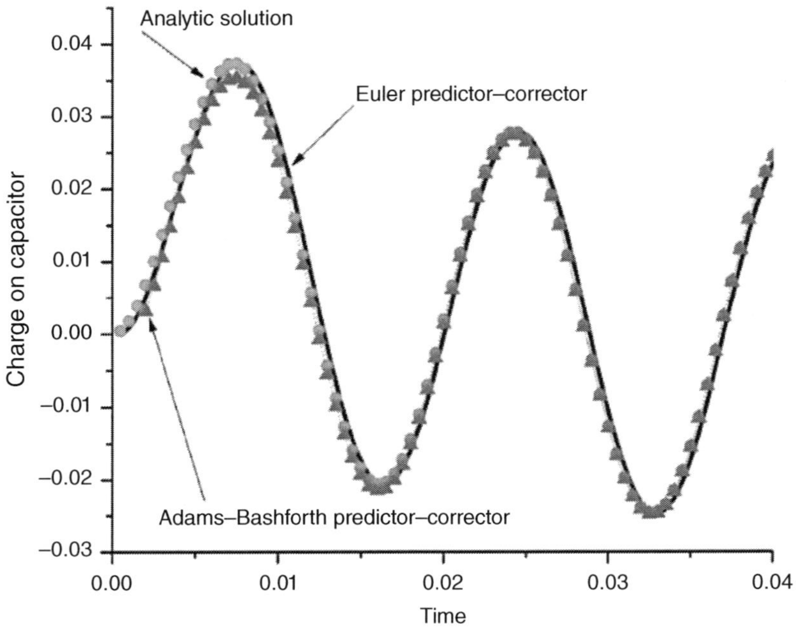

Figure 6.7 Charge on the capacitor as computed with the analytic solution (filled circles), the Euler predictor–corrector (solid line), and the Adams–Bashforth predictor–corrector with correction applied just once (filled triangles).

We use partial fraction expansion to find the analytic solution for this problem:

$$q(t) = 0.02478\exp(-100t) - 0.02478\cos(120\pi t) + 0.006574\ \sin(120\pi t). \tag{6.69}$$

By the time $t = 0.05$ s, the contribution from the *leading exponential term* is just 1.67×10^{-4} and at $t = 0.10$ s, merely 1.125×10^{-6}. We will use the Euler method to get Adams–Bashforth started—recognizing that any lack of accuracy in those initial points will be reflected by our solution. We will also solve eq. (6.67) with the Euler predictor–corrector and include the analytic solution eq. (6.69) for comparison.

The results shown in Figure 6.7 confirm an observation we made earlier in this section: The Adams–Bashforth (*AB*) *four-point* method will give excellent results *if* the initial calculations required for startup are *accurate*. Of course, this means that the coding required for solution of ODEs using *AB* will be a bit more complicated than we see with self-starting methods. Please note that other Adams–Bashforth methods (including two- and three-point algorithms) are common. The two-point method is easily developed using the Taylor series expansion, and it requires only one point to the left of the interval:

$$y_{n+1} = y_n + \frac{h}{2}\left(3y'_n - y'_{n-1}\right). \tag{6.70}$$

Of course, with equal step sizes eq. (6.70) will not provide the same quality of outcome as eq. (6.64).

6.8 Split Boundary Conditions

A situation that occurs regularly in problems concerning momentum, heat, and mass transfer in boundary layers is split boundary conditions; e.g., some variables (or derivatives) might be known at $x = 0$, but others at some different value of the independent variable. Consider two-dimensional (boundary layer) flow past a flat surface. The model developed by Prandtl and Blasius consists of the third-order nonlinear ODE:

$$\frac{d^3f}{d\eta^3} + \frac{1}{2}f\frac{d^2f}{d\eta^2} = 0, \tag{6.71}$$

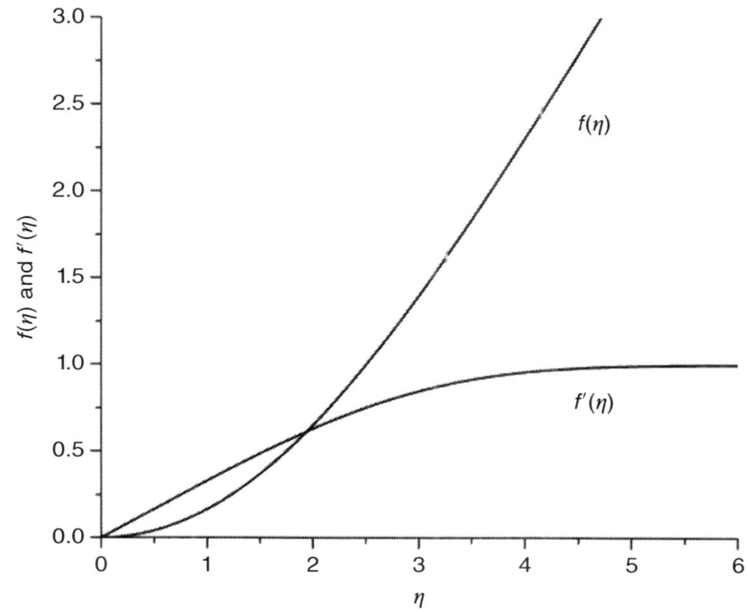

Figure 6.8 Solution of the Blasius equation for flow in the boundary layer on a flat plate. Note how $f'(\eta)$ approaches 1.0 asymptotically as η becomes large.

where $\eta = y\left(\frac{V}{\nu x}\right)^{1/2}$, $v_x = Vf'$, and $v_y = \frac{1}{2}\sqrt{\frac{\nu V}{x}}(\eta f' - f)$. The kinematic viscosity of the fluid is ν, and V is the fluid velocity outside the boundary layer. The plate surface corresponds to $\eta = 0$, where both v_x and v_y must be zero. Consequently, both $f(0)$ and $f'(0) = 0$. *However*, the required third boundary condition comes from the fact that v_x must correspond to the external potential flow outside the boundary layer (where the velocity is V): As $\eta \to \infty$, $v_x \to V$; since $f' = v_x/V$, it is clear that needed condition is $f'(\eta \to \infty) = 1$. Now consider application of the Runge–Kutta method to this equation. Because RK is self-starting, we must have values for $f(0)$, $f'(0)$, and $f''(0)$ in order to solve the boundary-value problem, eq. (6.71). Of course, we can guess the value of the second derivative, integrate across the boundary layer, find the result for f', and then adjust $f''(0)$ accordingly. And the reader may want to use this approach to find that $f''(0) = 0.33206$. The results for this problem are given below in Figure 6.8 to facilitate verification. The inefficient process employed here is equivalent to converting a boundary-value problem into *multiple* initial value problems; the process will take some time, and naturally one might wonder if there is a line of attack that would diminish our workload.

Hamming (1973) provided a useful example that illustrates one possible approach for split boundary conditions using a finite-difference approximation for the second derivative. Suppose we have the following second-order ODE:

$$\frac{d^2y}{dx^2} = y + x\frac{dy}{dx} = (x - x^2)y \text{ with } y(0) = 2. \tag{6.72}$$

We use the second-order central difference approximation for the left-hand side, resulting in

$$\frac{y_{n+1} - 2y_n + y_{n-1}}{h^2} \cong y_n + x_n. \tag{6.73}$$

Now suppose we break the interval (0–1) into four pieces—this gives us five nodal points with $n = 0$ and $n = 4$ corresponding to the ends of the interval. We apply the approximation at the three interior nodes, obtaining:

$$\frac{y_2 - 2y_1 + y_0}{(1/4)^2} = y_1 + \frac{1}{4} \tag{6.74}$$

$$\frac{y_3 - 2y_2 + y_1}{(1/4)^2} = y_2 + \frac{1}{2} \tag{6.75}$$

$$\frac{y_4 - 2y_3 + y_2}{(1/4)^2} = y_3 + \frac{3}{4} \tag{6.76}$$

Of course, $y_0 = 0$ and $y_4 = 1$; we have three equations and three unknown nodal values. It is easy to show that the interior nodal values (y_1, y_2, and y_3) are 0.180229, 0.387348, and 0.649927, respectively. How do these compare with the analytic solution (which is $y = 0.850918(e^{+x} - e^{-x}) - x$)? The three corresponding values for y_n are 0.179905, 0.386819, and 0.649448. The results are encouraging, and even more accurate results could be obtained by simply increasing the number of interior nodal points. But the equation we used for this example eq. (6.72) certainly lent itself to the process we carried out, so we might want to think about some alternatives that are more broadly applicable. To underscore the point, the reader may want to consider how "easily" a finite-difference technique could be applied to the Blasius eq. (6.71).

Shooting methods have been developed precisely to deal with this situation; let us consider the more general differential equation,

$$\frac{d^2y}{dx^2} = f\left(x, y, \frac{dy}{dx}\right), \tag{6.77}$$

along with the boundary values $y(x=a) = y_a$ and $y(x=b) = y_b$. Now we contemplate the corresponding *initial value* problem:

$$\frac{d^2y}{dx^2} = f\left(x, y, \frac{dy}{dx}\right) \text{ with } y(x = a) = y_a, \text{ and } dy/dx = s \text{ at } x = a. \tag{6.78}$$

It is clear that we would find the solution we want for the *boundary-value* problem if we could identify a zero for the function:

$$F = y(b, s) - y_b. \tag{6.79}$$

Naturally, each value selected for s will yield a corresponding F. Identifying the particular value we want can be done many different ways including an iterative scheme using Newton's method of tangents (or Newton–Raphson). The superscripts refer to the iterate number in eq. (6.80):

$$s^{n+1} = s^n - \frac{F(s^n)}{F'(s^n)}. \tag{6.80}$$

Stoer and Bulirsch (1993) point out that the derivative of F appearing in the denominator is often replaced by a first-order forward difference approximation. They add that Δs must be chosen carefully—a value either too large or too small may either produce a derivative that is not sufficiently accurate, or possibly lead to convergence problems. Moreover, as Hanna and Sandall (1995) note, it is clear that we now have two sources of error: We have error associated with the estimate of s, and we have error associated with the numerical solution of the ODE itself.

We can better appreciate the shooting method through an example. Suppose we have the following second-order ODE:

$$\frac{d^2y}{dx^2} + 0.06667\frac{dy}{dx} + 0.11111y = 0, \tag{6.81}$$

with $y(0) = 2$ and $y(3) = 10/3$. We plan to use a Runge–Kutta method, but we do not know $\left(\frac{dy}{dx}\right)_{x=0}$. Let our initial guess for s be 0.4 (with $\Delta s = 0.1$) and proceed:

$y'(0)$	$y(3)$	$F = y(3) - 10/3$
0.4	2.05325	−1.28008
0.5	2.28208	−1.05125

These values allow us to obtain a new estimate for s:

$$s^{n+1} = 0.4 - \frac{-1.28008}{2.2883} = 0.959402. \tag{6.82}$$

Now we return to our Runge–Kutta procedure, but this time with $y'(0) = 0.959402$. The result is $y(3) = 3.333318$, which is pretty close to the correct value of 10/3. We could then refine our estimates if greater precision was required:

$y'(0)$	$y(3)$	$F = y(3) - 10/3$
0.959	3.332397	−0.000936
0.960	3.334686	+0.001353

The improved estimate for s is 0.959409, which yields $y(3) = 3.333333$. We have now seen the ease with which a shooting method can be employed; this will be far more efficient than solving *many* initial value problems in an effort to identify $y'(0)$ or $y''(0)$.

6.9 Finite-Difference Methods

We demonstrated in the previous section how finite-difference approximations could be conveniently used to solve certain types of ODEs with split boundary conditions. We would like to further explore this topic, providing some amplification, and treating a problem of some practical importance. Let us consider the steady-state conduction in a homogeneous medium in which thermal energy is absorbed (possibly by an endothermic chemical reaction). We will also assume that the energy absorption is directly proportional to local temperature; therefore,

$$\frac{d^2T}{dx^2} - \beta T = 0. \tag{6.83}$$

The ends of the material are maintained at different temperatures: $T(x = 0) = T_0$, and $T(x = L) = T_L$. The continuum is discretized, i.e., we place n equally spaced nodes in the interior resulting in $n + 1$ intervals of length h (Δx), along with $n + 2$ nodes (including the boundaries). If we use a second-order central difference for the second derivative, our *algebraic* approximation for the differential equation appears:

$$\frac{T_{n+1} - 2T_n + T_{n-1}}{h^2} \cong \beta T_n. \tag{6.84}$$

It is important to remember that we have introduced *truncation* error with this approximation; Taylor's series expansions used to construct this finite difference have been truncated and, accordingly, we have introduced an error that is of the same order as the neglected terms. Note that the coefficients for the three nodal points are:

$n-1$	n	$n \pm 1$
1	$-h^2\beta - 2$	1

When we apply this pattern to all of the interior nodes (the *interior* nodes begin with $n = 1$), we find:

$n = 0$	$n = 1$	$n = 2$	$n = 3$	$n = 4$
1	$-h^2\beta - 2$	1	0	0
0	1	$-h^2\beta - 2$	1	0
0	0	1	$-h^2\beta - 2$	1
0	0	0	1	$-h^2\beta - 2$
0	0	0	0	1 … etc.

Please remember that the first column ($n = 0$) corresponds to the *left-hand boundary where the temperature is specified.* We see that the coefficient matrix is sparse—it contains many zeroes. Furthermore, it follows the *tridiagonal* form; this structure arises so frequently in applied mathematics that very efficient procedures for solving such sets of equations have been developed.

Let us give our conduction problem some quantitative definition and apply this technique to its solution: We have a slab of material 10 cm wide, with one edge ($x = 0$) maintained at 100° and the other at 200°. Let $\beta = 0.3\,\text{cm}^{-2}$ and let $h = 1$ cm, so we have nine interior nodes (we must solve a set of nine simultaneous algebraic equations). The analytic solution is just:

$$T = 99.1655 \exp(-\sqrt{0.3}x) + 0.8345 \exp(+\sqrt{0.3}x), \tag{6.85}$$

which allows for easy comparison. The results for the finite-difference computation are provided in Figure 6.9 along with the analytic solution; the largest discrepancy between the two is a little worse than 3%. We could easily cut the interval in half (to 0.5 cm) if we wanted improved accuracy. Indeed, if we do this the worst-case deviation is reduced to less than about 0.9%.

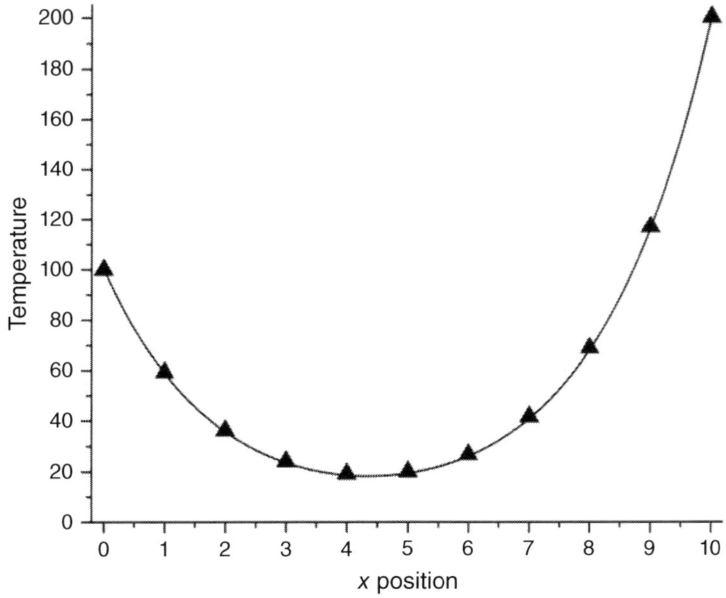

Figure 6.9 Comparison of the analytic solution (solid curve) for steady-state conduction in a slab with absorption of thermal energy with the finite-difference computation (filled triangles) using just nine interior nodes.

6.10 Stiff Differential Equations

A serious complication can arise in the numerical solution of ODEs when the equation(s) is *stiff*. When we say that an ODE (or a system of ODEs) is stiff, we are recognizing that the equations have widely varying characteristic values—they may differ by several orders of magnitude. In such cases, the stability of the calculation may become an issue. To examine the full range of system behavior we may be forced to use a *very small* step size; yet the important (ultimate) outcome may take a *very long time* to develop. This is exactly the case for the following hypothetical system where the characteristic times differ by three orders of magnitude:

$$\frac{dy_1}{dt} = -y_1 \text{ and } \frac{dy_2}{dt} = -1{,}000y_2, \text{ where both } y_1(0) = 1 \text{ and } y_2(0) = 1. \tag{6.86}$$

Of course, both have elementary exponential solutions, making the difficulty quite apparent; for $t = 1/4$, we have $y_1 = \exp(-t)$=0.7788 and $y_2 = \exp(-1{,}000t) = e^{-250}$. Such a parametric disparity can be very difficult to reconcile. Furthermore, the techniques we have discussed so far in this chapter may not work very well, or perhaps not at all. We will illustrate this point with the following example. Suppose we wish to solve the following test equation given by Hanna and Sandall (1995):

$$\frac{dy}{dx} = 50(x^2 - y) \text{ with } y(0) = 0. \tag{6.87}$$

Our objective is to find y for $x = 2$; we select the third-order Runge–Kutta method we introduced previously, and we will begin by using $h = 0.1$, and then successively halving h.

h	$y\,(x = 2)$
0.1	-2.78×10^{27}
0.05	-1.515×10^{35}
0.025	-1.783×10^{33}
0.0125	−1.601
0.00625	3.9208
0.003125	3.92075
0.0015625	3.92072
0.0010	3.9208+

The exact value for $y(2)$ is 3.9208. Though we were able to obtain something very close to that value by making h sufficiently small, the behavior we discovered above is disconcerting! It would be nice to be able to anticipate this problem and be able to assess its severity. One way to approach this difficulty is to define a "stiffness ratio" as follows:

$$S = |\lambda|(x_f - x_i), \tag{6.88}$$

where x_f and x_i are the end and initial points of the integration, respectively. We can obtain a *local* estimate for λ from:

$$\lambda = \frac{dy/dx}{y}. \tag{6.89}$$

In this example, $dy/dx = 50(x^2 - y)$ and for $x = 2$, $y = 3.9208$; thus,

$$\lambda \cong \frac{50(4 - 3.9208)}{3.9208} = 1.01, \tag{6.90}$$

resulting in $S = 2$. There are two important points to remember here: First, this is a very modest value for S, and second, for many nonlinear problems of interest, the value of S will vary considerably between the initial- and end-points of the integration. In such cases, a *single-point* evaluation of S may not be very informative.

We can illustrate this difficulty with a deceptively simple, but nonlinear, first-order ODE:

$$\frac{dy}{dt} = 3y + 4y^2, \tag{6.91}$$

with the initial condition $y(0) = 0.670$. Suppose we are interested in $y(t)$ for $0 \le t \le 0.25$: An analytic solution can be found, and the reader is invited to demonstrate that $y(t = 0.25) = 665.324$. Be forewarned, however, if you try to confirm this numerically you may run into trouble. Using Euler predictor–corrector and double precision, we find:

Δt	10^{-3}	10^{-4}	10^{-5}
$y(t = 0.25)$	331.2	641.64	660.51

Even the last entry in the table has an error of about 0.7%, so it is appropriate for us to wonder what has happened here. At least we have the information we need to evaluate λ at the end-points, $t = 0$ and $t = 0.25$; for the former we get 5.68, and for the latter, 2,664.3. Therefore, we can calculate S at both end-points resulting in 1.42 and 666.1—the stiffness ratio varies enormously over the interval of integration, suggesting that eq. (6.91) might be more challenging that it appears and that our observation regarding a single-point evaluation of S was well founded.

6.10.1 Gear's method

Gear's method is one of the best known techniques for dealing with systems of stiff differential equations. In fact, Gear devised an example problem (unsteady, batch, chemical kinetics) that has been used extensively to test ODE solvers; it consists of three simultaneous equations:

$$\frac{dy_1}{dt} = -0.013y_1 - 1,000y_1y_3 \tag{6.92}$$

$$\frac{dy_2}{dt} = -2,500y_2y_3 \tag{6.93}$$

$$\frac{dy_3}{dt} = -0.013y_1 - 1,000y_1y_3 - 2,500y_2y_3, \tag{6.94}$$

with $y_1(0) = 1$, $y_2(0) = 1$, and $y_3(0) = 0$. The range of integration is from $t = 0$ to $t = 50$. Suppose we begin our exploration of this example using the fourth-order Runge–Kutta procedure that we described previously. We begin by setting $h = 0.002$; in order to reach $t = 50$, about 100,000 k-function evaluations would be necessary. However, we never get to that point since the calculation self-destructs almost immediately. Even with $h = 0.001$ we find negative (unphysical) values for y_3. In order for the fourth-order RK scheme (with fixed step size) to work, it is necessary that h be less than about 0.0005. The results from such a computation are shown in Figure 6.10.

Gear (1971) developed a fourth-order backward differentiation formula (BDF) method specifically to deal with stiff ODEs, and it has been widely applied in this context. BDF methods will be discussed in detail in the next section.

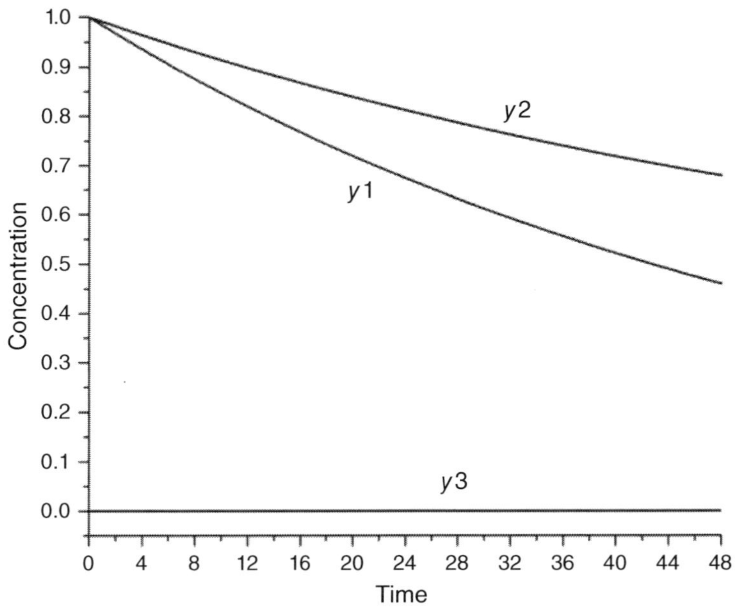

Figure 6.10 Computed results for Gear's example (a chemical kinetics problem). The concentration for the third species, y_3, is very small, but *not* zero.

6.11 BDF (Backward Differentiation Formula) Methods

BDF methods have been widely used for the solution of stiff differential equations, particularly since the publication of Gear's book in 1971. We begin our discussion of these techniques with an elementary illustration. Suppose that the differential equation to be solved is

$$\frac{dy}{dx} = f(x, y). \tag{6.95}$$

We formulate an Euler algorithm (but one that is *implicit*):

$$y_{i+1} = y_i + h\frac{dy_{i+1}}{dx} = y_i + hf(x_{i+1}, y_{i+1}). \tag{6.96}$$

What distinguishes this technique from those we have seen previously is that the function value that we wish to calculate (y_{i+1}) appears on *both sides of the equation*. Moreover, many of the problems for which this method is appropriate will be nonlinear; therefore, we will almost certainly need to solve nonlinear algebraic equations. But the enhanced stability of such methods for stiff ODEs will make the extra pain worthwhile. Let us apply this technique to a specific example. Suppose we have

$$\frac{dy}{dx} = 0.01xy^2 \text{ with } y(x = 1) = 3. \tag{6.97}$$

We want to know how y behaves as $x \to 8$. The analytic solution is simply:

$$y = \frac{1}{0.338333 - 0.005x^2}, \tag{6.98}$$

so $y(x = 8) = 54.5464$. We will now use the implicit Euler method starting with $h = 1/2$ to see how well the technique performs for this example eq. (6.97); we must keep in mind, though, that this is a first-order algorithm and accuracy may be poor.

h	*y* (*x* = 8)
0.5	134.696
0.25	271.482

h	$y(x = 8)$
0.125	466.908
0.0625	431.449
0.03125	109.204
0.015625	69.176
0.007813	60.645
0.003906	57.370
0.001953	55.908
0.000977	55.211
0.000488	54.876
0.000244	54.652
0.000122	54.540

By making h sufficiently small, we were able to obtain a reasonable result. However, it is clear that a higher-order BDF method is needed here. The generalized algorithm for order, n, can be written:

$$a_0 y_{i+1} + a_1 y_i + \cdots + a_n y_{i-n+1} = hf(x_{i+1}, y_{i+1}). \tag{6.99}$$

The coefficients required for this equation have been compiled by Lee and Schiesser (2004), and their table is presented in modified form below.

n	a_0	a_1	a_2	a_3	a_4
1	1	−1			
2	3/2	−2	1/2		
3	11/6	−3	3/2	−1/3	
4	25/12	−4	3	−4/3	1/4
5	137/60	−5	−10/3	5/4	−1/5

Therefore, if we wished to employ the third-order BDF method we would use:

$$\frac{11}{6} y_{i+1} - 3y_i + \frac{3}{2} y_{i-1} - \frac{1}{3} y_{i-2} = hf(x_{i+1}, y_{i+1}). \tag{6.100}$$

Since this algorithm requires two previous values for y, it is not self-starting, and some initial calculations will have to made with another method.

Let's underscore the merits of the third-order BDF using the test equation we examined in Section 6.10, $y' = 3y + 4y^2$, with $y(0) = 0.670$. Recall that $y(t = 0.25) = 665.324$, and that the stiffness ratio varied substantially over the interval of integration. We initially set $\Delta t = 0.0025$ (which we know from our previous discussion to be a *very large* value) and employ both the Euler predictor–corrector and the third-order BDF; we will then cut Δt in half (to 0.00125) and repeat:

	Euler PC	Third-order BDF
$\Delta t = 0.0025$	$y(t = 0.25) = 171.34$	$y(t = 0.25) = 345.48$
$\Delta t = 0.00125$	$y(t = 0.25) = 287.10$	$y(t = 0.25) = 620.10$

You can see that the BDF scheme has provided much improved estimates even though we employed very large Δt's—in fact, for $\Delta t = 0.00125$, the error at the end-point has been reduced to about 6.8%.

6.12 Bulirsch–Stoer Method

We previously introduced the Richardson extrapolation, and we saw what a powerful tool it could be. Bulirsch and Stoer recognized that it might be used as the basis for an extremely efficient method for solving ODEs; the technique they developed has garnered some very enthusiastic advocates (see, for example, page 563 in Press et al., 1989). For problems in which it is essential that the computational effort be minimized, Bulirsch–Stoer is definitely worth consideration.

The principal idea is a simple one: Suppose we have a differential equation $dy/dx = f(x,y)$ we wish to solve from some initial $x = 0$, to $x = L$. The function y is computed using a mid-point method across a *large* interval initially using two steps. The calculation is carried out again, but this time using four steps. We now have a function $f(h)$ from the two different values of h that can be fit to an analytic form. An extrapolation is then attempted to $h = 0$ (corresponding to an infinite number of steps); if the estimated error is deemed satisfactory, we proceed. If not, we increase the number of steps (to 6, then 8, then 12, etc.), and after each increase we try to the extrapolation to $h = 0$ again. The usual sequence of step number is 2, 4, 6, 8, 12, 16, 24, 32, 48, 64, 96, etc. Press et al. note that a more *expected sequence*, where we double the number of steps with each trial, will cause h to become too small too rapidly.

We will treat an elementary example to demonstrate the Bulirsch–Stoer method. Consider the differential equation,

$$\frac{dy}{dx} = xy^2, \text{ with } y(x = 0) = 2. \tag{6.101}$$

We are interested in $y(x = 0.975)$; since the analytic solution is $y = \frac{2}{1-x^2}$, $y(0.975) = 40.5063$. Let us use the modified mid-point method (which is sometimes referred to as Gragg's method) with $h = H/n_s$, where H is the interval we wish to cover (which is 0.975) and n_s is even (we start with $n_s = 4$).

$$y_{ep} = 6.159404$$

We repeat the process, but with eight steps rather than four:

$$y_{ep} = 10.258197$$

Now we use the two end-point (*ep*) estimates for a polynomial extrapolation:

$$y(x = 0.975) \cong \frac{4y_8 - y_4}{3} = \frac{(4)(10.258197) - (6.159404)}{3} = 11.6245. \tag{6.102}$$

It is clear that we are not there yet. We continue, but this time with 16, 32, 64, 128, … steps (we will double the number of steps each time to make use of the polynomial extrapolation immediately above). The resulting succession of estimates takes the form:

$$18.4863, 26.9801, 34.5736, 38.8291, 40.2004, 40.4649, \text{and } 40.5014.$$

Now notice how the error for our estimate has diminished through this sequence:

$$54.36\%, 33.39\%, 14.64\%, 4.14\%, 0.755\%, 0.102\%, \text{and } 0.012\%.$$

The very last estimate is from the 512- and 1,024-step pair. As you can see, this gets us within about 0.012% of the correct value. We would need about 2,000 steps with a third-order Runge–Kutta scheme to get within the same error window.

The Bulirsch–Stoer method is also available with Mathcad™ (*bulstoer*), and we illustrate its application using eq. (6.101). Note that there are seven parameters in the parameter list (unlike *rkfixed* which has only five).

Mathcad™ implementation of Bulirsch–Stoer method for eq. (6.101):

Solve $dy/dx = xy^2$ with $y(x = 0) = 2.$

$y_0 := 2$

$D(x, y) := x(y_0)^2$

$Z := \text{bulstoer}(y, 0, 0.975, 0.00001, D, 10, 0.000001)$

$$Z = \begin{pmatrix} 0 & 2 \\ 9.75\times10^{-3} & 2 \\ 0.107 & 2.023 \\ 0.61 & 3.183 \\ 0.705 & 3.973 \\ 0.817 & 6.006 \\ 0.927 & 14.29 \\ 0.975 & 40.503 \end{pmatrix}$$

The fourth parameter controls the accuracy of the solution, and in this case we have set its value to 1×10^{-5}; note that the value reported for $y(x = 0.975)$ is 40.503 which corresponds to an error of about 0.008%.

6.13 Phase Space

Among the problems that arise when an analyst must work with a nonlinear differential equation (or model) is the fact that he/she may not have any idea what form the dynamic behavior of the system will take. This makes it difficult to interpret a numerical result and extremely difficult to detect errors when they occur. Furthermore, an output stream of numbers appearing on screen (or written to a file) does not provide much feedback—it might be nearly impossible to detect a periodicity, or a lack of periodic behavior from such output. This is exactly the kind of situation where a phase-space analysis can be useful. Our strategy is to construct a *system trajectory* by cross-plotting dependent variables and removing time from the visualization of results. For systems of one, two, or three dimensions this is straightforward as we shall see.

Consider the simple sinusoid, $y(t) = A \sin \omega t$, with the derivative, $y'(t) = \omega A \cos \omega t$. Of course, if we plot $y(t)$ we know exactly what to expect. But suppose instead that we plot corresponding values of $y(t)$ and the derivative $y'(t)$ as time evolves; we will see a *closed* system trajectory in the form of what is called a *limit cycle*. We will select $A = 5$ and $\omega = 1$ for this illustration.

If the amplitude of the sinusoid was increasing with time (unstable behavior) then the form we would see in Figure 6.11 would be an outward-directed spiral. If the sinusoid was decaying, for example, if $y(t) = A \exp(-\beta t) \sin(\omega t)$, then we would expect an inward-directed spiral that would approach 0,0 on the phase plane as time became large. Of course, what we really want to know is what phase-space analysis can do for us in the context of challenging nonlinear differential equations.

6.13.1 Identifying a Strange Attractor

In 1963 Edward Lorenz published an extremely important paper entitled, "Deterministic Nonperiodic Flow" in the *Journal of Atmospheric Sciences*. Lorenz set out to develop the simplest possible model for atmospheric phenomena, accounting for the intensity of convective motion (X), the temperature difference between rising and falling currents (Y), and deviation of the vertical temperature profile from linearity (Z). The resulting set of ODEs can be written:

$$\frac{dX}{dt} = \Pr(Y - X) \quad \frac{dY}{dt} = -XZ + rX - Y \quad \text{and} \quad \frac{dZ}{dt} = XY - bZ. \tag{6.103a,b,c}$$

We will assume the following values for the parameters that appear in eq. (6.103a,b,c): $\Pr = 10$, $r = 28$, and $b = 8/3$. For initial conditions (X,Y,Z) we select (0,1,0), then we will obtain the projected (on the Y-Z plane) system trajectory by numerical solution of the differential equations. The two-dimensional cut (from the three dimensional system) shown in Figure 6.12 is a "portrait" of a *strange attractor*. For a simple mechanical system that oscillates with decaying amplitude, the phase-space trajectory (2D) as we observed previously will be an inward spiral—this is characteristic of dissipative systems. The point in phase space to which the trajectory is drawn is called an "attractor." If a frictionless system oscillates with constant amplitude, the phase-space portrait will be an ellipse (a limit cycle as we saw in Figure 6.11); such systems are said to be conservative because the phase "volume" remains constant.

What the two-dimensional cut provided in Figure 6.12 *cannot* reveal is that no point in phase space is *ever* revisited. Thus, the Lorenz model—though fully deterministic—is nonperiodic. The implications are staggering, and it would be far more difficult to comprehend the behavior of this system if we were trying to do so from a table of numbers, say $Y(t)$. This is a case where phase-space analysis is invaluable.

6.13.2 The Binet Equation

A *central force* problem is one in which an entity, or particle, is either attracted toward or repelled from a central point in space (which we take to be the origin). The Binet equation (Jacques Philippe Marie Binet, 1786–1856) can be used to determine the shape of an orbit in a central force problem given a force law of some particular form. Gravity is an example of a central force, and its magnitude at a point depends upon the distance, r, to the origin. Thus, the Binet differential equation can be used to *estimate* the elliptical orbits of objects about the sun for example. Alternatively, given orbital measurements one can deduce the form of $f(r)$:

$$f(r) = f\left(\frac{1}{u}\right) = -mh^2u^2\left(\frac{d^2u}{d\theta^2} + u\right). \tag{6.104}$$

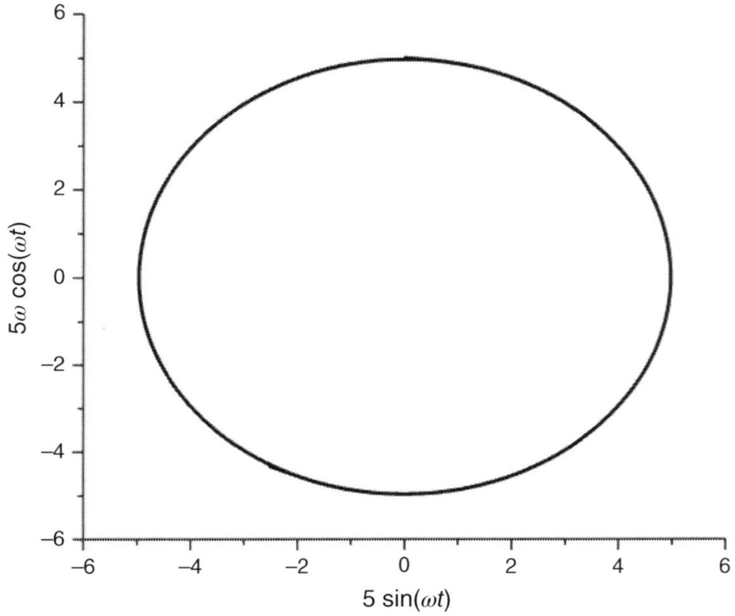

Figure 6.11 The system trajectory for this sinusoid is a closed path, or limit cycle. The amplitude of the oscillation is neither increasing nor decreasing. Motion on this limit cycle is clockwise.

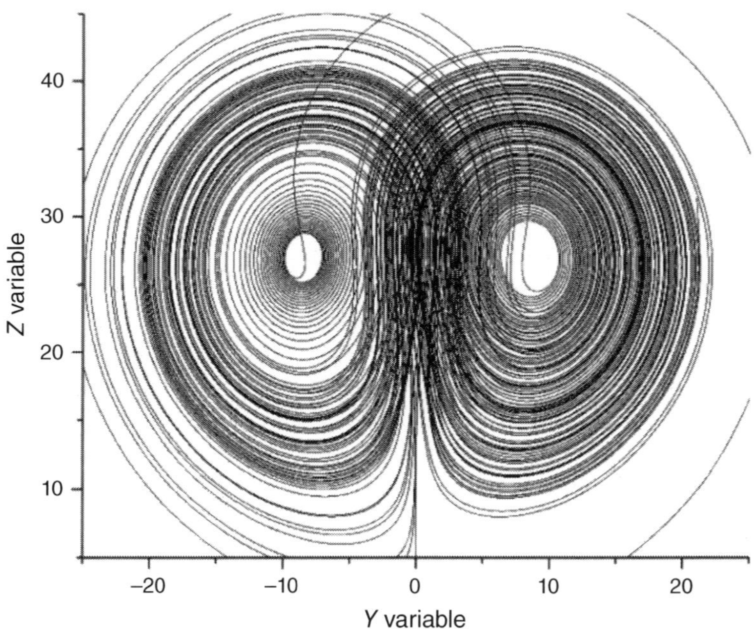

Figure 6.12 A two-dimensional (*Y*-*Z* plane) portrait of a *strange attractor* from the Lorenz model with $r = 28$.

Here m and h are the mass and specific angular momentum, and u is the reciprocal of r. In the Kepler problem, f is taken to be proportional to $-u^2$ (that is, the Newtonian gravitational force follows the inverse square law) resulting in a linear differential equation:

$$\frac{d^2u}{d\theta^2} + u = \frac{k}{mh^2}. \tag{6.105}$$

In this case the solution of the Binet equation can be found using techniques described in Chapter 5; the complementary function is obtained from the homogeneous equation, $(D^2 + 1)u = 0$, which when factored yields $(D + i)(D - i)u = 0$. And therefore, $CF = c_1 \exp(-i\theta) + c_2 \exp(+i\theta)$. The particular integral can be determined by inspection by setting the second

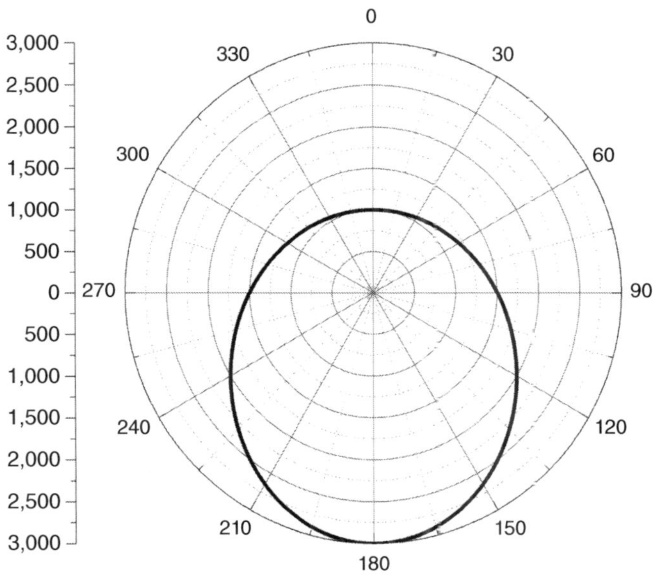

Figure 6.13 Polar plot of a solution to the Kepler problem with $f(r) = -ku^2$ and $\beta = 1/2$.

derivative to zero: $PI = \frac{k}{mh^2}$. We use the Euler relation to write the solution in terms of sin and cos, or equivalently:

$$u = C_1 \cos(\theta - \phi) + \frac{k}{mh^2}. \tag{6.106}$$

If $\phi = 0$, the elliptical orbit will exhibit minimum distance to the origin (as visualized in the x-y plane) on the x-axis where $\theta = 0$. The maximum distance from the origin will occur at $\theta = \pi$ (180°) of course. We can alter the orientation of the ellipse by including an offset or phase angle; that is, $\phi \neq 0$. Since $u = 1/r$, we invert eq. (6.106) with $\phi = 0$: $r = \frac{mh^2/k}{\beta \cos\theta + 1}$. If $\beta = 0$, r is fixed and the orbit is circular; if $\beta < 1$ the traveled pathway is elliptical. Figure 6.13 illustrates a polar plot of a typical solution of the linear Binet equation with $\beta = 1/2$; the orbit is elliptical and one of the foci is located at the origin.

What we have seen to this point is entirely familiar; in the Kepler version the Binet equation is just another second-order linear differential equation with constant coefficients. However, the Binet equation can produce a spectrum of behaviors depending upon what is chosen for the left-hand side of eq. (6.104); for example, the reader might consider $u'' + u = c_1/\exp(c_2\theta)$ which can—and no surprise here—produce an unstable orbit (for positive c_2) as shown in Figure 6.14. Note that the initial value for r was 1,000.

The Binet equation becomes more interesting when the force law assumes some form resulting in a nonlinear ODE. This is what we would like to consider next.

Let us examine the nonlinear equation:

$$\frac{d^2u}{d\theta^2} + u = c_1/\sqrt[4]{u}. \tag{6.107}$$

We set $u(0) = a$ and $u'(0) = 0$ and solve eq. (6.107) numerically (the reader is invited to try this). The result is affected by the choice for c_1 of course, but one outcome appears in Figure 6.15.

The results shown in Figure 6.15, a polar plot of $r(\theta)$, reveal an orbital behavior in which the apogee and the perigee (or in the case of an orbit around the sun, aphelion, and perihelion) are *the same in each cycle*. However, the computed pathway exhibits *precession*, i.e., rotation of the orbital path just as observed in our own solar system (both the orbital migration and the eccentricity are unrealistically large in this example). We have one dependent variable, so our construction of a phase-space portrait will entail cross-plotting u and its derivative, u'. What should we expect to see? The function, $u(\theta)$, is purely periodic, so the system will exhibit a limit cycle, a closed elliptical path in phase space.

The orbital precession observed in our solar system is due to the gravitational pull on each body by the other planets, and Newtonian gravitational theory provides reasonable estimates of this effect for most of the planets. However, a discrepancy (a perihelion advance) for the orbit of Mercury was noted by the French astronomer Le Verrier around 1859. The perihelion shift problem for the innermost planet was examined by Einstein, and the relativistic Binet equation (derived for

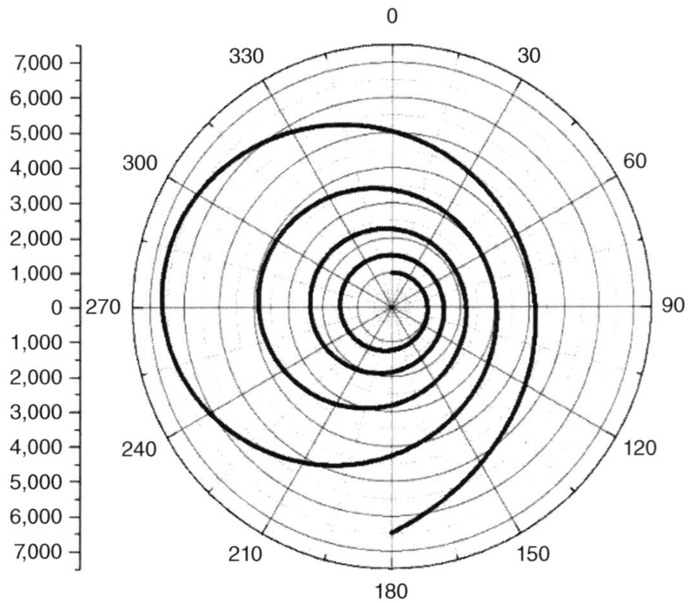

Figure 6.14 Numerical solution of the linear ODE, $u'' + u = c_1/\exp(c_2\theta)$, presented as $r(\theta)$.

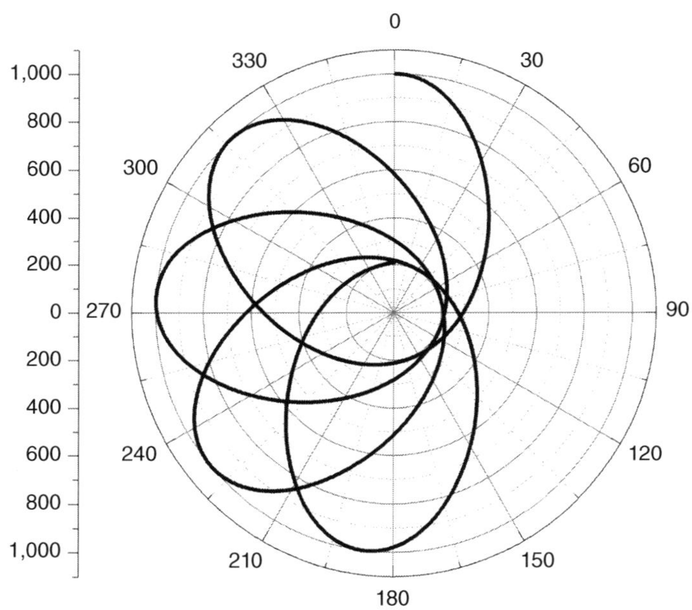

Figure 6.15 Numerical solution of a nonlinear Binet equation, $u'' + u = c_1/u^{1/4}$. The plot is $1/u$ (which is r) as a function of θ, and the observed path is an exaggerated facsimile of *apsidal precession.*

Schwarzschild coordinates) is:

$$\frac{d^2u}{d\theta^2} + u = \frac{R_s c^2}{2h^2} + \frac{3R_s}{2}u^2 = a + bu^2. \tag{6.108}$$

It is worth noting that precession in our solar system was first recorded more than 2,000 years ago when Hipparchus made observations of the moon orbiting the earth. But in the last 200 years meticulous measurements of the orbit of Mercury about the sun revealed that observed value was 43 s of arc per century *greater* than the value predicted by Newton's gravitational theory. Einstein was elated when his calculation of the relativistic perihelion advance (caused by curvature of space–time as Mercury dips into the sun's gravity well) produced the missing 43 s of arc per century—providing a quantitative validation of general relativity.

Similar ODEs—with the inclusion of a first derivative term—arose in Rayleigh's study of vibrations where the linear oscillator ODE,

$$\frac{d^2u}{d\theta^2} + u = k\frac{du}{d\theta} \tag{6.109}$$

led him to consider the equation:

$$\frac{d^2u}{d\theta^2} + u = c_1\frac{du}{d\theta} + c_2\left(\frac{du}{d\theta}\right)^3. \tag{6.110}$$

A phase-space portrait of the latter, eq. (6.110), is shown in Figure 6.16 where both c_1 and c_2 are negative.

Rayleigh found the case where c_1 and c_2 have different signs to be particularly interesting, and the reader may want to spend some time to examine some possible outcomes. Figure 6.17 depicts a situation where the numerical solution initially suggests unstable behavior; however, it evolves into a purely periodic motion producing a limit cycle in the phase plane.

What happens if we move the initial value of $u'(0)$ so that we start outside the limit cycle in Figure 6.17? If $u'(0) = 2.5$, the system trajectory spirals in toward the limit cycle in the clockwise direction. By changing the values of c_1 and c_2 to -0.275 and $+0.050795$, respectively, we can obtain unstable behavior (after a lengthy period of behavior that appears to be periodic).

We conclude our discussion of the utility of phase space with a related example adopted from LaSalle (*Proceedings of Symposia in Applied Mathematics, Volume XIII, Hydrodynamic Instability* 1962). Consider the deceptively simple (but non-linear) ODE:

$$\frac{d^2y}{dt^2} + a\frac{dy}{dt} + 2y + 3y^2 = 0, \tag{6.111}$$

with $a = ½$. We will solve this ODE numerically beginning with $y(0) = -½$ and $y'(0) = 0$. This system is stable and the phase-space portrait reveals an inward spiral as shown in Figure 6.18.

The LaSalle model eq. (6.111) becomes especially interesting when we change the initial condition to explore the threshold of instability; for example, suppose we first take $y(0) = +0.562866$ and then $y(0) = +0.562867$ (we maintain the zero initial value for the derivative both times) and solve eq. (6.111) for large values for t. In these cases the phase-plane portraits are very different from those depicted in Figure 6.18; in fact, the dynamic behavior in the second case is unstable. We will explore this model further in a student exercise at the end of the chapter.

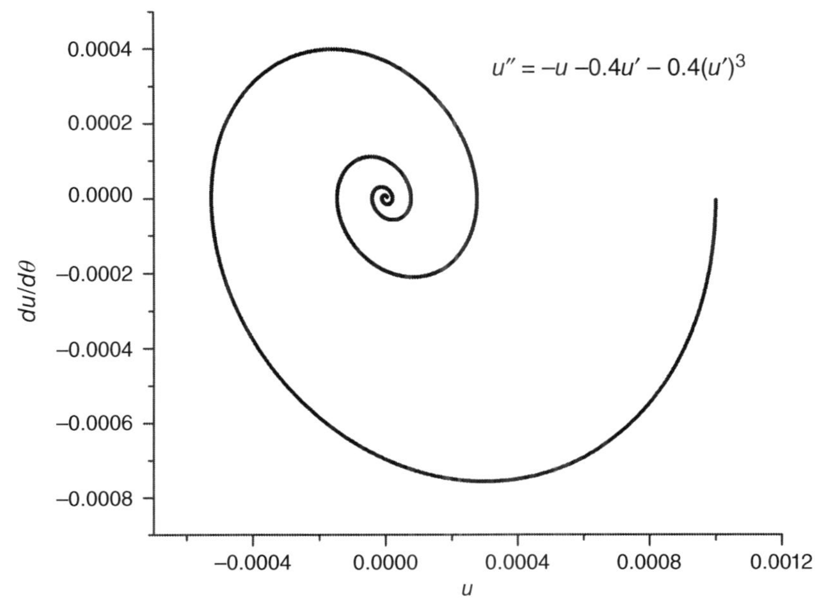

Figure 6.16 An inward-directed spiral obtained from eq. (6.109), $u'' + u = c_1u' + c_2(u')^3$.

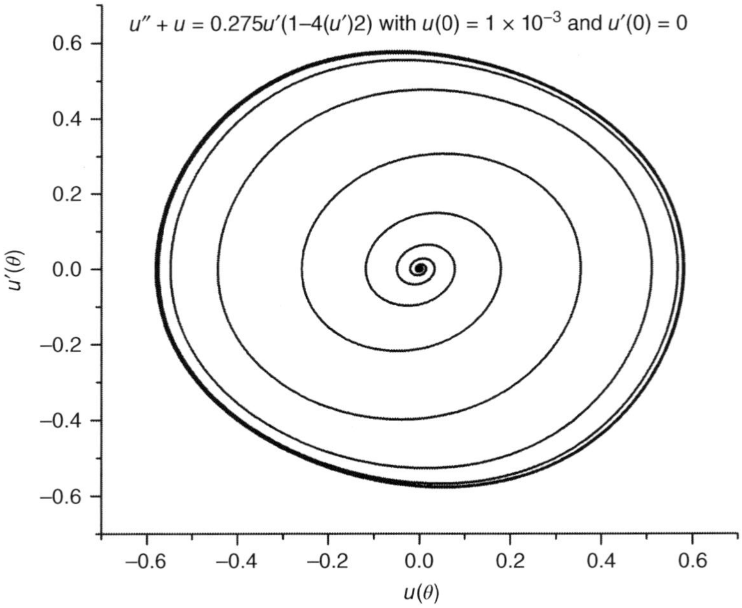

Figure 6.17 Phase-space portrait of the behavior of the ODE, $u'' + u = 0.275u'(1 - 4(u')^2)$, with $c_1 = 0.275$ and $c_2 = -1.10$. The outward-directed spiral converges upon the observed limit cycle.

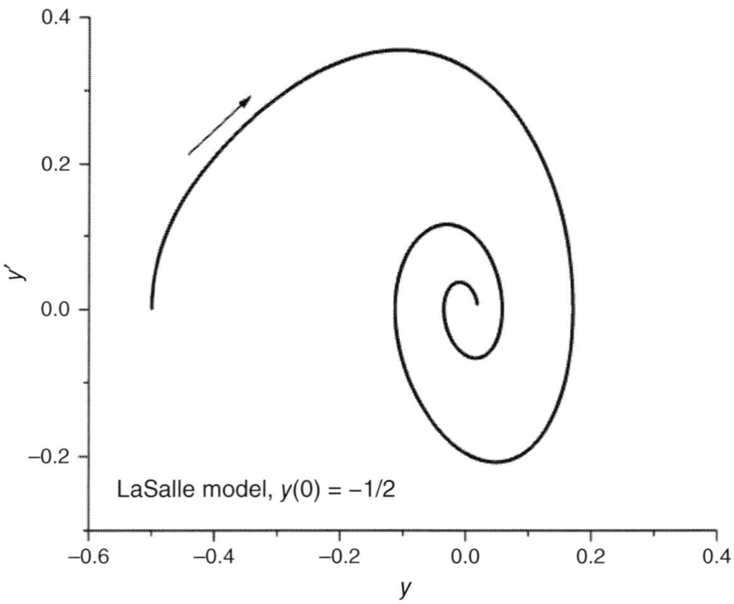

Figure 6.18 Phase plane for the LaSalle model with $y(0) = -½$. Note that the system trajectory is approaching the equilibrium point located at $y = 0$.

6.14 Summary

We have said very little about the error inherent in the numerical solutions of differential equations. This is because the general topic is too broad and too complex for the space available here. But there are some elementary observations we should make. For example, we need to draw the reader's attention to the difference between roundoff error and truncation error. Roundoff error is hardware-dependent, i.e., it is a consequence of the precision of the computing device being used. For example, a calculator might report 2/3 as 0.666667. Truncation error, on the other hand, is the result of neglected terms in the construction of the algorithm. For example, if we truncate a Taylor series expansion, the resulting error will be of the order of the first neglected term.

And there is another aspect to this discussion that is crucial: numerical stability which is sometimes referred to in the literature as *weak* stability). We have primarily talked about choosing the step size in terms of the accuracy of the resulting solution. And often—particularly for non-demanding problems—that is entirely appropriate. In the case of certain ODEs, however, step size is chosen not for accuracy, but to ensure *stability* of the computation. It is this feature of the numerical solution of ODEs that we wish to discuss briefly here. For additional detail consult Lee and Schiesser (2004) and for greater mathematical rigor, see Hairer and Wanner (1996).

Consider the elementary differential equation:

$$\frac{dy}{dx} = -\beta y. \tag{6.112}$$

Of course the solution has exponential form, $\frac{y}{y_0} = \exp(-\beta x)$. Now suppose we wished to use the Euler method for this equation beginning from $x = 0$; the first step would look like this:

$$y(\Delta x) \cong -\Delta x\beta y(0) + y(0) = y(0)(1 - \Delta x\beta), \tag{6.113}$$

and the second step:

$$y(2\Delta x) \cong -\Delta x\beta y(\Delta x) + y(\Delta x) = y(\Delta x)(1 - \Delta x\beta). \tag{6.114}$$

Note that the right-hand side of the latter equation can be written equivalently as

$$y(2\Delta x) = y(0)(1 - \Delta x\beta)^2, \tag{6.115}$$

and in general for n-steps:

$$y(n\Delta x) = y(0)(1 - \Delta x\beta)^n. \tag{6.116}$$

If Δx is selected such that the product, $\Delta x\beta = 2$, then the procedure results in oscillation between $-y(0)$ and $+y(0)$. Even worse, if $\Delta x\beta > 2$, the oscillations grow in amplitude; e.g., for $\Delta x\beta = 4$, we get the sequence: $-3y(0)$, $+9y(0)$, $-27y(0)$, etc. On the other hand, if $\Delta x\beta < 1$, we at least get qualitatively correct behavior where successive steps reveal diminishing y which is in accord with the *decaying* exponential solution. Setting $\Delta x\beta = 4/5$ produces the sequence: 1/5, 1/25, 1/125, 1/625, … each multiplied by $y(0)$ of course. It is clear, therefore, that the explicit Euler method will require that $|\Delta x\beta| < 2$ although for the sake of accuracy the product $\Delta x\beta$ will usually be much smaller than this.

In the case of explicit Runge–Kutta methods a stability assessment can be carried out in a similar manner. We take the model equation (sometimes referred to as the Dahlquist test equation):

$$\frac{dy}{dt} = \lambda y \text{ with } y(0) = y_0, \tag{6.117}$$

and let $z = h\lambda$. The solution for the test equation is exponential, and for a single step forward we write it as $\exp(\lambda t) = e^z$. Therefore a stability function can be written (for the fourth-order RK procedure in this case):

$$1 + \frac{z}{1!} + \frac{z^2}{2!} + \frac{z^3}{3!} + \frac{z^4}{4!}. \tag{6.118}$$

For stability of the numerical calculation it is necessary that $\left|\frac{y_{i+1}}{y_i}\right| \leq 1$, so the stability function given above *is set equal to* 1. This determines the limiting value for z (or $h\lambda$), which is −2.785 for the fourth-order RK. Lambert (1973) shows that for the third-order RK procedure the limiting value is −2.51. The implication of course is that the explicit Runge–Kutta methods are only *slightly more stable* than the Euler technique—i.e., the *limit* for the magnitude of $h\lambda$ is only a bit larger. Complete stability domains are given graphically in Hairer and Wanner (1996) which permit easy comparison of different solution techniques. Because the elementary explicit techniques have limited stability domains, they are not suitable for stiff problems; in these cases one must turn to an implicit method (like BDF) as we noted previously.

Although BDF methods are not self-starting, their enhanced stability has made them very popular for the solution of stiff ODEs. Lee and Schiesser (2004) summarize the stability properties of BDF methods in Chapter 1 of their book, and they note that BDF algorithms through sixth order are unconditionally stable along the negative real axis—clearly superior in this regard to the Euler and Runge–Kutta techniques considered earlier. Lee and Schiesseer also point out that there are many very high-quality codes available for the solution of ODEs, both within commercial software packages and from the public domain. For the latter, a good starting point is the Netlib Repository at http://www.netlib.org/index.html.

Problems

6.1 Show that the analytic solution for the ODE, $\frac{dy}{dx} = x - \frac{y}{x}$ is $y = \frac{x^2}{3} + \frac{C}{x}$. Then, given that $y(x = 2) = 2$, use the modified Euler method and find $y(x = 4)$.

6.2 Consider the second-order differential equation,

$$16\frac{d^2y}{dt^2} + 5\frac{dy}{dt} + y = x(t).$$

Initially, both y and dy/dt are zero. If $x(t) = 1/(1+t)$, what is the maximum value attained by y, and when does that occur?

6.3 A tank with a capacity of 1,900 gal (the tank diameter is 7 ft) is initially half-full. Water enters the tank at a constant rate of 85 gpm; it also drains from the bottom of the tank through a round orifice. The velocity through the orifice is given (approximately) by Torricelli's theorem: $V_0 = \sqrt{2gh}$, where h is the depth of water above the hole. The area of that orifice is 7.854×10^{-5} ft^2. Will the tank overflow? If so, when? If not, when will the maximum depth be attained? Repeat your calculation with an orifice area of 0.01 ft^2; what is the maximum depth in this case?

Answer part two: 5.589 ft.

6.4 We want to revisit eq. (6.17) from Section 6.2:

$$\frac{dy}{dx} = \frac{1}{\sqrt{x + \frac{1}{8}}}\left(y^2 - \frac{1}{9}y\right) \text{ with } y(0) = 1 \text{ and } 0 \le x \le 0.653.$$

You will recall that the Euler predictor–corrector method struggled with this first-order ODE, leaving us quite uncertain about the actual value for $y(x = 0.653)$. One could continue to halve Δx, and then try to extrapolate the results to $\Delta x \to 0$, but we looked at other methods of solution in this chapter that may be better suited to the problem. If you have access to *Mathcad*™ or similar software, you can quickly compare RK_{fixed}, RK_{adaptive}, *Bulirsch–Stoer*, etc. for this equation. For example, if we implement Bulirsch–Stoer with *Mathcad*:

$$y_0 := 1$$

$$D(x, y) := \frac{\left(y^2 - \frac{y}{9}\right)}{(x + 0.125)^{0.5}}$$

$$Z := \text{bulstoer}(y, 0, 0.653, 0.00001, D, 40, 1e-06)$$

$$Z = \begin{pmatrix} 0 & 1 \\ 0.653 & 325.733 \end{pmatrix}$$

Explore this ODE using more advanced methods, such as an adaptive Runge–Kutta or Bulirsch–Stoer. Is $y(x = 0.653) = 325.73$ a definitive result?

6.5 We want to examine the dynamic behavior of two populations in conflict, coyotes and rabbits (this is a subset of Volterra's problem but with heredity neglected). We will initiate the calculation with 100 rabbits (n_2) and 3 coyotes (n_1). Assume the governing equations are:

$$\frac{dn_1}{dt} = an_1n_2 - bn_1 \text{ and } \frac{dn_2}{dt} = cn_2{}^2 - dn_1n_2$$

Solve the two simultaneous equations and prepare a phase-space portrait by cross-plotting n_1 and n_2. The values for the constants a,b,c,d are 0.3, 0.2, 0.25, 0.7, respectively. Is extinction of a species possible in this problem? If heredity factors are added to this problem, exactly how will the nature of the problem be changed?

6.6 In a misguided attempt to celebrate the new year, Eric fires his 9 mm pistol vertically into the air. The 115 grain bullet leaves the barrel with an initial velocity of 1,225 ft/s. Assume the drag acting upon the projectile is given by $F = AKf$, where $A = \pi R^2$, $K = ½\rho V^2$, and the drag coefficient, f, is assumed to be constant at 0.5. When will the bullet strike the ground, and what will the velocity be at impact? Is it possible that the returning projectile could be lethal?

6.7 The driven pendulum has been the focus of intensive investigation because of the possibility of chaotic behavior. Begin your analysis with the elementary, linearized damped pendulum with the equation of motion:

$$\frac{d^2\theta}{dt^2} + \frac{d\theta}{dt} + \theta = 0.$$

Solve this equation (giving the pendulum an initial disturbance) and prepare a plot of the system trajectory by cross-plotting θ with its derivative. How does this system behave dynamically? Now, assume the three governing equations for the *driven pendulum* have the form:

$$\frac{d\omega}{dt} = -\frac{\omega}{q} - \sin\theta + g\cos\phi$$

$$\frac{d\theta}{dt} = \omega$$

$$\frac{d\phi}{dt} = \omega_D$$

We want to solve this set of equations using $q = 2$, $\omega_D = 2/3$, and $g = 1.0$, and once again we would like to construct the system trajectory (using ω and θ) as we suspect that chaotic behavior may be possible. If you need further help with this problem, see Baker and Gollub, *Chaotic Dynamics: An Introduction*, Cambridge University Press (1990).

6.8 Three-body problems are particularly interesting because they can exhibit chaotic behavior. Begin this problem by consulting the work of Durham et al. (NASA-CR-57430, November 1964) entitled: *Study of Methods for the Numerical Solution of Ordinary Differential Equations*. On page 104 they describe a model for a restricted three-body problem (earth–moon–satellite):

$$\frac{d^2x}{dt^2} = x + 2\frac{dy}{dt} - \mu'\frac{x+\mu}{((x+\mu)^2+y^2)^{3/2}} - \mu\frac{x-\mu'}{((x-\mu')^2+y^2)^{3/2}}$$

and

$$\frac{d^2y}{dt^2} = y - 2\frac{dx}{dt} - \mu'\frac{y}{((x+\mu)^2+y^2)^{3/2}} - \mu\frac{y}{((x-\mu')^2+y^2)^{3/2}},$$

where $\mu' = 1-\mu$. We want to solve this fourth-order system. The initial conditions are $x(0) = 0.994$, $y(0) = 0$, $dx/dt = 0$ for $t = 0$, and $dy/dt = -2.03173262955733683566$ for $t = 0$. The parameter, μ is 0.012277471. The period for the motion is 11.12434033726608513507 and the time step will need to be very small (and probably variable). *Please note: In this restricted three-body problem, the earth, moon, and satellite all lie in the x-y plane and the earth and moon are on the x-axis. The mass of the satellite is considered negligible compared to the two massive bodies. You can expect pronounced sensitivity to the initial conditions.*

6.9 Hiemenz stagnation flow is governed by the nonlinear, third-order ODE:

$$\nu f''' = f'^2 - ff'' - a^2,$$

where ν is the kinematic viscosity of the fluid and a is the strength of the potential flow approaching the flat surface. Assume the fluid is water such that $\nu = 0.01$ and that $a = 5$. Find a solution for this problem given that $f(0) = 0$, $f'(0) = 0$, and $f'(\infty) = a$.

6.10 One form of Bessel's differential equation is:

$$\frac{d^2y}{dx^2} + \frac{1}{x}\frac{dy}{dx} - k^2y = 0.$$

This might describe, for example, the disappearance of a reactant species in an infinitely long cylinder (catalyst pellet). It is obvious that there is a regular singular point at $x = 0$. The analytic solution for this problem is:

$$y = C_1 I_0(kx) + C_2 K_0(kx)$$

We want to solve this equation numerically, integrating from $x = 0$ to $x = 1$ given that the concentration is finite at the center and equal to a surface value (say, 1) at $x = 1$. Take $k = 2.325$.

6.11 Consider the differential equation, $\frac{dy}{dx} = 3x + y^2$. We know that $y(x = 0) = 1$, and we want to find an estimate of $y(x = 0.1)$. In particular, Ayres (*Differential Equations*, Schaum's Outline Series, Mc-Graw-Hill, 1952) gives a value for this problem of 1.12725, and want to know how accurate this value is.

6.12 A steel cable with a per-unit-length weight of W (lb/ft) hangs between two supports. The deflection is described by the differential equation,

$$\frac{d^2y}{dx^2} = \frac{W}{h}\sqrt{1+\left(\frac{dy}{dx}\right)^2}.$$

The shape assumed by a suspended cable is a *catenary* which can be described with the hyperbolic cosine (*cosh*). At each support, of course, $y(x = -L \text{ and } x = L) = 0$. Solve the differential equation numerically, assuming $W/h = 2$ and $L = 10$.

6.13 The Leveque equation describes the temperature distribution for an inlet flow past a heated wall. The x-axis corresponds to the wall and the y-coordinate extends into the fluid phase. By assuming that the velocity distribution in proximity to the stationary wall is linear: $v_x = cy$, the following differential equation can be developed:

$$\frac{d^2T}{d\eta^2} + 3\eta^2\frac{dT}{d\eta} = 0 \text{ where } \eta = y\left(\frac{c}{9\alpha x}\right)^{1/3}.$$

Of course, α is the thermal diffusivity of the fluid (assume that it is water). Given that $T(\eta = 0) = T_w$, the temperature of the heated wall, and that $T(\eta \to \infty) = T_\infty$, solve this second-order equation numerically and compare your results with the analytic solution for $\eta = 0.3, 0.5, 0.7, 0.9$, and 1.10. Let $T_w = 50°$ and $T_\infty = 25°$.

6.14 The Rayleigh–Plesset equation describes the oscillatory behavior of the interface of a disturbed spherical gas bubble immersed in a liquid:

$$\frac{P_i - P_\infty}{\rho} = R\frac{d^2R}{dt^2} + \frac{3}{2}\left(\frac{dR}{dt}\right)^2 + \frac{4\nu}{R}\frac{dR}{dt} + \frac{2\sigma}{\rho R}.$$

R, of course, is the bubble radius. This differential equation is notoriously stiff, and the term that includes the kinematic viscosity (ν) is often small and therefore frequently neglected. We are interested in the case in which the spherical gas bubble is subjected to an instantaneous increase in external pressure. This should result in compression and rebound—repeatedly. Use Borotnikova and Soloukhin (A Calculation of the Pulsations of Gas Bubbles in an Incompressible Liquid Subject to a Periodically Varying Pressure, *Soviet Physics—Acoustics*, 10:28, 1964) as a guide, and solve this *dimensionless* (and simplified) Rayleigh–Plesset equation:

$$\frac{d^2R}{dt^2} = -\frac{3}{2R}\left(\frac{dR}{dt}\right)^2 + \frac{\mu}{R}(R^{3\gamma} - 30).$$

The initial value for R is 1 and $\gamma = 1.4$; the increased pressure is 30 and $\mu = 0.837 \times 10^{-6}$. *Please note it will be necessary to use an adaptive step size as you approach the compression limit and the beginning of the rebound.*

6.15 Use the Bulirsch–Stoer method to solve the differential equation,

$$\frac{dy}{dx} = x^2y^3$$

with $y(x = 0) = 3$, and find estimates for $y(x = 0.5475)$.

6.16 Phase-space analysis is a powerful tool for exploring the dynamic behavior of complex systems (with a limited number of dimensions). It will allow us to assess stability and identify any attractors should they exist. We wish to reconsider the nonlinear LaSalle model from Section 6.13,

$$\frac{d^2y}{dt^2} + a\frac{dy}{dt} + 2y + 3y^2 = 0,$$

with $a = ½$. There is an obvious equilibrium point at $y = 0$, obtained by setting the time derivatives equal to zero. There is another located at $y = -⅔$. Solve this equation numerically using different initial values for $y(t = 0)$, letting

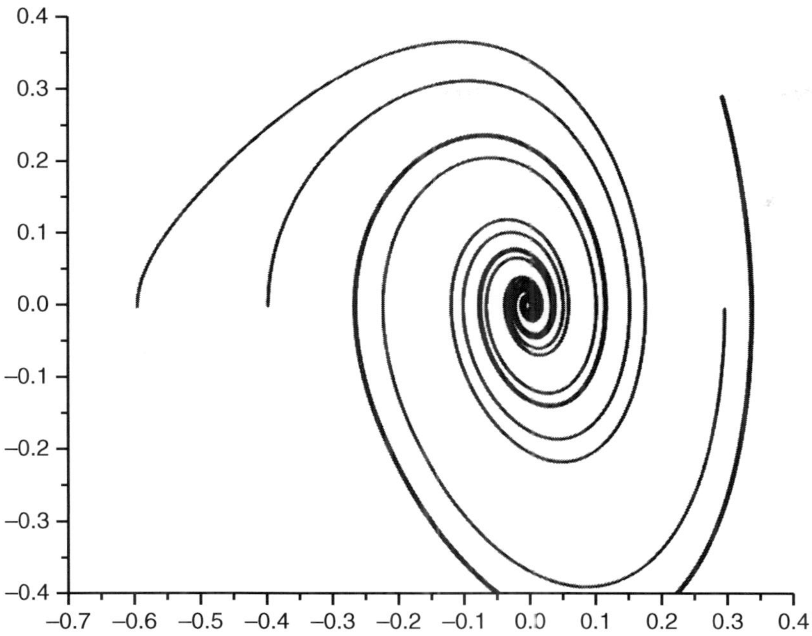

Figure P6.16 A few trajectories plotted for the LaSalle model.

$y'(t = 0) = 0$ in every case. Draw a sufficient number of trajectories on the phase plane such that the delineation between stable and unstable regions is apparent. We are particularly interested in trajectories near the second equilibrium point (at $y = -2/3$). Can you find any trajectory that actually terminates at $y = -2/3$? How would you describe that point? What is a saddle point? Place the dependent variable, y, on the horizontal axis and derivative, y', on the vertical axis. Note that many of the trajectories can be expected to spiral inward toward the point (0,0); this is illustrated in Figure P6.16.

6.17 The Chebyshev ODE is:

$$(1-x^2)\frac{d^2y}{dx^2} - x\frac{dy}{dx} + n^2y = 0.$$

We are interested in the case where $n = 5$ such that the last term on the left-hand side is $25y$. We know that $y(x = -1) = -1$ and $y(x = +1) = +1$. We want to solve this equation *numerically* on the interval $-1 \le x \le +1$, and we will pretend that *we do not know* that analytic solution is $T_5(x)$. One of the requirements of this problem is that we need to obtain a very accurate estimate for $\left.\frac{dy}{dx}\right|_{x=-1}$. Use the method of your choice and prepare a plot of your results for $y(x)$.

6.18 Schmitz and Amundson (*Chemical Engineering Science*, 18:265, 1963) studied the behavior of a reacting system consisting of two immiscible liquid phases (α and β) that are fed continuously to a stirred-tank reactor (CSTR). A first-order, irreversible chemical reaction occurs in both phases with the rate constants, k_α and k_β. The temperatures of the two phases are equal, and the components are distributed according to an equilibrium law. Because the heat generation function (for the exothermic case) is sigmoidal in shape with respect to temperature (T), and the heat removal rate is linear with T, it is possible that the two intersect three times (that is, there could be a maximum of three "steady-state" temperatures). Naturally this condition would only occur if the slope and intercept of the heat removal line were injudiciously chosen (a reactor designer would certainly try to avoid such a regime for safety reasons). Schmitz and Amundson solved three simultaneous (nonlinear) ODEs that formed the model for this system (their eqs. 18, 19, and 20). The three equations describe the dynamic behavior of the reduced total concentration, the reduced temperature, and the volume fraction for the α-phase. The parametric values of interest to us are provided as *Case d, Table 1*, page 280 in the Schmitz–Amundson paper; they are reproduced here:

$$k_\alpha = \exp(23.1 - 63.5/T')$$

$$k_\beta = \exp(22.1 - 51.5/T')$$

$$v_\beta/v_\alpha = C_\beta/C_\alpha = 1$$

$$K_A = 0.2$$

$$\varepsilon = 1$$

$$A_{\beta 0}/A_{\alpha 0} = 0.2$$

$$Uv_\alpha/C_\alpha V = 6.37 \times 10^{-3}$$

$$\tau = \exp(6)$$

$$T'_{\alpha 0} = T'_{\beta 0} = 1.85$$

$$\rho_s = 1$$

Solve the Schmitz–Amundson model numerically, recognizing that the oscillatory behavior expected for this case may require a procedure with variable step size. Prepare a phase-space portrait of the system by cross-plotting reduced mole fraction and reduced temperature. Verify the limit-cycle behavior presented in the original paper.

6.19 In Chapter 1 we contemplated a problem in which recently acquired video revealed the flight characteristics of a falcon stalking and attacking a flying crow. This is a classic curve of pursuit problem, but in three dimensions of course. In a planar (*x,y*) problem *if* the prey travels a straight line in the *y*-direction, starting from $x = a$, then the trajectory of the pursuer is described by the differential equation,

$$1 + \left(\frac{dy}{dx}\right)^2 = k^2(a - x)^2\left(\frac{d^2y}{dx^2}\right)^2,$$

where k is the ratio of velocities (velocity of the pursuer)/(velocity of the prey). Solve this equation numerically and plot the path taken by the pursuer given that $k = 1.5$ and $a = 10$. The prey starts from the *x*-axis and the pursuer starts from the origin. What happens to the solution if $k = 1$?

6.20 Suppose an SR-71 (the Lockheed reconnaissance aircraft that won the Collier Trophy in 1963) is flying a photographic intel mission over hostile territory. The "Blackbird" flies a straight course at 2,045 mph at a constant altitude of 89,500 ft. As the plane passes directly overhead, an anti-aircraft installation fires a ground-to-air missile that accelerates rapidly to Mach 5 (take this to be 3,600 mph); the missile maintains that speed until its fuel is exhausted (after 38 s). Can an intercept occur? If so, how far down-range does it occur? If not, how close does the missile get to the SR-71? This is another *curve of pursuit* problem; see problem 6.19 above and also *Introduction to Nonlinear Differential and Integral Equations* (Chapter 5, Section 7) by H. T. Davis, Dover Publications (1962).

6.21 Previously we looked at the behavior of the driven pendulum modeled as a third-order system (Problem 6.7). One of the earliest studies of nonlinear oscillators was the work of Van der Pol in the 1920's. In dimensionless form, Van der Pol's model is:

$$\frac{d^2\theta}{dt^2} - (\varepsilon - \theta^2)\frac{d\theta}{dt} + \theta = 0,$$

and he arrived at this form by assuming that the dissipative (friction) term would be a function of the amplitude of the oscillation. We want to solve this equation numerically using two different starting points: $(\theta, \theta') = (-0.1,0)$ and $(-0.1,2.8)$. Determine the phase-plane portrait of the Van der Pol system by cross-plotting θ with θ', and use both $\varepsilon = 0.4$ and $\varepsilon = 0.1$. Do all of the trajectories end up at the same limit cycle?

6.22 The Legendre differential equation figures prominently in mathematical physics; it arises when the Laplace equation in spherical coordinates is solved using the product method (separation of variables). The process is illustrated in the next chapter, Section 7.6. The resulting ODE is:

$$(1 - x^2)\frac{d^2y}{dx^2} - 2x\frac{dy}{dx} + p(p + 1)y = 0.$$

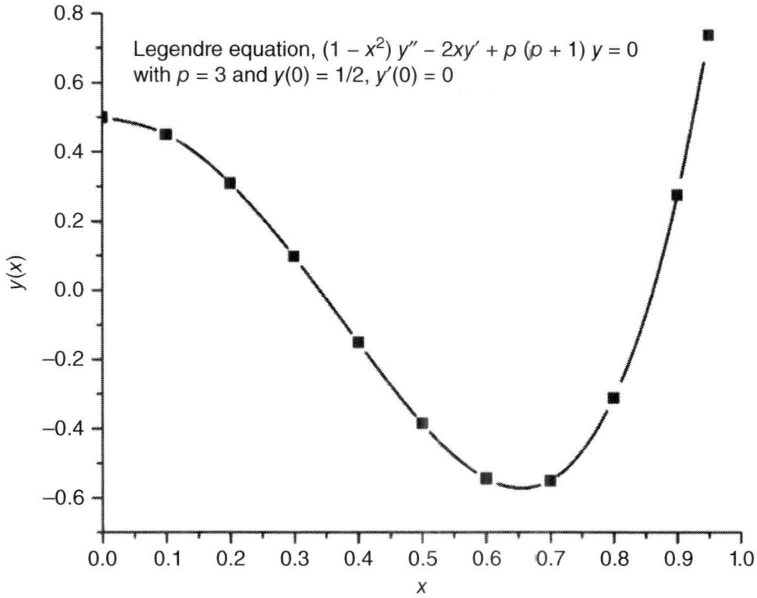

Figure P6.22 Approximation using Legendre polynomial, P_4.

Note that $x = 1$ is a regular singular point. This equation can be solved by integration in series as shown by Ayres (*Schaum's Outline Series: Theory and Problems of Differential Equations*). The result for positive integer values of p is the sequence of Legendre polynomials (a table is provided in Chapter 4); for example, $P_4 = \frac{35}{8}x^4 - \frac{30}{8}x^2 + \frac{3}{8}$. Suppose we are interested in solving this ODE with $p = 3$, $y(0) = 1/2$, and $y'(0) = 0$, and in particular we want to know if the solution can be reasonably approximated by $y \approx a_4P_4$. We recognize that if $a_4 = 4/3$, then both $y(0)$ and $y'(0)$ will have the correct values. The behavior of this approximation is shown in the accompanying figure:

Solve this Legendre ODE numerically and compare your results to the approximation illustrated in Figure P6.22. Does a_4P_4 roughly represent $y(x)$ on the interval 0–1?

6.23 R. E. Kidder published a study of transient flow of a gas through porous media in the *Journal of Applied Mechanics*, 27:329 (1957). He developed a nonlinear ODE:

$$\sqrt{1 - \alpha y}\frac{d^2y}{dx^2} + 2x\frac{dy}{dx} = 0,$$

with $y(0) = 1$ and $y(x \to \infty) = 0$. Solve this equation numerically for $\alpha = 0.75$, and identify the correct value for $y'(x = 0)$.

Answer: −1.23976039

6.24 Georg Duffing initiated a study of nonlinear oscillations in the early twentieth century stemming from his interest in vibrations occurring in reciprocating engines. Assisted by his eldest daughter, his theoretical and experimental investigations were compiled in a book, *Beitrag zur Bestimmung der Formveränderung gekröpfter Kurbelwellen*, in 1918. Duffing's nonlinear ODE,

$$y'' + y + ry^3 = K\sin(\Omega t),$$

has attracted the attention of mathematicians and physicists for 100 years due to the fact that it can exhibit chaotic behavior. In some cases, the solution for this equation may appear to yield a stable oscillation, only to devolve into unstable behavior at some advanced time, t. For example, if we take $r = -1/6$, $K = 1/4$, and $\Omega = 0.752045$, with $y(t = 0) = 0$ and $y'(t = 0) = 0$, and solve the ODE (to a sufficiently large t) we get the portrait in phase space shown in Figure P6.24.

The system's behavior is clearly unstable for these parametric values. This raises a number of intriguing questions, and foremost among them might be identifying values for K (holding all other parameters constant) that correspond to the stability threshold. Begin your study with $K = 1/10$, and generate an appropriate trajectory by cross-plotting dy/dt and $y(t)$. Then, increase K and repeat until you identify the critical value.

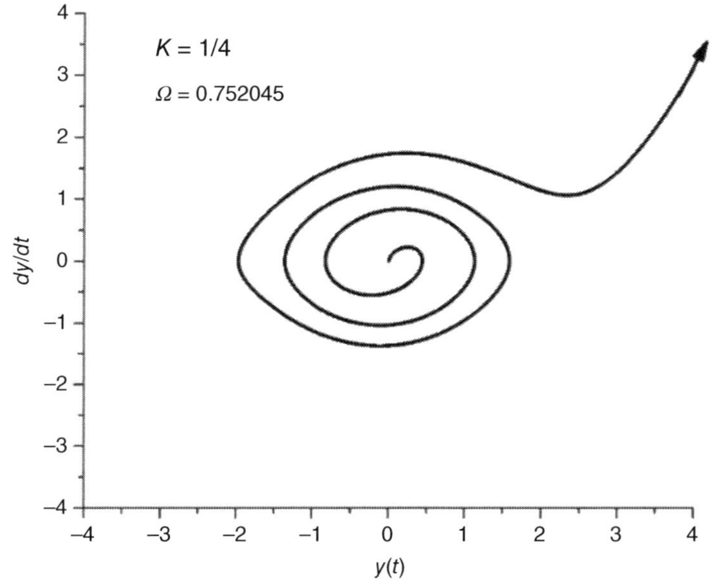

Figure P6.24 Phase plane for Duffing's equation with $r = -1/6$, $K = 1/4$, and $\Omega = 0.752045$.

6.25 A fascinating variation of problem 6.24 can be formulated by making a couple of changes to Duffing's equation; the most important is the addition of a damping term, i.e., a term proportional to $\frac{dy}{dt}$:

$$y'' = y - \frac{1}{4}y^3 - \frac{1}{10}y' + K\sin(\Omega t). \tag{6.119}$$

We want to solve this equation with $y(0) = 0$ and $y'(0) = 0$, letting $K = 2.5$ and $\Omega = 2$. The phase-plane portrait of the system's behavior will be much more interesting and you should explore its construction.

Verify this unusual behavior by solving Duffing's eq. (6.119) numerically, then generate a *Poincaré section* by repeating the phase-plane construction but this time *plotting only points at intervals of 2π*. Think about the implication of this process; if a function is purely periodic, like say $\sin(x)$, then the Poincaré section would consist of a *single point*. However, in this (chaotic) case the figure generated will be much more complex—an example is provided in Figure P6.25, one visualization of a *strange attractor*.

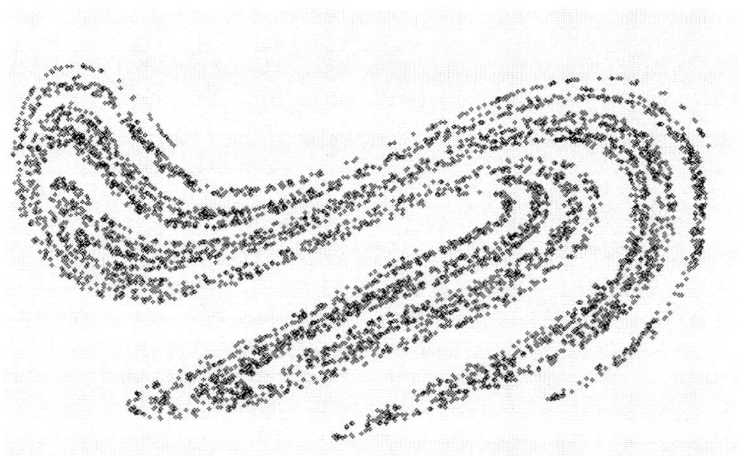

Figure P6.25 Poincaré section for the Duffing equation.

6.26 *The Taylor series approximation:* Consider the nonlinear ODE,

$$\frac{dy}{dx} + y^2 = 2,$$

with y(0) = ½ and $0 \leq x \leq 1$. This is a Riccati equation, and the analytic solution can be found with relative ease by separation:

$$y = \sqrt{2}\tan h[\sqrt{2}(x + 0.261276)].$$

Therefore, $y(x)$ increases with increasing x, but with diminishing slope. The analytic solution for the ODE will provide a basis for comparison with the results of this exercise.

We want to examine a method for finding an approximate solution for this nonlinear ODE, and we will assess whether the desired accuracy can be obtained. We begin by differentiating the ODE successively:

$$\begin{aligned} y' + y^2 &= 2 \\ y'' + 2yy' &= 0 \\ y''' + 2y'y' + 2yy'' &= 0 \\ y'''' + 6y'y'' + 2yy''' &= 0, \text{etc.} \end{aligned}$$

Next, we write a Taylor series expansion about the point $x_{0:}$

$$y(x) = y(x_0) + y'(x_0)(x - x_0) + y''(x_0)\frac{(x - x_0)^2}{2} + y'''(x_0)\frac{(x - x_0)^3}{6} + y''''(x_0)\frac{(x - x_0)^4}{24} + \ldots$$

If this series converges, it provides an approximate solution for this and perhaps other, more difficult ODEs; let's explore this idea further. Since $y(0) = 1/2$, we find $y'(0) = 7/4$, $y''(0) = -7/4$, $y'''(0) = -35/8$, etc. Now suppose we retain terms up to the third derivative in the Taylor series, the result is:

x	0.2	0.4	0.6	0.8	0.9
$y(x)$	0.80917	1.01333	1.0775	0.96667	0.83469

If we compare these results with the analytic solution, we see that the error is only about 2% at $x = 0.4$, but 9% at $x = 0.6$ and 36% at $x = 0.9$. If the Taylor series approximation is to be useful for this Riccati equation, we will need to add more terms. Take a few moments to add the fourth derivative term to the series and quantify the improvement to the approximation; compare your results with the table below for verification.

x	0.2	0.4	0.6	0.8	0.9
$y(x)$	0.80981	1.02360	1.12948	1.13093	1.09781

The discrepancy at $x = 0.9$ is still excessive. How many terms must we retain to keep the error below 2% for the last entry in the table ($x = 0.9$)?

There is one other matter of note with respect to this Riccati equation: As x becomes large, y must be asymptotic to $\sqrt{2}$ or 1.4142. Inspecting the results for the two cases tabulated above, we conclude that obtaining the correct limiting behavior may be difficult. Can the Taylor series expansion mimic $y(x)$ for large values of x?

Answer: 6

6.27 Emden's differential equation,

$$\frac{d^2y}{dx^2} + \frac{2}{x}\frac{dy}{dx} + y^n = 0,$$

was developed as a model of the thermal behavior of a spherical body of gas, and it is used to estimate the density and temperature of stars. If we solve this equation for $n = 2$ with $y(0) = 1$ and $y'(0) = 0$ we obtain the result shown in Figure P6.27.

Of particular interest is the value of x corresponding to $y(x) = 0$, and the BAAS (*British Association for the Advancement of Science*) has reported that this occurs at $x = 4.35287$ for $n = 2$. Solve Emden's equation numerically and assess the accuracy of this value. Then, solve the equation again but for $n = 3$, and obtain the values of x that yield $y(x) = 0$ (*there are several*).

Answer: The first crossing occurs at $x \approx 6.897$.

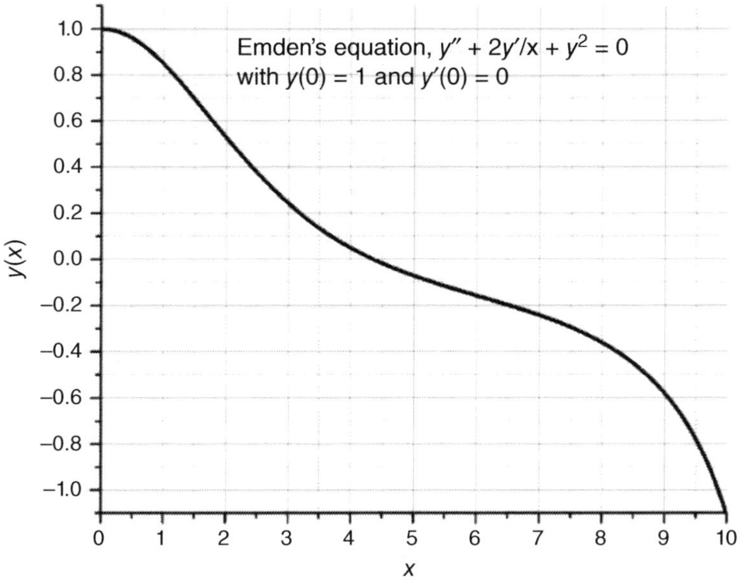

Figure P6.27 Numerical solution of Emden's equation with $n = 2$.

6.28 The second-order ODE,

$$x^2\frac{d^2y}{dx^2} + 3x\frac{dy}{dx} + 5y = \frac{1}{x}$$

is an example of a Cauchy equation. It is a linear ODE with variable coefficients and a singular point at $x = 0$. Solve this equation numerically for $\frac{1}{2} \leq x \leq 10$ with $y(x = ½) = 2$ and $y'(x = ½) = 8$. Find the locations of the maximum, the minimum, and the values of x for which $y = 0$.

Answer: The maximum and minimum occur at $x = 0.6673$ *and* 3.431*, respectively.*

6.29 Among the many well-known problems used to test ODE solvers' capability for stiff equations is the three-equation chemical kinetics model of H. H. Robertson (*The Solution of a Set of Reaction Rate Equations*, in *Numerical Analysis: An Introduction*, J. Walsh, Editor, Academic Press, 1967):

$$x' = -0.04x + 1 \times 10^4 yz$$
$$y' = 0.04x - 1 \times 10^4 yz - 3 \times 10^7 y^2$$
$$z' = 3 \times 10^7 y^2$$

with $x(0) = 1$ and $y(0) = z(0) = 0$. Some features of this stiff system are apparent: Species x will slowly disappear and species z will accumulate. But what does the second equation tell us about species y? The rate constants for the chemical destruction of y are *very* large, so although y is being produced by the decomposition of x, it will be rapidly converted to z. The source of our difficulty here is clear. Find the time when y attains its maximum value, and find the time when the concentrations for x and z are equal.

Answer: Species y attains maximum concentration when $t = 0.00848$.

6.30 Marine structures such as piers, docks, pilings and pipelines are exposed to dynamic forces from waves and in cases where the wave frequency is comparable to the natural frequency of the structure, amplification of the structure's response is possible. Consider a cylindrical piling subjected to a periodic flow, with the wave action represented by

$$U = U_0 + U_m \cos(\omega t).$$

The equation of motion for the structure (where x is the in-line displacement) has been given by Blevins (*Flow-Induced Vibration, Second Edition*, Krieger Publishing, 1994):

$$(m + \rho AC_A)\frac{d^2x}{dt^2} + 2m\zeta\omega_A\frac{dx}{dt} + kx = \rho AC_m\frac{du}{dt} + \frac{1}{2}\rho\left|U - \frac{dx}{dt}\right|\left(U - \frac{dx}{dt}\right)DC_D.$$

Note the similarity of the *left-hand side* to eq. (6.30) in Section 6.3; however, the model we are considering here is nonlinear by virtue of the product $\frac{dx}{dt}\frac{dx}{dt}$ appearing in the drag term on the far-right. At the far left of our model you will see the product ρAC_A; this is the *added mass* and it represents the weight of the fluid displaced by the structure (per unit length). This term would be negligible in air, but in water where ρ is about 62.4 lb_m/ft^3 or 1,000 kg/m^3 it may be significant. Let the wave frequency be 0.1 Hz so that $\omega = 0.628$ rad/s. Using cgs units, we take $D = 10$ cm, $\rho = 1$ g/cm^3, and m to be 100 g/cm of length. $A = \pi D^2/4 = 78.54$ cm^2, and $C_D = C_m = 1.5$. Assume the damping coefficient, ζ, is small (say $\leq 1/2$), and that $U_0 = 1$ and $U_m = 0.25$. Find the dynamic response of the structure assuming both x and dx/dt are zero initially; how much time is required for the piling to reach its steady oscillating state (within 2% of ultimate peaks and troughs)? The table provided below illustrates the impact of the damping coefficient upon the oscillatory motion of the piling; it contains the values for displacement, x, at successive peaks and troughs. Notice how a decreasing value for ζ affects both the duration of the transient behavior and the amplitude of the oscillation.

$\zeta = 2$	$\zeta = 1$	$\zeta = 1/2$	$\zeta = 1/10$
0.0139	0.0186	0.022	0.0262
−0.058	−0.0866	−0.113	−0.146
0.201	0.301	0.405	0.559
−0.023	−0.125	−0.269	−0.542
0.202	0.303	0.467	0.860
−0.023	−0.124	−0.292	−0.749

6.31 You may recall our discussion in Chapter 1 regarding engineering catastrophes. Often models for phenomena related to disasters are developed after the fact, when there is urgency to prevent additional property damage or loss of life. Such is the sad story of the Lockheed L-188 (Electra) turboprop airliner; it was designed to carry just under 100 passengers on short-to-intermediate routes at speeds around 370 mph. It immediately appealed to a number of airlines servicing smaller markets where the runway lengths were often not sufficient to handle first-generation jet airliners such as the Boeing 707 (or Dash 80). The L-188 could take off in 5,000 ft or less (on its first flight it was airborne in 1,800 ft), while the 707 required a 7,500 to 10,000 ft take off run (depending upon version, gross weight, temperature, and density-altitude). And on shorter routes the difference in cruising speed was of little consequence. The first flight of the L-188 occurred in December 1957, but the airplane suffered two major disintegrations (each involving the complete loss of one wing) after its introduction to commercial service; the first occurred on September 29, 1959 (34 deaths) and the second on March 17, 1960 (63 deaths). Passenger air service was still in its infancy and of course the resulting loss of public confidence was a liability for the entire industry—thus there was much pressure to figure out exactly what had happened and why. The L-188 was powered by four Allison 501 turbine engines, each capable of producing 3,750 hp at 13,820 rpm, and they were equipped with Aeroproducts square-tipped propellers (diameter 13.5 ft). Of course the turbine engines could not turn the propellers directly: Sonic velocity is about 1,100 ft/s and this means that the angular velocity of a 13.5 ft diameter propeller had to be *less* than about 163 rad/s, whereas 13,820 rpm is 1,447 rad/s. Therefore the engines were connected to a gearbox (by a drive shaft) to reduce the rotational speed. There is a cutaway model of this arrangement at the National Air and Space Museum should you be curious as to the actual layout (look for Allison T56-A1 online). However, such an engine–gearbox–propeller combination has some important implications, one of those is a nacelle with a long overhang and another is the fact is that the important masses (engine, drive shaft, gearbox, propeller) are distributed, and not concentrated near the center of the engine. Additionally, it turned out that the L-188 was susceptible to crack development that affected the integrity and rigidity of the engine mount. Propeller-driven aircraft can experience a displacement of the center of thrust when the loads on the propeller are not angularly symmetric. This produces a circular oscillation called whirl mode where the entire assembly (propeller, engine, and nacelle) begins to move very similarly to the precession (wobble) of a spinning top. When the engine mount has been compromised the whirl-mode oscillation can be severe, and if the frequency of the wobble should happen to match the frequency of the wing the result could be structural flutter. Furthermore, if the excitation is amplified, the structural limits could be exceeded and the wing might separate from the airframe. This is exactly what happened in the incidents in 1959 and 1960. *There is an* *intriguing film (L-553) of the flutter testing of a 1/8 scale model of the L-188 (Electra) available online; the work was conducted at NASA's Langley Research Center.* We are going to explore a simplified model of this phenomenon using Houbolt and Reed (*Propeller-Nacelle Whirl Flutter, Journal of the Aerospace Sciences, March 1962)* as our starting point. The coupled, second-order ODEs are

$$\frac{d^2\psi}{dt^2} + c_1\frac{d\psi}{dt} + c_2\psi + c_3\Omega\theta = F_y$$

and

$$\frac{d^2\theta}{dt^2} + b_1\frac{d\theta}{dt} + b_2\theta - b_3\Omega\frac{d\psi}{dt} = F_z,$$

where ψ is yaw angle and θ is pitch angle. By yaw we mean left and right, and by pitch up and down; Ω is the rotational speed of the propeller. Once again we note that these two equations are partially similar to something we have seen previously: Section 6.3, eq. (6.30). Use the following values for the parameters and determine whether the whirl-mode oscillation is unstable.

$c_1 = 2$	$b_1 = 2$	$F_y = (1/12)\sin(435\,t)$
$c_2 = 2$	$b_2 = 2$	$F_z = (1/14)\sin(435\,t)$
$c_3 = 1/20$	$b_3 = 1/30$	$\Omega = 145$

Please note: We observed in the problem statement that one of the difficulties characteristic of turboprop aircraft is the distribution of masses due to the engine–drive shaft–gearbox–propeller arrangement. A more complete model would account for the locations of the important masses and the fact that the coefficients of stiffness would be different for each mass. In fact, realistic models result in a matrix with the different masses, their spatial positions, and the rotational inertias accounted for, and the complexity of such models increases rapidly with efforts to represent the engine–nacelle–wing system accurately. For an introduction to more accurate modeling efforts, see P. Hylton, Analyzing a Simple Prop-Whirl-Flutter Model Using Modern Analysis Tools, Proceedings of the 2006 IJME-INTERTECH Conference.

6.32 The spectacular collapse of the Tacoma Narrows Bridge in November 1940 is an example of structural failure familiar to nearly every student of engineering physics. But the oft-cited cause of the disaster—resonance resulting from periodic vortex shedding—is incorrect as the frequency of vortex shedding did not match the fundamental frequency of the bridge. In recent years other explanations of the abrupt transition from the benign longitudinal wave (in which the bridge deck remained horizontal while vibrating vertically) to the destructive torsional oscillation have emerged. Green and Unruh (The failure of the Tacoma Bridge: A physical model, *American Journal of Physics*, 74:706, 2006) observed that the Barnet Elliott film reveals a large vortex moving across the bridge surface during torsional oscillation, creating variations in local pressure that reinforced and amplified the twisting motion. They simulated vorticity transport across the bridge deck to obtain a qualitative match with frames from the Elliott film. They also noted that a negative damping coefficient had been identified as a cause of the failure as early as 1941, although the actual cause of negative damping was unexplained. Let us consider the equation of motion,

$$M\frac{d^2Y}{dt^2} + 2M\zeta\frac{dY}{dt} + Ky = A\sin(\omega t).$$

We modify this ODE by assuming that the damping coefficient, ζ, *varies with time* as

$$\zeta = 0.55 - \cos(t),$$

assuring that it will be negative intermittently. We want to explore the consequences of this choice numerically with the modified equation:

$$\frac{d^2Y}{dt^2} = -2(0.55 - \cos(t))\frac{dY}{dt} - KY + 2\sin(2.666t).$$

Let $K = 0.239$ and take $Y(0) = Y'(0) = 0$. Solve the ODE out to $t = 80$ and plot $Y(t)$ from your results. Are the computed oscillations periodic?

6.33 In Chapter 5 problems 5.24 and 5.25 dealt with analytic solutions for flow in channels with non-parallel walls; the first employed the boundary-layer simplification and the final problem (5.25) required solution using elliptic functions. In the present exercise we will explore a related problem numerically; however, we now consider a channel with *divergent* walls. We recognize that flow of a viscous fluid in a *divergent channel* means that the average fluid velocity will decrease while the pressure at the wall will increase. That is, we are moving the fluid in the *positive r*-direction, but the pressure gradient near the wall will push the slow-moving fluid in the *negative r*-direction, potentially creating a region with reverse flow (this is, of course, an indicator of *separation*). As before, the governing ODE is

$$F'' + F^2 + +4F + k = 0 \text{ where } v_r = \frac{\nu}{r}F(\theta).$$

Assume that the included angle is 10°, with the lower wall located at −5° (−0.08727 rad) and the upper wall at +5° (+0.08727 rad); thus, the center of the channel corresponds to $\theta = 0$. We have zero velocity at the walls and maximum velocity at the centerline; therefore, $F(\theta = -5^\circ) = 0$ and $F'(\theta = 0) = 0$ (the flow is symmetric about the centerline). Let the centerline velocity at a radial position, r, be u_0, and then define a Reynolds number for the flow as $Re = \frac{u_0 r}{\nu}$ (ν is the kinematic viscosity of the fluid). We know from work carried out by Millsaps and Pohlhausen (Thermal distribution in Jeffery-Hamel flows between nonparallel plane walls, *Journal Aeronaut. Sci.*, 20, 187–196, 1953) that for divergent walls with a 10° included angle and a Reynolds number of 5,000, reverse flow *will occur* near the wall. The parameter, k, appearing in the ODE is the radial pressure gradient at the wall:

$$k = -\frac{1}{\rho}\frac{\partial p}{\partial r}\left(\frac{r^3}{\nu^2}\right),$$

and we note that it may have a large negative value since $\frac{\partial p}{\partial r}$ will be positive in this flow. We will let $r = 2$, $u_0 = 25$, and $\nu = 0.01$ such that $Re = 5{,}000$. Solve the second-order ODE numerically and verify that reverse flow will occur in this scenario. Obviously, if you select a forward-marching technique like Runge–Kutta, you will have to find the correct value for $F'(\theta = -0.08727)$. Let $k = -5.44 \times 10^6$; determine the location of maximum reverse (negative) velocity. What is the value of F on the centerline?

An aside: The existence of reverse flow near the walls in this case is borne out by experiment; however, in the laboratory this flow is not normally symmetric. Usually, separation will occur at one wall with the flow remaining attached at the other.

6.34 You may recall that in problem 2.2 we developed a force balance for a spacecraft subjected to gravity from both the earth and the moon. In that problem we were trying to identify the distance from the earth for which the two gravitational forces were essentially equal. Because the mass of the moon is so much less than that of the earth, the point of balance occurs close to the moon, at about 90% of r_{tot} (the total center-to-center distance, r_{tot}, is about 3.825×10^8 m). Let m_e, m_m, and m_{sp} be the masses of the earth, moon, and spacecraft, respectively. Our interest here is the dynamic behavior of a spacecraft traveling from the earth to the moon. The distance between the earth and the spacecraft is r_{esp} and an approximate force balance is:

$$m_{sp}\frac{d^2 r_{esp}}{dt^2} = -\frac{G m_e m_{sp}}{r_{esp}^2} + \frac{G m_m m_{sp}}{r_{msp}^2}.$$

We are neglecting the drag experienced by the craft as it moves through the earth's atmosphere. Of course, the distance between the moon and the spacecraft is $r_{msp} = r_{tot} - r_{esp}$. The numerical values of the necessary constants are:

$$G = 6.674 \times 10^{-11} \quad m_e = 5.972 \times 10^{24} \quad m_m = 7.348 \times 10^{22} \text{ and } r_{tot} = 3.825 \times 10^8.$$

The initial value for r_{esp} is about 6.44×10^6 m (the spacecraft starts its journey from the earth's surface), and we will assume that its initial velocity is 11,100 m/s (about 24,607 mph). Assuming the spacecraft travels in a *straight line*[1] from the earth to the moon, what is the minimum velocity of the craft, when and where does it occur, and what is its velocity when it arrives near the moon's surface?

1 The actual trajectory of a flight to the moon is called a trans-lunar injection (TLI), and it obviously cannot be a straight line since the moon is in motion around the earth. A typical lunar mission begins from earth orbit where the craft's velocity is around 8,000 m/s; an additional impulse is required to achieve escape velocity. Thus, the real distance traveled by the spacecraft will be much larger than 3.8×10^8 m, and the travel time will be significantly greater than the value obtained from our simplified model.

References

Bailey, H.E. (1969). Numerical Integration of the Equations Governing the One-Dimensional Flow of a Chemically Reactive Gas. *Physics of Fluids* 12: 2292.

Cash, J.R. and Karp, A.H. (1990). A Variable Order Runge-Kutta Method for Initial Value Problems with Rapidly Varying Right-Hand Sides. *ACM Transactions on Mathematical Software* 16: 201.

Chapra, S.C. and Canale, R.P. (2002). *Numerical Methods for Engineers*, 4the. Boston: McGraw-Hill.

Ferrell, J.E., Tsai, T.Y.C., and Yang, Q. (2011). Modeling the Cell Cycle: Why Do Certain Circuits Oscillate? *Cell* 144: 874.

Gear, C.W. (1971). *Numerical Initial Value Problems in Ordinary Differential Equations*. Englewood Cliffs, NJ: Prentice-Hall.

Hairer, E. and Wanner, G. (1996). *Solving Ordinary Differential Equations II*, 2nd revisede. Berlin: Springer.

Hamming, R.W. (1973). *Numerical Methods for Scientists and Engineers*, 2nde. New York: McGraw-Hill.

Hanna, O.T. and Sandall, O.C. (1995). *Computational Methods in Chemical Engineering*. Upper Saddle River, NJ: Prentice Hall PTR.

James, M.L., Smith, G.M., and Wolford, J.C. (1977). *Applied Numerical Methods for Digital Computation*, 2nde. New York: Harper & Row.

Lambert, J.D. (1973). *Computational Methods in Ordinary Differential Equations*. London: John Wiley & Sons.

LaSalle, J. P. Asymptotic Stability Criteria. *Proceedings of Symposia in Applied Mathematics, Volume XIII, Hydrodynamic Instability*. American Mathematical Society, Providence, RI (1962).

Lee, H.J. and Schiesser, W.E. (2004). *Ordinary and Partial Differential Equation Routines in C, C++, Fortran, Java, Maple, and MATLAB*. Boca Raton, FL: Chapman and Hall.

Lorenz, E. (1963). Deterministic Nonperiodic Flow. *Journal of the Atmospheric Sciences* 20: 130.

Press, W.H., Flannery, B.P., Teukolsky, S.A., and Vetterling, W.T. (1989). *Numerical Recipes: The Art of Scientific Computing*. Cambridge: Cambridge University Press.

Schmitz, R.A. and Amundson, N.R. (1963). An Analysis of Chemical Reactor Stability and Control-Va: Two-phase systems in physical equilibrium. *Chemical Engineering Science* 18: 265.

Southworth, R.W. and Deleeuw, S.L. (1965). *Digital Computation and Numerical Methods*. New York: McGraw-Hill.

Stoer, J. and Bulirsch, R. (1993). *Introduction to Numerical Analysis*, 2nde. New York: Springer-Verlag.

7

Analytic Solution of Partial Differential Equations

7.1 Introduction

Many of the phenomena that are of interest to us in engineering and the applied sciences are modeled with partial differential equations (PDEs). Fluid flow, heat transfer, and mass transfer are prime examples, but problems in gravitation, electrostatics, and quantum theory all give rise to similar equations. The purpose of this chapter is to provide the reader with some basic skills enabling him/her to find analytic solutions for many commonly encountered partial differential equations.

Several valuable references will be provided as we move through this material, but at the outset we want to point out that there are two uniquely important monographs devoted to the analytic solution of partial differential equations: J. Crank's, *The Mathematics of Diffusion, Second Edition* (1975), and *Conduction of Heat in Solids, Second Edition* (1959) by H. S. Carslaw and J. C. Jaeger. These two books are known to nearly every worker in applied mathematics. Both are incredibly useful as guides to the solution of practical problems where diffusional (molecular) transport processes are dominant. Practitioners in this field are often heard to say, "…..I found a similar problem in Crank…." or "…..I verified my solution with Carslaw and Jaeger." Anyone wishing to become adept with the subject matter of this chapter simply must own both of these books.

We need to make one other observation before we get started: Our main focus in this chapter is upon *linear* partial differential equations. Although many important physical phenomena are inherently nonlinear, the theory for nonlinear PDEs is quite undeveloped. This is one reason why so little is known about the Navier–Stokes equation (the partial differential equation governing fluid flow). Despite 175 years of intense effort by the world's foremost mathematicians and physicists, only about 75 solutions for this PDE have been found. The universe of practical flow scenarios is immeasurably vast, so 75 is an astonishingly small number.

7.2 Classification of Partial Differential Equations and Boundary Conditions

We have to be able to recognize and classify partial differential equations in order to attack them successfully; a book like Powers (1979) can be a valuable ally in this effort. Consider the generalized second-order partial differential equation where ϕ is the dependent variable and x and y are arbitrary independent variables:

$$A\frac{\partial^2\phi}{\partial x^2}+B\frac{\partial^2\phi}{\partial x\partial y}+C\frac{\partial^2\phi}{\partial y^2}+D\frac{\partial\phi}{\partial x}+E\frac{\partial\phi}{\partial y}+F\phi+G=0. \tag{7.1}$$

A, B, C, D, E, F, and G can be functions of x and y, but *not* of ϕ. This *linear* partial differential equation can be classified as follows:

$B^2-4\,AC<0$ *elliptic*

$B^2-4\,AC=0$ *parabolic*

$B^2-4\,AC>0$ *hyperbolic*

Problem Solving in Engineering: Analytical Mathematics and Numerical Analysis, First Edition. Larry A. Glasgow.

Companion website: www.wiley.com/go/glasgow/problemsolvingengineering

For illustration, we look at the "heat" equation (one-dimensional transient conduction):

$$\frac{\partial T}{\partial t} = \alpha \frac{\partial^2 T}{\partial y^2} \tag{7.2}$$

You can see that $A = \alpha$, $B = 0$, and $C = 0$; the equation is parabolic. Compare this with the governing (Laplace) equation for two-dimensional potential flow (ψ is the stream function):

$$\frac{\partial^2 \psi}{\partial x^2} + \frac{\partial^2 \psi}{\partial y^2} = 0. \tag{7.3}$$

In this case $A = 1$ and $C = 1$ while $B = 0$; the equation is elliptic. Next, we consider a vibrating string (the wave equation):

$$\frac{\partial^2 u}{\partial t^2} = s^2 \frac{\partial^2 u}{\partial y^2} \tag{7.4}$$

Note that $A = 1$ and $C = -s^2$; therefore, $-4AC > 0$ and eq. (7.4) is hyperbolic. In applied mathematics, transient problems with molecular transport only (heat or diffusion equations) will have parabolic character. Equilibrium problems such as steady-state diffusion, conduction, or viscous flow in a duct will be elliptic in nature (phenomena governed by Laplace- or Poisson-type partial differential equations). We will see numerous examples of both in this chapter. Hyperbolic equations are common in quantum mechanics, problems with vibrations, and high-speed, compressible flows, e.g., inviscid supersonic flow about an airfoil. The Navier–Stokes equation that has been the focus of much attention by physicists and mathematicians over the last 175 years is of mixed character.

The three common types of boundary conditions encountered in applied mathematics are Dirichlet, Neumann, and Robin's. For Dirichlet boundary conditions (or conditions of the first kind) the field variable is specified at the boundary. Two examples: In a conduction problem the temperature at a surface might be fixed (at $y = 0$, $T = T_0$); alternatively, in a viscous fluid-flow problem, the velocity at a stationary duct wall would be zero (for a Newtonian fluid). A condition of the first kind can also be written as a function of time, at $y = 0$, $T = f(t)$.

For Neumann conditions (or boundary conditions of the second kind) the flux is specified, e.g., for a conduction problem with an insulated wall located at $y = 0$, $\left(\frac{\partial T}{\partial y}\right)_{y=0} = 0$. Of course this gradient could also be written as a function of time.

A Robin's-type boundary condition (or condition of the third kind) results from equating the fluxes; for example consider the solid–fluid interface in a heat transfer problem. On the solid side heat is transferred by conduction (Fourier's law), but on the fluid side of the interface we might have mixed heat transfer processes approximately described by Newton's "law" of cooling:

$$-k\left(\frac{\partial T}{\partial y}\right)_{y=0} = h(T_0 - T_f). \tag{7.5}$$

We hasten to add that the heat transfer coefficient, h, that appears in (7.5) is an empirical quantity. The numerical value of h is known only for a small number of cases, usually those in which molecular transport is dominant. Thus, the use of a Robin-type boundary condition usually means that an additional unknown has been brought into the problem. An analogous relationship can be used for mass transfer at interfaces: $-D_{AB}\left(\frac{\partial C_A}{\partial y}\right)_{y=0} = K(C_{A0} - C_{A\infty})$. And again, the mass transfer coefficient, K, is unknown and would generally have to be estimated.

A critical observation with regard to these boundary conditions is that all three kinds are *linear* with respect to the dependent variable. We also should point out that there are other types of boundary conditions, the ones we described above are merely the most common. In neutron diffusion for example, we might extrapolate the neutron flux to zero at some distance *outside* the reactor. We might also employ an *albedo* boundary condition in which the ratio of the neutron fluxes in and out is specified.

7.3 Fourier Series

In his prize-winning work submitted to the Paris Academy in 1811, and later published within *Theorie analytique de la Chaleur* (1822), Fourier claimed that an arbitrary function, $f(x)$, could be represented by the trigonometric series,

$$f(x) = a_0 + (a_1 \cos x + b_1 \sin x) + (a_2 \cos 2x + b_2 \sin 2x) + \cdots \tag{7.6}$$

The constants that appear in this series are given by

$$a_0 = \frac{1}{2\pi}\int_{-\pi}^{+\pi} f(x)dx \tag{7.7}$$

$$a_n = \frac{1}{\pi}\int_{-\pi}^{+\pi} f(x)\cos nx dx \tag{7.8}$$

and

$$b_n = \frac{1}{\pi}\int_{-\pi}^{+\pi} f(x)\sin nx dx. \tag{7.9}$$

The general idea had surfaced earlier; prior to Fourier's work (in fact, in the eighteenth century) a number of prominent mathematicians worked on the vibrating string problem. Carslaw (1950) records that Bernoulli obtained a solution (for a string starting from rest) in the form of a trigonometric series. Euler responded to this work by noting that if Bernoulli was correct, then an arbitrary function of a single variable could be represented by an infinite series of sines (of integer multiples of the independent variable). Euler did not believe this was possible; he observed that sine was both a periodic function and one that was odd (i.e., $\sin(-\theta) = -\sin\theta$). If the function that was being represented did not have the same characteristics, how could it be obtained from sine?

Let us assume we are interested in a function, $f(x)$ for $(-\pi \le x \le +\pi)$:

$$f(x) = \frac{4}{\pi}x + 4 \quad \text{for} -\pi \le x \le -\pi/2 \tag{7.10}$$

$$f(x) = -\frac{4}{\pi}x \quad \text{for} -\pi/2 \le x \le +\pi/2 \tag{7.11}$$

and

$$f(x) = \frac{4}{\pi}x - 4 \quad \text{for} +\pi/2 \le x \le +\pi. \tag{7.12}$$

The function's behavior (a triangular or sawtooth wave) is illustrated in Figure 7.1.

We would like to know if this function can be represented in the manner suggested by Fourier, and if so, how do the coefficients actually behave? Figure 7.2 shows the approximations for $f(x)$ using 3, 5, 10, 20, and 90 terms in the series. Though minor discrepancies are apparent, the results are similar to the function illustrated in Figure 7.1.

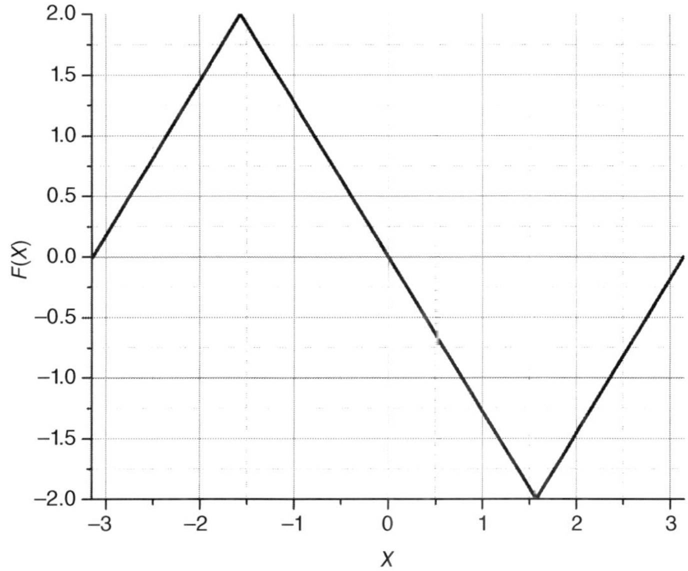

Figure 7.1 A triangular wave on the interval $(-\pi \le x \le +\pi)$.

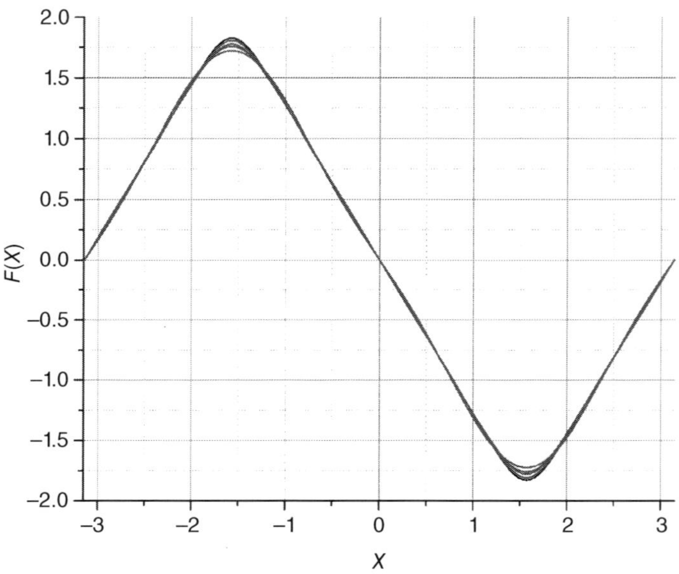

Figure 7.2 Representation of the triangular wave with Fourier's series technique. Many of the essential characteristics of $f(x)$ are reproduced reasonably well with just three terms in the series.

If this is your first exposure to Fourier series and their application to boundary value problems, then Spiegel's (1974) book can be an extremely useful tool and learning guide. The basic idea with Fourier series is that we use superposition to construct a representation of a periodic function using combinations of the oscillating functions, sine and cosine. Since many boundary value problems require us to expand a function into trigonometric series, we can expect Fourier series to prove extremely useful. If you are skeptical about using sine and cosine in this way, you are in good company. Korner (1989) notes that both Laplace and Lagrange initially had doubts about Fourier's development; part of their concern was a consequence of Fourier's lack of rigor.

Consider a function, $f(x)$, defined over an interval, $-L<x<+L$. The Fourier series corresponding to this function is, by definition,

$$f(x) = \frac{A_0}{2} + \sum_{n=1}^{\infty} A_n \cos\frac{n\pi x}{L} + B_n \sin\frac{n\pi x}{L}. \tag{7.13}$$

One concern that students new to Fourier series typically have is exactly how this expression will prove to be of value. After all, if this equation is to be used to reconstruct the function, $f(x)$, then we might need to know a very large number of A_n's and B_n's. The effort required appears formidable—until we think about orthogonality of the functions sine and cosine. Specifically, consider that

$$\int_{-\pi}^{\pi} \sin nx dx = 0 \tag{7.14}$$

$$\int_{-\pi}^{\pi} \sin nx \cos mx dx = 0 \tag{7.15}$$

$$\int_{-\pi}^{\pi} \sin nx \sin mx dx = 0 \ \text{ if } n \neq m, \ \text{ and } \pi \text{ if } n = m \tag{7.16}$$

$$\int_{-\pi}^{\pi} \cos nx \cos mx dx = 0 \ \text{ if } n \neq m, \ \text{ and } \pi \text{ if } n = m \neq 0. \tag{7.17}$$

These relationships suggest the following approach: Multiply the expression for $f(x)$ by $\sin(mx)dx$ and integrate from $-\pi$ to $+\pi$, so that

$$\int_{-\pi}^{\pi} f(x)\sin mxdx = \int_{-\pi}^{\pi} \frac{A_0}{2}\sin mxdx + \int_{-\pi}^{\pi} (A_n \cos nx + B_n \sin nx)\sin mxdx. \tag{7.18}$$

It follows immediately that for a function $f(x)$ of period, 2π,

$$B_n = \frac{1}{\pi}\int_{-\pi}^{\pi} f(x)\sin nxdx. \tag{7.19}$$

More generally for a function $f(x)$ defined over the interval, $-L < x < +L$, we have:

$$B_n = \frac{1}{L}\int_{-L}^{+L} f(x)\sin\frac{n\pi x}{L}dx. \tag{7.20}$$

Of course, the A_n's can be determined analogously.

In many of the problems of interest to us, the series solutions we obtain may only involve either sine or cosine terms. We refer to such cases as half-range Fourier sine (or cosine) series, and often our attention in such problems is focused upon just half of the interval, that is, from $x = 0$ to $x = +L$. In case of the half-range series,

$$\text{Cosine} : A_n = \frac{2}{L}\int_{0}^{L} f(x)\cos\frac{n\pi x}{L}dx \tag{7.21}$$

and

$$\text{Sine} : B_n = \frac{2}{L}\int_{0}^{L} f(x)\sin\frac{n\pi x}{L}dx \tag{7.22}$$

An obvious question of concern to us is whether an arbitrary function that is piecewise continuous over some interval 0 to L can be represented successfully in this way (by "successfully" we mean that we can obtain sufficient accuracy using a reasonable number of terms). Recall that in Figure 7.2, we saw that a triangular waveform could be very easily represented with just a few Fourier series terms. Now let us consider a function formed by two straight lines; represented by $f(x) = 3x$ from $x = 0$ to $x = 3$, and then $f(x) = 18-3x$ from $x = 3$ to $x = 6$. Since this function is defined only from $x = 0$ to $x = L$, and since the form requires an odd function representation:

$$f(x) = \sum_{n=1}^{\infty} B_n \sin\frac{n\pi x}{L}. \tag{7.23}$$

The necessary coefficients are determined directly from the integral(s):

$$B_n = \frac{2}{L}\left[\int_{0}^{3} 3x\sin\frac{n\pi x}{L}dx + \int_{3}^{6} (18-3x)\sin\frac{n\pi x}{L}dx\right]. \tag{7.24}$$

We will compute the B_n's, and then see how well the series represents $f(x)$ using successively, 2, 5, 10, 20, and 40 terms. The first seven B_n's are: 7.2951, -2.6408×10^{-5}, -0.8106, 2.3685×10^{-15}, 0.2918, -8.8027×10^{-6}, and -0.1489, and we note that that the even terms are actually zero.

As you can see from Figure 7.3, the approximation obtained with $n = 40$ is very good, with $f(x = 3)$ only 1% below the correct value (8.9088 as opposed to $f(3) = 9.0$).

7.3.1 A Preview of the Utility of Fourier Series

We want to explore a problem that will make it very clear why Fourier's work is so useful to us in our efforts to solve partial differential equations, and we will preface this example with an observation made by Lord Kelvin: "*Fourier's theorem is not only one of the most beautiful results of modern analysis, but it is said to furnish an indispensable instrument in the treatment of nearly every recondite question in modern physics.*" Let us see why Kelvin was so enthusiastic.

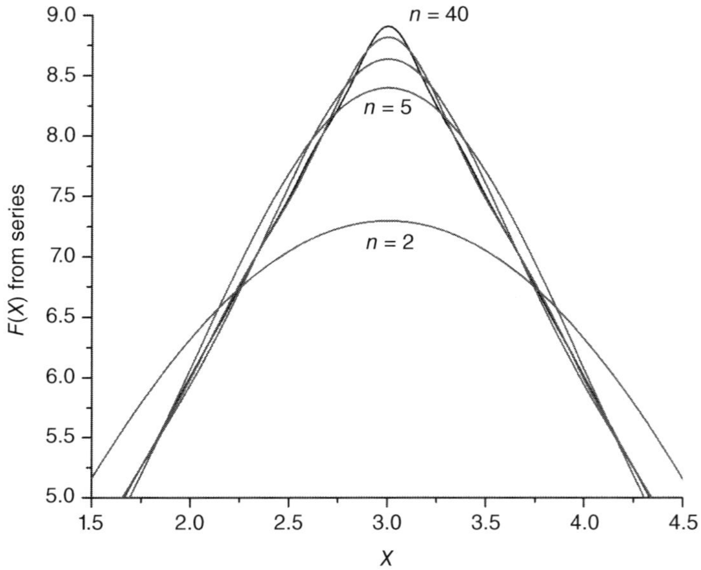

Figure 7.3 Reconstruction of triangular function using 2, 5, 10, 20, and 40 terms in the Fourier sine series. Please note that both axes have been abridged to better show the differences between the results.

We will assume that we have a slab of homogeneous material that extends in the x-direction such that $0 \leq x \leq 3$. This slab has an initial temperature distribution given by

$$T = f(x) = 6x - 2x^2, \tag{7.25}$$

which yields a temperature of zero (°C) at both ends and 4.5° at the center ($x = 1.5$). At $t = 0$, the temperature at both ends is instantaneously raised to 10 °C. The evolution of temperature in the slab is governed by a parabolic partial differential equation,

$$\frac{\partial T}{\partial t} = \alpha \frac{\partial^2 T}{\partial x^2}, \tag{7.26}$$

which we will modify by defining a new dependent variable, $\theta = T - 10$. Of course this means that θ will be zero at both ends—which proves to be very convenient. α that appears in eq. (7.26) is the thermal diffusivity, $\alpha = \frac{k}{\rho C_p}$. A solution for this problem has the form:

$$\theta = C_1 \exp(-\alpha \lambda^2 t)[A \sin \lambda x + B \cos \lambda x]. \tag{7.27}$$

It is easy to show that in order to satisfy the boundary conditions, we must have $B = 0$ and $\lambda_n = \frac{n\pi}{3}$. Consequently the solution we seek has the form:

$$\theta = \sum_{n=1}^{\infty} A_n \exp\left(-\alpha \frac{n^2\pi^2}{9} t\right) \sin \frac{n\pi x}{3}. \tag{7.28}$$

Our initial condition ($t = 0$) corresponds to

$$6x - 2x^2 - 10 = \sum_{n=1}^{\infty} A_n \sin \frac{n\pi x}{3}. \tag{7.29}$$

Of course this is a Fourier sine series, and we know that the needed coefficients are determined by integration:

$$A_n = \frac{2}{3} \int_0^3 (6x - 2x^2 - 10) \sin \frac{n\pi x}{3} dx. \tag{7.30}$$

Notice that the solution that we developed, eq. (7.28), will have a lot of exponential damping when t becomes *large*. Conversely, we should anticipate that for very small times the infinite series may converge *very slowly*! To further explore this "worst-case" scenario, we will find a few A_n's by integration (for even n's the coefficients are zero).

n	A_n
1	−8.08812
3	−4.07208
5	−2.50927
7	−1.80530
9	−1.40824
11	−1.15388
13	−0.97716
15	−0.84729
17	−0.74784
19	−0.66925
21	−0.60559
23	−0.55296
25	−0.50874
27	−0.47106
29	−0.43856

We see immediately that this sequence of coefficients is not diminishing rapidly—which may not be a good sign for convergence of the infinite series particularly if t is small. Since we have all of the pieces in place, we will compute the initial temperature distribution using increasing values of n (we start with $n = 50$ and go up to $n = 700$).

This example, illustrated in Figure 7.4, has taught us an important lesson: Convergence of the Fourier series to the initial temperature distribution is abysmally slow—we needed hundreds of terms in order to get a good approximation. However, we must remember that for modestly larger times, the exponential damping in the infinite series will greatly improve convergence; so much so that for intermediate t's we might only need *one or two terms* to get satisfactory results. We will illustrate this behavior by using the solution eq. (7.28) to calculate the temperature at the center ($x = 1.5$) for times ranging from 1/32 (0.03125) to 10. Recall that the initial temperature at $x = 1.5$ was 4.5 as given by eq. (7.25). We will arbitrarily choose a thermal diffusivity that corresponds to a material like lead, $\alpha = 1/5$.

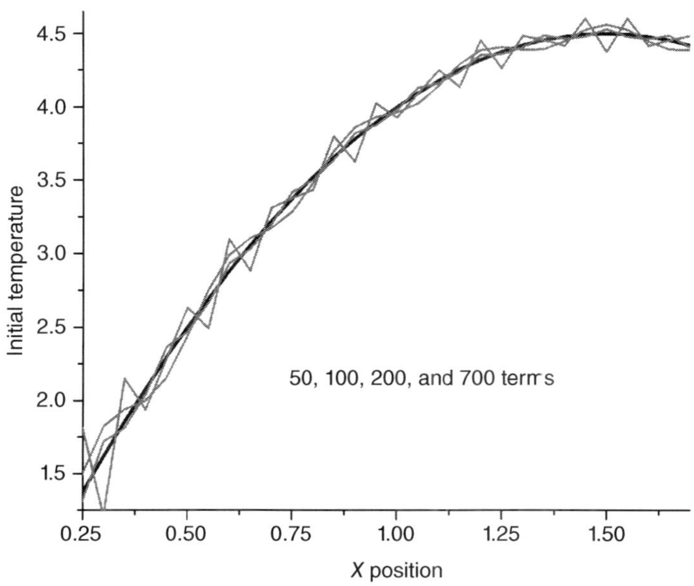

Figure 7.4 Computation of the initial temperature distribution in the slab using 50, 100, 200, and 700 terms. Both axes have been truncated to better reveal the behavior of the Fourier series (which is especially bad near the ends of the slab for small n).

Time (s)	Temperature for $x = 1.5$	Number of Terms Required in Series
$^1/_{32}$	4.475	19
$^1/_{16}$	4.450	15
$^1/_8$	4.400	12
$^1/_4$	4.300	10
$^1/_2$	4.116	7
1	4.060	5
2	4.862	4
4	6.638	2
6	7.831	2
8	8.601	2
10	9.098	1

Note that the temperature in the center of the slab *decreases* at first; this is a consequence of the shape of the initial temperature distribution and the fact that thermal energy is transferred downhill, in the direction of decreasing temperature. These results show that by the time we get to $t = 4$ s, we only need two terms in the series to get an acceptable value for T. This is important because the usefulness of infinite series solutions for PDEs is directly tied to the number of terms required to get a suitably accurate result. If *many* terms are required, an analyst cannot justify the computational burden.

7.4 The Product Method (Separation of Variables)

We begin by noting that if a PDE, say $\frac{\partial \phi}{\partial t} = \kappa \frac{\partial^2 \phi}{\partial y^2}$, with accompanying initial and boundary conditions, is to provide a suitable model of a physically real phenomenon, it must be properly posed. This means that a solution must exist, that the solution is unique, and that it exhibits continuity. We interpret the latter to mean that small changes in the independent variables (t and y in this case produce modest differences in the dependent variable, $\phi(y,t)$. The problems that we are about to examine will meet these criteria.

The product method is a technique by which certain partial differential equations can be solved analytically. As the name implies, the method is based upon equating the dependent variable to a product of functions of the independent variables. If the separation is successful, the result will be ordinary differential equations for which familiar methods of solution may be employed. There are, however, important restrictions upon the applicability of this technique with respect to the form of the differential equation, the shape of the boundary, and the nature of the boundary conditions.

It is obvious that the partial differential equation itself must be separable; i.e., it must be linear and it must not have any cross-derivatives. For example, the equation,

$$A\left(\frac{\partial \phi}{\partial x}\right)^2 + B\frac{\partial^2 \phi}{\partial x \partial y} + \cdots \tag{7.31}$$

would violate both of these restrictions. Furthermore, the equation must be homogeneous, or of a form that can be rendered homogeneous through suitable transformation. Thus, a Poisson-type partial differential equation might be handled successfully if the constant can be removed through a change in the dependent variable. We should also note that it is not necessary that an equation have constant coefficients. Consider the wave (hyperbolic) equation:

$$\frac{\partial^2 \phi}{\partial t^2} - s^2(x)\frac{\partial^2 \phi}{\partial x^2} = 0. \tag{7.32}$$

We propose $\phi = f(t)g(x)$, which results in $f''g = s^2 f g''$, and then: $\frac{f''}{f} = s^2 \frac{g''}{g} = -\lambda^2$.
So the two ordinary differential equations are simply:

$$f'' + \lambda^2 f = 0 \quad \text{and} \quad s^2(x)g'' + \lambda^2 g = 0. \tag{7.33}$$

For the problem types discussed above, the boundary conditions must correspond to constant values of x and y. Thus, $\phi(x = 0) = \phi_0$ or $\phi(y = B) = \phi_B$ would be satisfactory, but $\phi = \phi_0$ for $y/x = 2$ would not. Also, boundary conditions applied at say, $x = B$, cannot include any partial derivatives involving y. Similarly, a boundary condition written for $y = A$ cannot include partial derivatives with respect to x.

The preceding discussion may make it seem as though the applicability of the product method is severely limited, and it is certainly true that there are many partial differential equations of interest to us that cannot be solved with this technique. However, we should not be too hasty to discount separation. There are many important problems in applied mathematics that can be solved in this manner, and we will find examples in fluid flow, heat transfer, diffusion, and wave phenomena, among others. Moreover, we have more than two centuries of work in this branch of mathematics to draw upon, so usefully complete examples abound.

Before we begin to examine applications of this technique, a final word of caution: The solutions we are about to construct (and use) give the impression of being exact. As Weinberger (1965) notes, we truncate these infinite series solutions as a practical matter. This means that only a finite number of terms will be used to construct a Fourier series, for example, and therefore we will be generating an *approximation* to the solution. In many cases, it will be a very close approximation, but keep in mind that our results will not be *exactly correct*.

7.5 Parabolic Equations

The parabolic PDEs that we are most likely to see will involve transient molecular transport (by viscous friction, conduction, or diffusion) in one or more spatial directions. The general procedure we will employ will be similar in every case: We perform the separation, solve the resulting ordinary differential equations, use the boundary conditions to simplify the result and identify the constant of separation, and finally, use the initial condition (with Fourier theorem or orthogonality) to identify the proper values for the leading coefficient in the infinite series. The process is straightforward and usually quite transparent. There are, however, a few subtle issues that can complicate our analysis. We will try to illustrate some of these in the examples that follow.

7.5.1 Implementing Different Boundary Conditions

Let us begin by examining transient conduction of heat in a *finite* slab of material; the governing parabolic PDE is: $\frac{\partial T}{\partial t} = \alpha \frac{\partial^2 T}{\partial y^2}$. We let this object extend from $y = 0$ to $y = 1$, and we can have either a uniform initial temperature or a temperature distribution that can be written as a function of y. At $t = 0$, both faces are instantaneously heated to some new temperature, T_s. We define a dimensionless temperature, $\theta = \frac{T-T_s}{T_i-T_s}$, and let $\theta = f(y)g(t)$. The product method yields

$$g' = -\alpha\lambda^2 g \quad \text{and} \quad f'' + \lambda^2 f = 0. \tag{7.34}$$

As expected, we get

$$g = C_1 \exp(-\alpha\lambda^2 t) \quad \text{and} \quad f = A \sin \lambda y + B \cos \lambda y. \tag{7.35}$$

B must be zero since $\theta(y = 0,t) = 0$, and $\sin(\lambda) = 0$ because $\theta(y = 1,t) = 0$. Of course sine is periodic so an infinite number of λ's will satisfy the second of this pair of boundary conditions. Consequently, we use *superposition*, combining all of the possible solutions for this linear PDE and expressing the result as an infinite series:

$$\theta = \sum_{n=1}^{\infty} A_n \exp\left(-\alpha\lambda_n{}^2 t\right) \sin \lambda_n y, \quad \text{where } \lambda_n = n\pi. \tag{7.36}$$

If we have a uniform initial temperature, T_i, then application of the initial condition results in:

$$1 = \sum_{n=1}^{\infty} A_n \sin \lambda_n y, \tag{7.37}$$

a half-range Fourier sine series. By definition,

$$A_n = \frac{2}{L}\int_0^L f(y) \sin \frac{n\pi y}{L} dy, \tag{7.38}$$

Table 7.1 Illustration of Infinite Series Convergence for Small t's.

Term no.	$t = 0.001$	$t = 0.005$	$t = 0.025$	$t = 0.125$	$t = 0.625$
1	1.271981	1.266969	1.242205	1.12546	0.6870893
3	0.851322	0.8609938	0.9023096	0.9856378	0.6854422
5	1.099763	1.086086	1.039727	0.9972914	0.6854423
7	0.926459	0.9432634	0.9854355	0.9968604	0.6854423
9	1.05706	1.038121	1.004608	0.9968669	0.6854423
11	0.954341	0.9744126	0.9987616	0.9968669	0.6854423
13	1.037236	1.01695	1.000275	0.9968669	0.6854423
15	0.969256	0.9889856	0.9999457	0.9968669	0.6854423
17	1.025566	1.006978	1.000006	0.9968669	0.6854423
19	0.97864	0.9956936	0.9999966	0.9968669	0.6854423
21	1.017874	1.002573	0.9999977	0.9968669	0.6854423
23	0.985031	0.9985044	0.9999976	0.9968669	0.6854423
25	1.012515	1.000835	0.9999976	0.9968669	0.6854423
27	0.98955	0.9995433	0.9999976	0.9968669	0.6854423
29	1.008694	1.000235	0.9999976	0.9968669	0.6854423
31	0.992785	0.9998772	0.9999976	0.9968669	0.6854423
33	1.005956	1.000056	0.9999976	0.9968669	0.6854423
35	0.995097	0.9999698	0.9999976	0.9968669	0.6854423
37	1.004008	1.00001	0.9999976	0.9968669	0.6854423
39	0.996732	0.9999919	0.9999976	0.9968669	0.6854423
41	1.002642	0.9999996	0.9999976	0.9968669	0.6854423
43	0.997868	0.9999964	0.9999976	0.9968669	0.6854423

but for our case $L = 1$ and the function, $f(y)$, is also 1. The integral above is zero for even n and equal to $4/(n\pi)$ for $n = 1,3,5,\ldots$ With this example we have a good opportunity to examine the convergence of the infinite series solution. Let $y = 1/2$, $\alpha = 0.1$, and let t range from 0.001 to 0.625 by repeated factors of 5. We shall examine the series for n's from 1 to 43; see Table 7.1. Note that for small t's the series does not converge quickly. However, for $t = 0.125$ we need only five terms and at $t = 0.625$, only three. The results should not be surprising. For very small t's, the temperature profile is virtually half a cycle of a square wave.

Now suppose we have a Neumann condition at one boundary; in particular, let us assume we have an insulated boundary located at $y = L$, such that $\left(\frac{\partial T}{\partial y}\right)_{y=L} = 0$. The reader may wish to show using eq. (7.36) as our starting point, that $\cos\lambda_n L = 0$. Therefore, $\lambda_n = \frac{\pi}{2L}, \frac{3\pi}{2L}, \frac{5\pi}{2L}$, etc. Once again we have discovered a Fourier series problem since the separation constants are integer multiples of *pi* (π). Let us demonstrate how this modified problem can be solved; the reader can gain valuable experience by verifying the process shown below. The initial steps are exactly the same, so our product solution is $\theta = A\exp(-\alpha\lambda^2 t)\sin(\lambda y)$.

Since the right-hand boundary is insulated, $\left(\frac{\partial \theta}{\partial y}\right)_{y=L} = 0$, and therefore $\cos(\lambda L) = 0$ resulting in $\lambda_n = \frac{(2n-1)\pi}{2L}$. Our solution is now written

$$\theta = \sum_{n=1}^{\infty} A_n \exp\left(-\alpha\lambda_n{}^2 t\right)\sin(\lambda_n y). \tag{7.39}$$

The uniform initial temperature in the slab is T_i so $\theta(y, t = 0) = 1$, and the leading coefficients can be determined:

$$A_n = \frac{2}{L}\int_0^L (1)\sin(\lambda_n y)dy = \frac{4}{(2n-1)\pi}. \tag{7.40}$$

We will fix $y = L/2$, $\alpha = 0.1$, $L = 1$, and $t = 1$; we construct a table to show the successive terms in the series solution.

n	λ_n	A_n	θ_n
1	$\pi/(2L)$	$4/\pi$	0.7035
2	$3\pi/(2L)$	$4/(3\pi)$	0.0326
3	$5\pi/(2L)$	$4/(5\pi)$	−0.000377
4	$7\pi/(2L)$	$4/(7\pi)$	-7.22×10^{-7}

The summation of the θ_n's is: 0.7357.

We also want to address the issues that arise when one of the boundary conditions for our slab of material is of the Robin's type (the third kind). All of the initial steps in the problem are the same as in the previous case, but we make one adjustment by defining a new dependent variable, $\theta = T - T_\infty$, where T_∞ is the temperature of the surroundings at large distance. The situation we are describing would correspond to the case where one end of the slab (at $y = L$) loses heat to the surroundings, and we assume that this process can be described by Newton's law of cooling. Since the first part of this problem is common to what we saw previously, we can start with:

$$\theta = A \exp(-\alpha\lambda^2 t) \sin \lambda y. \tag{7.41}$$

But this time, at the end of the slab where $y = L$, we equate the fluxes:

$$-k \left.\frac{\partial\theta}{\partial y}\right|_{y=L} = h(T_{y=L} - T_\infty). \tag{7.42}$$

Consequently, we find

$$-kA\lambda \exp(-\alpha\lambda^2 t) \cos \lambda L = hA \exp(-\alpha\lambda^2 t) \sin \lambda L, \tag{7.43}$$

and this is equivalent to $-\frac{k}{hL}\lambda L = \tan \lambda L$, which we write as

$$\lambda L \cot \lambda L + \frac{hL}{k} = 0. \tag{7.44}$$

This transcendental equation arises frequently in applied mathematics, and you may recognize the dimensionless grouping, hL/k, as the Biot number (or modulus). We can find a few roots for this equation in Carslaw and Jaeger (1959), and a table of values (with eight roots) is provided below. Eq. (7.44) has been rewritten as $x\cot(x) + C = 0$, and the negative values of C come about for a sphere that loses heat to the surroundings through a Robin's-type boundary condition.

$C = -1$	$C = -0.8$	$C = -0.6$	$C = -0.4$	$C = -0.2$	$C = 0$
0.0000	0.7593	1.0528	1.2644	1.4320	1.5708
4.4934	4.5379	4.5822	4.6261	4.6696	4.7124
7.7253	7.7511	7.7770	7.8028	7.8284	7.8540
10.9041	10.9225	10.9408	10.9591	10.9774	10.9956
14.0662	14.0804	14.0946	14.1088	14.1230	14.1372
17.2208	17.2324	17.2440	17.2556	17.2672	17.2788
20.3713	20.3811	20.3909	20.4007	20.4106	20.4204
23.5195	23.5280	23.5365	23.5450	23.5535	23.5619

0.2	0.4	0.6	0.8	1.0	2.0
1.6887	1.7906	1.8798	1.9586	2.0288	2.2889
4.7544	4.7956	4.8358	4.8750	4.9132	5.0870
7.8794	7.9045	7.9295	7.9542	7.9787	8.0962
11.0137	11.0318	11.0498	11.0677	11.0855	11.1727
14.1513	14.1654	14.1795	14.1935	14.2074	14.2764
17.2903	17.3019	17.3134	17.3249	17.3364	17.3932
20.4301	20.4399	20.4497	20.4594	20.4692	20.5175
23.5704	23.5789	23.5874	23.5958	23.6043	23.6463
4	**6**	**8**	**10**	**20**	**40**
2.5704	2.7165	2.8044	2.8628	2.9930	3.0651
5.3540	5.5378	5.6669	5.7606	5.9921	6.1311
8.3029	8.4703	8.6031	8.7083	9.0018	9.1987
11.3348	11.4773	11.5993	11.7027	12.0250	12.2688
14.4080	14.5288	14.6374	14.7335	15.0625	15.3417
17.5034	17.6072	17.7032	17.7908	18.1136	18.4180
20.6120	20.7024	20.7877	20.8672	21.1772	21.4980
23.7289	23.8088	23.8851	23.9574	24.2516	24.5817

Now, suppose for our example $hL/k = 1/2$, the first eight values for the product λL are: 1.8366, 4.8158, 7.9171, 11.0409, 14.1724, 17.3076, 20.4448, and 23.5831.

Notice the *spacing* between the consecutive pairs: 2.9792, 3.1013, 3.1238, 3.1315, etc. It is now clear that values of the constant of separation, λ, are *not* integer multiples of *pi*, and this means that this case is *not a Fourier series problem.* So although our solution has the form

$$\theta = \sum_{n=1}^{\infty} A_n \exp\left(-\alpha \lambda_n{}^2 t\right) \sin \lambda_n y, \tag{7.45}$$

the coefficients (A_n's) must be determined using orthogonality as we will now demonstrate. At $t = 0$, $\theta = \theta_0$, so

$$\theta_0 = \sum_{n=1}^{\infty} A_n \sin \lambda_n y. \tag{7.46}$$

We multiply both sides of the equation by $\sin\lambda_m y dy$ and integrate from $y = 0$ to $y = L$, making use of the fact that

$$\int_0^L \sin \lambda_n y \sin \lambda_m y dy = 0 \text{ unless } n = m. \tag{7.47}$$

Therefore, the needed coefficients are obtained from the quotient of integrals:

$$A_n = \frac{\int_0^L \theta_0 \sin \lambda_n y dy}{\int_0^L \sin^2 \lambda_n y dy} = \frac{-\frac{\theta_0}{\lambda_n}(\cos \lambda_n L - 1)}{\frac{L}{2} - \frac{1}{4\lambda_n} \sin 2\lambda_n L}. \tag{7.48}$$

You may want to verify that for $hL/k = 1/2$ and $\theta_0 = 50$, and $L = 1$, $A_1 = 60.414$, $A_2 = 18.233$, and $A_3 = 13.322$; use these values with eq. (7.46) to see if $\theta_0 = 50$. With just these three terms you should find 53.45.

7.5.2 Diffusion in a Plane Sheet

In the previous examples we considered a parabolic PDE with a domain extending from $y = 0$ to $y = L$, subject to different boundary conditions; we will now explore a problem with symmetry. Imagine we have a porous slab of material like a membrane with a thickness $2w$ but with the origin placed at the center such that $-w \leq x \leq +w$. We will assume that the membrane has some constant and uniform initial concentration of solute, C_0; at $t = 0$ the concentration of solute at both planar surfaces is changed to C_W. Our interest is the dynamic behavior of C in the interior of the membrane where

$$\frac{\partial C}{\partial t} = D\frac{\partial^2 C}{\partial x^2}. \tag{7.49}$$

We now set $C = f(x)g(t)$ and introduce this product into the PDE to yield two ordinary differential equations:

$$\frac{dg}{dt} = -D\lambda^2 g \quad \text{and} \quad \frac{d^2 f}{dx^2} + \lambda^2 f = 0. \tag{7.50}$$

Of course we knew that the separation constant must be negative since the first equation of this pair yields a solution that is exponential in time. We can write down the solutions by inspection producing

$$C = c_1 \exp(-D\lambda^2 t)[A\sin\lambda x + B\cos\lambda x]. \tag{7.51}$$

Since the concentration distributions must be symmetric about the origin, we recognize that $\left.\frac{\partial C}{\partial x}\right|_{x=0} = 0$; that is, the solution must be constructed with even functions so $A = 0$:

$$C = B\exp(-D\lambda^2 t)\cos\lambda x. \tag{7.52}$$

It is convenient to let $C^* = C - C_W$ so that $\cos\lambda x = 0$ for $x = \pm w$. This of course means that $\lambda_n = \frac{(2n-1)\pi}{2w}$, and therefore

$$C - C_W = \sum_{n=1}^{\infty} B_n \exp\left(-D\lambda_n^2 t\right)\cos\lambda_n x. \tag{7.53}$$

We apply the initial condition: at $t = 0$, $C = C_0$ for $-w \leq x \leq +w$, so that

$$C_0 - C_W = \sum_{n=1}^{\infty} B_n \cos\frac{(2n-1)\pi}{2w}x. \tag{7.54}$$

For our half-range Fourier series problem,

$$B_n = \frac{2}{w}\int_0^w (C_0 - C_W)\cos\frac{(2n-1)\pi}{2w}x\,dx, \tag{7.55}$$

and the solution we are seeking is

$$C - C_W = \sum_{n=1}^{\infty} \frac{4(C_0 - C_W)}{(2n-1)\pi}\sin\frac{(2n-1)\pi}{2}\exp\left(-D\frac{(2n-1)^2\pi^2}{4w^2}t\right)\cos\frac{(2n-1)\pi}{2w}x. \tag{7.56}$$

This is an important result which is applicable to problems in both mass *and* heat transfer; for the latter of course, T replaces C and the binary diffusivity, D, is replaced by the thermal diffusivity, α. Eq. (7.56) is presented in a convenient graphical form in Figure 7.5, and we will illustrate its use with two examples: Suppose we have a block of cast iron for which $\alpha = 0.12\,\text{cm}^2/\text{s}$. Let the initial temperature be 20 °C and the face temperatures both be 200 °C. The thickness of the slab, $2w$, is 20 cm. When will the temperature at $x/w = 0.5$ become 60 °C? We find $\frac{60-20}{200-20} = \frac{40}{180} = 0.222$. Therefore, for $x/w = 0.5$, we find $\frac{\alpha t}{w^2} \cong 0.09$ and $t \cong (.09)(100)/(0.12) = 75$ s.

Now assume we have a slab of porous material initially saturated with a solute species such that for $-w \leq x \leq +w$, $C(x, t = 0) = 1$. At $t = 0$ the slab is introduced into a solvent bath of very large volume such that both faces are instantaneously changed to $C = 0$. If $D = 1 \times 10^{-5}$ and $w = 5$ cm, when will the center concentration fall to 50% of its initial value? From Figure 7.5 we find $\frac{Dt}{w^2} \cong 0.38+$; therefore, the required concentration in the center of the slab will be attained in 955,000 s, or about 11 days.

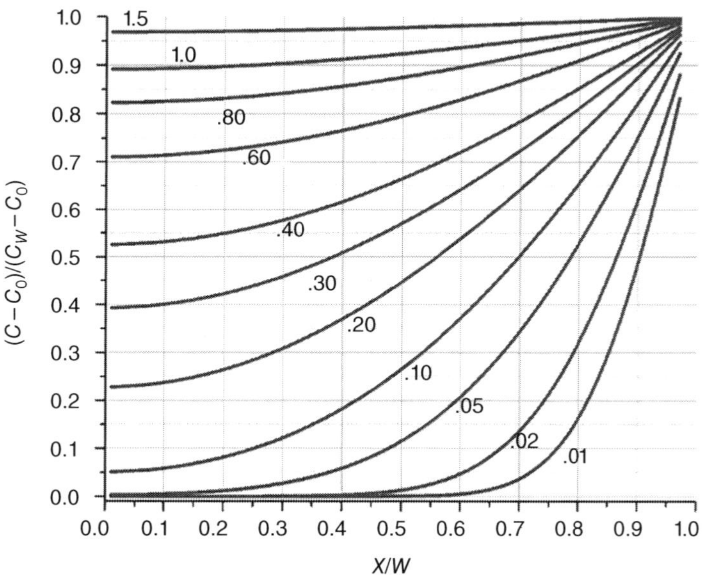

Figure 7.5 Concentration distributions for diffusion in a plane sheet of thickness, $2w$, as computed from the infinite series solution. The labels on the curves represent values of the quotient, $\frac{Dt}{w^2}$. C_0 is the initial concentration in the sheet, and C_W is the concentration at both faces, $x = \pm w$, imposed instantaneously at $t = 0$.

7.5.3 Cylindrical Coordinates

It is important that we be able to deal with analogous problems for "infinite" ($L \gg d$) cylinders too. Suppose we begin with a cylinder that has some initial temperature (or initial temperature distribution). At $t = 0$, the surface at $r = R$ is instantly cooled or heated; we wish to determine how the temperature of the medium for $0 < r < R$ responds to this change. For this case the governing PDE is:

$$\frac{\partial T}{\partial t} = \alpha \left[\frac{\partial^2 T}{\partial r^2} + \frac{1}{r}\frac{\partial T}{\partial r} \right]. \tag{7.57}$$

Eq. (7.57) fits the criteria that we had established for separation, so we propose $T = f(r)g(t)$; this results in

$$\frac{g'}{\alpha g} = \frac{f'' + \frac{1}{r}f'}{f} = -\lambda^2. \tag{7.58}$$

We know that the constant of separation must be negative so that the solution remains bounded as $t \to \infty$. The two ordinary differential equations (the second one is a form of Bessel's differential equation), with their solutions are shown immediately below:

$$g' = -\alpha\lambda^2 g \quad \text{and} \quad f'' + \frac{1}{r}f' + \lambda^2 f = 0 \tag{7.59}$$

resulting in

$$g = C_1 \exp(-\alpha\lambda^2 t) \quad \text{and} \quad f = AJ_0(\lambda r) + BY_0(\lambda r). \tag{7.60}$$

For all problems of this type that involve a solid cylindrical medium that extends from $r = 0$ to $r = R$, we can immediately simplify since T must be finite at the center; recall that $Y_0(0) = -\infty$ as we saw in Chapter 5, so consequently $B = 0$:

$$T = A\exp(-\alpha\lambda^2 t)J_0(\lambda r). \tag{7.61}$$

As noted above, we assume that the *surface* temperature is changed instantaneously to a new value, T_s. Furthermore, we define a new dependent variable, $\theta = T - T_s$, which means that $\theta(r = R, t) = 0$. This will be satisfied as long as $J_0(\lambda R) = 0$; this equation has an infinite number of roots and the first few λR's are: 2.40483, 5.52008, 8.65373, 11.79153, 14.93092, etc. Therefore,

$$\theta = \sum_{n=1}^{\infty} A_n \exp\left(-\alpha {\lambda_n}^2 t\right) J_0(\lambda_n r). \tag{7.62}$$

We are ready to apply the initial condition for the interior of the cylinder; typically that would be $T(r,t=0)=$ constant, or $T(r,t=0)=f(r)$. We start with the constant case and utilize orthogonality (readers unfamiliar with this process for Bessel functions may find Section 7.5 of Carslaw and Jaeger, 1959, or Chapter 10 in Spiegel, 1971, to be quite helpful): We begin with the initial condition

$$\theta_0 = \sum_{n=1}^{\infty} A_n J_0(\lambda_n r), \tag{7.63}$$

and multiply both sides by $rJ_0(\lambda_m r)dr$, then integrate from $r = 0$ to R. Please pay particular attention to the inclusion of the weighting function, r. Consequently,

$$A_n = \frac{\int_0^R \theta_0 r J_0(\lambda_n r)dr}{\int_0^R r J_0{}^2(\lambda_n r)dr}. \tag{7.64}$$

The integral in the denominator is known, and you should consult eq. (7.69) for confirmation: $\frac{R^2}{2}J_1{}^2(\lambda_n R)$. So therefore

$$A_n = \frac{2\theta_0}{R^2 J_1{}^2(\lambda_n R)} \int_0^R r J_0(\lambda_n r)dr. \tag{7.65}$$

Completing the problem, we find

$$\theta = \frac{2\theta_0}{R}\sum_{n=1}^{\infty} \exp\left(-\alpha\lambda_n{}^2 t\right) \frac{J_0(\lambda_n r)}{\lambda_n J_1(\lambda_n R)}. \tag{7.66}$$

This example serves as an important reminder: the zeros of J_0 figure prominently in many practical problems in which a radially directed flux occurs in cylindrical coordinates; furthermore, we often need to obtain the corresponding values for J_1. The first 20 pairs of these values are given here.

First 20 zeros of J_0 (x) and corresponding values for J_1 (x)

x	$J_1(x)$
2.404826	+0.519147
5.520078	−0.340265
8.653728	+0.271447
11.791534	−0.232115
14.930918	+0.206546
18.071064	−0.187729
21.211637	+0.173266
24.352472	−0.161702
27.493479	+0.152181
30.634606	−0.144166
33.775820	+0.137297
36.917098	−0.131325
40.058426	+0.126069
43.199792	−0.121399
46.341188	+0.117211
49.482610	−0.113429
52.624052	+0.109991
55.765511	−0.106848
58.906984	+0.103960
62.048469	−0.101293

In cases such as the present example, we will need to use orthogonality in conjunction with the initial condition in order to determine the leading coefficients for the infinite series solution. This process is predicated upon the fact that

$$\int_0^R rJ_0(\lambda_m r)J_0(\lambda_n r)dr = 0 \ \ unless\ m = n. \tag{7.67}$$

Two integrals arise commonly in this process:

$$\int_0^R rJ_0(\lambda_n r)dr = \frac{R}{\lambda_n}J_1(\lambda_n R) \tag{7.68}$$

and

$$\int_0^R rJ_0^{\,2}(\lambda_n r)dr = \frac{R^2}{2}J_1^{\,2}(\lambda_n R)\ \ \textbf{\textit{if}}\, J_0(\lambda_n R) = 0. \tag{7.69}$$

We should look at a specific example to better understand how well the solution will work for us. Suppose we have a long ABS (acrylonitrile butadiene styrene) plastic rod exactly 2 cm in diameter. The thermal diffusivity (α) for ABS is about 0.00108 cm^2/s, and we will assume our interest is for $r = 1/2$ cm and $t = 50$ s (we have purposefully selected a time that is neither short nor long). We obtain:

$$\frac{\theta}{\theta_0} = \frac{2}{R}[0.39267 + 0.017297 - 0.002659\cdots]$$

and since $R = 1$, this corresponds to $\frac{\theta}{\theta_0} \cong 0.81$. The reader should also be aware that the solution of this infinitely long cylinder problem is so important in practical situations that it has been presented graphically throughout the literature of heat and mass transfer, including Carslaw and Jaeger (1959) and Glasgow (2010). A similar graph is reproduced here as Figure 7.6; we can use it to corroborate our results for the ABS plastic rod as calculated from the series solution above. We have $\frac{\alpha t}{R^2} = \frac{(0.00108)(50)}{(1)^2} = 0.054$ and $\frac{r}{R} = 0.5$; using these values we find that $\frac{T-T_i}{T_b-T_i} \cong 0.2$. Consequently, $\frac{\theta}{\theta_0} \cong 0.8$.

A related problem, but one that is a bit more difficult, arises when a boundary condition of the third kind must be used at the cylinder surface where $r = R$. Although the discussion that follows concerns heat transfer to a cylinder, keep in mind that the analysis is the same for the equivalent mass transfer problem. Suppose we have heat transfer from the fluid phase to a long, solid cylinder with relatively large conductivity. Because k (or α) is large, much of the thermal energy that arrives

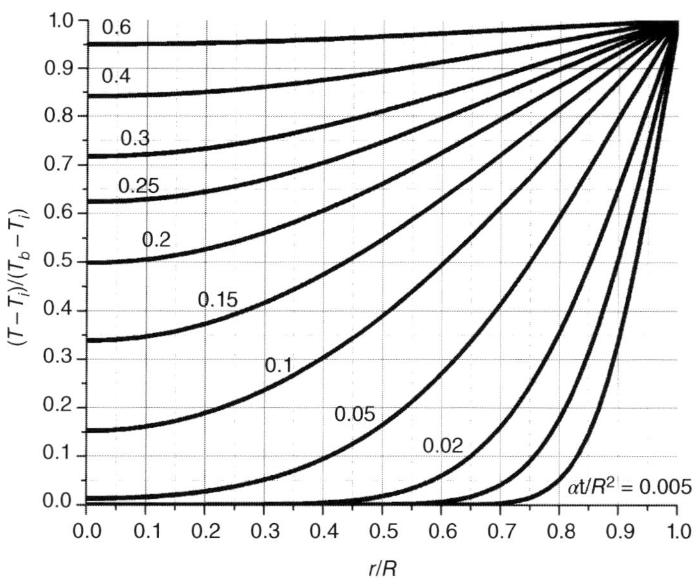

Figure 7.6 Solution for heat transfer to a long, solid cylinder when the surface acquires a new temperature instantaneously. Note that the centerline temperature attains 50% of the total change when $\alpha t/R^2 = 0.2$. For example, for a PFTE cylinder with $R = 5$ cm, $\alpha t/R^2 = 1.24 \times 10^{-7} t/(0.05)^2$, so a 50% change will require 4,032 s (1.12 hr).

at the interface will be conducted readily into the interior of the cylinder. Consequently, the surface temperature will *not* acquire the fluid temperature *instantaneously*. It is appropriate to think of this behavior in terms of the relative resistances: A small Biot number (hR/k) indicates that the bulk of the resistance to heat transfer lies in the fluid phase. The initial steps for this problem are the same as before leading to eq. (7.61), so: $T = A\exp(-\alpha\lambda^2 t)J_0(\lambda r)$. However, at $r = R$, a Robin's-type boundary condition must be used:

$$-k\left.\frac{\partial T}{\partial r}\right|_{r=R} = h(T|_{r=R} - T_\infty). \tag{7.70}$$

It is convenient to define $\theta = T - T_\infty$, so that when we introduce eqs. (7.61) into (7.70):

$$\lambda R J_1(\lambda R) - \frac{hR}{k} J_0(\lambda R) = 0. \tag{7.71}$$

Naturally, the roots of this transcendental equation depend upon the value of the Biot number. In a typical application of the kind we are discussing, hR/k might be about 0.7 for which the $\lambda_n R$'s are: 1.0873, 4.0085, 7.1143, 10.2419, 13.3761, 16.5131, etc. The roots for this transcendental eq (7.71) are needed frequently for the solution of transport problems in engineering, so it may be useful to provide an abbreviated table here.

Biot(hR/k)	$\lambda_1 R$	$\lambda_2 R$	$\lambda_3 R$	$\lambda_4 R$	$\lambda_5 R$
0.01	0.1412	3.8343	7.0170	10.1745	13.3244
0.02	0.1995	3.8369	7.0184	10.1754	13.3252
0.05	0.3401	3.8443	7.0225	10.1784	13.3274
0.1	0.4417	3.8577	7.0298	10.1833	13.3312
0.2	0.6170	3.8835	7.0440	10.1931	13.3387
0.5	0.9408	3.9594	7.0864	10.2225	13.3611
0.7	1.0873	4.0085	7.1143	10.2419	13.3761
1	1.2558	4.0795	7.1558	10.2710	13.3984
2	1.5994	4.2910	7.2884	10.3658	13.4719
5	1.9898	4.7131	7.6177	10.6223	13.6786
10	2.1795	5.0332	7.9569	10.9363	13.9580
20	2.2880	5.2568	8.2534	11.2677	14.2983
100	2.3809	5.4652	8.5678	11.6747	14.7834

Of course we now know that the solution for the modified problem must be written as

$$\theta = \sum_{n=1}^{\infty} A_n \exp\left(-\alpha\lambda_n{}^2 t\right) J_0(\lambda_n r), \tag{7.72}$$

and the coefficients (A_n's) are determined by application of the initial condition and orthogonality as before, with one important difference: *The values for the separation constant come from the transcendental eq. (7.71) rather than from the zeros of J_0, so the required integration produces a different result.* This is a commonly encountered situation in both heat and mass transport problems. When a boundary condition of the third kind results in the transcendental equation, $\lambda J_0{}'(\lambda R) + CJ_0(\lambda R) = 0$, which is the form shown in eq. (7.71), we use it to identify the infinite set of λ_n's. Application of the initial condition typically yields $f(r) = \sum_{n=1}^{\infty} A_n J_0(\lambda_n r)$; the leading coefficients for the series are then determined from:

$$A_n = \frac{2\lambda_n{}^2}{R^2(C^2+\lambda_n{}^2)J_0{}^2(\lambda_n R)}\int_0^R r f(r) J_0(\lambda_n r)dr$$

Again, this is described clearly in Chapter VII in Carslaw and Jaeger (1959) and also in Chapter 10 in Spiegel (1971); the consequence for this case is

$$A_n = \frac{2\lambda_n R J_1(\lambda_n R)}{\left(\frac{h^2R^2}{k^2} + \lambda_n{}^2 R^2\right) J_0{}^2(\lambda_n R)}. \tag{7.73}$$

Let us take $k = 0.09$, $R = 1.27$, and $h = 0.0496$ (all cgs units), such that $hR/k = 0.7$. Then using (38) we find $\lambda_1 = 0.8561$, $\lambda_2 = 3.1563$, and by eq. (7.73), $A_1 = 1.1522$ and $A_2 = -0.21197$.

As a practical matter we should emphasize the differences between the two preceding examples. In the first case the cylinder surface acquired the temperature of the surroundings instantaneously; that is, the main resistance to heat transfer was located *in the solid material*. For the second case we set $hR/k = 0.7$ which placed the principal resistance *in the fluid phase*. Let's illustrate the effects of this change with a table of results for $r = R/2$ and $t = 1$ s. We will select $T_i = 0°$ and $T_b = 75°$ with $\alpha = 0.25$.

Bi = *hR/k*	***T*(*r* = *R*/2, *t* = 1 s)**
0.7	8.56
7	31.45
70	40.79
700	41.90

Please note that if we assume the surface temperature of the cylinder acquires the fluid phase value instantaneously, and use Figure 7.6 with $\alpha = 0.25$, $T(r = R/2, t = 1) = 42°$. The reader may want to obtain some practice with these solutions by confirming the results given above.

Our task becomes a bit more difficult in cases where the dependent variable has a more complicated initial distribution. Consider mass transfer in the interior of a long, porous cylinder where the initial distribution of solute is second degree with respect to r.

Transient diffusion in a porous cylindrical solid is described by the parabolic partial differential equation,

$$\frac{\partial C}{\partial t} = D\left[\frac{\partial^2 C}{\partial r^2} + \frac{1}{r}\frac{\partial C}{\partial r}\right]. \tag{7.74}$$

We take $0 \le r \le 1$ and let $C(r, t = 0) = C_0(1 - r^2)$; therefore, the initial concentration is C_0 at the center of the cylinder and zero at the outer surface where $r = 1$. At $t = 0$ the concentration at the surface is instantaneously elevated to a new value, C_S; thus, $C(r = 1, t = 0) = C_S$. Naturally as t becomes very large the concentration of solute in the interior of the cylinder will asymptotically approach C_S for all r positions. We can apply separation of variables, noting that the concentration must be finite at the center, to obtain the familiar result, $C = C_S + A\exp(-D\lambda^2 t)J_0(\lambda r)$. Since $C(r = R = 1, t) = C_S$, we must require $J_0(\lambda R) = 0$ and thus

$$C - C_S = \sum_{n=1}^{\infty} A_n \exp\left(-D\lambda_n{}^2 t\right) J_0(\lambda_n r). \tag{7.75}$$

The initial condition is of course the parabolic distribution cited above, and for simplicity we will rewrite the left-hand side of eq. (7.75) as $a - br^2$:

$$a - br^2 = \sum_{n=1}^{\infty} A_n J_0(\lambda_n r), \tag{7.76}$$

where $a = C_0 - C_S$ and $b = C_0$. The leading coefficients must be determined using orthogonality, so we multiply both sides of eq. (7.76) by $rJ_0(\lambda_m r)dr$, set $m = n$, and integrate from 0 to R. Thus A_n is given by:

$$A_n = \frac{a\int_0^R rJ_0(\lambda_n r)dr - b\int_0^R r^3 J_0(\lambda_n r)dr}{\int_0^R rJ_0{}^2(\lambda_n r)dr}. \tag{7.77}$$

We have seen two of these integrals previously, but the second integral in the numerator is unfamiliar:

$$b\int_0^R r^3 J_0(\lambda_n r)dr = b\left[\frac{R^3}{\lambda_n}J_1(\lambda_n R) - \frac{2R^2}{\lambda_n{}^2}J_2(\lambda_n R)\right]. \tag{7.78}$$

Now let $R = 1$, $C_0 = 1$, and $C_S = ½$, so that $a = ½$ and $b = 1$; the solution we are seeking becomes

$$C - C_S = \sum_{n=1}^{\infty} \exp\left(-D\lambda_n^2 t\right) J_0(\lambda_n r) \frac{\frac{(1-2R^2)}{\lambda_n} J_1(\lambda_n R) + \frac{4}{\lambda_n^2} J_2(\lambda_n R)}{J_1^2(\lambda_n R)}. \tag{7.79}$$

As we have emphasized previously, whether or not this analytic solution is useful depends entirely upon how rapidly the series converges. We will explore this further by looking at $n = 1$ through 4:

n	λ_n	$J_1(\lambda_n)$	$J_2(\lambda_n)$
1	2.4048	0.5191	0.432
2	5.5201	−0.3403	−0.123
3	8.6537	0.2714	0.063
4	11.7915	−0.2321	−0.039

We will use just the first four terms in the series, fixing $r = ½$ and $D = 1 \times 10^{-4}$, but varying t; the numerical solution has also been computed to help us with our assessment.

t (s)	$C_{analytic}$	$C_{numerical}$
100	0.7210	0.7103
400	0.6512	0.6508
1,000	0.6127	0.6124
4,000	0.5204	0.5206

These results reaffirm an observation we have made repeatedly in the context of parabolic partial differential equations: If t is sufficiently large an infinite series solution may converge rapidly enough to be valuable, but if t is small, the level of work required may be daunting. In the case of diffusion within this porous cylindrical solid, if t is much less than 100 s, one would not want to expend the effort required with eq. (7.79).

7.5.4 The Annulus or Hollow Cylinder

A cylindrical geometry of immense practical importance is the hollow cylinder (annulus), and problems of this type are more difficult because we cannot take advantage of the simplification at the center where $r = 0$. Assume we have mass transfer (diffusion) in a porous annular solid with the inner surface located at $r = R_1$ and the outer surface located at $r = R_2$. We require that $L >> (R_2 - R_1)$ so that z-direction transport can be neglected; the governing equation is eq. (7.74), $\frac{\partial C}{\partial t} = D\left[\frac{\partial^2 C}{\partial r^2} + \frac{1}{r}\frac{\partial C}{\partial r}\right]$. We employ the product method just as in the preceding examples, with the same preliminary result obtained before, $C = c_1 \exp(-D\lambda^2 t)[AJ_0(\lambda r) + BY_0(\lambda r)]$. But this time $BY_0(\lambda r)$ *cannot* be eliminated as it was in the previous case since we are only interested in the region from R_1 to R_2. Now suppose the porous annular solid is initially saturated with some solute such that $C = 1$; at $t = 0$ this hollow cylinder is immersed in an agitated solvent bath where the prevailing C is much reduced. Let this solute concentration near the inner and outer surfaces be represented by C_1, and define a new dependent variable, $C^* = C - C_1$ which can be introduced into eq. (7.74) without change. In order to satisfy the boundary conditions at the two surfaces it is necessary that:

$$0 = AJ_0(\lambda R_1) + BY_0(\lambda R_1) \quad \text{and} \quad 0 = AJ_0(\lambda R_2) + BY_0(\lambda R_2). \tag{7.80}$$

Therefore, the required values for the separation constant are obtained from the positive roots of

$$J_0(\lambda R_2)Y_0(\lambda R_1) = J_0(\lambda R_1)Y_0(\lambda R_2). \tag{7.81}$$

Now let us assume that $R_2 = 3$ and $R_1 = 1$ so that $R_2/R_1 = 3$; the first seven values of λ are 1.5485, 3.1291, 4.7038, 6.2767, 7.8487, 9.4204, and 10.9918. Eq. (7.81) often arises in problems with a radially directed flux in cylindrical coordinates where the center ($r = 0$) is not part of the range for variable, r. Therefore, we should compile some roots for this transcendental equation, $J_0(b)Y_0(\kappa b) - J_0(\kappa b)Y_0(b) = 0$:

κ	b_1	b_2	b_3	b_4	b_5
1.2	15.7014	31.4126	47.1217	62.8302	78.5385
1.5	6.2702	12.5598	18.8451	25.1294	31.4133
2.0	3.1230	6.2734	9.4182	12.5614	15.7040
2.5	2.0732	4.1773	6.2754	8.3717	10.4672
3.0	1.5485	3.1291	4.7038	6.2767	7.8487
3.5	1.2339	2.5002	3.7608	5.0196	6.2776
4.0	1.0244	2.0809	3.1322	4.1816	5.2301
4.5	0.8750	1.7816	2.6831	3.5830	4.4820
5.0	0.7632	1.5571	2.3464	3.1340	3.9208

Proceeding with our example, $A = \frac{-BY_0(\lambda R_1)}{J_0(\lambda R_1)}$, and we find

$$C^* = \sum_{n=1}^{\infty} B_n \exp\left(-D\lambda_n^2 t\right) \frac{1}{J_0(\lambda_n R_1)} [Y_0(\lambda_n r)J_0(\lambda_n R_1) - Y_0(\lambda_n R_1)J_0(\lambda_n r)] \tag{7.82}$$

It remains for us to find the B_n's using the initial condition; we have taken $C(r, t = 0) = 1$ and we will assume that $C_1 = 0$. We define a *cylinder function* (convenient shorthand notation):

$$Z_0 = Y_0(\lambda_n r)J_0(\lambda_n R_1) - Y_0(\lambda_n R_1)J_0(\lambda_n r) \tag{7.83}$$

Please note that eq. (7.83) automatically satisfies the boundary conditions at both R_1 and R_2. Consequently, the initial condition is

$$1 = \sum_{n=1}^{\infty} B_n \frac{Z_0(\lambda_n r)}{J_0(\lambda_n R_1)}. \tag{7.84}$$

The needed coefficients are obtained using orthogonality just as we have seen previously; that is,

$$\int_{R_1}^{R_2} rZ_0(\lambda_n r)Z_0(\lambda_m r)dr = 0 \quad \text{unless } m = n. \tag{7.85}$$

As an exercise the student may wish to demonstrate that $B_1 = 2.166$. *Suggestion:* If you decide to try this, a software package like Mathcad™ will make things much easier!

Some results are included here to assist the student with his/her exploration of this problem. The table immediately below provides the time required for the concentration in the middle of the annular wall ($r = 2$) to fall to the specified value, given $D = 0.001$:

$C(r = 2)$	Time Required
0.95	100.5
0.9	131.7
0.85	159.9
0.8	187.6
0.7	245.7
0.6	310.8
0.5	387.2
0.4	480.3
0.3	600.5
0.2	769.8
0.1	1059.2
0.075	1179.3

7.5.5 Spherical Coordinates

To this point we have said nothing about parabolic equations arising in spherical geometries. Let us now consider transient conduction in a spherical entity; thermal energy is transferred only in the r-direction:

$$\rho C_p \frac{\partial T}{\partial t} = k\left[\frac{1}{r^2}\frac{\partial}{\partial r}\left(r^2\frac{\partial T}{\partial r}\right)\right]. \tag{7.86}$$

We rewrite the equation,

$$\frac{\partial T}{\partial t} = \alpha\left[\frac{\partial^2 T}{\partial r^2} + \frac{2}{r}\frac{\partial T}{\partial r}\right]. \tag{7.87}$$

This PDE eq. (7.87) can be separated resulting in the two ordinary differential equations:

$$g' = -\alpha\lambda^2 g \quad \text{and} \quad f'' + \frac{2}{r}f' + \lambda^2 f = 0.$$

Let's focus our attention on the latter which is a form of Bessel's equation; if we refer to Chapter 5, we find $a = 2$, $b = 0$, $c = 0$, and $s = 1$. Since

$$n = \frac{1}{2}\sqrt{\left(\frac{1-a}{2}\right)^2 - c} = -\frac{1}{2}$$

the solution for this equation will be Bessel functions of ½-order:

$$f = \frac{C_1}{\sqrt{r}}J_{\frac{1}{2}}(\lambda r) + \frac{C_2}{\sqrt{r}}J_{-\frac{1}{2}}(\lambda r). \tag{7.88}$$

But these ½-order Bessel functions can be written in terms of sine and cosine:

$$J_{\frac{1}{2}}(x) = \sqrt{\frac{2}{\pi r}}\sin(x) \quad \text{and} \quad J_{-\frac{1}{2}}(x) = \sqrt{\frac{2}{\pi r}}\cos(x). \tag{7.89}$$

We can arrive at exactly the same place much more directly (and ultimately much more usefully) by noting that it is advantageous to let $T = \theta/r$ in eq. (7.87). This variable change results in the familiar "slab" equation: $\frac{\partial\theta}{\partial t} = \alpha\frac{\partial^2\theta}{\partial r^2}$, for which we know:

$$\theta = C_1\exp(-\alpha\lambda^2 t)[A\sin\lambda r + B\cos\lambda r]. \tag{7.90}$$

Naturally, we need only to divide by r to return to $T(r,t)$. This means that a great many problems of this type in spherical coordinates can be solved by simply adapting appropriate results from solutions worked out for slab problems.

We will illustrate a typical solution procedure for spheres with a detailed example: A solid sphere of radius 3 has a uniform initial temperature of 30°. At $t = 0$, the surface of the sphere is instantaneously heated to 80°. Of course, we already know that $T = \frac{C_1}{r}\exp(-\alpha\lambda^2 t)[A\sin\lambda r + B\cos\lambda r]$. Since T must be finite at the center, we require $B = 0$. If we now define a dimensionless temperature, $\theta = \frac{T-80}{30-80}$, then $\theta = 0$ for $r = R$, and

$$\theta = \sum_{n=1}^{\infty}\frac{A_n}{r}\exp\left(-\alpha{\lambda_n}^2 t\right)\sin\lambda_n r. \tag{7.91}$$

The constants of separation are integer multiples of *pi* (π), $\lambda_n = \frac{n\pi}{R}$, and we use the initial condition to find: $1 = \sum_{n=1}^{\infty}\frac{A_n}{r}\sin\lambda_n r$. This is a Fourier series problem, so we can determine the leading coefficients by $A_n = \frac{2}{R}\int_0^R r\sin\lambda_n r\,dr = -\frac{2R}{n\pi}\cos n\pi$. We note that $\cos(n\pi)$ will be -1 for odd n and $+1$ for even n. A complete solution is now at hand and we will replace R with its value, 3:

$$\frac{T-80}{30-80} = -6\sum_{n=1}^{\infty}\frac{\cos n\pi}{n\pi}\exp\left(-\alpha\frac{n^2\pi^2}{9}t\right)\frac{\sin\frac{n\pi r}{3}}{r}. \tag{7.92}$$

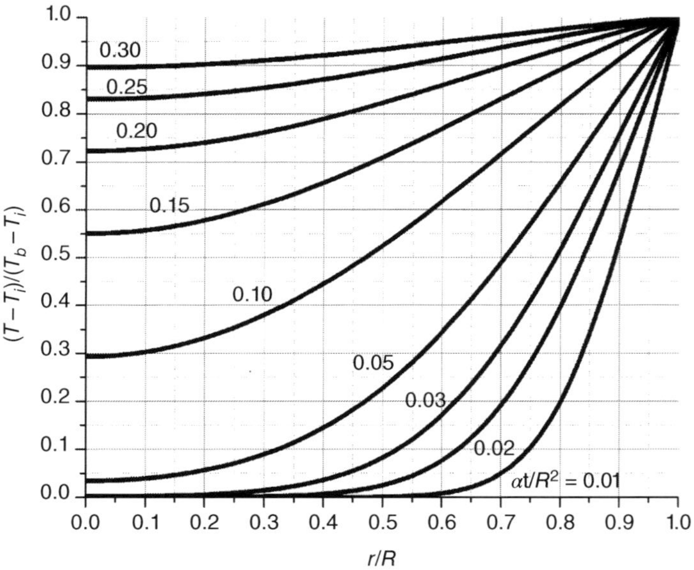

Figure 7.7 Solution for a sphere, initially at a uniform temperature, T_i. At $t = 0$ the temperature at the surface of the sphere is instantaneously elevated to T_b. We have selected $r/R = 0.5$ for our example, which corresponds to the vertical line in the middle of the figure.

We will set $\alpha = 0.005$, $r = 1.5$, vary t from 1 to 1000, and use 50 terms in the series:

t	1	10	100	400	1000
$T(r/R = 0.5)$	29.99999	30.00021	43.36091	72.89833	79.7354

Because this is such an important practical problem, the solution is often presented graphically in the form shown in Figure 7.7. Please observe that the dimensionless temperature used in the figure is different than the definition we used above; it is convenient to have $\theta \to 0$ as t becomes large in the analytic solution, whereas in Figure 7.7, $\theta \to 1$ as t becomes large.

We can use this graphical presentation to confirm our analytic results. For $t = 400$, we find $\frac{\alpha t}{R^2} = 0.222$, which means that $\frac{T-30}{80-30} \cong 0.855$, yielding a temperature of 72.7° (very close to the result we obtained above using the infinite series solution). For additional practice the reader might want to try this scenario: A spherical watermelon with an initial temperature of 90 °F is placed in an ice-water bath at 32 °F. If $\alpha = 0.0015\,\text{cm}^2/\text{s}$, and if $R = 15\,\text{cm}$, when will the center temperature fall to 45 °F? The center of the watermelon corresponds to the left-hand side of Figure 7.7, and we find that $\frac{\alpha t}{R^2} \cong 0.23$. Consequently, the time required should be about 34,500 s, or 9.5 hr.

There is an important variation of the preceding problem that merits discussion. Suppose we have a porous sphere that is initially saturated with some volatile organic compound (VOC); at $t = 0$ the sphere is exposed to an atmosphere that leads to evaporation of the VOC. The fluid (solvent) phase is of great extent such that the concentration at large distance from the surface of the sphere is effectively zero. The governing equation is

$$\frac{\partial C}{\partial t} = D\left[\frac{\partial^2 C}{\partial r^2} + \frac{2}{r}\frac{\partial C}{\partial r}\right], \tag{7.93}$$

and as before we employ the transformation, $\phi = rC$, which yields the slab equation, $\frac{\partial \phi}{\partial t} = D\frac{\partial^2 \phi}{\partial r^2}$. The solution is the same as eq. (7.90) above: $\phi = C_1 \exp(-D\lambda^2 t)[A \sin \lambda r + B \cos \lambda r]$. The concentration must be finite at the center, so $B = 0$ (remember we must divide by r because of the transformation). Loss of the VOC occurs at the surface (at $r = R$) where we equate the fluxes:

$$-D\left.\frac{\partial C}{\partial r}\right|_{r=R} = K(C_{r=R} - C_\infty). \tag{7.94}$$

The reader should show that this Robin's-type boundary condition at the sphere surface leads to the transcendental equation: $\lambda R \cot \lambda R - 1 = -\frac{KR}{D}$. You will recall that we saw this equation previously and noted its importance. In the table provided below, we set $B = -1 + \frac{KR}{D}$; we also reiterate that in the case of spheres, B can be *negative*. So for example, if $B = -1$, we are saying that the overwhelming resistance to mass transfer is in the fluid phase. That is, $K \ll D$.

B	$(\lambda R)_{n=1}$	$n = 2$	$n = 3$	$n = 4$	$n = 5$	$n = 6$
−1	0	4.4934	7.7253	10.9041	14.0662	17.2208
−0.5	1.1656	4.6042	7.7899	10.9499	14.1017	17.2498
0	1.5708	4.7124	7.8540	10.9956	14.1372	17.2788
0.5	1.8366	4.8158	7.9171	11.0409	14.1724	17.3076
1	2.0288	4.9132	7.9787	11.0856	14.2075	17.3364
5	2.6537	5.4544	8.3914	11.4086	14.4699	17.5562
10	2.8628	5.7606	8.7083	11.7027	14.7335	17.7908
50	3.0801	6.1606	9.2420	12.3247	15.4090	18.4953

Consequently,

$$C^* = C - C_\infty = \sum_{n=1}^{\infty} \frac{A_n}{r} \exp\left(-D\lambda_n^2 t\right) \sin \lambda_n r. \tag{7.95}$$

We set $\frac{KR}{D} = \frac{1}{2}$ which means $B = -1/2$, yielding the following set of (the first 12) λR's:

n	1	2	3	4	5	6	7	8
λR	1.1656	4.6042	7.7899	10.9499	14.1017	17.2498	20.3958	23.5407
n			9	10	11	12		
λR			26.6848	29.8284	32.9716	36.1145		

Of course we see that this is not a Fourier series problem, so we must use orthogonality with the initial condition to identify the correct set of A_n's. We will take $R = 1$, $D = 0.001$, and $K = 0.0005$ and let the initial concentration in the interior of the sphere be 1 (uniformly). You may wish to verify that the first few values for A_n are: 0.9576, 0.1741, 0.0726, 0.0522, 0.035, 0.029, 0.022, 0.0196, 0.0162, and so on. We should expect some convergence problems for small t's, and this is why we compiled the first 12 values for λR above. Our concern is borne out by the results provided in the following table using fixed radial position, $r = 1/2$. The results from the truncated infinite series are compared with concentrations computed numerically.

r	t	C Analytic	C Numerical
1/2	50	1.069	0.984
"	100	0.951	0.935
"	200	0.807	0.822
"	400	0.612	0.631
"	800	0.356	0.369
"	1,600	0.120	0.126
"	3,200	0.014	0.015

The discrepancies between the two C-values shown in pairs here are usually less than about 5% except at small t.

7.5.6 Multiple Spatial Variables

We also need to point out that Fourier series solutions can be extended to parabolic problems with *multiple* spatial variables. Suppose we have a slab of material that extends from $x = 0$ to $x = L$ and from $y = 0$ to $y = H$ that has some initial distribution of temperature in the interior, $T(x,y)$. At $t = 0$, all four edges are instantaneously changed to a new temperature which we take to be zero for convenience. The governing equation is

$$\frac{\partial T}{\partial t} = \alpha \left[\frac{\partial^2 T}{\partial x^2} + \frac{\partial^2 T}{\partial y^2}\right], \tag{7.96}$$

and we let $T = f(x)g(y)h(t)$. After dividing by $\alpha\, fgh$, we find

$$\frac{h'}{\alpha h} = \frac{f''}{f} + \frac{g''}{g} = -\lambda^2. \tag{7.97}$$

Of course, we immediately see that $h = C_1 \exp(-\alpha\lambda^2 t)$, and that

$$\frac{f''}{f} = -\lambda^2 - \frac{g''}{g}, \tag{7.98}$$

which gives us a function of x on the left and a function of y on the right. We use the familiar argument and conclude that both sides must be equal to a constant:

$$\frac{f''}{f} = -\lambda^2 - \frac{g''}{g} = -\eta^2. \tag{7.99}$$

Accordingly, we find

$$f = A\sin\eta x + B\cos\eta x \quad \text{and} \quad g = C\sin\sqrt{\lambda^2-\eta^2}y + D\cos\sqrt{\lambda^2-\eta^2}y. \tag{7.100}$$

We arrange the dependent variable, T, such that $T = 0$ for both $x = 0$ and $y = 0$; therefore, $B = D = 0$ leaving us with:

$$T = A\exp(-\alpha\lambda^2 t)\sin\eta x\sin\sqrt{\lambda^2-\eta^2}y. \tag{7.101}$$

When $x = L$, $T = 0$, so $\sin(\eta L) = 0$, resulting in $\eta = m\pi/L$. Similarly, when $y = H$, $T = 0$, so $\sin\sqrt{\lambda^2-\eta^2}H = 0$, requiring that $\sqrt{\lambda^2-\eta^2} = \frac{n\pi}{H}$. The solution for our problem can be written:

$$T = \sum_{m=1}^{\infty}\sum_{n=1}^{\infty} A_{mn}\exp(-\alpha\lambda^2 t)\sin(\eta x)\sin\sqrt{\lambda^2-\eta^2}y, \tag{7.102}$$

with λ and η determined as shown above. Now we will assume that the slab has an initial distribution of temperature, $T(x,y,t = 0) = T_0 xy$ (with $T_0 = 100$), and let $L = H = 1$. This means that the maximum *initial* temperature is in the upper right-hand corner, and for small t's we should see something similar to the contours shown in Figure 7.8.

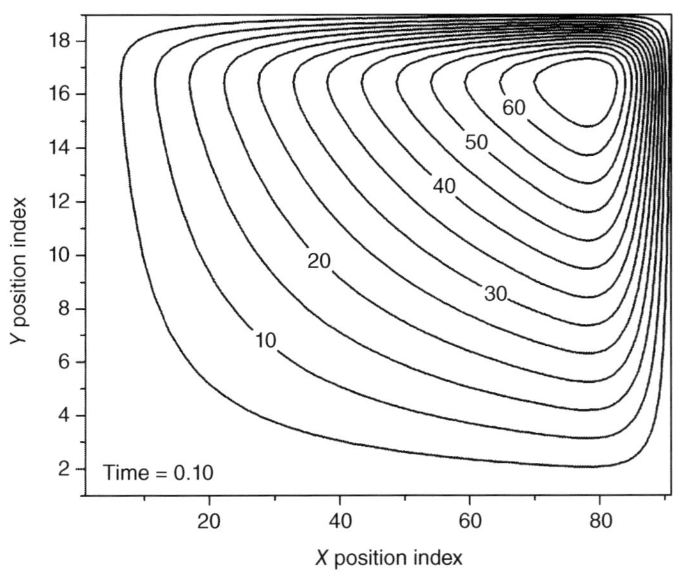

Figure 7.8 Typical temperature distribution in slab at $t = 0.10$ given an initial distribution of $T(x,y,t = 0) = 100xy$; at $t = 0$, all four edges are reduced to $T = 0$.

The coefficients for our double Fourier series are determined from

$$A_{mn} = 400\int_0^1\int_0^1 xy \sin m\pi x \sin n\pi y dxdy. \tag{7.103}$$

A few of the computed coefficients are provided immediately below to allow the student to further explore this situation.

(m,n) =

(1,1)	40.5283	**(1,2)**	−20.268	**(1,3)**	13.5125
(2,2)	10.1321	**(2,3)**	−6.7571	**(2,4)**	5.0673
(3,3)	4.5031	**(3,4)**	−3.3780		
(4,4)	2.5330	**(4,5)**	−2.0268		
(5,5)	1.6211	**(5,6)**	−1.3513		
(6,6)	1.1258				

We want to offer a final word regarding the solution of parabolic PDEs: You may recall that at the beginning of this chapter we mentioned the possibility of time-varying boundary conditions. For example, we can envision a heat (or mass) transfer problem in which a surface condition (or flux) changes with time; as an illustration we might think about the diurnal variation of solar radiation upon an exterior surface. Another possibility is that the concentration of solute in a solvent might increase from zero and approach some maximum value (maybe its solubility) asymptotically. Such problems, when posed correctly, can be solved with *Duhamel's theorem*; the solution is constructed from the *fundamental* solution obtained for *constant* surface conditions. An illustration for the interested reader will be provided (along with a numerical analysis) in the next chapter; a very useful discussion also appears in Carslaw and Jaeger (1959) in Section 1.14.

7.6 Elliptic Equations

Elliptic (or often, *potential*) equations apply to equilibrium phenomena. Familiar situations include steady-state conduction (of thermal energy) in a slab and viscous flow in a duct—these are examples of the Laplace and Poisson equations, respectively. We should keep in mind, however, there are many other applications for *potential* equations including gravitation, electrostatics, and ideal (inviscid) fluid flow. Let us illustrate some of the issues we will encounter while using molecular conduction and diffusion as our example phenomena.

7.6.1 Rectangular Coordinates

We have a two-dimensional slab of material with three sides maintained at some fixed temperature and the upper (top) surface maintained at a different (elevated) temperature (this is exactly the same problem as *slow* viscous flow in a rectangular duct in which motion is driven by an upper surface sliding in the z-, or axial, direction). Since the dependent variable (T) is specified everywhere on the boundary, this is an example of a *Dirichlet problem*. We will place the origin at the lower left-hand corner. The slab extends from $x = 0$ to $x = L$ and from $y = 0$ to $y = H$. The governing equation in this case is:

$$\frac{\partial^2 T}{\partial x^2} + \frac{\partial^2 T}{\partial y^2} = 0. \tag{7.104}$$

As has been our practice, we take: $T = f(x)g(y)$. This leads directly to two ordinary differential equations:

$$f'' + \lambda^2 f = 0 \quad \text{and} \quad g'' - \lambda^2 g = 0. \tag{7.105}$$

The solutions for the two differential equations are:

$$f = A\sin\lambda x + B\cos\lambda x \tag{7.106}$$

and

$$g = C\sinh\lambda y + D\cosh\lambda y. \tag{7.107}$$

We now define the dependent variable T as the difference between T and the temperature of the two sides and the bottom; this gives us $T = 0$ on the left, the right, and the bottom. Since we placed the origin in the lower-left corner, we will use odd functions to build our solution. Thus:

$$T = A\sin\lambda x \sinh\lambda y. \tag{7.108}$$

Note how the boundary condition(s) for $x = 0$ and $y = 0$ are satisfied. Of course, T must also be zero when $x = L$; consequently, it is necessary for $\sin(\lambda L) = 0$ which means $\lambda = n\pi/L$. The solution we seek is therefore:

$$T = \sum_{n=1}^{\infty} A_n \sin\frac{n\pi x}{L} \sinh\frac{n\pi y}{L}. \tag{7.109}$$

It remains for us to identify the A_n's, and we use the boundary condition at the top for this purpose. Let us assume that $T = 100°$ for $y = H$; of course this leads to the Fourier sine series:

$$\frac{100}{\sinh\frac{n\pi H}{L}} = \sum_{n=1}^{\infty} A_n \sin\frac{n\pi x}{L}, \tag{7.110}$$

and by definition,

$$A_n = \frac{200}{L\sinh\frac{n\pi H}{L}} \int_0^L \sin\frac{n\pi x}{L} dx. \tag{7.111}$$

The integral is -2 for odd n (which must be multiplied by $-L/n\pi$) and zero for even n; therefore A_1, A_3, and A_5 are 11.025, 0.00685, and 7.7×10^{-6}, respectively. Alternatively,

$$T = \frac{400}{\pi} \sum_{n=1,3,5,\ldots}^{\infty} \frac{\sin\frac{n\pi x}{L} \sinh\frac{n\pi y}{L}}{n\sinh\frac{n\pi H}{L}}. \tag{7.112}$$

A contour plot of this result, given $L = 10$ and $H = 10$, follows in Figure 7.9. Note that the infinite series in eq. (7.112), converges rapidly. If we fix $(x,y) = (6,6)$, with $n = 1$, we obtain $T = 11.025 \sin\left(\frac{6\pi}{10}\right) \sinh\left(\frac{6\pi}{10}\right) = 33.73$, which is confirmed by the following contour plot.

Now, let us consider how the solution will be impacted if the right-hand boundary has a Neumann boundary condition; for example, it may be insulated such that $\left(\frac{\partial T}{\partial x}\right)_{x=L} = 0$. The preliminary steps are exactly the same, leading us to eq. (7.108): $T = A\sin\lambda x \sinh\lambda y$. The Neumann condition for the right-hand boundary will require that

$$0 = A\lambda\cos\lambda L \sinh\lambda y, \tag{7.113}$$

and therefore, $\lambda = \pi/2L, 3\pi/2L, 5\pi/2L$, etc. Applying the condition at the top of the slab leads to a Fourier sine series just as before; the A_n's are determined by definition and the integration results in:

$$T = \frac{400}{\pi} \sum_{n=0,1,2,3,\ldots}^{\infty} \frac{\sin\frac{(1+2n)\pi x}{2L} \sinh\frac{(1+2n)\pi y}{2L}}{(1+2n)\sinh\frac{(1+2n)\pi H}{2L}}. \tag{7.114}$$

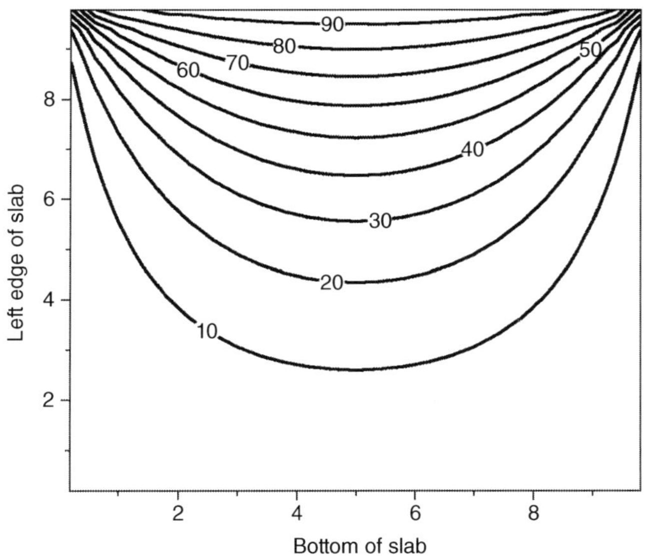

Figure 7.9 Temperature distribution in a two-dimensional slab with the top maintained at 100° and the other three sides at $T = 0°$.

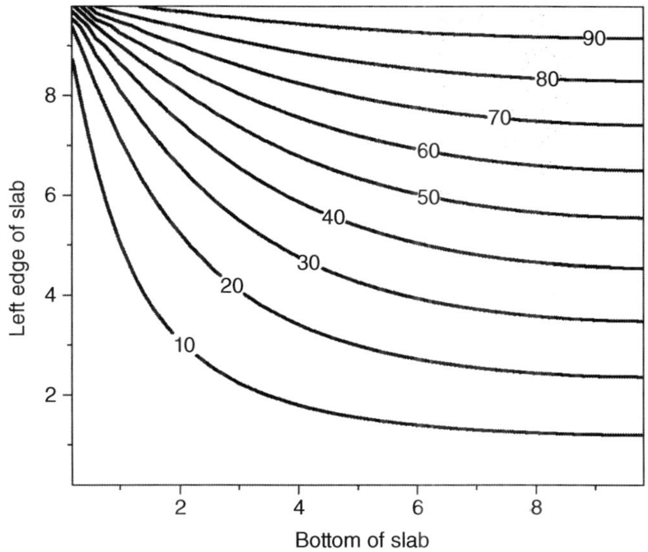

Figure 7.10 Temperature distribution in a two-dimensional slab with the right-hand side insulated, the top edge maintained at 100°, and the left-hand side and bottom at 0°.

Of course in this case we have blocked heat transfer through the right-hand side, so we can expect the temperature contours to be perpendicular to that edge; this is illustrated in Figure 7.10.

We encounter a new level of difficulty when we tackle slab problems with Dirichlet boundary conditions that lack symmetry, i.e., the dependent variable may differ from one side to the other. To illustrate we will consider steady diffusion in porous slab for which $0 \leq x \leq L$ and $0 \leq y \leq H$; the governing elliptic equation is

$$\frac{\partial^2 C}{\partial x^2} + \frac{\partial^2 C}{\partial y^2} = 0. \tag{7.115}$$

The origin is in the lower left-hand corner and we take $C = 0$ on the (vertical) sides, but $C = 1/4$ for the bottom edge and $C = 1/2$ for the top. As usual, we set $C = f(x)g(y)$ to obtain two ordinary differential equations, $f'' + \lambda f = 0$ and $g'' - \lambda g = 0$. These familiar differential equations have the solutions:

$$f = A \sin \lambda x + B \cos \lambda x \quad \text{and} \quad g = c_1 \sinh \lambda y + c_2 \cosh \lambda y. \tag{7.116}$$

Since $C = 0$ for both $x = 0$ and $x = L$, we note that $B = 0$ and $\sin\lambda L = 0$. Therefore we know $\lambda = \frac{n\pi}{L}$ and

$$C = \sum_{n=1}^{\infty} A_n \sin \lambda_n x[c_{1n} \sinh \lambda_n y + c_{2n} \cosh \lambda_n y]. \tag{7.117}$$

The concentration across the bottom, $C(x,y = 0)$ is simply ¼, which yields a familiar Fourier series ($\cosh(0) = 1$, of course):

$$\frac{1}{4} = \sum_{n=1}^{\infty} c_{2n} \sin \lambda_n x, \tag{7.118}$$

and consequently,

$$c_{2n} = \frac{2}{L}\int_0^L \frac{1}{4} \sin \lambda_n x dx = \frac{1}{n\pi}. \tag{7.119}$$

The concentration across the top of the slab, $C(x,y = H)$ is ½, so

$$\frac{1}{2} = \sum_{n=1}^{\infty} [c_{1n} \sinh \lambda_n H + c_{2n} \cosh \lambda_n H] \sin \lambda_n x. \tag{7.120}$$

This is also a Fourier series which provides us with a means for determining c_{1n}; since

$$c_{1n}\sinh\lambda_n H + c_{2n}\cosh\lambda_n H = \frac{2}{L}\int_0^L \frac{1}{2}\sin\lambda_n x dx, \tag{7.121}$$

we have

$$c_{1n} = \frac{\frac{2}{L}\int_0^L \frac{1}{2}\sin\lambda_n x dx - c_{2n}\cosh\lambda_n H}{\sinh\lambda_n H}. \tag{7.122}$$

As we observed above, the c_{2n}'s are known and the first few values are:

n	1	3	5	7	9
c_{2n}	0.3183	0.1061	0.0637	0.0455	0.0354

It is a valuable exercise for the reader to complete the analytic solution and determine concentrations in the porous slab; a few values along the vertical centerline of the slab are given here so you can verify your work. The solution is also provided in graphical form in Figure 7.11 allowing the reader to examine other spatial positions as well.

y/H	0.2	0.3	0.4	0.5	0.6	0.7	0.8	0.9
$C(x = L/2, y)$	0.2484	0.2537	0.2657	0.2856	0.3140	0.3508	0.3945	0.4392

In our previous examples of solutions for the Laplace equation for steady molecular transport in two-dimensional slabs, we placed the origin at the lower left-hand corner. However, there can be definite advantages to placing it at the center; for example, in cases with symmetry, the solution can be built from even functions. And it could also facilitate adapting a solution to a problem in a spherical geometry. Consider a slab extending from $x = -L$ to $x = +L$, and from $y = -H$ to $y = +H$. We maintain constant temperature of 80° for the top and bottom, and 0° for the left- and right-hand sides. All of the initial steps in the solution procedure are the same as used previously except that the solution must be constructed using cos and cosh (that is, we need even functions in both directions):

$$T = B\cos\lambda x\cosh\lambda y. \tag{7.123}$$

Since $T = 0$ for $x = L$, it is necessary that $\lambda L = \pi/2$, $3\pi/2$, $5\pi/2$, etc. Therefore,

$$T = \sum_{n=0}^{\infty} B_n\cos\lambda_n x\cosh\lambda_n y \tag{7.124}$$

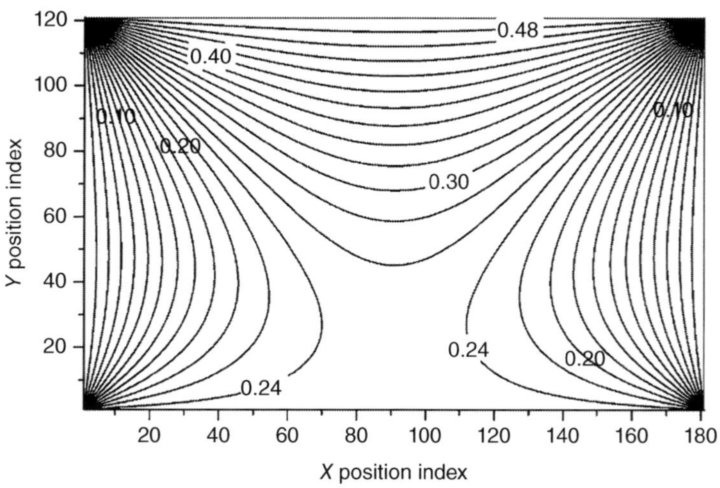

Figure 7.11 Concentration distribution in a rectangular slab where the top is fixed at $C = 1/2$ and the bottom at $C = 1/4$. The vertical centerline is located at an index position of 91. For these computed results, $L/H = 3/2$, and at the position $1/2L$ and $2/3H$, $C = 0.337$.

where

$$\lambda_n = \frac{(1+2n)\pi}{2L}. \tag{7.125}$$

Of course for $y = H$, $T = 80°$, so

$$80 = \sum_{n=0}^{\infty} B_n \cos\frac{(1+2n)\pi x}{2L} \cosh\frac{(1+2n)\pi H}{2L}. \tag{7.126}$$

The leading coefficients can now be determined using Fourier's theorem, resulting in

$$B_n = \frac{320}{(1+2n)\pi \cosh\frac{(1+2n)\pi H}{2L}} \sin\frac{(1+2n)\pi}{2}, \tag{7.127}$$

and the distribution of T in the slab appears as shown in Figure 7.12.

We want to illustrate another slab problem with several rather interesting features. Suppose we have steady conduction in a slab that extends from $x = 0$ to $x = L$, and from $y = 0$ to $y = H$. Once again we place the origin in the lower left-hand corner. Both the left edge and the top surface are insulated so at $x = 0$, $\partial T/\partial x = 0$ and at $y = H$, $\partial T/\partial y = 0$. The right edge ($x = L$) loses heat to the surroundings, so a Robin's-type boundary condition will be applied. The bottom of the slab has a temperature distribution: at $y = 0$, $T = f(x)$, and we will specify $f(x)$ later.

As in our previous examples,

$$T = (A \sin \lambda x + B \cos \lambda x)(C \sinh \lambda y + D \cosh \lambda y). \tag{7.128}$$

Because of the Neumann condition at the left edge of the slab $A = 0$. Things appear a bit more difficult with respect to the y-surfaces (or edges), but there is an easy fix: We take

$$T = B \cos \lambda x \cosh \lambda(H - y). \tag{7.129}$$

Now we use the Robin's-type condition at the right-hand edge, and you may want to verify that it results in the transcendental equation:

$$\lambda L \tan \lambda L = \frac{hL}{k}. \tag{7.130}$$

For our purposes we will take the Biot modulus to be 1 and also let $H = L = 1$. The first ten roots are shown here *along with* the B_n's which will be determined using eq. (7.131) below:

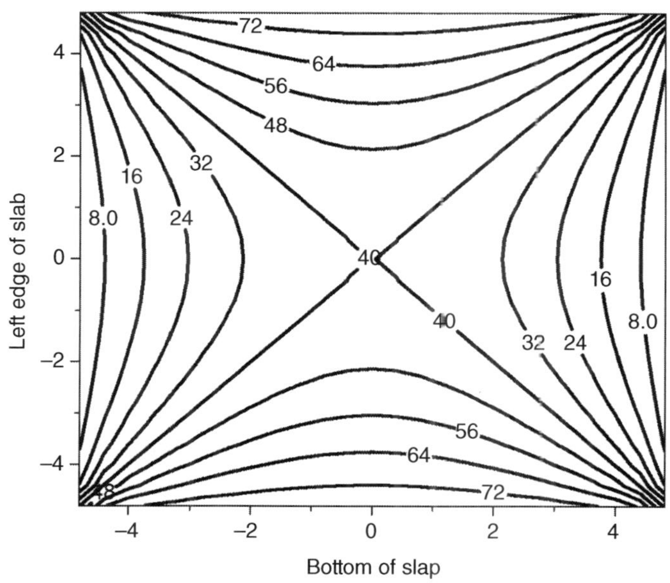

Figure 7.12 Temperature distribution in a slab with symmetric (left–right, and top-bottom) Dirichlet boundary conditions: 80° top and bottom, and 0° for the left-hand and right-hand sides.

n	λ	B_n
1	0.8603	117.681
2	3.4256	−3.9488
3	6.4373	0.02904
4	9.5293	-1.1756×10^{-3}
5	12.6453	$+1.4677\times10^{-5}$
6	15.7713	-7.3618×10^{-7}
7	18.9024	$+1.1305\times10^{-8}$
8	22.0365	-5.6184×10^{-10}
9	25.1725	$+1.0054\times10^{-11}$
10	28.3096	-4.033×10^{-13}

Of course this is another example where we must use orthogonality—this is not a Fourier series problem! We now select $f(x) = 100 + 100x$; therefore, the temperature across the bottom of the slab varies from 100° to 200°. Consequently, we can determine the B_n's from:

$$B_n = \frac{\int_0^1 (100 + 100x)\cos\lambda_n x dx}{\cosh\lambda_n H \int_0^1 \cos^2\lambda_n x dx}. \tag{7.131}$$

An immediate question is raised: Is it possible that the first ten roots of the transcendental equation, along with the accompanying ten values for B_n, accurately portray the solution? If so, then we should be able to extract $f(x)$ for the bottom where $y = 0$ using just B_1 through B_{10}. We will examine this behavior in Figure 7.13.

The transcendental eq. (7.130) occurs so regularly in applied mathematics that it is useful to have ready access to roots. The first 10 roots of $x\tan(x) = C$ have been computed for different values of C (ranging from 0.001 to 100) using Newton–Raphson, and these results are presented in tabular form below.

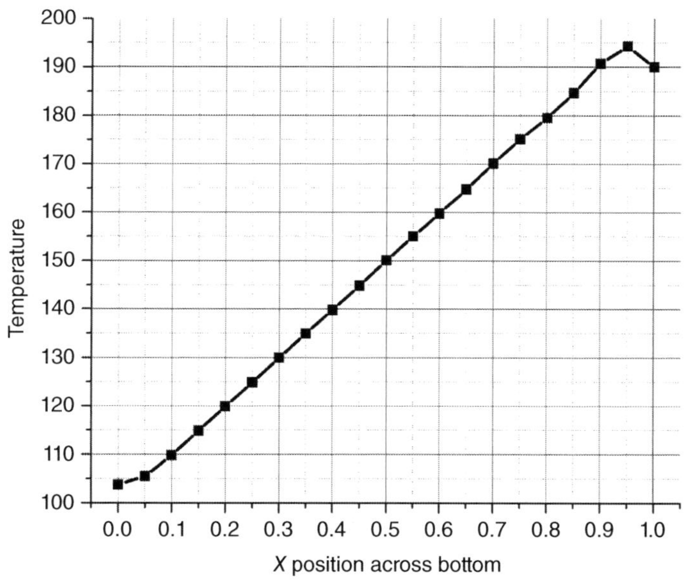

Figure 7.13 The temperature should vary across the bottom from 100° at $x = 0$ to 200° at $x = 1$. Therefore, at $x = 0.3$, $T = 130°$ and for $x = 0.75$, $T = 175°$, etc. Clearly, the intermediate values are reasonably accurate, but the truncated series (with ten terms) does not work so well at the ends. Additional terms will be necessary if we wish to improve our solution across the bottom.

$C = 0.001$	0.002	0.004	0.006	0.008	0.010	0.020
0.03162	0.04471	0.06320	0.07738	0.08932	0.09983	0.14095
3.14191	3.14223	3.14287	3.14350	3.14414	3.14477	3.14795
6.28334	6.28350	6.28382	6.28414	6.28446	6.28478	6.28637
9.42488	9.42499	9.42520	9.42541	9.42563	9.42584	9.42690
12.56645	12.56653	12.56669	12.56685	12.56701	12.56717	12.56796
15.70803	15.70809	15.70822	15.70835	15.70847	15.70860	15.70924
18.84961	18.84966	18.84977	18.84987	18.84998	18.85009	18.85062
21.99119	21.99124	21.99133	21.99142	21.99151	21.99160	21.99206
25.13278	25.13282	25.13290	25.13298	25.13306	25.13314	25.13354
28.27437	28.27440	28.27448	28.27455	28.27462	28.27469	28.27504
$C = 0.040$	**0.060**	**0.080**	**0.1**	**0.2**	**0.4**	**0.6**
0.19868	0.24253	0.27913	0.31105	0.43284	0.59324	0.70507
3.15427	3.16057	3.16685	3.17310	3.20393	3.26355	3.32037
6.28955	6.29272	6.29589	6.29906	6.31485	6.34613	6.37700
9.42902	9.43114	9.43326	9.43538	9.44595	9.46700	9.48793
12.56955	12.57114	12.57273	12.57432	12.58226	12.59811	12.61390
15.71051	15.71178	15.71305	15.71433	15.72069	15.73338	15.74605
18.85168	18.85274	18.85380	18.85486	18.86016	18.87075	18.88132
21.99297	21.99388	21.99479	21.99570	22.00024	22.00932	22.01839
25.13433	25.13513	25.13592	25.13672	25.14070	25.14865	25.15659
28.27575	28.27645	28.27716	28.27787	28.28141	28.28847	28.29554
$C = 0.8$	**1**	**2**	**4**	**6**	**8**	**10**
0.79103	0.86033	1.07687	1.26459	1.34955	1.39782	1.42887
3.37438	3.42562	3.64360	3.93516	4.11162	4.22636	4.30580
6.40740	6.43730	6.57833	6.81401	6.99236	7.12628	7.22811
9.50871	9.52933	9.62956	9.81188	9.96667	10.09492	10.20026
12.62963	12.64529	12.72230	12.86776	12.99881	13.11413	13.21419
15.75869	15.77129	15.83361	15.95363	16.06540	16.16746	16.25936
18.89188	18.90241	18.95468	19.05646	19.15314	19.24354	19.32703
22.02745	22.03650	22.08148	22.16965	22.25450	22.33509	22.41085
25.16452	25.17245	25.21190	25.28961	25.36502	25.43744	25.50638
28.30259	28.30964	28.34478	28.41419	28.48196	28.54756	28.61058
$C = 20$	**40**	**60**	**80**	**100**		
1.49613	1.53250	1.54505	1.55141	1.55525		
4.49148	4.59794	4.63529	4.65428	4.66577		
7.49541	7.66466	7.72592	7.75732	7.77637		
10.51167	10.73341	10.81720	10.86064	10.88713		
13.54198	13.80484	13.90937	13.96435	13.99809		
16.58640	16.87944	17.00262	17.06855	17.10931		
19.64394	19.95755	20.09715	20.17334	20.22083		
22.71311	23.03937	23.19308	23.27878	23.33272		
25.79232	26.12497	26.29056	26.38496	26.44501		
28.88002	29.21432	29.38965	29.49194	29.55774		

We will conclude our discussion of *elliptic equations in rectangular coordinates* with a Poisson-type example; the solution procedure will lend itself to a variety of problems, including heat transfer with constant thermal energy production and pressure-driven viscous flow in a duct. Suppose we have slab of material that extends in the x-direction from $x = -a$ to $x = +a$ and in the y-direction from $y = -b$ to $y = +b$. We have steady conduction accompanied by thermal energy production throughout the interior. The production occurs at a *constant rate* such that

$$k\left[\frac{\partial^2 T}{\partial x^2} + \frac{\partial^2 T}{\partial y^2}\right] + P = 0. \tag{7.132}$$

We will take the temperature at each edge of the slab to be zero. We referred above to the similarity of this problem to steady viscous flow in a duct which is governed by the equation:

$$\mu\left[\frac{\partial^2 v_z}{\partial x^2} + \frac{\partial^2 v_z}{\partial y^2}\right] - \frac{dp}{dz} = 0. \tag{7.133}$$

Indeed, with the no-slip (zero velocity) condition at the walls, this is *exactly* the same problem. In either case, if we can eliminate the inhomogeneity, we may be able to solve this problem just as we have in previous examples in this section. Let us now look at the conduction problem with *constant production* of thermal energy. Consider the effect of adding $-\frac{P}{2k}x^2$ to the usual product of functions of x and y; the result is that $-P/k$ appears on both sides of the equation, eliminating the problem caused by the production term. Consequently,

$$T = -\frac{P}{2k}x^2 + B\cos\lambda x\cosh\lambda y. \tag{7.134}$$

Since $T = 0$ for $x = a$, we conclude that the solution must be written

$$T = \frac{P}{2k}(a^2 - x^2) + \sum_{n=0}^{\infty} B_n \cos\lambda_n x\cosh\lambda_n y, \tag{7.135}$$

where the λ_n's come from $\cos(\lambda A) = 0$, of course. We select $P/k = 100$ and let $a = b = 1$; therefore,

$$\lambda = \frac{(2n+1)\pi}{2} \quad \text{for } n = 0,1,2,3,\ldots. \tag{7.136}$$

This is a Fourier series problem, and the B_n's are determined from

$$B_n = \frac{-100}{\cosh\lambda_n}\int_0^1 (1 - x^2)\cos\lambda_n x dx. \tag{7.137}$$

Of course the maximum temperature will occur in the center of the slab as shown in Figure 7.14.

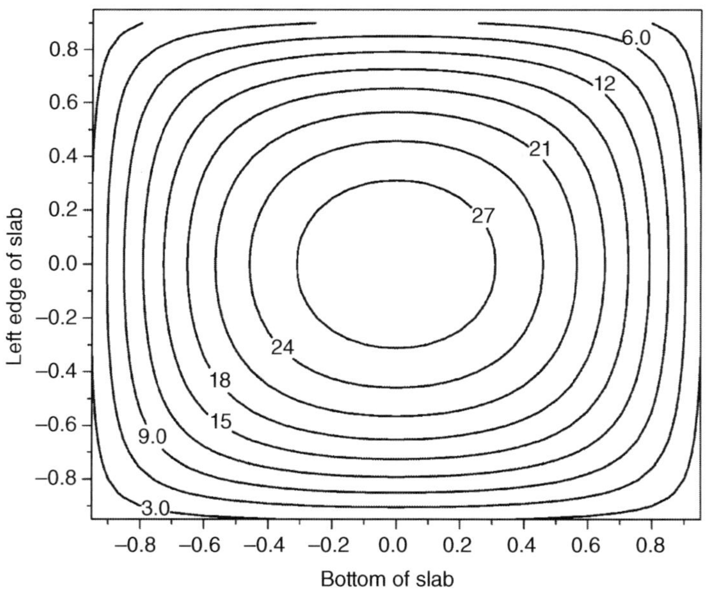

Figure 7.14 Temperature contours in a two-dimensional slab with constant production throughout the interior and the edges maintained at 0°, as computed from eqs. (7.135), (7.136), and (7.137).

7.6.2 Elliptic Equations in Cylindrical Coordinates

The same procedures we employed for the elliptic equations above can be used in *cylindrical coordinates* as well. For example, consider a solid cylinder with diameter, $2R$, that extends from $z = 0$ to $z = L$. The curved surface and the flat, circular end at $z = L$ are always maintained at $T = 0$. The end located at $z = 0$ is maintained at $T = T_0$ for all time. The governing equation for this case is:

$$\frac{\partial^2 T}{\partial r^2} + \frac{1}{r}\frac{\partial T}{\partial r} + \frac{\partial^2 T}{\partial z^2} = 0. \tag{7.138}$$

We take $T = f(r)g(z)$, which results in two ordinary differential equations:

$$f'' + \frac{1}{r}f' + \lambda^2 f = 0 \quad \text{and} \quad g'' - \lambda^2 g = 0. \tag{7.139}$$

The solutions for these two equations are

$$f = AJ_0(\lambda r) + BY_0(\lambda r) \quad \text{and} \quad g = C\sinh \lambda z + D\cosh \lambda z. \tag{7.140}$$

Since the temperature must be finite at the center, $B = 0$. We can accommodate the boundary condition at the end of the cylinder at $z = L$, by taking

$$T = AJ_0(\lambda r)\sinh \lambda(L - z). \tag{7.141}$$

Furthermore, since $T(r = R) = 0$, $J_0(\lambda R) = 0$ (recall that the zeroes for J_0 were given in Section 7.5), and

$$T = \sum_{n=1}^{\infty} A_n J_0(\lambda_n r)\sinh \lambda_n(L - z). \tag{7.142}$$

We must get $T = T_0$ for $z = 0$, so utilizing orthogonality:

$$\frac{T_0}{\sinh \lambda_n L}\int_0^R rJ_0(\lambda_n r)dr = A_n\int_0^R rJ_0{}^2(\lambda_n r)dr. \tag{7.143}$$

The coefficients are therefore given by

$$A_n = \frac{2T_0}{\lambda_n R J_1(\lambda_n R)\sinh \lambda_n L}. \tag{7.144}$$

We will choose $R = 1$, $L = 2$, and $T_0 = 200°$ so we can examine the behavior of this solution; we fix $r = 1/2$ and use increasing values for z, beginning with $z = 1/16$:

z	1/16	1/8	1/4	1/2	1	3/2
$T°$	184.7	158.9	117.6	64.4	19.2	5.3

Now we will explore a variation of the previous problem, a right-circular cylinder with diameter $2R$ that extends from $z = 0$ to $z = L$. But in this case the ends of cylinder will be maintained at $T = 0$ and the curved (convex) surface at T_S:

$$T(r, z = 0) = 0$$

$$T(r, z = L) = 0$$

$$T(r = R, z) = T_S$$

The governing elliptic equation is exactly the same as eq. (7.138), $\frac{\partial^2 T}{\partial r^2} + \frac{1}{r}\frac{\partial T}{\partial r} + \frac{\partial^2 T}{\partial z^2} = 0$, and the separation proceeds just as before:

$$\frac{f'' + \frac{1}{r}f'}{f} = -\frac{g''}{g} = +\lambda^2. \tag{7.145}$$

But we immediately notice that the sign of the constant of separation has been changed. This is an important lesson: with elliptic PDEs it is not always immediately apparent if the constant should be positive or negative. Let us think about this for a moment. If we choose a *negative* constant, then the second of the pair of ordinary differential equations becomes

$$g'' - \lambda^2 g = 0 \text{ with the solution, } g = C_1\sinh(\lambda z) + C_2\cosh(\lambda z).$$

The difficulty is clear: What do we do about the boundary condition at $z = L$? By selecting a positive constant we obtain the pair of second-order ODEs:
$f'' + \frac{1}{r}f' - \lambda^2 f = 0$ and $g'' + \lambda^2 g = 0$ with the solutions:

$$f = AI_0(\lambda r) + BK_0(\lambda r) \quad \text{and} \quad g = C_1 \sin(\lambda z) + C_2 \cos(\lambda z). \tag{7.146}$$

Naturally T must be finite at $r = 0$, so $B = 0$; and g must disappear for $z = 0$, so $C_2 = 0$ as well. This means that

$$T = AI_0(\lambda r)\sin(\lambda z) \tag{7.147}$$

and since $T = 0$ for $z = L$, we must have $sin(\lambda L) = 0$ such that $\lambda_n = \frac{n\pi}{L}$. The solution we seek is

$$T = \sum_{n=1}^{\infty} A_n I_0(\lambda_n r)\sin(\lambda_n z). \tag{7.148}$$

Finally, we take care of the boundary condition at the curved surface, $r = R$, resulting in a familiar Fourier series problem:

$$\frac{T_S}{I_0(\lambda_n R)} = \sum_{n=1}^{\infty} A_n \sin(\lambda_n z). \tag{7.149}$$

The Fourier coefficients are determined by integration:

$$A_n = \frac{2}{L}\int_0^L \frac{T_S}{I_0(\lambda_n R)} \sin(\lambda_n z)dz, \tag{7.150}$$

and the even values are zero of course. The solution for this problem is then:

$$T(r,z) = \frac{4T_S}{\pi}\sum_{n=1}^{\infty} \frac{I_0[(2n-1)\pi r/L]\sin\frac{(2n-1)\pi z}{L}}{(2n-1)I_0[(2n-1)\pi R/L]}. \tag{7.151}$$

For practice the reader may wish to examine the following scenario: First, given $R = 2$, $L = 3$, and $T_S = 100°$, show that the first three values for A_n (1,3,5) are 52.246, 0.4872, and 0.00578, respectively. Then, set $r = 1.5$ and $z = 2.7$ and calculate $T(r,z)$, which should result in 38.218°. We note—merely for confirmation—that the numerical solution to this problem yields 38.22° at $r = 1.5$, $z = 2.7$. A contour plot of the solution is provided in Figure 7.15.

Now we will consider a squat cylinder (with diameter 50% greater than the cylinder's length) with its ends maintained at some constant temperature, and its curved (convex) surface at zero. This situation is a kind of "reciprocal" of the preceding example. The governing equation is once again just eq. (7.138); $\frac{\partial^2 T}{\partial r^2} + \frac{1}{r}\frac{\partial T}{\partial r} + \frac{\partial^2 T}{\partial z^2} = 0$. But in this case we will place the origin at the *center* of the cylinder such that $-H \le z \le +H$ to take advantage of the vertical symmetry. We apply the product method noting that we have carried out this separation before. However, the essential point to recognize in this case is that $T(r = R,z) = 0$, so the two ordinary differential equations we obtain must have solutions:

$$f = AJ_0(\lambda r) + BY_0(\lambda r) \quad \text{and} \quad g = c_1 \sinh(\lambda z) + c_2 \cosh(\lambda z). \tag{7.152}$$

Of course the temperature will be finite at $r = 0$, so $B = 0$, and we note that we need to represent the z-variation with an even function due to the cited vertical symmetry:

$$T = AJ_0(\lambda r)\cosh(\lambda z). \tag{7.153}$$

Since T must disappear at $r = R$, we know $J_0(\lambda R) = 0$, and therefore

$$T = \sum_{n=1}^{\infty} A_n J_0(\lambda_n r)\cosh(\lambda_n z). \tag{7.154}$$

Now suppose the ends of the cylinder ($z = \pm H$) are maintained at 100°:

$$100 = \sum_{n=1}^{\infty} A_n J_0(\lambda_n r)\cosh(\lambda_n H). \tag{7.155}$$

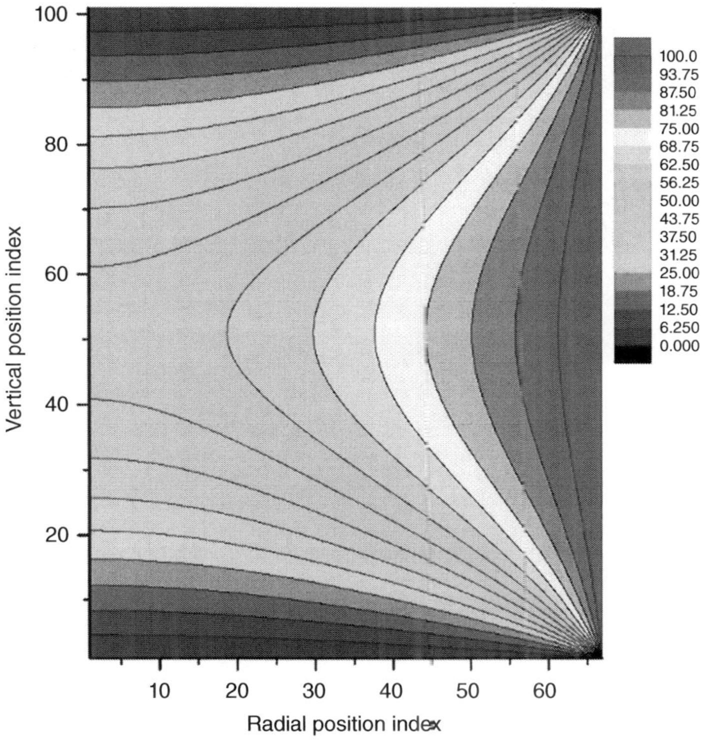

Figure 7.15 Temperature distribution in a circular cylinder with ends maintained at $T = 0°$ and the curved (convex) surface at 100°.

We use orthogonality to determine the leading coefficients just as we have done previously:

$$A_n = \frac{\frac{100}{\cosh(\lambda_n H)} \int_0^R rJ_0(\lambda_n r)dr}{\int_0^R rJ_0^{\,2}(\lambda_n r)dr}, \tag{7.156}$$

resulting in $A_n = \frac{\frac{200}{\cosh(\lambda_n H)}}{\lambda_n R J_1(\lambda_n R)}$. Here is where the table provided in Section 7.5.3 is particularly useful as it allows us to get the necessary values for $J_1\,(\lambda_n R)$. Try setting $H = 2$ and $R = 3$; verify that the correct values for A_1 through A_6 are: 61.978, −5.367, 0.532, −0.056, 0.0062, and −0.00069. In this case the analytic solution converges rapidly enough to be of value to us; for example, we can estimate the temperature at $r = 2.25$ and $z = 1.25$, $T(2.25,1.25) \approx 45.5°$. The numerical solution (shown in Figure 7.16) yields 45.1° at this point, a difference that amounts to just 0.89%. For $r = 1$ and $z = 1$, the analytic solution produces $T(1,1) \approx 64.14°$ and the numerical solution indicates 64.7°; once again, the difference is less than 1%. A graph illustrating the solution for this problem is provided in Figure 7.16, allowing the reader to work through this example and then check his/her results.

We will now turn our attention to an elliptic equation in cylindrical coordinates in which the dependent variable, T, is a function of r and θ. Suppose we have a thin circular disk for which the flat faces are very well insulated. The temperature of the circular edge is maintained at T_1 for $0° \le \theta \le 180°$ and T_2 for $180° \le \theta \le 360°$. Under steady-state conditions the temperature distribution in the disk is governed by

$$\frac{\partial^2 T}{\partial r^2} + \frac{1}{r}\frac{\partial T}{\partial r} + \frac{1}{r^2}\frac{\partial^2 T}{\partial \theta^2} = 0. \tag{7.157}$$

We will let $T = f(r)g(\theta)$ and divide the result by the product, fg:

$$\frac{f''g + \frac{1}{r}f'g + \frac{1}{r^2}fg''}{fg} = 0. \tag{7.158}$$

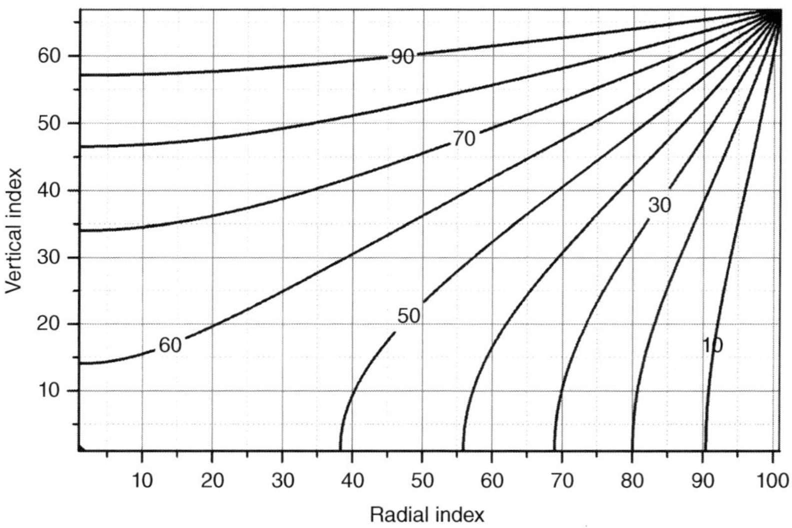

Figure 7.16 Solution for steady conduction in a squat cylinder with the flat, circular ends maintained at 100° and the curved surface at $r = R$ maintained at 0°. The upper half of the cylinder is shown here and the left-hand edge corresponds to $r = 0$.

This permits separation and yields two ordinary differential equations, so in familiar fashion:

$$\frac{r^2 f''}{f} + \frac{rf'}{f} = -\frac{g''}{g} = +\lambda^2. \tag{7.159}$$

The simpler of the two equations is $g'' + \lambda^2 g = 0$ with the solution $g = A\cos\lambda\theta + B\sin\lambda\theta$. The more challenging of the pair is a Cauchy–Euler (or simply Euler) differential equation:

$$r^2 f'' + rf' - \lambda^2 f = 0. \tag{7.160}$$

We will let $f = r^p$ such that $f' = pr^{p-1}$ and $f'' = p(p-1)r^{p-2}$, with the consequence that r^p is common to all three terms and can be divided out resulting in $p^2 - \lambda^2 = 0$. Since $p = \pm\lambda$, we have

$$f = C_1 r^{\lambda} + C_2 r^{-\lambda}. \tag{7.161}$$

Of course T must be finite at the center of the disk where $r = 0$, so $C_2 = 0$. Furthermore, since T has a period of 2π, the parameter λ (the separation constant) must be an integer, n. Thus:

$$T(r,\theta) = \sum_{n=0}^{\infty} r^n [A_n \cos n\theta + B_n \sin n\theta]. \tag{7.162}$$

Recall from the beginning of this chapter that if a function, $f(x)$, has a period $2L$, then the Fourier expansion for $f(x)$ is given by

$$f(x) = \frac{A_0}{2} + \sum_{n=1}^{\infty}\left(A_n \cos\frac{n\pi x}{L} + B_n \sin\frac{n\pi x}{L}\right) \tag{7.163}$$

where

$$A_n = \frac{1}{L}\int_{-L}^{+L} f(x)\cos\frac{n\pi x}{L}dx \quad \text{and} \quad B_n = \frac{1}{L}\int_{-L}^{+L} f(x)\sin\frac{n\pi x}{L}dx. \tag{7.164}$$

Note what happens for $n = 0$: If $f(x) = C_1$ for x between 0 and L, and if $f(x) = C_2$ for x between $-L$ and 0, then $A_0 = C_1 + C_2$. Therefore, $\frac{A_0}{2}$ is the *average* value over the period and in our case that would be $\frac{T_1+T_2}{2}$; you may want to consult Chapter 2 in Spiegel (1974). It is left to the reader to show that $A_n = 0$ if $n \neq 0$. For all $n > 0$,

$$B_n = \frac{1}{\pi}\int_0^{2\pi} T(r = R, \theta)\sin n\theta d\theta. \tag{7.165}$$

We assume that the disk has *unit radius* such that $R = 1$, and break the integral into two parts to accommodate the two different edge temperatures:

$$B_n = \frac{1}{\pi}\int_0^{\pi} T_1 \sin n\theta d\theta + \frac{1}{\pi}\int_{\pi}^{2\pi} T_2 \sin n\theta d\theta = \frac{T_1 - T_2}{n\pi}(1 - \cos n\pi). \tag{7.166}$$

Consequently, our solution has the form

$$T(r,\theta) = \frac{T_1 + T_2}{2} + \sum_{n=1}^{\infty} \frac{(T_1 - T_2)(1 - \cos n\pi)}{n\pi} r^n \sin n\theta. \tag{7.167}$$

We should explore this expression a little further particularly since we anticipate difficulties with series convergence as $r \to 1$. We will take $T_1 = 300°$ and $T_2 = 100°$ and use 50 terms in the series:

θ	$T(r = 1/2,\theta)$	$T(r = 3/4,\theta)$	$T(r = 15/16,\theta)$	$T(r = 31/32,\theta)$
$\pi/8$	230.04	258.54	289.2	294
$\pi/4$	248.13	275.09	294.2	297
$\pi/2$	259.03	281.93	295.9	298
$3\pi/4$	248.13	275.09	294.2	297
π	200	200	200	200
$5\pi/4$	151.87	124.91	105.8	103
$3\pi/2$	140.97	118.07	104.1	102
$7\pi/4$	151.87	124.91	105.8	103

We emphasize that the values in the last column in this table (for $r = 31/32$) are not very accurate. The temperature distribution that results from these calculations is shown in Figure 7.17.

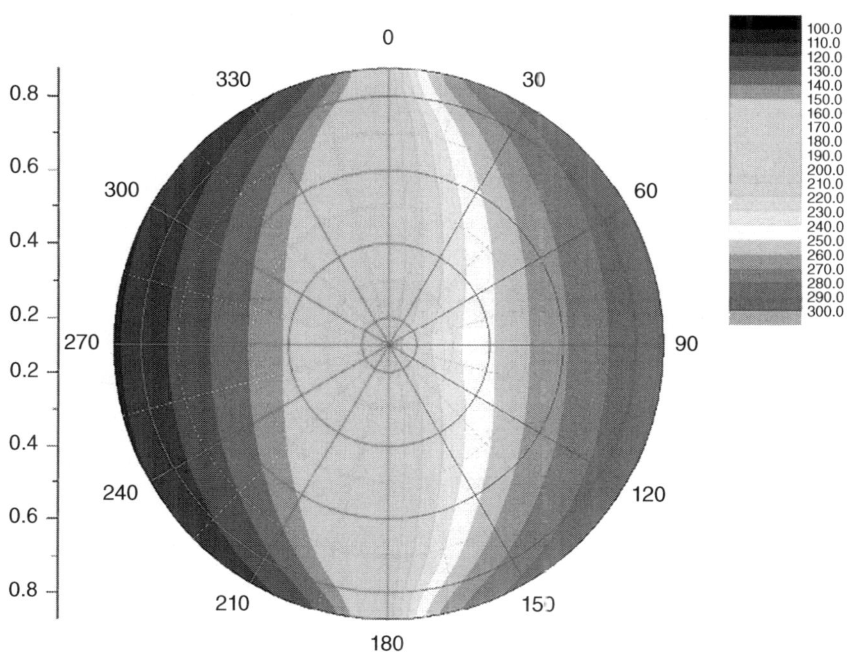

Figure 7.17 Temperature distribution in a flat disk with half of the edge maintained at 300° and half at 100°. $T(r, \theta)$ was computed from the analytic solution using $n = 50$.

7.6.3 Elliptic Equations in Spherical Coordinates

We would also like to explore application of separation of variables to an elliptic equation in spherical coordinates, and we should expect to encounter some additional challenges. Consider:

$$\frac{\partial}{\partial r}\left(r^2\frac{\partial \psi}{\partial r}\right)+\frac{1}{\sin\theta}\frac{\partial}{\partial \theta}\left(\sin\theta\frac{\partial \psi}{\partial \theta}\right)=0. \tag{7.168}$$

Now let $\psi = f(r)g(\theta)$ and divide by the product fg, yielding two ordinary differential equations

$$\frac{r^2f''+2rf'}{f}=-\frac{1}{\sin\theta}\frac{\cos\theta g'+\sin\theta g''}{g}=-\lambda^2. \tag{7.169}$$

The first of the pair is $r^2f''+2rf'+\lambda^2f=0$, and we set $f=r^p$ (as we saw previously) which results in $f'=pr^{p-1}$ and $f''=p(p-1)r^{p-2}$. Just as before, r^p is common to all three terms and can be divided out resulting in $p^2+p+\lambda^2=0$. The roots of this quadratic are $-\frac{1}{2}\pm\sqrt{\frac{1}{4}-\lambda^2}$, and we now define $n=-\frac{1}{2}+\sqrt{\frac{1}{4}-\lambda^2}$, which will be used to replace λ^2, i.e., $n^2+n=-\lambda^2$ or $\lambda^2=-n(n+1)$.

The second ordinary differential equation resulting from separation is $g''+\frac{\cos\theta}{\sin\theta}g'-\lambda^2g=0$, but we will replace λ^2 and rewrite the equation in a more useful form:

$$\frac{d}{d\theta}\left[\sin\theta\frac{dg}{d\theta}\right]+n(n+1)\sin\theta g=0; \tag{7.170}$$

this a second-order equation with variable coefficients. You may remember that we pointed out in Chapter 5 that there is no general procedure that can guarantee success with such equations. In this instance we will employ a transformation with the hope that we can reduce eq. (7.170) to some familiar form for which the solution is known. Let us try the following: set $\cos\theta=x$, such that $\frac{d\theta}{dx}=-\frac{1}{\sin\theta}$ and therefore, $g'=\frac{dg}{d\theta}=-\sin\theta\frac{dg}{dx}$, resulting in

$$\frac{d}{d\theta}\left[-\sin^2\theta\frac{dg}{dx}\right]+n(n+1)\sin\theta g=0. \tag{7.171}$$

Of course, $\sin^2\theta=1-\cos^2\theta=1-x^2$. We also replace $\frac{d}{d\theta}$ with $\frac{d}{dx}$, remembering to multiply by $\frac{dx}{d\theta}$ which is $-\sin\theta$. Finally, we divide out the $\sin\theta$, and obtain

$$(1-x^2)\frac{d^2g}{dx^2}-2x\frac{dg}{dx}+n(n+1)g=0. \tag{7.172}$$

This second-order ODE is referred to as the associated Legendre differential equation, and its solution can be written in terms of Legendre functions of the first and second kinds:

$$g=c_1P_n(x)+c_2Q_n(x). \tag{7.173}$$

If n is an integer, then $P_n(x)$ reduces to the Legendre polynomials which are tabulated in Chapter 4 and illustrated in Figure 7.23. At least, in principle, we have found a solution for this elliptic partial differential equation:

$$\psi=f(r)g(\theta)=\left(Ar^n+\frac{B}{r^{n+1}}\right)[c_1P_n(\cos\theta)+c_2Q_n(\cos\theta)]. \tag{7.174}$$

We note that if the region of interest in the sphere includes the center, then of course $B=0$. It is reasonable to ask about the circumstances under which this solution might be of value. Consider the variation of temperature in the earth's crust arising from the decreasing surface temperature as one moves from the equator to the north pole; Carslaw and Jaeger (1959) observe in Section 9.12 of their book that the temperature distribution in this case is given by:

$$\psi=\sum_{n=0}^{\infty}A_n\left(\frac{r}{R}\right)^nP_n(\cos\theta). \tag{7.175}$$

7.6.4 Neutron Diffusion

Before we leave our discussion of elliptic partial differential equations we should spend a little time discussing neutron diffusion models. We preface these deliberations with a description of the simplifications we are employing to make these models tractable. First, we assume *one group* theory; i.e., all of the neutrons are of a single energy. This is not realistic as a broad range of neutron energies exists in nuclear power operations, and the conservation equation is typically subdivided by integration over finite energy intervals. This process results in a series of coupled elliptic partial differential equations that must generally be solved iteratively. Such problems are routine in nuclear reactor design, and the interested reader is referred to Chapter 3 in Wachspress (1966). Second, we assume that the neutron current is related to the gradient of the neutron flux, ϕ, by Fick's law:

$$J = -D_{ij}\nabla\phi. \tag{7.176}$$

Validation for this relation is taken from the fact that neutrons tend to be transported from regions of high density to regions where they are fewer in number. This is a heuristic argument, but one that leads to familiar elliptic PDEs that can be solved with relative ease. Note that the parameter, D_{ij}, is a second-order tensor—neutron diffusivity is in general, directional. We however will assume that D is single-valued. A usefully complete neutron balance for a finite and homogeneous reactor must account for accumulation, diffusion, absorption, and include a source term; therefore:

$$\frac{\partial\phi}{\partial t} + D\nabla^2\phi = -\sum_a \phi + S, \tag{7.177}$$

where the source term is normally written as: $S = k\sum_f \phi$. By assuming steady state and presuming that the absorption and source terms are linear in ϕ (for the latter, we are assuming that the neutrons produced by chain reaction are directly proportional to the flux), the one-group diffusion model can be written as:

$$\nabla^2\phi = -B^2\phi. \tag{7.178}$$

This is an elliptic (Helmholtz) PDE. B is referred to as the *geometric buckling,* and it is an indicator of how the neutron flux is distorted by leakage in a finite reactor; in a small nuclear reactor the neutron flux will have greater curvature, so B^2 will be larger.

Let us now consider a three-dimensional "reactor" in the form of a cube with

$$-\frac{L}{2} \le x \le \frac{L}{2}, -\frac{L}{2} \le y \le \frac{L}{2}, \quad \text{and} \quad -\frac{L}{2} \le z \le \frac{L}{2}.$$

The appropriate model in this case will take the form:

$$\frac{\partial^2\phi}{\partial x^2} + \frac{\partial^2\phi}{\partial y^2} + \frac{\partial^2\phi}{\partial z^2} = -B^2\phi. \tag{7.179}$$

In familiar fashion we take $\phi = f(x)g(y)h(z)$ and obtain three ordinary differential equations, each with a sinusoidal solution; for example, $\frac{f''}{f} = -B^2 - \frac{g''}{g} - \frac{h''}{h} = -\lambda^2$. Therefore,

$$f = C_1\sin(\lambda x) + C_2\cos(\lambda x). \tag{7.180}$$

However, we placed the origin at the center of the cube, so only an even function can reflect the inherent symmetry; we will also require that the neutron flux disappear at the edge of the cube where $x = L/2$ (we note that this is physically incorrect as ϕ will approach zero only as $x \to \infty$), so: $f = C_2\cos\left(\frac{n\pi x}{L}\right)$ with analogous results for g and h. A negative neutron flux has no physical meaning, so only the fundamental solution ($n = 1$) is of interest and $\lambda = \pi/L$ resulting in $B^2 = 3(\pi/L)^2$ for a cubic reactor with the origin placed at the center.

An analogous development is possible in the case of a cylindrical reactor of finite length:

$$\frac{\partial^2\phi}{\partial r^2} + \frac{1}{r}\frac{\partial\phi}{\partial r} + \frac{\partial^2\phi}{\partial z^2} = -B^2\phi, \tag{7.181}$$

where $0 \le r \le R$ and $-L/2 \le z \le +L/2$. This is another example of a Helmholtz equation, and it can be separated in the usual way, by setting $\phi = f(r)g(z)$, and resulting in

$$\frac{f'' + \frac{1}{r}f'}{f} = -B^2 - \frac{g''}{g} = -\lambda^2. \tag{7.182}$$

For Bessel's differential equation we obtain $f = C_1J_0(\lambda r) + C_2Y_0(\lambda r)$, and of course C_2 must be zero since we cannot have an infinite flux at the center. We proceed similarly with the ODE for g, obtaining: $g = a_1\sin(\eta z) + a_2\cos(\eta z)$. As noted, the neutron flux must be finite at the center of the cylinder, and it must also disappear at both the curved surface ($r = R$) and the (vertical) ends where $z = \pm L/2$:

$$\phi = C_1J_0(\lambda r)\cos(\eta z). \tag{7.183}$$

Therefore, $\cos\left(\frac{\eta L}{2}\right) = 0$ so that $\eta = n\pi/L$. We observed previously that only the fundamental mode is physically tolerable, so the solution must have the form

$$\phi = C_1 J_0\left(\frac{2.4048r}{R}\right)\cos\left(\frac{\pi z}{L}\right). \tag{7.184}$$

You will recognize the constant (2.4048) in the numerator of the argument for the zero-order Bessel function of the first kind, J_0; it corresponds to the first zero and consequently, ϕ disappears when $r = R$. In the case of the finite cylindrical reactor the buckling assumes the form: $B^2 = (2.4048/R)^2 + (\pi/L)^2$.

The same steps can be carried out for a spherical geometry as well, but if we assume there is neither polar nor azimuthal variation in ϕ, then the governing equation is an ODE:

$$\frac{d^2\phi}{dr^2} + \frac{2}{r}\frac{d\phi}{dr} = -B^2\phi, \tag{7.185}$$

with the solution

$$\phi = \frac{A_1}{r}\sin\frac{\pi r}{R}. \tag{7.186}$$

Of course, this means that the buckling for the spherical reactor is $B^2 = \pi^2/R^2$.

7.7 Application to Hyperbolic Equations

7.7.1 The Vibrating String Problem

We pointed out at the beginning of this chapter that hyperbolic equations are usually associated with wave-type phenomena. Separation of variables can be applied to many of these problems and we can illustrate this with the vibrating "string." It is a fascinating sidelight to note that vibrating string problems were solved by Euler and Bernoulli, among others, in the *middle of the eighteenth century*. Thus we have 260 years' worth of experience with elementary hyperbolic problems to draw upon.

Assume our string extends from $x = 0$ to $x = L$ and that both ends are fixed such that $\phi(x = 0) = 0$ and $\phi(x = L) = 0$ for all t. The string is given an initial displacement $\phi(x, t = 0) = s(x)$ where the function $s(x)$ is specified. It may also have a distribution of initial velocity, $\frac{\partial\phi}{\partial t}(x, t = 0) = v(x)$. The equation of interest is:

$$\frac{\partial^2\phi}{\partial x^2} = \frac{1}{c^2}\frac{\partial^2\phi}{\partial t^2}. \tag{7.187}$$

As usual, we take $\phi = f(x)g(t)$, and of course, this hyperbolic PDE meets all of the criteria for separation. Substitution and division by the product, fg, results in:

$$\frac{f''g = \frac{1}{c^2}fg''}{fg}. \tag{7.188}$$

Consequently we find two second-order, ordinary differential equations:

$$f'' + \lambda^2 f = 0 \quad \text{and} \quad g'' + c^2\lambda^2 g = 0. \tag{7.189}$$

Of course, $f = A\sin\lambda x + B\cos\lambda x$, and since f must vanish for both $x = 0$ and $x = L$, we note $B = 0$ and $\lambda = \frac{n\pi}{L}$. We can also see by inspection that

$$g = C\sin c\lambda t + D\cos c\lambda t. \tag{7.190}$$

Therefore,

$$\phi = \sum_{n=1}^{\infty}\sin\lambda_n x[C_n\sin c\lambda_n t + D_n\cos c\lambda_n t]. \tag{7.191}$$

For $t = 0$, the string has some initial displacement: $\phi(x, t = 0) = f(x)$, and consequently,

$$f(x) = \sum_{n=1}^{\infty} D_n\sin\lambda_n x. \tag{7.192}$$

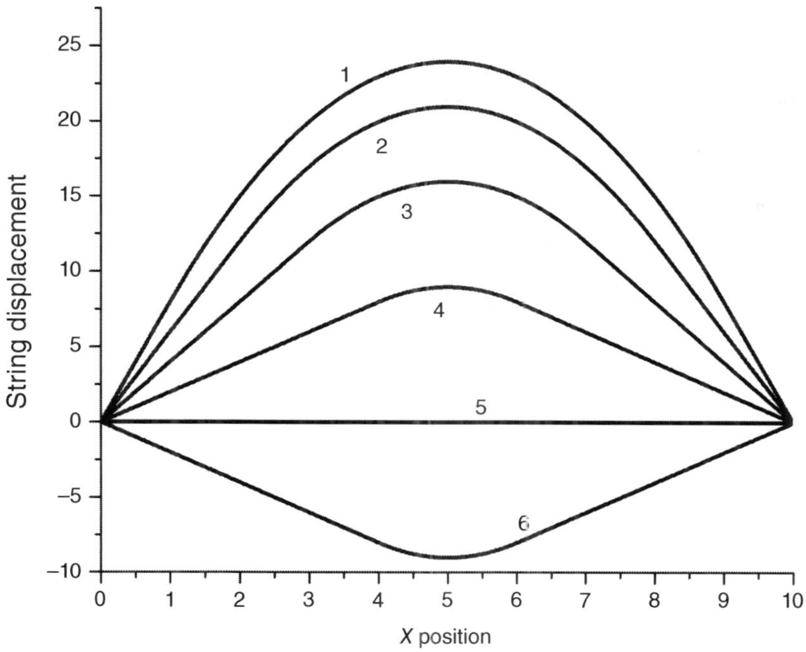

Figure 7.18 String displacement for t's of 1, 2, 3, 4, 5, and 6 from solution of the hyperbolic PDE, eq. (7.187). The initial displacement was $f(x) = 10x - x^2$.

This is merely a Fourier sine series, so

$$D_n = \frac{2}{L}\int_0^L f(x)\sin\frac{n\pi x}{L}dx. \tag{7.193}$$

We can also accommodate a distribution of initial velocity by differentiating, $\frac{\partial \phi}{\partial t}$. Of course, *if the initial velocity is zero*, then $C_n = 0$, and the solution can be written

$$\phi = \sum_{n=1}^{\infty}\frac{2}{L}\int_0^L f(x)\sin\frac{n\pi x}{L}dx \cdot \sin\frac{n\pi x}{L}\cos\frac{cn\pi t}{L}. \tag{7.194}$$

To illustrate what this solution will produce, we take $L = 10$, $c = 1$, and $f(x) = 10x - x^2$; some results are shown in Figure 7.18.

7.7.2 Membranes, Drums, and Chains

Now suppose we have a rectangular *membrane* that spans the gap between fixed supports located at $x = 0$ and $x = L$ and also at $y = 0$ and $y = H$. The appropriate equation is:

$$\frac{\partial^2\phi}{\partial t^2} = s^2\left[\frac{\partial^2\phi}{\partial x^2} + \frac{\partial^2\phi}{\partial y^2}\right] \tag{7.195}$$

We are interested in the response of the membrane to some initial displacement, which may be a function of both x and y, but the initial velocity of the membrane, $\frac{\partial \phi}{\partial t}$, is zero. We begin by letting $\phi = f(x)g(y)h(t)$, and this results in:

$$\frac{h''}{s^2 h} = \frac{f''}{f} + \frac{g''}{g} = -\lambda^2. \tag{7.196}$$

Consequently, $h'' + s^2\lambda^2 h = 0$ and $h = A\sin s\lambda t + B\cos s\lambda t$. The balance of the equation is rewritten so that

$$\frac{f''}{f} = -\frac{g''}{g} - \lambda^2, \tag{7.197}$$

and we note that the left-hand side is a function of x only. Therefore a second constant, η, is introduced such that

$$f'' + \eta^2 f = 0 \quad \text{and} \quad \frac{g''}{g} + \lambda^2 = \eta^2 \text{ or } g'' + (\lambda^2 - \eta^2)g = 0. \tag{7.198}$$

The solutions for the two additional equations are

$$f = C \sin \eta x + D \cos \eta x \tag{7.199}$$

and

$$g = E \sin \sqrt{\lambda^2 - \eta^2} y + F \cos \sqrt{\lambda^2 - \eta^2} y. \tag{7.200}$$

We now have:

$$\phi = (A \sin s\lambda t + B \cos s\lambda t)(C \sin \eta x + D \cos \eta x)(E \sin \sqrt{\lambda^2 - \eta^2} y + F \cos \sqrt{\lambda^2 - \eta^2} y), \tag{7.201}$$

but because the membrane is anchored at the edges, we must have $\phi = 0$ for both $x = 0$ and $y = 0$. This will require that $D = F = 0$. We also know that the initial velocity of the membrane is zero, i.e., $\frac{\partial \phi}{\partial t} = 0$ for $t = 0$; therefore, $A = 0$ as well. What is left is a bit more tractable:

$$\phi = B \cos s\lambda t \sin \eta x \sin \sqrt{\lambda^2 - \eta^2} y. \tag{7.202}$$

Now we have to ensure that the other two (supported) edges are fixed, so $\phi = 0$ for both $x = L$ and $y = H$. For the former, $\sin(\eta L) = 0$ so $\eta = \frac{m\pi}{L}$, and for the latter:
$\sqrt{\lambda^2 - \eta^2} = \frac{n\pi}{H}$. Therefore: $\lambda^2 = \frac{n^2\pi^2}{H^2} + \frac{m^2\pi^2}{L^2}$, and if we take $L = H = 1$, then

$$\phi = \sum_{m=1}^{\infty}\sum_{n=1}^{\infty} B_{mn} \cos(s\pi\sqrt{n^2 + m^2} t) \sin(m\pi x) \sin(n\pi y). \tag{7.203}$$

The membrane has some initial displacement, so for $t = 0$, $\phi = \phi_0(x, y)$:

$$\phi_0(x, y) = \sum_{m=1}^{\infty}\sum_{n=1}^{\infty} B_{mn} \sin(m\pi x) \sin(n\pi y), \tag{7.204}$$

which allows us to determine the needed coefficients:

$$B_{mn} = 4\int_0^1\int_0^1 \phi_0(x, y) \sin(m\pi x) \sin(n\pi y) dx dy. \tag{7.205}$$

Depending upon the form of the function, $\phi_0(x, y)$, it may be easier to evaluate the double integral eq. (7.205) numerically. Algorithms for this purpose are discussed in Chapter 4.

Hyperbolic PDEs in cylindrical coordinates can often be handled in the same way. For example, consider a circular drum with radius, R. At $t = 0$ the drum skin is given some initial displacement, and our interest is how the skin reacts to this disturbance. Our starting point, assuming angular symmetry, is:

$$\frac{\partial^2 \phi}{\partial t^2} = c^2 \left[\frac{\partial^2 \phi}{\partial r^2} + \frac{1}{r}\frac{\partial \phi}{\partial r}\right]. \tag{7.206}$$

We presume that $\phi = f(r)g(t)$, and this substitution produces two ordinary differential equations:

$$g'' + c^2\lambda^2 g = 0 \quad \text{and} \quad f'' + \frac{1}{r}f' + \lambda^2 f = 0, \tag{7.207}$$

with the solutions, $g = C_1 \sin(c\lambda t) + C_2 \cos(c\lambda t)$ and $f = AJ_0(\lambda r) + BY_0(\lambda r)$. Since there is no initial velocity (only displacement), and since the drum skin response must be finite at the center, we have $C_1 = 0$ and $B = 0$:

$$\phi = A \cos(c\lambda t) J_0(\lambda r). \tag{7.208}$$

The drum skin is fixed rigidly at the rim where $r = R$, so $\phi(r = R, t) = 0$; therefore, we note $J_0(\lambda R) = 0$, and thus,

$$\phi = \sum_{n=1}^{\infty} A_n \cos(c\lambda_n t) J_0(\lambda_n r). \tag{7.209}$$

The initial condition—some type of displacement—is $\phi(r, t = 0) = \psi(r)$, so

$$\psi(r) = \sum_{n=1}^{\infty} A_n J_0(\lambda_n r), \tag{7.210}$$

and using orthogonality we find

$$A_n = \frac{\int_0^R \psi(r) r J_0(\lambda_n r) dr}{\int_0^R r J_0^2(\lambda_n r) dr}. \tag{7.211}$$

The integral in the denominator is one we have encountered previously, and we know it is simply $\frac{R^2}{2} J_1^2(\lambda_n R)$. Thus, the solution we are seeking is

$$\phi = \sum_{n=1}^{\infty} \frac{\cos(c\lambda_n t) J_0(\lambda_n r)}{\frac{R^2}{2} J_1^2(\lambda_n R)} \int_0^R \psi(r) r J_0(\lambda_n r) dr. \tag{7.212}$$

Now we take $0 \leq r \leq 1$ and assume the initial displacement is just $\psi(r) = \psi_0(1 - r^2)$. The integral to the far right in eq. (7.212) is then

$$\psi_0 \left[\int_0^R r J_0(\lambda_n r) dr - \int_0^R r^3 J_0(\lambda_n r) dr \right], \tag{7.213}$$

which yields $\psi_0 \left[\frac{2R^2}{\lambda_n^2} J_2(\lambda_n R) \right]$ (since we are requiring $R = 1$). We will now evaluate this infinite series solution to see how useful it will be. We set $c = 1$, $\psi_0 = 1$, and fix $r = ½$ so that $\psi(r) = ¾$; for the first five terms we have:

n	λ_n	$J_1(\lambda_n)$	$J_0(\lambda_n/2)$	$J_2(\lambda_n)$
1	2.4048	+0.5191	+0.670	+0.432
2	5.5201	−0.3403	−0.168	−0.123
3	8.6537	+0.2714	−0.356	+0.063
4	11.7915	−0.2321	+0.121	−0.039
5	14.9309	+0.2065	+0.271	+0.028

At short times we should get something very close to the initial displacement, which of course was $\psi = ¾$ at $r = ½$. For $t = 0.01$, the first five terms in the infinite series are: 0.7427, 0.0234, −0.0162, −0.0025, and 0.00316, yielding $\phi(r = 0.5, t = 0.01) = 0.7506$. The oscillation of the drum skin is driven by $\cos(c\lambda_n t)$, and using just five terms in the series we see the following behavior at $r = ½$:

t	0.01	0.25	0.5	0.75	1.0	1.25	1.5	1.75	2	3
Σ	0.751	0.626	0.251	−0.197	−0.526	−0.710	−0.692	−0.366	0.073	0.421

Next we turn our attention to a heavy rope or chain of length, L, secured at the top end where $x = 0$ but hanging freely where $x = L$. If this chain is given some initial displacement (ϕ_0) at the free end, and then released, it will oscillate (swinging back and forth under the influence of gravity). If the initial displacement is modest, then the governing hyperbolic PDE is:

$$\frac{\partial^2 \phi}{\partial t^2} = g \left(x \frac{\partial^2 \phi}{\partial x^2} + \frac{\partial \phi}{\partial x} \right). \tag{7.214}$$

We can employ separation for this problem by setting $\phi = f(x)h(t)$, resulting in the two ordinary differential equations,

$$h'' + \lambda^2 h = 0 \quad \text{and} \quad x f'' + f' + \frac{\lambda^2 f}{g} = 0. \tag{7.215}$$

For the former, $h = c_1 \sin(\lambda t) + c_2 \cos(\lambda t)$, but the second ODE is a bit more interesting. We can put it in standard form by multiplying by x, and then we can compare the result with Bessel's equation in Chapter 5, noting that $a = 1$, $b = 0$, $c = 0$, $d = \lambda^2$, and $s = 1/2$. Consequently,

$$f = AJ_0\left(2\lambda\sqrt{\frac{x}{g}}\right) + BY_0\left(2\lambda\sqrt{\frac{x}{g}}\right). \tag{7.216}$$

Since there is no initial velocity, only displacement, $c_1 = 0$, and since the oscillations must be finite for all x, $B = 0$ as well. We can conveniently rewrite our solution as

$$\phi = A\cos(\lambda t)J_0\left(2\lambda\sqrt{\frac{L-x}{g}}\right), \tag{7.217}$$

noting that $J_0\left(2\lambda\sqrt{\frac{L}{g}}\right) = 0$ (there can be no lateral displacement at the top where $x = 0$), thus resulting in the infinite series,

$$\phi = \sum_{n=1}^{\infty} A_n \cos(\lambda_n t)J_0\left(2\lambda_n\sqrt{\frac{L-x}{g}}\right). \tag{7.218}$$

The leading coefficients are determined using the initial displacement and orthogonality:

$$A_n = \frac{\int_0^L \phi_0(x)xJ_0\left(2\lambda_n\sqrt{\frac{L-x}{g}}\right)dx}{\int_0^L xJ_0{}^2\left(2\lambda_n\sqrt{\frac{L-x}{g}}\right)dx}. \tag{7.219}$$

It is a useful exercise for the student to determine some values for A_n; let $L = 50$ cm, $g = 980$ cm/s, and set $\phi_0(x) = ax^2$ with $a = 0.002$ which provides an initial horizontal displacement of 5 cm for the free end of the chain. You should find that $\lambda_1 = 5.3232$ and $A_1 \approx 4.53$.

We must emphasize that the above model, with its analytic solution, represents a highly simplified case for pendulum motion. After all, the PDE was linear, the displacement was small, and the system was not dissipative. Driven pendulums suspended on springs provide a far more interesting area of study, in some cases producing chaotic motions that may reveal (in phase space) strange attractors. Of course such systems cannot be analyzed using separation of variables. The interested reader is referred to Baker and Gollub (1990).

7.7.3 The Schrödinger Equation

As part of our discussion of hyperbolic partial differential equations, we want to consider one of the most important developments of twentieth century physics. Erwin Schrödinger (1926) developed the equation that bears his name that, assuming it can be solved, yields the wave function for a system of interest. The Schrödinger equation revolutionized our thinking about particles and waves at small scales, and it made it very clear that classical Newtonian mechanics was wrong—only very slightly wrong at macroscopic scales to be sure—but very wrong at atomic scales. When Schrödinger was able to demonstrate that his model confirmed the discrete electron energy levels for hydrogen atoms that Niels Bohr had predicted more than a decade earlier, the proof was at hand. From that point on, quantum mechanics rapidly expanded our understanding of how atoms and particles behave.

Waves or particles or both? Thomas Young's dual-slit experiment with light 200 years ago revealed interference patterns, a sure sign that light was wave-like in its behavior (contrary to Newton's corpuscular theory, the idea that light consisted of small particles). By the late nineteenth and early twentieth centuries, evidence had begun to accumulate suggesting that there were serious problems with classical mechanics. Atkins (1978) reviews how quantum mechanics resolved some of the phenomena known to be problematic, including: the thermal properties of solids at low temperatures, the UV catastrophe, the photoelectric effect, and atomic and molecular spectra. The latter are particularly persuasive since the spectra reveal that a molecule can only absorb and emit light at specific, discrete frequencies. Then in the 1920s Louis de Broglie suggested that in addition to light (photons) other types of particles (like neutrons and electrons) would have a wavelength related to their momentum. And sure enough, it was discovered that interference patterns were generated by those particles as well; wave-like behavior was demonstrated in a variety of classic experiments, including the dual slit mentioned earlier and the diffraction of electrons from a crystal lattice (the Davisson–Germer experiment was a definitive moment in twentieth century physics).

If a particle like an electron exhibits wave-like behavior, then it would seem that the simple wave eq. (7.187) we considered previously might have an important connection to quantum mechanics. We will begin our exploration of this topic by rewriting eq. (7.187):

$$\frac{\partial^2\phi}{\partial x^2} = \frac{1}{c^2}\frac{\partial^2\phi}{\partial t^2}, \tag{7.220}$$

and then setting $\phi = \psi(x)g(t)$ but with $g = \sin(2\pi ft)$. The result, after we divide by the product of $\psi(x)g(t)$, is

$$\frac{\psi''}{\psi} = -\frac{4\pi^2 f^2}{c^2}\psi \quad \text{or} \quad \frac{d^2\psi}{dx^2} + \frac{4\pi^2 f^2}{c^2}\psi = 0. \tag{7.221}$$

We will refer to ψ as the wave function. Now we assume that the total energy (E) of a particle is the sum of kinetic and potential (U) contributions: $E = \frac{1}{2}mv^2 + U$. We will let the momentum (p) of a particle be represented by $p = mv$, so that $v^2 = \frac{p^2}{m^2}$, and thus: $E = \frac{1}{2m}p^2 + U$. The wavelength, λ, is related to the momentum by Planck's constant: $\lambda = h/p$ (this is known as de Broglie's relationship). Since velocity can be written as the product of frequency and wavelength, $v = f\lambda$, we find

$$f^2 = 2m(E - U)\frac{v^2}{h^2}. \tag{7.222}$$

The characteristic velocity in eq. (7.221), c, is replaced by v, so we obtain:

$$\frac{d^2\psi}{dx^2} + \frac{8\pi^2 m}{h^2}(E - U)\psi = 0. \tag{7.223}$$

This is the *time-independent Schrödinger equation* for one spatial dimension. We can generalize for three-dimensional problems:

$$\nabla^2\psi + \frac{8\pi^2 m}{h^2}(E - U)\psi = 0. \tag{7.224}$$

It is common practice in quantum mechanics to replace h with $\hbar$; we note that $\hbar = \frac{h}{2\pi}$, such that $h^2 = 4\pi^2\hbar^2$. We will restrict our attention momentarily to the one-dimensional case for which E is a positive constant, but $U = 0$. Let us assume that this one-dimensional "box" extends from $x = 0$ to $x = L$, and at these ends, an infinite potential barrier is in place. Accordingly, we have

$$\frac{d^2\psi}{dx^2} + \beta\psi = 0, \quad \text{where } \beta = \frac{8\pi^2 mE}{h^2}. \tag{7.225}$$

Therefore, $(D^2 + \beta)\psi = 0$ such that $(D + i\sqrt{\beta})(D - i\sqrt{\beta}) = 0$, and consequently,

$$\psi = A\sin\sqrt{\beta}x + B\cos\sqrt{\beta}x. \tag{7.226}$$

We require that $\psi = 0$ at both $x = 0$ and $x = L$; from the former, $B = 0$, and from the latter,

$$\sin\sqrt{\beta}L = 0. \tag{7.227}$$

Thus, $\sqrt{\beta}L = n\pi$ and $E_n = \frac{n^2h^2}{8mL^2}$. n is a *quantum number*, and the expression for E_n gives us the discrete, allowable energy states. This in turn means that the behavior of the wave function is described by $\sin\left(\frac{n\pi x}{L}\right)$; when $n = 1$ we get a half-wave over the interval $0 \le x \le L$, when $n = 2$ we get a complete cycle, and for $n = 3$, 1½ cycles, etc.

Before we leave our example of the one-dimensional box, we should make note of an important consequence of quantum mechanics. Let us rewrite the solution for this case in a fully equivalent form:

$$\psi = C_1\exp\left(\frac{2\pi ix}{h}\sqrt{2m(E - U)}\right). \tag{7.228}$$

We now imagine that our particle exits the one-dimensional box and enters the region where the potential barrier is very large (*but not infinite*). Under these conditions, $U > E$ and $2m(E - U)$ is negative. If we factor out the i resulting from the $\sqrt{(-1)}$, we find

$$\psi = C_1\exp\left(-\frac{2\pi x}{h}\sqrt{2m(U - E)}\right). \tag{7.229}$$

This is a representation of the wave function *inside* the potential barrier, and it is important because Max Born recognized that the probability of finding the particle of interest at a particular position, x, was: $|\psi(x)|^2$. This means that based upon eq. (7.229), there would be a small *but finite* probability that the particle in question could be found *outside* the box. This is called *tunneling*, and while it clearly does not apply to a marble sealed in a tin can, it does apply to electrons if they are confined by a finite potential barrier.

Now we turn our attention to the particle in a two-dimensional enclosure, but we let the potential well be circular such that $U = 0$ for $0 \le r \le R$, but $U = \infty$ for $r > R$. For this case,

$$\frac{1}{r}\frac{\partial}{\partial r}\left(r\frac{\partial \psi}{\partial r}\right) + \frac{1}{r^2}\frac{\partial^2 \psi}{\partial \theta^2} + \frac{8\pi^2 m}{h^2}E\psi = 0. \tag{7.230}$$

Let us again try the product method, setting $\psi = f(r)g(\theta)$; the result is:

$$-\frac{r^2}{f}\left(f'' + \frac{1}{r}f'\right) - \frac{8\pi^2 mE}{h^2}r^2 = \frac{g''}{g} = -\eta^2. \tag{7.231}$$

And in the familiar pattern we obtain two ordinary differential equations,

$$g'' + \eta^2 g = 0 \tag{7.232}$$

with

$$f'' + \frac{1}{r}f' + \left(\frac{8\pi^2 mE}{h^2} - \frac{\eta^2}{r^2}\right)f = 0. \tag{7.233}$$

The solution for the first member of this pair is $g = A\sin\eta\theta + B\cos\eta\theta$. We simply choose to have $g = 0$ at $\theta = 0$ (think of the prime meridian at Greenwich) such that $B = 0$. And since $g(0°)$ must be the same as $g(360°)$, it is clear that we can have only integers for η (integer multiples of π). Now we rewrite eq. (7.233) in a more useful form (standard form for Bessel's differential equation):

$$r^2\frac{d^2f}{dr^2} + r\frac{df}{dr} + \left(\frac{8\pi^2 mE}{h^2}r^2 - \eta^2\right)f = 0. \tag{7.234}$$

The solution for eq. (7.234) can be written:

$$f = C_1 J_\eta\left(\sqrt{\frac{8\pi^2 mE}{h^2}}r\right) + C_2 Y_\eta\left(\sqrt{\frac{8\pi^2 mE}{h^2}}r\right). \tag{7.235}$$

However the wave function must be finite at the center (at $r = 0$) so $C_2 = 0$; furthermore, the wave function must be zero at $r = R$, and consequently,

$$J_\eta\left(\sqrt{\frac{8\pi^2 mE}{h^2}}R\right) = 0. \tag{7.236}$$

This constrains E to a series of distinct values (as expected), and we illustrate with the case for which $\eta = 1$. Since the zeros for J_1 occur at 3.83171, 7.01559, 10.17347, 13.32369, 16.47063, and so on., we find

$$\frac{8\pi^2 mE_1}{h^2} = \left(\frac{3.83171}{R}\right)^2, \tag{7.237}$$

or

$$E_1 = \left(\frac{3.83171}{R}\right)^2\frac{h^2}{8\pi^2 m}. \tag{7.238}$$

We have a complete description of the behavior of the wave function inside the circular potential well. An obvious extension of this problem is to make the box three dimensional by allowing the cylinder to have some finite height in the z-direction. Wolfram™ has a very nice demonstration project that illustrates the behavior of the wave function for this case.

In the two previous examples we employed infinitely deep potential wells for elementary one- and two-dimensional problems. We now move to a more realistic particle-in-a-box scenario in three dimensions; we want to develop a model for the hydrogen atom with one electron (or a *hydrogen-like* atom), so our "box" will actually be a spherical shell. But in this case the potential energy of the electron will be $U = -\frac{e^2}{r}$, such that $U(r = 0) = -\infty$, and $U(r \to \infty) = 0$. The partial differential equation for the wave function will be

$$\frac{1}{r^2}\frac{\partial}{\partial r}\left(r^2\frac{\partial \psi}{\partial r}\right) + \frac{1}{r^2\sin\theta}\frac{\partial}{\partial \theta}\left(\sin\theta\frac{\partial \psi}{\partial \theta}\right) + \frac{1}{r^2\sin^2\theta}\frac{\partial^2 \psi}{\partial \phi^2} + \frac{8\pi^2 m}{h^2}\left(E + \frac{e^2}{r}\right)\psi = 0. \tag{7.239}$$

Note that this is a linear PDE and once again we have a candidate for separation; we will apply the product method and see if we can obtain radial and angular components. We will let $\psi(r, \theta, \phi) = \beta(r)\cdot\gamma(\theta, \phi)$ to begin. After a little work we divide by the product of $\beta\gamma$ to obtain:

$$\frac{r^2\frac{d^2\beta}{dr^2} + 2r\frac{d\beta}{dr}}{\beta} + \frac{1}{\gamma\sin\theta}\frac{\partial}{\partial\theta}\left(\sin\theta\frac{\partial\gamma}{\partial\theta}\right) + \frac{1}{\gamma\sin^2\theta}\frac{\partial^2\gamma}{\partial\phi^2} + \frac{8\pi^2 m}{h^2}(Er^2 + e^2 r) = 0. \tag{7.240}$$

The radial and angular portions can be taken to opposite sides and set equal to a constant of separation, say, $-\eta^2$. The $\beta(r)$ result is:

$$r^2\frac{d^2\beta}{dr^2} + 2r\frac{d\beta}{dr} + \frac{8\pi m}{h^2}(Er^2 + e^2 r)\beta = \eta^2\beta. \tag{7.241}$$

It is standard procedure to replace the constant of separation (η^2) in eq. (7.241) with $\ell(\ell+1)$; this is done for reasons that have been explained by Wieder (1973, page 135) and the rationale will become apparent shortly. Now we can also separate the remaining equation for $\gamma(\theta, \phi)$ by setting $\gamma = C(\theta)D(\phi)$. Of course, another constant of separation arises, and we let it be $-\alpha^2$. The polar (θ) part produces

$$\sin^2\theta\frac{d^2C}{d\theta^2} + \sin\theta\cos\theta\frac{dC}{d\theta} + (\sin^2\theta\cdot\ell(\ell+1) - \alpha^2)C = 0, \tag{7.242}$$

and the azimuth (ϕ) part yields

$$\frac{d^2D}{d\phi^2} + \alpha^2 D = 0. \tag{7.243}$$

It is clear that eq. (7.243) is the easiest of the trio, and we recognize

$$D = a_1\exp(-i\alpha\phi) + a_2\exp(+i\alpha\phi). \tag{7.244}$$

Let us turn our attention to eq. (7.242); we let $z = \cos\theta$ such that $\frac{dC}{d\theta} = \frac{dC}{dx}\frac{dx}{d\theta}$ where $\frac{dx}{d\theta} = -\sin\theta$; we also replace $C(\theta)$ with $U(z)$. The result is

$$(1 - z^2)\frac{d^2U}{dz^2} - 2z\frac{dU}{dz} + \left(\ell(\ell+1) - \frac{\alpha^2}{1 - z^2}\right)U = 0, \tag{7.245}$$

which is an *associated Legendre differential equation*, with a solution that can be written in terms of Legendre polynomials. The product of the solutions for eqs. (7.243) and (7.245) is central to quantum mechanics, and it is referred to as a spherical harmonic (presented in normalized form):

$$Y_\ell^\alpha(\theta, \phi) = \varepsilon\left[\frac{(2\ell+1)}{4\pi}\frac{(\ell - |\alpha|)!}{(\ell + |\alpha|)!}\right]^{1/2}\exp(+i\alpha\phi)P_\ell^\alpha(\cos\theta), \tag{7.246}$$

where $\varepsilon = (-1)^\alpha$ for $\alpha \geq 0$. The P_ℓ^α's are associated Legendre functions obtained by differentiation of the Legendre polynomials; e.g., $P_\ell^\alpha = (1 - z^2)^{\alpha/2}\frac{d^\alpha P_\ell}{dz^\alpha}$. Legendre polynomials were introduced in Chapter 4 in our discussion of numerical quadrature, and the first five Legendre polynomials are provided for the reader near the end of this chapter. For the constants of separation appearing in eq. (7.246), $\ell = 0, 1, 2, \ldots.$ and $\alpha = 0, \pm 1, \pm 2, \ldots. \pm \ell$. Lastly we consider the equation for the radial portion of the solution, eq. (7.241). If we divide by r^2 and restrict our attention to "large" values of r, then we get an equation that will be valid asymptotically (with the electron moved very far from the nucleus):

$$\frac{d^2\beta}{dr^2} + \frac{8\pi mE}{h^2}\beta \approx 0. \tag{7.247}$$

Using differential operator notation again, $\left(D^2 + \frac{8\pi mE}{h^2}\right)\beta = (D + i\kappa)(D - i\kappa)\beta = 0$ with $\kappa = \sqrt{\frac{8\pi mE}{h^2}}$. Therefore,

$$\beta = a_1\exp\left(-i\sqrt{\frac{8\pi mE}{h^2}}r\right) + a_2\exp\left(+i\sqrt{\frac{8\pi mE}{h^2}}r\right). \tag{7.248}$$

Obviously, we also need to know how the wave function behaves nearer to the nucleus. We can achieve a little simplification of eq. (7.247) by setting $\beta = \Omega/r$, which results in

$$\frac{d^2\Omega}{dr^2} + \frac{8\pi m}{h^2}\left(E + \frac{e^2}{r}\right)\Omega = \frac{\ell(\ell+1)}{r^2}\Omega. \tag{7.249}$$

This equation can be solved for certain cases, for example, for the infinite spherical potential well, for which one obtains spherical Bessel and Neumann functions. However, in the instance of the hydrogen atom with its Coulomb potential, the equation is solved by expansion in a power series. This is made a little easier by the fact that we know something about the asymptotic radial behavior of the wave function. In principle at least, we can obtain an analytic solution for a particle in a spherical "box" which is the product of the three solutions, subject to any simplifying restrictions imposed. You should make note of the fact that three quantum numbers have appeared quite naturally in the solution procedure; they are referred to as the *principal*, *azimuthal (often called orbital)*, and *magnetic* quantum numbers.

7.7.4 Telegrapher's Equations

The development of the telegraph in the nineteenth century allowed rapid communication between cities, and eventually, between continents. However, attempts to construct and operate a trans-Atlantic cable in the middle of the century were plagued by problems, not the least of which was signal degradation occurring due to inductance and leakage. William Thomson (Lord Kelvin) had also predicted that speed problems expected from the *law of the squares*[1] might result in signal overlap, making it impossible to send messages at a normal telegrapher's pace (around 30 words per minute). And, to much disappointment, it was discovered that Thomson was correct, sending a *single word* required about 10 min. E. O. W. Whitehouse (an English surgeon who became Chief Electrician for the Atlantic Telegraph Company) thought that merely increasing the voltage would cure the problem but that attempt burned up the cable. Of course, signals of different frequencies also travel at different velocities, resulting in distortion and rendering interpretation at the far end of the cable difficult or even impossible. In 1876 Oliver Heaviside published an investigation of signal transmission in wires and presented what has become known as the telegrapher's equations, one for voltage, V, and one for current, I:

$$\frac{\partial V}{\partial x} = -L\frac{\partial I}{\partial t} - RI, \tag{7.250}$$

and

$$\frac{\partial I}{\partial x} = -C\frac{\partial V}{\partial t} - GV. \tag{7.251}$$

R is the distributed resistance, L is the distributed inductance, C is the capacitance, and G is the conductance of the dielectric (for a vacuum $G = 0$). If we differentiate the first equation with respect to x and the second with respect to t, we can combine them to eliminate I:

$$\frac{\partial^2 V}{\partial x^2} - LC\frac{\partial^2 V}{\partial t^2} = (RC + GL)\frac{\partial V}{\partial t} + GRV. \tag{7.252}$$

You will note that the left-hand side of this equation is familiar from previous discussions in this section. The terms on the right-hand side are responsible for signal attenuation and dispersion; *if* $\omega L >> R$ *and if* $\omega C >> G$, then the transmission line is *lossless* and eq. (7.252) can be rewritten as

$$\frac{\partial^2 V}{\partial t^2} - s^2\frac{\partial^2 V}{\partial x^2} = 0, \tag{7.253}$$

where $s = 1/\sqrt{LC}$. This, of course, is just the wave equation considered previously and we have already discovered the types of behavior we can expect from it; for a line with no losses a wave form would be transmitted without degradation. Real transmission lines, however, exhibit signal attenuation and dispersion, and we should consider exactly how the additional terms

1 The law of the squares was based upon an analogy with the conduction equation as applied to a metal conductor where the temperature at one end would be instantaneously raised: $RC\frac{\partial V}{\partial t} = \frac{\partial^2 V}{\partial x^2}$. Using this equation, Thomson estimated that the time required for the current to reach its maximum following application of a voltage impulse would be: $t = \frac{1}{6}RCx^2$. Thus Thomson concluded that for very long cables it would be necessary for R to be small (of course this would require much more copper), and additional insulation would be needed to reduce C (much more gutta percha). Since these changes resulted in a more expensive cable, Thomson's ideas met with considerable resistance.

in eq. (7.252) affect the solution. The PDE is linear and looks like a candidate for separation; letting $V = f(x)g(t)$,

$$f''g - LCfg'' = (RC + GL)fg' + GRfg. \tag{7.254}$$

We divide by $LCfg$, and group terms in the usual manner:

$$\frac{1}{LC}\frac{f''}{f} = \frac{g''}{g} + \left(\frac{R}{L} + \frac{G}{C}\right)\frac{g'}{g} + \frac{GR}{LC}. \tag{7.255}$$

Now we have a function of x on the left and a function of t on the right; as noted we set $s^2 = 1/(LC)$ and obtain:

$$\frac{f''}{f} = \frac{1}{s^2}\left[\frac{g''}{g} + \left(\frac{R}{L} + \frac{G}{C}\right)\frac{g'}{g} + \frac{GR}{LC}\right] = -\lambda^2. \tag{7.256}$$

The solution of the equation for f ($f'' - \lambda^2 f = 0$) is by inspection: $f = A_1 \sin(\lambda x) + A_2 \cos(\lambda x)$. We require $V(x = 0) = 0$ and $V(x = \ell) = 0$, and therefore $A_2 = 0$ and $\lambda_n = n\pi/\ell$. Consequently,

$$f = A_1 \sin(\lambda_n x). \tag{7.257}$$

For the second ODE we obtain

$$g'' + \left(\frac{R}{L} + \frac{G}{C}\right)g' + \left(\frac{GR}{LC}\right)g = -s^2\lambda^2 g. \tag{7.258}$$

Using linear differential operator notation, this equation is written as

$$(D^2 + \alpha D + \beta)g = 0 \tag{7.259}$$

where
$\alpha = \frac{R}{L} + \frac{G}{C}$ and $\beta = \frac{GR}{LC} + s^2\lambda^2$. The roots for eq. (7.259) will be

$$\frac{-\alpha \pm \sqrt{\alpha^2 - 4\beta}}{2} = \frac{-\frac{R}{L} - \frac{G}{C} \pm \sqrt{\left(\frac{R}{L} + \frac{G}{C}\right)^2 - 4\left(\frac{GR}{LC} + s^2\lambda^2\right)}}{2}, \tag{7.260}$$

and therefore the nature of the solution depends upon the sign of the discriminant, $\alpha^2 - 4\beta$; since s in cables is so large, typically about 50% of the speed of light, this difference will be negative resulting in

$$g = c_1 e^{(-a-ib)t} + c_2 e^{(-a+ib)t}. \tag{7.261}$$

We use the Euler relation to rewrite this result in a more useful form:

$$g = c_1 e^{-at}(\cos(bt) + i\sin(bt)) + c_2 e^{-at}(\cos(bt) - i\sin(bt)).$$

This linear combination of sine and cosine can be written compactly as $ce^{-at}\cos(bt - \phi)$, and therefore

$$V = \sum_{n=1}^{\infty} A_n \sin\frac{n\pi x}{\ell} e^{-at}\cos(b_n t - \phi_n). \tag{7.262}$$

For an initial condition we assign some form for $V(x, t = 0)$, for example, a pulse or partial wave, $p(x)$, located at a particular x-position. This wave form is initially stationary, so $\frac{\partial V}{\partial t} = 0$ for $t = 0$. The leading coefficients can now be determined by integration since eq. (7.262) is just a half-range Fourier sine series:

$$A_n = \frac{2}{\ell\cos(\phi)}\int_0^{\ell} p(x)\sin\frac{n\pi x}{\ell}dx.$$

We noted previously that Heaviside realized that signals transmitted over cables will have different wavelengths and travel at different velocities. Thus, signal overlap at the receiving end of a long cable could render a message undecipherable. To combat this problem Heaviside determined that the condition, $\frac{G}{C} = \frac{R}{L}$, should be enforced to minimize distortion. This worked well enough at the end of the nineteenth century, but the conductance, G, in modern cables is quite small such that in long lines $\frac{G}{C} \ll \frac{R}{L}$ making the balance unachievable. Repeaters (devices that receive signals and retransmit them) are now widely used in telephony, telecommunications, radio broadcasting, and so on; in long cables digital repeaters are regularly spaced to maintain readable signals.

7.8 Applications of the Laplace Transform

You may recall from our previous discussion in Chapter 5 that the formal definition of the Laplace transform of a function of time, $f(t)$, is

$$L\{f(t)\} = F(s) = \int_0^{\infty} e^{-st} f(t) dt. \tag{7.263}$$

The effect of course is that a continuous function of time is transformed to the s-plane. In our present context, the characteristic of the Laplace transform that is most important concerns time derivatives. Let $df/dt = f'(t)$:

$$L\{f'(t)\} = sL\{f(t)\} - f(t = 0). \tag{7.264}$$

In other words, we replace the derivative with *multiplication* by s and subtract off the initial condition. It is obviously advantageous to formulate the problem in terms of "deviation" variables such that the initial value of f (for $t = 0$) is *zero*. Let us now see exactly what this will accomplish for us when applied to a parabolic PDE.

Assume we have a transient problem with molecular transport in one spatial direction in a semi-infinite medium such that

$$\frac{\partial \phi}{\partial t} = \kappa \frac{\partial^2 \phi}{\partial x^2}. \tag{7.265}$$

Our boundary conditions will have the form: $\phi(x = 0, t) = 2$ and $\phi(x \rightarrow \infty, t) = 0$, with the initial condition, $\phi(x, t = 0) = 0$. Naturally, this means that for positive t, ϕ (whatever that is) will flow—actually diffuse—into the medium from the left-hand boundary where $x = 0$. Rearranging the equation and applying the Laplace transform yields the *subsidiary* equation:

$$\frac{d^2\phi(s)}{dx^2} - \frac{1}{\kappa} s\phi(s) = 0. \tag{7.266}$$

If this ordinary differential equation can be solved (with the corresponding boundary conditions), then we will obtain the *Laplace transform of the solution* of the PDE. If we can successfully invert that transform, we will find the solution we seek. An exponential expression is found for the ODE:

$$\phi(s) = C_1 \exp\left(-\sqrt{\frac{s}{\kappa}}x\right) + C_2 \exp\left(+\sqrt{\frac{s}{\kappa}}x\right) \tag{7.267}$$

Of course, this result cannot be unbounded in the x-direction, so $C_2 = 0$. We also transform the boundary condition: for $x = 0$, $\phi(s) = \frac{2}{s}$, and therefore,

$$\phi(s) = \frac{2}{s} \exp\left(-\sqrt{\frac{s}{\kappa}}x\right). \tag{7.268}$$

In this case we can turn immediately to a table of Laplace transforms (see the table in Chapter 5 or Section 29 in Abramowitz and Stegun, 1972), finding the pair:

$$\frac{1}{s}\exp(-k\sqrt{s}) \quad erfc\left(\frac{k}{2\sqrt{t}}\right) \tag{7.269}$$

Accordingly, the analytic result we seek is

$$\phi = 2erfc\left(\frac{x}{\sqrt{4\kappa t}}\right). \tag{7.270}$$

Let us illustrate this process again using the very same partial differential equation (applied to the molecular transport of thermal energy), but with a more difficult boundary condition; our semi-infinite slab extends in the x-direction away from the interface located at $x = 0$. This time the end of the slab at $x = 0$ is exposed to a fluid maintained at some elevated temperature, so we write

$$-k\left.\frac{\partial T}{\partial x}\right|_{x=0} = h(T_\infty - T|_{x=0}). \tag{7.271}$$

We divide by $-k$ and transform the boundary condition at $x = 0$:

$$\frac{\partial T(s)}{\partial x} - \frac{h}{k}T(s) = -\frac{hT_\infty}{k}\frac{1}{s}. \tag{7.272}$$

The subsidiary equation is precisely the same as before:

$$\frac{d^2T(s)}{dx^2} - \frac{s}{\alpha}T(s) = 0, \tag{7.273}$$

so

$$T(s) = C_1 \exp\left(-\sqrt{\frac{s}{\alpha}}x\right). \tag{7.274}$$

Of course, $C_2 = 0$ since the transform must remain bounded. We differentiate $T(s)$ with respect to x and set $x = 0$; the transformed boundary condition is then used to find C_1:

$$C_1 = \frac{\frac{hT_\infty}{k}\frac{1}{s}}{\frac{h}{k} + \sqrt{\frac{s}{\alpha}}}. \tag{7.275}$$

Therefore,

$$T(s) = \frac{\frac{hT_\infty}{k}\frac{1}{s}}{\frac{h}{k} + \sqrt{\frac{s}{\alpha}}} \exp\left(-\sqrt{\frac{s}{\alpha}}x\right) = \frac{HT_\infty}{s(H+Q)}\exp(-Qx), \tag{7.276}$$

where $H = h/k$ and $Q = \sqrt{(s/\alpha)}$. Once again this is a form that we can find in a suitable table of Laplace transforms, and returning to the time domain we can write

$$\frac{T}{T_\infty} = erfc\frac{x}{\sqrt{4\alpha t}} - \exp\left(\frac{hx}{k} + \frac{h^2\alpha t}{k^2}\right) erfc\left(\frac{x}{\sqrt{4\alpha t}} + \frac{h}{k}\sqrt{\alpha t}\right). \tag{7.277}$$

The utility of the Laplace transform for these transient conduction–diffusion problems in semi-infinite slabs is apparent. But can the technique also be applied to more difficult problems? The answer is a qualified yes, and we demonstrate this with a problem in which heat is transferred in the z-direction in a cylindrical rod with loss to a fluid surrounding the rod's surface. Heat flow is initiated by raising the temperature of the end of the rod at $z = 0$. The conductivity of the metal rod is large, so that the bulk of the *resistance* to heat flow is on the *fluid side* of the interface; therefore, we assume that the temperature of the solid rod does not vary (much) in the transverse, or, r-direction. An *approximate* model for this process is

$$\rho C_p\frac{\partial T}{\partial t} = k\frac{\partial^2 T}{\partial z^2} - \frac{2h}{R}(T - T_\infty), \tag{7.278}$$

where $\frac{2h}{R}(T - T_\infty)$ accounts for loss at the surface of the rod. We define a new dependent variable, $\theta = T - T_\infty$, and divide by ρC_p, obtaining

$$\frac{\partial\theta}{\partial t} = \alpha\frac{\partial^2\theta}{\partial z^2} - \frac{2h}{\rho C_p R}\theta, \tag{7.279}$$

with the following boundary and initial conditions:

$$\theta(z = 0, t) = \theta_0, \theta(z \to \infty, t) = 0, \text{and } \theta(z, t = 0) = 0.$$

Note that the second boundary condition given here is an *idealization*; for physically real situations the medium that extends in the z-direction will certainly be of finite extent. However, if t is small, the medium may effectively appear to be very "deep" in the z-direction. The subsidiary equation is

$$\frac{d^2\theta(s)}{dz^2} - \frac{2h}{kR}\theta(s) - \frac{s}{\alpha}\theta(s) = 0. \tag{7.280}$$

Once again the solution for this equation can be found in a suitable table of Laplace transforms (e.g., Abramowitz and Stegun, 1972), allowing us to return directly to the time domain:

$$\theta = \frac{1}{2}\left[\exp\left(\sqrt{\frac{2h}{kR}}z\right) erfc\left(\frac{z}{\sqrt{4\alpha t}} + \sqrt{\frac{2h}{\rho C_p R}t}\right) + \exp\left(-\sqrt{\frac{2h}{kR}}z\right) erfc\left(\frac{z}{\sqrt{4\alpha t}} - \sqrt{\frac{2h}{\rho C_p R}t}\right)\right]. \tag{7.281}$$

The reader may recognize that this is *exactly the same problem* as absorption into a quiescent liquid accompanied by a first-order, irreversible chemical reaction.

Now, suppose the solution of the subsidiary equation *cannot* be found in a table of transforms; is there any recourse? One possibility is through application of the *inversion theorem* which requires contour integration. The procedure is described by Carslaw and Jaeger (1959) in Chapter XII, and they provide an illustrative example in Section 12.6. In some cases, it is also possible that the solution of the subsidiary equation can be expanded into a series whose *individual* terms can be found in the table of transforms. For example, consider the quotient, $\cosh(bx)/\cosh(bL)$; we rewrite the hyperbolic functions so that: $\cosh(bx)/\cosh(bL) = \frac{e^{bx}+e^{-bx}}{e^{bL}+e^{-bL}} = \frac{e^{bx}+e^{-bx}}{e^{bL}(1+e^{-2bL})}$, then use the binomial theorem to expand this into the series: $(e^{-b(L-x)} + e^{-b(L+x)})\sum_{n=0}^{\infty}(-1)^n e^{-2nbL}$. This technique can be useful *if* the resulting series converges rapidly enough. Many of the details of this procedure are provided by Carslaw and Jaeger (C&J) in *Operational Methods in Applied Mathematics* (1963). For example, one of the cases they consider is a viscous fluid, initially at rest, contained between parallel planar surfaces. At $t = 0$ a uniform force (F) is applied to the fluid resulting motion in the x-direction:

$$\frac{\partial V}{\partial t} = F + \nu\frac{\partial^2 V}{\partial y^2} \tag{7.282}$$

with $-h \leq y \leq +h$. Applying the Laplace transform,

$$sV(s) = \frac{F}{s} + \nu\frac{d^2V(s)}{dy^2}. \tag{7.283}$$

Of course, $V(s) = 0$ for $y = \pm h$. The solution of the subsidiary equation is:

$$V(s) = \frac{F}{s^2}\left[1 - \frac{\cosh\sqrt{s/\nu}y}{\cosh\sqrt{s/\nu}h}\right]. \tag{7.284}$$

The solution is given by Carslaw and Jaeger (C&J) in Section 75, Chapter VIII; using the inversion theorem and contour integration they obtain the result:

$$V = \frac{F(h^2 - y^2)}{2\nu} - \frac{16h^2F}{\nu\pi^3}\sum_{n=0}^{\infty}\frac{(-1)^n}{(2n+1)^3}\cos\frac{(2n+1)y}{2h}\exp\left[-\frac{\nu\pi^2}{h^2}\left(n+\frac{1}{2}\right)^2 t\right]. \tag{7.285}$$

Students interested in learning more about application of the inversion theorem and contour integration for similar problems will find examples from conduction, mechanical vibrations, hydrodynamics, electromagnetics, and diffusion in C&J.

7.9 Approximate Solution Techniques

Occasionally we will encounter a problem for which the techniques described previously will not work, and an alternative numerical solution is either undesirable or simply not useful for the analyst's purpose. In such cases we may be forced to seek an approximate analytic solution. We do have options in these circumstances, and we will describe a couple of useful approaches here.

Many of the methods that are available to us for this purpose have the same underlying theme: We choose a suitable polynomial that either automatically satisfies the boundary conditions or can easily be made to satisfy them. We then "adjust" the polynomial by determining values for the coefficients that—*in some sense*—give us the best possible performance. We can begin to think about this in the following way: Suppose we have a differential equation,

$$D(\phi) = 0 \text{ where } \phi = f(y) \quad \text{and} \quad a \leq y \leq b. \tag{7.286}$$

We propose a trial function:

$$\phi_{trial} = \phi_0 + \sum_{i=1}^{n} c_i\phi_i(y). \tag{7.287}$$

We define the *residual*, R, as

$$R = D(\phi_{trial}). \tag{7.288}$$

If we could somehow force $R = 0$ for all y between a and b, we would have *the* solution! Of course, that really is not the objective; our aim is to find an analytic approximation that is reasonably accurate and cost effective from the standpoint of time invested. Thus, we will settle for a compromise.

We can illustrate some of the principal ideas with a simple steady-state example from conduction. Imagine a slab of type 347 stainless steel for which one face is maintained at 0 °F and the other at 1,000 °F. Over this temperature range, the thermal conductivity of 347 increases (almost linearly) by more than 60%. We let $k = a + bT$ and note that in rectangular coordinates,

$$\frac{d}{dy}\left[k(T)\frac{dT}{dy}\right] = 0. \tag{7.289}$$

Therefore, the nonlinear differential equation of interest is:

$$(a + bT)\frac{d^2T}{dy^2} + b\left(\frac{dT}{dy}\right)^2 = 0. \tag{7.290}$$

Our boundary conditions for this problem are:

$$\text{at } y = 0, T = 0\ °\text{F}, \quad \text{and}$$

$$\text{at } y = h, T = 1{,}000\ °\text{F}.$$

For convenience, we set $h = 1$ ft, and we arbitrarily propose

$$T = \sum C_n y^n, \text{ such that}$$

$$T = C_0 + C_1 y + C_2 y^2 + C_3 y^3 + \cdots \tag{7.291}$$

If we set $C_0 = 0$, the boundary condition at $y = 0$ is automatically satisfied. We form the residual (R) by truncating eq. (7.291) and substituting the result into eq. (7.290):

$$[a + b(C_1 y + C_2 y^2 + C_3 y^3)](2C_2 + 6C_3 y) + b(C_1 + 2C_2 y + 3C_3 y^2)^2 = R. \tag{7.292}$$

Our task now is to choose values for C_1, C_2, and C_3 that result in the smallest possible value for R. This minimization of R can take several different forms; for example, if we employ a weighting function, $W(y)$, and write

$$\int_0^h W(y) R\,dy = 0, \tag{7.293}$$

we have the method of weighted residuals (MWR). Finlayson (1980) points out that if we use the Dirac delta function for $W(y)$, then we are employing a simple collocation scheme where the residual will be zero at a few select points.

If we force the residual to be zero at the end points and also require eq. (7.291) to satisfy the boundary condition at $y = h$, then we have the three simultaneous algebraic equations:

$$2aC_2 + bC_1^2 = 0 \tag{7.294}$$

$$[a + b(C_1 + C_2 + C_3)](2C_2 + 6C_3) + b(C_1 + C_2 + C_3)^2 = 0 \tag{7.295}$$

and

$$1{,}000 - C_1 - C_2 - C_3 = 0. \tag{7.296}$$

A solution is found by successive substitution:

$$C_1 = 1641.434, \quad C_2 = -920.838, \quad \text{and} \quad C_3 = 279.40.$$

We will also use a fourth-order Runge–Kutta scheme to solve eq. (7.290) numerically for comparison, and both solutions are shown in Figure 7.19.

It is essential that we note exactly what occurred here: We set the residual, R, to zero only at the *end points* of the interval, and this was done strictly for convenience. We cannot generally expect to obtain quantitatively useful results this way.

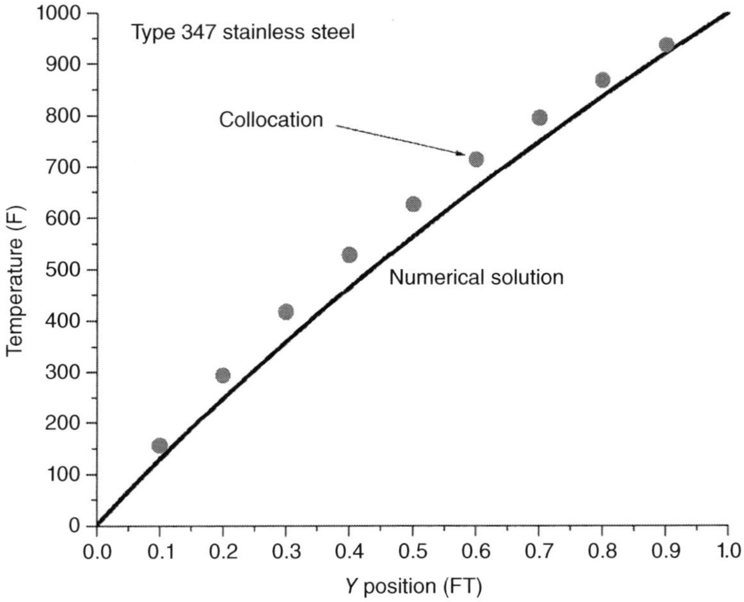

Figure 7.19 Comparison of the exact numerical solution with the collocation result which was obtained by requiring that $R = 0$ at both $y = 0$ and $y = h$. Although the approximate solution exhibits similar behavior, it is rough quantitatively.

7.9.1 Galerkin MWR Applied to a PDE

Let us look at an improved variant of the process described above and apply it to a transient conduction (or diffusion) problem with a temperature (or concentration)-dependent diffusivity. Such an equation might appear as

$$\frac{\partial \phi}{\partial t} = \frac{\partial}{\partial y}\left(\kappa(\phi)\frac{\partial \phi}{\partial y}\right). \tag{7.297}$$

We take $\kappa = 1 + \phi$ for simplicity, which results in the nonlinear partial differential equation,

$$\frac{\partial \phi}{\partial t} = (1+\phi)\frac{\partial^2 \phi}{\partial y^2} + \left(\frac{\partial \phi}{\partial y}\right)^2. \tag{7.298}$$

Our boundary and initial conditions are: $\phi(y = 0, t) = 1$, $\phi(y = 1, t) = 0$, and $\phi(y, t = 0) = 0$. In other words, we have a medium that initially has uniform (or zero) concentration or temperature; we elevate the concentration (or temperature) at the front face ($y = 0$) and diffusion into the medium commences. Finlayson (1980) points out that the method of weighted residuals (MWR) is well suited to this type of problem. We will use the Galerkin technique (named after the Russian mathematician, Boris G. Galerkin) and begin by taking

$$\phi = 1 + b(t)y + c(t)y^2. \tag{7.299}$$

Of course the boundary condition at the far face ($y = 1$) must be satisfied, so

$$-1 = b(t) + c(t) \text{ or } c(t) = -1 - b(t). \tag{7.300}$$

We use this to eliminate $c(t)$ from eq. (7.299) resulting in trial function,

$$\phi = 1 - y^2 + b(t)y(1-y). \tag{7.301}$$

Now we take the original PDE and multiply by the weighting function (which in the Galerkin MWR is taken from the basis, or trial, functions) and integrate from $y = 0$ to $y = 1$. The left-hand side becomes

$$\int_0^1 y(1-y)\frac{\partial \phi}{\partial t}dy, \tag{7.302}$$

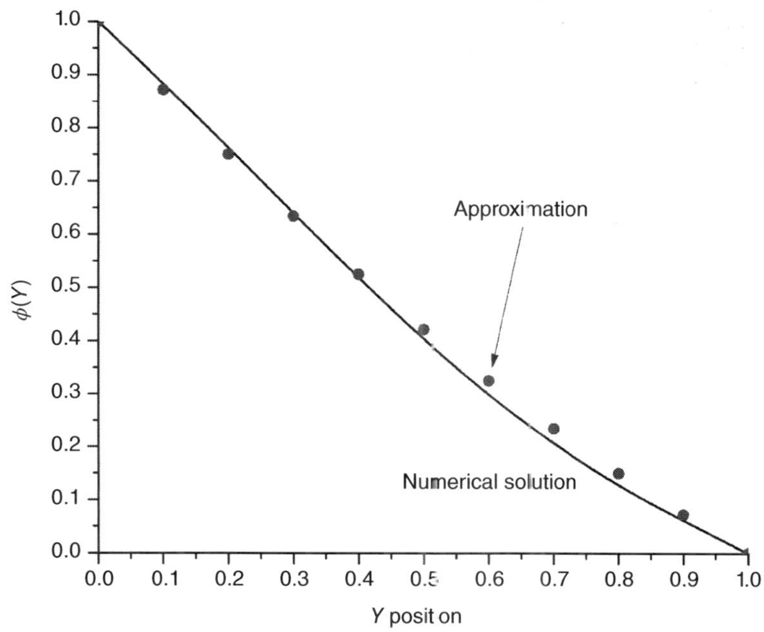

Figure 7.20 Comparison of the numerical solution (solid curve) of the nonlinear PDE for $t = 0.10$ with the approximation obtained from Galerkin MWR (filled circles).

where $\frac{\partial \phi}{\partial t} = b'y(1-y)$. Therefore this integral is simply:

$$\int_0^1 \frac{db}{dt} y^2(1-y)^2 dy = \frac{1}{30}\frac{db}{dt}. \tag{7.303}$$

For the right-hand side of the equation we differentiate the trial function as indicated in eq. (7.303), multiply by $y(1-y)dy$, and obtain:

$$\int_0^1 \{(2 - y^2 + b(y - y^2))(-2 - 2b) + (b - 2y(1 + b))^2\} y(1 - y)dy \tag{7.304}$$

Equating the results of this integration with eq. (7.303) yields a first-order, ordinary differential equation for $b(t)$: $\frac{db}{dt} = -11 - 17b - b^2$. Although this equation could be integrated to produce an analytic solution, it is certainly easier to evaluate $b(t)$ numerically. Finlayson (1980) used the initial value (for b) of -2; however, advanced knowledge of the numerical solution makes it possible to choose a "better" value in terms of the quality of the approximation at advanced times: We will employ $b(t = 0) = -3.325$. This results in $b(t = 0.10) = -1.3125$, and a comparison of the results from the Galerkin method with the actual numerical solution is shown in Figure 7.20.

This relatively simple approach to the solution of a nonlinear partial differential equation has yielded acceptable results requiring determination of only one unknown function of time, $b(t)$. Naturally, the approximation could be improved by simply continuing the expansion eq. (7.299), but at the risk of defeating the whole purpose; remember, our objective is to *quickly* find a suitable analytic approximation for $\phi(y)$.

7.9.2 The Rayleigh–Ritz Method

At the beginning of the twentieth century, Walther Ritz devised a method for approximating eigenfunctions based upon the minimization of certain integrals. This technique, commonly referred to as the Rayleigh–Ritz method, can be used to solve boundary value problems governed by elliptic partial differential equations. The core of the procedure involves application of the *Dirichlet principle*, which concerns identification of a function that minimizes the integral:

$$I = \iiint \left(\phi_x^{\ 2} + \phi_y^{\ 2} + \phi_z^{\ 2}\right) dxdydz. \tag{7.305}$$

It can be written for two dimensions (which will be of direct use to us) as

$$I = \iint |grad\phi|^2 dxdy. \tag{7.306}$$

Let us illustrate how this method works with an example adapted from Chapter XII in Weinberger (1965). Consider the elliptic partial differential equation,

$$\frac{\partial^2\phi}{\partial x^2} + \frac{\partial^2\phi}{\partial y^2} = 0, \tag{7.307}$$

defined over a triangular region for which $x > 0$, $y > 0$, and $x + 2y < 2$. For the bottom of the triangle,

$$\phi(x, y = 0) = x(2 - x), \tag{7.308}$$

which means that the maximum value at the bottom boundary occurs at $x = 1$: $\phi(x = 1, y = 0) = 1$. For the left-hand edge,

$$\phi(x = 0, y) = 0. \tag{7.309}$$

For the hypotenuse,

$$\phi(x = 2 - 2y, y) = 0. \tag{7.310}$$

Our plan is to select trial functions that satisfy the boundary conditions such that

$$\phi \cong \phi_0 + a_1\phi_1 + a_2\phi_2 + \cdots \tag{7.311}$$

We hope to identify the constants, a_1, etc., that give us the *best possible* approximation. This is to be achieved by finding values that give

$$\iint |grad(\phi_{trial} - \phi)|^2 dxdy \tag{7.312}$$

the smallest obtainable value. We will now truncate our approximation, eq. (7.311), and demonstrate how this actually works. We use Dirichlet's principle to formulate an algebraic equation in a_1 (a quadratic in a_1) that can be differentiated and set equal to zero (please keep in mind that if we retained additional trial functions, we would obtain a set of equations for a_1, a_2, etc. by setting the *partial derivatives* equal to zero):

$$\iint |grad(\phi_0 + a_1\phi_1)|^2 dxdy = \iint |grad\phi_0|^2 dxdy + 2a_1 \iint grad\phi_0 \cdot grad\phi_1 dxdy + a_1{}^2 \iint |grad\phi_1|^2 dxdy \tag{7.313}$$

Consequently, the optimal value for the coefficient, a_1, is:

$$a_1 = \frac{\iint grad\phi_0 \cdot grad\phi_1 dxdy}{\iint |grad\phi_1|^2 dxdy}. \tag{7.314}$$

The trial functions must satisfy the boundary conditions, and we will first try: $\phi_0 = x(2 - x - 2y)$ and $\phi_1 = xy(2 - x - 2y)$. Therefore, we will differentiate with respect to x, then y, for the numerator:

$$\frac{\partial\phi_0}{\partial x} = 2(1 - x - y) \quad \text{and} \quad \frac{\partial\phi_0}{\partial y} = -2x \tag{7.315}$$

and

$$\frac{\partial\phi_1}{\partial x} = 2y(1 - x - y) \quad \text{and} \quad \frac{\partial\phi_1}{\partial y} = x(2 - x - 4y). \tag{7.316}$$

The double integral in the numerator is then:

$$\int_0^1 \int_0^{2-2y} (4y(1 - x - y)^2 - 2x^2(2 - x - 4y)dxdy, \tag{7.317}$$

and the form for the denominator is:

$$\int_0^1 \int_0^{2-2y} 4y^2(1 - x - y)^2 + x^2(2 - x - 4y)^2 dxdy. \tag{7.318}$$

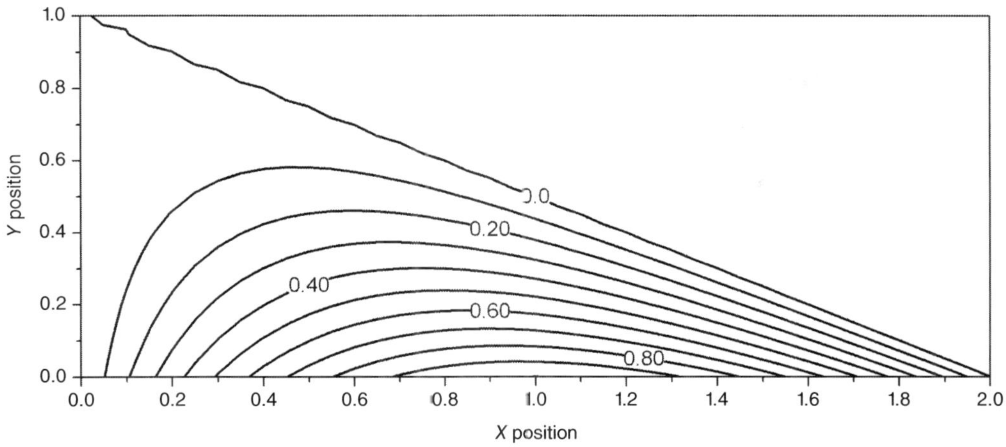

Figure 7.21 Numerical solution for the elliptic partial differential equation used for the Rayleigh–Ritz example.

The resulting quotient is −3/5, resulting in

$$\phi \cong x(2 - x - 2y) - \frac{3}{5}xy(2 - x - 2y) = x\left(1 - \frac{3}{5}y\right)(2 - x - 2y). \tag{7.319}$$

The elliptic PDE given by eq. (7.307) was also solved numerically so that the quality of the Rayleigh–Ritz approximation could be better assessed, and these numerical results are shown in Figure 7.21.

Now we will compute several values from the approximate solution eq. (7.319) for comparison:

x	*Y*	*ϕ(x,y)*
0.1	0.8	0.0156
0.4	0.6	0.1024
0.6	0.3	0.3936
0.8	0.2	0.5632
1.0	0.2	0.5280
1.2	0.2	0.4224
1.4	0.1	0.5264
1.7	0.06	0.2950

You will immediately note that the Rayleigh–Ritz approximation has produced reasonable results; for most of the points provided above the error is less than 10%, and in many cases it is only about 2%. In particular, if we look at $(x,y) = (0.6,0.3)$, the numerical solution yields about 0.4, and for $(x,y) = (1.0,0.2)$ it produces about 0.53 (the approximate solution has an error that is less than 0.4%). We will explore the Rayleigh–Ritz method more thoroughly in Chapter 11.

7.9.3 Collocation

You may have noticed that in the introductory example for this section—the solution of eq. (7.290)—a number of quite arbitrary choices were made; these include the polynomial itself and the location of the points where we forced the residual to be zero. A critical question concerns the placement of the collocation points—an equidistant or haphazard siting is likely to be less than optimal. Therefore, we should contemplate changes to the procedure that may improve the outcome. Suppose we begin by selecting a polynomial that *automatically satisfies the boundary conditions*. In addition, if we use *orthogonal* polynomials, and place the collocation points at the roots of one or more of the terms, we will significantly decrease the burden placed on the analyst. We are now describing what Villadsen and Stewart (1967) called *interior collocation*.

We can illustrate our first improvement with a nonlinear ODE example from fluid mechanics. Suppose we have a non-Newtonian fluid in a rectangular duct, subjected to a constant pressure gradient. If the fluid exhibits power-law

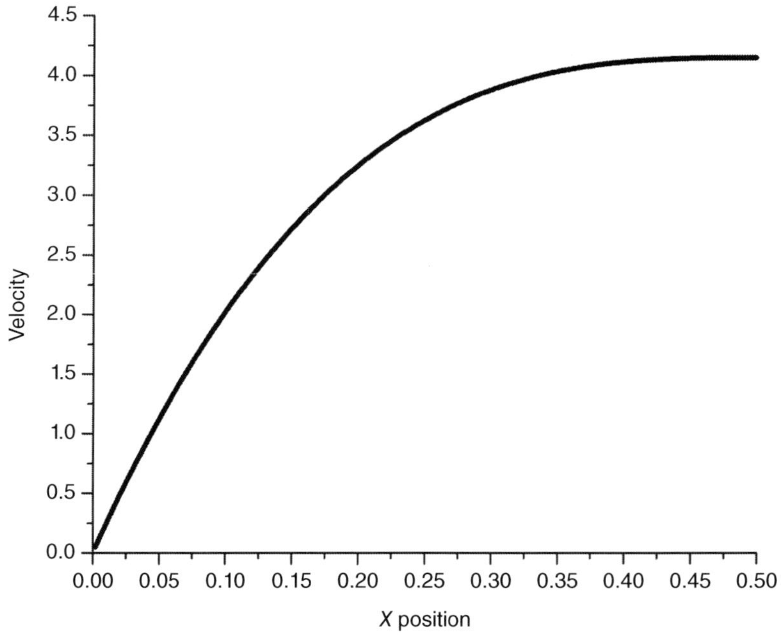

Figure 7.22 Exact numerical solution for non-Newtonian flow through a rectangular duct with $C_0 = -20$.

behavior, then one of the possibilities is

$$\frac{d^2v_x}{dy^2} = -C_0\sqrt{\frac{dv_x}{dy}}. \tag{7.320}$$

The boundary conditions are:

at $y = 0, v_x = 0$ and

at $y = 1, v_x = 0$.

We can avoid any difficulties caused by the sign change on the velocity gradient by noting that at $y = 1/2$, $dv_x/dy = 0$. For this example we choose the polynomial

$$v_x = c_1(y - y^2) + c_2(y - y^2)^2 + c_3(y - y^2)^3 + \cdots \tag{7.321}$$

The conditions at $y = 0$ and $y = 1/2$ are automatically satisfied. We will select $C_0 = -20$ and find the exact numerical solution (provided in Figure 7.22), so we have a basis for comparison.

The reader should complete this example and compare his/her result with the computed profile above. Note that it is necessary for $c_1 = 24.91347$ (the reader should confirm this); this value results in an excellent approximation. How many terms must one retain in the assumed polynomial in order to get accurate results? If we terminate the polynomial with the c_2-term and require the residual to be zero only at $y = 1/4$, we actually find that $c_1 = 46.52397$ and $c_2 = -21.68451$. Although the resulting shape is correct, this solution is unacceptable because the centerline velocity is roughly twice the correct value. It is clear that we should contemplate further improvements for this technique.

Polynomials are said to be orthogonal on the interval (a,b) with respect to the weighting function, $W(x)$, if

$$\int_a^b W(x)P_n(x)P_m(x)dx = 0, \quad \text{where } n \neq m. \tag{7.322}$$

Let us consider the first few Legendre polynomials on the interval $(-1,1)$ for problems that *lack symmetry*. We want to explore how orthogonality may work to our advantage.

$$P_0 = 1 \; P_1 = x \; P_2 = \frac{1}{2}(3x^2 - 1) \; P_3 = \frac{1}{2}(5x^3 - 3x) \; P_4 = \frac{1}{8}(35x^4 - 30x^2 + 3)$$

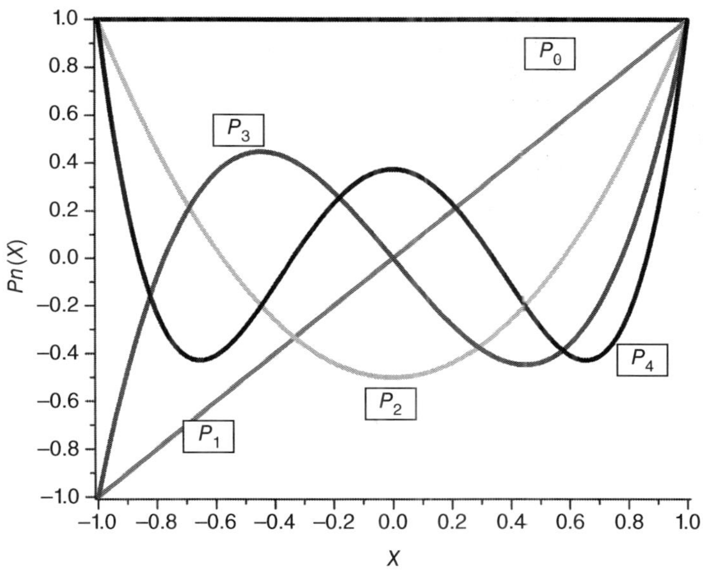

Figure 7.23 Legendre polynomials, P_0 through P_4, on the interval −1 to 1.

You may want to confirm, for example, that

$$\int_{-1}^{+1} P_1(x)P_2(x)dx = \frac{1}{2}\left[\frac{3}{4}x^4 - \frac{1}{2}x^2\right]_{-1}^{+1} = 0. \tag{7.323}$$

The first five Legendre polynomials are shown in Figure 7.23.

Note that if we were to locate collocation points at $x = \pm 1/\sqrt{3}$, then $P_2 = 0$. Similarly, for $x = \pm\sqrt{(3/5)}$, $P_3 = 0$. A further improvement can be obtained by making the dependent variables the functional values *at the collocation points* rather than the coefficients appearing in the polynomial representation. This modified procedure was described by Villadsen and Stewart (1967), Villadsen and Michelsen (1978), and it is also explained very clearly by Finlayson (1980) on pages 73–74.

Let us now suppose that we have a boundary-value problem *with symmetry* about the centerline where

$$\frac{d^2\phi}{dx^2} + f(x,\phi) = 0 \quad \text{and} \quad f(x,\phi) = \phi/(1+x^2) \tag{7.324}$$

The independent variable, x, extends from −1 to 1, and the field variable, ϕ, has a set value at the end points, $\phi(\pm 1) = 1$. Naturally, at the centerline, $\frac{d\phi}{dx} = 0$. Accordingly, we propose

$$\phi = \phi(\pm 1) + (1 - x^2)\sum C_n P_n, \tag{7.325}$$

where the P_n's are Jacobi polynomials for a slab:

$n = 0$	1	
$n = 1$	$(1-5x^2)$	±0.447214
$n = 2$	$(1-14x^2+21x^4)$	±0.2852315, ±0.7650555
$n = 3$	$(1-27x^2+99x^4-85.8x^6)$	±0.209299, ±0.5917, ±0.87174

At this point, eq. (7.325) is substituted into eq. (7.324) to form the residual. We can solve this set of equations for the coefficients (the C_n's), or we can develop an alternative set of equations written in terms of the functional values (ϕ_n's) at the collocation points. The reader is encouraged to try both approaches for this example. If we use *only* P_0 and P_1, the maximum error will be about 10%. Here are some exact values to use for verification: $x = \pm 0.7$, $\phi = 1.3232 : x = \pm 0.4$, $\phi = 1.5658 : x = \pm 0.1$, $\phi = 1.6880$. You may also find it useful to know that $\phi'(x = -1) = 1.170198$ and $\phi(x = 0) = 1.69648$.

7.9.4 Orthogonal Collocation for Partial Differential Equations

Orthogonal collocation has also been used to solve elliptic partial differential equations of the form:

$$\frac{\partial^2 \phi}{\partial x^2} + \frac{\partial^2 \phi}{\partial y^2} = f(x,y), \tag{7.326}$$

on the unit square, $x(0,1)$ and $y(0,1)$. Examples of the method's application are provided by Villadsen and Stewart (1967), Houstis (1978), and Prenter and Russell (1976). Please note that an elliptic equation for any rectangular region $x(a,b)$ and $y(c,d)$, can be mapped into the unit square by employing the transformation,

$$x \to \frac{x-a}{b-a} \quad \text{and} \quad y \to \frac{y-c}{d-c}.$$

This broadens the applicability of the technique considerably. Now, let us suppose for illustration that eq. (7.326) has a solution given by

$$\phi = 3e^x e^y (x - x^2)(y - y^2), \tag{7.327}$$

which can be plotted to yield the results shown in Figure 7.24:

Prenter and Russell (1976) solved this problem using bicubic Hermite polynomials, and their results indicate very favorable performance relative to the Ritz–Galerkin method. Furthermore, in some cases the use of collocation with Hermite polynomials has outperformed solution of elliptic equations by the finite difference method. Section 22 in Abramowitz and Stegun (1972) is a good starting point for the reader interested in the use of Hermite polynomials.

In an example provided by Villadsen and Stewart (1967), the Poisson equation,

$$\frac{\partial^2 \phi}{\partial x^2} + \frac{\partial^2 \phi}{\partial y^2} = -1, \tag{7.328}$$

(for Poiseuille flow through a duct) was solved on the square $(-1 \le x \le +1)$, $(-1 \le y \le +1)$ by taking

$$\phi = (1 - x^2)(1 - y^2) \sum \sum A_{ij} P_i(x^2) P_j(y^2) \tag{7.329}$$

If the expansion is limited to the Jacobi polynomial, $P_1 = (1 - 5x^2)$, and the collocation point is placed at $(x_1, y_1) = (0.447214, 0.447214)$, then

$$\phi \cong \frac{5}{16}(1 - x^2)(1 - y^2). \tag{7.330}$$

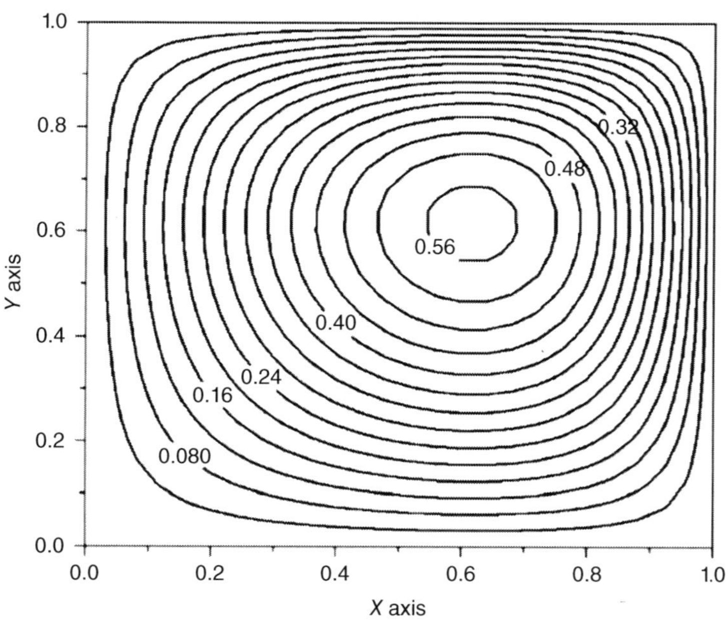

Figure 7.24 Solution for the elliptic partial differential equation, $\frac{\partial^2 \phi}{\partial x^2} + \frac{\partial^2 \phi}{\partial y^2} = 6xye^x e^y(xy + x + y - 3)$.

This truncated approximation is surprisingly good; it yields a center point velocity of 0.3125, whereas a highly accurate numerical solution produces 0.29466. Thus, the value of ϕ from eq. (7.330) at the center of the duct is about 6% too large. Villadsen and Stewart refined this rough solution by including $P_2 = (1-14x^2+21x^4)$ in the expansion with the three collocations points located at $(x,y) \rightarrow (0.2852315, 0.2852315)$, $(0.7650555, 0.2852315)$, and $(0.7650555, 0.7650555)$. The improved result was:

$$\phi \cong (1-x^2)(1-y^2)[0.31625 - 0.013125(1-5x^2+1-5y^2) + 0.00492(1-5x^2)(1-5y^2)] \quad (7.331)$$

The value of ϕ at the center obtained from eq. (7.331) is 0.308 which is a modest improvement over the initial truncation eq. (7.330). It is fair for us to observe that the obvious symmetry in this example lends itself to an approximate solution by collocation. We know that viscous flow in a rectangular duct must be approximately parabolic in both directions and, in fact, one could construct a reasonable solution from many different combinations of *even* functions. For example, take a few moments to consider

$$\phi \cong a_1(1.54308 - \cosh(x))(1.54308 - \cosh(y)) + . \quad (7.332)$$

Set $a_1 = 1$ and try a few (x,y) pairs in a truncated eq. (7.332); are the results reasonable?

Several collocation schemes for elliptic partial differential equations are available through a FORTRAN-based system called ELLPACK. The development of this software was initiated in 1976, and the effort was coordinated by John Rice of Purdue (see Rice and Boisvert, 1985). Support for the project came from NSF, DOE, and ONR; collocation modules include COLLOCATION, HERMITE COLLOCATION, and INTERIOR COLLOCATION. See the ELLPACK Home Page for recent developments of this software. ELLPACK allows a user with a minimal knowledge of FORTRAN to solve elliptic partial differential equations rapidly; even more importantly, the analyst can compare different solution techniques for accuracy and computational speed.

7.10 The Cauchy–Riemann Equations, Conformal Mapping, and Solutions for the Laplace Equation

Earlier in this chapter we discussed the solution of elliptic partial differential equations using separation of variables. We now want to illustrate a very different approach that can be applied to a limited class of elliptic PDEs; we will restrict our attention to the two-dimensional Laplace equation. Functions that satisfy this PDE are said to be harmonic, and it is important for us to remember that only *conservative* fields can be represented by the two-dimensional Laplace equation.

Our initial focus in this section is the function,

$$w = f(z) = \phi(x, y) + i\psi(x, y). \quad (7.333)$$

If w is *analytic* in a region denoted by R, and if ϕ and ψ are related by the *Cauchy–Riemann equations* (see eq. (7.336) given below), then the real and imaginary parts are solutions for the two-dimensional Laplace equation. Indeed, if these conditions are met, then *any* analytic function $f(z)$ is the solution for *some problem* governed by the Laplace equation. Under these circumstances we refer to the $f(z)$'s as solutions for *potential* problems, and we will explore this technique using a topic familiar to students of hydrodynamics, ideal fluid flow.

Although we are using *ideal potential flow* as the framework for our discussion the technique we present here will be applicable to other types of problems as well including electrostatics, steady two-dimensional diffusion and heat conduction, as well as two-dimensional scattering of electromagnetic waves. Let us begin by clarifying exactly what we mean by an ideal flow: We stipulate that the *fluid* is inviscid and incompressible, and that the *flow* is irrotational; a useful mnemonic device in this context is to think of the *three I*'s (inviscid, incompressible, irrotational). For a two-dimensional ideal flow the velocity vector components can be obtained by differentiation of the *velocity potential*, ϕ, in the corresponding directions:

$$v_x = \frac{\partial \phi}{\partial x} \quad \text{and} \quad v_y = \frac{\partial \phi}{\partial y}. \quad (7.334)$$

Next we define the *stream function*, ψ, such that

$$v_x = -\frac{\partial \psi}{\partial y} \quad \text{and} \quad v_y = \frac{\partial \psi}{\partial x}. \quad (7.335)$$

We think of a streamline (a curve of constant ψ) as the path followed by a fluid particle. If streamlines are converging locally, then the flow is accelerating in that region; if the streamlines are diverging, the fluid velocity is decreasing. Evidently the velocity potential and the stream function are related:

$$\frac{\partial \phi}{\partial x} = -\frac{\partial \psi}{\partial y} \quad \text{and} \quad \frac{\partial \phi}{\partial y} = \frac{\partial \psi}{\partial x}. \tag{7.336}$$

These are the *Cauchy–Riemann equations*, and they *guarantee* that *any analytic function* of the complex variable, z, where $z = x + iy$, is the solution for *some* potential flow problem. That is, given a function of the complex variable, z, which we will write as $w = f(z)$, we have a mapping between the x-y plane and the $\phi - \psi$ plane; we need only equate the real and imaginary parts.

$$\phi + i\psi = f(z). \tag{7.337}$$

This branch of mathematics is known as conformal mapping due to the fact that angles are preserved; in the $\phi - \psi$ plane, velocity potential lines and streamlines intersect at right angles, just as lines of constant x and y do in the x-y plane. We specified an incompressible fluid for which $\nabla \cdot v = 0$, so if we differentiate eq. (7.334) appropriately, then

$$\frac{\partial^2 \phi}{\partial x^2} + \frac{\partial^2 \phi}{\partial y^2} = 0. \tag{7.338}$$

We also indicated that the flow was to be irrotational, which means that $\nabla \times v = 0$, and if we differentiate (7.335) accordingly, then

$$\frac{\partial^2 \psi}{\partial x^2} + \frac{\partial^2 \psi}{\partial y^2} = 0. \tag{7.339}$$

Thus, *both* the velocity potential and the stream function are governed by the Laplace equation and—most importantly—if $w(z)$ is a single-valued function of z over the region, R, and if it is differentiable at every point in that region, then every analytic $w(z)$ gives us a solution for the Laplace equation. But it is critical for us to emphasize: Had we included fluid friction (a *dissipative* process) in this study of moving fluids then the Laplace equation would no longer be applicable since we would not have a conservative field.

Let us illustrate the process we have in mind with an elementary example. We set

$$w = f(z) = z^2 + \frac{1}{z^2} = (x + iy)^2 + \frac{1}{(x + iy)^2} = \frac{x^2 - y^2}{x^4 + 2x^2y^2 + y^4} - \frac{2xyi}{x^4 + 2x^2y^2 + y^4}. \tag{7.340}$$

The function is clearly not analytic at $z = 0$, so we exclude that point from the region of interest. Therefore,

$$\phi = (x^2 - y^2)\left[1 + \frac{1}{x^4 + 2x^2y^2 + y^4}\right]$$

and

$$\psi = 2xy\left[1 - \frac{1}{x^4 + 2x^2y^2 + y^4}\right]. \tag{7.341}$$

So we have identified the velocity potential and the stream function for some potential flow, but we are faced with the immediate question: Exactly what problem governed by the Laplace equation is this a solution for? We will find out by plotting values of ψ for (x,y) pairs in the first quadrant, then we construct appropriate contours. The result is shown in Figure 7.25.

Notice that according to the definition of the stream function given by eq. (7.335), the flow in this case is right to left (the fluid enters the figure at the right-hand boundary and leaves through the upper surface).

We have found a solution for the Laplace equation by a backward process that is easy enough to execute, but might not be very useful under more general circumstances. We can of course simply write down more functions of z: $w = f(z)$ and identify the results by plotting $\psi(x,y)$. The reader is encouraged to explore this approach, and an interesting case (flow over a circular obstruction, or log) is given by

$$w = f(z) = \pi RU \coth\left(\frac{\pi R}{z}\right), \tag{7.342}$$

where R is the radius of the obstruction and U is the intensity of the approaching flow. Should you wish to try this, start by writing $\coth(x) = \frac{e^x + e^{-x}}{e^x - e^{-x}}$, and note that the polar form for a complex number is: $x + iy = r(\cos\theta + i\sin\theta)$. Of course, $r = \sqrt{x^2 + y^2}$.

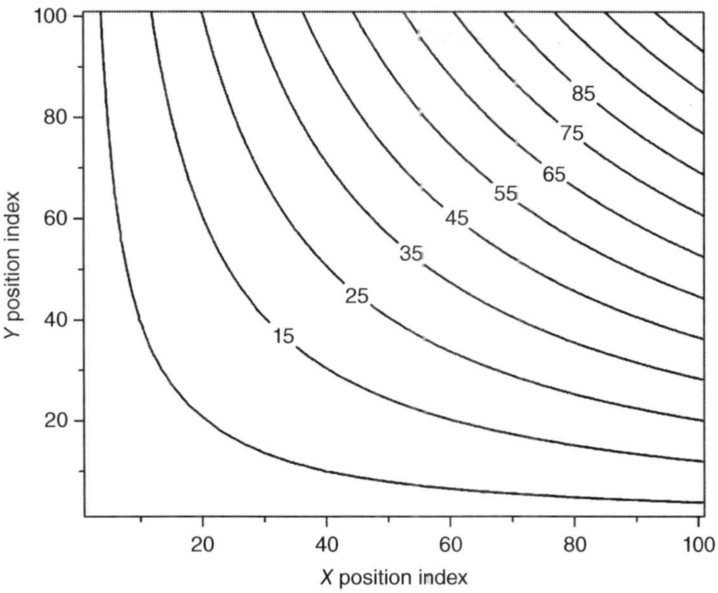

Figure 7.25 Plot of the stream function for the complex potential given by eq. (7.340).

Because the Laplace equation is linear, we can also use superposition to combine individual solutions, building complex potentials (or stream functions) for more complicated problems. For example, we could take horizontal potential flow around a cylinder, for which $\psi = V_\infty \sin\theta \left(r - \frac{R^2}{r}\right)$, and add to that the vortex, $\psi = K \ln r$. This combination will produce flow about a right circular cylinder with rotation and will result in a vertical lift being generated (the *Magnus effect*). If we combine a source with a uniform flow (both in polar form for convenience) we get $\psi = -\frac{K}{2\pi}\theta + Vr\sin\theta$, which is flow about a half-body. As you can see, one can obtain the velocity potential and the stream function for many situations of interest by merely combining elementary solutions.

The methodology employed in the earlier example and additionally recommended for a student exercise will certainly work, but it is not very useful when we are seeking a solution for the Laplace equation for a *particular* problem. Moreover, in many cases this backward, or indirect process, would just be needless duplication of effort. Many conformal mappings are known, and compilations exist that we can consult directly. One example is the book by H. Kober (*Dictionary of Conformal Representations*, Dover Publications, Inc., 1952). We can find, for example, an extensive collection of functions of the type, $w = f(z) = z^a$, where a is real, in *Part Two* of Kober's book. *Part Three* is devoted to exponential (and related) functions, e.g., $w = e^z$.

Let us explore the use of such a dictionary with an example: We will consider ideal flow into a channel, or alternatively, the potential field (actually equipotential lines) accompanying two charged plates of finite size separated by a distance, $2b$. This problem appears on pages 116 and 117 in Kober, Section 11.5, from which we find:

$$z = w + \exp(w) \text{ or } x + iy = \phi + i\psi + \exp(\phi - i\psi). \tag{7.343}$$

Since $e^{i\psi} = \cos\psi + i\sin\psi$, it is easy to show that

$$x = \ln\left(\frac{y-\psi}{\sin\psi}\right) + \left(\frac{y-\psi}{\tan\psi}\right). \tag{7.344}$$

The form of this equation suggests that we select a constant value for ψ, allow y to range through a plausible sequence of values, and compute the corresponding x-positions. This process will allow us to prepare an appropriate plot, which is provided here as Figure 7.26.

We can conclude that conformal mapping is an easy approach to the solution of a limited class of problems described by the Laplace equation. But as we pointed out previously, the indirect approach of writing down an analytic complex potential and then determining what problem is solved by it, is not very practical. There are a couple of techniques that can be applied more broadly. For example, let us consider the case in which we have a constant potential along some parametrically

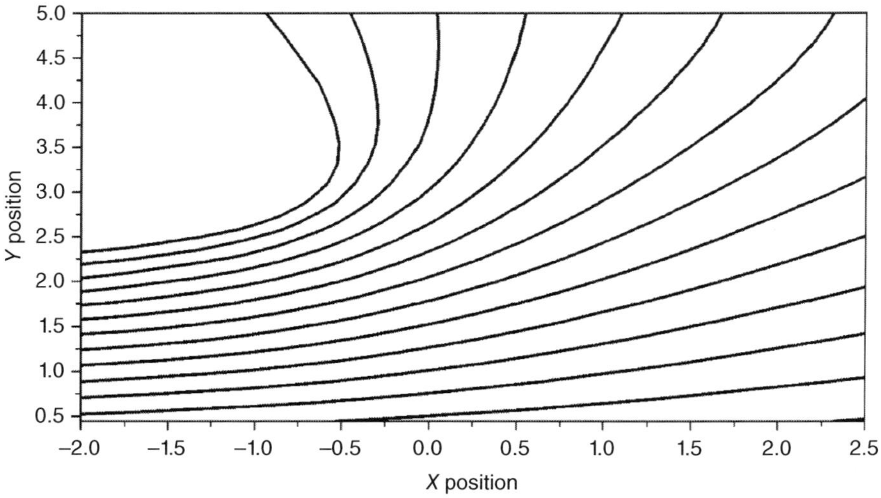

Figure 7.26 A partial construction of streamlines for a potential flow entering a channel or a canal. This view shows just the upper right-hand corner of the field.

defined curve. Specifically, suppose we know that the potential is constant over an ellipse for which $x = A\cos t$ and $y = B\sin t$; therefore,

$$z = x + iy = A\cos w + iB\sin w, \tag{7.345}$$

where $w = \phi + i\psi$. With a bit of work we can show:

$$A^2 = \frac{x^2}{\left[\cosh\psi + \frac{B}{A}\sinh\psi\right]^2} + \frac{y^2}{\left[-\sinh\psi + \frac{B}{A}\cosh\psi\right]^2}. \tag{7.346}$$

Thus the "streamlines" or equipotential lines are confocal ellipses, as expected. Another approach that can be quite useful when the potential is known on a *polygonal boundary* is the Schwarz–Christoffel formula; Smith (1953) provides several examples of its application and Bieberbach's (1953) discussion is helpful as well. The SC formula is:

$$w = \int_0^z \frac{dt}{\Pi(t - a_k)^{\alpha_k}} \text{ with } \Sigma\alpha_k = 2. \tag{7.347}$$

Thus, if the half-plane were to be mapped onto the interior of a triangle with *exterior* angles, $\alpha_1\pi$, $\alpha_2\pi$, and $\alpha_3\pi$, then:

$$w = \int_0^z \frac{dt}{(t - a_1)^{\alpha_1}(t - a_2)^{\alpha_2}(t - a_3)^{\alpha_3}}, \tag{7.348}$$

where all of the α's are greater than zero and their sum, $\alpha_1 + \alpha_2 + \alpha_3 = 2$. The a_1, a_2, etc. are the vertices mapped onto the real axis of the z-plane. We can look at an elementary case given by Lamb (1945) for which two finite points are chosen on the real axis at ± 1:

$$w = A\int \frac{dt}{\sqrt{t^2 - 1}} = A\cosh^{-1}(t) + B. \tag{7.349}$$

Lamb shows how this method can be used to model a Borda entrance (in two dimensions), and he provides several other interesting results obtained with the Schwarz–Christoffel formula in Chapter IV of *Hydrodynamics*.

7.11 Conclusion

Many important problems involving molecular (or diffusive) transport arising in engineering and the applied sciences can be solved using the product method or separation of variables. Moreover, extensive collections of these solutions exist (examples include Crank and Carslaw and Jaeger); frequently, an analyst can consult such resources and directly adapt an existing solution to their needs. This does not mean that every problem that can be solved using the product method, has been. There will always be variations that present a new challenge. But, once a student understands the technique, he/she will be much better able to assess what is possible and what is prohibitively difficult. In the more general case of nonlinear partial differential equations, one must either accept the limitations of an approximate analytic solution or proceed to a numerical simulation. The importance of the latter has grown rapidly—and pretty much in step with the expanding availability of computing power. The numerical solution of partial differential equations is the subject of the next chapter.

Problems

7.1 The function $f(x)$ is equal to +3 for $-2 \le x \le 0$ and −3 for $0 \le x \le 2$. We wish to reconstruct this function using a Fourier series, and since $f(x)$ is odd, we know that a half-range Fourier sine series is appropriate. Therefore

$$B_n = \frac{2}{L}\int_0^L f(x)\sin\frac{n\pi x}{L}dx,$$

where $L = 2$. We can determine the B_n's either analytically or numerically, and we see by inspection that all of the even coefficients are zero. Find (at the very least) the first 30 coefficients and compare your results with the abbreviated table given here:

$n = 1$	3	5	7	9	11
$B_n = -3.8196$	−1.2732	−0.7639	−0.5457	−0.4244	−0.3472

Of course, we want to know how many terms will be necessary to adequately represent $f(x)$, and we can anticipate that the end points will be problematic. Compare your results with the Figure P7.1 (note that the x-axis interval has been chosen to make this result look better than it actually is).

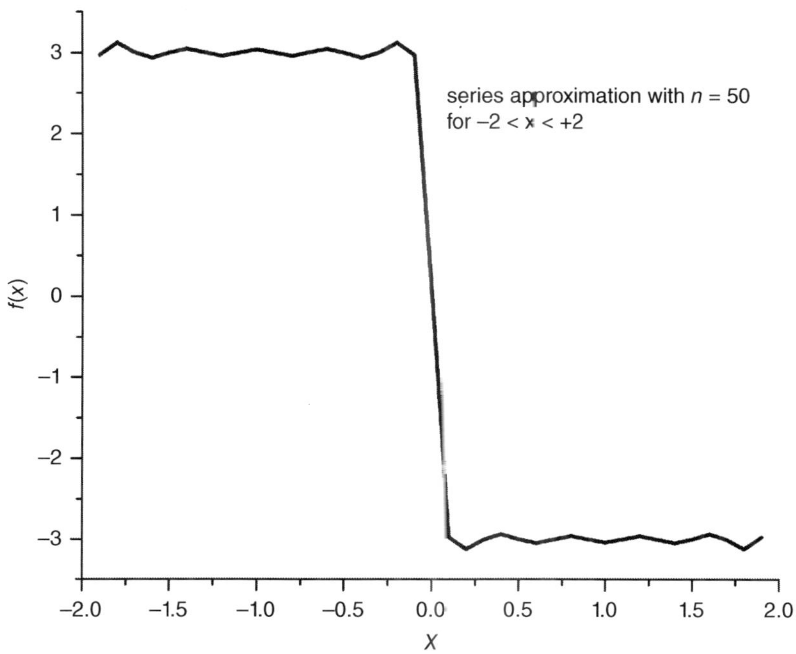

Figure P7.1 Fourier series approximation for $f(x)$ with $n = 50$ for $-2 < x < +2$.

7.2 Consider the even function, $f(x) = 16 - 2x^2$ for $-3 \leq x \leq +3$. We will use a Fourier cosine series to represent this function and consequently

$$A_n = \frac{2}{L}\int_0^L f(x)\cos\frac{n\pi x}{L}dx.$$

Since $L = 3$, we can easily show that $A_0 = 20$. Compute the A_n's and check your values with the numerically computed coefficients shown in this table:

n	0	1	2	3	4	5	6	7
A_n	19.999	7.296	−1.825	0.812	−0.458	0.293	−0.204	0.150

Now, using n's from zero to 50, approximate $f(x)$ and see how closely your results match the actual function. You should find that the agreement is excellent except at the upper end of the range of x:

X	0	0.5	1	1.5	2	2.5	3
$f(x)$	16.002	15.502	13.999	11.497	7.996	3.495	−1.917

Remember that in the case of this Fourier cosine series,

$$f(x) = \frac{A_0}{2} + \sum_{n=1}^{\infty} A_n \cos\frac{n\pi x}{L}.$$

7.3 Consider the polynomial: $f(x) = 5 - 5x + 2x^2 - \frac{1}{4}x^3$ over the range $0 \leq x \leq 6$. We want to know if a half-range Fourier sine series can adequately represent this function, and we are particularly interested in values of x near the end points. Reconstruct the function $f(x)$ and determine how many terms will be necessary for your infinite series to produce *values near the end points* that are within 5% of the correct values. Some results are shown in Figure P7.3 for the interior values of x using 100 terms in the series.

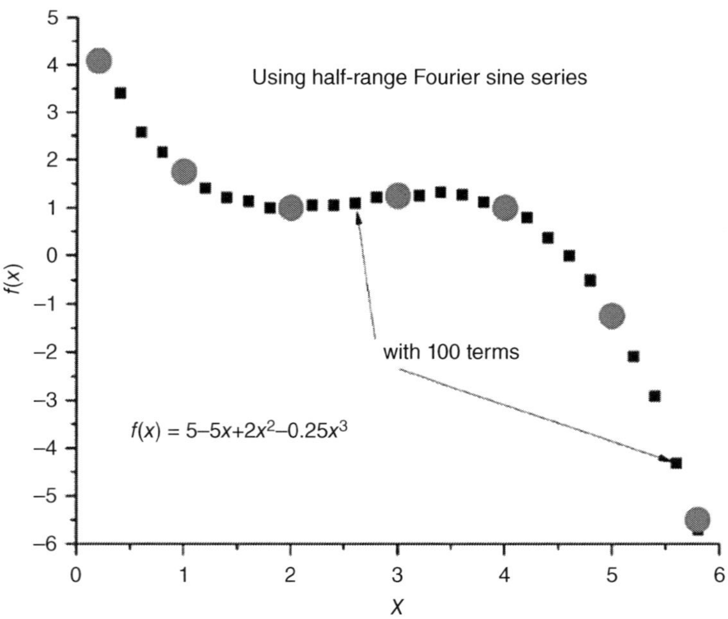

Figure P7.3 Comparison of $f(x)$ represented by a truncated Fourier sine series (small black squares) with correct values computed for select values of x (filled red circles). The worst agreement for the data shown here occurs at $x = 5.8$ where the value obtained from the truncated series is in error by about 4%. The fit is *considerably worse* for values of x around 5.95.

7.4 In the application of the product method to certain PDEs, some boundary conditions lead to transcendental equations with solutions that are not integer multiples of *pi*. In these cases the leading coefficients of the infinite series solution must be obtained through orthogonality. Consider the two functions:

$$f(x) = 5 + bx^2 \quad \text{and} \quad g(x) = x\sin(x).$$

We want to determine whether or not these two functions can be made orthogonal on the interval 0 to 2π through proper selection of the constant, b. That is, can an appropriate b be found such that

$$\int_0^{2\pi} (5 + bx^2)(x\sin x)dx = 0?$$

Note that a value that will work is located between 0 and −0.25; find an accurate value for b (you should get −0.149356). Is this answer unique? Will other values for b meet the required condition?

7.5 Solve the boundary value problem

$$\frac{\partial T}{\partial t} = \alpha \frac{\partial^2 T}{\partial y^2} \quad \text{for } 0 \le y \le 10$$

given $\alpha = 1$ and $T(y = 0,t) = 0$, $T(y = 10,t) = 45$, and $T(y,t = 0) = 20$.

7.6 Solve the boundary value problem

$$\frac{\partial T}{\partial t} = \alpha \frac{\partial^2 T}{\partial y^2} \quad \text{for } 0 \le y \le 10$$

given $\alpha = 2$ and $T(y = 0,t) = 20$, $T(y = 10,t) = 10$, and $T(y,t = 0) = 5 + 5y$.

7.7 Solve the boundary value problem

$$\frac{\partial T}{\partial t} = \alpha \frac{\partial^2 T}{\partial y^2} \quad \text{for } 0 \le y \le 10$$

given $\alpha = 1/4$ and $T(y = 0,t) = 10$, and for $y = 10$:

$$-k\left.\frac{\partial T}{\partial y}\right|_{y=10} = h(T|_{y=10} - T_{\infty}).$$

The initial condition is $T(y,t = 0) = 30$ and the Biot number, $hL/k = 1/4$.

7.8 Find the distribution of S over the annular region $R_1 \le r \le R_2$ where S is governed by the potential equation

$$\frac{\partial^2 S}{\partial r^2} + \frac{1}{r}\frac{\partial S}{\partial r} + \frac{1}{r^2}\frac{\partial^2 S}{\partial \theta^2} = 0.$$

The constant values of the edges are $S(r = R_1) = 100$ and $S(r = R_2) = 10$.

7.9 Repeat problem 7.8 but with $R_1 = 1$, $R_2 = 3$, $S(r = R_1) = 100$, and

$$-\kappa \frac{\partial S}{\partial r}_{r=R_2} = \beta(S_{r=R_2} - 10).$$

We know that $\beta/\kappa = 1/4$.

7.10 We are investigating a problem governed by the Poisson equation

$$\frac{\partial^2 \phi}{\partial x^2} + \frac{\partial^2 \phi}{\partial y^2} = -1.$$

Suppose $\phi(x, y) = a_0 + a_1x + a_2y + a_3x^2 + a_4xy + a_5y^2$. Find the relationships between coefficients in this polynomial.

7.11 The Laplace equation is applied to a rectangular region that extends from $x = 0$ to $x = L$ and from $y = 0$ to $y = H$

$$\frac{\partial^2 U}{\partial x^2} + \frac{\partial^2 U}{\partial y^2} = 0.$$

For the left and right sides ($x = 0$ and $x = L$), $U = 1$. For the top and bottom edges ($y = 0$ and $y = H$) the flux is zero; i.e., $\frac{\partial U}{\partial y} = 0$. Find the distribution, $U(x,y)$.

7.12 Find the distribution of temperature in a slab of material that extends in the x-direction: $0 \le x \le L$. The governing equation is

$$\frac{\partial T}{\partial t} = \alpha \frac{\partial^2 T}{\partial x^2}$$

given that $T(x = 0, t) = T_0$, $\frac{\partial T}{\partial x}(x = L, t) = 0$, and $T(x, t = 0) = T_i$.

7.13 Find the solution for $\frac{\partial \phi}{\partial t} = \kappa \left[\frac{\partial^2 \phi}{\partial r^2} + \frac{1}{r}\frac{\partial \phi}{\partial r}\right]$ for the annular region, $R_1 \le r \le R_2$, where $R_1 = 1$ and $R_2 = 4$. We know that $\phi(r = R_1, t) = 200$, $\phi(r, t = 0) = 10$, and

$$-\alpha \frac{\partial \phi}{\partial r} = \beta \phi(r = R_2).$$

7.14 Steady viscous flow in a duct is driven by a sliding upper surface, moving with constant velocity in the z-direction. By Newton's law of friction, the velocity at the other three walls will be zero, of course. The governing equation is

$$0 = \frac{\partial^2 V_z}{\partial x^2} + \frac{\partial^2 V_z}{\partial y^2} \quad \text{for } 0 \le x \le L \quad \text{and} \quad 0 \le y \le H.$$

For $x = 0$, $V = 0$, for $y = 0$, $V = 0$, for $x = L$, $V = 0$, and for $y = H$, $V = 10$. Let $H = L = 2$, and find $V\,(x,y)$.

7.15 Find $T(x,y)$ in a two-dimensional slab with the origin placed in the lower left-hand corner. The bottom, the top, and the left edges are all maintained at 50°. At the right-hand edge ($x = L$):

$$-k \left.\frac{\partial T}{\partial x}\right|_{x=L} = h(T|_{x=L} - T_\infty).$$

The temperature of the surroundings (T_∞) is zero, $H = 2$, $L = 1$, and $hL/k = 3/4$.

7.16 A long cylinder has a uniform initial temperature, $T_i = 75°$. At $t = 0$, the surface (at $r = R$) is rapidly cooled to 0°. Find $T(r,t)$.

$$\frac{\partial T}{\partial t} = \alpha \left[\frac{\partial^2 T}{\partial r^2} + \frac{1}{r}\frac{\partial T}{\partial r}\right].$$

The thermal diffusivity, α, is 2, and the cylinder radius is 2.

7.17 A long hollow cylinder has an initial temperature distribution to $T = f(r)$ for $R_1 \le r \le R_2$. For all positive t's, the surfaces at R_1 and R_2 are maintained at 0°. Find $T(r,t)$ for two different cases:

$$f(r) = T_0, \quad \text{and then } f(r) = 10 + 10\frac{(r - R_1)}{(R_2 - R_1)}.$$

7.18 Consider a diffusion tube that extends from $z = -L$ to $z = +L$. The concentration of the species of interest is governed by

$$\frac{\partial C}{\partial t} = D\frac{\partial^2 C}{\partial z^2}.$$

Of course the ends are impermeable, so for $z = \pm L$, $\frac{\partial C}{\partial z} = 0$. For the initial condition, $-L \le z \le 0$, $C = 1$ and $0 \le z \le +L$, $C = 0$. Find $C(z,t)$.

7.19 Given $k\left[\frac{\partial^2 T}{\partial x^2} + \frac{\partial^2 T}{\partial y^2}\right] + P = 0$, find $T(x,y)$ given $P/k = 200$. The origin is in the lower left-hand corner, and all four edges are maintained at 50°. The slab is square with $L = H = 2$.

7.20 A porous rectangular slab extends from $x = 0$ to $x = L$ and from $y = 0$ to $y = H$. Steady-state diffusion of species "A" is occurring in the slab with three sides (bottom, left, and right) all maintained at a constant dimensionless concentration of ¼. The concentration of "A" along the top of the slab (at $y = H$) varies with position as: $C = \frac{1}{4} + \frac{3}{4}\frac{x}{L}$. The governing elliptic partial differential equation is

$$\frac{\partial^2 C}{\partial x^2} + \frac{\partial^2 C}{\partial y^2} = 0.$$

Solve this problem using separation of variables and find the concentration distribution in the porous slab. Begin by redefining the concentration as C-¼ so that you have zero on three sides of the slab. Place the origin in the lower left-hand corner and confirm that the solution for this problem is

$$C = \frac{1}{4} + \sum_{n=1}^{\infty} A_n \sin\frac{n\pi x}{L} \sinh\frac{n\pi y}{L}.$$

Evaluate the coefficients by Fourier thecrem:

$$A_N = \frac{2}{L}\int_0^L \frac{\frac{3}{4}\frac{x}{L}}{\sinh\frac{n\pi H}{L}} \sin\frac{n\pi x}{L}\,dx.$$

Let $H = 5$ and $L = 8$; verify the entries in the following table:

x	y	C(x,y)
1	1	0.2660
2	2	0.3155
3	3	0.3993
4	4	0.5258
5	5	0.7277

7.21 Find the distribution of the variable, ϕ, in a circular disk given

$$\phi(r = R, 0 < \theta < \pi) = 1 \text{ and } \phi(r = R, -\pi < \theta < 0) = 0.$$

The governing equation is

$$\frac{1}{r}\frac{\partial}{\partial r}\left(r\frac{\partial \phi}{\partial r}\right) + \frac{1}{r^2}\frac{\partial^2 \phi}{\partial \theta^2} = 0.$$

7.22 A porous slab, with surfaces located at $x = \pm b$, is initially saturated with solvent. We wish to model the diffusion process within the slab for a drying problem where the loss of solvent at the surface(s) is described by $-D\frac{\partial C}{\partial x}\Big|_{x=b} = K(C_{x=b} - C_\infty)$. The governing equation is $\frac{\partial C}{\partial t} = D\frac{\partial^2 C}{\partial x^2}$. Find $C(x,t)$, and then find an expression for the total amount of solvent lost from the slab over a time, t. Remember: The slab loses solvent from *both* surfaces.

7.23 A taut string is secured (fixed and stationary) at $y = 0$ and $y = 9$. At a point corresponding to $y = 3$, the string is displaced ½ unit in the transverse direction and then released. Find $u(y,t)$ using the product method given:

$$\frac{\partial^2 u}{\partial y^2} = \frac{1}{c^2}\frac{\partial^2 u}{\partial t^2}.$$

7.24 A drum head is a membrane (or skin) stretched over a circular mounting rim of radius, R. The displacement of the membrane in response to an initial forcing function is governed by

$$\frac{\partial^2 \phi}{\partial t^2} = c^2 \left[\frac{\partial^2 \phi}{\partial r^2} + \frac{1}{r}\frac{\partial \phi}{\partial r} + \frac{1}{r^2}\frac{\partial^2 \phi}{\partial \theta^2}\right]$$

Use the product method to find a solution for this problem given that $\phi(r = R, \theta, t) = 0$, $\frac{\partial \phi}{\partial t}(r, \theta, t = 0) = 0$, and $\phi(r, \theta, t = 0) = F(r)$.

7.25 A solid sphere of radius, $R = 3$ cm, has an initial temperature of 45 °C. At $t = 0$, the sphere is plunged into a large, cooled bath maintained at 0 °C. If the heat transfer coefficient between the fluid phase and the surface of the sphere is 0.02 cal/(cm^2 s °C), and if the thermal diffusivity, α, is 0.9 cm^2/s, find *T(r,t)* and plot the temperature distributions at $t = 4$, 10, 20, and 40 s. The density and heat capacity of the material are 1.74 and 0.24, respectively (cgs units). The governing equation for this problem is:

$$\rho C_p \frac{\partial T}{\partial t} = k\left[\frac{1}{r^2}\frac{\partial}{\partial r}\left(r^2 \frac{\partial T}{\partial r}\right)\right].$$

7.26 A circular disk of radius, R, is well insulated on the flat, circular faces (top and bottom sides). At $t = 0$, the edge of the disk is rapidly cooled to 0° and heat flows in the r-direction toward the rim:

$$\frac{\partial T}{\partial t} = \alpha \left[\frac{\partial^2 T}{\partial r^2} + \frac{1}{r}\frac{\partial T}{\partial r}\right].$$

Find *T(r,t)* given $T(r, t = 0) = T_0(1 - r^2)$, and $R = 1$.

7.27 Use the Galerkin MWR technique to find an approximate solution for

$$\frac{\partial \phi}{\partial t} = \frac{\partial}{\partial y}\left[\kappa \frac{\partial \phi}{\partial y}\right] \quad \text{for } 0 \le y \le 1,$$

given $\kappa = (\phi_0 + 4\phi)^{-1}$. The boundary and initial conditions are: $\phi(y = 0, t) = 1$, $\phi(y = 1, t) = 0$, and $\phi(y, t = 0) = 0$. Figure P7.27 contains some numerical results for this problem, and it is provided below to assist the student with their work.

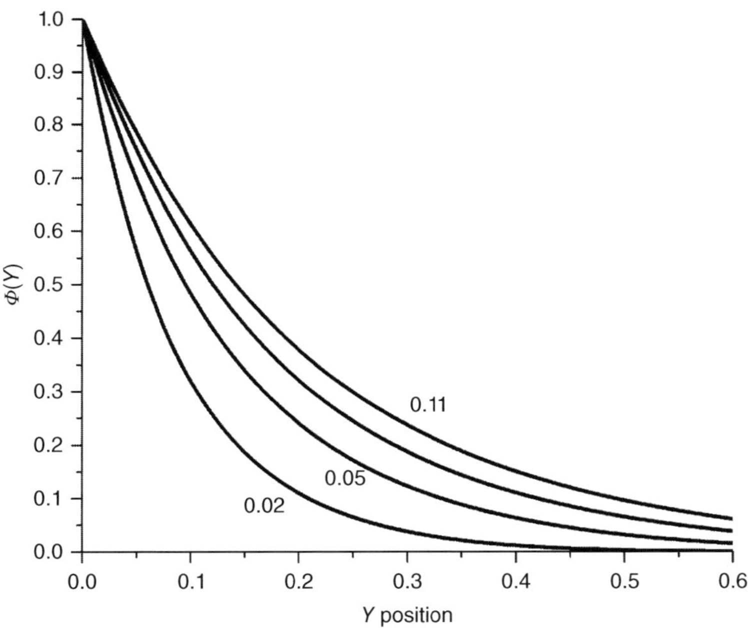

Figure P7.27 Computed numerical solutions for t's of 0.02. 0.05, 0.08, and 0.11. Note that the horizontal axis has been truncated to better show the time evolution.

7.28 We have a long, solid cylinder in which thermal energy is produced at a uniform rate (per unit volume) corresponding to S. The surface of the cylindrical solid will be maintained at $T = 0$ for all time, t. The governing equation has the form

$$\frac{\partial T}{\partial t} = \alpha \left[\frac{1}{r}\frac{\partial}{\partial r}\left(r\frac{\partial T}{\partial r}\right)\right] + \frac{S}{\rho C_p}.$$

We want to find the analytic solution for this problem. Begin by verifying that the steady-state solution for this situation is

$$T_{ss} = \frac{S}{4k}(R^2 - r^2).$$

Then, let the dependent variable, T, be written as the sum of steady-state and transient parts: $T = T_{ss} + T_1$. Use this sum to eliminate the inhomogeneity and demonstrate that

$$\frac{\partial T_1}{\partial t} = \alpha\left[\frac{\partial^2 T_1}{\partial r^2} + \frac{1}{r}\frac{\partial T_1}{\partial r}\right].$$

Now complete your solution and check your result with Section 7.9 in Carslaw and Jaeger (1959).

7.29 Consider a solid sphere with a radius of 3 (three), in which thermal energy production occurs at a constant rate

$$\rho C_p \frac{\partial T}{\partial t} = k\left[\frac{\partial^2 T}{\partial r^2} + \frac{2}{r}\frac{\partial T}{\partial r}\right] + S_0.$$

The source term, S_0, will have units of cal/(cm^{3}sec). We want to solve this problem analytically (and later numerically) and then plot our solutions. Two substitutions are required in order to find a solution by separation of variables. The first, which is already familiar to us, is $T = \phi/r$. However, this step leaves the inhomogeneity which must be dealt with by setting $\phi = \theta - \frac{S_0 r^3}{6k}$. Prove that the steady-state solution is $T = \frac{S_0}{6k}(R^2 - r^2)$, and more usefully, $\theta = \frac{S_0}{6k}R^2 r$, leading to:

$$\theta = \frac{S_0}{6k}R^2 r + A\exp(-\alpha\lambda^2 t)\sin(\lambda r).$$

Show that $\sin(\lambda R) = 0$ such that $\lambda_n = \frac{n\pi}{R}$. Finally, use the initial condition, $T(r,t = 0) = 0$, to find the A_n's. The parametric values we will employ are:

$$k = 0.05 \; \alpha = 0.05 \; R = 3 \text{ and } S_0 = 10$$

We are particularly interested in values of the parameter, $\frac{\alpha t}{R^2}$ =0.055, 0.111, and 0.222. Find the temperature profiles for each and plot them on the same figure. Use the dimensionless group for the dependent variable, $\frac{6k}{S_0}\frac{T}{R^2}$. Assume that that the initial temperature of the sphere is 0 (zero) and that the surface is maintained at zero for all time. Show that the temperature at the center of the sphere at $t = 20$ s is 179.27° and at $t = 40$ s, 259.28°

7.30 Suppose we have a finite cylinder (of length, L) which is at some initial temperature, T_i. At $t = 0$, the end of the cylinder (at $z = 0$) is instantaneously heated to T_0. The governing equation is:

$$\rho C_p \frac{\partial T}{\partial t} = k\left[\frac{1}{r}\frac{\partial}{\partial r}\left(r\frac{\partial T}{\partial r}\right) + \frac{\partial^2 T}{\partial z^2}\right].$$

The surface of the cylinder, at $r = R$, loses heat to the surroundings such that

$$-k\left(\frac{\partial T}{\partial r}\right)_{r=R} = h(T_R - T_\infty),$$

and we will assume that $T_\infty = 0$. The far end of the cylinder, at $z = L$, is maintained at $T = 0$ for all time. Begin by finding the *steady-state temperature distribution* in the cylinder, and check your result with Section 8.3 in Carslaw and Jaeger (1959). Then, explore the transient problem and find an analytic solution if you can.

7.31 A viscous fluid, which is initially at rest, lies upon a planar surface corresponding to the x-axis. The fluid extends very far in the vertical (or y-) direction. At $t = 0$, the planar wall begins to oscillate such that $V_x(y = 0) = V_0\cos(\omega t)$.

Find the velocity distribution in the fluid assuming that the flow is governed by

$$\frac{\partial V_x}{\partial t} = \nu \frac{\partial^2 V_x}{\partial y^2}.$$

This scenario is referred to as *Stokes' second problem*, and you may want to consider use of the Laplace transform.

7.32 We have a potential field in spherical coordinates with ϕ symmetry such that

$$\frac{1}{r^2}\frac{\partial}{\partial r}\left(r^2\frac{\partial \psi}{\partial r}\right) + \frac{1}{r^2 \sin\theta}\frac{\partial}{\partial \theta}\left(\sin\theta \frac{\partial \psi}{\partial \theta}\right) = 0.$$

We want to find an analytic solution for this problem using Section 7.6 as a guide. Begin by proposing $\psi = f(r)g(\theta)$ and carry out the separation. The equation for f will be of the Cauchy type, and the equation for g can be transformed into Legendre's differential equation such that the solution can be written in terms of Legendre polynomials. You may also want to consult Chapter 12 in Spiegel (*Advanced Mathematics for Engineers and Scientists*, McGraw-Hill, 1971) for assistance. Express the solution as the product of f and g, and assume that ψ must be finite at the center. You should obtain something that looks like eq. (7.174) in Section 7.6. Describe in detail how you will find values for the A_n's in the infinite series solution.

7.33 A porous sphere with $d = 2$ cm is initially saturated with solute (species "A") such that the interior concentration is uniform: $C(0 \le r \le R) = 1$. At $t = 0$ the sphere is immersed in a very large volume of stirred solvent that contains *no* "A." If $D = 0.001$ cm^2/s, when will $C(r = ½)$ fall to 10% of its initial value? Next, assume that the solvent phase is quiescent (i.e., it is offering significant resistance to mass transfer) such that

$$-D\frac{\partial C}{\partial r}\bigg|_{r=R} = K(C_{r=R} - C_\infty),$$

with $C_\infty = 0$. Set $\frac{KR}{D} = 1$; thus, $K = 0.001$ cm/s. At what time will half (½) of the total amount of "A" initially present in the porous sphere be removed? *Note that the answer for the first part of this problem is about* 250 s.

7.34 Consider the simple parabolic PDE, $\frac{\partial \phi}{\partial t} = \kappa \frac{\partial^2 \phi}{\partial y^2}$, with the boundary conditions, $\phi(0, t) = 0$ and $\phi(\pi, t) = 0$. The initial condition is: $\phi(y, 0) = 2\sin 3y$ and $0 \le y \le \pi$. Solve this problem using the product method and then describe in detail the consequences if the initial condition contained multiple sinusoids. Can you think of a physical situation where such might occur?

7.35 A porous right-circular cylinder is initially saturated with a solute species such that everywhere in the interior, $C = 1$. At $t = 0$ the cylinder will be immersed in a solvent bath of large volume, and the solute will begin to leave the cylinder by diffusion. The cylinder is *squat* however, and in particular, $0 \le z \le 1$ and $0 \le r \le 1$. This means that $L/d = ½$, and consequently the concentration in the interior will be governed by

$$\frac{\partial C}{\partial t} = D\left[\frac{\partial^2 C}{\partial r^2} + \frac{1}{r}\frac{\partial C}{\partial r} + \frac{\partial^2 C}{\partial z^2}\right].$$

We will solve this problem using the product method. Begin by setting: $C = f(r)g(z)h(t)$, and showing that

$$\frac{h'}{Dh} = \frac{f''}{f} + \frac{1}{r}\frac{f'}{f} + \frac{g''}{g} = -\lambda^2.$$

Therefore, we immediately observe that $h = c_1 \exp(-D\lambda^2 t)$, and move on to the rest of the equation:

$$\frac{f''}{f} + \frac{1}{r}\frac{f'}{f} = -\lambda^2 - \frac{g''}{g} = -\beta^2.$$

In the case of f, we see Bessel's differential equation and thus $f = AJ_0(\beta r) + BY_0(\beta r)$. Combine the two constants by defining a new γ^2 and show that

$$C = c_1 \exp(-D\lambda^2 t)[AJ_0(\beta r) + BY_0(\beta r)][c_2 \sin \gamma z + c_3 \cos \gamma z].$$

Now require C to be finite at the center where $r = 0$, and recognize that the function's behavior with respect to z must be odd, *not even*. Employ two boundary conditions, by noting that $C(r = 1, z, t) = 0$, and that $C(r, z = 1, t) = 0$

and prove that

$$C = \sum_{k=1}^{\infty}\sum_{n=1}^{\infty} A_{kn} \exp\left(-D\left(\beta_n^2 + k^2\pi^2\right)t\right) J_0(\beta_n r) \sin k\pi z.$$

Use the initial condition to write:

$$1 = \sum_{k=1}^{\infty}\left[\sum_{n=1}^{\infty} A_{kn} J_0(\beta_n r)\right] \sin k\pi z.$$

Treat the product in the interior brackets as a constant and note that the above equation is merely a Fourier sine series—therefore the sequence of values for the product in the interior brackets is simply the Fourier coefficients. Finally, complete the solution by determining values for A_{kn} using orthogonality. Verify your solution by comparison with the following table of approximate results using $D = 0.001$:

$C(r = ½, z = ½)$	Time (s)
0.9	26.7
0.7	45.1
0.5	66.5
0.3	98.4
0.1	167.9
0.05	212.1
0.01	315.3

7.36 A flat annular disk (a shape like a washer) extends from R_1 to R_2. The flat surfaces are insulated, and the *initial* temperature distribution is given by:

$$T(r, t = 0) = \beta(r - R_1)(R_2 - r).$$

The two circular edges are maintained at $T = 0$ for all time, t. Heat transfer in the disk will be governed by

$$\frac{\partial T}{\partial t} = \alpha\left[\frac{\partial^2 T}{\partial r^2} + \frac{1}{r}\frac{\partial T}{\partial r}\right]$$

with the solution,

$$T = c_1 \exp(-\alpha\lambda^2 t)[AJ_0(\lambda r) + BY_0(\lambda r)].$$

Since the temperature must be zero at both edges, it will be necessary to have

$$Y_0(\lambda R_2)J_0(\lambda R_1) = Y_0(\lambda R_1)J_0(\lambda R_2).$$

Of course there are an infinite number of solutions for this equation; we will set $R_1 = 2$ and $R_2 = 3$, and consequently the first three values for λ are 3.1353, 6.2795, and 9.4225. Determine the evolution of the temperature distribution in the disk, given that $\beta = 800$ and $\alpha = 0.0022$. When will the temperature at the central point, $r = 2.5$, fall to 18°? Some data are provided in the following table to allow you to check your analytic solution.

Time, $t = 1$	5	10	15	20	30	50
$T(r = 2.5) = 196.5$	182.4	165.1	148.8	133.7	107.8	70.0

7.37 Let's examine unsteady axial flow in an annulus; initially, a viscous fluid is at rest in the annular space between two concentric cylinders. At $t = 0$, the outer cylinder begins to slide with constant velocity (V_0) in the $+z$ direction. The velocity of the fluid in the annulus is governed by the partial differential equation

$$\frac{\partial v_z}{\partial t} = \nu\left[\frac{\partial^2 v_z}{\partial r^2} + \frac{1}{r}\frac{\partial v_z}{\partial r}\right]$$

This equation is a candidate for separation so we take $v_z = f(r)g(t)$ which results in

$$\frac{g'}{\nu g} = \frac{f'' + \frac{1}{r}f'}{f} = -\lambda^2.$$

The resulting two ordinary differential equations have the solutions

$$g = c_1 \exp(-\nu\lambda^2 t) \text{ and } f = AJ_0(\lambda r) + BY_0(\lambda r).$$

Verify that the *steady-state* velocity distribution is $v_z = \frac{V_0}{\ln(R_2/R_1)} \ln\left(\frac{r}{R_1}\right)$, and combine this expression with the product of f and g. Note that B is *not* zero in this problem, but v_z must disappear on both sides of the annulus, at $r = R_1$ and $r = R_2$. Given that $R_1 = 1$, $R_2 = 2.5$, $V_0 = 10$, and $\nu = 0.01$ (all cgs), assume that our interest is the velocity at $r = 2$, i.e., $v_Z(r = 2,t)$. When will the velocity at this position exceed 4 cm/s? 6 cm/s? 7 cm/s? For the last of these you should obtain about 55.96 s.

7.38 A viscous fluid is at rest in a cylindrical tube. At $t = 0$, a constant pressure gradient is applied to the fluid, and it begins to move in the positive z-direction. The governing equation for this situation is:

$$\rho\frac{\partial v_z}{\partial t} = -\frac{\partial p}{\partial z} + \mu\left[\frac{\partial^2 v_z}{\partial r^2} + \frac{1}{r}\frac{\partial v_z}{\partial r}\right].$$

We want to solve this PDE and reveal the dynamic behavior of the fluid. However, we note that the equation is not homogeneous due to the presence of the pressure gradient. Therefore, our first task is to eliminate dp/dz. Begin your analysis by showing that the steady-state solution is the familiar parabolic velocity profile:

$$V = \frac{1}{4\mu}\frac{dp}{dz}(r^2 - R^2).$$

Then, let $v_z = V + v_1$, where v_1 is the transient part of the solution. Substitute this combination into the parabolic PDE to obtain: $\frac{\partial v_1}{\partial t} = \nu\left[\frac{\partial^2 v_1}{\partial r^2} + \frac{1}{r}\frac{\partial v_1}{\partial r}\right]$, and use the product method to find: $v_1 = A\exp(-\nu\lambda^2 t)J_0(\lambda r)$. Of course Y_0 has been removed from the solution because v_1 must be finite at the center of the tube where $r = 0$. Therefore

$$v_z = \frac{1}{4\mu}\frac{dp}{dz}(r^2 - R^2) + A\exp(-\nu\lambda^2 t)J_0(\lambda r),$$

and application of the "no-slip" condition at $r = R$ will require $J_0(\lambda R) = 0$, resulting in

$$v_z = \frac{1}{4\mu}\frac{dp}{dz}(r^2 - R^2) + \sum_{n=1}^{\infty} A_n \exp\left(-\nu{\lambda_n}^2 t\right) J_0(\lambda_n r).$$

Use orthogonality to evaluate the A_n's and then confirm your results using the following parameters:

$$R = 1 \ \mu = 0.01 \ \nu = 0.01 \ \rho = 1 \ dp/dz = -0.8.$$

A few select points for centerline velocity are provided here:

t (s)	10	20	60
V (cm/s)	7.7	13	19.2

Note that the evolution of this velocity distribution is slow—even after 100 s the centerline velocity is 0.4% below its ultimate value.

7.39 Solve the PDE, $\frac{\partial \phi}{\partial t} = \frac{\partial^2 \phi}{\partial y^2} - a^2\phi$, using separation of variables. We can think of this as a model of the diffusional (or molecular) transport of ϕ with a loss term that is linear in the dependent variable. We are given the following:

$$\phi(y = 0, t) = 10 \quad \phi(y = L, t) = 2 \quad \phi(y, t = 0) = 0.$$

Begin by finding the steady-state solution, and then show that

$$\phi = c_1 \exp(-ay) + c_2 \exp(+ay) + \sum_{n=1}^{\infty} A_n \exp\left(-{\lambda_n}^2 t\right) \sin(\beta_n y).$$

Now let $L = 5$ and $a = 1/2$, evaluate c_1 and c_2, and find the coefficients that cause this series to converge to the correct values. See if you obtain $A_1 = -4.677$ and $A_2 = -2.198$. Since $\beta_n = \frac{n\pi}{L}$, this is a Fourier series problem. Check your results with the small table provided here (40 terms were used for the series solution):

t	*ϕ(y = L/2,t) Numerical*	*ϕ(y = L/2,t) Analytic*
0.5	0.1355	0.1348
1	0.777	0.7756
2	1.891	1.890
5	2.991a	2.991
10	3.1699	3.1698

7.40 Steady diffusion is occurring in a rectangular slab for which $0 \le x \le L$ and $0 \le y \le H$. The governing elliptic equation is:

$$\frac{\partial^2 C}{\partial x^2} + \frac{\partial^2 C}{\partial y^2} = 0$$

with $C = 0$ for $x = 0$ and $x = L$ and $y = H$, but at $y = 0$, $C = f_0(x)$. By separation of variables show that the solution for this problem is

$$C = \sum_{n=1}^{\infty} A_n \sin(\lambda_n x) \sinh(\lambda_n (H - y)).$$

Then, with $L = 10$ and $H = 8$, set $f_0(x) = ax - bx^6$ where $b = 1.7172 \times 10^{-6}$ and $a = 0.1712$. This means that the maximum value of C (which is 1) occurs at $x = 6.9883$. Show that $A_1 = 0.15007$, $A_2 = -0.00379$, and $A_3 = 0.000114$, then verify the values for $C(x,y)$ provided in the following table.

x	5.25	5.25	5.25	5.25	5.25
y	0	1	3	5	7
C(x,y)	0.8657	0.6504	0.3449	0.1638	0.0481

7.41 This problem makes use of the results of exercise 5.26. Consider a long, solid cylinder with an initial temperature distribution given by

$$T(r) = 100 - 50r + 30r^3$$

with $0 \le r \le 2$. Accordingly, this initial $T(r)$ behaves as indicated here:

r	0	0.5	0.7454	1.0	1.5	2.0
T(r)	100	78.75	75.155	80	126.25	240

At $t = 0$ the temperature at the surface ($r = 2$) is instantaneously changed to 0°, and the dynamic behavior of the temperature distribution is governed by:

$$\frac{\partial T}{\partial t} = \alpha \left[\frac{\partial^2 T}{\partial r^2} + \frac{1}{r}\frac{\partial T}{\partial r} \right].$$

We want to solve this problem analytically, and one of the integrals arising from application of the initial condition is

$$\int_0^2 (100r - 50r^2 + 30r^4) J_0(\lambda r) dr.$$

The surface boundary condition requires that $J_0(\lambda R) = 0$; thus, since $R = 2$, $\lambda_1 = 1.2024$, $\lambda_2 = 2.7605$, etc. Of course, the solution for this problem has the familiar form:

$$T = \sum_{n=1}^{\infty} A_n \exp\left(-\alpha \lambda_n^2 t\right) J_0(\lambda_n r).$$

The challenging part of this exercise involves the determination of the leading coefficients, the A_n's, that produce a series converging to the desired result. And a critical component of the process will be the ability of the series to reproduce the initial temperature distribution in the cylinder. One suspects this will not be trivial. For $t = 0$,

$$100 - 50r + 30r^3 = \sum_{n=1}^{\infty} A_n J_0(\lambda_n r)$$

and therefore,

$$A_n = \frac{\int_0^2 (100r - 50r^2 + 30r^4) J_0(\lambda_n r) dr}{\int_0^2 r J_0^2(\lambda_n r) dr}.$$

Find a suitable set of coefficients capable of reproducing the correct $T(r)$ for very small t. Using *Mathcad* with λ_1 for example:

$$\frac{\int_0^2 (100r - 50r^2 + 30r^4) J_0(1.2024r) dr}{\int_0^2 r J_0^2(1.2024r) dr} =$$

This yields $A_1 = 157.611$. We want to know if the infinite series—with a reasonable number of terms—can produce the correct behavior of $T(r, t)$ for small t's.

References

Abramowitz, M. and Stegun, I.A. (1972). *Handbook of Mathematical Functions*. New York: Dover Publications.

Atkins, P.W. (1978). *Physical Chemistry*. San Francisco: W. H. Freeman and Company.

Baker, G.L. and Gollub, J.P. (1990). *Chaotic Dynamics: An Introduction*. Cambridge: Cambridge University Press.

Bieberbach, L. (1953). *Conformal Mapping*. New York: Chelsea Publishing Company.

Carslaw, H.S. (1950). *An Introduction to the Theory of Fourier's Series and Integrals*, revised 3rd edition. New York: Dover Publications.

Carslaw, H.S. and Jaeger, J.C. (1959). *Conduction of Heat in Solids*, 2nde. Oxford: Oxford Clarendon Press.

Carslaw, H.S. and Jaeger, J.C. (1963). *Operational Methods in Applied Mathematics*. New York: Dover Publications.

Crank, J. (1975). *The Mathematics of Diffusion*, 2nde. Oxford: Oxford Clarendon Press.

Finlayson, B.A. (1980). *Nonlinear Analysis in Chemical Engineering*. New York: McGraw-Hill.

Glasgow, L.A. (2010). *Transport Phenomena: An Introduction to Advanced Topics*. New York: John Wiley & Sons.

Houstis, E.N. (1978). Collocation Methods for Linear Elliptic Problems. *BIT* 16: 301.

Kober, H. (1952). *Dictionary of Conformal Representations*. New York: Dover Publications.

Korner, T.W. (1989). *Fourier Analysis*. Cambridge: Cambridge University Press.

Lamb, H. (1945). *Hydrodynamics*, 6the. New York: Dover Publications.

Powers, D.L. (1979). *Boundary Value Problems*, 2nde. New York: Academic Press.

Prenter, P.M. and Russell, R.D. (1976). Orthogonal Collocation for Elliptic Partial Differential Equations. *SIAM Journal of Numerical Analysis* 13: 923.

Rice, J.R. and Boisvert, R.F. (1985). *Sovling Elliptic Problems Using ELLPACK*. New York: Springer-Verlag.

Schrödinger, E. (1926). Quantisierung als Eigenwertproblem. *Annalen der Physik* 79: 489.

Smith, L.P. (1953). *Mathematical Methods for Scientists and Engineers*. New York: Dover Publications.

Spiegel, M.R. (1971). *Advanced Mathematics for Engineers and Scientists*. Schaum's Outline Series. New York: McGraw-Hill.

Spiegel, M.R. (1974). *Theory and Problems of Fourier Analysis*. Schaum's Outline Series. New York: McGraw-Hill.

Villadsen, J. and Michelsen, M.L. (1978). *Solution of Differential Equation Models by Polynomial Approximation*. Englewood Cliffs, NJ: Prentice-Hall.

Villadsen, J. and Stewart, W.E. (1967). Solution of Boundary-Value Problems by Orthogonal Collocation. *Chemical Engineering Science* 22: 1483.

Wachspress, E. (1966). *Iterative Solution of Elliptic Systems and Applications to the Neutron Diffusion Equation of Reactor Physics*. Englewood Cliffs, NJ: Prentice-Hall.

Weinberger, H.F. (1965). *A First Course in Partial Differential Equations*. New York: John Wiley & Sons.

Wieder, S. (1973). *The Foundations of Quantum Theory*. New York: Academic Press.

8

Numerical Solution of Partial Differential Equations

8.1 Introduction

Our usual approach will involve discretization of partial differential equations, followed by solution of the resulting algebraic equations. Discretization is key to both finite difference methods (FDM) and finite element methods (FEM). The two approaches require the same level of numerical effort, but the latter is particularly useful for problems involving irregular shapes and boundaries (an introduction to FEM will be provided in Chapter 11). On the other hand, finite difference methods are much less software dependent, and for simple problems FDM solutions can be obtained with a broad spectrum of hardware–software combinations, even through use of commonplace tools like spreadsheet programs. Thus, the analyst can solve many important practical problems without commercial modeling software, without high-level language proficiency, without compiler experience, and without mesh generation and refinement.

We should anticipate that when we solve a partial differential equation numerically we may not obtain a completely accurate solution. Of course, we expect discrepancies arising from both roundoff and truncation, and a common view is that we are solving the given PDE with some acceptable level of error. There is a second viewpoint that is useful in the context of certain computations, and it reveals a more insidious problem that we need to recognize: *When we discretize a partial differential equation we are actually creating a PDE that may have additional terms; i.e., we end up with an equation that is not the original model for the phenomenon of interest. Clearly, we need to understand how those additional terms affect the solution.* We will give a very brief introduction to this topic here, but the interested reader should consult Chapter 6 in Anderson (1995) for detail. Consider the fragmentary equation for a transient problem with convective transport:

$$\frac{\partial \phi}{\partial t} + V\frac{\partial \phi}{\partial x} = \dots \tag{8.1}$$

One possible discretization can be written

$$\frac{\phi_{i,j+1} - \phi_{i,j}}{\Delta t} + V\frac{\phi_{i,j} - \phi_{i-1,j}}{\Delta x} \cong \dots \tag{8.2}$$

where the index $j+1$ refers to the new time step, $t + \Delta t$. Please note that the x-direction gradient of ϕ is written in the *upwind* (backward) form; the need for this particular difference will be explained later. If we now expand $\phi_{i,j+1}$ and $\phi_{i-1,j}$ in Taylor series:

$$\phi_{i,j+1} = \phi_{i,j} + \left(\frac{\partial \phi}{\partial t}\right)_{i,j}\Delta t + \left(\frac{\partial^2 \phi}{\partial t^2}\right)_{i,j}\frac{(\Delta t)^2}{2} + \left(\frac{\partial^3 \phi}{\partial t^3}\right)_{i,j}\frac{(\Delta t)^3}{6} + \cdots \tag{8.3}$$

$$\phi_{i-1,j} = \phi_{i,j} - \left(\frac{\partial \phi}{\partial x}\right)_{i,j}\Delta x + \left(\frac{\partial^2 \phi}{\partial x^2}\right)_{i,j}\frac{(\Delta x)^2}{2} - \left(\frac{\partial^3 \phi}{\partial x^3}\right)_{i,j}\frac{(\Delta x)^3}{6} + \cdots \tag{8.4}$$

and then substitute the results into the original equation, eq. (8.1), we recover the original terms *but with the addition of new ones*. Anderson shows through a process of differentiation and subtraction that the time derivatives that appear in "new" terms in the equation can be replaced by derivatives with respect to x ultimately resulting in:

$$\frac{\partial \phi}{\partial t} + V\frac{\partial \phi}{\partial x} = A\frac{\partial^2 \phi}{\partial x^2} + B\frac{\partial^3 \phi}{\partial x^3} + \cdots \tag{8.5}$$

The original equation is recovered on the left-hand side, but new derivatives appear on the right. The even derivatives are *dissipative*, and in computational fluid dynamics (CFD) they are referred to as *artificial viscosity*; they exert a stabilizing

Problem Solving in Engineering: Analytical Mathematics and Numerical Analysis, First Edition. Larry A. Glasgow.

Companion website: www.wiley.com/go/glasgow/problemsolvingengineering

influence upon the computation. The odd derivatives are *dispersive*, and they can create distortions and in some cases destabilize a computation. Let us emphasize the essential point of this discussion: The discretization process we employ can produce additional terms in the PDE. We may, in fact, be solving a partial differential equation that differs from the actual model eq. (8.1) of the phenomenon of interest. Though this sounds ominous it may be beneficial in particular circumstances; artificial viscosity, for example, can be used intentionally to make an unstable computational scheme stable. But to be absolutely clear, if we pursue this course (rendering an unstable computation stable by adding artificial viscosity), we are adopting the viewpoint that finding *some* kind of numerical solution is better than not finding one at all.

8.2 Finite Difference Approximations for Derivatives

Finite difference approximations allow us to develop algebraic representations for differential equations. Consider the following Taylor series expansions:

$$y(x+h) = y(x) + hy'(x) + \frac{h^2}{2}y''(x) + \frac{h^3}{6}y'''(x) + \cdots \tag{8.6}$$

and

$$y(x-h) = y(x) - hy'(x) + \frac{h^2}{2}y''(x) - \frac{h^3}{6}y'''(x) + \cdots \tag{8.7}$$

When we add the two equations together we obtain:

$$y(x+h) + y(x-h) = 2y(x) + h^2 y''(x) + f(h^4) + \cdots$$

If we discard all of the terms involving h^4 (and up), we get:

$$y''(x) \cong \frac{y(x+h) - 2y(x) + y(x-h)}{h^2}. \tag{8.8}$$

This second-order, central difference approximation for the second derivative has a leading error on the order of h^2. If h is small, this approximation should be good. For example, let

$$y = x\sin x, \text{ thus } \frac{dy}{dx} = \sin x + x\cos x, \text{ and } \frac{d^2y}{dx^2} = 2\cos x - x\sin x.$$

Now let $x = 0.3$:

$$y = 0.088656, \frac{dy}{dx} = 0.582121, \text{ and } \frac{d^2y}{dx^2} = 1.822017;$$

then choose $h = 0.01$:

$$\frac{d^2y}{dx^2} \cong \frac{0.094568 - 2(0.088656) + 0.082926}{(0.01)^2} = 1.820$$

This is about 0.11% less than the analytic value for the second derivative. By simply combining Taylor series expansions, we can build any number of approximations and for derivatives of any order. Furthermore, these approximations can be forward, backward, centered, or skewed. Some of the more useful forms are compiled for you below. Note that **F** $\Rightarrow$ forward, **C** $\Rightarrow$ central, **B** $\Rightarrow$ backward, and h is convenient shorthand for Δx.

First order:

$$\mathbf{F} \quad y_i' = \frac{1}{h}(y_{i+1} - y_i) \tag{8.9}$$

$$\mathbf{B} \quad y_i' = \frac{1}{h}(y_i - y_{i-1}) \tag{8.10}$$

Second order:

$$\mathbf{F} \quad y_i' = \frac{1}{2h}(-3y_i + 4y_{i+1} - y_{i+2}) \tag{8.11}$$

$$y_i'' = \frac{1}{h^2}(y_i - 2y_{i+1} + y_{i+2}) \tag{8.12}$$

C
$$y_i' = \frac{1}{2h}(y_{i+1} - y_{i-1}) \tag{8.13}$$

$$y_i'' = \frac{1}{h^2}(y_{i+1} - 2y_i + y_{i-1}) \tag{8.14}$$

B
$$y_i' = \frac{1}{2h}(3y_i - 4y_{i-1} + y_{i-2}) \tag{8.15}$$

$$y_i'' = \frac{1}{h^2}(y_i - 2y_{i-1} + y_{i-2}) \tag{8.16}$$

Third order:

F
$$y_i' = \frac{1}{6h}(2y_{i+3} - 9y_{i+2} + 18y_{i+1} - 11y_i) \tag{8.17}$$

$$y_i'' = \frac{1}{h^2}(-y_{i+3} + 4y_{i+2} - 5y_{i+1} + 2y_i) \tag{8.18}$$

$$y_i''' = \frac{1}{h^3}(y_{i+3} - 3y_{i+2} + 3y_{i+1} - y_i) \tag{8.19}$$

B
$$y_i' = \frac{1}{6h}(11y_i - 18y_{i-1} + 9y_{i-2} - 2y_{i-3}) \tag{8.20}$$

$$y_i'' = \frac{1}{h^2}(2y_i - 5y_{i-1} + 4y_{i-2} - y_{i-3}) \tag{8.21}$$

$$y_i''' = \frac{1}{h^3}(y_i - 3y_{i-1} + 3y_{i-2} - y_{i-3}) \tag{8.22}$$

Fourth order:

F
$$y_i' = \frac{1}{12h}(-3y_{i+4} + 16y_{i+3} - 36y_{i+2} + 48y_{i+1} - 25y_i) \tag{8.23}$$

$$y_i'' = \frac{1}{12h^2}(11y_{i+4} - 56y_{i+3} + 114y_{i+2} - 104y_{i+1} + 35y_i) \tag{8.24}$$

$$y_i''' = \frac{1}{2h^3}(-3y_{i+4} + 14y_{i+3} - 24y_{i+2} + 18y_{i+1} - 5y_i) \tag{8.25}$$

$$y_i'''' = \frac{1}{h^4}(y_{i+4} - 4y_{i+3} + 6y_{i+2} - 4y_{i+1} + y_i) \tag{8.26}$$

C
$$y_i' = \frac{1}{12h}(-y_{i+2} + 8y_{i+1} - 8y_{i-1} + y_{i-2}) \tag{8.27}$$

$$y_i'' = \frac{1}{12h^2}(-y_{i+2} + 16y_{i+1} - 30y_i + 16y_{i-1} - y_{i-2}) \tag{8.28}$$

$$y_i''' = \frac{1}{2h^3}(y_{i+2} - 2y_{i+1} + 2y_{i-1} - y_{i-2}) \tag{8.29}$$

$$y_i'''' = \frac{1}{h^4}(y_{i+2} - 4y_{i+1} + 6y_i - 4y_{i-1} + y_{i-2}) \tag{8.30}$$

B
$$y_i' = \frac{1}{12h}(25y_i - 48y_{i-1} + 36y_{i-2} - 16y_{i-3} + 3y_{i-4}) \tag{8.31}$$

$$y_i'' = \frac{1}{12h^2}(35y_i - 104y_{i-1} + 114y_{i-2} - 56y_{i-3} + 11y_{i-4}) \tag{8.32}$$

$$y_i''' = \frac{1}{2h^3}(5y_i - 18y_{i-1} + 24y_{i-2} - 14y_{i-3} + 3y_{i-4}) \tag{8.33}$$

$$y_i'''' = \frac{1}{h^4}(y_i - 4y_{i-1} + 6y_{i-2} - 4y_{i-3} + y_{i-4}) \tag{8.34}$$

8.3 Boundary Conditions

The simplest type of boundary condition is the Dirichlet BC where the field variable is specified at the boundary. For a fluid flow problem this could take the form of $v_z(r = R) = 0$; in mass transfer we might have $C(y = y_0) = C_0$. These are simple enough to implement in our numerical solution schemes, but we will see that things are a bit more difficult when fluxes are involved.

Consider a conduction problem in a slab for which the right-hand boundary is insulated, thus $q_x = 0$; this is an example of a Neumann boundary condition $\left(\frac{\partial T}{\partial x} = 0\right)$. Let the nodal point on the boundary be represented by the index, n, and let the temperatures for $n-2$ and $n-1$ be 50° and 45°, respectively. We can determine the temperature at the boundary by setting the derivative equal to zero. However, if we use a first-order, backward difference in this situation:

$n-2$	$n-1$	n
50°	45°	?°

then $T_n = 45°$, a result that is clearly unphysical because the temperature "profile" on this row has a discontinuity in slope. One alternative is to employ eq. (8.15) from Section 8.2:

$$T_n = \frac{1}{3}(-50 + 4(45)) = 43.333°. \tag{8.35}$$

Of course, a third- or fourth-order backward difference could be used as well; if we go with third order and set the temperature at $n-3$ to 56°, we find $T_n = 42.909°$.

We should also examine the use of a Robin's-type boundary condition (a boundary condition of the third kind) for a solid–fluid interface:

$$-k_s \frac{\partial T}{\partial x} = h_f(T_n - T_\infty). \tag{8.36}$$

Let the Biot modulus, $Bi = \frac{\Delta x h_f}{k_s}$; then, *one possible expression for* T_n *is:*

$$T_n = \frac{2BiT_\infty + 4T_{n-1} - T_{n-2}}{3 + 2Bi}. \tag{8.37}$$

If we select $Bi = 1$ and $T_\infty = 20°$ and use the temperatures given above for the $n-1$ and $n-2$ positions, then:

$$T_n = \frac{2(20) + 4(45) - 50}{5} = 34°. \tag{8.38}$$

Note how eq. (8.37) is affected when Bi is very low—the result is exactly the same as eq. (8.35)! Of course a very small Bi places the main resistance to heat transfer on the fluid side of the interface.

These are not the only types of boundary conditions that one might encounter—they are merely the most common. In neutron diffusion for example, one might have a non-zero neutron flux at the boundary that would be extrapolated to zero at some distance *outside* the reactor. Another possibility is the *albedo* boundary condition (where albedo, β, is defined as the ratio of entering and exiting neutron currents) applied at a surface, $r = R$:

$$D\nabla\phi + \frac{1}{2}\left(\frac{1-\beta}{1+\beta}\right)\phi = 0. \tag{8.39}$$

If the albedo is zero, there are no neutrons entering at the boundary; if the albedo is one (1), the reactor medium is bounded by a reflector.

8.4 Elliptic Partial Differential Equations

Our main focus in this section is upon Laplace- and Poisson-type elliptic partial differential equations that apply to equilibrium phenomena. Examples include steady-state conduction in a slab,

$$\frac{\partial^2 T}{\partial x^2} + \frac{\partial^2 T}{\partial y^2} = 0, \tag{8.40}$$

steady viscous flow in the z-direction in a duct,

$$\frac{\partial^2 V_z}{\partial x^2} + \frac{\partial^2 V_z}{\partial y^2} = \frac{1}{\mu}\frac{dp}{dz}, \tag{8.41}$$

and the Laplacian of the stream function for two-dimensional potential flow:

$$\frac{\partial^2 \psi}{\partial x^2} + \frac{\partial^2 \psi}{\partial y^2} = 0. \tag{8.42}$$

We should begin this part of our discussion by looking at a solution procedure for eq. (8.40). Suppose we have a square slab of material with prescribed temperatures on all four edges (400° across the top, and 100° for both sides and the bottom); we wish to find the interior temperature distribution, $T(x, y)$. We discretize the slab using $\Delta x = \Delta y$ (a square mesh) with five nodes in each direction. Since the boundary temperatures are known, we have nine interior nodes where the temperature must be determined:

400	400	400	400	400
100				100
100				100
100				100
100	100	100	100	100

If we approximate eq. (8.40) with second-order central differences, we find

$$T_{i+1,j} + T_{i-1,j} + T_{i,j+1} + T_{i,j-1} - 4T_{i,j} = 0. \tag{8.43}$$

Thus, we have an elementary problem in which we must find the solution for nine simultaneous, linear algebraic equations. This can be accomplished many different ways, and we choose to employ Crout's (also known as Cholesky's) method which was described in detail in Chapter 2. The reader may wish to verify that the solution for the given problem is:

400	400	400	400	400
100	228.6	258	228.6	100
100	156.3	175	156.3	100
100	121.4	129.5	121.4	100
100	100	100	100	100

This raises an interesting question: How accurate is this solution? For example, is the temperature at the center of the slab really 175°? Since this is a problem for which the analytic solution is known (see Section 7.6), we can test the given result using the infinite series very easily. To four decimal places, the center temperature from the analytic solution is 174.9995°. This shows that our approximate solution for the center (175°) is unusually accurate; we can also find the actual temperatures immediately above and below this point to get a clearer picture of the overall quality of the solution. One node above the center the analytic solution produces 262.158° (as opposed to 258°) and one node below, we find 128.624° (as opposed to 129.5°). Our numerical solution is surprisingly close considering the coarse discretization that was employed.

Many elliptic PDEs can be solved in this manner, and since the coefficient matrix is usually sparse, such problems can be solved very efficiently. Although our slab example used only nine interior nodes, much larger problems can be solved in the same way. Some care must be exercised in such cases, however, because roundoff error can accumulate and corrupt the solution. If very large sets of simultaneous equations are to be solved using an elimination method it may be necessary to either use greater precision in the calculations, or alternatively, to incorporate *error equations* into the procedure. We will give a brief sketch of this process here, but the interested reader may want to consult Chapter 3 in James et al. (1977).

Suppose we have a set of simultaneous equations,

$$a_{11}X_1 + a_{12}X_2 + a_{13}X_3 + \ldots = C_1$$

$$a_{21}X_1 + a_{22}X_2 + \ldots \text{etc.}$$

When we solve these equations we obtain a set of *approximate* values for the X_n's that we will represent like this: Y_1, Y_2, etc. We take these approximate values back to the set of equations and compute the new constants, i.e.,

$$a_{11}Y_1 + a_{12}Y_2 + a_{13}Y_3 + \ldots = D_1. \tag{8.44}$$

If roundoff errors have been generated, then $D_1 \neq C_1$, $D_2 \neq C_2$, etc. Now we presume that the desired values for the X_n's can be obtained by *adding a correction to the approximate solution*: $X_1 = Y_1 + \Delta X_1$, $X_2 = Y_2 + \Delta X_2$, etc. The correction expressions are substituted into the original algebraic equations replacing the unknown X_n's:

$$a_{11}(Y_1 + \Delta X_1) + a_{12}(Y_2 + \Delta X_2) + a_{13}(Y_3 + \Delta X_3) + \ldots = C_1. \tag{8.45}$$

Next we subtract the set of equations obtained with the initial estimates, resulting in

$$\begin{aligned} &a_{11}\Delta X_1 + a_{12}\Delta X_2 + a_{13}\Delta X_3 + \ldots = C_1 - D_1 \\ &a_{21}\Delta X_1 + a_{22}\Delta X_2 + a_{23}\Delta X_3 + \ldots = C_2 - D_2, \text{etc.} \end{aligned} \tag{8.46}$$

The solution of this set of equations produces the corrections that are added to the original estimates: $X_1 = Y_1 + \Delta X_1$, $X_2 = Y_2 + \Delta X_2$, …. This process can be repeated any number of times should that prove necessary.

8.4.1 An Iterative Numerical Procedure: Gauss–Seidel

There are alternative solution techniques that can be applied to elliptic PDEs and we will now examine a straightforward iterative scheme; let us consider laminar flow in a rectangular duct for this example. By using the second-order central difference approximations for the second derivatives (where the *i*- and *j*-indices represent the *x*- and *y*-directions, respectively), eq. (8.42) can be written:

$$\frac{1}{\mu}\frac{dp}{dz} \cong \frac{V_{i+1,j} - 2V_{i,j} + V_{i-1,j}}{(\Delta x)^2} + \frac{V_{i,j+1} - 2V_{i,j} + V_{i,j-1}}{(\Delta y)^2} \tag{8.47}$$

If the discretization employs a square mesh ($\Delta x = \Delta y$), then we can isolate the term with the largest numerical coefficient, with the convenient result

$$V_{i,j} \approx \frac{1}{4}\left[V_{i+1,j} + V_{i-1,j} + V_{i,j+1} + V_{i,j-1} - \frac{(\Delta x)^2}{\mu}\frac{dp}{dz}\right]. \tag{8.48}$$

Please note that the *z*-direction subscript has been dropped from velocity to minimize clutter. This approximation is the basis for a simple Gauss–Seidel iterative computational scheme for the solution of such problems. In this case, of course, the velocity is zero on the boundaries, so we merely apply the algorithm to all of the interior points, row by row. The newly computed values are employed as soon as they become available (which distinguishes Gauss–Seidel from the Jacobi iterative method). As an example, consider the case of laminar flow in a rectangular duct 8 cm wide and 4 cm high; the pressure gradient is −3 dynes/cm^2/cm and the viscosity is 0.04 g/(cm s). All of the nodal velocities will be initialized to zero to start the computation. For the specified pressure gradient the centerline (maximum) velocity will be about 139 cm/s. The computed velocity distribution is shown in Figure 8.1 as a contour plot.

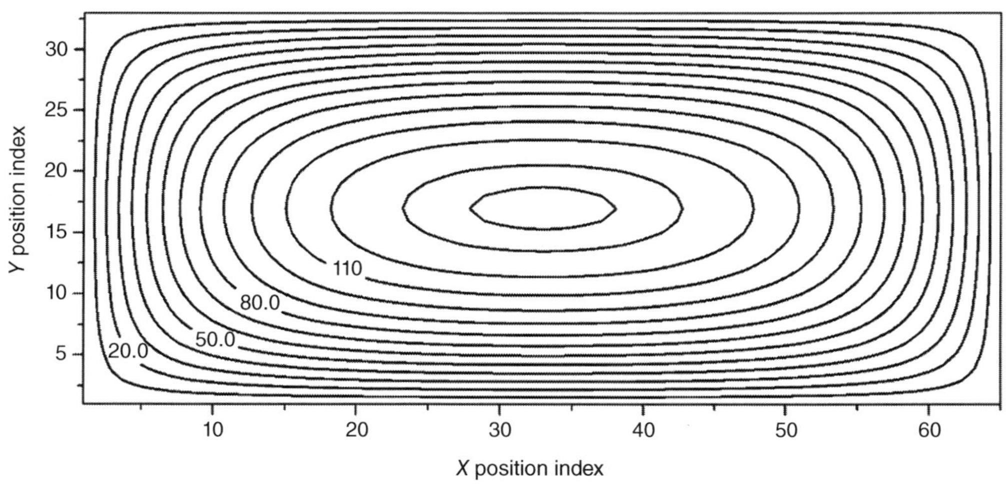

Figure 8.1 Velocity distribution in a rectangular duct computed with the Gauss–Seidel iterative method. The duct measures 8 cm by 4 cm with $dp/dz = -3$ dynes/cm^2/cm.

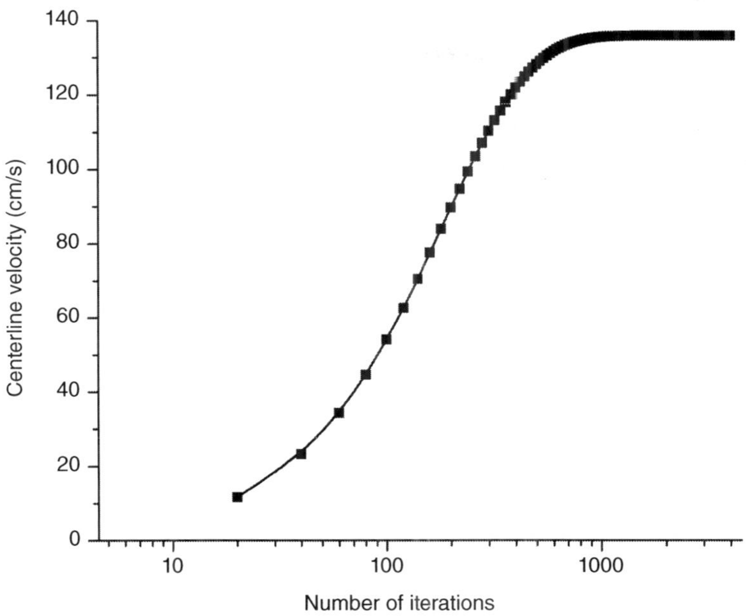

Figure 8.2 Centerline velocity as a function of the number of iterations for the solution of the Poisson equation for laminar flow in a rectangular duct as approximated by eq. (8.48).

In a computation of this type, a key issue is the number of iterations required to attain convergence. For the example shown here we can monitor the evolution of centerline velocity during the calculations as illustrated in Figure 8.2; keep in mind that we initialized all of the interior nodes at zero velocity. We could certainly improve the speed with which convergence is attained by starting the computation with a better initial estimate, i.e., providing a more suitable distribution for $V_{i,j}$.

Note that a reasonably accurate value is obtained with about 1,000 iterations, and after 3,000 iterations the third decimal place is essentially fixed.

8.4.2 Improving the Rate of Convergence with SOR

The rate of convergence of iterative solutions can be accelerated significantly through use of the *extrapolated Liebmann method* (also known as successive over-relaxation, or SOR). In this technique, the change that *would have been produced* by a single Gauss–Seidel iteration is increased through use of an accelerating factor which is usually denoted by ω. SOR can be implemented easily in the previous example by a slight modification of eq. (8.48):

$$V_{i,j}^{(new)} \approx V_{i,j} + \frac{1}{4}\omega\left[V_{i+1,j} + V_{i-1,j} + V_{i,j+1} + V_{i,j-1} - 4V_{i,j} - \frac{(\Delta x)^2}{\mu}\frac{dp}{dz}\right] \tag{8.49}$$

The $V_{i,j}$'s appearing on the right-hand side of eq. (8.49) are from the latest available calculations, of course. You can see immediately that if $\omega = 1$, this is identically the Gauss–Seidel algorithm. For *over-relaxation*, ω will have a value between 1 and 2; the rate of convergence is very sensitive to the value of the acceleration parameter. Please see Smith (1965) for additional discussion. Frankel (1950) has shown that for large rectangular domains such as that used in our example,

$$\omega_{opt} \approx 2 - \sqrt{2}\pi\left(\frac{1}{p^2} + \frac{1}{q^2}\right)^{1/2}, \tag{8.50}$$

where p and q are the number of nodal points used in the x- and y-directions, respectively. For our case, $p = 65$ and $q = 33$, so $\omega_{\text{opt}} \approx 1.85$. The consequences of a poor choice are shown clearly in Figure 8.3 where the number of iterations required to achieve a desired degree of convergence is reported. While it is apparent that SOR can significantly reduce the computational effort required to solve elliptic partial differential equations, the acceleration parameter, ω, must be chosen carefully to obtain the greatest possible benefit. We should make one other observation regarding ω: In the iterative solution of *nonlinear* PDEs, stability can sometimes be maintained by using under-relaxation, i.e., by setting $\omega < 1$.

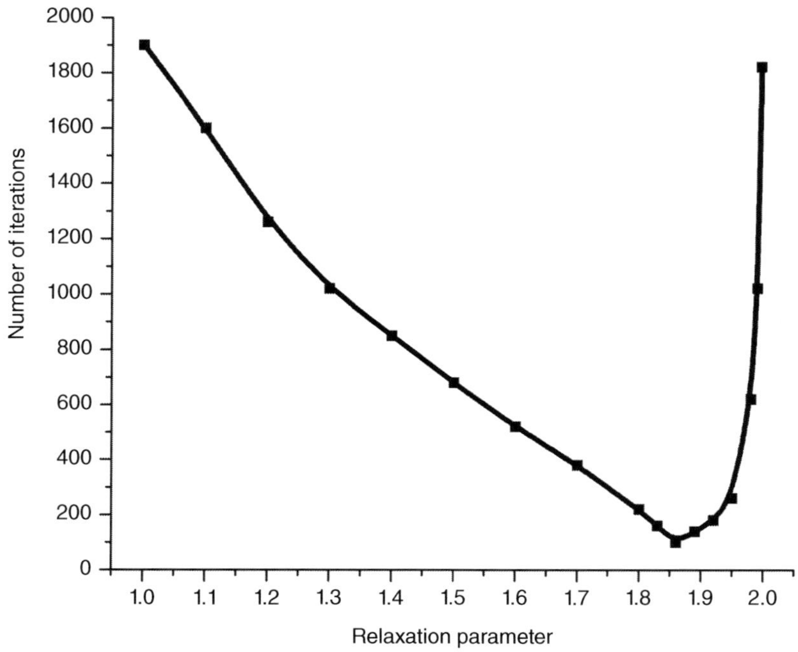

Figure 8.3 Number of iterations required to achieve $\varepsilon = 2 \times 10^{-7}$ as a function of ω. A Poisson-type equation for laminar flow in a rectangular duct is being solved and the minimum is located at about $\omega \approx 1.86$.

We will now illustrate the application of SOR with a very detailed example. Assume we have a mild steel slab; the left-hand side is maintained at 1,000°, the bottom at 500°, and the top is insulated. The right-hand side loses heat to the surroundings according to:

$$-k\left.\frac{\partial T}{\partial x}\right|_{x=L} = h(T|_{x=L} - T_\infty). \tag{8.51}$$

This is a steady-state problem, so the temperature in the interior of the slab is simply governed by

$$\frac{\partial^2 T}{\partial x^2} + \frac{\partial^2 T}{\partial y^2} = 0. \tag{8.52}$$

A typical program structure might appear as shown below:

```
        #COMPILE EXE
#DIM ALL
              REM *** Illustration of SOR computation for heat transfer in a steel slab
              REM *** Left-hand boundary maintained at 1000, bottom at 500, top is
insulated, and right side loses heat
                GLOBAL L,HT,h,k,dx,dy,w,Tsur AS SINGLE
                GLOBAL i,j,iter AS SINGLE
FUNCTION PBMAIN
                DIM T(251,201) AS SINGLE
                dx=0.1:dy=0.1:k=0.108:h=0.0203:Tsur=100:w=1.98
              REM *** initialize temperature
                  FOR j=1 TO 201
                      T(1,j)=1000
                      NEXT j
                  FOR i=2 TO 251
                      T(i,1)=500
                      NEXT i
                  iter=0
   50 REM *** continue
          FOR j=2 TO 200
```

```
        FOR i=2 TO 250
            T(i,j)=T(i,j)+w/4*(T(i+1,j)+T(i-1,j)+T(i,j+1)+T(i,j-1)-4*T(i,j))
                  NEXT i:NEXT j
      REM *** top boundary (insulated)
              FOR i=2 TO 250
                  T(i,201)=(4*T(i,200)-T(i,199))/3
                  NEXT i
      REM *** right-hand boundary-Robin's type BC
              FOR j=2 TO 200
                  T(251,j)=(-h*Tsur+k/(2*dx)*(-4*T(250,j)+T(249,j)))/(-3*k/(2*dx)-h)
                  NEXT j
                    iter=iter+1
                        PRINT iter,T(150,100)
                    IF iter>6000 THEN 200 ELSE 50
  200 REM *** continue
          OPEN "c:STslab1.dat" FOR OUTPUT AS #1
                  FOR j=1 TO 201
                      FOR i=1 TO 251
                          WRITE#1,i,j,T(i,j)
                              NEXT i:NEXT j
          CLOSE:END
```

Results obtained from this program are shown in the contour plot provided in Figure 8.4.

As we indicated previously, the progress of such a computation (that is, the rate at which one obtains a satisfactory solution) is very sensitive to the value selected for the relaxation parameter, ω. We will illustrate this by changing ω and monitoring the value of $T(150,100)$ at exactly 500 iterations:

ω	$T(150,100)$
1.0	0.0122
1.1	0.0672
1.2	0.2837
1.3	0.9799
1.4	2.912

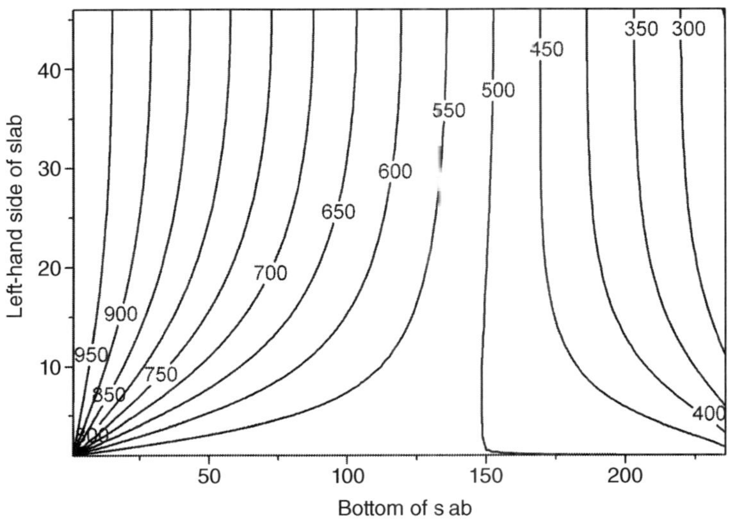

Figure 8.4 Isotherms computed for a mild steel slab with the left-hand side maintained at 1,000° and the bottom at 500°. The top surface is insulated and the right-hand side loses thermal energy to the surroundings according to Newton's law of cooling.

ω	T(150,100)
1.5	7.774
1.6	19.39
1.7	46.86
1.8	112.55
1.85	174.57
1.90	270.81
1.95	418.41
1.98	513.84
1.99	523.79

The initial value for all of the interior nodes was 0°. Note that the correct temperature at this location, T(150,100), is 523.8°.

Although we have purposefully tried to minimize connecting our discussions to specific computational software, the student needs to be aware that there are *many* commercial packages that have capabilities for elementary partial differential equations. We will illustrate one such option here, using Mathcad™. Suppose we have a two-dimensional (square) slab of material with the edges (T,B,L,R) maintained at the following temperatures: 200, 10, 50, and 200°. The temperature in the slab will be governed by the Laplace equation, i.e., eq. (8.52):

$$\frac{\partial^2 T}{\partial x^2} + \frac{\partial^2 T}{\partial y^2} = 0.$$

Our discretization for this PDE is exactly the same as eq. (8.43), of course:

$$T_{i+1,j} + T_{i-1,j} + T_{i,j+1} + T_{i,j-1} - 4T_{i,j} = 0.$$

The Mathcad™ function we will employ is *relax(a,b,c,d,e,f,u,rjac)*. Note that a is the matrix of coefficients on $(i+1,j)$, and all of them are 1's. b is the matrix of coefficients for $(i-1,j)$, and again, these values are all 1's. c is the matrix of coefficients on $(i,j+1)$, all 1's, and d is the matrix of coefficients for $(i,j-1)$, also 1's. The matrix of coefficients for the central temperature (i,j) in the pattern is e, and of course those values are all -4. f would correspond to the source term, if one were present. In our case, all of the f's are zero. The matrix u contains the constant boundary temperatures, and estimates for the interior nodes. *rjac* is a constant between 0 and 1 that affects the rate of convergence of the relaxation algorithm. For the simple problem we are considering here, the Mathcad™ procedure is not much affected by the choice of *rjac*. Using 25 interior nodes (so the coefficient matrices are all 7×7), we obtain the following result:

$$\text{relax}(a, b, c, d, e, f, u, 0.5) = \begin{pmatrix} 200 & 200 & 200 & 200 & 200 & 200 & 200 \\ 50 & 123.747 & 153.013 & 168.152 & 179.073 & 189.353 & 200 \\ 50 & 91.977 & 120.151 & 140.523 & 158.788 & 178.341 & 200 \\ 50 & 74.009 & 95.093 & 115 & 137.214 & 165.221 & 200 \\ 50 & 58.967 & 71.212 & 87.169 & 109.848 & 145.33 & 200 \\ 50 & 40.646 & 43.619 & 52.617 & 69.679 & 106.252 & 200 \\ 50 & 10 & 10 & 10 & 10 & 10 & 200 \end{pmatrix}$$

Naturally, the first question one should ask concerns the reliability of this computation: How accurate is it? To address this, we will refine the mesh and compute the interior temperatures ourselves using one of the algorithms we have already discussed. Using 13 nodes in each direction (and single precision), and reporting only every other node for ease of comparison with the above results, we obtain:

200	200	200	200	200	200	200
50	123.77609	153.79272	168.72809	179.47177	189.59695	200
50	91.29428	120.20753	140.88583	159.26443	178.76828	200
50	73.57849	94.815605	114.99994	137.57727	165.87845	200
50	58.76750	70.73545	86.72105	109.79236	146.24150	200
50	40.40302	42.99232	51.81481	68.67136	106.22386	200
50	10	10	10	10	10	200

These data indicate that the Mathcad™ solution obtained with *relax* is reasonably accurate, even with the coarse discretization we employed. The largest discrepancies between the two sets of results are on the order of 1.5%, and those errors appear on the bottom interior row. Of course we can use exactly the same discretization (employing 13 × 13 matrices) in Mathcad™, and when we do so we get the following results on the diagonal (starting in the lower left-hand corner and proceeding toward the upper right): 50, 40.403, 70.737, 115.002, 159.266, 189.597, and 200. These numbers are nearly identical with our do-it-yourself computation—the largest discrepancy is smaller than 0.002%.

We will use another example, and this one in cylindrical coordinates, to emphasize some of the points made regarding iterative solution of elliptic partial differential equations. Suppose we have a right-circular cylinder of radius, R, and length, L; one end of the cylinder is maintained at elevated temperature and the other surfaces are fixed at $T = 0$. Since we have angular symmetry with heat flow in the r- and z-directions, the governing equation for this situation is:

$$\frac{\partial^2 T}{\partial r^2} + \frac{1}{r}\frac{\partial T}{\partial r} + \frac{\partial^2 T}{\partial z^2} = 0. \tag{8.53}$$

We use the subscript, i, to represent the r-direction and j to represent z. Our finite difference approximation for this equation is

$$\frac{T_{i+1,j} - 2T_{i,j} + T_{i-1,j}}{(\Delta r)^2} + \frac{1}{r}\frac{T_{i+1,j} - T_{i-1,j}}{2\Delta r} + \frac{T_{i,j+1} - 2T_{i,j} + T_{i,j-1}}{(\Delta z)^2} \cong 0. \tag{8.54}$$

Letting $\Delta r = \Delta z$, the algorithm for the Gauss–Seidel iteration is simply

$$T_{i,j} = \frac{1}{4}\left[\left(1 + \frac{\Delta r}{2r}\right) T_{i+1,j} + \left(1 - \frac{\Delta r}{2r}\right) T_{i-1,j} + T_{i,j+1} + T_{i,j-1}\right]. \tag{8.55}$$

For successive over-relaxation, the above equation is modified slightly:

$$T_{i,j} = T_{i,j} + \frac{\omega}{4}\left[\left(1 + \frac{\Delta r}{2r}\right) T_{i+1,j} + \left(1 - \frac{\Delta r}{2r}\right) T_{i-1,j} + T_{i,j+1} + T_{i,j-1} - 4T_{i,j}\right]. \tag{8.56}$$

Again, note that if we set $\omega = 1$, we get the exact form of the preceding equation, eq. (8.55). Now we give the problem definite form by selecting $R = 5$, $L = 15$, $\Delta r = 0.083333$, with $T(r, z = 0) = 100$. These choices result in 10,561 interior nodes (all initialized at $T = 0$), and the centerline values for T are obtained by symmetry since $\left.\frac{\partial T}{\partial r}\right|_{r=0} = 0$. We will monitor the value of $T(30,90)$, the approximate center of the computational domain, and employ both Gauss–Seidel and successive over-relaxation (SOR) using $\omega = 1.9$. Our main interest is how rapidly the estimated temperature approaches the correct value which is 3.1225; this is shown in Figure 8.5.

For this elliptic partial differential equation, Gauss–Seidel requires more than 10,000 iterations to achieve a result that SOR accomplishes in 400.

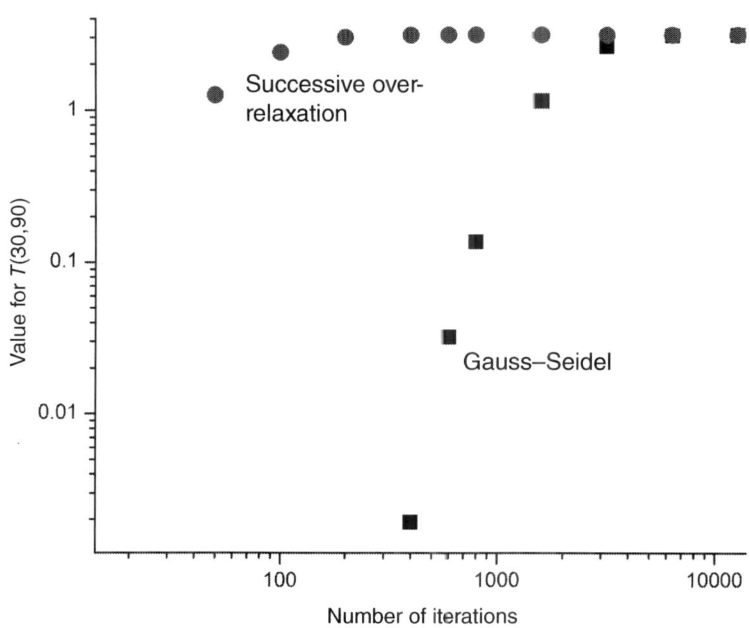

Figure 8.5 Approach of the node $T(30,90)$ to the correct temperature, 3.1225, using both Gauss–Seidel and successive over-relaxation with $\omega = 1.9$.

8.5 Parabolic Partial Differential Equations

8.5.1 An Elementary, Explicit Numerical Procedure

Suppose we have a viscous fluid that extends far in the y-direction, initially at rest near a plane wall that is set in motion with velocity, V_0, at time, $t = 0$; thus, $V_x(y = 0, t) = V_0$. Letting $V = V_x/V_0$,

$$\frac{\partial V}{\partial t} = \nu \frac{\partial^2 V}{\partial y^2}. \tag{8.57}$$

This scenario is known as Stokes' first problem, and the analytic solution is just

$$\frac{V_x}{V_0} = 1 - erf\left(\frac{y}{\sqrt{4\nu t}}\right),$$

where the error function, $erf(\eta) = \frac{2}{\sqrt{\pi}} \int_0^\eta \exp(-\eta^2) d\eta$:

η	0	0.1	0.2	0.4	0.8	1.6	3.2
$erf(\eta)$	0.00	0.1125	0.2227	0.4284	0.7421	0.9764	1.0000

An *explicit* algorithm is easily developed for eq. (8.57); use of a first-order forward difference for the time derivative followed by isolation of the V value on the new time step results in:

$$V_{i,j+1} = \frac{\Delta t \nu}{(\Delta y)^2} [V_{i+1,j} - 2V_{i,j} + V_{i-1,j}] + V_{i,j}. \tag{8.58}$$

Therefore we can march forward in time, computing new V's on each spatial row as we go. Eq. (8.58) is attractive because of its simplicity; it is easy to understand and easy to execute, but it poses a potential problem. To ensure stability, it is necessary that

$$\frac{\Delta t \nu}{(\Delta y)^2} \leq \frac{1}{2}. \tag{8.59}$$

We will illustrate this using eq. (8.58) by choosing $\nu = 0.05\ \text{cm}^2/\text{s}$, $\Delta y = 0.1$ cm, and $\Delta t = 0.12$ s; of course, this guarantees that we are over the limit of ½ (actually 0.6). We can put the calculation into a table and monitor the evolution of the nodal velocities, which will reveal the consequence of our choices. Since the analytic solution for this problem is known, we have a convenient comparison available (Table 8.1).

The problem we see immediately above is easy to resolve. We change our parametric choices to yield: $\frac{\Delta t \nu}{(\Delta y)^2} = 0.4$, and repeat the calculation.

Table 8.1 Explicit Computation with Unstable Parametric Choice(s).

t	$i = 1$	$i = 2$	$i = 3$	$i = 4$	$i = 5$	$i = 6$	$i = 7$
0	1	0	0	0	0	0	0
Δt	1	0.6	0	0	0	0	0
$2\Delta t$	1	0.48	0.36	0	0	0	0
$3\Delta t$	1	0.72	0.216	0.216	0	0	0
$3\Delta t$	1	0.5856	0.5184	0.0864	0.1296	0	0
$4\Delta t$	1	0.7939	0.2995	0.3715	0.0259	0.0777	0
$5\Delta t$	1	0.6209	0.6394	0.1210	0.2644	0	0.0467
$6\Delta t$	1	0.8594	0.3173	0.5181	0.0197	0.1866	−0.0093
$7\Delta t$	1	0.6185	0.7630	0.0986	0.4189	−0.0311	0.1306

Table 8.2 Explicit Computation with Stable Parametric Choice(s).

t	$i = 1$	$i = 2$	$i = 3$	$i = 4$	$i = 5$	$i = 6$	$i = 7$
0	1	0	0	0	0	0	0
Δt	1	0.4	0	0	0	0	0
$2\Delta t$	1	0.48	0.16	0	0	0	0
$3\Delta t$	1	0.56	0.224	0.064	0	0	0
$4\Delta t$	1	0.6016	0.2944	0.1024	0.0256	0	0
$5\Delta t$	1	0.6381	0.3405	0.1485	0.0461	0.0102	0
$6\Delta t$	1	0.6638	0.3872	0.1843	0.0727	0.0205	0.0041
$7\Delta t$	1	0.6859	0.4158	0.2190	0.0965	0.0348	0.0090

This is an important lesson. If we need good spatial resolution, Δy will be small and Δt will need to be *very small,* perhaps prohibitively small. Fortunately, we do have options that will work well for this type of problem. Before we consider them, however, we will look specifically at the entry in Table 8.2 for $i = 4$ and $t = 7\Delta t$ (which is 0.2190); the analytic solution for this particular point is $1-erf(0.8964) = 0.205$, so the discrepancy produced by the explicit computation amounts to a little less than 7%. Though larger than we would like, this would still be satisfactory for many applications.

This explicit method can be used to solve more difficult problems as well. Consider a fluid flowing through a cylindrical tube with the provision for heat to be added at a constant rate at the wall. This is a steady-state problem, so there is no time derivative but we can treat $\frac{\partial T}{\partial z}$ in precisely the same way. The governing equation is

$$V_z \frac{\partial T}{\partial z} = \alpha \left[\frac{\partial^2 T}{\partial r^2} + \frac{1}{r}\frac{\partial T}{\partial r} \right]. \tag{8.60}$$

We are concerned with laminar flow through the tube, so $V_z = V_{\max}\left[1 - \frac{r^2}{R^2}\right]$ where $V_{\max}$ is the centerline ($r = 0$) velocity; therefore,

$$\frac{\partial T}{\partial z} = \frac{\alpha}{V_{\max}\left(1 - \frac{r^2}{R^2}\right)} \left[\frac{\partial^2 T}{\partial r^2} + \frac{1}{r}\frac{\partial T}{\partial r} \right]. \tag{8.61}$$

Although $V_z(r = R) = 0$, we do not need to worry about dividing by zero since our computational procedure will not visit the node at the wall. Using the index i for the r-position, and j for z, our finite difference approximation for this equation is written:

$$\frac{T_{i,j+1} - T_{i,j}}{\Delta z} \cong \frac{\alpha}{V_{\max}\left(1 - \frac{r^2}{R^2}\right)} \left[\frac{T_{i+1,j} - 2T_{i,j} + T_{i-1,j}}{(\Delta r)^2} + \frac{1}{r}\frac{T_{i+1,j} - T_{i-1,j}}{2\Delta r} \right]. \tag{8.62}$$

There are two features of the discretized equation we should note: First, $T_{i,j+1}$ appears only on the left-hand side of the equation so it can be easily isolated, and second, a second-order central difference has been used for $\frac{\partial T}{\partial r}$. The algorithm will allow us to march downstream in the z-direction, computing new $T(r)$'s as we go. Of course we must be able to compute the wall temperature, $T(r = R)$ at each z-position, and this will be obtained from our requirement that the flux of thermal energy into the fluid be constant. Using Fourier's law, we write

$$q_0 = -k \left.\frac{\partial T}{\partial r}\right|_{r=R}, \tag{8.63}$$

where k is the thermal conductivity of the fluid. Since the flux is constant we can use a second-order backward difference and write

$$-\frac{q_0}{k} = \left.\frac{\partial T}{\partial r}\right|_{r=R} \cong \frac{3T_{i,j} - 4T_{i-1,j} + T_{i-2,j}}{2\Delta r} \tag{8.64}$$

where the index, i, corresponds to the wall position. The location where this relation is implemented in the code below is labeled: ◀*wall temperature*. The centerline temperature is obtained by symmetry since $\left.\frac{\partial T}{\partial r}\right|_{r=0} = 0$ and this line in the code is labeled: ◀*centerline temperature*. We assume that the fluid enters the heated section at some uniform temperature (10°), and since heat is added at a constant rate, the average (or bulk fluid) temperature, T_b, must increase linearly as the fluid moves downstream. We define the *heat transfer coefficient*, h, by writing

$$q_0 = h(T_{r=R} - T_b), \tag{8.65}$$

and this means that if we compute the bulk fluid (or mixing-cup) temperature we can obtain values for $h(z)$. But we must remember that since thermal energy is being transferred in the *negative* r-direction (toward the center of the tube), q_0 is negative by convention.

One of the nice things about this particular problem is that the analytic solution is known and as $z \to \infty$, the heat transfer coefficient must assume a value such that in dimensionless form, $Nu = \frac{hd}{k} = 4.3636$ (this grouping is known as the Nusselt number). In our initial discussion of the explicit method we observed that stability was a principal concern; in this second example we will focus upon accuracy by exploring whether the computed limiting value of Nu approaches 4.3636 as it should.

We will need to determine the bulk fluid temperature in order to find h and Nu, and this must be done by integration since both the temperature and the velocity vary with respect to r-position:

$$T_b = \frac{\int_0^R 2\pi r V_z(r) T(r) dr}{\pi R^2 \langle V_z \rangle}. \tag{8.66}$$

$\langle V_z \rangle$ is the average velocity and for laminar flow in a tube, $\langle V_z \rangle = \frac{1}{2} V_{\max}$. Now we are in position to assemble the necessary logic to solve this problem explicitly, and the required structure appears below.

```
#COMPILE EXE
#DIM ALL
   GLOBAL vz,rpos,dr,alpha,kcon,rho,cp,vmax,RR,q0,dtdr,d2tdr2,dz,zpos,zlimit AS
SINGLE
   GLOBAL nn,tbar,nusselt,sum,pi,h AS SINGLE
   GLOBAL i AS INTEGER
FUNCTION PBMAIN
   DIM t(56,2) AS SINGLE
     RR=1:dr=RR/55:alpha=0.001445:cp=1:rho=1:kcon=alpha*rho*cp:vmax=10:q0=-0.02:dz=
0.02
     zlimit=300:pi=3.1416
  REM *** initialize temps everywhere
     FOR i=1 TO 55
      T(i,1)=10
      NEXT i
       T(56,1)=1/3*(-q0/kcon*2*dr+4*T(55,1)-T(54,1))
  100 REM *** prepare for forward marching in z-direction
       FOR i=2 TO 55
        d2tdr2=(T(i+1,1)-2*T(i,1)+T(i-1,1))/dr^2
        dtdr=(T(i+1,1)-T(i-1,1))/(2*dr)
          rpos=(i-1)*dr
          vz=vmax*(1-rpos^2/RR^2)
          T(i,2)=dz*alpha/vz*(d2tdr2+1/rpos*dtdr)+T(i,1)
          NEXT i
  REM *** handle symmetry at center of tube and wall temp
```

```
        T(1,2)=1/3*(4*T(2,2)-T(3,2))                    ◀centerline temperature
        T(56,2)=1/3*(-q0/kcon*2*dr+4*T(55,2)-T(54,2))   ◀wall temperature
    zpos=zpos+dz
      PRINT zpos,T(55,2),T(52,2)
REM *** swap z-position values
    FOR i=1 TO 56
     T(i,1)=T(i,2)
        NEXT i
     IF zpos>zlimit THEN 200 ELSE 100
200 REM *** continue
REM *** find the average bulk fluid temperature
        sum=0
         FOR i=2 TO 55
           rpos=(i-1)*dr
           vz=vmax*(1-rpos^2/RR^2)
           sum=sum+2*pi*rpos*vz*T(i,1)*dr
             NEXT i
              tbar=sum/(pi*RR^2*vmax/2)
              PRINT tbar
REM *** find h and the nusselt number
        h=-q0/(T(56,1)-TBAR)
          nusselt=h*2*RR/kcon
          PRINT nusselt
       INPUT "Shall we continue execution?";nn
       IF nn>0 THEN END
```

This program was used to compute $Nu(z)$ for comparison with the known limiting behavior.

The computations reveal that $Nu = 4.382$ at $z/R = 1250$, which is about 0.4% bigger than the correct limiting value. Observe that for the parameters employed in this program,

$$\frac{\alpha(\Delta z)}{(\Delta r)^2} = \frac{(0.001445)(0.02)}{(.01818)^2} = 0.0874. \tag{8.67}$$

The behavior evinced by Figure 8.6 is an excellent result, and it should allay any fears we might have regarding the reliability of computations performed with the explicit technique. We see that it is possible to obtain accurate answers for important practical problems using this method.

For a concluding illustration of the implementation of the elementary explicit method, let us look at a common type of problem in spherical coordinates. A porous adsorbent particle is placed in a well-stirred aqueous solution bearing a contaminant species that is to be removed. The concentration of the contaminant in the spherical particle during the sorption process is governed by the parabolic partial differential equation,

$$\frac{\partial C}{\partial t} = D\left[\frac{\partial^2 C}{\partial r^2} + \frac{2}{r}\frac{\partial C}{\partial r}\right]. \tag{8.68}$$

The initial concentration in the adsorbent sphere is zero and the solution volume is large enough that the contaminant concentration at the surface of the particle is constant, $C(r = R,t) = 1$. Letting the subscript i refer to r-position and j to time, we arrange a finite difference approximation for the PDE for explicit solution:

$$C_{i,j+1} = \Delta tD\left[\frac{C_{i+1,j} - 2C_{i,j} + C_{i-1,j}}{(\Delta r)^2} + \frac{2}{r}\frac{C_{i+1,j} - C_{i-1,j}}{2\Delta r}\right] + C_{i,j}. \tag{8.69}$$

We are also interested in the total uptake of contaminant by the spherical particle which can be determined by integration:

$$M_t = \int_0^t 4\pi R^2 N_{r=R}dt, \tag{8.70}$$

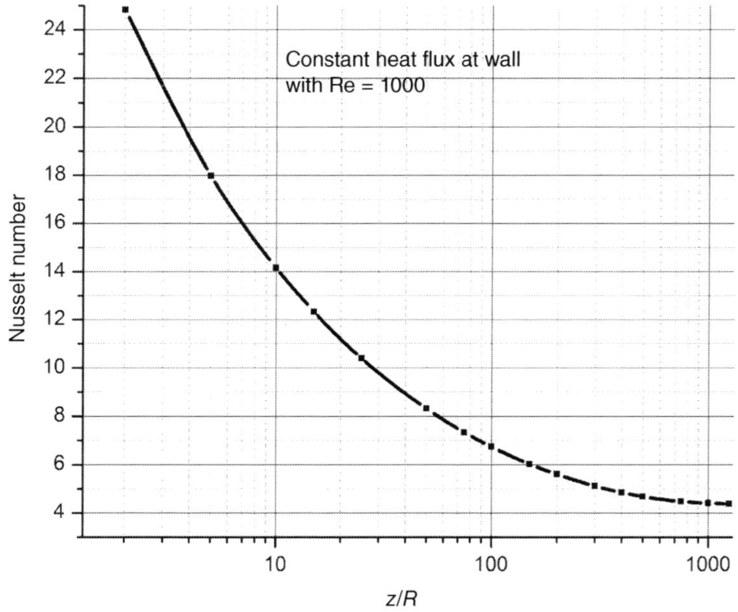

Figure 8.6 Computed Nusselt number as a function of *z/R*. The Reynolds number, Re, is the ratio of inertial and viscous forces and for flow in a tube, $\mathrm{Re} = \frac{d\langle V_z\rangle}{\nu}$.

where the contaminant flux is given by $N = -D\frac{\partial C}{\partial r}$. Again, we observe that this flux is negative since transport is occurring in the negative *r*-direction. The analytic solution for $C(r,t)$ is readily determined as we saw in the previous chapter:

$$C = 1 + \sum_{n=1}^{\infty} \frac{A_n}{r} \exp\left(-D\lambda_n^2 t\right) \sin \lambda_n r \tag{8.71}$$

where $\lambda_n = \frac{n\pi}{R}$ and $A_n = -\frac{2}{R}\int_0^R r \sin \frac{n\pi r}{R} dr$ since $C(r,t = 0) = 0$. We take $R = 0.2\,\text{cm}$, $D = 1\times 10^{-4}\ \text{cm}^2/\text{s}$, $\Delta t = 0.02\,\text{s}$, and $\Delta r = 0.2/50 = 0.004\,\text{cm}$; consequently, $\frac{\Delta t D}{(\Delta r)^2} = 0.125$. Note that the chosen value of the diffusivity, *D*, is quite large for an aqueous system—this was done solely to decrease the time required for the sorptive process. Finally, we recognize that M_t will approach some ultimate value (M_∞) as the available sorptive sites in the interior of the particle are occupied; we would like to explore the dynamics of that approach to M_∞ as part of our solution procedure.

We can perform the explicit computation and compare the results with the analytic solution. The main functional piece of the scheme we will employ is simply:

```
FOR i=2 TO 50
        rpos=(i-1)*dr
                dcdr=(C(i+1,1)-C(i-1,1))/(2*dr)
                d2cdr2=(C(i+1,1)-2*C(i,1)+C(i-1,1))/dr^2
                        C(i,2)=dt*D*(d2cdr2+2/rpos*dcdr)+C(i,1)
                        NEXT i
```

You will observe that second-order central differences are being used for both first and second derivatives on the right-hand side of eq. (8.68). Of course the concentration must be finite at the center of the sphere where we have symmetry, so we will set $\left.\frac{\partial C}{\partial r}\right|_{r=0} = 0$. The numerical and analytic solutions are compared in Figure 8.7.

Due to the small particle size and the large value of the diffusivity, *D*, the sorption of contaminant is rapid. The total uptake is determined by integration, and this is easily incorporated into the explicit calculation. The results show that the sorptive occupancy of the spherical particle (fractional approach to saturation) is at 50% in just 12 s, 90% in about 74 s, and virtually complete by 400 s. This is an excellent practice problem for the student new to numerical solution of partial differential equations, so he/she may want to confirm the results provided here.

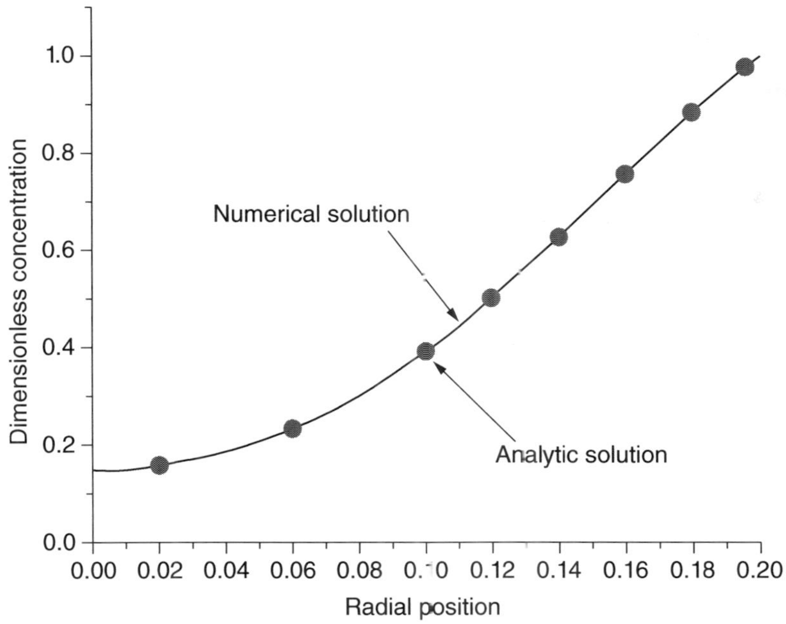

Figure 8.7 Concentration of contaminant in the interior of the porous spherical adsorbent particle at $t = 30$ s. The explicitly computed results are shown as a solid curve and select points from the analytic solution are provided for comparison. The agreement is excellent.

8.5.2 Du Fort–Frankel Scheme

In our initial discussion of the elementary explicit technique we observed that the stability requirement was: $\frac{\Delta t \nu}{(\Delta y)^2} \leq \frac{1}{2}$. We will discuss the origin of this condition momentarily, but it is worthwhile to question whether or not some other finite difference representation of the parabolic PDE might permit a direct, but less constrained computation. After all, we have an enormous variety of difference approximations including forward, backward, centered, and skewed, first order, second order, and third order. Could a parabolic PDE be approximated in some other way that would ensure stability and perhaps even improve accuracy while still providing an elementary computation? Consider a region (a slab) that extends from $x = 0$ to $x = L$, where

$$\frac{\partial U}{\partial t} = \kappa \frac{\partial^2 U}{\partial x^2}. \tag{8.72}$$

The dependent variable, U, is initially zero for all values of x, but at $t = 0$, U is elevated to 500 at both ends of the slab. The analytic solution for this problem is straightforward, and the reader may wish to show that

$$U = 500 - \sum_{n=1,3,5,}^{\infty} \frac{2000}{n\pi} \exp\left(-\kappa \frac{n^2\pi^2}{L^2} t\right) \sin \frac{n\pi x}{L}. \tag{8.73}$$

This will allow us to assess whether or not the proposed scheme has merit. Now we write an alternative difference representation for eq. (8.72); let the index j represent x-position and let m represent time:

$$\frac{U_{j,m+1} - U_{j,m-1}}{2\Delta t} \cong \frac{\kappa}{(\Delta x)^2}[U_{j+1,m} - (U_{j,m+1} + U_{j,m-1}) + U_{j-1,m}]. \tag{8.74}$$

Notice that a second-order, central difference has been used for the accumulation term (the time derivative), and that the expected $2U_{j,m}$ on the right-hand side has been replaced by the average of the old $(m-1)$ and new $(m+1)$ U values. Now let $B = \frac{2\Delta t \kappa}{(\Delta x)^2}$ and rearrange eq. (8.74):

$$U_{j,m+1}[1 + B] = B[U_{j+1,m} - U_{j,m-1} + U_{j-1,m}] - U_{j,m-1}. \tag{8.75}$$

The new time-step value $(m + 1)$ appears only on the left-hand side, so eq. (8.75) is explicit—this is the *Du Fort–Frankel* scheme. However, the algorithm does require the central value for U on the *old* time step $(m-1)$. If we can figure out how to get the computation started, we should be able to calculate new U's everywhere in the interior of the slab. We will set

$L = 10$, $\kappa = 0.125$, $\Delta t = 0.0025$, and $\Delta x = 10/400 = 0.025$, and therefore $B = \frac{(2)(0.0025)(0.125)}{(0.025)^2} = 1.0$. We will look at the dynamic behavior of U in the center of the slab using both the analytic solution and eq. (8.75):

x Position	Time	*U* Analytic	*U* Du Fort–Frankel
5	5	0.00766	0.00766
5	10	1.565	1.564
5	25	45.500	45.49
5	50	157.28	157.26
5	100	314.61	314.40
5	200	446.01	445.64
5	300	484.28	483.91
5	400	495.42	495.06
5	500	498.67	498.30

At $t = 500$, the discrepancy amounts to about 0.074%; the Du Fort–Frankel method has given us exceptionally good results for this example.

8.5.3 Von Neumann Stability Analysis

How does one determine if his or her proposed computational scheme will actually work? We will answer this question by revisiting eq. (8.72), and writing it in a very familiar finite difference form:

$$U_{j,m+1} = U_{j,m} + C\left(U_{j+1,m} - 2U_{j,m} + U_{j-1,m}\right), \tag{8.76}$$

where $C = \frac{\kappa \Delta t}{(\Delta x)^2}$. Now we define an error, ε, as the difference between the computed numerical solution and the actual solution of eq. (8.76) with no round-off error (with unlimited numerical precision). Since the error must also satisfy eq. (8.76), we note

$$\varepsilon_{j,m+1} = \varepsilon_{j,m} + C(\varepsilon_{j+1,m} - 2\varepsilon_{j,m} + \varepsilon_{j-1,m}). \tag{8.77}$$

The underlying idea here is that the error, which is a function of both x-position and time, can be represented as a Fourier series. If we assume that the Fourier coefficients depend only upon time and that the error will either grow or be damped exponentially, then

$$\varepsilon(x,t) = \sum_{n=1}^{N} \exp(at)\exp(ik_n x), \tag{8.78}$$

where N is the number of discrete intervals; for $0 \le x \le L$, $N = L/\Delta x$. We substitute this expansion into eq. (8.77):

$$e^{a(t+\Delta t)}e^{ik_n x} = e^{at}e^{ik_n x} + C\left(e^{at}e^{ik_n(x+\Delta x)} - 2e^{at}e^{ik_n x} + e^{at}e^{ik_n(x-\Delta x)}\right) \tag{8.79}$$

and divide by $e^{at}e^{ik_n x}$:

$$e^{a\Delta t} = 1 + C\left(e^{ik_n \Delta x} - 2 + e^{-ik_n \Delta x}\right). \tag{8.80}$$

Since $e^{ik_n \Delta x} = \cos k_n \Delta x + i \sin k_n \Delta x$ and $e^{-ik_n \Delta x} = -\cos k_n \Delta x - i \sin k_n \Delta x$, we find

$$e^{a\Delta t} = 1 + C(2\cos k_n \Delta x - 2), \tag{8.81}$$

but we know from a double-angle formula that $2(\cos\theta - 1) = -4\sin^2\frac{\theta}{2}$ and consequently,

$$e^{a\Delta t} = 1 + C\left(-4\sin^2\frac{k_n \Delta x}{2}\right). \tag{8.82}$$

We cannot allow the error to grow exponentially, so

$$\left|1 - 4C\sin^2\frac{k_n \Delta x}{2}\right| \le 1. \tag{8.83}$$

The *largest* value that can be taken on by $\sin^2\theta$ is 1; thus, $4C \le 2$, and the stability threshold for the elementary explicit technique is:

$$\frac{\kappa \Delta t}{(\Delta x)^2} \le \frac{1}{2}. \tag{8.84}$$

Let us make perfectly clear what John von Neumann's stability analysis provides for us: It is a Fourier technique that produces a stability threshold for finite difference solutions of *linear* partial differential equations. If we apply this method to the Du Fort–Frankel scheme described previously, we find that it is *stable*. However, do not be misled by this statement; we cannot simply use *any* Δt in a stable computational scheme and expect to get satisfactory results! We will underscore this point in the next section.

8.5.4 The Crank–Nicolson Method

Consider a transient diffusion problem in two spatial dimensions (rectangular coordinates):

$$\frac{\partial C}{\partial t} = D\left[\frac{\partial^2 C}{\partial x^2} + \frac{\partial^2 C}{\partial y^2}\right], \tag{8.85}$$

where C is the molar concentration of the species of interest and D is the diffusivity.

If we were to solve this problem using the explicit approach described in the previous section, we would have to choose Δt such that

$$\Delta t D\left(\frac{1}{(\Delta x)^2} + \frac{1}{(\Delta y)^2}\right) \le \frac{1}{2}. \tag{8.86}$$

If the problem required enhanced spatial resolution, then the time-step size, Δt, would need to be very small, and the required computational effort might be excessive (particularly in view of the large time required for the concentration field in many diffusion problems to develop, for example in many liquids, $D \approx 10^{-5}$ cm^2/s). Suppose we choose $\Delta x = \Delta y = 1/20$ and $D = 1/10$; then, $80 \cdot \Delta t \le \frac{1}{2}$ and Δt must be less than 0.00625 if we are performing the computation explicitly. Fortunately there are alternatives and the Crank–Nicolson method is one option.

In the Crank–Nicolson approach, a first-order, forward difference is used for the time derivative, and the second derivatives (the molecular transport terms) are written twice, once on the present time-step row, t, and once for $t + \Delta t$. The arithmetic average of the two values is used in the computation. Let the i, j, and k indices correspond to x, y, and t, respectively. The scheme can be written out:

$$\frac{C_{i,j,k+1} - C_{i,j,k}}{\Delta t} \cong \frac{D}{2}\left[\frac{C_{i+1,j,k} - 2C_{i,j,k} + C_{i-1,j,k}}{(\Delta x)^2} + \frac{C_{i,j+1,k} - 2C_{i,j,k} + C_{i,j-1,k}}{(\Delta y)^2} + \right.$$
$$\left. + \frac{C_{i+1,j,k+1} - 2C_{i,j,k+1} + C_{i-1,j,k+1}}{(\Delta x)^2} + \frac{C_{i,j+1,k+1} - 2C_{i,j,k+1} + C_{i,j-1,k+1}}{(\Delta y)^2}\right]. \tag{8.87}$$

Of course, this algorithm is implicit, which means that a set of simultaneous algebraic equations must be solved to advance to the new time-step row, $k+1$, i.e., $t + \Delta t$. Note that the computational pattern involves five points: the central node, i,j, then left and right, and up and down. The Crank–Nicolson method is stable for *any* value of Δt. We employ a square mesh so that $\Delta x = \Delta y$ and isolate the $k+1$ values on the left-hand side of the equation:

$$C_{i,j,k+1}\left[\frac{1}{\Delta t} + \frac{2D}{(\Delta x)^2}\right] - \frac{D}{2(\Delta x)^2}(C_{i+1,j,k+1} + C_{i-1,j,k+1} + C_{i,j+1,k+1} + C_{i,j-1,k+1})$$
$$= \frac{D}{2(\Delta x)^2}\left[C_{i+1,j,k} + C_{i-1,j,k} + C_{i,j+1,k} + C_{i,j-1,k} - 4C_{i,j,k}\right] + \frac{C_{i,j,k}}{\Delta t}. \tag{8.88}$$

An attractive feature of this approach is that the coefficients for the computational pattern on the new time-step row are simply:

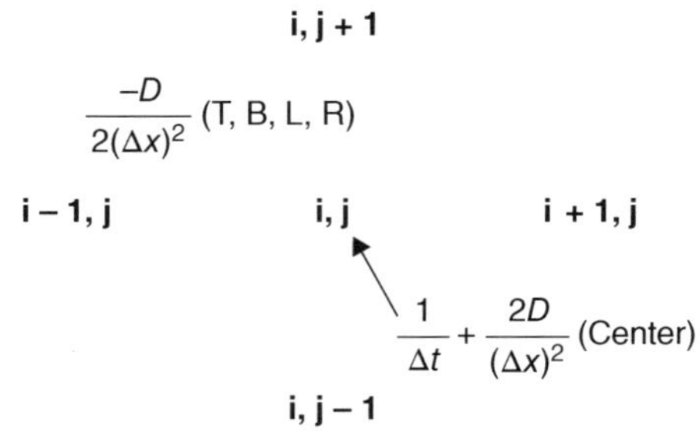

Let us illustrate the advantages offered by Crank–Nicolson with an example. Suppose we have a slab of material with a thermal diffusivity (α) of 0.03 cm^2/s, which is roughly characteristic of minerals like fluorite and quartz. The slab measures 6 cm by 6 cm, and it has an initial temperature of 0°. At $t = 0$, the temperature of the left-hand side is instantaneously raised to 1,000° and the top edge to 600°. The other two edges are maintained at 0° for all time. In this case, of course, the governing equation is $\frac{\partial T}{\partial t} = \alpha \left[\frac{\partial^2 T}{\partial x^2} + \frac{\partial^2 T}{\partial y^2} \right]$, which is completely analogous to eq. (8.85). We are interested in the temperature distribution in the slab at $t = 50$ s. We first compute the result explicitly, using $\Delta t = 0.01$ s which corresponds to 5,000 time steps. The results are shown in the array immediately below.

600	600	600	600	600	600	600
1,000	644.62	461.64	374.39	324.39	245.17	0
1,000	607.64	348.74	217.09	153.31	96.47	0
1,000	570.37	280.35	133.69	70.72	36.25	0
1,000	525.29	231.49	91.31	36.06	14.12	0
1,000	404.95	154.94	53.92	17.69	5.51	0
1,000	0	0	0	0	0	0

Now we carry out the computation a second time, but we use Crank–Nicolson with $\Delta t = 50$ s; i.e., we employ just *one time step*! We should not expect the two sets of results to compare favorably.

600	600	600	600	600	600	600
1,000	831.73	520.05	430.11	389.99	314.98	0
1,000	715.85	311.76	183.87	134.90	89.89	0
1,000	674.36	242.93	103.88	55.70	29.52	0
1,000	637.82	205.64	71.53	28.78	11.86	0
1,000	521.68	144.47	43.21	14.42	4.92	0
1,000	0	0	0	0	0	0

By no means is this acceptable. But remember that we have *replaced 5,000 time steps* (explicit) *with just one* (Crank–Nicolson). If we reduce the total time by a factor of 10, that is, we carry out the calculations to $t = 5$ s using both the explicit technique with $\Delta t = 0.01$ s and a single 5 s step with Crank–Nicolson, the typical discrepancy is just a few percent. And, if we drop down to 2 s to compare 200 time steps (explicit) to just one (Crank–Nicolson) we find that the typical difference for values in the first interior column (at $t = 2$ s) is less than 0.5%; this is illustrated below.

Explicit	Crank–Nicolson
88.39	88.21
57.79	57.52
56.87	56.67
56.82	56.61
55.25	55.13

Because Crank–Nicolson is so easy to use in one spatial dimension, the reader is encouraged to try applying the method to the following slab example. The initial temperature of the semi-infinite slab is zero; at $t = 0$ the temperature of the front face is elevated to 500°. Given a thermal diffusivity of 0.12 cm^2/s, we compute the temperature distribution in the slab using both the analytic solution, $T = 500 erfc\left(\frac{y}{\sqrt{4\alpha t}}\right)$, and Crank–Nicolson with a single time step. After 16 s, the temperature profiles appear as shown here:

y (cm)	0	1	2	3	4	5	6	7
T (analytic)	500	303.5	154.2	65.9	23.9	7.5	2.1	0.5
T (CN)	500	375	140.6	52.7	19.8	7.4	2.8	1.0

Again, the reader should note that the Crank–Nicolson calculation employed *just one* 16 s time step; he/she might also consider repeating the calculation but with $\Delta t = 1$ s.

8.5.5 Alternating-Direction Implicit (ADI) Method

The Peaceman-Rachford (1955), or alternating direction implicit (ADI), method can be particularly useful for the types of parabolic partial differential equations we have been discussing, and it is more efficient than Crank–Nicolson. Let the indices i, j, and k represent x, y, and t, respectively. We will use transient conduction in two spatial dimensions for our example:

$$\frac{\partial T}{\partial t} = \alpha \left[\frac{\partial^2 T}{\partial x^2} + \frac{\partial^2 T}{\partial y^2}\right]. \tag{8.89}$$

The first half of the ADI algorithm is used to advance to the $k+1$ time step:

$$\frac{T_{i,j,k+1} - T_{i,j,k}}{\alpha \Delta t} = \frac{T_{i+1,j,k+1} - 2T_{i,j,k+1} + T_{i-1,j,k+1}}{(\Delta x)^2} + \frac{T_{i,j+1,k} - 2T_{i,j,k} + T_{i,j-1,k}}{(\Delta y)^2}, \tag{8.90}$$

and the second half takes us to $k+2$:

$$\frac{T_{i,j,k+2} - T_{i,j,k+1}}{\alpha \Delta t} = \frac{T_{i+1,j,k+1} - 2T_{i,j,k+1} + T_{i-1,j,k+1}}{(\Delta x)^2} + \frac{T_{i,j+1,k+2} - 2T_{i,j,k+2} + T_{i,j-1,k+2}}{(\Delta y)^2}. \tag{8.91}$$

Note that neither step can be repeated unilaterally. Let us examine a simple application. A two-dimensional slab of material is at a uniform initial temperature of 100°. At $t = 0$, one face (the bottom) is instantaneously heated to 400°. Let $\Delta x = \Delta y = 1$, as well as $\alpha = 1$ and $\Delta t = 1/8$. We rewrite eq. (8.90) isolating the $k+1$ terms on the right-hand side:

$$-T_{i,j+1,k} + \left(2 - \frac{(\Delta x)^2}{\alpha \Delta t}\right) T_{i,j,k} - T_{i,j-1,k} = T_{i-1,j,k+1} - \left(2 + \frac{(\Delta x)^2}{\alpha \Delta t}\right) T_{i,j,k+1} + T_{i-1,j,k+1}. \tag{8.92}$$

Now we will illustrate the process with a simple square slab; the top, left, and right sides are all maintained at 100°. The bottom will be set to 400°. The nine interior nodes are initialized at 100°.

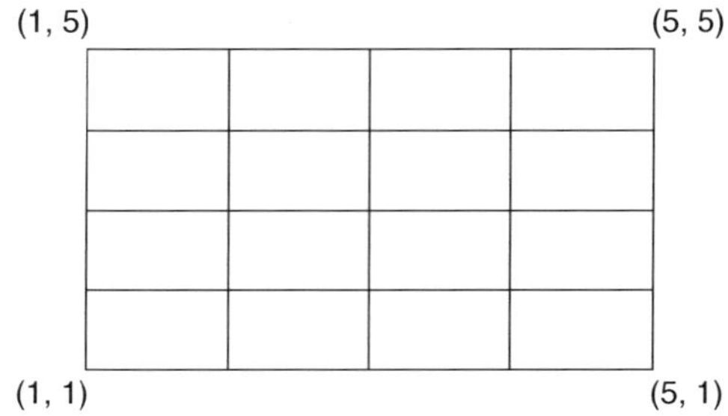

We apply eq. (8.90) at the interior points, row by row; the first horizontal sweep results in:

100	100	100
100	100	100
133.67	136.73	133.67

for the nine interior points. Now we recast eq. (8.91) for application to the columns in order to advance to the $k+2$ time step:

$$-T_{i+1,j,k+1}+\left(2-\frac{(\Delta x)^2}{\alpha\Delta t}\right)T_{i,j,k+1}-T_{i-1,j,k+1}=T_{i,j+1,k+2}-\left(2+\frac{(\Delta x)^2}{\alpha\Delta t}\right)T_{i,j,k+2}+T_{i,j-1,k+2}. \tag{8.93}$$

We solve the simultaneous equations that result from applying this algorithm to the columns, and obtain:

100.55	100.6	100.55
105.5	106	105.5
154.42	159.37	154.42

If the total number of equations is modest, then a direct elimination scheme can be used for solution. The coefficient matrix follows the tridiagonal pattern (with 1, −10, 1, for the selected parameters), so the process is easy to automate. Smith (1965) states that for rectangular regions the ADI method requires about 25 *times less work* than an explicit computation. Carrying out the procedure to $t = 1.75$ yields:

114.91	120.25	114.91
146.35	161.01	146.35
221.06	247.42	221.06

for the interior nodes. Chung (2002) notes that this scheme is unconditionally stable which makes it very attractive for problems in which the time evolution is slow; we can employ a *very* large Δt relative to the elementary explicit technique and still obtain acceptable accuracy.

Now we will look at a detailed example accompanied by a complete program structure that highlights some of the important advantages of ADI. We will consider a square geometry where the dependent variable, U, is governed by

$$\frac{\partial U}{\partial t}=\kappa\left[\frac{\partial^2 U}{\partial x^2}+\frac{\partial^2 U}{\partial y^2}\right]. \tag{8.94}$$

Let the height and width of the domain be 5 by 5, and let the molecular transport coefficient (or diffusivity), $\kappa = 1/10$. We will take the initial value for U to be zero everywhere in the interior of the square slab, but at $t = 0$, the bottom edge will instantaneously jump to $U = 500$. Our interest of course is the evolution of $U(x,y,t)$ in the interior of the slab. We will employ a fairly coarse discretization with 19 interior mesh points in both the x- and y-directions, which means that we will have to solve 361 simultaneous algebraic equations (NEQ = 361) for both the horizontal and vertical sweeps. The coefficient matrix is sparse since each equation will involve just three unknown nodal values—the tridiagonal form. There are many routines that can be used to solve these algebraic equations, and we will adapt Crout's method from Chapter 2 for this purpose. Consequently, our main task in setting up the program logic is to assign the proper values to the coefficient matrix for each sweep. Note that the total number of mesh points with the boundaries included will be 441 (21 by 21); this appears in the code as NX = 21. The unknowns, the nodal values of U, will be numbered consecutively starting with the first (bottom) interior row, and these values will be $X(1)$ through $X(19)$ in Crout's routine; this means that $X(1)$ corresponds to $U_{i,j} = U(2,2)$ since the position indices start with 1,1 in the lower right-hand corner.

```
#COMPILE EXE
#DIM ALL
  REM *** ADI program with Crout's method for solution of algebraic equations
    GLOBAL I,J,K,M,N,NEQ AS INTEGER
    GLOBAL SUM,DX,DY,KAPPA,DT,CMID,CLEFT,CRIGHT,RHS,ZZ AS SINGLE
    GLOBAL II,JM1,K,IP1,IM1,JJ,NN,L,NX AS INTEGER
    GLOBAL TT,CUP,CDOWN AS SINGLE
  FUNCTION PBMAIN
  DIM A(361,362) AS SINGLE
  DIM X(361) AS SINGLE
  DIM U(21,21,3) AS SINGLE
     DX=0.25:DY=0.25:DT=0.05:KAPPA=1/10:NEQ=361:NX=21
  REM *** initialize all values of dependent variable U
      FOR J=1 TO NX
       FOR I=1 TO NX
        U(I,J,1)=0
          NEXT I:NEXT J
           FOR I=1 TO NX
            U(I,1,1)=500
            NEXT I
  100 REM *** continue
  REM *** horizontal sweep starting with bottom row, first return coefficient matrix
to zero
               GOSUB 300
               FOR J=2 TO NX-1
                FOR I=2 TO NX-1
                    N=(J-2)*(NX-2)+(I-1)
                  CMID=1+2*KAPPA*DT/DX^2
                  CRIGHT=-DT*KAPPA/DX^2
                  CLEFT=CRIGHT
                  RHS=U(I,J,1)*(1-2*KAPPA*DT/DY^2)+KAPPA*DT/DY^2*(U(I,J+1,1)+
U(I,J-1,1))
                    A(N,N)=CMID
                    A(N,N+1)=CRIGHT
                    A(N,N-1)=CLEFT
                    A(N,NEQ+1)=RHS
                      IF I=2 THEN A(N,N-1)=0
                      IF I=(NX-1) THEN A(N,N+1)=0
                         NEXT I:NEXT J
                         GOSUB 500
             FOR J=2 TO NX-1
               FOR I=2 TO NX-1
                 N=(J-2)*(NX-2)+(I-1)
                 U(I,J,2)=X(N)
                   NEXT I:NEXT J
                    PRINT U(2,2,2),U(3,2,2),U(4,2,2),U(5,2,2),U(6,2,2)
                    TT=TT+DT
```

```
REM *** prepare and carry out vertical sweep, swap time values first
                   FOR J=2 TO NX-1
                    FOR I=2 TO NX-1
                     U(I,J,1)=U(I,J,2)
                       NEXT I:NEXT J
           GOSUB 300
           FOR I=2 TO NX-1
             FOR J=2 TO NX-1
               N=(J-2)*(NX-2)+(I-1)
             CMID=1+2*KAPPA*DT/DY^2
             CUP=-DT*KAPPA/DY^2
             CDOWN=CUP
             RHS=U(I,J,1)*(1-2*KAPPA*DT/DX^2)+KAPPA*DT/DX^2*(U(I+1,J,1)+U(I-1,J,1))
                A(N,N)=CMID
                A(N,NEQ+1)=RHS
                  IF N>(NX-2) THEN A(N,N-(NX-2))=CDOWN
                  IF N<(NEQ-(NX-2)) THEN A(N,N+(NX-2))=CUP
                     NEXT J:NEXT I
                     GOSUB 500
          FOR J=2 TO NX-1
            FOR I=2 TO NX-1
              N=(J-2)*(NX-2)+(I-1)
              U(I,J,2)=X(N)
                NEXT I:NEXT J
                 TT=TT+DT
                   FOR J=2 TO NX-1
                     FOR I=2 TO NX-1
                       U(I,J,1)=U(I,J,2)
                         NEXT I:NEXT J
             IF TT<5 THEN 100 ELSE 200
200 REM *** continue
                  OPEN "c:squarADI.dat" FOR OUTPUT AS #1
                   FOR I=1 TO NX
                     FOR J=1 TO NX
                       WRITE#1,I,J,U(I,J,1)
                        NEXT J:NEXT I
                     INPUT "Shall we continue?";zz
                     IF zz>0 THEN CLOSE
                     END
300 REM *** reset matrix to all zeros
       FOR J=1 TO NEQ+1
       FOR I=1 TO NEQ
         A(I,J)=0
         NEXT I:NEXT J
         RETURN

500 REM *** Crout's method of matrix decomposition
     M=NEQ+1
     FOR J=2 TO M
     A(1,J)=A(1,J)/A(1,1):NEXT J
   FOR I=2 TO NEQ
       J=I
```

```
         FOR II=J TO NEQ
         SUM=0
         JM1=J-1
           FOR K=1 TO JM1
           SUM=SUM+A(II,K)*A(K,J):NEXT K
     A(II,J)=A(II,J)-SUM:NEXT II
     IP1=I+1
        FOR JJ=IP1 TO M
        SUM=0
        IM1=I-1
          FOR K=1 TO IM1
          SUM=SUM+A(I,K)*A(K,JJ):NEXT K
     A(I,JJ)=(A(I,JJ)-SUM)/A(I,I):NEXT JJ
     NEXT I
   X(NEQ)=A(NEQ,NEQ+1)
   L=NEQ-1
     FOR NN=1 TO L
     SUM=0
     I=NEQ-NN
     IP1=I+1
        FOR J=IP1 TO NEQ
        SUM=SUM+A(I,J)*X(J)
        NEXT J
     X(I)=A(I,M)-SUM:NEXT NN

         RETURN
END FUNCTION
```

The development of $U(x,y,t)$ is illustrated by Figure 8.8; for this computation, $\kappa=5/10$ and the time variable, $tt = 10$. We can expect U to be symmetric about the vertical centerline of the slab.

8.5.6 Three Spatial Dimensions

Naturally, the solution techniques for parabolic PDEs that we have discussed in this section can be extended to three dimensions as we shall now demonstrate. Consider a cube of solid material, measuring 10 cm on each side, initially at some uniform temperature, T_i. At $t = 0$ the temperatures of the four vertical faces are instantaneously changed to elevated values. In particular, the front face will be 400°, the right-hand face 200°, the left-hand face 1,000°, and the back 600°. The bottom of the cube is insulated, and the top horizontal surface will lose thermal energy to the surroundings by Newton's law of cooling. A sketch of the arrangement appears in Figure 8.9.

The governing equation for this case is just

$$\frac{\partial T}{\partial t} = \alpha\left[\frac{\partial^2 T}{\partial x^2} + \frac{\partial^2 T}{\partial y^2} + \frac{\partial^2 T}{\partial z^2}\right]. \tag{8.95}$$

We will take the thermal conductivity (k) of the medium to be 0.075 cal/(cm s °C) and discretize the equation letting Δx, Δy, and Δz all be 0.16667 cm. Accordingly, the number of interior mesh points will be 205,379. We will use a first-order forward difference for the time derivative and second-order, central differences for the conduction terms:

$$\frac{T_{i,j,k,2} - T_{i,j,k,1}}{\Delta t} \cong \frac{\alpha}{(\Delta x)^2}\left[T_{i+1,j,k,1} + T_{i-1,j,k,1} + T_{i,j+1,k,1} + T_{i,j-1,k,1} + T_{i,j,k+1,1} + T_{i,j,k-1,1} - 6T_{i,j,k,1}\right]. \tag{8.96}$$

Our intent is to solve the equation explicitly by forward marching in time. We will employ just two values for the time index, 1 and 2, corresponding to the old and new time steps (this is done to minimize storage requirements). We take the thermal diffusivity, α, to be 0.088 cm^2/s (therefore $\rho C_p = 0.852$ cal/cm^3 °C) and the time step, Δt, to be 0.01 s, resulting in:

$$\Delta t\alpha\left[\frac{1}{(\Delta x)^2} + \frac{1}{(\Delta y)^2} + \frac{1}{(\Delta z)^2}\right] = 0.095, \tag{8.97}$$

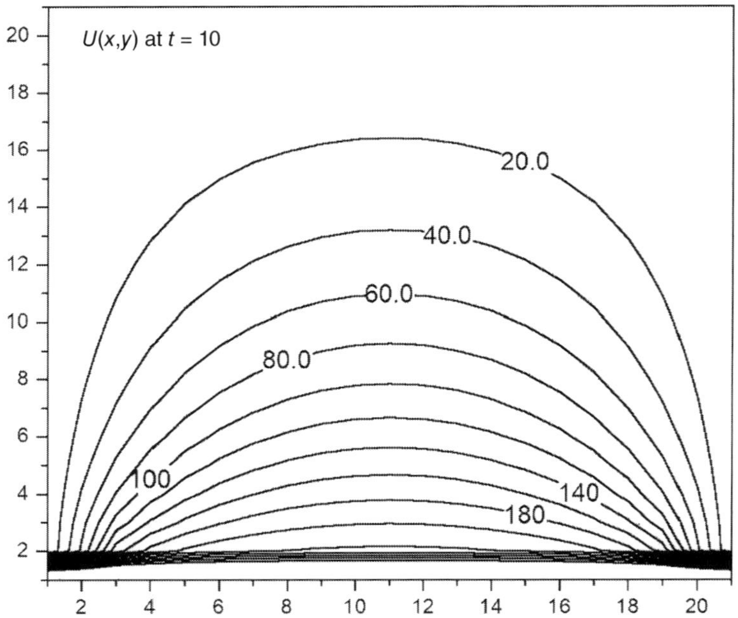

Figure 8.8 Development of $U(x,y)$ in a square domain for time (variable *TT* in program) of 10 and $\kappa = 5/10$ (the value for κ was increased here to speed the evolution of U). The bottom boundary was instantaneously elevated to 500 at $t = 0$, and the sides and top were maintained at 0 for all t. The spatial index for the computation extends from 1 to 21 for both horizontal and vertical directions.

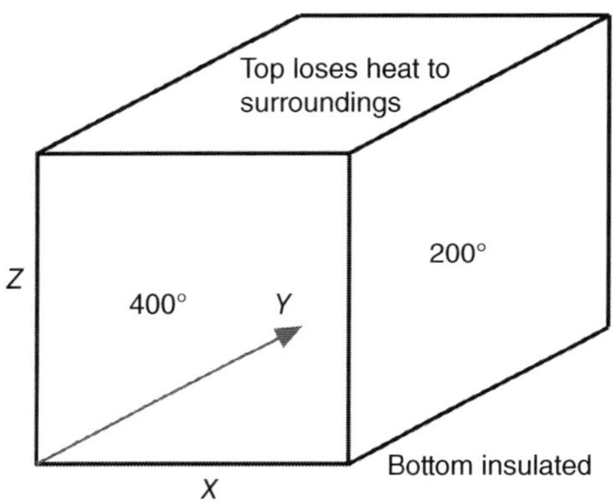

Figure 8.9 Cube of material with four vertical sides maintained at different temperatures for all $t > 0$.

which is much less than the limit for stability (recall that the limit is ½). We can get a sense of how $T(x,y,z,t)$ develops by looking at the *top surface* of the cube at $t = 7.5$, 15, 30, and 60 s; these results are shown in Figure 8.10.

The sequence of contour plots shown in Figure 8.10 reveals the speed with which thermal energy is conveyed throughout the cube. Although the explicit method was used to solve this problem, the execution time was not prohibitively long—despite the fact that each time step required approximately 212,000 calculations. Since Δt was 0.01 s, about 1.27×10^9 calculations were required to reach $t = 60$ s. The ease with which this problem was solved suggests that many momentum, heat, and mass transfer problems involving three spatial dimensions can be handled exactly this way.

8.6 Hyperbolic Partial Differential Equations

Perhaps the best-known example of a hyperbolic PDE is the "wave" equation; for one spatial dimension it can be written:

$$\frac{\partial^2 u}{\partial t^2} = c^2 \frac{\partial^2 u}{\partial x^2}. \tag{8.98}$$

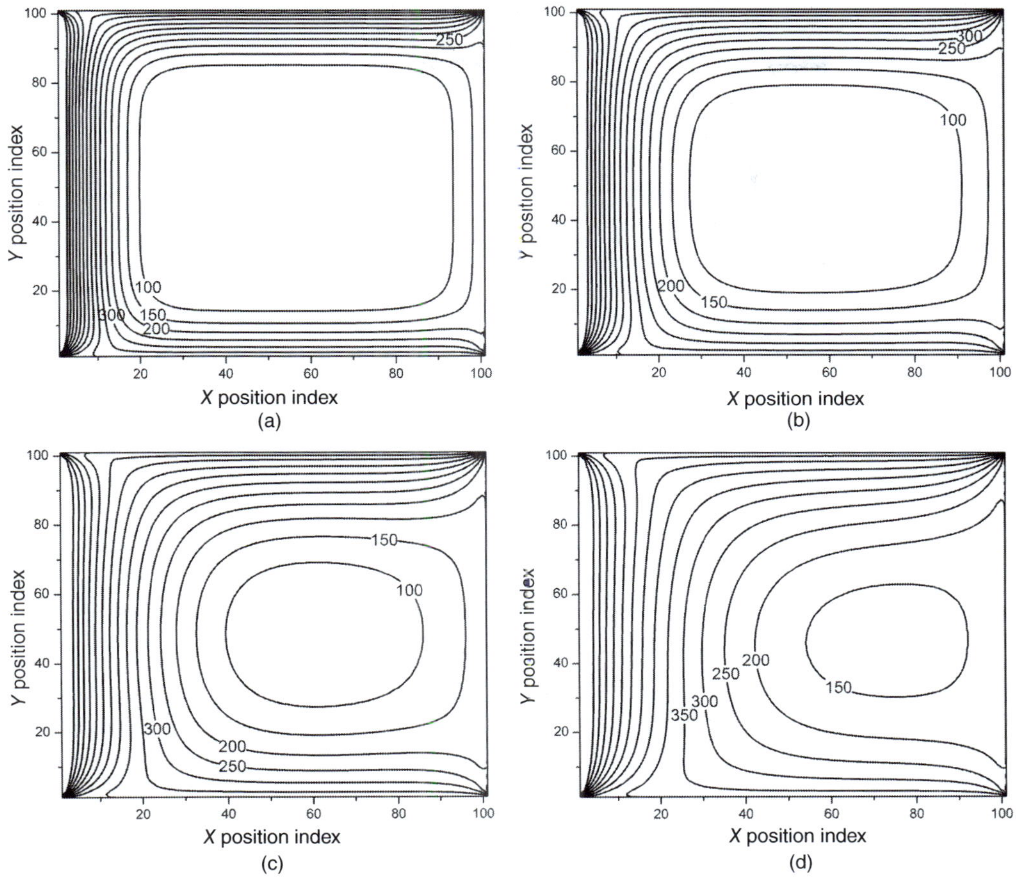

Figure 8.10 Evolution of the temperature distribution *on the top surface* of the cube; the four contour plots correspond to 7.5, 15, 30, and 60 s (top-to-bottom). For the two-dimensional top-view shown here, the right-hand edge is maintained at 200°, the left-hand side at 1,000°, the bottom at 400°, and the top at 600°.

Of course our immediate thought with respect to a physical interpretation might center upon a vibrating string. But wave-like behavior can be found for many different phenomena, including electrical and magnetic fields and even nerve impulses (for the latter, the interested reader should explore the *FitzHugh–Nagumo model*).

Because the wave equation has been around since the middle of the eighteenth century, much is known about its solutions. In fact, the reader is encouraged to apply the variable transformation,

$$r = x + ct \quad \text{and} \quad s = x - ct, \tag{8.99}$$

to eq. (8.98) to reproduce d'Alembert's solution process from 1747. This approach is of particular interest to us because it represents a special case of a technique we wish to discuss, the *method of characteristics*. The name of this technique arises from the fact that at every point in the x–t plane, two characteristic directions can be identified for which *ordinary* differential equations can be used to "solve" eq. (8.98) in a step-wise process.

Before we begin that discussion we will illustrate several important points using an extremely simple first-order, "constant coefficient advection" (first-order wave equation) model:

$$\frac{\partial \phi}{\partial t} + 3\frac{\partial \phi}{\partial x} = 0. \tag{8.100}$$

We let $\phi(x, t = 0) = 0$ and introduce the disturbance (at $x = 0$) that will propagate in the x-direction. We can use this particular model to highlight some of the problems that one may encounter with hyperbolic PDEs. First, we will introduce a finite-duration impulse (with finite amplitude) and solve eq. (8.100) numerically using an explicit approach. Since the "velocity" in the x-direction is "3" we will solve the equation for specific times of 1/3, 2/3, …, 5/3; thus, the advected disturbance should be centered at 1, 2, …, 5.

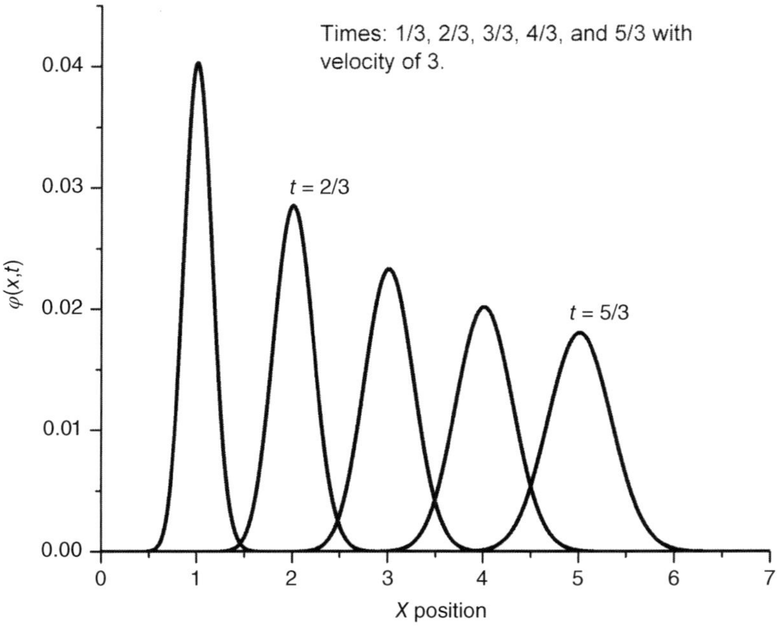

Figure 8.11 Propagation of a finite impulse in the x-direction due to a constant velocity of 3.

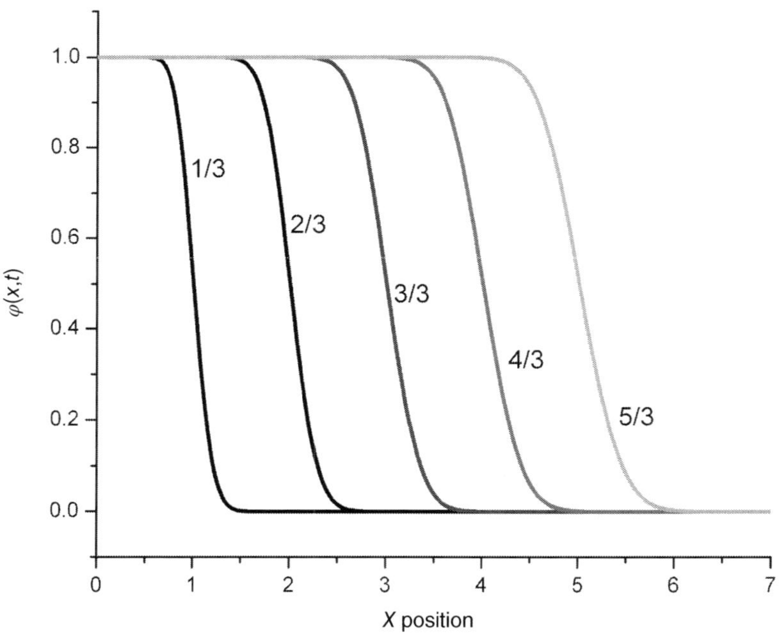

Figure 8.12 Advection of a unit step in the x-direction due to the constant velocity, 3.

The result depicted in Figure 8.11 is probably not what you expected; note the decrease in amplitude and the broadening of the distribution. We can make the nature of the problem even clearer by inputting a unit step change at $x = 0$, letting the "sharp-edged" step be carried along in the positive x-direction (Figure 8.12).

It may be apparent to you that this result also fails to meet expectations; for a homogeneous wave equation in one dimension, the shape of the traveling wave should not change! We will now demonstrate what *should have* transpired. We first do this using a familiar technique, the Laplace transform, which will eliminate the time derivative. Applying the transform and solving the subsidiary equation results in

$$\phi(s) = C_1 \exp\left(-\frac{sx}{3}\right). \tag{8.101}$$

The unit step is put in at the left-hand boundary; that is, at $x = 0$, $\phi(s) = \frac{1}{s}$. Therefore,

$$\phi(s) = \frac{1}{s}\exp\left(-\frac{sx}{3}\right). \tag{8.102}$$

We can invert directly by consulting a table of transforms. Letting $k = x/3$, we find

$$\phi = 0 \text{ for } 0 < t < k \quad \text{and} \quad \phi = 1 \text{ if } t > k.$$

In other words, the unit step disturbance propagates downstream *unchanged* according to the analytic solution—exactly like the behavior of an idealized plug-flow tubular reactor (PFTR). Clearly the solution technique used in our first attempt with eq. (8.100) does not work; one that has been much used in the solution of hyperbolic PDEs is the method of characteristics. The reader may find additional detail helpful; consultation with Smith (1965) or Sarra (2003) is recommended.

8.6.1 The Method of Characteristics

We direct our attention toward a particular curve given by $(x(s), t(s))$. We let the derivatives dt/ds and dx/ds be equated to the coefficients on the $\frac{\partial\phi}{\partial t}$ and $\frac{\partial\phi}{\partial x}$ terms in eq. (8 100). Please note what the consequence of this action is; therefore,

$$\frac{dt}{ds} = 1 \quad \text{and} \quad \frac{dx}{ds} = 3. \tag{8.103}$$

For the latter we find $x = 3s + C_1$; for $s = 0$ we have $x = x_0$, and thus: $x_0 = x - 3s$. From the former of this pair we find $t = s + C_2$ where $C_2 = 0$; i.e., for this type of problem there is only *one* characteristic equation to solve. Therefore, $x_0 = x - 3t$ and $\phi = f(x - 3t)$; we have identified the transformation from (x,t) to (x_0,s).

We will now look at an example that permits us to more fully gauge the usefulness of the method of characteristics. Consider the behavior of an ideal string suspended between supports located at $x = 0$ and $x = L$, where $L = 2$. The velocity of propagation will be taken as 1 (i.e., $c = 1$) therefore:

$$\frac{\partial^2 u}{\partial t^2} = \frac{\partial^2 u}{\partial x^2}. \tag{8.104}$$

The initial shape of the string is specified, $f(x,0) = \sin(\pi x)$, and the initial velocity is zero. The analytic solution for this case is known, and we will want to make use of it:

$$u(x,t) = \sum_{n=1}^{\infty}\left[\int_0^2 \sin\pi x \sin\frac{n\pi x}{2}dx\right]\sin\frac{n\pi x}{2}\cos\frac{n\pi t}{2}. \tag{8.105}$$

Some results from this equation are presented graphically in Figure 8.13 for specific x-positions corresponding to: $x = \frac{1}{16}, \frac{1}{8}, \frac{1}{4}, \frac{1}{2}$.

We begin with values taken from a "curve" along which the $u(x,t)$'s are known. The slopes of the "characteristic" directions are obtained from the roots of the quadratic equation:

$$\left(\frac{dt}{dx}\right)^2 - 1 = 0, \tag{8.106}$$

And, of course, these values are +1 and −1. We use these slopes to extrapolate from known positions, P and Q, to a new position R. The initial points for this example are selected from Figure 8.13: Let $(x_P,t_P) = (1/8, 0.2)$ and $(x_Q,t_Q) = (1/4, 0.2)$. The new point is identified from the linear approximations,

$$t_R - t_P = +1(x_R - x_P) \quad \text{and} \quad t_R - t_Q = -1(x_R - x_Q). \tag{8.107}$$

The solution for the simultaneous eq. (8.107) is $x_R = 0.1875$ and $t_R = 0.2625$. We now use the differential relationships along the characteristics to obtain new estimates for $p = \frac{\partial u}{\partial x}$ and $q = \frac{\partial u}{\partial t}$. These slopes (for positions P and Q) in turn allow us to estimate the change in the dependent variable, u; we use the average of the initial and projected slopes to compute this change (and hence, the new value for the displacement, u). It is easy to show that

$$q_R = \frac{1}{2}\left[p_Q - p_P + q_P + q_Q\right].$$

For the points selected from Figure 8.13, $p_P = 2.12$, $p_Q = 0.9$, $q_P = -0.714$, and $q_Q = -1.438$. Therefore, $q_R = -1.686$ and $p_R = 1.148$. Since the change in the dependent variable, u, is just: $du = pdx + qdt$, we find that

$$u_R - u_P = \frac{1}{2}(p_P + p_R)(x_R - x_P) + \frac{1}{2}(q_P + q_R)(t_R - t_P),$$

which yields $u_R = 0.3371$. The reader should turn immediately to the results presented in Figure 8.13 and estimate the value of u at this new point, "R"—the value is approximately 0.34. In this case the linear extrapolation combined with use of the average slopes has produced a very good estimate.

Smith (1965) shows how this estimate for the dependent variable u can be subsequently refined by using the average slopes to improve the coordinates of the projected position—which are in turn used to get improved slopes, etc. In this manner very accurate solutions for hyperbolic PDEs can be obtained through iteration; *however*, the technique is not the easiest to automate and for that reason is probably not used as commonly as it once was. There is a finite difference method for solution of some hyperbolic PDEs that is extremely easy to implement, and it is described in the next section.

8.6.2 The Leapfrog Method

Let us now return to the familiar wave equation with one spatial dimension:

$$\frac{\partial^2 u}{\partial t^2} = c^2 \frac{\partial^2 u}{\partial x^2} \tag{8.108}$$

and formulate one possible finite-difference approximation for it:

$$\frac{u_{i,j+1} - 2u_{i,j} + u_{i,j-1}}{(\Delta t)^2} \cong c^2 \frac{u_{i+1,j} - 2u_{i,j} + u_{i-1,j}}{(\Delta x)^2}. \tag{8.109}$$

We now isolate the value on the new time-step row:

$$u_{i,j+1} \cong \frac{c^2(\Delta t)^2}{(\Delta x)^2}[u_{i+1,j} - 2u_{i,j} + u_{i-1,j}] + 2u_{i,j} - u_{i,j-1}. \tag{8.110}$$

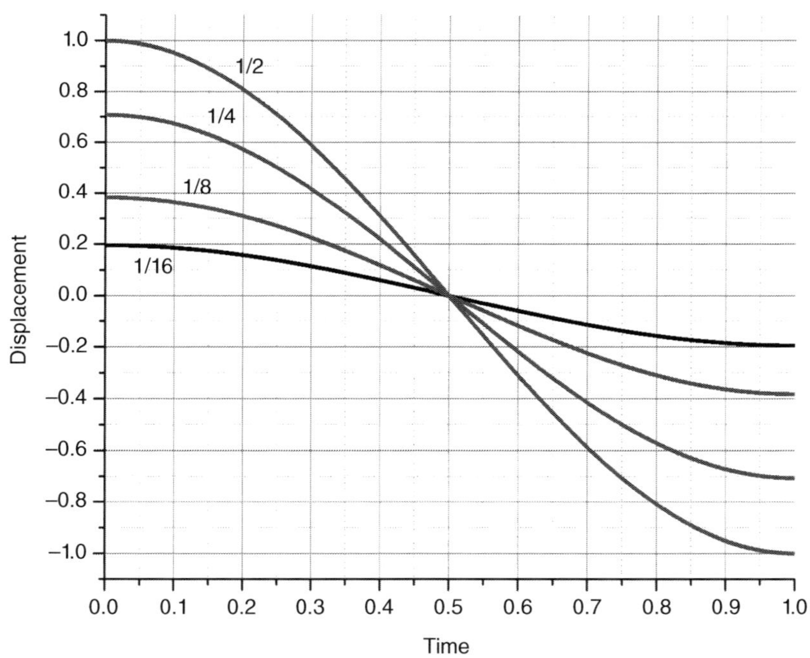

Figure 8.13 Analytic solution for the wave eq. (8.104) for specific x-positions, 1/16, 1/8, 1/4, and 1/2.

Since c has dimensions of velocity, l/t, it is clear that the quotient, $\frac{c^2(\Delta t)^2}{(\Delta x)^2}$, is dimensionless. In fact, it is the Courant number, Co, squared of course. Notice what happens if we select $Co = 1$; the finite difference approximation is now simply:

$$u_{i,j+1} \cong u_{i+1,j} + u_{i-1,j} - u_{i,j-1}. \tag{8.111}$$

This very compact expression forms the basis for what is called the "leapfrog" method, and it will allow us to solve certain wave-equation problems. You may notice, however, that the algorithm requires values for *two previous* time steps; it is not self-starting.

Now let us suppose that we have initial values for both position and velocity such that

$$u(x,0) = f(x) \quad \text{and} \quad \frac{\partial u}{\partial t}(x,0) = g(x). \tag{8.112}$$

We let $u_{i,1} = f_i$ and since

$$\frac{\partial u}{\partial t} \cong \frac{u_{i,j+1} - u_{i,j-1}}{2\Delta t},$$

if we set $j = 1$, then this derivative is just

$$\frac{u_{i,2} - u_{i,0}}{2\Delta t} = g_i. \tag{8.113}$$

We take this result back to the leapfrog algorithm and (letting $j = 1$) isolate $u_{i,2}$. Therefore,

$$u_{i,2} = \frac{1}{2}[f_{i+1} + f_{i-1} + 2\Delta t g_i]. \tag{8.114}$$

This allows us to get the computation started.

We will now illustrate how this works with a typical example. Suppose we have a "string" stretched between supports located at $x = 0$ and $x = L$. The "string" is perfectly elastic and is under great tension such that the gravitational force is unimportant. The displacement (deflection) of the string is described by the wave equation:

$$\frac{\partial^2 u}{\partial t^2} = c^2 \frac{\partial^2 u}{\partial x^2}. \tag{8.115}$$

The string has an initial displacement and an initial velocity given by $u(x,0) = f(x)$ and $\left.\frac{\partial u}{\partial t}\right|_{t=0} = g(x)$, respectively. We will take $L = 10$, an initial velocity of zero, but an initial deflection described by $u = x$ for $0 < x < 1$ and $u = 1 - \frac{(x-1)}{9}$ for $1 < x < 10$. We will use the leapfrog method to compute the string's displacement as a function of time. Note that the initial deflection propagates to the right (and down) as illustrated in Figure 8.14.

8.6.3 Lax–Wendroff Method

Consider the pair of constant coefficient advection equations (also referred to as first-order convection equations):

$$\frac{\partial \phi}{\partial t} + v\frac{\partial \psi}{\partial x} = 0 \quad \text{and} \quad \frac{\partial \psi}{\partial t} + v\frac{\partial \phi}{\partial x} = 0, \tag{8.116 a and b}$$

where v is the velocity at which a disturbance is propagated. If we differentiate the first of this pair with respect to t, and the second with respect to x, we can eliminate $\frac{\partial^2 \psi}{\partial x \partial t}$ and obtain the familiar wave equation,

$$\frac{\partial^2 \phi}{\partial t^2} = v^2 \frac{\partial^2 \phi}{\partial x^2}. \tag{8.117}$$

The Lax–Wendroff method is an explicit technique by which we can obtain the solution for eq. (8.117) by solving the pair of eq. (8.116 a and b). As noted previously, we use the Courant number, $Co = \frac{v\Delta t}{\Delta x}$, to consolidate constants. For the first of the pair (8.116) we write

$$\phi_{i,j+1} = \phi_{i,j} - \frac{Co}{2}(\psi_{i+1,j} - \psi_{i-1,j}) + \frac{Co^2}{2}(\phi_{i+1,j} - 2\phi_{i,j} + \phi_{i-1,j}) \tag{8.118}$$

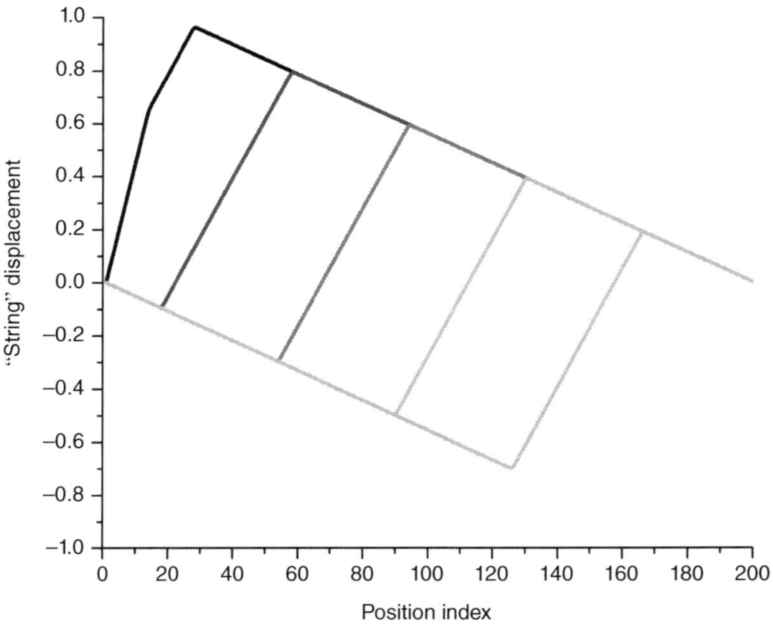

Figure 8.14 Computed string displacement for times of 0.06, 0.36, 0.72, 1.08, and 1.44; we can render these t's dimensionless using ct/L and the corresponding values are 0.03, 0.18, 0.36, 0.54, and 0.72. The position index of 200 corresponds to $L = 10$.

and for the second:

$$\psi_{i,j+1} = \psi_{i,j} - \frac{Co}{2}(\phi_{i+1,j} - \phi_{i-1,j}) + \frac{Co^2}{2}(\psi_{i+1,j} - 2\psi_{i,j} + \psi_{i-1,j}). \tag{8.119}$$

The subscript i refers to spatial position (x) and the subscript j is the time index; thus, $j+1$ implies $t+\Delta t$. Of course the attractive feature of eqs. (8.118) and (8.119) is that the procedure is fully explicit; values of ϕ and ψ on the new time-step row appear only on the left-hand side of the equations. You should recognize that the far-right terms in both eqs. (8.118) and (8.119) are second-order central difference approximations for the second (spatial) derivatives, and you might wonder how that came about. Suppose we expand ϕ using a Taylor series:

$$\phi(t+\Delta t) \cong \phi(t) + \frac{\partial \phi}{\partial t}\Delta t + \frac{1}{2}\frac{\partial^2 \phi}{\partial t^2}(\Delta t)^2 + \ldots \tag{8.120}$$

We know from eq. (8.116 a and b) that $\frac{\partial \phi}{\partial t} = -v\frac{\partial \psi}{\partial x}$ and $\frac{\partial^2 \phi}{\partial t^2} = -v\frac{\partial^2 \psi}{\partial x \partial t} = -v\frac{\partial^2 \psi}{\partial t \partial x}$; but by eq. (8.116 a and b) we have $\frac{\partial \psi}{\partial t} = -v\frac{\partial \phi}{\partial x}$ and thus $\frac{\partial^2 \psi}{\partial t \partial x} = -v\frac{\partial^2 \phi}{\partial x^2}$. Now, both the first and second derivatives in eq. (8.120) can be replaced by appropriate spatial derivatives of ψ and ϕ:

$$\phi(t+\Delta t) \cong \phi(t) - v\Delta t\frac{\partial \psi}{\partial x} + \frac{1}{2}v^2(\Delta t)^2\frac{\partial^2 \phi}{\partial x^2} + \ldots \tag{8.121}$$

The algorithm is completed by introducing second-order central differences for the spatial derivatives: $\frac{\partial \psi}{\partial x} \cong \frac{\psi_{i+1}-\psi_{i-1}}{2\Delta x}$ and $\frac{\partial^2 \phi}{\partial x^2} \cong \frac{\phi_{i+1}-2\phi_i+\phi_{i-1}}{(\Delta x)^2}$. Let's illustrate how easily this scheme can be implemented. We will take the velocity of propagation (v) to be 300 m/s, $\Delta x = 0.02$ m, and $\Delta t = 0.00001$ s such that the Courant number, $Co = 0.15$. The medium of interest will extend from $x = 0$ to $x = 50$ m, and we will position an initial disturbance at the center:

$$\phi_0 = A\exp\left(-\frac{1}{3}(x-25)^2\right), \tag{8.122}$$

and allow ψ to be zero everywhere initially. The amplitude, A, of this "spike" will be two (2), and therefore $\phi_0(x=23) = 0.5272$, $\phi_0(x=21) = 0.00966$, and $\phi_0(x=19) = 1.2\times10^{-5}$. The logic for this procedure is straightforward:

```
#COMPILE EXE
#DIM ALL
  REM *** Lax-Wendroff solution for wave equation via solution
  REM *** of two first-order convection equations
   GLOBAL co,dx,x,dt,tt,v,zz AS SINGLE
        GLOBAL i AS INTEGER
FUNCTION PBMAIN
   DIM phi(2500,2) AS SINGLE
   DIM psi(2500,2) AS SINGLE
       dx=0.02:dt=0.00001:v=300:co=v*dt/dx
       OPEN "c:LAXwendP01.dat" FOR OUTPUT AS #1
   REM *** initialize phi
           FOR i=1 TO 2500
             x=(i-1)*dx
               phi(i,1)=2*EXP(-0.3333*(x-25)^2)
                  NEXT i
   100 REM *** explicit calculation
        FOR i=2 TO 2499
       phi(i,2)=phi(i,1)-co/2*(psi(i+1,1)-psi(i-1,1))+co^2/2*(phi(i+1,1)-2*phi(i,1)+
phi(i-1,1))
       psi(i,2)=psi(i,1)-co/2*(phi(i+1,1)-phi(i-1,1))+co^2/2*(psi(i+1,1)-2*psi(i,1)+
psi(i-1,1))
                NEXT i
                 tt=tt+dt
                    PRINT tt,phi(1250,2)
                FOR i=2 TO 2499
                  phi(i,1)=phi(i,2)
                     psi(i,1)=psi(i,2)
                       NEXT i
                        IF tt>0.01 THEN 200 ELSE 100
  200 REM *** continue
          FOR i=1 TO 2500
            WRITE#1,i,phi(i,1)
            NEXT i
               INPUT "end program now?";zz
               IF zz>0 THEN CLOSE:   END
```

You will notice that the time subscript in this elementary program has only two values (1 and 2), corresponding to the old and time new step rows, respectively. This is done simply to minimize the storage requirement for the arrays. How will the initial disturbance, centered at $x = 25\,\mathrm{m}$, be propagated in this computation? The behavior is shown in Figure 8.15.

In the preceding discussion we examined the use of the Lax–Wendroff method for the solution of the constant coefficient (velocity) advection equations, eq. (8.116 a and b, and we demonstrated that this pair was equivalent to the familiar one-dimensional wave eq. (8.117). Recall that the wave eq. (8.117) is *not* dissipative—therefore an initial disturbance will propagate through the medium *unchanged*, implying that a wave transmitted along a taut string for example will *not* decay in amplitude. Of course, this *lossless* wave transmission is unphysical. If you are curious about the addition of diffusion to such problems, please see the discussion of eq. (8.126) in Section 8.8.

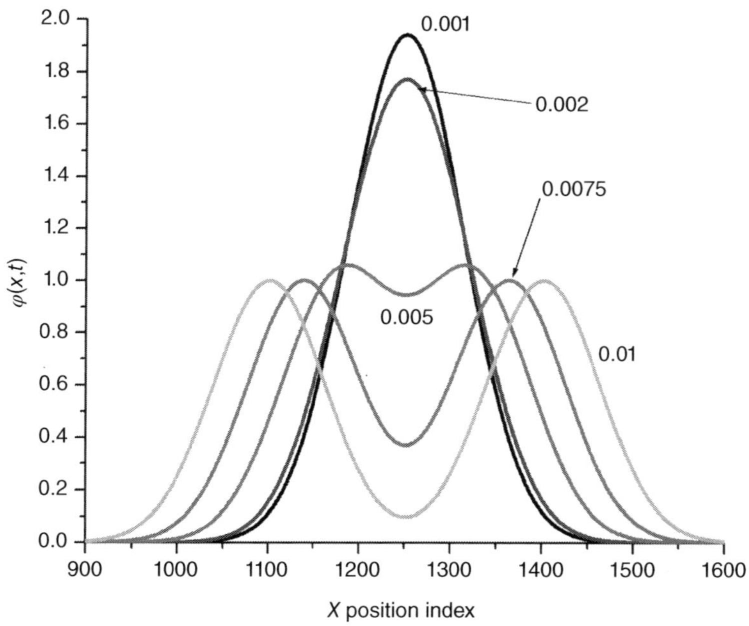

Figure 8.15 Propagation of an initial disturbance centered at $x = 25$ m, corresponding to a position index of 1,250. $\phi(x,t)$'s are provided for 0.001, 0.002, 0.005, 0.0075, and 0.01 s. Since $v = 300$ m/s, the disturbance will reach the boundaries of the computation in just 0.083 s.

8.7 Problems with Moving Boundaries

An extremely important group of practical problems concerns the change of phase that may result when liquid and solid phases are in contact at a temperature at or very near the melting point. In such cases heat transfer across the interface may result in solidification or melting, depending upon the magnitudes of the fluxes of thermal energy. Let us assume that the interface is located at x_0, with the solid phase (to the left) denoted by the subscript "1" and the liquid phase (to the right) by "2." The liquid phase is initially at some elevated uniform temperature, and the solid phase, where $x < x_0$, is at a temperature just below the melting point. Heat will be transferred from right to left in this scenario and, as melting occurs, the interface will move to the left, toward smaller values of x. The relative size of the fluxes at the interface and the latent heat of fusion will determine the speed with which the interface moves (dX/dt):

$$k_1 \left.\frac{\partial T_1}{\partial x}\right|_{x=x_0} - k_2 \left.\frac{\partial T_2}{\partial x}\right|_{x=x_0} = \Delta H_f \rho \frac{dX}{dt}. \tag{8.123}$$

Heat transfer in both phases is assumed to occur solely by conduction, and therefore we have two parabolic partial differential equations to solve,

$$\frac{\partial T_1}{\partial t} = \alpha_1 \frac{\partial^2 T_1}{\partial x^2} \quad \text{and} \quad \frac{\partial T_2}{\partial t} = \alpha_2 \frac{\partial^2 T_2}{\partial x^2}. \tag{8.124}$$

It is important to recognize that we have neglected buoyancy in the liquid phase, and we have assumed that the thermal diffusivities are constant; these simplifications may be questionable if the temperature differences in a particular problem are large.

Now suppose we have solid copper at 1,084 °C ($\alpha_1 = 1.14$ cm^2/s) brought instantaneously into contact with a large pool of liquid copper at 1,500 °C ($\alpha_2 = 0.38$ cm^2/s). The melting point of copper is 1,085 °C and the latent heat of fusion, ΔH_f, is 13.59 kJ/mol or 51.11 cal/g. As heat is transferred to the left, toward smaller values of x, melting will occur and the interface will move to the left. The driving force for heat transfer is very large initially, but it will rapidly diminish as the sharp shoulder for $T(x)$ is rounded off. The one-dimensional conduction eq. (8.124) are easily solved explicitly, and the fluxes at the interface are compared according to eq. (8.123), but there is a difficulty we should anticipate: We have discretized the

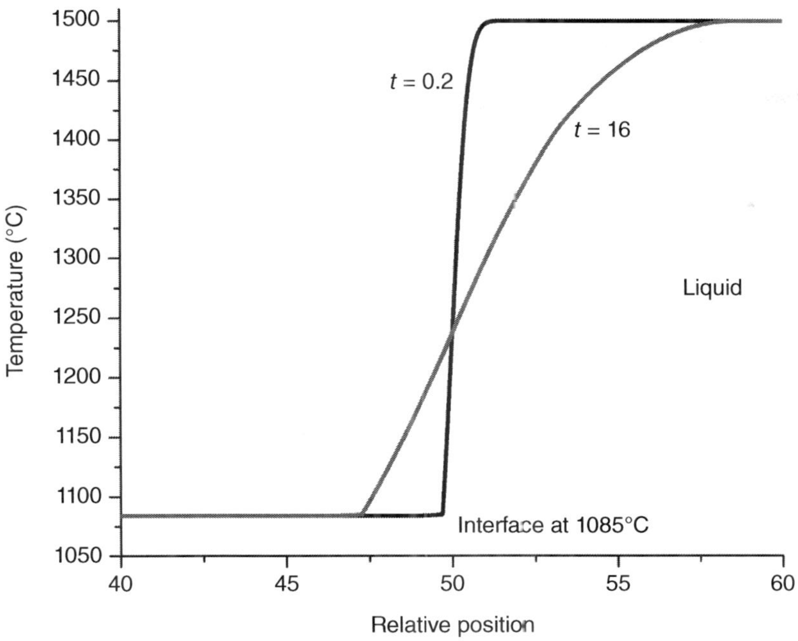

Figure 8.16 Change in the temperature distribution near the interface between solid (left) and liquid (right) copper.

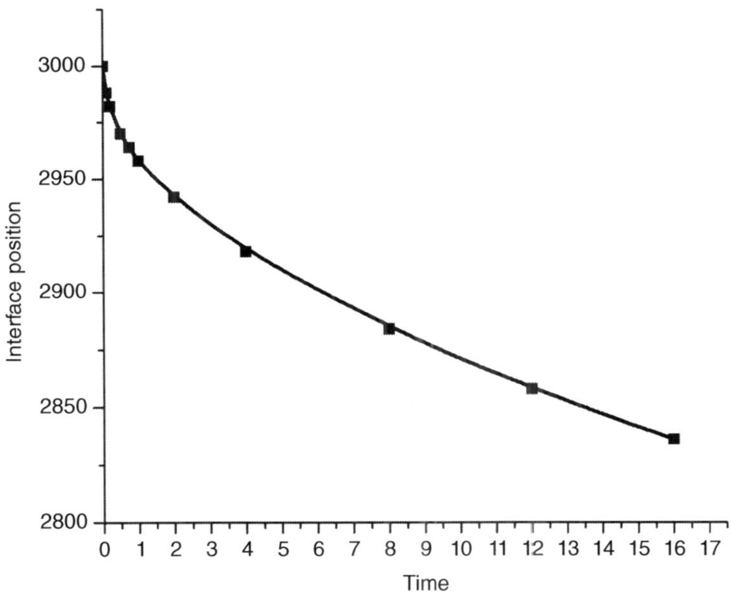

Figure 8.17 Movement of the interface over the first 16 s of heat transfer. Each integer unit for interface position corresponds to 0.01667 cm, so the interface has moved 2.72 cm in 16 s.

continuum requiring that the motion of the interface be interpreted in terms of integer values of Δx. As the thermal driving force decreases, the motion of the interface over a single time step may only be a small fraction of 1 (one) Δx. It will therefore be necessary for our numerical scheme to accumulate these increments until the sum exceeds Δx and then the interface can be moved to $n-1$. Figure 8.16 illustrates the change in the temperature distribution over the first 16 s of melting.

The velocity of the solid–liquid boundary to the left diminishes as the thermal driving force decreases. This is apparent in the change in slope shown in (Figure 8.17).

8.8 Elementary Problems with Convective Transport

Our focus in this section concerns problems in which convection is important; that is, problems where momentum, heat, mass, etc. are transported by virtue of a nonzero velocity vector component. Examples of the terms of interest are

$$\rho\left(v_x\frac{\partial v_x}{\partial x}+v_y\frac{\partial v_x}{\partial y}+v_z\frac{\partial v_x}{\partial z}\right)=\ldots \text{ for } x\text{-momentum,} \tag{8.125 a,b,c}$$

$$\rho C_p\left(v_x\frac{\partial T}{\partial x}+v_y\frac{\partial T}{\partial y}+v_z\frac{\partial T}{\partial z}\right)=\ldots \text{ for thermal energy,}$$

and

$$\left(v_x\frac{\partial C_A}{\partial x}+v_y\frac{\partial C_A}{\partial y}+v_z\frac{\partial C_A}{\partial z}\right)=\ldots \text{ for mass (concentration of species “A”)}$$

Inclusion of such terms in a model represents a serious escalation in difficulty as we shall see. However, a limited class of such problems can be solved readily using techniques that are now familiar to us.

Imagine a situation in which a chemical species with concentration, C, is carried in the z-direction by fluid motion. Furthermore, assume that any mixing that occurs—possibly as a result of turbulence—can be represented as though it were diffusive in character. One model for this phenomenon can be written as:

$$\frac{\partial C}{\partial t}+V\frac{\partial C}{\partial z}=D\frac{\partial^2 C}{\partial z^2}. \tag{8.126}$$

This is an *axial dispersion model* in which the velocity (V) in the z-direction is taken to be constant. Let the index i represent z-position and j represent time; one possible discretization for this equation is

$$\frac{C_{i,j+1}-C_{i,j}}{\Delta t}\cong D\frac{C_{i+1,j}-2C_{i,j}+C_{i-1,j}}{(\Delta z)^2}-V\frac{C_{i,j}-C_{i-1,j}}{\Delta z}. \tag{8.127}$$

Please note that an *upwind* difference has been used in the convective transport term. The significance of this choice will be discussed in the next section—for now we will simply operate on the assumption that this is appropriate. If we multiply by Δt and add $C_{i,j}$ to both sides of the equation we obtain an elementary explicit algorithm:

$$C_{i,j+1}\cong \Delta t\left[D\frac{C_{i+1,j}-2C_{i,j}+C_{i-1,j}}{(\Delta z)^2}-V\frac{C_{i,j}-C_{i-1,j}}{\Delta z}\right]+C_{i,j}. \tag{8.128}$$

The reader might wonder why we have not chosen a “better” approximation for the first derivative, $\frac{\partial C}{\partial z}$. For example, we could select a second-order central difference where

$$\frac{\partial C}{\partial z}\cong\frac{C_{i+1,j}-C_{i-1,j}}{2\Delta z}. \tag{8.129}$$

Please consider trying this change yourself; it is worth taking a little time to discover that it will not work. The result of the computation is unphysical.

Now assume that some finite pulse is put into the flow at the entrance where $z=0$; our intent is to model the behavior of this pulse as it is carried downstream by the flow. However, there is an obvious limitation to our modeling approach: If the length of the test section is L, then at a time corresponding to L/V (actually a little before) the tracer pulse will reach the outflow boundary. Naturally, an outflow boundary condition of $C(z=L,t)=0$ would be violated. This a common, recurring problem in the computational solution of transport problems—an outflow condition is necessary, but specification of the wrong one will constrain the solution and produce an incorrect result. An obvious “remedy” is to stop the computation before the tracer pulse arrives at the far end of the computational domain. Let us explore how this very simple dispersion model performs; we set $V=2$, $D=1/4$, $\Delta z=0.2$, and $\Delta t=0.005$. The inlet concentration is set to 2 for $0<t<0.4$, and then it reverts to zero. The evolution of the input pulse is illustrated by Figure 8.18.

Since the fluid velocity is 2, the successive peaks are centered at axial positions corresponding to $2t$, of course. The dispersion is causing the attenuation of the initial pulse height and the broadening of the distributions; note that by $t=50$, the tracer is covering axial positions from 80 to 120, which is double the width seen at $t=10$. The significance of the dispersion model illustrated in the above example is that eq. (8.126) allows us to characterize the performance

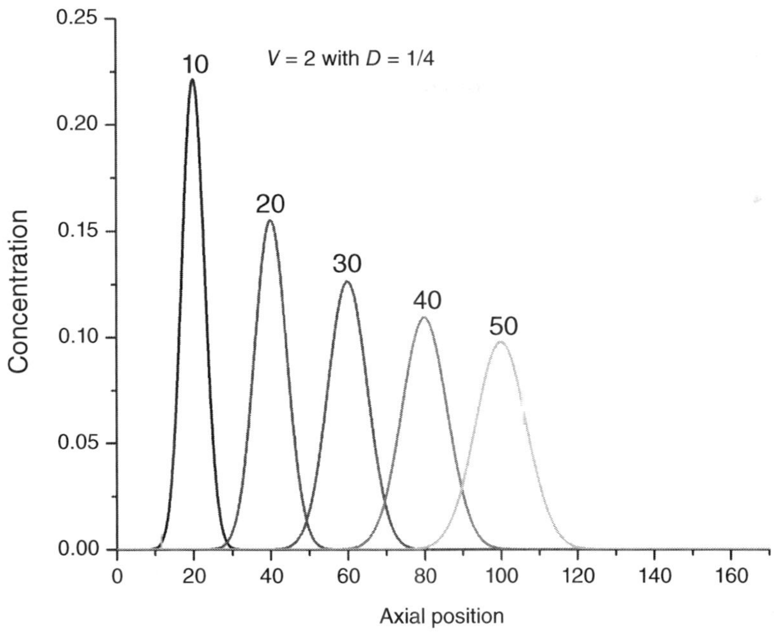

Figure 8.18 Dispersion model results for a flow reactor with $Pe = VL/D = 6{,}400$ at times corresponding to 10, 20, 30, 40, and 50. Note that *Pe* is a kind of *Peclet* number for mass transfer.

of a flow reactor. By monitoring an inert tracer injected into the reactor, we can evaluate V and D (and thus the *Peclet* number). These results can be used in turn through the inclusion of chemical kinetics to predict performance (that is, the expected conversion). The model itself, eq. (8.126), can be augmented to account for radial dispersion by incorporating two dispersion coefficients:

$$\frac{\partial C}{\partial t} + V\frac{\partial C}{\partial z} = \frac{D_R}{r}\frac{\partial}{\partial r}\left(r\frac{\partial C}{\partial r}\right) + D_L\frac{\partial^2 C}{\partial z^2}. \tag{8.130}$$

We turn now to an example of convective heat transport. Consider laminar flow in a duct (formed by parallel walls); the lower wall is located at $y = 0$ and the upper wall at $y = B$. The flow is fully developed, and the fluid enters at a uniform (low) temperature. The heated walls are maintained at a higher temperature and heat transfer to the fluid occurs as the fluid moves in the x-direction. This is a steady-state problem with the velocity distribution given by:

$$v_x = \frac{1}{2\mu}\frac{dp}{dx}(y^2 - By). \tag{8.131}$$

The appropriate energy equation for this case is

$$\rho C_p v_x \frac{\partial T}{\partial x} = k\left[\frac{\partial^2 T}{\partial x^2} + \frac{\partial^2 T}{\partial y^2}\right]. \tag{8.132}$$

Note that conduction terms in both the transverse and the flow directions have been included. The latter can be neglected if the product of the *Reynolds* and *Prandtl* numbers (RePr, where $\text{Re} = \frac{<v>B}{\nu}$ and $\text{Pr} = \frac{\mu C_p}{k} = \frac{\nu}{\alpha}$) is greater than about 100, and for many liquids this is likely to be the case. Therefore,

$$\frac{\partial T}{\partial x} = \frac{\alpha}{A(y^2 - By)}\frac{\partial^2 T}{\partial y^2}, \tag{8.133}$$

where $A = \frac{1}{2\mu}\frac{dp}{dx}$. Such problems provide us with an opportunity for simple explicit solution by forward marching in the x-direction; we discretize the equation so that:

$$T_{i+1,j} \cong \frac{\Delta x \alpha}{A(y^2 - By)}\left[\frac{T_{i,j+1} - 2T_{i,j} + T_{i,j-1}}{(\Delta y)^2}\right] + T_{i,j}. \tag{8.134}$$

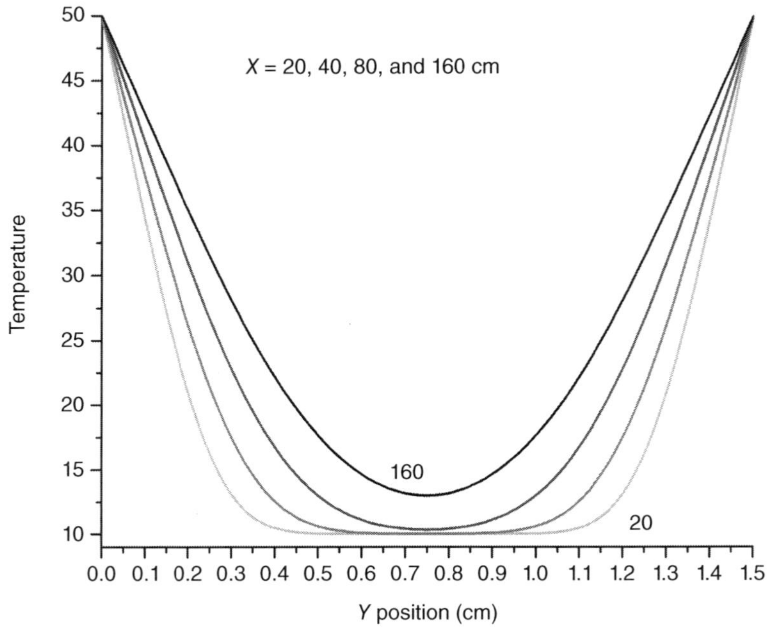

Figure 8.19 Heat transfer to fully developed laminar flow between parallel walls, 1.5 cm apart. Temperature distributions are shown for *x*-positions of 20, 40, 80, and 160 cm. The walls are maintained at 50 °C, and the fluid (water) enters the heated section with a uniform temperature of 10 °C. The fluid properties are taken as constant although that is not entirely realistic since the viscosity of water at 50 °C is only about 0.55 cp.

We now have an algorithm that allows us to march downstream, computing new temperatures as we go, given the initial uniform inlet temperature distribution. Let us illustrate how this is going to work with an example; we select an inlet temperature of 10 °C and maintain the parallel walls at 50 °C. We take the fluid to be water with $\alpha = 0.00147\ \text{cm}^2/\text{s}$ and $\text{Pr} = 6.8$. The pressure gradient, dp/dx, is set to $-0.1778\ \text{dyne/cm}^2/\text{cm}$. We expect the temperature distribution to evolve slowly—after all, we are relying upon molecular conduction to carry the thermal energy into the interior of the fluid. In fact, Figure 8.19 shows that the bulk fluid temperature is only 21.2 °C at $x = 160$ cm; even by $x = 400$ cm, the bulk fluid temperature is only 29.94 °C.

In the last example of this section we would like to treat the case of mass transfer in the annular space between concentric cylinders. The fluid contained within is initially at rest, but at $t = 0$ the outer cylinder (located at $r = R_2$) begins to rotate with constant angular velocity, ω. Of course this means that the tangential velocity vector component at the wall is $v_\theta(r = R_2) = \omega R_2$. We place a small spot of inert tracer adjacent to the inner wall; we are interested in how this material is transported as the velocity distribution develops. The two governing equations are:

$$\frac{\partial v_\theta}{\partial t} = v\left[\frac{\partial^2 v_\theta}{\partial r^2} + \frac{1}{r}\frac{\partial v_\theta}{\partial r} - \frac{v_\theta}{r^2}\right] \tag{8.135}$$

and

$$\frac{\partial C}{\partial t} + \frac{v_\theta}{r}\frac{\partial C}{\partial \theta} = D\left[\frac{\partial^2 C}{\partial r^2} + \frac{1}{r}\frac{\partial C}{\partial r}\right]. \tag{8.136}$$

Notice that we have omitted molecular (diffusional) transport in the θ-direction as it should be small relative to convective transport once the velocity distribution begins to develop. Our solution strategy is as follows: we will solve the discretized version of eq. (8.135) explicitly by forward marching in time. Each time we compute a new velocity distribution we will use $v_\theta(r, t)$ in the discretized version of eq. (8.136) to calculate C over the entire array of r and θ positions: $R_1 \le r \le R_2$ and $0 \le \theta \le 2\pi$. Of course, we must use upwind differencing for the convective transport term $\left(\frac{v_\theta}{r}\frac{\partial C}{\partial \theta}\right)$ since we are using the explicit technique. A complete pass through *both* equations corresponds to a time step. We merely repeat the process until the desired final time is attained. An elementary code for the computational process is included below.

```
#COMPILE EXE
#DIM ALL
  REM *** Convective transport example
   GLOBAL dt,dth,dr,R1,R2,kvis,D,w,d2vdr2,dvdr,rpos AS SINGLE
   GLOBAL tt,zz,d2cdr2,dcdr,dcdth AS SINGLE
   GLOBAL i,j AS INTEGER
FUNCTION PBMAIN
   DIM v(91,2) AS SINGLE
   DIM C(91,91,2) AS SINGLE
     dr=0.03333:dth=0.069813:kvis=0.02:D=0.001:w=3:R1=5:R2=8
     dt=0.001
  REM *** initialize velocity and cencentration fields
    FOR i=1 TO 90
      v(i,1)=0
      NEXT i
       v(91,1)=w*R2
      FOR i=1 TO 90
        FOR j=1 TO 90
          c(i,j,1)=0
          NEXT j:NEXT i
           FOR i=1 TO 5
             FOR j=1 TO 5
               c(i,j,1)=1
               NEXT j:NEXT i
        tt=0
  100 REM *** continue
  REM *** compute revised velocities
     FOR i=2 TO 90
       rpos=R1+(i-1)*dr
       d2vdr2=(v(i+1,1)-2*v(i,1)+v(i-1,1))/dr^2
       dvdr=(v(i+1,1)-v(i-1,1))/(2*dr)
       v(i,2)=dt*kvis*(d2vdr2+1/rpos*dvdr-v(i,1)/rpos^2)+v(i,1)
       NEXT i
        FOR i=2 TO 90
          v(i,1)=v(i,2)
          NEXT i
  REM *** begin computation for concentration field
      FOR j=2 TO 90
        FOR i=2 TO 90
          d2cdr2=(c(i+1,j,1)-2*c(i,j,1)+c(i-1,j,1))/dr^2
          dcdr=(c(i+1,j,1)-c(i-1,j,1))/(2*dr)
          dcdth=(c(i,j,1)-c(i,j-1,1))/dth
            rpos=R1+(i-1)*dr
       c(i,j,2)=dt*D*(d2cdr2+1/rpos*dcdr)-v(i,1)/rpos*dcdth*dt+c(i,j,1)
                NEXT i:NEXT j
                 FOR j=2 TO 90
                   FOR i=2 TO 90
                    c(i,j,1)=c(i,j.2)
                    NEXT i:NEXT j
                    FOR i=2 TO 90
```

```
                        c(i,91,1)=c(i,90,1)
                        c(i,1,1)=c(i,91,1)
                        NEXT i
        tt=tt+dt
             PRINT c(6,6,1),c(7,7,1),c(8,8,1),c(9,9,1)
               IF tt<55 THEN 100 ELSE 300
   300 REM *** continue
         OPEN "c:MTcylind.dat" FOR OUTPUT AS #1
            FOR j=1 TO 91
              FOR i=1 TO 91
                WRITE#1,i,j,c(i,j,1)
                NEXT i:NEXT j
            CLOSE:END
```

Figure 8.20 illustrates the progress of the tracer (contaminant) plume as the velocity distribution develops. Remember, the outer cylinder is put in motion at $t = 0$ with $v_\theta(r = R_2) = \omega R_2$. The inner cylinder is fixed (stationary), so at very small t mass transfer occurs mainly by molecular diffusion.

In the examples of convective transport provided above, we were able to find solutions with familiar, elementary numerical procedures. Of course, in all cases we had only one non-zero component of the velocity vector. The situation for transport involving two-dimensional flows is more difficult, but a very useful procedure for such problems is described in the next section.

8.9 A Numerical Procedure for Two-Dimensional Flow and Transport Problems

We now will describe a very powerful technique that can be used to solve a host of important flow and transport problems in *two dimensions*. We begin by writing down the governing equations for the motion of an incompressible fluid:

$$\frac{\partial v_x}{\partial t} + v_x\frac{\partial v_x}{\partial x} + v_y\frac{\partial v_x}{\partial y} = -\frac{1}{\rho}\frac{\partial p}{\partial x} + \nu\left[\frac{\partial^2 v_x}{\partial x^2} + \frac{\partial^2 v_x}{\partial y}\right] \tag{8.137}$$

$$\frac{\partial v_y}{\partial t} + v_x\frac{\partial v_y}{\partial x} + v_y\frac{\partial v_y}{\partial y} = -\frac{1}{\rho}\frac{\partial p}{\partial y} + \nu\left[\frac{\partial^2 v_y}{\partial x^2} + \frac{\partial^2 v_y}{\partial y^2}\right] \tag{8.138}$$

and

$$\frac{\partial v_x}{\partial x} + \frac{\partial v_y}{\partial y} = 0. \tag{8.139}$$

These three equations must be solved simultaneously for the general problem we are contemplating. The main difficulty in such cases—and one that plagues computational fluid dynamics—is the determination of $p(x,y,t)$. As the flow field evolves, $p(x,y,t)$ must change to ensure that continuity is satisfied ($\nabla \cdot v = 0$). In the approach we are about to describe, the problem of $p(x,y,t)$ is circumvented.

We now cross-differentiate eqs. (8.137) and (8.138), resulting in:

$$\frac{\partial^2 v_x}{\partial t\partial y} + \frac{\partial v_x}{\partial y}\frac{\partial v_x}{\partial x} + v_x\frac{\partial^2 v_x}{\partial x\partial y} + \frac{\partial v_y}{\partial y}\frac{\partial v_x}{\partial y} + v_y\frac{\partial^2 v_x}{\partial y^2} = -\frac{1}{\rho}\frac{\partial^2 p}{\partial x\partial y} + \nu\left[\frac{\partial^3 v_x}{\partial x^2\partial y} + \frac{\partial^3 v_x}{\partial y^3}\right] \tag{8.140}$$

and

$$\frac{\partial^2 v_y}{\partial t\partial x} + \frac{\partial v_x}{\partial x}\frac{\partial v_y}{\partial x} + v_x\frac{\partial^2 v_y}{\partial x^2} + \frac{\partial v_y}{\partial x}\frac{\partial v_y}{\partial y} + v_y\frac{\partial^2 v_y}{\partial y\partial x} = -\frac{1}{\rho}\frac{\partial^2 p}{\partial y\partial x} + \nu\left[\frac{\partial^3 v_y}{\partial x^3} + \frac{\partial^3 v_y}{\partial y^2\partial x}\right]. \tag{8.141}$$

The *vorticity vector component*, which for two-dimensional flow in the x–y plane is a measure of rotation about the z-axis, is defined by:

$$\omega_z = \left(\frac{\partial v_y}{\partial x} - \frac{\partial v_x}{\partial y}\right). \tag{8.142}$$

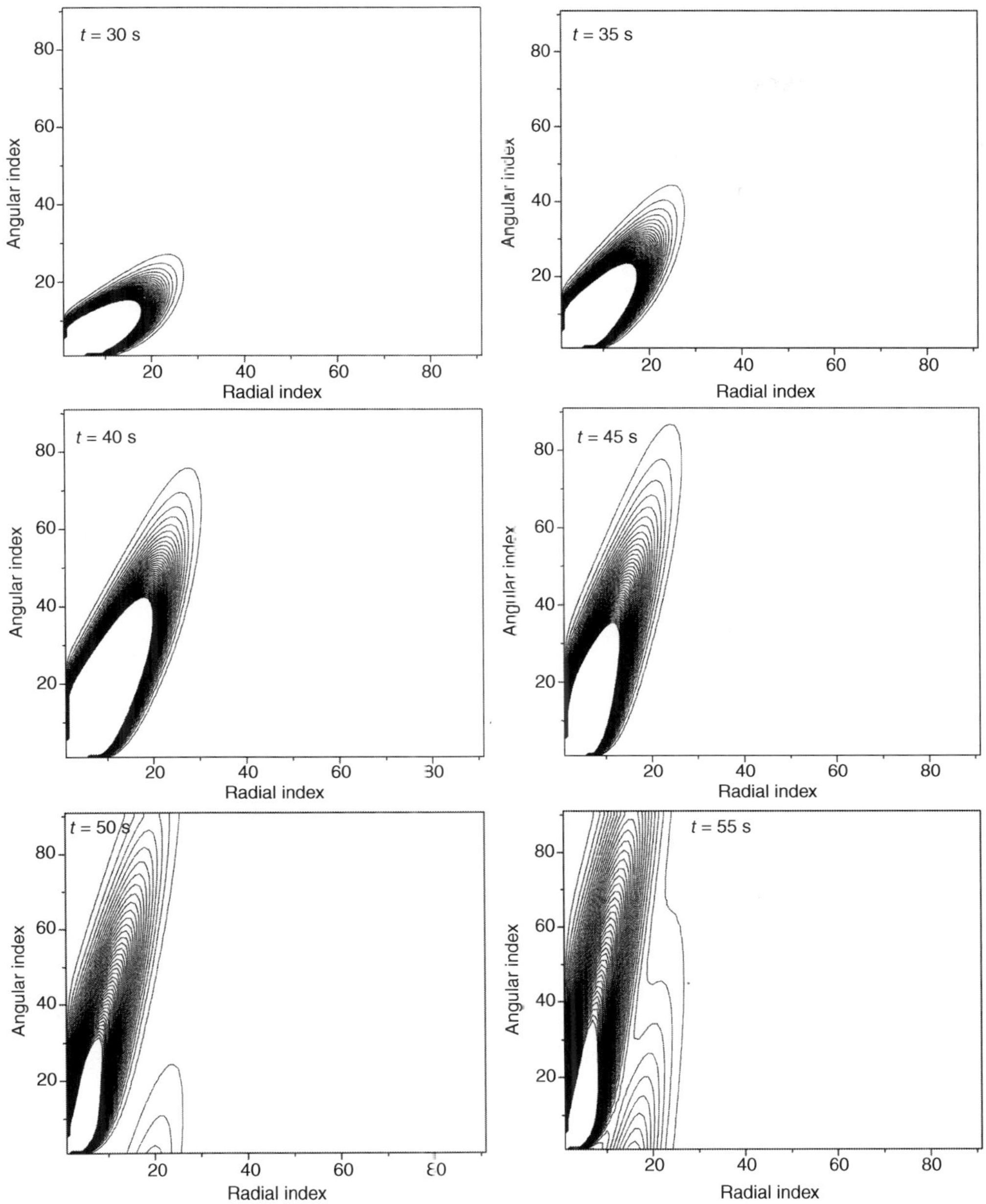

Figure 8.20 Transport of the tracer plume between concentric cylinders. The inner cylinder is at rest and the outer cylinder begins to rotate with constant angular velocity at $t = 0$.

Note what happens when we subtract eq. (8.140) from eq. (8.141) employing the vorticity definition:

$$\frac{\partial \omega}{\partial t} + v_x \frac{\partial \omega}{\partial x} + v_y \frac{\partial \omega}{\partial y} = \nu \left[\frac{\partial^2 \omega}{\partial x^2} + \frac{\partial^2 \omega}{\partial y^2} \right]. \tag{8.143}$$

This is the *vorticity transport equation*. We have eliminated pressure, and if we now introduce the stream function,

$$v_x = \frac{\partial \psi}{\partial y} \quad \text{and} \quad v_y = -\frac{\partial \psi}{\partial x}, \tag{8.144}$$

we can guarantee that continuity will automatically be satisfied and we can use the vorticity definition, eq. (8.142), to obtain the relationship between ω and ψ:

$$-\omega = \frac{\partial^2 \psi}{\partial x^2} + \frac{\partial^2 \psi}{\partial y^2}. \tag{8.145}$$

One of the reasons that the vorticity approach is so powerful is that vorticity is only created at the boundaries of the flow—not in the interior! A solution procedure suggests itself: We use the vorticity distribution to get the stream function, use the stream function to get the velocity vector components, and then use the new velocities to get an updated vorticity distribution, and repeat. The numerical processes required are already familiar to us, but we must emphasize: *this approach can be used to solve partial differential equations that represent some very complex situations*. Consequently, we are going to devote time and effort to explaining and illustrating just what can be accomplished with vorticity transport.

8.9.1 Vorticity Transport at Low Reynolds Numbers

We want to gain some experience with the application of vorticity transport, and we will begin with an elementary example that permits some significant simplifications. Consider a square cavity that contains a viscous fluid initially at rest. At $t = 0$, the upper surface (the lid) begins to slide across the top of the cavity at some constant velocity, V_{top}. For this introductory example, we *assume* that the *Reynolds* number is small enough such that the convective transport of vorticity can be neglected; that is, we need only concern ourselves with viscous (or molecular) transport:

$$\frac{\partial \omega}{\partial t} = \nu \left[\frac{\partial^2 \omega}{\partial x^2} + \frac{\partial^2 \omega}{\partial y^2} \right]. \tag{8.146}$$

This is a parabolic partial differential equation and we can solve it explicitly by forward marching in time. The vorticity and the stream function (ψ) are related by eq. (8.145):

$$-\omega = \frac{\partial^2 \psi}{\partial x^2} + \frac{\partial^2 \psi}{\partial y^2};$$

this is an elliptic (Poisson-type) partial differential equation and we can solve it iteratively using Gauss–Seidel or preferably, successive over-relaxation (SOR). We now will construct the necessary logic to solve this problem, in detail, and step by step. We place the origin in the lower left-hand corner of the cavity.

1) We assign an arbitrary value to the stream function on the *entire* boundary of the cavity; we choose $\psi = 0$ for convenience.
2) Since vorticity is neither created nor destroyed in the interior of a homogeneous fluid, we realize that the motion of the upper surface will create velocity gradients, and hence, vorticity. We must adjust the stream function near the top of the cavity to reflect the constant velocity, V_{top}. Let the subscript j indicate the top (lid) position; we must assign the correct value to ψ for the uppermost *interior* row:

$$\psi_{i,j-1} = \frac{1}{18}(11\psi_{i,j} + 9\psi_{i,j-2} - 2\psi_{i,j-3} - 6V_{top}\Delta y). \tag{8.147}$$

You may recognize that we have used a third-order backward difference to obtain ψ on the top *interior* ($j-1$) row (the boundary of course has been specified, $\psi_{i,j} = 0$).
3) We can now compute vorticity across the top of the cavity, for example,

$$\omega_{i,j} = -\frac{1}{(\Delta y)^2}(\psi_{i,j} - 2\psi_{i,j-1} + \psi_{i,j-2}). \tag{8.148}$$

This expression comes directly from the fact that we only have the x-component of velocity at the very top of the cavity:

$$\omega = -\frac{\partial v_x}{\partial y} = -\frac{\partial^2 \psi}{\partial y^2}. \tag{8.149}$$

The same reasoning applies at the bottom too, but in that case we will use a second-order forward difference approximation. One must also make sure that the no-slip condition is enforced at the bottom and the vertical sides of the cavity in anticipation that we will want to examine larger Reynolds numbers at some point. For the bottom this means $\frac{\partial \psi}{\partial y} = 0$, and for the side walls, $\frac{\partial \psi}{\partial x} = 0$.
4) We calculate the vorticity on both vertical walls, recognizing that only the y-component of velocity need be taken into account. Let the subscript i represent the index for the left vertical wall:

$$\omega_{i,j} = -\frac{1}{(\Delta x)^2}(\psi_{i,j} - 2\psi_{i+1,j} + \psi_{i+2,j}), \tag{8.150}$$

where $\psi_{i,j} = 0$. We treat the right-hand vertical wall in exactly the same manner, but we employ a second-order *backward* difference.

5) Now that we have obtained vorticity values on all four cavity walls, we can compute updated vorticity in the *interior* of the cavity explicitly:

$$\omega_{i,j,k+1} = \nu\Delta t\left[\frac{\omega_{i+1,j,k} - 2\omega_{i,j,k} + \omega_{i-1,j,k}}{(\Delta x)^2} + \frac{\omega_{i,j+1,k} - 2\omega_{i,j,k} + \omega_{i,j-1,k}}{(\Delta y)^2}\right] + \omega_{i,j,k}. \tag{8.151}$$

We only need two values for the time subscript, k, since we will discard the old vorticity values once the new ones are computed.

6) We use the updated vorticity distribution to calculate the stream function iteratively everywhere in the interior of the cavity:

$$\psi_{i,j} = \psi_{i,j} + \frac{W_{SOR}}{4}(\psi_{i+1,j} + \psi_{i-1,j} + \psi_{i,j-1} + \psi_{i,j-1} - 4\psi_{i,j} + (\Delta x)^2\omega_{i,j,k+1}). \tag{8.152}$$

W_{SOR} is the relaxation (accelerating) factor, and for a problem like this it should have a value of roughly 1.8 to 1.9 (note that we have changed our notation for the relaxation factor to avoid possible confusion with vorticity, ω). The actual number of iterations required here will depend upon other parameters selected for a particular problem, but it is normally modest, possibly on the order of 100. This completes the sequence of calculations for one time step; we have advanced from t to $t + \Delta t$. We now return to 2) and repeat the process.

We construct a code written to execute this logic:

```
#COMPILE EXE
#DIM ALL
  REM *** slow viscous flow in cavity, Re=5
     GLOBAL L,H,dt,tt,kvisc,dx,dy,Vtop,d2wdx2,d2wdy2,vx,wsor,cc AS SINGLE
   GLOBAL i,j,iter AS INTEGER
FUNCTION PBMAIN
    DIM w(81,81,2) AS SINGLE
    DIM psi(81,81) AS SINGLE
       L=5:dx=L/80:dy=dx:kvisc=0.05:Vtop=1/20:wsor=1.9:dt=0.01
       OPEN "c:PSIcavitySLO.dat" FOR OUTPUT AS #1
         FOR j=1 TO 81
           FOR i=1 TO 81
             psi(i,j)=0
             NEXT i:NEXT j
  100 REM *** fix stream function and vorticity across top and bottom of cavity
        FOR i=2 TO 80
          psi(i,80)=1/18*(11*psi(i,81)+9*psi(i,79)-2*psi(i,78)-6*Vtop*dy)
          psi(i,2)=1/18*(11*psi(i,1)+9*psi(i,3)-2*psi(i,4))
          w(i,81,1)=-1/dy^2*(psi(i,81)-2*psi(i,80)+psi(i,79))
          w(i,1,1)=-1/dy^2*(psi(i,1)-2*psi(i,2)+psi(i,3))
         NEXT i
  REM *** get vorticity on two vertical walls
         FOR j=2 TO 80
         psi(2,j)=1/18*(11*psi(1,j)+9*psi(3,j)-2*psi(4,j))
         psi(80,j)=1/18*(11*psi(81,j)+9*psi(79,j)-2*psi(78,j))
         w(1,j,1)=-1/dx^2*(psi(1,j)-2*psi(2,j)+psi(3,j))
         w(81,j,1)=-1/dx^2*(psi(81,j)-2*psi(80,j)+psi(79,j))
           NEXT j
  REM *** explicit computation in interior
         FOR j=2 TO 80
         FOR i=2 TO 80
           d2wdx2=1/dx^2*(w(i+1,j,1)-2*w(i,j,1)+w(i-1,j,1))
           d2wdy2=1/dy^2*(w(i,j+1,1)-2*w(i,j,1)+w(i,j-1,1))
             w(i,j,2)=dt*kvisc*(d2wdx2+d2wdy2)+w(i,j,1)
```

```
            NEXT i:NEXT j
  400 REM *** calculate new psi distribution from poisson equation iteratively
        FOR j=2 TO 80
          FOR i=2 TO 80
            psi(i,j)=psi(i,j)+wsor/4*(psi(i+1,j)+psi(i-1,j)+psi(i,j+1)+psi(i,j-1)-
4*psi(i,j)+dx^2*w(i,j,2))
            w(i,j,1)=w(i,j,2)
            NEXT i:NEXT j
            iter=iter+1
               IF iter>100 THEN 500 ELSE 400
  500 REM *** continue
      iter=0
      tt=tt+dt
        PRINT tt,psi(71,41)
         IF tt>10 THEN 600 ELSE 100
  600 REM *** continue
        FOR j=1 TO 81
          FOR i=1 TO 81
            WRITE#1,i,j,psi(i,j),w(i,j,1)
            NEXT i:NEXT j
             CLOSE
             OPEN "c:VxCAVITYcl.dat" FOR OUTPUT AS #1
               FOR j=2 TO 80
                 vx=1/(2*dy)*(psi(41,j+1)-psi(41,j-1))
                   WRITE#1,j,vx
                 NEXT j
        INPUT "end program now?";cc
        IF cc>0 THEN CLOSE
        END
END FUNCTION
```

There are a few observations we should make regarding this implementation of our computational scheme. First, the cavity's dimensions are 5×5 and the kinematic viscosity of the fluid is 0.05. Therefore, a top-plate velocity of 1/20 yields a *Reynolds* number of 5. This is certainly pushing our initial assumption that the vorticity diffuses through the cavity—there must be some minimal convective transport too. Second, notice that we are reassigning the time-indexed vorticity values (to limit the time-storage requirements to two, old and new) inside an existing nested loop. This saves about four lines of code at the cost of slower execution. Finally, at the very bottom of the program you will notice that we are computing the x-component of velocity on the vertical centerline of the enclosure. This will allow us to see where the largest negative velocity occurs, and of course it will reveal that the velocity at the top is +0.05 and at the bottom, 0, as expected. Now we should ask ourselves: what do we expect to obtain from this solution? The flow is occurring in an enclosure, driven by a sliding top surface; clearly the fluid must recirculate, so the streamlines should reveal a series of nested, closed paths. The results of the computation are provided in Figure 8.21. We can also assess the validity of our solution by consulting the examples of flow visualization for creeping flow in cavities provided by Taneda (Journal of the Physical Society of Japan, 46:1935, 1979); figures 17(b) and 18 from Taneda's paper are relevant although obtained for very low Re. We should also note that the cited flows were driven by a viscous fluid *flowing across the top* of the cavity rather than a sliding top surface. As a result, some slight sag of the streamlines is evident at the top of the square cavity, and the sag is more pronounced for shallow, wide cavities.

Notice that the streamlines are closer together (vertically) near the top; this indicates higher velocity. They are farthest apart in the bottom corners where the velocity vector components are much smaller. Should you decide to try this problem yourself—and I encourage you to do so—make note of how the "center" of the recirculation is concentrated very near the upper sliding surface at small time, t. It will migrate downward toward the position you see above as t increases. We should also anticipate what the ultimate distribution of vorticity looks like, and we should expect to see greater magnitudes in

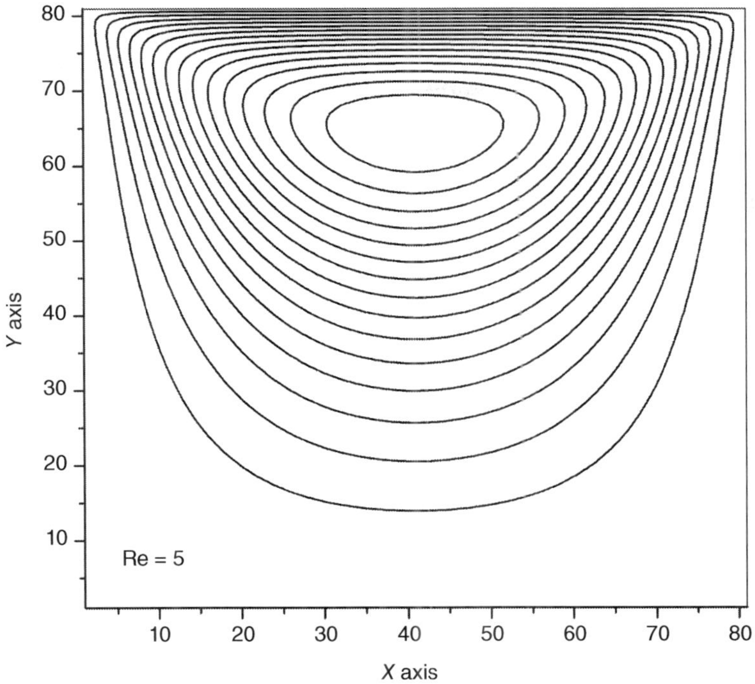

Figure 8.21 Computed streamlines for recirculating flow in a square cavity. The fluid motion is driven by a sliding upper surface and the Reynolds number is 5.

regions of the cavity where the velocity gradients are largest; of course, this will correspond to the top corners and the along the sliding lid.

Let us emphasize a crucial point here: We have solved this problem—viscous flow in a square cavity—using nothing but elementary procedures that are already familiar to us. In doing so, we have gained confidence that we can now tackle more complicated two-dimensional flow scenarios with an expectation of success.

8.9.2 The Deep Cavity at Large Reynolds Numbers

With this in mind, we will move forward with a similar physical situation, but employing two specific changes that will produce a problem that is more difficult but far more interesting. We will change the aspect ratio of the cavity, increasing the depth, H, such that H is twice the width, W. Furthermore, we increase the Reynolds number to 2,000 (a 400-fold increase); we will not be able to use eq. (8.146) in this case since the transport of vorticity by the fluid motion will be very important. Our full model, when (partially) discretized, will appear as:

$$\frac{\omega_{i,j,k+1}-\omega_{i,j,k}}{\Delta t}=\nu\left[\frac{\omega_{i+1,j,k}-2\omega_{i,j,k}+\omega_{i-1,j,k}}{(\Delta x)^2}+\frac{\omega_{i,j+1,k}-2\omega_{i,j,k}+\omega_{i,j-1,k}}{(\Delta y)^2}\right]-v_{x_{i,j,k}}\frac{\partial\omega}{\partial x}-v_{y_{i,j,k}}\frac{\partial\omega}{\partial y} \tag{8.153}$$

Notice that the convective transport terms have been moved to the right-hand side, but the derivatives of vorticity with respect to x and y have not been specified. This is because those derivatives must be discretized in *upwind form* if we are to employ our explicit procedure to find vorticity on the new time-step row. Consequently, we must test the velocity vector components at each point, i,j,k, and write the derivative appropriately. For example, if v_x is greater than zero, then $\frac{\partial\omega}{\partial x}\cong\frac{\omega_{i,j,k}-\omega_{i-1,j,k}}{\Delta x}$. Alternatively, if the x-component of velocity is negative (in an area of reverse flow), then $\frac{\partial\omega}{\partial x}\cong\frac{\omega_{i+1,j,k}-\omega_{i,j,k}}{\Delta x}$. The convective transport of vorticity in the y-direction must be handled similarly, of course. Once the vorticity distribution is updated, we use it to compute the revised stream function (ψ) everywhere in the interior of the cavity. The velocity vector components are then computed from the stream function:

$$v_{x_{i,j}}=\frac{\psi_{i,j+1,}-\psi_{i,j-1}}{2\Delta y}\quad\text{and}\quad v_{y_{i,j}}=-\frac{\psi_{i+1,j}-\psi_{i-1,j}}{2\Delta x}. \tag{8.154}$$

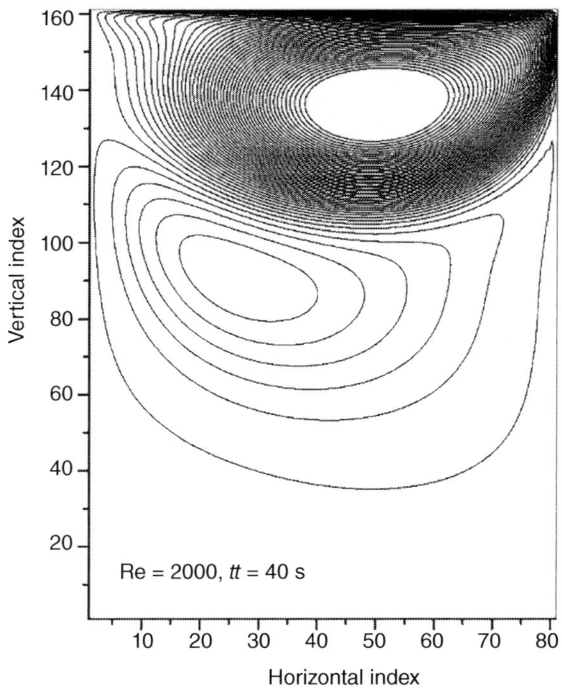

Figure 8.22 Flow in a deep cavity for which $H = 2$ W and Re = 2,000; this is the flow field 40 s after initiation of motion at the top surface. A third vortex will appear at the bottom of the cavity, but it is very weak and not revealed by the scaling employed here.

We have dropped the time subscript to minimize clutter, but we recognize that these are the *new* values of velocity. Let us summarize the changes that are necessary in order to properly modify the previous computational scheme which we employed to model the diffusional transport of vorticity through a square cavity:

1) The aspect ratio of the cavity must be changed.
2) The velocity of the upper sliding surface must be increased to reflect the larger Reynolds number.
3) The full vorticity transport equation must be used to advance $\omega_{i,j,k}$ to the new time-step row, and the convective transport terms must employ upwind differencing. Consequently, the velocity vector components must be tested at every interior point.
4) Once the stream function distribution has been obtained from the updated vorticity, the velocities v_x and v_y must be computed at every interior point.

We can anticipate some significant changes from the low-*Re* flow field we saw previously. Because of the deep cavity, multiple vortices may appear. And, due to the large velocity of the upper sliding surface, the uppermost vortex will not be centered in the cavity as it was previously; this is illustrated in Figure 8.22.

The vorticity distribution will assume a very different form (compared to the low Re case) too; if you examine the streamlines in Figure 8.22, you will note that the velocity gradients are very large in the upper right-hand corner. As a result, we would expect to see a large concentration of vorticity in that area.

8.9.3 Adding Heat and Mass Transfer to Vorticity Transport Models

You will recall that we previously observed that we have tools in hand that permit us to solve some far more complex—and more interesting—problems. Let us move farther toward that objective by writing both the energy equation and the continuity equation (for species "*A*") assuming that we have some arbitrary two-dimensional flow.

$$\frac{\partial T}{\partial t} + v_x \frac{\partial T}{\partial x} + v_y \frac{\partial T}{\partial y} = \alpha \left[\frac{\partial^2 T}{\partial x^2} + \frac{\partial^2 T}{\partial y^2} \right]. \tag{8.155}$$

and

$$\frac{\partial C_A}{\partial t} + v_x \frac{\partial C_A}{\partial x} + v_y \frac{\partial C_A}{\partial y} = D_{AB} \left[\frac{\partial^2 C_A}{\partial x^2} + \frac{\partial^2 C_A}{\partial y^2} \right]. \tag{8.156}$$

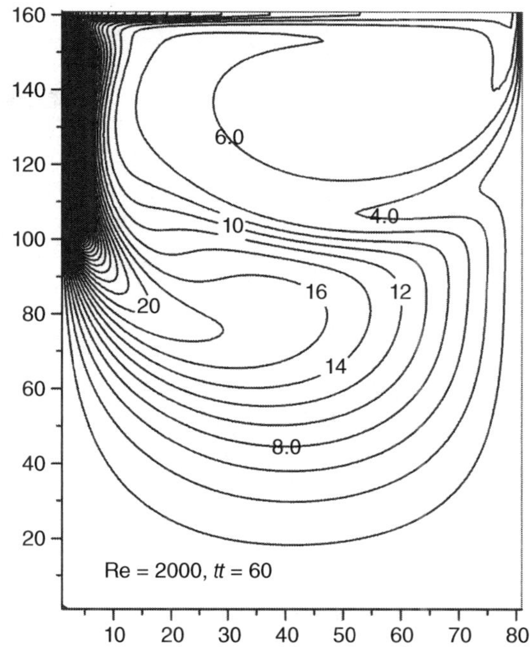

Figure 8.23 Isotherms for the deep cavity with the heated panel (upper portion of left-hand vertical wall) set to 75°, Re = 2,000, Pr = 1, and $t = 60$ s. Notice how the vortices are wrapping the thermal energy around and transporting it throughout the cavity.

We see that both of these equations have *exactly the same form* as the vorticity transport equation; therefore, the explicit scheme we used for vorticity transport can also be employed to solve two-dimensional heat and mass transfer problems. An identical logic structure will work as long as we remember to use upwind differencing on the convective transport terms. With this in mind, we now add heat transfer to the deep cavity problem.

Let the upper (sliding) surface be insulated, and let the upper portion of the left-hand vertical wall be maintained at elevated temperature. All other wall sections will be maintained at 0° and the fluid (again, initially at rest) will have a uniform initial temperature of 0° also. We discretize eq. (8.155) in exactly the same manner as the vorticity transport equation; this allow us to update both $\omega(x,y,t)$ and $T(x,y,t)$ using common code structure. We fix the temperature of the heated wall section at 75°, set Re = 2,000, Pr = 1, and carry out the computation until $t = 60$ s The resulting isotherms appear in Figure 8.23.

8.9.4 Adding Buoyancy to Vorticity Transport

There is one more aspect to this heat transfer problem that warrants discussion: Large local temperature variations result in buoyancy due to the decrease in fluid density. For example, the density of water is 1,000 kg/m^3 at 0° but only 975 kg/m^3 at 75 °C. Consequently, the coefficient of volumetric expansion, β, is about 0.0004 °C^{-1} at 75°. For ideal gases you may recall that $\beta = 1/T$; thus, if $T = 400$ K, $\beta = 0.0025$ K^{-1}. In the situation illustrated by Figure 8.23, the heated upper left-hand wall will promote fluid motion in the $+y$-direction, increasing the size and intensity of the upper vortex. We can extend our previous model to include this generated buoyancy using the Boussinesq approximation (which is appended to the vertical component of the equation of fluid motion). The buoyant force is the product of the fluid density, the acceleration of gravity, the coefficient of volumetric expansion, and the characteristic temperature difference. Using values appropriate for the example heat transfer problem considered previously, we find:

$$\rho g \beta \Delta T = (1)(980)(0.0004)(75) = 29.4 \text{ dynes/cm}^3 \text{ in cgs units.}$$

A dyne is a very small unit of force, so this value of 29.4 may strike you as too small to be of any significance. Let us explore this further; first, we need to determine exactly how buoyancy enters the computational picture. Clearly it will make its appearance in the vorticity transport equation—but perhaps not in the way you imagine. Consider a two-dimensional flow field for which the x-axis is horizontal and the y-axis is vertical: The horizontal component of the equation of motion is just as it appears in (8.137), but the y-component is amended as follows

$$\frac{\partial v_y}{\partial t} + v_x \frac{\partial v_y}{\partial x} + v_y \frac{\partial v_y}{\partial y} = -\frac{1}{\rho}\frac{\partial p}{\partial y} + \nu \left[\frac{\partial^2 v_y}{\partial x^2} + \frac{\partial^2 v_y}{\partial y^2}\right] + g\beta(T - T_\infty) \tag{8.157}$$

(remember that we have divided by the fluid density, ρ). In order to develop the modified vorticity transport equation, we cross-differentiate the two equations of motion and then subtract the x-component from the y-component. As a result of the differentiation, the Boussinesq approximation (which was appended to eq. (8.157) immediately above) takes the form $g\beta\frac{\partial T}{\partial x}$. This is the added source term that appears in the vorticity transport equation:

$$\frac{\partial \omega}{\partial t} + v_x \frac{\partial \omega}{\partial x} + v_y \frac{\partial \omega}{\partial y} = \nu \left[\frac{\partial^2 \omega}{\partial x^2} + \frac{\partial^2 \omega}{\partial y^2}\right] + g\beta\frac{\partial T}{\partial x}. \tag{8.158}$$

We should reflect upon this for a moment. Did you expect to see a *derivative with respect to x* in the equation? Remember that vorticity resulting from fluid motion in the x–y plane is a direct measure of rotation about the z-axis. Buoyancy—for our choice of coordinates—will create fluid velocity in the $+y$-direction. If rotation is to be produced about the z-axis, then there must be some variation in buoyancy *horizontally*.

The deep cavity example that we have been studying can be used to demonstrate the effects of buoyancy upon the flow field. The cavity dimensions will remain the same, and the *Prandtl* number will still be 1 (one). We will decrease the velocity of the sliding top surface so that Re = 500. The *Grashof* number is used to characterize problems in heat transfer in which buoyancy is thought to be important:

$$Gr = \frac{g\beta\Delta T L^3}{\nu^2} = \frac{(980)(0.0001)(25)(5)^3}{(0.05)^2} = 2.45 \times 10^6. \tag{8.159}$$

Though this may seem large to you, this is a modest value for Gr; 10^9 is not unusual for a heated surface immersed in a fluid. We want to compare two cases: in the first, we will heat the lower half of the *left-hand* wall, maintaining the heated panel at a constant 25 °C. The lid will be insulated, and all other surfaces in the cavity will have constant $T = 0°$. For case two, we will move the heated wall section to the lower half of the *right-hand* wall. We should anticipate some significant differences; since the cavity lid is sliding left to right, buoyancy generated at the left-hand wall will tend to increase the strength of the upper vortex. If we heat the right-hand wall, buoyancy will act in opposition to the motion created by the moving lid as shown in Figure 8.24.

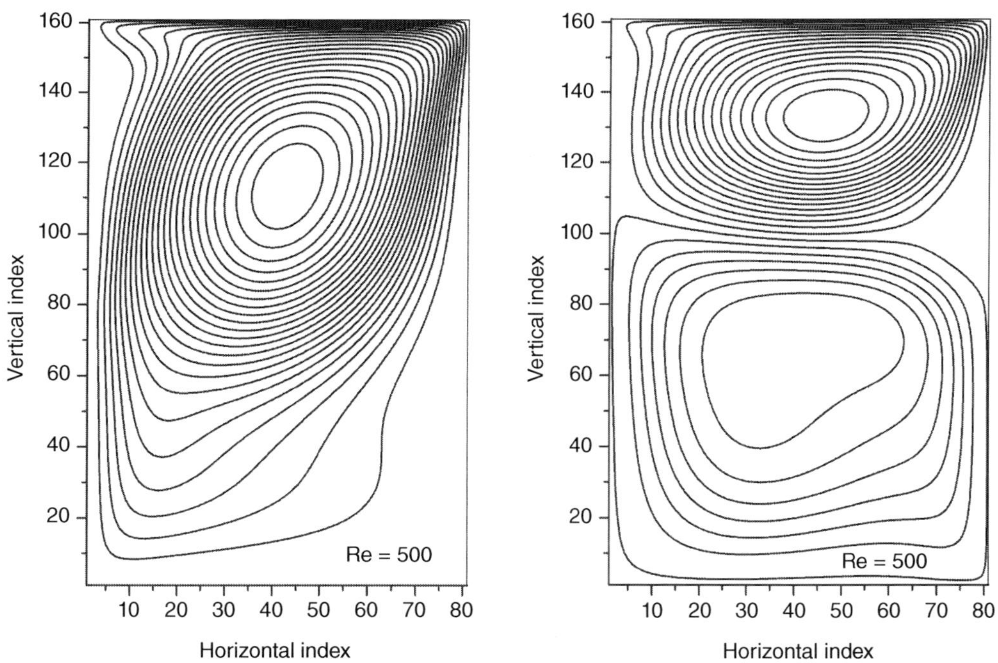

Figure 8.24 Streamlines illustrating the effects of buoyancy upon solution of the vorticity transport equation for lid-driven flow in a deep cavity with Re = 500, Pr = 1, and Gr = 2.45×10^6. On the left, the heated panel is the lower half of the *left-hand* wall; on the right, the heated section is the lower half of the *right-hand* vertical wall.

If we heat the left-hand wall buoyancy contributes to the size and strength of the upper vortex, and alters its position such that it reaches the bottom of the cavity. If we heat the right-hand wall buoyancy strengthens bottom vortex and inhibits the size of the upper vortex.

8.9.5 Vorticity Transport and the Rayleigh–Bénard Scenario

There is a closely related flow phenomenon that can be dealt with easily using the tools we now have in our grasp: natural convection in enclosures, or *Rayleigh–Bénard* problems. The very same computational procedure that we employed in the examples above can be used with just two modifications: The aspect ratio of the cavity (or enclosure) will be changed, and the sliding motion of the upper surface will be stopped. In this new scenario, a viscous fluid, initially at rest and at uniform temperature, is confined within the rectangular enclosure. At $t = 0$ the temperature of the *bottom* surface is elevated slightly, initiating *buoyancy-driven fluid motion*. The warmer fluid rises and the cooler fluid falls, setting up a pattern of recirculation. We assume that the flow is two-dimensional and that the Boussinesq approximation adequately describes the buoyancy effect; we saw previously that the latter entails adding a term, $\rho g\beta\Delta T$, to the y-component of the Navier–Stokes equation. After dividing by the fluid density, ρ, cross-differentiating, and then subtracting the x-component from the y-component, we found that the vorticity transport equation had been appended with a source term, $g\beta\frac{\partial T}{\partial x}$. The meaning here, as we discussed above, is that vorticity will be created when buoyancy varies in the horizontal direction. We should examine the consequences of this modification to vorticity transport. We choose a rectangular enclosure in which the (horizontal) width is *four times* larger than the height and all four surfaces are stationary. We will take $\mathrm{Pr} = 1$, and fix the *Grashof* number,

$$\mathrm{Gr} = \frac{g\beta\Delta TL^3}{\nu^2} = 6,125.$$

The fluid has an initial (uniform) temperature of 0°; at $t = 0$ the temperature of the bottom surface is instantaneously elevated. Since the side walls are maintained at 0° for all t, it is clear that the initial impetus for vorticity creation will occur at the bottom corners where $\frac{\partial T}{\partial x}$ may assume relatively large values. This phenomenon is clearly illustrated in the two-time sequence shown in Figure 8.25 where flow development is depicted at $t = 4$ and 40 s.

Of course, it is entirely appropriate to wonder about the physical fidelity of this Rayleigh–Bénard simulation. Do the convections rolls illustrated in Figure 8.25 bear any relation to real buoyancy-driven flows in enclosures? Fortunately this

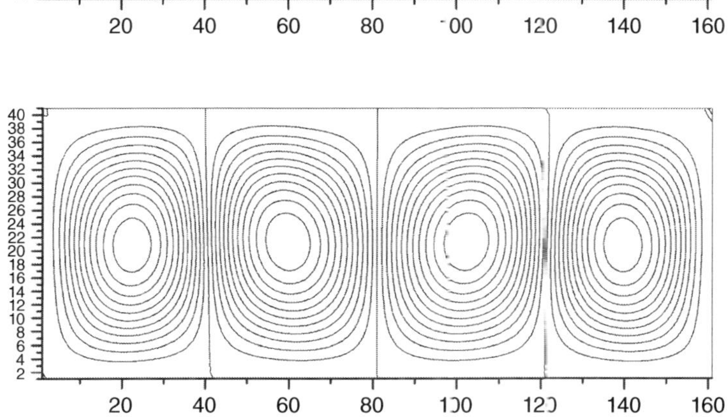

Figure 8.25 Development of the convection rolls in a two-dimensional Rayleigh–Bénard problem. At $t = 0$ the temperature of the bottom surface is elevated and buoyancy-driven fluid motion is initiated. For these calculations, $\mathrm{Pr} = 1$ and $\mathrm{Gr} = 6{,}125$. The illustrated sequence is for $t = 4$ and 40 s.

is an area of heat transfer research that has seen decades of activity, and we can obtain comparisons from the literature that will reveal whether or not our model has meaning. Van Dyke (1982) assembled interferograms (in figure 139, page 82) from the work of Oertel and Kirchartz (1979); these experimental data show marked similarity to the results of our computations, so our modeling procedure does have utility in terms of representing natural convection in enclosures.

In the previous example, the ultimate outcome was partly determined by geometrical constraints imposed by the width and height of the enclosure; the aspect ratio was ¼, so four convection rolls of equal size would exactly fill the available space. What happens if we remove that constraint and increase the thermal driving force consecutively? Will the vorticity transport computation prove to be robust under far more demanding circumstances? To address these questions we will turn our attention to a deep cavity containing a viscous fluid initially at rest. We will raise the temperature of the bottom surface at $t = 0$, maintain the vertical walls at 0°, and provide an insulated (and stationary) lid. This is a problem that has attracted both experimental and computational attention (particularly for three-dimensional enclosures) in recent years because the flow exhibits transitions as the driving force is increased; first from steady state to oscillatory flow, and then from unsteady flow to chaos. The transitions are usually characterized in terms of critical values of the *Rayleigh* number, where $Ra = GrPr$ and the reader is referred to Xia and Murthy (2002) for a readable introduction to this type of study. Of course our purpose is not to determine the route to chaos in buoyancy-driven flows, but rather to test the capabilities of our elementary vorticity transport scheme. We should expect to find some kind of limit—a threshold (in terms of thermal driving force) at which our computational procedure may fail to produce results.

We select a deep cavity for which the aspect ratio (depth/width) is two. The fluid contained within is initially at rest and at a uniform temperature of 0°. At $t = 0$ we elevate the temperature of the cavity *bottom* just as we did in the initial *Rayleigh–Bénard* example. We will set the *Prandtl* number to 1 such that Ra = Gr. As we have discussed, vorticity will initially be created in the bottom corners of the enclosure where $\frac{\partial T}{\partial x}$ (part of the Boussinesq source term) will *not be zero*. Let us begin a sequence of computations starting with a modest value for the *Grashof* number (49,000); our plan is to carry out the computations to a fixed time (say, $t = 5$ s) and then increase Gr and repeat. This will provide us with a series of "snapshots" of the developing flow field, and with a clear picture of the limitations of our implementation of vorticity transport. The actual driving force for the developing flow will be determined by the product of the coefficient of volumetric expansion and the characteristic temperature difference, that is, $\beta\Delta T$. It is worth noting that Xia and Murthy found that the first critical *Rayleigh* number was 25,450 for a rectangular (three-dimensional) enclosure with an aspect ratio of two; below $Ra = 25{,}450$ the fluid was found to remain at rest. Of course our simulation is two dimensional, so we should not expect anything more than qualitative correlation with three-dimensional studies. In this series of computations we start with a modest thermal driving force using Gr = 49,000; the *Prandtl* number was fixed at Pr = 1 and the calculations were stopped at $t = 5$ s. For smaller *Grashof* numbers the vortices that arise from heating the cavity floor are symmetric (left and right), and of course the initial impetus for fluid motion arises at the bottom corners. However, as we increase Gr we should expect a very different pattern of flow evolution. In fact, at much larger *Grashof* numbers the flow soon after initiation becomes far more complex; Figure 8.26 illustrates the flow field for $Gr = 6 \times 10^6$ five seconds after initiation.

The stream function "snapshot" provided in Figure 8.26 shows that the flow development under large thermal driving force conditions is dynamically complex; this is radically different than the recirculations generated at much small Gr. In fact, at Gr = 49,000 and $t = 5$ s we obtain two corner vortices (with symmetry), and we find that a steady-state flow condition will ultimately be attained. In contrast, for large driving forces (e.g., Gr = 6,000,000) the vorticity transport model reveals a complex and dramatically transient flow that—at least at $t = 5$ s—certainly appears chaotic.

A final observation regarding our modeling efforts for the driven cavity and Rayleigh–Bénard phenomena: Our computational modeling cannot reproduce certain aspects of the transient behavior seen in real flows. For example, for a buoyancy-driven flow the sense of the initial rotation is always the same in our procedure; this is at odds with a carefully crafted Rayleigh–Bénard *experiment* where localized heating may result in either clockwise or counter-clockwise rotation of the first convection roll. This provides emphasis for an observation we made previously: When we discretize a partial differential equation we are, in fact, making changes to the model itself. The gross results may be realistic, but the effects of altering the PDE, coupled with the computational pattern employed, may effectively disguise some of the nuances seen in the actual physical phenomenon. In this connection, we are also hostage to the resolution of our mesh; Moffatt (1964) predicted the existence of a sequence of small transient vortices appearing particularly proximate to interior corners. In the case of lid-driven flows in rectangular cavities, this means that *small*, counter-rotating vortices should appear in the bottom corners; yet in the examples of this type of flow given previously in this chapter, no evidence of such vortices is seen. This is because our procedure, coupled with the mesh size employed, is incapable of resolving these very small eddies. Biswas and Kalita (2016) demonstrated that they *can be captured* if the grid size adjacent to the critical areas is sufficiently refined; see

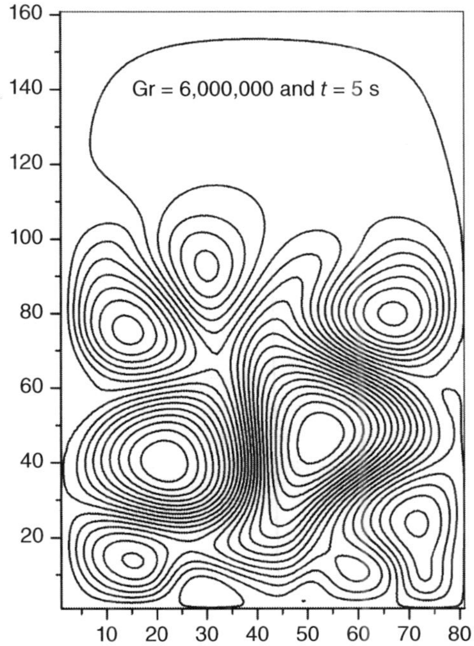

Figure 8.26 Application of vorticity transport to buoyancy-driven flow in a deep enclosure (aspect ratio 2) with heated bottom surface. In this image a large thermal driving force ($Gr = 6 \times 10^6$) was used and the computation was stopped at $t = 5$ s.

Figures 2,3, and 4 in their paper. In a square cavity with Re = 100, they successively refined their grid size (culminating with 321×321) and reported that the intensities of the first three Moffatt vortices decreased by four or five orders of magnitude. The very small scales and the extraordinarily low intensities of these flow features make it unlikely for our procedure to capture them without significant modification of the computational structure.

8.9.6 Vorticity Transport in More Difficult Geometries

In the preceding applications of the vorticity transport algorithm we dealt exclusively with rectangular cavities; in these cases the explicit updating of vorticity could proceed systematically from the bottom interior row (say, j), scanning horizontally (the x- or i-direction), then moving to $j+1$ and repeating the process. We must note, however, that a similar strategy can be applied to more complicated geometries; for example, we could place some type of obstruction in a channel, with the flow moving past it from left to right. Such a situation imposes a couple of additional challenges for the analyst: First, the horizontal scan (at all of the affected j-positions) must recognize the obstruction and skip past it of course. Second, the intrusion of the object in the channel will create additional vorticity-generating surfaces, so appropriate logic will need to be employed to accurately calculate vorticity upon them.

We should illustrate the modified computational process with an example; we place a triangular wedge upon the floor of a rectangular channel. At $t = 0$ flow is initiated in the channel (from left-to right) and the fluid is required move up the leading (inclined) surface of the wedge, flow over the vertex, and then decelerate as the fluid reoccupies the entire channel height. For a physical analogue we could think of a turbulence- or heat transfer-promoting strip placed in a channel. Since we have gained some practice with heat transfer occurring in combination with vorticity transport, we also allow the wedge to be heated to see the effect of the obstruction upon heat transfer in the channel. We will, however, initially neglect buoyancy for this example.

There is one more feature of this sort of problem that requires our attention. In the cavity flow examples we studied above, the moving fluid was completely enclosed; that is, the entire boundary was a streamline (for which we arbitrarily set $\psi = 0$). But in the channel flow we consider now, there are both inflow and outflow boundaries—the fluid enters the channel from the left-hand side and leaves on the right. Our computational domain will necessarily be finite (although we could circumvent this difficulty with a coordinate transformation). Consequently, when we fix the distribution of ψ on the outflow boundary we have constrained the flow in an unphysical manner. Think of it this way: If the wedge generates a vortex—and it will—then growth of the area of recirculation may result in the vortex interacting with the outflow boundary where we have already specified $\psi(x = L,y)$. What are our options? We could employ a computational domain that is very extensive at the cost of large storage requirements and much slower execution, or we might set the derivatives (with respect to x) equal to zero on the outflow boundary, effectively stopping further evolution of T and ψ in the flow direction. Let us examine the consequences of the latter.

An extensive series of computations was carried out for a heated triangular wedge placed upon a channel floor. The channel height was 10 cm and its length was 27.5 cm; the wedge was a 90-45-45 (isosceles) triangle with a height of 4 cm, and therefore it occupied 40% of the channel height. Reynolds numbers ranging from 80 to 320 were investigated and the streamlines and isotherms were plotted and as one might expect for this limited computational domain, the effects produced by the heated wedge did reach the outflow boundary. Furthermore, this situation was worsened by larger Reynolds numbers, increased ΔT, and the inclusion of buoyancy in the calculations. Buoyancy was initially neglected by simply setting $g\beta\Delta T = 0$ even though the temperature difference of 50° was relatively large. The calculations demonstrated that the thermal "plume" reached the outflow boundary (at short times) for $\text{Pr} = 1$, $\text{Re} = 160$, and $g\beta\Delta T = 0$, and the thermal wake of the heated wedge was enveloped by the vortex.

Although the ΔT employed in this example was large the coefficient of volumetric expansion, β, was initially set to zero (0) so that the contribution of the buoyancy source term to the vorticity transport equation was neglected. It is reasonable for us to wonder if the thermal "plume" and the vortex may be altered significantly by a non-zero β. You may recall that we looked at the product, $\rho g\beta\Delta T$, in our initial discussion of buoyancy and decided that 29.4 dynes/cm^3, although small, was probably not insignificant. Let us now include buoyancy in this study of flow over a heated wedge so we can better understand the implications for our modeling procedure. We use the same parameters as before, except we will examine increasing values of $g\beta\Delta T$ (instead of zero).

The results prove that buoyancy plays a very significant role in the application of vorticity transport to nonisothermal flow problems. We can better gauge the importance of this source term in the vorticity transport equation by examining the impact of the product $g\beta\Delta T$ upon the thermal plumes trailing the heated wedge. We will fix $\text{Pr} = 1$, the *Reynolds* number at 80 and stop each computation exactly 10 s after initiation of flow in the channel. We begin with $g\beta\Delta T = 0.2$, where the role of buoyancy in the production of vorticity is slight, and then employ successive increases until $g\beta\Delta T = 5$. For the latter, the role of buoyancy in the computation is profound; see Figure 8.27.

8.9.7 Flow in the Entrance of a Duct Formed by Parallel Planes

In the previous section we described difficulties posed by specification of an outflow boundary condition. Let us now demonstrate in a precise way why this is so problematic. Consider a two-dimensional rectangular duct with one horizontal wall located at $y = 0$ (where $\psi = 0$) and the other at $y = b$. At $t = 0$ a viscous fluid begins to move through the channel; it enters at $x = 0$ with a uniform velocity, V_{in}. This means that the stream function at the inlet, $\psi(x = 0,y)$ varies linearly such that

$$\psi_{inlet} = (j-1)\Delta y V_{in}, \tag{8.160}$$

where j is the y-position index ($1 \leq j \leq 41$). Consequently for the top horizontal wall,

$$\psi_{y=b} = V_{in} b. \tag{8.161}$$

Since this is a pressure-driven flow, we find that the velocity distribution for fully developed conditions is

$$v_x = \frac{1}{2\mu}\frac{dp}{dx}(y^2 - by). \tag{8.162}$$

Naturally our interest is how the flow transitions from the uniform inlet condition, V_{in}, to the parabolic velocity profile $v_x(y)$ given by eq. (8.162). Since the fluid is incompressible, the volumetric flow rate cannot change with x-position, so:

$$Q = V_{in} b = \int_0^b \frac{1}{2\mu}\frac{dp}{dx}(y^2 - by)dy = -\frac{1}{12\mu}\frac{dp}{dx}b^3. \tag{8.163}$$

Therefore, for *fully developed flow* it will be necessary that the pressure gradient satisfy

$$\frac{dp}{dx} = \frac{-12\mu V_{in}}{b^2}. \tag{8.164}$$

Thus we can determine the stream function distribution for the fully developed (outlet) flow condition; since

$$v_x = \frac{\partial\psi}{\partial y} = \frac{1}{2\mu}\frac{dp}{dx}(y^2 - by),$$

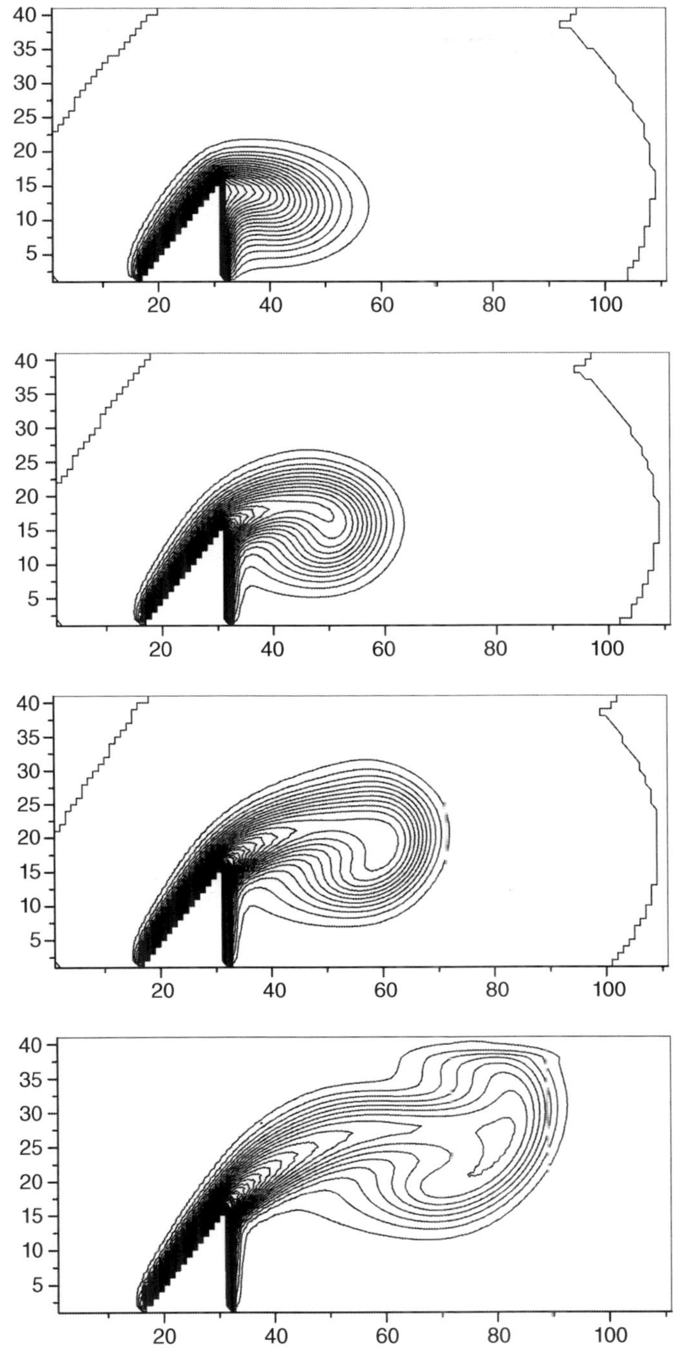

Figure 8.27 Thermal plumes behind wedge with Pr = 1, Re = 80, and $t = 10$ s (after initiation of flow). From top to bottom, the product $g\beta\Delta T$ is 0.2, 1, 2, and 5 (cm/s^2). The role of buoyancy in the production of vorticity is insignificant for the 0.2 case (top) and absolutely crucial at the bottom where $g\beta\Delta T = 5$.

we find

$$\psi(x \to \infty, y) = \frac{1}{2\mu}\frac{dp}{dx}\left(\frac{y^3}{3} - \frac{by^2}{2}\right) + C_1. \tag{8.165}$$

We selected $\psi = 0$ for $y = 0$, so the constant of integration is zero.

Now we are in position to apply the vorticity transport equation to this entrance flow problem. We begin by fixing the length of the duct, L, so that $L = 20b$ (at the same time suspecting that this will be an insufficient length for this situation). Accordingly, the stream function distribution at $x = L$ is given by eq. (8.165). The fluid is initially at rest so that $v_x = v_y = 0$

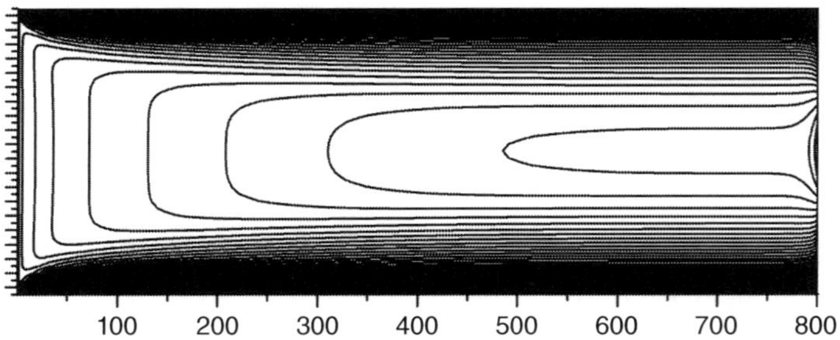

Figure 8.28 Contours of $v_x(x,y)$ for Re = 900 and t = 18 s. The vertical height of the duct has been exaggerated to facilitate interpretation of the results.

everywhere. Then, at $t = 0+$, we allow the uniform inlet velocity to begin to increase as

$$V_{in} = V(1 - \exp(-at)). \tag{8.166}$$

We perform the vorticity transport computation out to $t = 18$ s; with $a = 1/2$, this time will yield 99.99% of the ultimate uniform inlet velocity. We choose b and the kinematic viscosity, ν, so that for the ultimate condition, $\mathrm{Re} = \frac{bV_{in}}{\nu} = 900$. We will prepare a contour plot of the x-component of the velocity vector from our computational work.

What we see in Figure 8.28 is exactly the expected outcome of the computation—remember, we fixed the outflow boundary condition so that we *had to get* the parabolic profile at $x = L$. That is, our scheme for solution of this problem *required* that the entrance length, L_e, be 20 times the duct height: $L_e = 20b$. Is this close to being correct? For laminar flows the entrance length is Reynolds number-dependent; Shimomukai and Kanda (2006) summarized the results obtained by several researchers and compared them to their own computations. If we take the entrance length to correspond to the point where the centerline velocity attains 98% of its ultimate value, then approximately,

$$0.034 \cong \frac{x}{b\mathrm{Re}}.$$

For the example computation illustrated by Figure 8.28, $b = 3$ cm and Re = 900, and consequently, $L_e \approx 92$ cm. This means that we set the outflow boundary condition at a position ($x = 60$ cm) corresponding to about *two-thirds* of the actual entrance length; we have *constrained the solution in an unphysical way*. Our results may be of use qualitatively, but perhaps not quantitatively. For problems of similar type where accuracy is critical there are two options available to us: *greatly* expand the computational domain in the x-direction, or define a new variable, e.g., $\eta = 1 - \frac{1}{1+cx}$ so that $0 \le \eta \le 1$ as $0 \le x \le \infty$. The reader should explore the price to be paid should we choose the latter course of action; for example, consider the derivative of vorticity with respect to x:

$$\frac{\partial \omega}{\partial x} = \frac{\partial \omega}{\partial \eta}\frac{\partial \eta}{\partial x} = c(1-\eta)^2\frac{\partial \omega}{\partial \eta}. \tag{8.167}$$

8.9.8 Free Convection from Horizontal Ducts and the Effect of Pr

We previously described the difficulties created by flow scenarios for which there are inflow and outflow boundaries. Let us consider a buoyancy-driven flow created by a heated horizontal duct located in an "infinite" fluid medium. The heat loss that occurs in such cases is of significance to the process and utility industries where both pipes and rectangular ducts are routinely exposed to ambient conditions, and the resulting thermal energy transfer can adversely affect process efficiency. Prior to the advent of digital computers such problems were handled either by the development of correlations from experimental data or by boundary-layer approximations that facilitated analytic or numerical solutions. An example of the latter is Hermann's (1954) treatment of free convection from a horizontal cylinder in which the usual boundary-layer simplifications were introduced to reduce the governing partial differential equations to a set of ordinary differential equations (a third-order equation for the stream function and a second-order equation for temperature). The most difficult aspect of Hermann's approach to this problem was the accommodation of curvature in the usual boundary-layer coordinates, where x is measured along the object's surface and y is the normal coordinate extending out into the fluid.

Our purpose here, however, is to see if we can adapt the vorticity transport procedure to this class of problem. There are some critical issues that will impact our modeling—of greatest importance is the fact that our computational domain must necessarily be finite, whereas the physical phenomenon we are representing normally occurs in an infinite fluid medium—or one that is nearly so. We begin by selecting a rectangular duct measuring 6 cm in width and 3 cm in height. We assume there is a vertical plane of symmetry that bisects the box such that portion appearing in our computations measures 3 × 3. We arbitrarily select $\psi = 0$ for the plane of symmetry and the box surface, noting that $\frac{\partial v_y}{\partial x} = 0$ along that vertical centerline. The right-hand computational boundary for this problem is placed at $x = 12$ cm (four times the half-width of the heated box), and we emphasize that this is *not* a "wall" but a slip boundary which is more reflective of the physical situation we are trying to model. The computational "engine" we employ is precisely the one we have used repeatedly in earlier examples. It consists of the logic sequence below: *w(i,j,k)* is vorticity, *T(i,j,k)* is temperature, *vx(i,j)* and *vy(i,j)* are the *x*- and *y*-components of velocity, and *psi(i,j)* is the stream function:

```
700 REM *** subroutine for vorticity and temperature update
            d2wdx2=1/dx^2*(w(i+1,j,1)-2*w(i,j,1)+w(i-1,j,1))
            d2wdy2=1/dy^2*(w(i,j+1,1)-2*w(i,j,1)+w(i,j-1,1))
            d2tdx2=1/dx^2*(T(i+1,j,1)-2*T(i,j,1)+T(i-1,j,1))
            d2tdy2=1/dy^2*(T(i,j+1,1)-2*T(i,j,1)+T(i,j-1,1))
        IF vx(i,j)>0 THEN dwdx=(w(i,j,1)-w(i-1,j,1))/dx ELSE dwdx=(w(i+1,j,1)-
w(i,j,1))/dx
        IF vy(i,j)>0 THEN dwdy=(w(i,j,1)-w(i,j-1,1))/dx ELSE dwdy=(w(i,j+1,1)-
w(i,j,1))/dx
        IF vx(i,j)>0 THEN dtdx=(T(i,j,1)-T(i-1,j,1))/dx ELSE dtdx=(T(i+1,j,1)-
T(i,j,1))/dx
        IF vy(i,j)>0 THEN dtdy=(T(i,j,1)-T(i,j-1,1))/dx ELSE dtdy=(T(i,j+1,1)-
T(i,j,1))/dx
           w(i,j,2)=dt*kvisc*(d2wdx2+d2wdy2)-dt*(vx(i,j)*dwdx+vy(i,j)*dwdy)+dt*g*
beta*(T(i+1,j,1)-T(i-1,j,1))/(2*dx)+w(i,j,1)
           T(i,j,2)=dt*alpha*(d2tdx2+d2tdy2)-dt*(vx(i,j)*dtdx+vy(i,j)*dtdy)+
T(i,j,1)
                RETURN

 800 REM *** SOR subroutine for psi determination
          psi(i,j)=psi(i,j)+wsor/4*(psi(i+1,j)+psi(i-1,j)+psi(i,j+1)+psi(i,j-1)-4*
psi(i,j)+dx^2*w(i,j,2))
            RETURN

 900 REM *** calculate velocities in interior
          vx(i,j)=1/(2*dy)*(psi(i,j+1)-psi(i,j-1))
          vy(i,j)=-1/(2*dx)*(psi(i+1,j)-psi(i-1,j))
            RETURN

1000 REM *** swap time values
        w(i,j,1)=w(i,j,2)
        T(i,j,1)=T(i,j,2)
        RETURN
```

The outflow (upper) boundary is located at a vertical position corresponding to 8 times the heated duct height. The fluid is initially stationary and the temperature of the medium is a uniform 0°. At $t = 0$, the temperature of the horizontal duct is elevated and buoyancy-driven fluid motion is initiated. For the result illustrated in Figure 8.29, Gr = 40,000 and Pr = 1/3 and the elapsed time is $t = 18$ s.

We should briefly reconsider eq. (8.155), the energy equation in two spatial dimensions; in the case of the *x*-direction transport of thermal energy; for explicit solution, the convective transport term, $v_x \frac{\partial T}{\partial x}$, is taken to the right-hand side and the equation is discretized. We multiply by Δt and observe that a dimensionless quotient has now appeared: $\frac{\Delta t v_x}{\Delta x}$. This is the Courant number, *Co*, that we saw in our earlier discussion of the leapfrog method applied to hyperbolic PDEs; its value *must be* between 0 and 1 if the elementary explicit scheme is to work successfully. Furthermore, the temperature gradient or

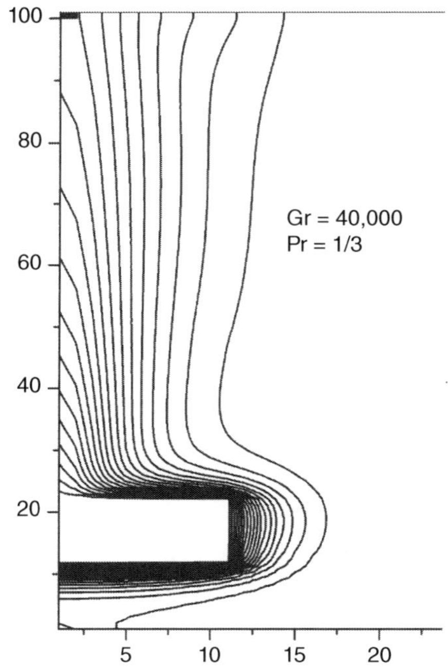

Figure 8.29 Thermal plume rising from a heated rectangular duct in free convection. In this illustration Gr = 40,000 and Pr = 1/3 (thus, Ra = GrPr = 13,333). You may recall that Pr $= \frac{\nu}{\alpha} = \frac{\mu C_p}{k}$, the quotient of the kinematic viscosity and the thermal diffusivity. Therefore, a smaller Prandtl number enhances the molecular transport of thermal energy and thickens the thermal wake (laterally) of the heated duct; conversely, larger values of Pr reduce the horizontal size of the plume. The thermal plume reaches the outflow boundary about 12 s after startup which means that average buoyancy-driven fluid velocity is about 2 cm/s. The horizontal scaling has been expanded in this example.

derivative, $\frac{\partial T}{\partial x}$, must be written in the "upwind" form so that a disturbance is only *advected in the direction of the fluid motion* (by the term "advection" we mean transport by fluid motion in a particular direction). Let us illustrate by assuming that the velocity in the x-direction can be taken to be a positive constant, V, then the first two terms in eq. (8.155) are written as:

$$\frac{\partial T}{\partial t} = -V\frac{\partial T}{\partial x} + \cdots \text{ which become } \frac{T_{i,j+1} - T_{i,j}}{\Delta t} \cong -V\frac{T_{i,j} - T_{i-1,j}}{\Delta x} + \cdots \tag{8.168}$$

and therefore,

$$T_{i,j+1} \cong (1 - Co)T_{i,j} + CoT_{i-1,j} \cdots \tag{8.169}$$

The power of the vorticity transport technique for two-dimensional problems with fluid motion and heat transfer may now be apparent to you, but let us extend it once more for emphasis. Recall that the transport of a chemical species in two dimensions is governed by the continuity eq. (8.156):

$$\frac{\partial C}{\partial t} + v_x\frac{\partial C}{\partial x} + v_y\frac{\partial C}{\partial y} = D\left[\frac{\partial^2 C}{\partial x^2} + \frac{\partial^2 C}{\partial y^2}\right].$$

We noted previously that this equation has *exactly the same form* as the vorticity transport equation and the two-dimensional energy equation. Thus, the explicit procedure employed before will work here as well. We now have the capability for solving two-dimensional problems involving the transport of momentum, heat, and mass, even when all three are occurring simultaneously. The only restrictive requirement is that the Reynolds number be low enough to guarantee that we have a highly ordered (laminar) flow. In a recent paper, Nikbakhti and Rahimi (2012) have solved eqs. (8.144), (8.155), and (8.156) simultaneously for *double-diffusive* natural convection in a rectangular cavity; the convective circulation was driven by a thermally active wall *combined with* a patch of elevated concentration. This means that the local fluid density was affected by both thermal expansion and by variations in solute concentration, and both are accounted for by using the Boussinesq approximation: $\rho = \rho_0[1 - \beta_T(T - T_c) + \beta_s(C - C_1)]$. Note that the effects of heat and mass transfer are opposed. The authors presented streamlines, isotherms, and concentration contours for different locations of the thermally active surface and some of the resulting circulation patterns are fascinating. For example, when the heated sections are placed at the top and bottom of the cavity, two strong (convection) cells arise, in the upper-left and the lower-right, with an off-set horizontally. Under these conditions, the local velocity near the horizontal centerline is very low.

8.10 MacCormack's Method

MacCormack's method (1969) for transient phenomena allows the analyst to solve time-dependent partial differential equations explicitly and for a couple of decades it was one of the most popular techniques for the solution of high-speed *compressible* flow problems. For simplicity let us contemplate a two-dimensional flow in which the density varies (this means we have a compressible fluid). We have two components of the Navier Stokes equation and continuity (conservation of mass). The reader may note an inconsistency here: We said the fluid density would vary, yet we are using the constant-density Navier–Stokes equation. We will resolve this momentarily. For now, let us arrange the equations to isolate the time derivatives:

$$\frac{\partial v_x}{\partial t} = -v_x\frac{\partial v_x}{\partial x} - v_y\frac{\partial v_x}{\partial y} - \frac{1}{\rho}\frac{\partial p}{\partial x} + \nu\left[\frac{\partial^2 v_x}{\partial x^2} + \frac{\partial^2 v_x}{\partial y^2}\right] \tag{8.170}$$

$$\frac{\partial v_y}{\partial t} = -v_x\frac{\partial v_y}{\partial x} - v_y\frac{\partial v_y}{\partial y} - \frac{1}{\rho}\frac{\partial p}{\partial y} + \nu\left[\frac{\partial^2 v_y}{\partial x^2} + \frac{\partial^2 v_y}{\partial y^2}\right] \tag{8.171}$$

and

$$\frac{\partial \rho}{\partial t} = -\frac{\partial}{\partial x}(\rho v_x) - \frac{\partial}{\partial y}(\rho v_y). \tag{8.172}$$

An equation of state is used to relate pressure (p) to the fluid density (ρ) in compressible flow problems, closing the system of equations. MacCormack's approach is a predictor–corrector scheme in which the first estimates for the time derivatives are obtained with forward differences for the inertial terms and central differences for the viscous terms. We let $u = v_x$ and $v = v_y$ to reduce clutter, and we take the i and j indices to represent the x- and y-directions, respectively. Thus, using the x-component as our example and *omitting pressure*:

$$\frac{\partial u}{\partial t} = -u_{i,j}\frac{u_{i+1,j} - u_{i,j}}{\Delta x} - v_{i,j}\frac{u_{i,j+1} - u_{i,j}}{\Delta y} + \nu\left[\frac{u_{i+1,j} - 2u_{i,j} + u_{i-1,j}}{(\Delta x)^2} + \frac{u_{i,j+1} - 2u_{i,j} + u_{i,j-1}}{(\Delta y)^2}\right]. \tag{8.173}$$

The computed values for the time derivatives are used to predict new values for all of the dependent variables; we illustrate this with u:

$$u_{i,j}^{new} = u_{i,j}^{old} + \left(\frac{\partial u}{\partial t}\right)\Delta t. \tag{8.174}$$

The *new* (predicted) values for the dependent variables are used to find a revised estimate for the time derivatives. But in this second step, *backward* differences are used for the inertial terms:

$$\left(\frac{\partial u}{\partial t}\right)^{revised} = -u_{i,j}\frac{u_{i,j} - u_{i-1,j}}{\Delta x} - v_{i,j}\frac{u_{i,j} - u_{i,j-1}}{\Delta y} + \nu\left[\frac{u_{i+1,j} - 2u_{i,j} + u_{i-1,j}}{(\Delta x)^2} + \frac{u_{i,j+1} - 2u_{i,j} + u_{i,j-1}}{(\Delta y)^2}\right]. \tag{8.175}$$

Now the two computed time derivatives are averaged:

$$\left(\frac{\partial u}{\partial t}\right)_{average} = \frac{1}{2}\left[\frac{\partial u}{\partial t} + \left(\frac{\partial u}{\partial t}\right)^{revised}\right], \tag{8.176}$$

and this average value is used to compute the corrected value (for each dependent variable, of course):

$$u_{i,j}^{corrected} = u_{i,j} + \left(\frac{\partial u}{\partial t}\right)_{ave}\Delta t. \tag{8.177}$$

You can see immediately why this scheme became so popular—this explicit algorithm is simple enough that the required programming logic is fairly easy to implement. However, it *cannot be applied* to incompressible flow problems without modification since for incompressible fluids continuity simplifies to $\nabla \cdot v = 0$ (no time derivative). Fortunately, MacCormack's method has been adapted for such problems by Bernard (1986 and 1989). In this variation, the changes in velocity are determined exactly as shown immediately above *omitting pressure*. Then the pressure gradient is determined by solution of a Poisson equation:

$$\nabla^2 p = \rho\nabla \cdot \left[\nu\nabla^2 u - \frac{du}{dt}\right]. \tag{8.178}$$

Bernard (1989) developed a code (STREMR) for two-dimensional flows of incompressible fluids using this approach. This scheme was later extended to three-dimensional flows (MAC3D) with the ultimate intended application to turbulent flow problems. You can assess how well these modifications to MacCormack's method work by consulting the examples Bernard (1992) provides; these include channel flow, the driven cavity, flow off of a step, and flow around a circular cylinder.

We should explore an elementary example to give more clarity to this discussion: We will use Burgers' equation as the basis for a transparent application of MacCormack's explicit scheme:

$$\frac{\partial u}{\partial t} + u\frac{\partial u}{\partial x} = \kappa\frac{\partial^2 u}{\partial x^2}. \tag{8.179}$$

This PDE has frequently been used in the literature as a kind of surrogate for the Navier–Stokes equation as it includes accumulation, convective transport, and diffusion (molecular transport) terms. However, it is important to note that the equation does *not* exhibit the chaotic aspects associated with turbulence. For this simplified example we first confine our attention to the "inviscid" form of Burgers' equation by setting $\kappa = 0$:

$$\frac{\partial u}{\partial t} + u\frac{\partial u}{\partial x} = 0. \tag{8.180}$$

The dissipative term has been removed and although the equation certainly looks innocuous its fundamental character has been changed as it is now a hyperbolic PDE. We will apply MacCormack's method to two versions of this PDE; first we will consider a linearized version:

$$\frac{\partial u}{\partial t} + c\frac{\partial u}{\partial x} = 0, \tag{8.181}$$

where c is a constant. Note that this equation is completely analogous to eq. (8.172) in Section 8.6. We will assume an initial distribution for u and initiate the predictor–corrector scheme; up will be the predicted value and uc the corrected one. Let i correspond to x position and j correspond to time, t. For the predictor step we have

$$up_{i,j+1} = u_{i,j} - c\frac{\Delta t}{\Delta x}(u_{i+1,j} - u_{i,j}). \tag{8.182}$$

And for the corrector:

$$uc_{i,j+1} = \frac{1}{2}\left[u_{i,j} + up_{i,j+1} - c\frac{\Delta t}{\Delta x}(up_{i,j+1} - up_{i-1,j+1})\right]. \tag{8.183}$$

Notice how the predictor step utilizes a first-order *forward* difference for the convective transport term, whereas the corrector step uses an *upwind* difference on the first derivative of the predicted value for u (which we designated as up). The reason that MacCormack's method was dominant in computational fluid dynamics (especially for compressible flows) for decades is probably apparent to you—it is extremely easy to implement. Because it is fully explicit we can march forward in time, computing new values for the variable u as we go. We are going to try this, setting $c = 1$ and using an initial distribution of $u(x, t = 0) = 1/i^2$, where i is simply the x-position index (which ranges from 1 to 101) (Figure 8.30).

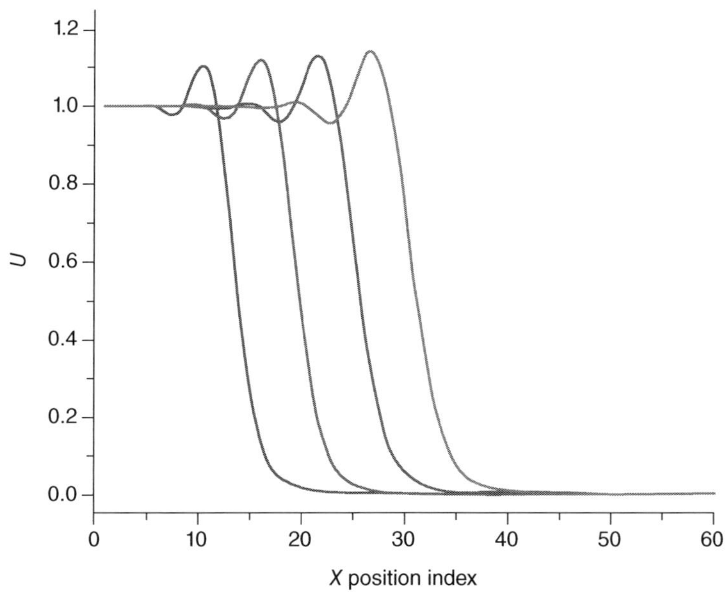

Figure 8.30 Solutions for the linearized "inviscid" Burgers' equation using MacCormack's method. Note the growing oscillations upstream of the transition. The curve farthest right corresponds to $50\Delta t$.

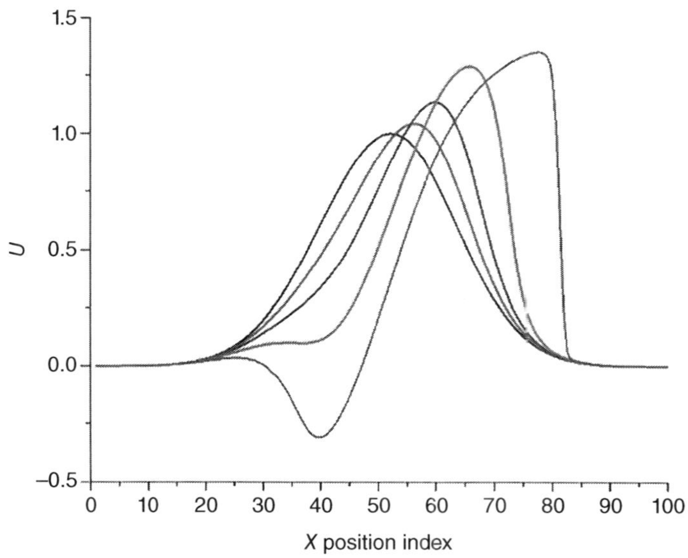

Figure 8.31 Solutions obtained with MacCormack's method for the nonlinear Burgers' equation where u was given an initial Gaussian distribution (black curve centered at an index value of 51). Time is increasing as we move successively to the right in this graph. Notice how the right-hand side of the wave is progressively steeper as time increases.

Now we return to the original PDE *without* the simplifying linearization; the nonlinear Burgers' equation is written in conservation form by setting $k = \frac{1}{2}u_{i,j}^2$ so that the predictor step appears as:

$$up_{i,j+1} = u_{i,j} - \frac{\Delta t}{\Delta x}(k_{i+1,j} - k_{i,j}). \tag{8.184}$$

In this nonlinear example the initial distribution of u is Gaussian, centered at an index value, $i = 50$. The numerical solution reveals the "breaking" phenomenon as t increases; the large-time solutions are approaching a discontinuity which is referred to as a *shock*. This behavior is evident in Figure 8.31.

It is a valuable exercise for the reader to contemplate what will happen in the example of above if $\kappa \neq 0$; i.e., how will the results shown in Figure 8.31 be changed if the dissipative term is added back to the computation? We recall that even terms (second, fourth, etc. spatial derivatives) have the effect of smoothing out dispersive events and rapid changes in the field variable. I recommend that you try this on the preceding example; set $\kappa = 0.04$, say, and perform MacCormack's procedure out to $t \approx 2.6$. You should see a damping effect on oscillatory behavior with diminished amplitudes and slopes. This will also serve as a reminder why it is sometimes advantageous to add artificial viscosity to the numerical solution of certain classes of PDEs.

MacCormack's method is an extremely powerful way to solve certain partial differential equations with time dependence. And although the technique has been around for more than four decades, it is still being applied to important practical problems and it has been focus of continued development efforts. The interested reader should see Selle et al. (An Unconditionally Stable MacCormack Method, 2007).

8.11 Adaptive Grids

At the beginning of this chapter we noted that many problems of significance could be solved without mesh (or grid) generation. Consequently, in our preceding discussion of the numerical solution of partial differential equations we normally used a square grid in which $\Delta x = \Delta y$ for two-dimensional problems. This meant that in most of the examples we considered the domain of interest was either rectangular, or regular is some way such that a square array of nodal points would coincide with the important boundaries of the problem. Obviously that will not always be the case, and we should expect to encounter problems involving objects placed in the field or walls or boundaries that exhibit curvature. In some cases such problems can be handled rather easily. Let us illustrate with an elementary two-dimensional case; we will assume that we have flow in a divergent duct as illustrated in Figure 8.32.

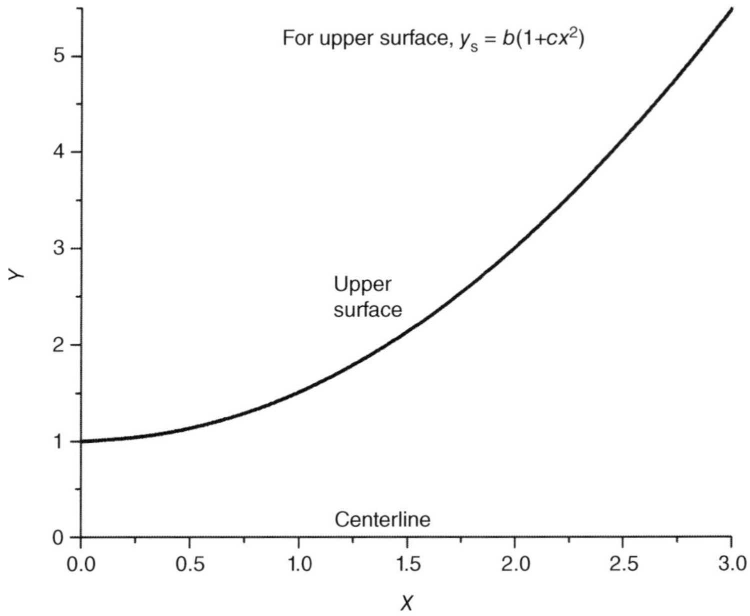

Figure 8.32 Half of a duct with increasing flow area. Fluid enters at the left and the origin has been placed on the centerline (the x-axis). The position of the upper surface is described by $y_s = b(1 + cx^2)$, so b is the half-width of the duct at the origin.

We will let

$$\eta = x \quad \text{and} \quad \phi = \frac{y}{b(1+cx^2)}. \tag{8.185}$$

Obviously, whenever y coincides with the upper surface of the duct, $y = y_s$ and $\phi = 1$. This transformation yields a rectangular grid in the *computational plane* since

$$0 \leq \eta \leq x \quad \text{and} \quad 0 \leq \phi \leq 1.$$

Now we will examine a fragment of the x-component of the Navier–Stokes equation:

$$\frac{\partial v_x}{\partial t} + v_x \frac{\partial v_x}{\partial x} + v_y \frac{\partial v_x}{\partial y} \cdots \tag{8.186}$$

Of course,

$$\frac{\partial v_x}{\partial x} = \frac{\partial v_x}{\partial \eta}, \quad \text{and} \quad \frac{\partial v_x}{\partial y} = \frac{\partial v_x}{\partial \phi}\frac{\partial \phi}{\partial y} = \frac{\partial v_x}{\partial \phi}\left(\frac{1}{b(1+cx^2)}\right). \tag{8.187}$$

What we have done in this case is *fit the coordinate system to the boundary*, and for some simple problems in which a duct wall does not correspond to our x-y grid this can be used effectively.

8.11.1 Von Mises Transformation

We observed previously that the finite difference method (FDM) is well suited to rectangular domains where grid points coincide with the surfaces of the objects under study. However, many problems of interest involve curved surfaces and, in such cases, body-fitted coordinates are much more appropriate. In the von Mises transformation we map a problem from the x-y plane to the x-ψ plane:

$$(x, y) \to (x, \psi) \quad \text{where} \quad y = y(x, \psi). \tag{8.188}$$

This provides a rectangular computational domain where specific values of ψ are the streamlines for the flow. Thus, solid boundaries in the flow field will have fixed values of ψ, and the inlet and outlet distributions are determined by specification of the flow. Although this transformation has been known for about 100 years, its use for machine computation of incompressible flows is more recent; the reason for the resurgence of interest in such an old technique is that it generates a rectangular computational domain to which FDM is readily applied. Much of the recent progress in this area was stimulated by the work of Barron (1989); more recently Alderson et al. (2007) sought to extend its application. We will briefly consider some of the details of the latter investigation.

Let us begin with a *steady-state* flow in two dimensions; in terms of the primitive variables we have three equations: the continuity equation for conservation of mass, and two components of the equation of motion. Recall that the stream function, ψ, automatically satisfies continuity since

$$v_x = \frac{\partial \psi}{\partial y} \quad \text{and} \quad v_y = -\frac{\partial \psi}{\partial x}. \tag{8.189}$$

By the familiar process of cross-differentiation and subtraction we can eliminate pressure from the equations of motion; we can then introduce vorticity since $\omega = \frac{\partial v_y}{\partial x} - \frac{\partial v_x}{\partial y}$ with the result:

$$\frac{\partial^2 \omega}{\partial x^2} + \frac{\partial^2 \omega}{\partial y^2} = \mathrm{Re}\left\{ \frac{\partial \psi}{\partial y}\frac{\partial \omega}{\partial x} - \frac{\partial \psi}{\partial x}\frac{\partial \omega}{\partial y} \right\}. \tag{8.190}$$

Rewriting the velocity vector components in terms of the stream function:

$$\frac{\partial^2 \psi}{\partial x^2} + \frac{\partial^2 \psi}{\partial y^2} = -\omega. \tag{8.191}$$

In this approach to steady, two-dimensional flow problems just two equations must be solved, eqs. (8.190) and (8.191). The boundary conditions for the stream function are of the Dirichlet type and of course, boundary values for vorticity must be computed.

Alderson et al. give the transformation of the vorticity-stream function formulation:

$$L_1(y) = \omega\left\{ \frac{\partial y}{\partial \psi} \right\}^3 \tag{8.192}$$

and

$$L_1(\omega) = \omega\frac{\partial \omega}{\partial \psi}\left\{ \frac{\partial y}{\partial \psi} \right\}^2 + \mathrm{Re}\frac{\partial y}{\partial \psi}\frac{\partial \omega}{\partial x}, \tag{8.193}$$

where the operator L_1 is

$$L_1 = \left\{ \frac{\partial y}{\partial \psi} \right\}^2 \frac{\partial^2}{\partial x^2} - 2\frac{\partial y}{\partial x}\frac{\partial y}{\partial \psi}\frac{\partial^2}{\partial x \partial \psi} + \left[1 + \left\{ \frac{\partial y}{\partial x} \right\}^2 \right] \frac{\partial^2}{\partial \psi^2}. \tag{8.194}$$

A significant simplification is possible for irrotational flows where $\omega = 0$; in this case, as shown by Barron, the governing second-order nonlinear elliptic partial differential equation is:

$$\left\{ \frac{\partial y}{\partial \psi} \right\}^2 \frac{\partial^2 y}{\partial x^2} - 2\frac{\partial y}{\partial x}\frac{\partial y}{\partial \psi}\frac{\partial^2 y}{\partial x \partial \psi} + \left[1 + \left\{ \frac{\partial y}{\partial x} \right\}^2 \right] \frac{\partial^2 y}{\partial \psi^2} = 0. \tag{8.195}$$

Barron recommends solving a linearized version of this equation iteratively; this is accomplished by setting the *coefficients* on the second derivatives equal to the values from the *previous* iteration. He provides examples of this computation for shapes including a (half) cylinder and a Joukowski airfoil where $y = \frac{1}{10}(1 - 2x)(1 - 4x^2)^{1/2}$. This function provides the values of the dependent variable on the x-axis for $-\frac{1}{2} \le x \le \frac{1}{2}$. Barron also describes how the analyst can increase the density of the computational grid in a region where larger values of the gradients are expected; this is achieved by employing stretching functions as described by Jones and Dickinson (1980).

The discretization of the partial differentiation eq. (8.195) is carried out as follows:

$$\begin{aligned}
\frac{\partial y}{\partial \psi} &= \left[\frac{y_{i,j+1} - y_{i,j-1}}{2\Delta\psi} \right]_{n-1} \\
\frac{\partial^2 y}{\partial x^2} &= \left[\frac{y_{i+1,j} - 2y_{i,j} + y_{i-1,j}}{(\Delta x)^2} \right]_{n} \\
\frac{\partial y}{\partial x} &= \left[\frac{y_{i+1,j} - y_{i-1,j}}{2\Delta x} \right]_{n-1} \\
\frac{\partial^2 y}{\partial x \partial \psi}: \ \beta &= \left[\frac{y_{i+1,j} - y_{i-1,j}}{2\Delta x} \right]_{n} \quad \text{then} \quad \left[\frac{\beta_{i,j+1} - \beta_{i,j-1}}{2\Delta\psi} \right]_{n} \\
\frac{\partial^2 y}{\partial \psi^2} &= \left[\frac{y_{i+1,j} - 2y_{i,j} + y_{i-1,j}}{(\Delta\psi)^2} \right]_{n}
\end{aligned} \tag{8.196a,b,c,d,e}$$

The subscript n corresponds to the present iteration and $n-1$ the previous iteration. By letting $\Delta x = \Delta\psi$ we can isolate $4y_{i,j}$ and solve the elliptic equation iteratively with Dirichlet boundary conditions; the computational domain in the x,ψ plane extends from $y = 0$ (with the inserted shape, $y = f(x)$) on the x-axis across the bottom with $y = \psi$ on the sides and top. On the sides of course, the values of y are determined by the inlet and outlet flow. If the inlet flow is uniform, then the dependent variable y would be obtained from a linear distribution of the stream function, $\psi = ay + b$.

8.11.2 Elliptic Grid Generation

Now suppose we have a transport process occurring where the size or shape of the domain changes—perhaps repeatedly. If the field variable has specified values on the boundaries, then an elliptic PDE will yield contours that will conform to the objects or intrusions. For example, consider the Laplace equation,

$$\frac{\partial^2\psi}{\partial x^2} + \frac{\partial^2\psi}{\partial y^2} = 0. \tag{8.197}$$

Let us examine the situation where a simple 2/3-cut step is placed in a rectangular domain. We will assign constant values to ψ at both the top and bottom, and vary ψ appropriately at the left and right boundaries (these might be inflow and outflow boundaries, respectively). We solve eq. (8.197) iteratively and plot the resulting contours in Figure 8.33.

Note how the contours conform smoothly to the intrusion projecting from the bottom boundary. This suggests that we might be able to obtain a coordinate system adapted to an arbitrary shape through solution of a Dirichlet problem. This is what we mean when we speak of *elliptic grid generation.*

A recurring scenario in fluid flow involves the situation where we have flow *around* an object—we can think of an airfoil, a bridge pier, or a heat exchanger tube. A hypothetical case is illustrated in Figure 8.34.

Of course, we could employ a square mesh and just interpolate near the boundaries, but if strong gradients exist near these surfaces (which is entirely likely) interpolation may lead to significant error. This is our motivation for finding a transformation that will yield coordinates that conform to the surface of the immersed object, corresponding to a rectangular grid in the computational plane. We will think of this process in the following way: Let the surface of the object correspond to the inner boundary—we know pairs of (x, y) values that correspond to this surface. For examples, if $x = 3/2$, $y = \pm 1.28906$, and if $x = 9/4$, $y = \pm 1.53809$. We take the outer surface to be some kind of enclosing curve where once again the (x,y) values are known. If the "ends" were also known, we would have a Dirichlet problem, and the connection between the x–y

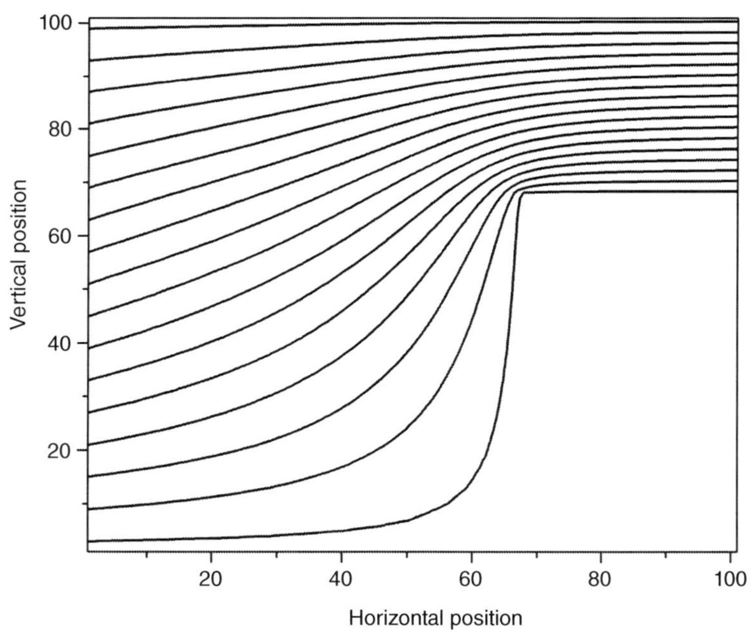

Figure 8.33 Contours plotted for $\psi(x,y)$ obtained from the solution of eq. (8.197).

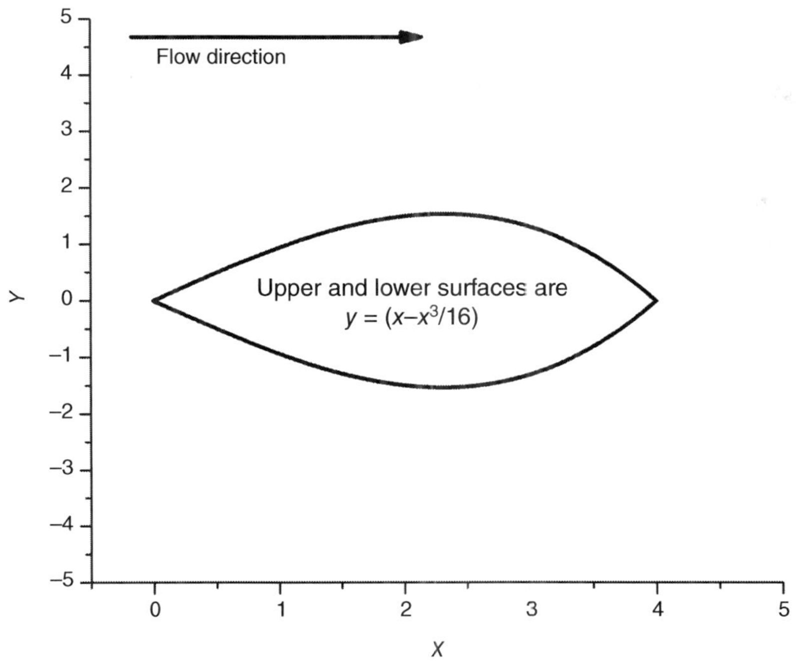

Figure 8.34 An object placed in a two-dimensional flow field. This is an instance where *elliptic grid generation* would be ideal. The upper and lower surfaces of the immersed object are given by $y = \pm\left(x - \frac{x^3}{16}\right)$.

(physical) field and the computational plane would be governed by two elliptic partial differential equations:

$$\frac{\partial^2\phi}{\partial x^2} + \frac{\partial^2\phi}{\partial y^2} = 0 \quad \text{and} \quad \frac{\partial^2\eta}{\partial x^2} + \frac{\partial^2\eta}{\partial y^2} = 0. \tag{8.198}$$

Unfortunately we must solve the *inverse* problem; we need to obtain the x–y positions corresponding to $\phi - \eta$ positions in the computational plane. When x and y are the dependent variables we find:

$$A\frac{\partial^2 x}{\partial\phi^2} - 2B\frac{\partial^2 x}{\partial\phi\partial\eta} + C\frac{\partial^2 x}{\partial\eta^2} = 0, \tag{8.199}$$

and

$$A\frac{\partial^2 y}{\partial\phi^2} - 2B\frac{\partial^2 y}{\partial\phi\partial\eta} + C\frac{\partial^2 y}{\partial\eta^2} = 0. \tag{8.200}$$

The parameters A, B, and C are:

$$A = \left(\frac{\partial x}{\partial\eta}\right)^2 + \left(\frac{\partial y}{\partial\eta}\right)^2 \tag{8.201}$$

$$B = \left(\frac{\partial x}{\partial\phi}\right)\left(\frac{\partial x}{\partial\eta}\right) + \left(\frac{\partial y}{\partial\phi}\right)\left(\frac{\partial y}{\partial\eta}\right), \tag{8.202}$$

and

$$C = \left(\frac{\partial x}{\partial\phi}\right)^2 + \left(\frac{\partial y}{\partial\phi}\right)^2. \tag{8.203}$$

In principle, if we know x and y on all four boundaries we can use eqs. (8.199) and (8.200) to determine the location of a mesh point in the physical (x–y) plane relative to a location in the computational (ϕ, η) field. However, if we base the mesh generation on the pair of elliptic equations, eq. (8.198), we often do not obtain sufficient computational points in regions where there are large gradients. Note in Figure 8.33, for example, how the contours are clustered near the step point (a convex region) but are sparsely distributed at the bottom of the step (a concave region). For this reason production (or source) terms are often added with the purpose of adjusting the shape of the computational mesh; e.g.,

$$\frac{\partial^2\phi}{\partial x^2} + \frac{\partial^2\phi}{\partial y^2} = P(\phi, \eta).$$

Now we will examine a discretization for eq. (8.199)—for simplicity we will omit the source term and we will take $\Delta\phi = \Delta\eta = 1$:

$$A(x_{i+1,j} - 2x_{i,j} + x_{i-1,j}) - \frac{B}{2}(x_{i+1,j+1} - x_{i-1,j+1} - x_{i+1,j-1} + x_{i-1,j-1}) + C(x_{i,j+1} - 2x_{i,j} + x_{i,j-1}) = 0 \tag{8.204}$$

with

$$A = \frac{1}{4}\left[(x_{i,j+1} - x_{i,j-1})^2 + (y_{i,j+1} - y_{i,j-1})^2\right], \tag{8.205}$$

$$B = \frac{1}{4}[(x_{i+1,j} - x_{i-1,j})(x_{i,j+1} - x_{i,j-1}) + (y_{i+1,j} - y_{i-1,j})(y_{i,j+1} - y_{i,j-1})] \tag{8.206}$$

$$C = \frac{1}{4}\left[(x_{i+1,j} - x_{i-1,j})^2 (y_{i+1,j} - y_{i-1,j})^2\right] \tag{8.207}$$

The equation for y, eq. (8.200) would be handled similarly of course. Functional forms for the source terms have been put forward in the literature, and Thompson et al. (1974) suggested that $P(\phi, \eta)$ might be expressed such that it was dependent upon velocity gradient or vorticity. The coordinate system would be accordingly time-dependent, and the generated mesh would automatically concentrate in regions where sharp changes in the field variables were occurring. Elliptic mesh generation is an important topic in modern computational fluid mechanics, and software designed specifically for this purpose has been incorporated into many commercial CFD packages. Anderson (1995) provides a nice introduction to this mesh generation and Chung (2002) offers more detail. The reader with greater interest in body-fitted coordinate systems may also find Thompson et al. (1974) useful.

8.12 Conclusion

Many problems of interest in engineering and the applied sciences are governed by partial differential equations, and only a very small number of these problems can be solved analytically. Most will require numerical solution. Consider, for example, a simple case of heat transfer in a finite cylinder for which $L/d = 3$. Thermal energy is produced throughout the interior of the cylinder, the ends are maintained at 100° (which is also the initial temperature of the object), and the curved surface loses heat to the surroundings (with $T_\infty = 70°$) according to Newton's law of cooling. Suppose the governing equation is determined to be

$$\rho C_p \frac{\partial T}{\partial t} = k\left[\frac{\partial^2 T}{\partial r^2} + \frac{1}{r}\frac{\partial T}{\partial r} + \frac{\partial^2 T}{\partial z^2}\right] + \beta T^{3/2}, \tag{8.208}$$

where β has dimensions of (energy)/[(volume)(time)($T^{3/2}$)]. The equation is nonlinear by virtue of the production term, making analytic solution improbable. Yet the equation is very easily discretized to yield:

$$\frac{T_{i,j,k+1} - T_{i,j,k}}{\Delta t} \cong \alpha\left[\frac{T_{i+1,j,k} - 2T_{i,j,k} + T_{i-1,j,k}}{(\Delta r)^2} + \frac{1}{r}\frac{T_{i+1,j,k} - T_{i-1,j,k}}{2\Delta r} + \frac{T_{i,j+1,k} - 2T_{i,j,k} + T_{i,j-1,k}}{(\Delta z)^2}\right] + \frac{\beta}{\rho C_p} T_{i,j,k}^{3/2} \tag{8.209}$$

We can solve this problem rapidly using the explicit technique described previously in this chapter; with the Biot modulus (hR/k) set equal to 0.596, we obtain the following sequence (Figure 8.35).

The solution for this problem required just 35 lines of code (with no particular effort to be efficiently compact), and the programming logic was devised in less than 15 min. What this example reveals is that even some rather formidable nonlinear PDEs can be solved by elementary means with no more computing power than that provided by ordinary personal computers. The range of problems that the analyst can solve in this way is broad, and the transformative power the computer has exerted upon routine solution of PDEs is obvious.

To provide emphasis for this last point, we want to provide the reader with an illustrative construction of a generalized elliptic PDE solver that uses SOR to handle Poisson and Laplace problems in rectangular coordinates. The code was written by the author, and it is designed to provide capability for fluid flow, heat transfer, and mass transfer with Dirichlet-, Neumann-, and Robin's-type boundary conditions. The governing equation has the form:

$$0 = \kappa\left[\frac{\partial^2 \phi}{\partial x^2} + \frac{\partial^2 \phi}{\partial y^2}\right] + P. \tag{8.210}$$

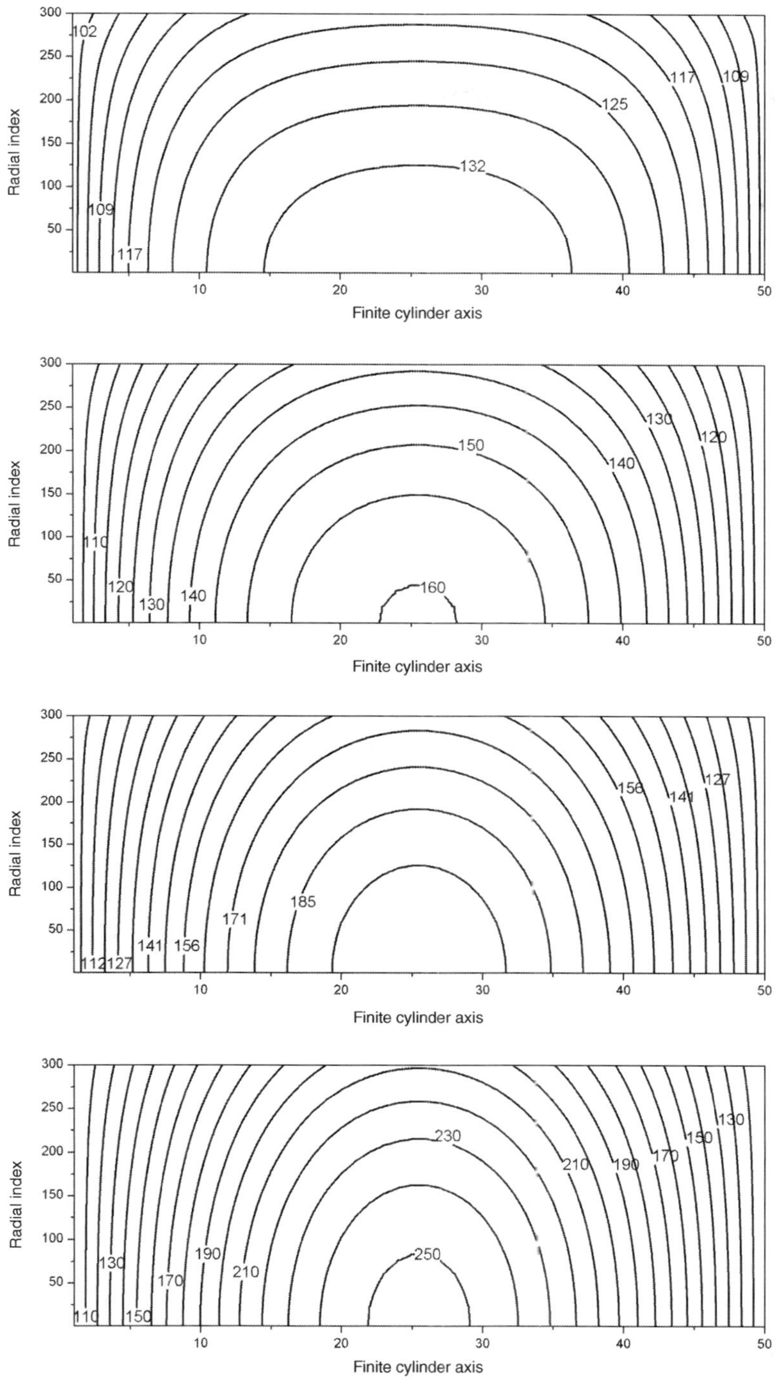

Figure 8.35 Results from the numerical solution of transient heat transfer in a finite cylinder with production. For these calculations, $L/d = 3$, Bi = 0.596, and $T_\infty = 70°$. The sequence shown is for times of 1, 2, 4, and 8 (top to bottom).

The user is queried for the length (L) and height (H) of the rectangular region and the source term, P, can be zero, a constant, or spatially variable, e.g.:

$$P = a + \frac{b}{c + (H - y)^2 + (L - x)^2}. \tag{8.211}$$

In discretized form eq. (8.210) appropriately cast for iterative solution appears as:

$$\phi_{i,j} = \phi_{i,j} + \frac{\omega}{\beta}\left[\frac{\phi_{i+1,j} + \phi_{i-1,j}}{(\Delta x)^2} + \frac{\phi_{i,j+1} + \phi_{i,j-1}}{(\Delta y)^2} + \frac{P_{i,j}}{\kappa} - (\beta)\phi_{i,j}\right] \tag{8.212}$$

where

$$\beta = \frac{2}{(\Delta x)^2} + \frac{2}{(\Delta y)^2}.$$

The rectangular region is divided into 101 nodes in each direction, resulting in 9,801 interior mesh points. The acceleration parameter for SOR, ω, is taken to be 1.85. An arbitrary measure of convergence is employed to terminate the computation (when the change in the *sixth* decimal place less than 1). First, we will apply the program to flow of an aqueous fluid in a square *microchannel*, 18 μm on each side. The applied pressure gradient will be (−)5,300 dynes/cm²/cm.

Next, we will apply the very same code to conduction of thermal energy in a slab with spatially variable production. In this case we choose $L = 10$ cm, $H = 4$ cm, and $k = 0.01$ cal/(cm s °C). The slab's bottom and left-hand side will be insulated, but the top and the right-hand side will lose thermal energy to the surroundings. The production function has its maximum value in the center of the slab.

The examples shown in Figures 8.36 and 8.37 are simply intended to illustrate how easy it is to solve a broad variety of elliptic PDEs using one basic code structure. The user merely selects the size of the rectangular domain, the numerical value of the transport coefficient, the boundary conditions, and the nature of the source term. Since the algorithm is not constrained to a square mesh, calculations can be made for a rectangular region of any aspect ratio.

Finally, we do want the reader to be aware of the existence of other useful numerical procedures for transport problems and computational fluid dynamics that have not been described here. The first of these is the *explicit Lax–Wendroff* technique (recall that we used this technique in Section 8.6 for the solution of hyperbolic PDEs) which is appropriate for forward marching in time. For the two-dimensional flow of an *inviscid* fluid we would write:

$$\frac{\partial \rho}{\partial t} = -\left(\rho\frac{\partial v_x}{\partial x} + v_x\frac{\partial \rho}{\partial x} + \rho\frac{\partial v_y}{\partial y} + v_y\frac{\partial \rho}{\partial y}\right) \tag{8.213}$$

$$\frac{\partial v_x}{\partial t} = -\left(v_x\frac{\partial v_x}{\partial x} + v_y\frac{\partial v_x}{\partial y} + \frac{1}{\rho}\frac{\partial p}{\partial x}\right), \tag{8.214}$$

and

$$\frac{\partial v_y}{\partial t} = -\left(v_x\frac{\partial v_y}{\partial x} + v_y\frac{\partial v_y}{\partial y} + \frac{1}{\rho}\frac{\partial p}{\partial y}\right). \tag{8.215}$$

Next, each of the dependent variables is expanded in a Taylor series to give the value of that quantity on the new time-step row; for example,

$$\rho_{i,j}^{t+\Delta t} = \rho_{i,j}^{t} + \left(\frac{\partial \rho}{\partial t}\right)_{i,j}^{t}\Delta t + \left(\frac{\partial^2 \rho}{\partial t^2}\right)_{i,j}^{t}\frac{(\Delta t)^2}{2} + \cdots \tag{8.216}$$

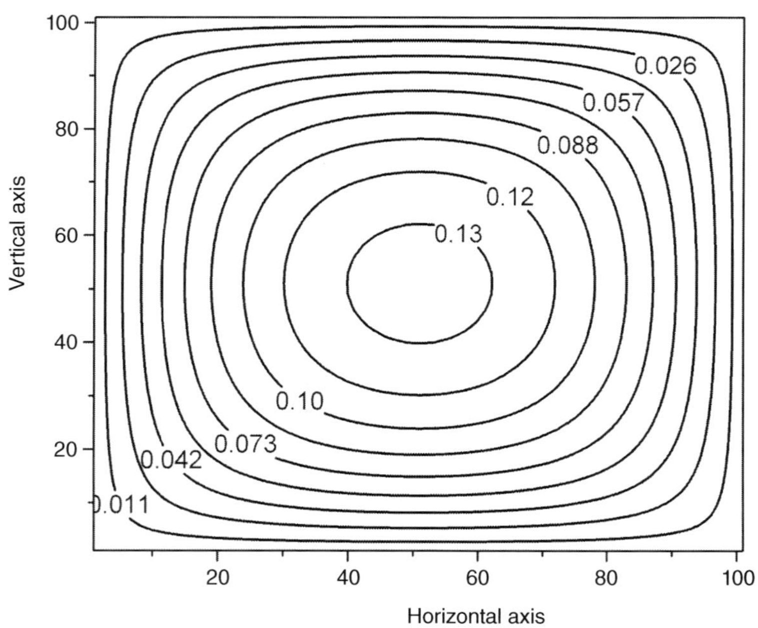

Figure 8.36 Velocity distribution in a square microchannel (18 μm on each side) resulting from a pressure gradient of −5,300 dynes/cm²/cm. The average velocity is a little less than 0.07 cm/s.

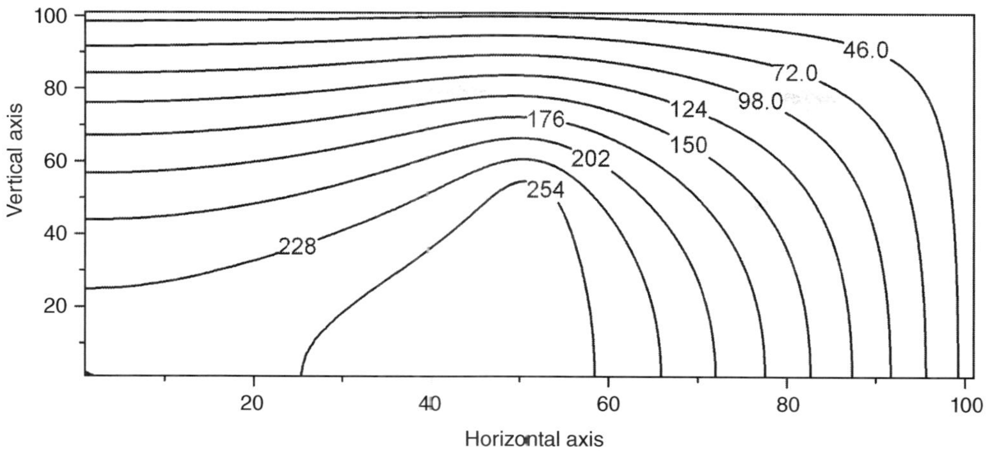

Figure 8.37 Temperature distribution in a two-dimensional slab with spatially variable production of thermal energy (with maximum at the center). The bottom and left-hand side are insulated, and the top and right-hand side lose thermal energy to the surroundings by Newton's law of cooling.

Note that the derivatives in eq. (8.216) are on the *previous* time-step row; if we could obtain values for these time derivatives, we could calculate $\rho_{i,j}$ on the new time-step row explicitly. The first derivative with respect to time is obtained directly from eq. (8.213), but with the *spatial derivatives on the right-hand side* rewritten as second-order central differences (all those quantities are known because they are from the previous time-step row). The estimate for the second derivative, $\left(\frac{\partial^2 \rho}{\partial t^2}\right)_{i,j}^{t}$, is obtained by differentiating eq. (8.213) with respect to time. Of course, this process generates mixed derivatives, such as $\left(\frac{\partial^2 v_x}{\partial x \partial t}\right)$, that are approximated by differentiating eq. (8.214) with respect to x, etc. Once again, all of the spatial derivatives are written as second-order, central differences. Although the algebra associated with Lax–Wendroff is complicated, the method is fully explicit (just like the related technique, MacCormack's method, discussed previously). For this reason it has been popular among practitioners of CFD. Both Chung (2002) and Anderson (1995) provide additional detail for the interested reader.

The second approach we want to mention is known as *SIMPLE* (semi-implicit method for pressure-linked equations), and it was devised by Patankar and Spalding and is very nicely explained by Patankar (1980). You may recall that we indicated previously that the major problem in CFD was computation of the pressure field (and you may also remember that we used the vorticity transport equation to circumvent this difficulty in two dimensions). In the SIMPLE procedure, the pressure field is estimated, the momentum equations are then solved to obtain the velocity vector components, the pressure field is corrected, and revised estimates are obtained for the velocities. The corrected pressure then serves as the initial estimate, and the entire process is repeated until convergence is obtained. Since its development in the 1970s, SIMPLE has been used by many fluid dynamicists, and it has been incorporated into a number of commercial CFD codes for both two- and three-dimensional flows. It has been found to be divergent in some applications, and under-relaxation has been used to cure that problem. Patankar's book is recommended reading for the student wishing to know more about SIMPLE.

We have shown in this chapter that a wide variety of phenomena governed by partial differential equations (PDEs) can be modeled successfully, and the solutions can be computed with relative ease. In many cases, even nonlinear phenomena can be dealt with using nothing more sophisticated than a personal computing device coupled with a spreadsheet. Most importantly, it only takes a little practice to become proficient at solving a broad spectrum of important practical problems.

Problems

8.1 Consider a slab of steel measuring 20 cm by 20 cm. The left-hand edge ($x = 0$) is maintained at 1,000° for all time and the top surface ($y = 20$ cm) is insulated. The bottom and the right-hand side lose thermal energy to the surroundings according to Newton's law of cooling. Using the right-hand side as an example, therefore, we write

$$-k\left.\frac{\partial T}{\partial x}\right|_{x=20} = h(T_{x=20} - T_{\infty})$$

Find (and prepare a contour plot illustrating) the equilibrium temperature distribution in the slab by solving

$$\frac{\partial^2 T}{\partial x^2} + \frac{\partial^2 T}{\partial y^2} = 0.$$

The thermal conductivity of mild steel is about 0.1 cal/(g cm °C) and the heat transfer coefficient (h) may be taken as 0.033 cal/(s cm^2 °C). The ambient temperature is 20°.

8.2 A solid copper cylinder with a radius of 6 cm (with very large L/d) has the upper half of its (curved) surface maintained at 300 °C. The lower half of the cylinder is embedded in a medium maintained at $T = 100$ °C. Find the steady-state temperature distribution in the copper cylinder. The thermal conductivity of copper is 0.93 cal/(g cm °C) and the governing equation is:

$$\frac{1}{r}\frac{\partial}{\partial r}\left(r\frac{\partial T}{\partial r}\right) + \frac{1}{r^2}\frac{\partial^2 T}{\partial \theta^2} = 0.$$

8.3 A slab of anisotropic material measuring 10 cm by 10 cm lies in the x–y plane. The thermal conductivity in the x-direction is 0.000834 cal/(cm s °C), but in the y-direction it is only 0.000361 cal/(cm s °C). If the left edge is maintained at 170 °C, and the right-hand side at 25 °C, find the temperature distribution in the slab. The top and bottom edges lose heat to the surroundings according to Newton's law of cooling; therefore, using the top for our example we have:

$$-k_y \left.\frac{\partial T}{\partial y}\right|_{y=H} = h(T_{y=H} - T_\infty).$$

The temperature of the surrounding air is 35 °C and the heat transfer coefficient, h, is 0.0001 cal/(cm^2 s °C). The governing equation for the interior of the slab is:

$$\frac{\partial}{\partial x}\left(k_x\frac{\partial T}{\partial x}\right) + \frac{\partial}{\partial y}\left(k_y\frac{\partial T}{\partial y}\right) = 0.$$

8.4 A viscous fluid is at rest in a square duct measuring 3 cm by 3 cm. At $t = 0$, a pressure gradient of 0.7 dyne/cm^2/cm is applied in the z-direction, and the fluid begins to move down the duct. The fluid's viscosity and density (μ and ρ) are 0.03 g/(cm s) and 1 g/cm^3, respectively. Solve the governing equation

$$\frac{\partial v_z}{\partial t} = -\frac{1}{\rho}\frac{dp}{dz} + \nu\left[\frac{\partial^2 v_z}{\partial x^2} + \frac{\partial^2 v_z}{\partial y^2}\right],$$

and find the velocity at the duct's center at 1, 10, and 100 s. Find the ultimate centerline velocity and the expected Reynolds number.

8.5 Consider the parabolic PDE given by:

$$\frac{\partial \phi}{\partial t} = 2\frac{\partial^2 \phi}{\partial x^2}.$$

Compute and plot a series of $\phi(x, t)$ curves such that the full range of dynamic behavior of this system is illuminated.

We are given that $0 \le x \le 4$, $\phi(0, t) = 0$, and $\phi(4, t) = 0$, with the initial condition, $\phi(x, 0) = 25x$. Of course, the analytic solution for this problem is just

$$\phi = \sum_{n=1}^{\infty} A_n \exp\left(-2\frac{n^2\pi^2}{L^2}t\right)\sin\left(\frac{n\pi x}{L}\right)$$

with $A_n = \frac{50}{L}\int_0^L x \sin\left(\frac{n\pi x}{L}\right) dx$, where $L = 4$.

8.6 A long cylindrical rod with $R = 1$ (and $d = 2$) is at an initial (uniform) temperature of zero. At $t = 0$, the surface of the rod is instantaneously heated to a constant 100°. Spiegel (*Advanced Mathematics for Engineers and Scientists,*

McGraw-Hill, 1971) provides the analytic solution for this problem:

$$T(r,t) = 100\left[1 - 2\sum_{n=1}^{\infty}\frac{J_0(\lambda_n r)}{\lambda_n R J_1(\lambda_n R)}\exp\left(-\alpha\lambda_n^2 t\right)\right]$$

The values for λ_n come from the roots of J_0, of course. Compute the numerical solution for this problem and compare your results at $t = 0.5$, 1.0, and 2.0 s with those obtained from the analytic solution above. The thermal diffusivity, α, has a value of 0.4. The governing PDE is:

$$\frac{\partial T}{\partial t} = \alpha\left[\frac{\partial^2 T}{\partial r^2} + \frac{1}{r}\frac{\partial T}{\partial r}\right].$$

8.7 An annular fin is placed on a pipe to help dissipate heat. The fin is constructed of mild steel, and it extends from $R_1 = 4$ cm to $R_2 = 8$ cm, with a thickness of 4 mm. At $r = R_1$ the pipe wall has a temperature of 200 °C for all time. The temperature in the steel fin is governed by:

$$\frac{\partial T}{\partial t} = \alpha\left[\frac{\partial^2 T}{\partial r^2} + \frac{1}{r}\frac{\partial T}{\partial r} + \frac{\partial^2 T}{\partial z^2}\right],$$

and the flat annular surfaces (denoted by s) lose heat to the surroundings according to

$$-k\left.\frac{\partial T}{\partial z}\right|_s = h(T_s - T_\infty).$$

We can assume that (nearly) no heat is lost through the outer cylindrical edge of the fin (at $r = R_2$). We are given that h, ρ, C_p, and α (*all in c.g.s. units*) are: 0.00678, 7.85, 0.118, and 0.12, respectively. We want to investigate the dynamic behavior of the temperature distribution in the fin if the temperature of the surroundings (T_∞) suddenly drops from 45 °C to 0 °C. Assume that this temperature had been 45 °C long enough that equilibrium had been established prior to the change.

8.8 Find the distribution of S over the annular region, $R_1 \le r \le R_2$, where S is governed by the equation

$$\frac{\partial^2 S}{\partial r^2} + \frac{1}{r}\frac{\partial S}{\partial r} + \frac{1}{r^2}\frac{\partial^2 S}{\partial \theta^2} = 0.$$

Let $S(r = R_1) = 100$, but $S(r = R_2) = 50 + 30\sin(\theta)$ where θ varies from 0 to 2π. The inner and outer radii are 1 and 10, respectively.

8.9 A solid cylindrical rod of radius, R, is immersed in a liquid such that the surface temperature of the rod is maintained constantly at 50°. The rod has an initial uniform temperature of 50°, but at $t = 0$ thermal energy begins to be produced inside the rod by a source term, S:

$$\rho C_p\frac{\partial T}{\partial t} = k\left[\frac{\partial^2 T}{\partial r^2} + \frac{1}{r}\frac{\partial T}{\partial r} + \frac{\partial^2 T}{\partial z^2}\right] + S.$$

The production of thermal energy in the interior follows:

$$S = \beta\left(\frac{1}{a + r^2}\right).$$

The rod is 50 cm long with $0 \le r \le 10$ cm. We know $\beta = 40$ and $a = 2$. The properties of the material (ρ, C_p, and k) are 7, 0.22, and 0.19, respectively. Explore the dynamic behavior of $T(r,t)$ over a time period sufficient to establish the full range of thermal behavior.

8.10 The thermal conductivity of chrome steel varies with temperature; between 0 and 500 °C, the relationship is approximately linear: $k = 62$–$0.049\,T$, W/(m°C). A slab of chrome steel 40 cm thick is at a uniform initial temperature of 500 °C. At $t = 0$, the front face is quickly cooled to 0 °C. The back face is insulated. Find the evolution of the temperature distribution in the slab by solving the equation:

$$\rho C_p\frac{\partial T}{\partial t} = \frac{\partial}{\partial x}\left(k(T)\frac{\partial T}{\partial x}\right).$$

The density of chrome steel is about 488 $\text{lb}_\text{m}/\text{ft}^3$ and the heat capacity is about 0.11 Btu/(lb_m °F).

8.11 A viscous fluid in an annulus (with $R_1 = 3$ cm and $R_2 = 5$ cm) is moving in the positive z-direction under the influence of a pressure gradient. The average velocity is 4 cm/s. At $t = 0$ the inner surface (which was stationary) begins moving in the *negative* z-direction at a constant velocity of −10 cm/s. When will the *net* flow be exactly zero? The fluid motion is governed by

$$\rho\frac{\partial V_z}{\partial t} = -\frac{\partial p}{\partial z} + \mu\left[\frac{1}{r}\frac{\partial}{\partial r}\left(r\frac{\partial V_z}{\partial r}\right)\right].$$

8.12 Consider water, initially at uniform temperature, flowing in the z-direction between parallel plates; for $z > 0$ both the upper and lower plates are maintained at a constant elevated temperature. The lower planar surface is located at $y = 0$ and the upper at $y = B$. For steady-state conditions, the governing equation is:

$$\rho C_p V_z\frac{\partial T}{\partial z} = k\left[\frac{\partial^2 T}{\partial y^2} + \frac{\partial^2 T}{\partial z^2}\right].$$

We will assume that the fluid properties are constant, the flow is fully developed, and that the centerline (maximum) velocity is V_{max}. We want to find the bulk fluid temperature and the Nusselt number as functions of z-position. Before we begin we should carefully consider the axial conduction term, $\partial^2 T/\partial z^2$. Under what circumstances can we expect this term to be negligible?

We will use the following specific values for our problem: $B = 3$ cm, $V_{max} = 5$ cm/s, $T_S = 150\,°$F, and $T_{in} = 50\,°$F. Assume that the fluid properties (for water) are constant, and use values corresponding to a temperature of 80 °F. Prepare a figure that shows both the bulk fluid temperature and the Nusselt number (Nu = hd/k) as functions of z-position. How long must the apparatus be if the bulk fluid temperature is to be 60, 70, 80, and 90 °F? Show that the bulk fluid temperature at $z = 600$ cm is 24.77 °C. Remember that the bulk fluid temperature must be determined by integration of the product, $V_z(y)T(y)$, over the flow area.

8.13 The FitzHugh–Nagumo model describes wave-like phenomena associated with nerve axons, and it consists of the pair of equations

$$\frac{\partial V}{\partial t} = -R + V - \frac{V^3}{3} + \frac{\partial^2 V}{\partial x^2} + I(t) \quad \text{and} \quad \frac{\partial R}{\partial t} = k(V - bR + a).$$

V is the membrane potential (output of the neuron), R is a restoring effect, and $I(t)$ is the input or forcing function. For certain parametric choices (a, b, k) this model can exhibit chaotic behavior. One variant of the model that appears frequently consists of the equations,

$$\frac{dV}{dt} = c\left(-R + V - \frac{1}{3}V^3 + I(t)\right) \quad \text{and} \quad \frac{dR}{dt} = V - bR + a.$$

Begin by solving this pair of *ordinary* differential equations with $a = 0.7$, $b = 0.8$, and $c = 10$, and let $I(t)$ be a simple periodic (sinusoidal) input with an amplitude of 1. Prepare a plot of $V(t)$ and $R(t)$ to reveal the phase-plane dynamics of this simplified system. What will the impact of the term, $\frac{\partial^2 V}{\partial x^2}$, be upon the solution of this system? Demonstrate the difference between the alternative models by solving the initial set of equations.

8.14 Solve the wave equation,

$$\frac{\partial^2 u}{\partial t^2} = c^2\frac{\partial^2 u}{\partial x^2},$$

numerically over the range $-10 < x < +10$. The initial distribution of displacement, $u(x, 0)$, is $u = \exp(-x^2)$, with zero initial velocity. Set the Courant number equal to 1 (one) and use the results shown in Figure P8.14 as a solution guide.

8.15 Solve the vorticity transport equation and the energy equation for the case in which a heated rectangular box is placed upon the floor of a channel. Allow the box height ($L = 1$ cm) to occupy half of the flow area of the channel. At $t = 0$ flow in the channel is initiated (left to right); we note that the severe constriction–expansion of flow area will result in strong vortex formation even at low Reynolds numbers. Our goal is to evaluate the importance of buoyancy in both flow development and rate of heat transfer. Recall that the Grashof number, $Gr = \frac{g\beta\Delta TL^3}{\nu^2}$, and the Rayleigh number, $Ra = GrPr$, can be used to assess the likelihood that buoyancy (and hence natural convection)

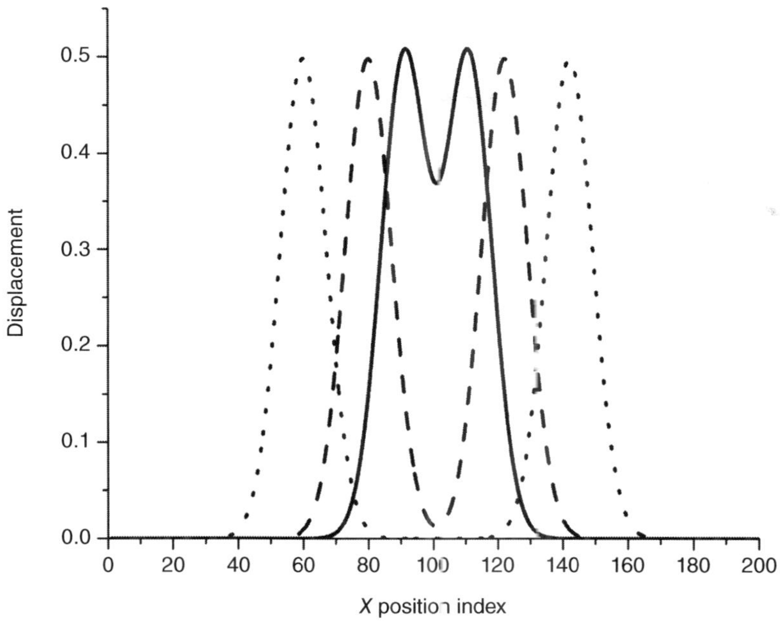

Figure P8.14 Note that the initial exponential distribution of displacement results in the formation of two peaks traveling in opposite directions. An x-index of 101 corresponds to the center of the interval (i.e., $x = 0$). Curves shown are for times of 0.015 (solid), 0.03 (dash), and 0.06 (dot) s.

will be important. For simplicity we will set Pr = 1, Re = 45, and Gr = 25,000, and allow the fluid to enter the channel at a uniform $T = 0°$ and set the box temperature to 20°. Of course we can easily remove the buoyancy effect from our calculations by simply setting $\beta = 0$. You may find it helpful to completely omit heat transfer initially and just concentrate on the flow problem. First, explore some computational outcomes (without buoyancy) for a dimensionless time, $t^* = \frac{\sqrt{\nu t}}{L} = 0.224$, looking only at the isotherms. Although the vortex has begun to form on the trailing edge of the rectangular obstruction, it is quite weak at this (short) time. The thermal plume shows no tendency to lift (of course since $\beta = 0$), and it is clearly being impacted by the strengthening circulation. Note that the bottom of the channel is insulated, $q_y|_{y=0} = 0$. Next, restore buoyancy and repeat the computations: Under these conditions, Pr = 1, Re = 45, Gr = 25,000, and $t^* = 0.224$, buoyancy is extremely important. We can expect the lifting effect created by the temperature difference will significantly alter the size, strength, and position of the vortex. Furthermore, it will be clear that a limited computational domain will constrain—and affect—your results. We have sacrificed physical fidelity for computational speed, and we would not want to rely upon this outcome for any critical purpose.

8.16 We have a two-dimensional slab that extends from $x = 0$ to $x = L$ and from $y = 0$ to $y = H$. Three sides of the slab are maintained at 100° and the bottom edge is insulated. Thermal energy is produced in the interior of the slab and the maximum rate of production occurs at the exact center. The governing equation is:

$$k\left[\frac{\partial^2 T}{\partial x^2} + \frac{\partial^2 T}{\partial y^2}\right] + \frac{A}{\left[(x-L/2)^2 + \frac{1}{4}\right]\left[(y-H/2)^2 + \frac{1}{4}\right]} = 0.$$

Compute the temperature distribution in the slab, $T(x,y)$, and plot the isotherms. Since the bottom is insulated, the isotherm contours should intersect the bottom at right angles.

Also, evaluate the rate of heat loss from the top edge of the slab. We know $L = H = 8$, $A = 0.085$, and $k = 0.005$.

8.17 Repeat the analysis of mass transport between concentric cylinders illustrated in Section 8.8, but allow the tangential component of velocity to oscillate such that:

$$v_\theta = (r = R_2, t) = \omega R_2 \left(0.5 + \sin\left(\frac{1}{9}t\right)\right).$$

How much impact does the oscillation of the outer cylinder have upon the radial development of the concentration plume? Keep in mind that your computational scheme must recognize the possibility of negative velocities during flow reversal and correct the convective terms accordingly.

8.18 Consider a cavity filled with a viscous fluid initially at rest. At $t = 0$, the upper surface of the cavity (a flat plate) begins to slide in the positive x-direction with constant velocity, V. Use the following equations to model the resulting flow:

$$\frac{\partial \omega}{\partial t} + v_x \frac{\partial \omega}{\partial x} + v_y \frac{\partial \omega}{\partial y} = \nu \left[\frac{\partial^2 \omega}{\partial x^2} + \frac{\partial^2 \omega}{\partial y^2} \right]$$

$$v_x = \frac{\partial \psi}{\partial y} v_y = -\frac{\partial \psi}{\partial x}$$

and

$$-\omega = \frac{\partial^2 \psi}{\partial x^2} + \frac{\partial^2 \psi}{\partial y^2}.$$

Solve this problem for a square cavity with unit width and depth and let $V = 1$. Place the origin in the lower left-hand corner and take the kinematic viscosity to be 0.10. The Reynolds number for this flow, therefore, will ultimately be $\mathrm{Re} = \frac{Vh}{\nu} = 10$. Prepare a series of plots of the stream function that illustrate the evolution of the flow. The student should note that *if the Reynolds number is sufficiently small*, this problem would simply require solution of the biharmonic equation, $\nabla^4 \psi = 0$; i.e., one could ignore the convective transport of vorticity (just as we did in the first vorticity transport example in this chapter). Chow (1979) provides details of the solution procedure for this creeping flow problem in Chapter 3 of his book, *An Introduction to Computational Fluid Mechanics*. If the upper surface is started impulsively at $t = 0$ with a velocity of 1, then at large times the analyst should obtain streamlines very similar to those shown in Figure 8.21 in Section 8.9.1.

8.19 An interesting variation of problem 8.18 is posed by a square cavity in which the top is a *free surface* (nearly zero momentum flux) and the bottom surface slides with constant velocity, V. Solve this modified problem with the same parameters employed immediately above in problem 8.18.

8.20 An even more interesting variation of problem 8.18 is the *deep* cavity with larger Reynolds number. This is a problem that has been intensively studied and solved with a variety of computational schemes. For example, in 2004 Maciej Matyka (*Computational Physics Section of Theoretical Physics, University of Wroclaw*) compared the performance of *SIMPLE*, *SIMPLE Revised*, and vorticity-stream function methods at Re = 300 using a coarse grid. Matyka provided an illustration (his Figure 10) of the deep cavity at Re = 100. Use a cavity depth that is twice the width, i.e., $h = 2w$. Allow the sliding upper surface to begin moving at $t = 0$ with a constant velocity, V, such that $\mathrm{Re} = \frac{wV}{\nu} = 3,000$, and compute the evolution of the vortices in the cavity. Keep in mind that with this kind of velocity you may encounter stability problems; the author found that with his program structure it was possible to go up to $\mathrm{Re} \cong 3,975$. Expected vortex evolution in deep cavity with sliding upper surface: At very small t's you should find just one vortex immediately beneath the sliding lid. As t increases the size of that vortex will diminish and a second region of recirculation will appear. At large t's a third vortex will appear at the bottom of the cavity; however, this vortex is weak, so you may have to change the scaling on your plotting software to completely reveal it.

8.21 We want to examine a solid-state device with three layers, but with production of thermal energy occurring in the intermediate layer (2). Each layer will have a different conductivity, but all three have the same thickness (0.667 cm). The device is 3 cm long (horizontally) and 2 cm high. The general form of the equation we must solve is:

$$k_n \left[\frac{\partial^2 T}{\partial x^2} + \frac{\partial^2 T}{\partial y^2} \right] + P_n = 0,$$

and the three thermal conductivities are 124/100, 52/100, and 130/100 for k_1 through k_3 (bottom to top), respectively. We are dividing by 100 to convert from W/(mK) to W/(cmK). The production term for the middle layer (2) has the constant value, 25 W per unit volume.

Assume that the bottom surface (the bottom of layer 1) is insulated, as is the left-hand side (vertical edge). The top loses thermal energy to the surroundings according to Newton's law of cooling: $h_{top}(T_{top} - T_{air})$. The right-hand edge also loses thermal energy to the surroundings, but the actual BC will vary with layer:

$$-k_n\left(\frac{\partial T}{\partial x}\right)_{x=L} = h_{edge}(T_{x=L} - T_{air}).$$

We are going to assume that $h_{top} = 0.75$ and that $h_{edge} = 1.5h_{top} = 1.125$. The air temperature will be taken as 25 °C. Find the temperature distribution in the interior of the device (temperature contours). Will this arrangement meet operating requirements if the thermal limit for the device is 60 °C? The results of a calculation for the case in which *both vertical sides and the bottom are insulated* is shown in Figure P8.21 as a guide.

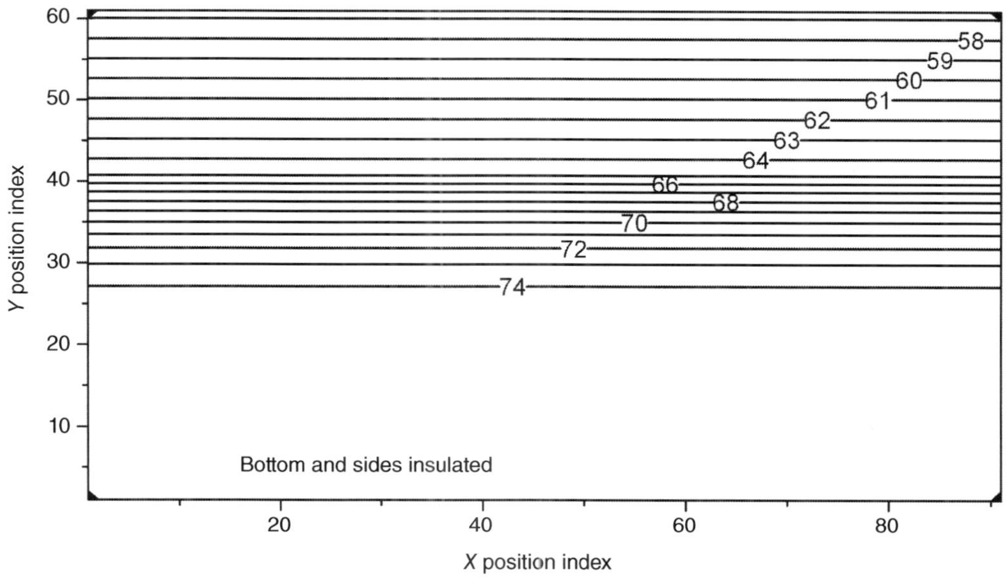

Figure P8.21 Temperature distribution in the three-layer device when the *bottom and both sides are insulated*. This arrangement results in the limiting operational temperature being exceeded.

8.22 Transport processes in porous media are sometimes simulated by placing impermeable obstructions in the migration pathway. Consider the situation illustrated in Figure P8.22, where an impermeable block has been placed in the center of the channel. The height of the block is 2 in., so it provides 33% occlusion of the pathway.

The contaminant, at a concentration of 1 (one), enters the region from the left where $z = 0$. The lower boundary is impermeable and the upper boundary is absorbing; i.e., the contaminant is instantaneously removed such that

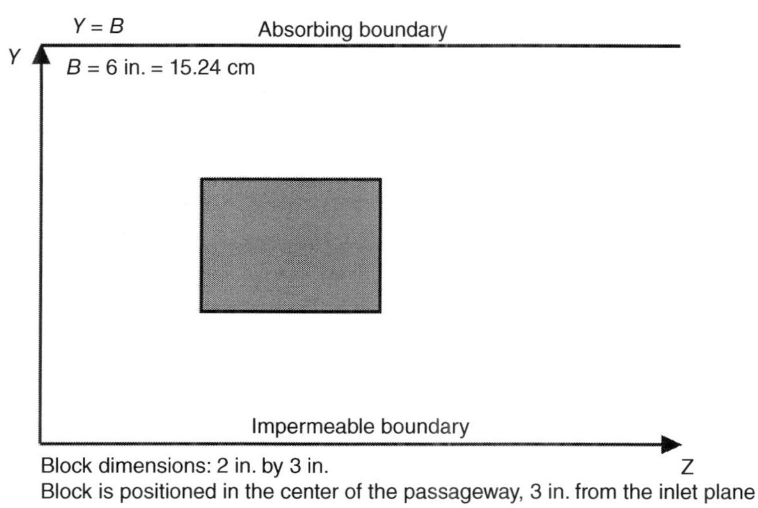

Figure P8.22 Porous medium with an impermeable obstruction placed in the migration path.

$C = 0$ for all t at the surface indicated by $y = B$. We assume that the mass transfer process is governed by

$$\frac{\partial C}{\partial t} = D\left[\frac{\partial^2 C}{\partial y^2} + \frac{\partial^2 C}{\partial z^2}\right].$$

Our interest is the total flow of contaminant past the trailing edge of the block where $z = 6$ in., or 15.24 cm. We will take $D = 2 \times 10^{-3}$ cm^2/s (a very large value, but this will shorten the computational time considerably). It takes almost 1 hr for the contaminant plume to reach the leading edge of the block and about 5 or 6 hr for it to reach the trailing edge. *Compare the rate at which contaminant migrates past the block at $z = 6$ in. with the rate of transport past that plane with the obstruction removed.*

8.23 Consider a porous, sorbent sphere placed in a well-agitated solution of limited volume; e.g., an activated carbon "particle" immersed in a beaker of water containing an organic contaminant. The contaminant (or solute) species ("A") is taken up by the sphere and the concentration of "A" in the liquid phase is depleted. The governing equation for transport in the sphere's interior is

$$\frac{\partial C_A}{\partial t} = D\left[\frac{\partial^2 C_A}{\partial r^2} + \frac{2}{r}\frac{\partial C_A}{\partial r}\right].$$

As we have seen previously, this equation can be transformed into an equivalent problem in a "slab" by setting $\phi = C_A r$. The total amount of "A" in solution initially is VC_{A0} and the rate at which "A" is removed from solution can be described by:

$$4\pi R^2 D \left.\frac{\partial C_A}{\partial r}\right|_{r=R}.$$

Therefore, the total amount removed over a time, t, can be obtained by integration of this equation. The *limiting case* is readily solved through use of the product method, and the transformation of our equation leads to:

$$\frac{\partial \phi}{\partial t} = D\frac{\partial^2 \phi}{\partial r^2},$$

which is a (familiar) candidate for separation of variables:

$$C_A = \frac{A}{r}\exp(-D\lambda^2 t)\sin \lambda r.$$

The cosine term has disappeared because the concentration of solute at the sphere's center must be finite, of course. It is convenient to switch to dimensionless concentration, where

$$C = \frac{C_A - C_{Ai}}{C_{As} - C_{Ai}}.$$

It is likely that the sphere contains no solute initially, so $C_{Ai} = 0$. When the solution volume is *unlimited* then we write:

$$C = 1 + \sum_{n=1}^{\infty} \frac{A_n}{r} \exp\left(-D\lambda_n^2 t\right) \sin \lambda_n r,$$

where $\lambda_n = \frac{n\pi}{R}$.

Complete the analytic solution started above for the limiting case and prepare a figure that shows M_t/M_∞ as a function of Dt/R^2. Note that M_t is the amount of solute taken up by the sphere through time, t. M_∞ is the amount taken up by the sphere after infinite time.

Solve the "limited solution" problem *numerically* and prepare similar curves for three cases: the portion of solute ultimately removed from solution (and taken up by the sphere) is 20%, 50%, and 80%.

Suppose spherical sorbent particles with $d = 4.8$ mm are used to remove benzyl alcohol from an initially saturated aqueous solution maintained at 17 °C (the solubility is 4 g per 100 g water). It is known that $D = 0.82 \times 10^{-6}$ cm^2/s. The number of particles employed is such that each is effectively surrounded by volume of liquid corresponding to 100 mm^3. Prepare a figure that shows the concentration of benzyl alcohol remaining in solution as a function of time, assuming the liquid phase is energetically stirred. Then, repeat this final part of the problem using a Robin's-type boundary condition at the surface with $KR/D = 0.2$.

8.24 Circumstances can arise in mass and heat transfer in which a concentration or temperature *front* can propagate through a medium. Such cases are referred to as *moving-boundary* problems, and examples of how they can occur include:

- the diffusivity experiences a sharp, discontinuous change at a particular concentration
- the diffusing species is immobilized at a limited number of available sites
- a chemical reaction occurs at a reactant interface and one or more *product* species migrate(s) through the medium
- heat flows through a medium in which a phase change occurs at a specific temperature (the latent heat effect may either release or absorb thermal energy).

Suppose we have a medium in which the diffusivity is 10^{-3} cm^2/s *if* the dimensionless concentration is *above* 0.5 but is 3 orders of magnitude *smaller* (10^{-6}) for concentrations below 0.5. At $t = 0$, the concentration at the front face where $x = 0$ is instantaneously elevated to 1; there is initially none of the diffusing species in the medium. Compute the migration of this species into the medium and plot the resulting concentration as a function of $\frac{x}{\sqrt{4D_1 t}}$, where D_1 is the elevated diffusivity (10^{-3}) so that the evolution of the concentration profile is evident. You can obtain a qualitative comparison for your results with Figure 13.7 in Crank (*The Mathematics of Diffusion*, 1975); however, in Crank's example mass transfer is completely shut off at the threshold concentration.

8.25 Refer to Figure 8.33 (regarding elliptic grid generation). Suppose that the paucity of contours at the base of the intrusion was preventing the analyst from accurately solving a particular problem. Develop and add an appropriate source or production term to eq. (8.197), Section 8.11 to rectify this difficulty and then demonstrate its effectiveness by computing the solution of the modified equation numerically and replotting the $\psi(x, y)$ contours.

8.26 A concrete drainage ditch is constructed in the form of a 90° "V," and it is to be used to carry 5,500 gpm of wastewater away from a plant. *Assume* that the flow is governed by:

$$0 \cong \mu \left[\frac{\partial^2 V_z}{\partial x^2} + \frac{\partial^2 V_z}{\partial y^2} \right] + \rho g \sin\theta.$$

The slope is 1 ft/1,000 ft such that $\sin\theta = 0.001$. By computation, determine the *depth* of water in the channel. Then, repeat the analysis but assume that the channel is a 60° "V." One concern with an open channel of this type is the possibility of particle deposition in the bottom of the "V," and if sedimentation occurs, the carrying capacity of the ditch could be reduced. Consequently particle "scour" is extremely important; make sure that your calculations also provide an estimate of the shear stress at the bottom of the channel.

8.27 Liquid flows through a partially filled large pipe, 1 m in diameter, under the influence of gravity. The depth of water in the pipe (measured vertically from the bottom) is $d/3$ or $0.667R$. The velocity distribution is (approximately) governed by

$$0 = \mu \left[\frac{1}{r}\frac{\partial}{\partial r}\left(r\frac{\partial V_z}{\partial r} \right) + \frac{1}{r^2}\frac{\partial^2 V_z}{\partial \theta^2} \right] + \rho g \sin\phi.$$

The angle of declination is small, so we will take $sin\phi = 0.00002$. The fluid flows over a section of coated pipe wall that is soluble, and we will take the concentration on the fluid side of the interface (immediately adjacent to the pipe wall) to be 1 (one). The continuity equation for the soluble species will be

$$V_z\frac{\partial C}{\partial z} = D\left[\frac{1}{r}\frac{\partial}{\partial r}\left(r\frac{\partial C}{\partial r} \right) + \frac{1}{r^2}\frac{\partial^2 C}{\partial \theta^2} \right].$$

The coated section of wall is 1 m long and $D = 0.02$ cm^2/s. Find the velocity distribution and the average fluid velocity and the concentration distribution 600 m downstream. Any contaminant that finds its way to the free surface disappears immediately (is volatilized). The viscosity, μ, is 0.04.

8.28 We want to revisit problem 7.29, uniform heat generation in a solid sphere. The governing equation was

$$\frac{1}{\alpha}\frac{\partial T}{\partial t} = \frac{1}{r^2}\frac{\partial}{\partial r}\left(r^2\frac{\partial T}{\partial r} \right) + \frac{S_0}{k}.$$

If we introduce a new variable such that $\phi = Tr$, we get

$$\frac{1}{\alpha}\frac{\partial \phi}{\partial t} = \frac{\partial^2 \phi}{\partial r^2} + \frac{S_0}{k} r.$$

By letting $\beta = \phi + \frac{S_0 r^3}{6k}$, we can transform the equation to the familiar slab form:

$$\frac{\partial \beta}{\partial t} = \alpha \frac{\partial^2 \beta}{\partial r^2},$$

with the equally well-known solution:

$$\beta = C_1 \exp(-\alpha\lambda^2 t)[A \sin \lambda r + B \cos \lambda r].$$

If the surface at $r = R$ is maintained at $T = 0$ for all time, then the analytic solution is:

$$T = \frac{S_0}{6k}(R^2 - r^2) + \frac{2S_0R^3}{k\pi^3 r}\sum_{n=1}^{\infty}\frac{(-1)^n}{n^3}\exp\left(-\frac{\alpha n^2\pi^2 t}{R^2}\right)\sin\frac{n\pi r}{R}.$$

We will begin this exercise by solving this problem *numerically* with the parametric values, $R = 2$, $k = 0.12$, $\alpha = 0.12$, and $S_0 = 0.05$ (all cgs units). Verify your work by comparing your solution with the values given here for $r = 0$ (center of sphere):

t	$6kT/(S_0R^3)$
3	0.51
5	0.72
10	0.94

Now we are prepared to solve a more interesting variation of this problem. Suppose that the generation of heat occurs only in a band (a certain range of r positions) in the interior of the sphere. In particular, assume that production only occurs for $0.75 < r < 1.50$; in this region S_0 is 0.065 (and zero elsewhere). Again, the surface is maintained at zero temperature for all time. Solve the modified problem and present your results in the same form as shown in the table above using the dimensionless variable, $6kT/(S_0 R^3)$; note the location of the maximum temperature and how its position changes with time.

A geophysical version of this problem was discussed by Louis Slichter, *Bull. Geol. Soc. Amer.*, 52:561 (1941), who pointed out that evidence exists that radioactive thermal energy production occurs in a relatively narrow band near the surface of the earth. The mathematical formulation of the problem begins on page 576 of his paper, and the most interesting feature of the work is the effect that different production distributions have upon *T(r)*. There is also an informative discussion of the relative effects of conduction and convection in the initial cooling of the earth.

8.29 In previous considerations of problems in cylindrical coordinates we focused upon transport in the r- and z-directions; of course, we can also have a transport process occurring in the angular (θ) direction. Suppose for example we have an infinite (large L/d) cylinder with unsteady heat flow occurring in both the radial and angular directions. The governing equation is then:

$$\frac{\partial T}{\partial t} = \alpha\left[\frac{\partial^2 T}{\partial r^2} + \frac{1}{r}\frac{\partial T}{\partial r} + \frac{1}{r^2}\frac{\partial^2 T}{\partial \theta^2}\right].$$

Assume that the cylinder is at a uniform initial temperature of 0°, but at $t = 0$ the surface temperature of the sector from 0 to $\pi/2$ (90°) is suddenly elevated to a temperature of 100°. The surface of the bottom half of the cylinder (from $\pi/2$ to $3\pi/2$) is covered with a good insulator such that.

$$\left.\frac{\partial T}{\partial r}\right|_{r=R} = 0.$$

The sector of the surface from $3\pi/2$ to 2π is maintained at $T = 0$ for all t, and an insulating sheet is inserted into the cylinder at the top so that it extends completely from $r = 0$ to $r = R$. Consequently, heat flow will occur mainly in the negative r-direction (toward the center of the cylinder) and in the positive (clockwise) θ-direction. Solve this problem numerically and provide results in polar coordinates that illustrate the angular progression of thermal energy.

There are several particularly interesting features of this problem. First, Carslaw and Jaeger (1959) point out in Section 7.12 that an analytic solution for this partial differential equation has the form.

$$\exp(-\alpha\lambda^2 t)J_n(\lambda r)[A_n \cos n\theta + B_n \sin n\theta].$$

However, turning this into something useful for our problem is a significant undertaking! Second, we should observe that we do not have symmetry at the center of the cylinder where $r = 0$; moreover, at the center θ is *not unique*. What are the consequences of this as the thermal energy approaches the center? Your numerical procedure will have to take this into account.

Take the radius if the cylinder to be $R = 4$, and the thermal diffusivity of solid cylinder to be $\alpha = 0.15$. The surface temperature of the sector from 0 to 90° should be set to 100°. The quarter of the surface from 270 to 360° will remain at 0° for all t. If you increase the thermal diffusivity, α, to a larger value (say 0.65), the clockwise migration of thermal energy is enhanced.

8.30 Consider a rectangular slab in the x–y plane that is 2 units wide ($0 \le x \le 2$) and 3 units high ($0 \le y \le 3$). The left-hand side and the top are insulated, while the right-hand side loses heat to the surroundings with the process described by Newton's law of cooling; take the temperature of the surroundings to be zero for convenience. Therefore:

$$-k\left.\frac{\partial T}{\partial x}\right|_{x=A} = hT|_{x=A}.$$

The bottom edge of the slab has a linear temperature distribution such that $T(x, y = 0) = 200 + \frac{100x}{2}$; i.e., the temperature across the bottom varies from 200 to 300°. Since this is an equilibrium problem, we have the elliptic partial differential equation,

$$\frac{\partial^2 T}{\partial x^2} + \frac{\partial^2 T}{\partial y^2} = 0.$$

We will solve this equation iteratively using successive over-relaxation (SOR). You may recognize that a very similar problem was dealt with analytically in Chapter 7, Section 7.6. Our purpose in this exercise is to compare the results of our computational solution with those obtained from the analytical result. We are interested in three specific points in the interior of the slab: $(x, y) = (1.6, 0.3), (1.6, 1.5), (1.6, 2.7)$. Set $k = 1$ and $h = 0.10$ such that $\frac{hA}{k} = 0.2$. How well do the results compare, and which solution method requires the least numerical work? Use this point temperature as a check for your procedure: $T(x = 1.6, y = 0.3)$ is between 258.5 and 258.65°. If the heat transfer coefficient were 10x larger, i.e., if $\frac{hA}{k} = 2$, how much would the solution(s) be affected?

8.31 A viscous liquid is initially at rest in a long, *half-filled* pipe. At $t = 0$ the pipe is inclined slightly and the fluid begins to move with its behavior set by the equation,

$$\frac{\partial v_z}{\partial t} = \nu\left[\frac{\partial^2 v_z}{\partial r^2} + \frac{1}{r}\frac{\partial v_z}{\partial r} + \frac{1}{r^2}\frac{\partial^2 v_z}{\partial \theta^2}\right] + g_z.$$

We are interested in the development of the average velocity in the half-filled pipe; we particularly want to know when the fluid will attain 85% of its ultimate average velocity. The pipe radius is 20 cm and the kinematic viscosity is 0.095. Set the driving force, $g_z = 0.02$. A partially completed table is provided below to allow you to assess your results.

Time	50	100	200	300	400	600
Ave Velocity	0.853	1.575	2.792	3.790	4.621	5.903

Note that because of the relatively large kinematic viscosity, the velocity distribution will develop very slowly. Since a vertical plane of symmetry passes through the center of the pipe, it will only be necessary to carry out calculations for the second quadrant, i.e., for theta ranging from 90 to 180°. Assume that no momentum is transferred across the free surface (the gas–liquid interface).

8.32 A spherical solid, initially at uniform temperature, $T = 0$, is exposed to a constant heat flux at its surface. We want to compute the evolution of the temperature distribution in the interior of the sphere and prepare a plot that shows the dimensionless temperature, $\frac{kT}{Rq_0} - \frac{3\alpha t}{R^2}$, as a function of r/R. The governing equation for this problem is

$$\frac{\partial T}{\partial t} = \alpha \left[\frac{\partial^2 T}{\partial r^2} + \frac{2}{r}\frac{\partial T}{\partial r} \right].$$

The parametric values we will employ are: $k = 0.12$, $R = 10$, $q_0 = 40$, and $\alpha = 0.12$. Compare your results with Figure P8.32. You can also compare with the graphical presentation of the analytic solution in Carslaw and Jaeger (*Conduction of Heat in Solids*) which appears on page 242 in Section 9.7.

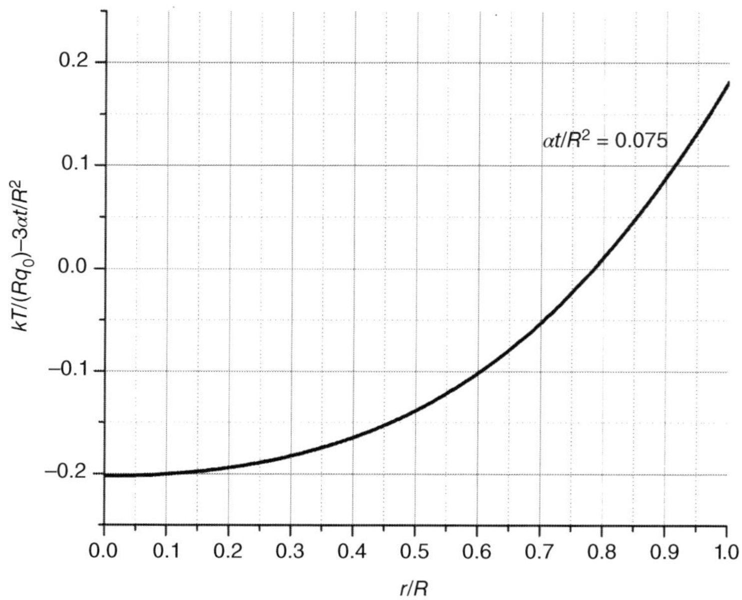

Figure P8.32 Dimensionless temperature as a function of radial position, r/R for $\frac{\alpha t}{R^2} = 0.075$.

8.33 Consider a hollow sphere with $R_1 \le r \le R_2$. The initial temperature distribution in this spherical shell is a function of r: $T_0(r - R_1)^2(r - R_2)^2 + 100$, where $T_0 = 1.5$. Assume that the inner surface is maintained at $T = 0°$ and the outer surface at $T(r = R_2) = 200°$ for all $t \ge 0$. Compute the approach of the temperature distribution to steady state and compare your numerical results with the analytic solution which is given on page 246 in Section 9.9 in Carslaw and Jaeger. $R_1 = 10$ and $R_2 = 20$ and $\alpha = 0.025$. The equation to be solved is:

$$\frac{\partial T}{\partial t} = \alpha \left[\frac{\partial^2 T}{\partial r^2} + \frac{2}{r}\frac{\partial T}{\partial r} \right].$$

8.34 We want to examine a nuclear fuel rod bundle in the shape of a cylinder where the production of thermal energy is *not* centered. We want to compute the steady-state temperature distribution in the bundle when there is a constant surface temperature maintained by the surrounding fluid, $T_S = 650°$. Although the bundle is obviously not homogeneous (uranium oxide and zirconium have very different thermal conductivities), we will assume that k is uniform at 0.25 cal/(cm s K). Take $R = 10$ cm and set $\Delta r = 10/60$ and $\Delta\theta = 2\pi/100$; therefore, our array for discretized T field will be $T(61,101)$, so for the indices: $1 \le i \le 61$ and $1 \le j \le 101$.

$$\frac{\partial^2 T}{\partial r^2} + \frac{2}{r}\frac{\partial T}{\partial r} + \frac{1}{r^2}\frac{\partial^2 T}{\partial \theta^2} + \frac{S_N}{k} = 0.$$

Let the source term be represented (using the position indices) by: $S_N = \frac{S_0}{1+(j-26)^2} + \frac{S_0}{1+(i-16)^2}$, where $S_0 = 215$. Your results should show a localized hot spot in quadrants 1 and 2 aligned with $\theta = 90°$.

8.35 Determination of the thermal conductivity is a significant problem in the evaluation of thermal energy transfer, particularly when the material or its composition is unknown. Suppose we have a small sphere that has been modified to allow monitoring of both the center and surface temperatures simultaneously. We immerse it in a heated fluid until the sample has attained some uniform, elevated temperature (200°). The specimen is removed from the heated bath such that it loses heat to its surroundings, probably through a combination of processes including conduction, radiation, and convection. The temperature in the interior of the sphere is governed by

$$\frac{\partial T}{\partial t} = \alpha \left[\frac{\partial^2 T}{\partial r^2} + \frac{2}{r}\frac{\partial T}{\partial r} \right]$$

and the boundary condition at the surface ($r = R$) is

$$-k \left. \frac{\partial T}{\partial r} \right|_{r=R} = h(T_{r=R} - T_\infty) \text{ where } T_\infty = 0°.$$

The conduction equation can be converted to slab form and solved using separation of variables yielding

$$T = \frac{c_1}{r} \exp(-\alpha \lambda^2 t)[A \sin \lambda r + B \cos \lambda r].$$

Of course, B must be zero since T is finite at the center of the sphere. Application of the boundary condition at the surface produces the transcendental equation,

$$\lambda R \cot \lambda R + \frac{hR}{k} = 1.$$

If $hR/k = ½$, then $\lambda_1 R = 1.1656$, $\lambda_2 R = 4.6042$, $\lambda_3 R = 7.7899$, etc.
And the solution, therefore, is:

$$T = \sum_{n=1}^{\infty} \frac{A_n}{r} \exp\left(-\alpha \lambda_n^2 t\right) \sin \lambda_n r.$$

We must use the initial condition with orthogonality to find the A_n's:

$$A_n = \frac{-\frac{T_0}{\lambda_n} \cos(\lambda_n R - 1)}{\int_0^R \frac{\sin^2 \lambda_n r}{r} dr}.$$

The real difficulty is a familiar one: Both k and h are unknown in this scenario. However, if t is large enough so that the infinite series can be accurately represented with a *single term*, and if we have data available for both the surface and center temperatures, then

$$\frac{T_S}{T_C} = \frac{\sin \lambda_1 R}{\lambda_1 R}.$$

We can subsequently use the transcendental equation to estimate the Biot modulus, hR/k. Try this using the accompanying data and then solve the problem numerically to determine how large t must be in order for one term to adequately represent the solution with $\alpha = 0.011$, $C_p = 0.21$, $k = 0.006$, and $\rho = 2.6$ (all cgs). The necessary data appear in Figure P8.35.

8.36 A very wide rectangular duct extends vertically from $y = 0$ to $y = H = 10$ cm, and it contains two immiscible fluids. Fluid "B" extends from $y = 0$ to $y = 5$ cm and has a kinematic viscosity of 0.07 cm^2/s, and fluid "A" occupies the space from $y = 5$ to $y = 10$ cm with a kinematic viscosity of 0.009 cm^2/s. At $t = 0$ a pressure gradient, dp/dz (0.007 g/(cm s^2)/cm), is imposed that causes both fluids to begin to move in the z-direction. Of course the transport of momentum must be continuous across the midpoint of the duct (the location of the interface). When will fluid "A" attain 50%, 75%, and 95% of its ultimate velocity? An illustration of the velocity distribution at $t = 1{,}000$ s is provided for your information. Make particular note of the point of inflection that occurs in this flow as it is an indicator of potential instability; see Figure P8.36.

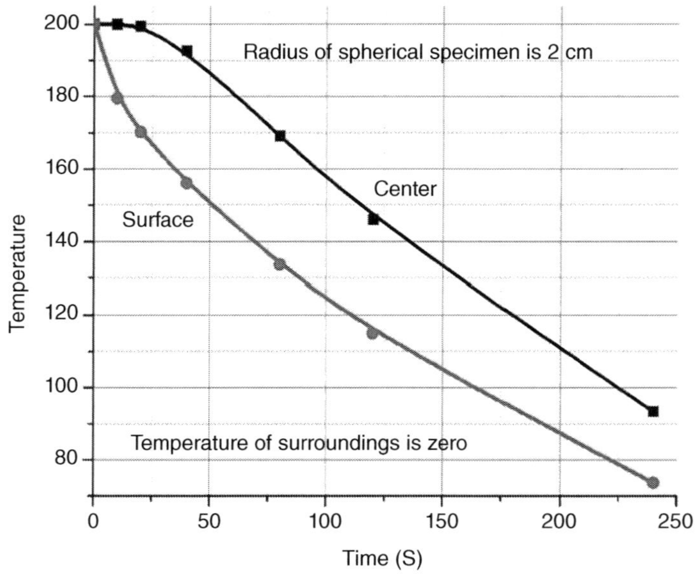

Figure P8.35 Surface and center temperatures for a spherical specimen with $R = 2$ cm.

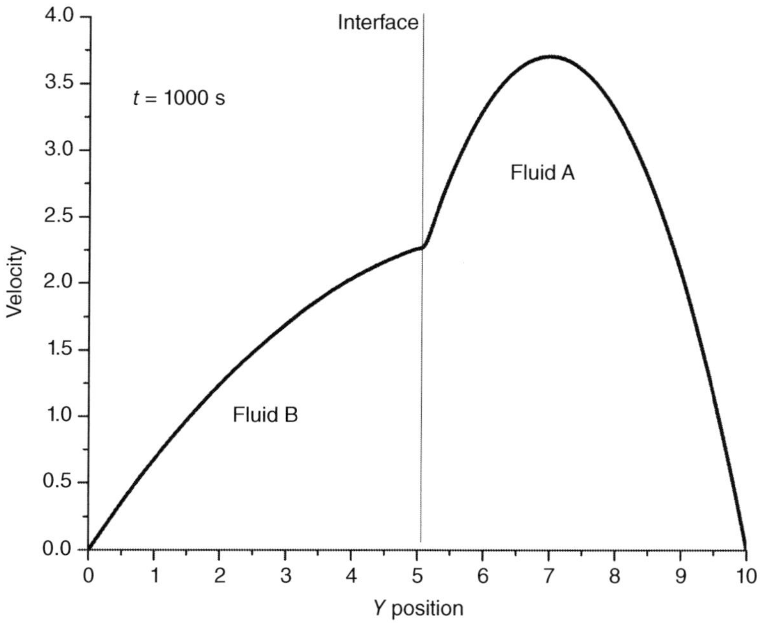

Figure P8.36 Typical velocity distribution for start-up flow through a duct containing immiscible fluids, $\frac{\nu_B}{\nu_A} = \frac{0.07}{0.009} = 7.78$.

8.37 A solid aluminum rod with $R = 2\,\text{cm}$ is initially at some uniform temperature, T. At $t = 0$ one end (at $z = 0$) is instantaneously heated to $T = 100°$ and heat then flows down the rod but it is also simultaneously lost to the surroundings from the surface. *If we neglect* the variation of T in the transverse (r-) direction, an approximate model can be written:

$$\frac{\partial T}{\partial t} = \alpha \frac{\partial^2 T}{\partial z^2} - \frac{2h}{\rho C_p R}(T_{r=R} - T_\infty)$$

We need to be absolutely clear about this approximation: We are saying that that there is much more resistance to heat transfer in the fluid phase than there is in the aluminum rod. Note that we can assess this presumption by looking at the dimensionless Biot number, $Bi = \frac{hR}{k}$. For our aluminum rod, taking the heat transfer coefficient, h, to be 0.002, we find

$$Bi = \frac{(0.002)(2)}{(0.48)} = 0.0083$$

We want to calculate the evolution of the temperature distribution for this case and of course your computations should reveal an increasing temperature along the length of the rod. Of course, we have assumed that $T \neq f(r)$ and this model is not strictly correct—there must be *some* variation in temperature in the r-direction. This raises an important question: How large must Bi be before we need to employ a more accurate model:

$$\frac{\partial T}{\partial t} = \alpha \left[\frac{\partial^2 T}{\partial r^2} + \frac{1}{r}\frac{\partial T}{\partial r} + \frac{\partial^2 T}{\partial z^2} \right]?$$

Solve this equation numerically employing larger values for the heat transfer coefficient, h. How large must the rate of heat loss from the surface be to make $T(r)$ important? The boundary condition at $r = R$ is:

$$-k\left.\frac{\partial T}{\partial r}\right|_{r=R} = h(T_{r=R} - T_\infty).$$

8.38 Imagine a squat cylindrical solid with diameter $2R$ that extends vertically from $z = 0$ to $z = L$. The curved surface ($r = R$) is maintained at $T = 0$ as is the top at $z = L$. The bottom, at $z = 0$, has a temperature distribution given by $F(r)$ for all time. We want to find the steady-state temperature distribution in the solid cylinder by computation. The equation to be solved is

$$\frac{\partial^2 T}{\partial r^2} + \frac{1}{r}\frac{\partial T}{\partial r} + \frac{\partial^2 T}{\partial z^2} = 0,$$

and of course there is no θ-dependence. This is a problem that can be solved analytically, and you may want to demonstrate that

$$T = AJ_0(\lambda r)\sinh\lambda(L - z),$$

where $J_0(\lambda R) = 0$ such that $T = \sum_{n=1}^{\infty} A_n J_0(\lambda_n r)\sinh\lambda_n(L - z)$. The difficult part of the process (as usual) is using orthogonality to find the leading coefficients that will reproduce $T(r, z = 0) = F(r)$. For our problem take $R = 4$ and $F(r) = 100 + \frac{400}{1+(2-r)^4}$; therefore, the temperatures at the outside edge of both the bottom and the bottom center will be 123.5°. Let $L = 4$. The temperature distribution, $F(r)$, across the bottom is given in this table:

R	0	0.5	1	1.5	2	2.5	3	3.5	4
T(r,z = 0)	123.5	166	300	477	500	477	300	166	123.5

8.39 One approach to determining the thermal properties of an unknown material is to heat the specimen to a uniform initial temperature and then immerse it in a calorimeter containing a specified volume of a known liquid. If the liquid is adequately stirred then the surface temperature of the solid will be virtually the same as the temperature of the calorimeter fluid. Suppose the specimen is a solid sphere with $R = 3$. The temperature in the interior of the sphere will be governed by

$$\frac{\partial T}{\partial t} = \alpha \left[\frac{\partial^2 T}{\partial r^2} + \frac{2}{r}\frac{\partial T}{\partial r} \right].$$

The temperature must be finite at the center of the sphere so as we have seen previously:

$$T = \frac{A}{r}\exp(-\alpha\lambda^2 t)\sin\lambda r.$$

The second boundary condition is the interesting one; the heat lost from the surface of the sphere is transferred to the calorimeter liquid:

$$-4\pi R^2 k\left.\frac{\partial T}{\partial r}\right|_{r=R} = M_f C_{Pf}\left.\frac{\partial T}{\partial t}\right|_{T(r=R)}.$$

Note that M_f and C_{Pf} are the mass and heat capacity of the *liquid* in the calorimeter, respectively. Therefore,

$$\tan \lambda R = \frac{\lambda R}{1 + \frac{M_f C_{pf} \alpha \lambda^2}{4\pi R k}}.$$

Assume that the calorimeter fluid is water with $M_f = 1{,}850$ and $C_{Pf} = 1$. The spherical specimen has the properties of granite: $\rho = 2.6$, $k = 0.006$, $C_p = 0.21$, and $\alpha = 0.011$ (all cgs). The initial temperature of the sphere is 90° and the water starts at 0°. Compute the evolution of the temperature distribution in the interior of the sphere (of course the surface temperature is of principal interest since it is the same as the temperature of the calorimeter liquid) and compare your results with the Figure P8.39.

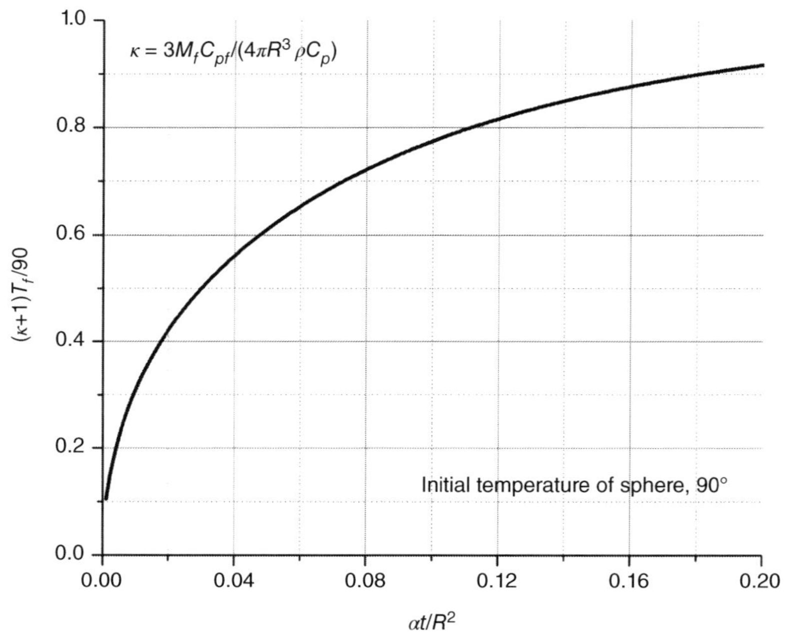

Figure P8.39 Temperature of the calorimeter liquid in dimensionless form. Note that for the given parameters, the thermal capacity of the liquid is much larger than that of the sphere: $\frac{M_f C_{Pf}}{M_s C_{Ps}} \cong 30$. Therefore, the temperature of the liquid will not change much.

8.40 *Should you decide to attack this problem from scratch, be forewarned: This is a difficult scenario requiring a bit more coding skill.* Consider a situation in which a refrigerated liquid is flowing through a horizontal pipe. At $t = 0$ an unexpected shutdown occurs and fluid motion stops completely. At the same instant the pipe's external wall—which is exposed to ambient conditions—is heated by incident solar radiation. The localized heating creates buoyancy and heat transfer by natural convection commences inside the pipe. We want to use vorticity transport to model this phenomenon and determine both expected flow patterns and the rate of heat transfer to the refrigerated fluid. We will look at two cases: one in which the pipe wall is heated uniformly, and one in which the incident solar radiation affects only half of the pipe's external surface. The former case (although for cooling rather than heating) has been considered by Ludovisi and Garza (*Proceedings of the ASME 2013 Power Conference*) who used a commercial code (Simcenter STAR-CCM+) to model the phenomenon. They present one example of a velocity distribution obtained for low *Rayleigh* number ($Ra = GrPr$) which you may consult for validation. Of course, with uniform wall temperature the ultimate internal condition will be uniform T and the cessation of buoyancy-driven fluid motion, so asymmetric heating is probably a more interesting subject for study. Let the pipe ID be 7.5 cm, Pr = 1, and Gr = 15,000 (thus $Ra = 15{,}000$, we start with fairly benign conditions). Assume the initial temperature of the refrigerated fluid is 0 °C and that the solar radiation heats the pipe wall to 10 °C instantaneously. Use vorticity transport to determine the evolution of the buoyancy-driven flow occurring inside the pipe and the rate of heat transfer to the fluid. Attempt the uniform heating case first! Some computational results for a more difficult case are shown in Figure P8.40 to provide you with qualitative guidance.

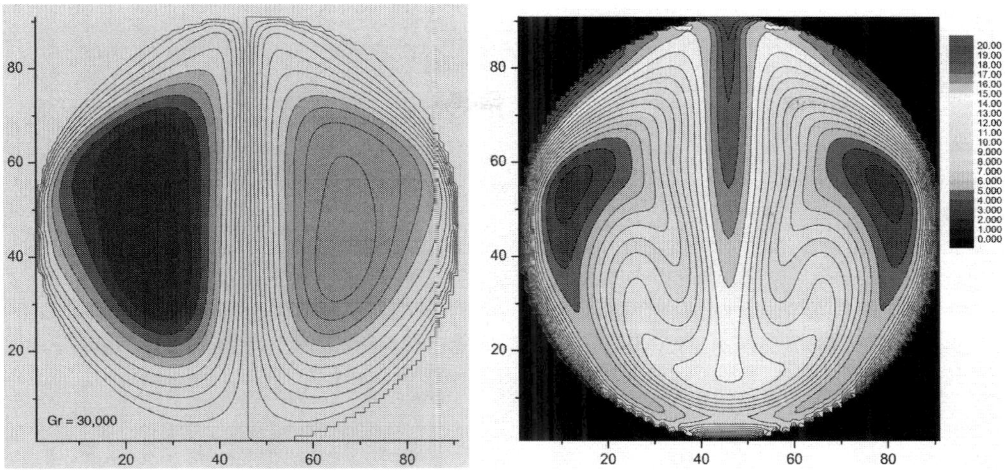

Figure P8.40 Streamlines and isotherms for natural convection inside a horizontal pipe (ID = 15 cm) with Gr = 30,000, Pr = 1. The temperature of the pipe wall (uniform heating) was instantaneously raised to 22.75 °C at $t = 0$, and the dimensionless time ($t^* = \frac{\sqrt{\nu t}}{d}$) corresponding to these results was 0.092. The results are dramatically changed in the case of *asymmetric heating* of the pipe wall (right-half hotter than left-half). Try this variation and contrast your results with symmetric case. Use Gr = 7,800.

An important consideration: Since we are using vorticity transport with rectangular coordinates, a square mesh will yield a pipe wall that *approximates* a cylindrical shape. Thus the "quality" of the approximation for the wall will depend upon the fineness of the mesh (you can see evidence of this problem on the right-hand side of the streamline plot shown immediately above). One of the unfortunate consequences of this roughly cylindrical wall can be variations in vorticity that lead to a flow exhibiting spurious localized accelerations/decelerations in fluid motion. Under difficult conditions this may produce completely anomalous results, particularly at advanced t.

8.41 We want to model viscous flow in a 30° V-shaped cavity with localized heating of the right-hand wall near the bottom. The fluid in the cavity is initially at rest, and at $t = 0$ the "lid" across the top of the cavity begins to slide from left to right at a constant velocity. At the same instant, the lower half of the right-hand wall of the cavity is slightly heated, creating a (local) buoyancy-driven flow in the $+y$ direction. Let the cavity be 14 cm deep, Pr = 1, Re = 200, and Gr = 30,000. The initial temperature of the fluid is 0°. Since the interior angle at the bottom is 30°, the cavity width, W, at the top should be about 7.5 cm.

$$\text{Re} = \frac{WV_{top}}{\nu} = 200 \quad \text{and} \quad Gr = \frac{g\beta\Delta TW^3}{\nu^2} = 30,000$$

At small values of t, the flow pattern (streamlines) should consist of an upper vortex driven by the sliding lid (rotating clockwise) and a lower vortex with opposite rotation driven by the heated lower wall.

8.42 A rectangular channel measures 6 cm in height and 10 cm in length (horizontally). A viscous fluid enters the box through the center 2 cm of the left-hand wall and leaves through an opening in the right-hand wall of the same size and located at the same position. This means that the bottom wall extends vertically upward 2 cm on each side and the top wall extends downward the same distance. The fluid flows through the left-hand opening with *uniform velocity* and exits on the right-hand side the same way. We are interested in both flow and heat transfer occurring in this channel. Since the flow area triples upon entrance to the box, we can expect to see areas of recirculation emerge at or near the corners. Place the origin in the lower left-hand corner and set the stream function on that solid boundary to zero. Allow the velocity of the fluid entering the box to be phased in, e.g., let $V_{inlet} = V\left(1 - \exp\left(-\frac{1}{2}t\right)\right)$ such that for $t = 6$ s, we will have attained about 95% of the ultimate velocity. Since the inlet and exit velocities are uniform, the stream function must increase linearly as one moves across the openings in the $+y$-direction. Begin your analysis by using vorticity transport to explore the transient flow problem *without* heat transfer. The streamlines for flow through the center of a rectangular channel with Re = 800 (based upon the ultimate velocity through the 2 cm opening) and $t = 6$ s will reveal vortices in all four corners as expected. But the vortices in the top corners are less developed due to the way the uniform velocity through the opening is increased gradually. Of course, the stream function, ψ, on the upper boundary of the channel is changing with each time step.

Now we want to add heat transfer to this problem by allowing the bottom surface to acquire some elevated temperature; in particular, we will set $T(x,y=0)=55\,°\text{C}$ and maintain the top and the left-hand side at 0°. The right-hand side will be insulated. Let the ultimate Reynolds number for the *channel entrance* be Re = 400, and set Pr = 1. Determine the rate at which thermal energy *leaves* the rectangular box as a function of time. Take ρ, C_p, and ν to be 1, 1, and 0.01, respectively (all cgs).

An interesting variation of problem 8.42 can be devised by inserting a baffle at the center of the rectangular channel, extending vertically from the floor with a height of 4.67 cm and a width of 0.4 cm. Let the Reynolds number (based upon the size of the channel entrance, 2 cm) be 500, and compute the flow field in the absence of heat transfer. Because the fluid velocity must increase significantly over the top of the baffle we can expect to see some large regions of recirculation.

8.43 One of the most studied problems in all of fluid mechanics is flow of a viscous fluid around a circular cylinder. This scenario has been used as a benchmark for measuring the performance of CFD software for decades. There are several reasons for this: the flow has been thoroughly documented experimentally, and the vortices that form on the trailing side of the body are of particular interest. At low Reynolds numbers the vortices occur in pairs, upper and lower, and they are fixed (remain attached to the backside of the cylinder). However, at slightly higher Reynolds numbers—greater than about 45—the vortices are shed from the cylinder alternately, top-bottom, top-bottom, etc. The shed vortices form a "vortex street," and the frequency at which they depart from the cylinder is normally represented with the *Strouhal* number, $St = df/V$, where f is the frequency of vortex shedding from *one side* of the cylinder. At Re = 1,000, St is about 0.20.

We want to use vorticity transport to model this flow, but we will do so *assuming* there is a horizontal plane of symmetry that bisects the cylinder, allowing us to just make computations for the upper half. This choice has implications with respect to the model's ability to reproduce the actual physical behavior for if vortices are shed alternately, we cannot expect to have upper–lower symmetry. Thus, we are imposing an unrealistic constraint upon the solution procedure, and naturally we will be interested in the consequences of this choice.

The challenges of using a square mesh to represent a curved surface were discussed previously. Of course this is crucial for this problem because we must be able to compute the vorticity upon the cylinder's surface. Here is an example of how one might identify the "surface" mesh points:

```
FOR j=2 TO 25
yc=j-1
xc=SQR(24^2-yc^2)
x=45-INT(xc)
xmin=x-1
xmax=45+INT(xc)+1
```

Vorticity can now be calculated on both sides of the cylinder, i.e., at the pair of points (*xmin,j*) and (*xmax,j*). The horizontal axis bisecting the circular cylinder corresponds to $j=1$ and the center of the cylinder (x-position) is $i=45$; therefore, when $j=2$, $xc=23.979$, $x=22$, $xmin=21$, and $xmax=69$. Notice that the radius, R, is set to $24\Delta x$ so that the cylinder diameter is divided into 48 internodal (mesh) spaces. Typical results obtained from the computation for Re = 80 reveal streamlines that are not in accord with the experimental data for this problem; at Re = 80 the vortices will certainly be shed alternately from the trailing half of the cylinder and Roshko's data (*NACA Report* 1191, 1954) show that $St \approx 0.18$ (with an observed range of $0.17 \leq St \leq 0.195$).

At Re = 80 you should find the lateral size of the vortex to be about one cylinder diameter, and if you perform the computation for Re = 150, nearly two diameters. The cylinder midpoint is located at an index of 45 on the horizontal axis where the cylinder extends from 21 to 69. Consequently, separation (of the boundary layer) occurs at an angle of about 118° as measured from the forward stagnation point. We can compare this with values cited throughout the literature—generally about 110°.

Solve this problem for both Re = 50 and Re = 100, assuming that the horizontal axis is a plane of symmetry, and evaluate the ratio: length of standing vortex divided by cylinder diameter.

8.44 A viscous fluid, containing a reactant species, flows between two parallel, catalytic walls. The lower wall is located at $y = 0$ and the upper wall at $y = b$, and the reaction upon the two surfaces is so rapid that the concentration of reactant there is virtually zero. The concentration of reactant in the entering fluid is uniform, $C = 1$, and the model for this process is

$$v_x \frac{\partial C}{\partial x} = D\frac{\partial^2 C}{\partial y^2}.$$

We neglect molecular transport in the flow direction, assuming that it will be insignificant. The laminar flow between the planar surfaces is fully developed and the velocity distribution is:

$$v_x = \frac{1}{2\mu}\frac{dp}{dx}(y^2 - by).$$

Determine the evolution of the concentration distribution numerically and present your results as a contour plot. We are given: $b = 3$, $D = 0.09$, $dp/dx = -0.088$, and $\mu = 0.01$, all with compatible units. At what x-position will the *average* concentration of reactant be reduced by 25%? 50%? 75%? (A 50% reduction requires $x \approx 52$.) Do not forget that the average concentration must be determined by integration of the product of v_x and C, since both vary in the transverse (y) direction.

An interesting comparison can be made between our numerical results from the first part of this problem and the analytic solution, *assuming plug flow* between the planar surfaces (v_x is taken as the constant, V). Verify that the analytic solution in this case is:

$$C = \sum_{n=1}^{\infty}\left(\frac{2}{b} - \frac{2}{n\pi}\cos(n\pi)\right)\exp\left(-\beta\frac{n^2\pi^2 x}{b^2}\right)\sin\left(\frac{n\pi y}{b}\right),$$

where $\beta = D/V$. Let the fixed velocity, V, correspond to the average velocity from the laminar flow case. Prepare a figure illustrating the evolution of the average concentration from part one, with the average concentration as determined from the analytic solution for the plug-flow case. The analytic solution indicates that the average concentration of reactant at $x = 80$ is approximately 0.247.

8.45 In problem 8.43 we considered flow past a circular cylinder by treating only half of the object; that is, we placed a horizontal plane of symmetry through the center of the cylinder and carried out our computations for the upper half only. We noted that the results thus obtained would be unphysical at larger Reynolds numbers because the generated vortices would remain attached to the cylinder surface. Such vortices are said to be fixed, or standing, and they remain so regardless of Re. In the present problem we are going to use vorticity transport to investigate uniform flow past a thin rectangular solid which we can think of as a facsimile of an impeller blade. Let the "blade" have a vertical height of 1.6 cm (normal to the uniform flow) and a thickness of 0.6 cm. Allow the uniform flow to accelerate past the blade, attaining an ultimate velocity corresponding to Re = 240. Compute the evolution of the flow field and determine the Stouhal number ($St = df/V$) for the fully developed flow condition. A stream function plot is provided in Figure P8.45 to assist you in evaluating your solution.

Finally, increase the flow rate such that the ultimate Reynolds number will be 500; will the Strouhal number change significantly?

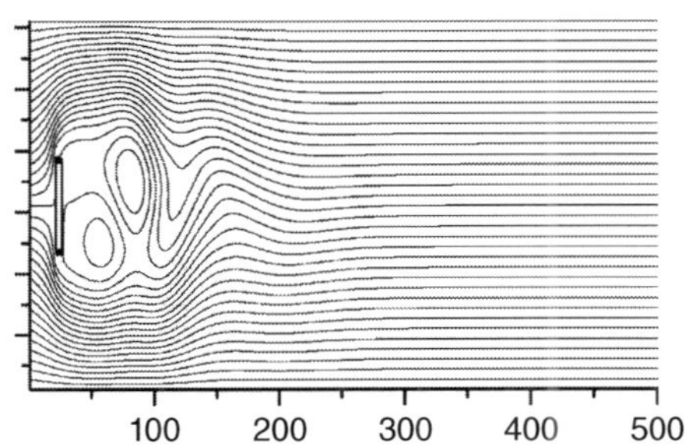

Figure P8.45 Vortex shedding from impeller "blade" at Re = 240. This image corresponds to a time of 10 s after flow initiation. A sequence of time "snapshots" will show the vortices being shed alternately from the top and bottom of the blade facsimile.

8.46 Viscous channel flow across the top of a rectangular cavity has been studied intensively both experimentally and computationally; an early computational study was carried out by Mehta and Lavan (Flow in a Two-Dimensional

Channel with a Rectangular Cavity, *NASA Contractor Report CR-1245*, January 1969). Their work was for steady-state conditions with the channel flow generated by a sliding upper wall employing Reynolds numbers of 1, 10, 100, and 500. They noted that the streamlines *sag into the cavity* at low Re, but *bulge upward* at large Re. This study is also of some historical interest: The program (provided in the report) was written in FORTRAN IV and executed on an IBM-360.

A more recent example is the numerical study of Yoshida and Watanabe (Numerical Simulation of Flow over an Open Cavity with Self-Sustained Oscillation Mode Switching, *Open Journal of Fluid Mechanics*, 6:361, 2016). They solved two components of the Navier–Stokes equation using a pressure correction method described by Armfield and Street (*International Journal for Numerical Methods in Fluids*, 38:255, 2002).

The exact nature of the flow depends upon the ratio of cavity length, L, and depth, D. If L/D is greater than 1.6, oscillations in the shear layer at the top of the cavity may occur. For flow initiation past a cavity with $L/D = 1$, the vorticity transport method reveals how the streamlines sag into the cavity at small t as shown in Figure P8.46.

It is the transient behavior of this flow that is particularly interesting. For small t's following startup, the streamlines sag into the cavity. At intermediate times the flow in the cavity becomes more complicated and this is the dynamic process we wish to explore. You can compare the results you obtain with figures presented by Yoshida and Watanabe, and (for large t's) Mehta and Lavan.

However, if $L/D > 1.6$, different modes of oscillation may occur in the shear layer; for $L/D = 1.7$ the oscillations occur at a Strouhal number ($St = f\delta/V$, *where f is the frequency of the dominant oscillation, δ is the momentum thickness of the boundary layer, and V is the freestream velocity above the cavity*) of about 0.018 for a Reynolds number (based upon the momentum thickness) of 190. The momentum thickness characterizes the loss of momentum in the boundary layer relative to the external flow: $\delta = \int_0^\infty \frac{v}{V}\left(1 - \frac{v}{V}\right) dy$.

It has been demonstrated that St decreases (from 0.018 to about 0.011) with increasing L/D *until* $L/D \approx 3.2$, where a jump or transition occurs to $St \approx 0.015$. This general behavior has been verified by both Knisely and Rockwell (*JFM*, 116:157, 1982) and Gharib and Roshko (*JFM*, 177:501, 1987).

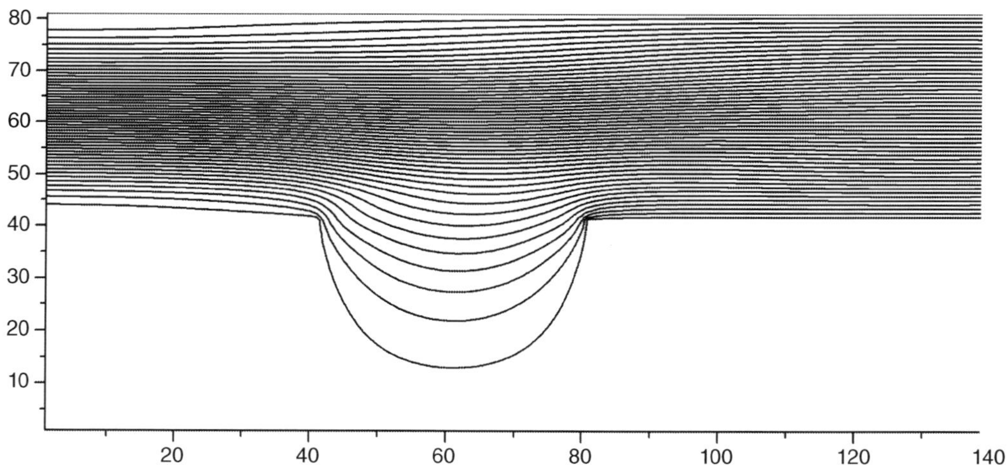

Figure P8.46 Startup flow across the top of a square ($L/D = 1$) cavity; computed streamlines are shown for short time.

8.47 A square cavity measuring 20 cm by 20 cm is filled with a viscous fluid initially at rest with a uniform temperature of 0°. A much smaller "box" measuring 3.73 by 3.73 cm is centered in the cavity, and at $t = 0$ its temperature is instantaneously elevated to some small value which is maintained for all t. We can expect the resulting buoyancy to produce rising fluid currents and perhaps patterns of recirculation within the large cavity. We will consider moderate thermal driving forces such that $Gr = \frac{g\beta\Delta T L^3}{\nu^2} = 1{,}000$ *to* $10{,}000$, and we will set $\Pr = 1$. We are interested in the evolution of flow within the cavity—recognizing that at small t, vorticity will be created along the vertical sides of the heated box, as well as the four corners. At larger times we can expect a buoyancy-driven plume to rise vertically from the box, turning as it approaches the upper surface of the large cavity, and producing a symmetric recirculation. We are particularly interested in determining whether or not small, weak vortices may be produced in the corners of the square cavity. Let the kinematic viscosity of the fluid be 0.05, and set $\beta = 0.0001$; this means that the temperature of heated (centered) box will be 0.4916° for Gr = 1,000. Interesting extensions of this problem might include the impact of different Prandtl numbers and the temporal change of the average temperature of fluid in the cavity.

8.48 Two circular disk electrodes are separated by a distance $2b$; one electrode is located at $z = -b$ and the other at $z = +b$. Each of the aluminum electrodes has radius, R. We want to determine the potential distribution between the two disk electrodes where the governing PDE is:

$$\frac{\partial^2 v}{\partial r^2} + \frac{1}{r}\frac{\partial v}{\partial r} + \frac{\partial^2 v}{\partial z^2} = 0 \tag{8.217}$$

for $-b \le z \le +b$ with $\frac{\partial v}{\partial r} = 0$ at $z = \pm b$ if $r > R$, and

$$\frac{\partial v}{\partial r} = \frac{S}{2\pi k R\sqrt{R^2 - r^2}} \quad \text{if } r < R.$$

This problem was solved by Riemann in 1855, and the analytic solution is

$$v = \frac{S}{\pi k b}\sum_{n=1}^{\infty} \sin\frac{n\pi}{2} \sin\frac{n\pi z}{2b} K_0\left(\frac{n\pi r}{2b}\right),$$

which is valid as long as $r > 0$ since $K_0(0)$ is infinite. There are several important points for us to note about this solution: The size of the circular disk electrodes is small relative to the extent of the medium in the r-direction, and since the electrode at $z = -b$ has some potential $-V_0$ and the potential of the electrode at $z = +b$ is $+V_0$, the solution must consist of odd functions (of course, $\sin(n\pi/2) = 0$ for all even n's) with respect to z-position. Therefore, at the mid-point where $z = 0$, we must have $v = 0$. We want to solve this problem numerically and compare our results with Riemann's solution, using the values $S = 1$, $b = 1$, $R = 0.5$, and $k = 0.48$.

We expect to solve eq. (8.217) iteratively, but we recognize there are other important issues to be considered when attempting to find a numerical solution:

1) The computational domain must be finite in the r-direction, whereas the analytic solution is for an infinite medium.
2) It is clear that $\partial v/\partial z$ at $z = +b$ will involve division by zero if $r = R$.

Our computational domain will extend from $0 \le z \le +b$ and $0 \le r \le mR$, so we must decide how large m should be to approximately represent the "infinite" medium—consider starting with $m = 3$. Finally, we note that we must have symmetry about the z-axis so $\frac{\partial v}{\partial z} = 0$ for $r = 0$.

We anticipate that the limitations we are imposing in our efforts to produce a numerical result will certainly affect the accord between analytic and numerical solutions.

8.49 The United States conducted the Castle Bravo test of a practical thermonuclear device on March 1, 1954, at Bikini Atoll. Although the fusion process itself is radiologically "clean" it must be initiated with a fission reaction, and consequently fission products with their radiological hazards are widely distributed in atmospheric and surface detonations. Fission products typically include about 200 isotopes of about 35 different elements and many are beta- and gamma-emitters. Moreover, when such explosions occur near the surface there is significant neutron activation of soil constituents resulting in additional radioactive contamination by isotopes with a broad spectrum of half-lives.

The predicted yield for Castle Bravo was 5 MT (megatons, a unit of explosive power equivalent to 1 million tons of TNT); it had been assumed that the lithium-7 isotope in the lithium deuteride fuel would be inert. It was not because exposure of lithium-7 to energetic neutrons can produce tritium, contributing significantly to the fusion yield. As a result, the actual power of the device was about 15 MT and it resulted in widespread contamination from radioactive fallout. After 96 hr a cumulative radiation exposure of 3,000 R (Roentgen, the quantity of gamma or X-radiation that will produce 1.61×10^{12} ion pairs when absorbed in one gram of air) extended out about 100 mi from Bikini. The initial reaction of U.S. officials (including Lewis Strauss, Chairman of the Atomic Energy Commission) to this nuclear catastrophe was inexcusable—several hundred Rongelese natives and the crew of a Japanese fishing boat were exposed to injurious levels of ionizing radiation. It was later discovered that females affected by this disaster were much more likely to develop reproductive cancers.

An acute radiation exposure of 500 rem (Roentgen equivalent man, 500 rem = 5 sievert) is *usually* lethal to humans. This threshold cannot be known with certainty since the effect of ionizing radiation upon biological tissues is stochastic—it is impossible to know precisely which cellular components will be directly impacted.

In the 1950s the U.S. began an extensive program of civil defense in which certain individuals with appropriate backgrounds were trained in matters related to nuclear war, including radioactive contamination by fallout. The AEC published a book that was widely distributed titled, *The Effects of Nuclear Weapons* (Samuel Glasstone, Editor, June 1957). Chapter 9 was devoted to estimation of fallout distributions resulting from both surface and atmospheric detonations. Figure 9.87 in Glasstone presented the idealized cumulative 36-hour exposure contours for Castle Bravo indicating that the lethal zone (exposure $\gtrsim$ 500 R) extended out about 150 mi from ground zero.

We want to perform numerical experiments with a simplified two-dimensional model of the dispersion of radioactivity resulting from a nuclear explosion. We will use cylindrical coordinates and assume the prevailing wind is west to east with east corresponding to $\theta = 0$:

$$\frac{\partial C}{\partial t} + V_r \frac{\partial C}{\partial r} = D\left[\frac{1}{r}\frac{\partial}{\partial r}\left(r\frac{\partial C}{\partial r}\right) + \frac{1}{r^2}\frac{\partial^2 C}{\partial \theta^2}\right]. \tag{8.218}$$

Assume C is the instantaneous dose rate and D is a dispersion coefficient (it will be quite large in order to crudely account for the effects of atmospheric turbulence). We will assume that

$$V_r = V_0 \exp(-b\theta^2),$$

with $b = 5.056$ which means that the velocity in the r-direction will be 25% of the centerline ($\theta = 0$) value when $\theta = \pm \pi/6$. At $\theta = \pm \pi/3$ $V_r(\theta)$ will be only 0.39% of the centerline value. This broadening of $V_r(\theta)$ will help mimic atmospheric dispersion. We will also assume $V_0 = 5{,}000$ m/hr (about 3.1 mph). Solve eq. (8.218) numerically recognizing that the convective transport term must be handled appropriately. Assume the initial contamination occupies about 280,000 m^2 centered about the origin; this is our ground zero (gz), where the initial value of C is $C_0 = 5{,}000$. In this ground zero area we let the dose rate decline exponentially with time:

$$C_{gz} = C_0 \exp\left(-\frac{t}{9}\right).$$

Thus, after 12 hr $C_{gz} \approx 26\%$ of C_0. This is a very crude treatment of what actually takes place: In a nuclear explosion the enormously intense *initial* radiation is mainly neutrons and gamma rays, but this diminishes rapidly over the first 60 s as the atomic cloud rises toward the stratosphere. The residual radiation from the explosion is due to fission products and unstable isotopes created by neutron capture; important neutron-activated soil constituents include Na^{24}, Ca^{41}, Mn^{56}, Al^{28}, and K^{42}, and (except for Al^{28}) these species have relatively short half-lives.

After 18 hr, what is the maximum value of C, and where is it located? A typical result for t = 13 hr indicates the contour for the highest dose rate (about 3,400) has reached a radial position of about 63,000 m (about 39 mi). Let $D = 675$ m^2/hr.

References

Alderson, T.L., Allan, F.M., and Hamdan, M.H. (2007). On the Universality of the von Mises Transformation. *International Journal of Applied Mathematics* 20: 109.

Anderson, J.D. (1995). *Computational Fluid Dynamics: The Basics with Applications*. New York: McGraw-Hill.

Barron, R.M. (1989). Computation of Incompressible Potential Flow Using von Mises Coordinates. *Mathematics and Computers in Simulation* 31: 177.

Bernard, R.S. (1986). *Discrete Solution of the Anelastic Equations for Mesoscale Modeling. Report 86/E/51*. Geesthacht: GKSS Forschungszentrum.

Bernard, R.S. (1989). *Explicit Numerical Algorithm for Modeling Incompressible Approach Flow. Technical Report REMR-HY-5*. Vicksburg: US Army Engineer Waterways Experiment Station.

Bernard, R.S. (1992). A MacCormack Scheme for Incompressible Flow. *Computers & Mathematics with Applications* 24: 151.

Biswas, S. and Kalita, J.C. (2016). Moffatt Vortices in the Lid-Driven Cavity Flow. *Journal of Physics: Conference Series* 759 (1): 01281.

Chow, C.Y. (1979). *An Introduction to Computational Fluid Mechanics*. New York: John Wiley & Sons.

Chung, T.J. (2002). *Computational Fluid Dynamics*. Cambridge: Cambridge University Press.

Frankel, S.P. (1950). Convergence Rates of Iterative Treatments of Partial Differential Equations. *Mathematical Tables and Other Aids to Computation* 4: 65.

Hermann, R. (1954). Heat Transfer by Free Convection from Horizontal Cylinders in Diatomic Gases. *NACA Technical Memorandum* 1366.

James, M.L., Smith, G.M., and Wolford, J.C. (1977). *Applied Numerical Methods for Digital Computation (with FORTRAN and CSMP)*, 2nde. New York: Harper and Row.

Jones, D. J. and Dickinson, R. G.A Description of the NAE Two-Dimensional Transonic Small Disturbance Computer Method. *NAE Laboratory Technical Report* LTR-HA-30 (1980).

Moffatt, H.K. (1964). Viscous and Resistive Eddies Near a Sharp Corner. *Journal of Fluid Mechanics* 18: 1.

Nikbakhti, R. and Rahimi, A.B. (2012). Double-Diffusive Natural Convection in a Rectangular Cavity with Partially Thermally Active Side Walls. *Journal of the Taiwan Institute of Chemical Engineers* 43: 535.

Oertel, H. and Kirchartz, K.R. (1979). *Recent Developments in Theoretical and Experimental Fluid Mechanics* (Muller, Roesner, and Schmidt, editors). Springer-Verlag.

Patankar, S.V. (1980). *Numerical Heat Transfer and Fluid Flow*. Washington, DC: Hemisphere Publishing.

Peaceman, D.W. and Rachford, H.H. (1955). The Numerical Solution of Parabolic and Elliptic Partial Differential Equations. *Journal of the Society for Industrial and Applied Mathematics* 3: 28.

Sarra, S.A. (2003). The Method of Characteristics and Conservation Laws. *The Journal of Online Mathematics and its Applications* 3: 1–16.

Selle, A., Fedkiw, R., Kim, B.M. et al. (2007). An Unconditionally Stable MacCormack Method. *Journal of Scientific Computing* 35: 350–371.

Shimomukai, K. and Kanda, H. (2006). Numeerical Study of Normal Pressure Distribution in Entrance Flow between Parallel Plates, 1. Finite Difference Calculations. *Electronic Transactions in Numerical Analysis* 23: 202.

Smith, G.D. (1965). *Numerical Solution of Partial Differential Equations*. Oxford: Oxford University Press.

Taneda, S. (1979). Visualiztion of Separating Stokes Flows. *Journal of the Physical Society of Japan* 46: 1935.

Thompson, J.F., Thames, F.C., and Mastin, C.W. (1974). Automatic Numerical Generation of Body-Fitted Curvilinear Coordinate Systems for Fields Contining Any Number of Arbitrary Two-Dimensional Bodies. *Journal of Computational Physics* 15: 299.

Van Dyke, M. (1982). *An Album of Fluid Motion*. Stanford, CA: Parabolic Press.

Xia, C. and Murthy, J.Y. (2002). Buoyancy-Driven Flow Transitions in Deep Cavities Heated from Below. *Transactions of the ASME* 124: 650.

9

Integro-Differential Equations

9.1 Introduction

As the name implies, integro-differential equations have both integral and derivative terms; often, the integral terms are present because the past behavior (the history) of the process affects the future outcome. This is in sharp contrast with models based solely upon differential equations (DEs) where the initial conditions, coupled with the DEs, dictate all future states of the system. As a somber example of the importance of process history, consider the world's first commercial jet airliner: The De Havilland Comet was a relatively small passenger aircraft with four jet engines buried in the wing roots. In 1952 the Comet flew its inaugural commercial flight (London to Johannesburg with intermediate stops), attracting much interest and fanfare everywhere it went. But merely two years later the aircraft was withdrawn from service because of two catastrophic airframe disintegrations at altitude. Pieces of the first failed airframe (registration G-ALYP) were recovered from the Mediterranean, and inspections revealed that metal fatigue (with subsequent explosive airframe failure) was the cause. To better understand what had happened, a Comet airframe was filled with water and subjected to pressurization–depressurization cycles (simulating a climb from sea level to 35,000 ft where the pressure is only 196 mmHg) at three-minute intervals. The fuselage failed dramatically after 1,800 cycles when it developed a rupture 2.4 m long from a crack propagating from a bolt hole near an escape hatch window frame.[1] The Comet 1's fate was dictated by several unfortunate design features: 1) the structure was very light due to the limited thrust (5,000 lb_f) of the Halford H.2 Ghost engines, 2) the fabrication utilized punch riveting as opposed to drilled, 3) the cut out in the cabin roof for the ADF antenna was square, and 4) the low efficiency of first-generation jet engines meant that flying at altitude (around 35,000 ft) was necessary to improve range (the atmospheric pressure at FL350 is only about 3.5 psi, so the interior–exterior pressure differential would be about $11 - 3.5 = 7.5\ lb_f$ psi). The Comet 1 never flew in commercial service again, and the importance of the *cumulative* aspect of process experiences (history) was recognized throughout the aviation industry.

Integro-differential equations (IDEs) arise in a variety of other contexts too; there are applications to process control and to diffusion along grain boundaries (Antipov and Gao, 2000), as well as modeling for neural networks (Jackiewicz et al., 2008), option prices (Cont and Voltchkova, 2005), and the spread of infectious diseases (Medlock and Kot, 2003). Only rarely can analytic solutions be found for such problems, so frequently the analyst must resort to numerical methods. It is worth noting that IDEs also figure prominently in the analysis of multiphase processes where countable entities such as bubbles, drops, and particles are borne by a fluid phase. Examples include solvent extraction/emulsification, flocculation, crystallization, sedimentation, and the operation of biochemical reactors. Because so many industrial processes involve countable entities, IDEs are critically important to process engineering and the applied sciences.

Let's begin our consideration of IDEs by exploring an elementary linear problem that can be solved analytically:

$$\frac{dy}{dt} + 2y + 5\int_0^t y(\tau)d\tau = 2, \tag{9.1}$$

with $y(0) = 0$. The easiest approach in a case like this is to make use of the Laplace transform:

$$sy(s) + 2y(s) + 5\frac{y(s)}{s} = \frac{2}{s}. \tag{9.2}$$

1 One of the principals in the Comet investigation was Anne (Pellew) Burns, a truly remarkable person. She was just the second woman to read Engineering Sciences at Oxford University where she worked with R. V. Southwell on convective fluid motions arising from localized heating. She flew on many of the Comet test flights during the investigation monitoring strain gauges—she later observed that they flew about at altitude just waiting for some kind of airframe failure to occur. She was also a champion glider pilot and an expert on clear-air turbulence.

Problem Solving in Engineering: Analytical Mathematics and Numerical Analysis, First Edition. Larry A. Glasgow.

Companion website: www.wiley.com/go/glasgow/problemsolvingengineering

Isolating $y(s)$ we find:

$$y(s) = \frac{2}{s^2 + 2s + 5}. \tag{9.3}$$

We can invert this transform by partial fraction expansion:

$$y(s) = \frac{A}{s + 1 + 2i} + \frac{B}{s + 1 - 2i}, \tag{9.4}$$

or we can complete the square in the denominator to obtain a form that may be found directly in tables of Laplace transforms:

$$\frac{2}{s^2 + 2s + 5} = \frac{2}{(s + 1)^2 + 4}. \tag{9.5}$$

If we choose the former approach, $A = i/2$ and $B = -i/2$; and we make use of the Euler relation to find

$$y(t) = \exp(-t)\sin(2t), \tag{9.6}$$

a decaying sinusoid. This raises an important question: Exactly how does the inclusion of the integral term in the original equation affect the solution? If we omit it, we get just the simple first-order ordinary differential equation (ODE), $dy/dt + 2y = 2$, with the solution $y(t) = 1 - \exp(-2\,t)$; thus, $y(t)$ will asymptotically approach 1 (one) as t becomes large. But suppose we change the value of the leading constant for the integral term (which was 5 in this example)—we will let it be 0.1, 1, and 10 successively, solving the IDE for each case. The results are shown in Figure 9.1.

For this example, increasing the impact of the integral term makes the solution increasingly oscillatory, while diminishing the amplitude of the response. It may be apparent to you that the nature of $y(t)$ can be assessed by looking at the roots of the polynomial $(as^2 + bs + c)$ in the denominator of eq. (9.3). In particular, if we look at $b^2 - 4ac$ we note that:

c	$b^2 - 4ac$
10	−36
5	−16
1	0
½	+2

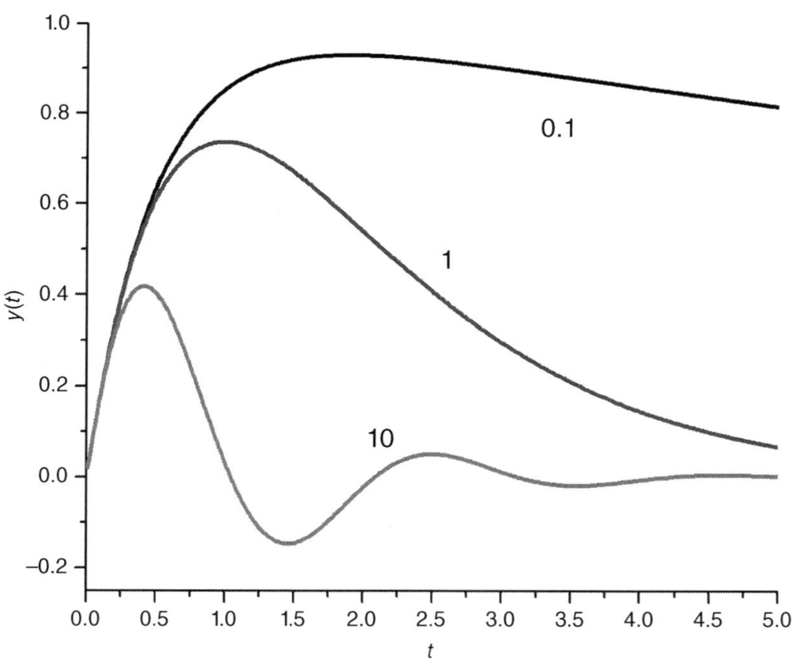

Figure 9.1 Solutions for the IDE, eq. (9.1), with increasing contributions from the integral term. This was accomplished by changing its prefactor successively from its original value (5) to 0.1, then 1, and finally 10.

Of course this means that $c = 1$ represents the threshold between oscillatory and exponential behavior; if we continue to decrease c (the constant in front of the integral term) we will approach the exponential limiting behavior noted above: $1 - \exp(-2\,t)$.

The conditions under which the Laplace transform becomes useful for the solution of IDEs are restrictive: We must have a linear equation with constant coefficients For this subset of IDEs the Laplace transform is an invaluable tool. You may find a brief review useful at this point, and if so, solve the following IDE using the Laplace transform (note the ramp function on the right-hand side):

$$\frac{dy}{dt} + 2y + 5\int_{0}^{t} y(\tau)d\tau = t, \tag{9.7}$$

and again we let $y(0) = 0$. Verify that

$$y(t) = \frac{1}{5} - \frac{1}{5}\exp(-t)\cos(2t) - \frac{1}{10}\exp(-t)\sin(2t), \tag{9.8}$$

such that

t	1/2	1	2	3	4	5
$y(t)$	0.083	0.197	0.228	0.192	0.199	0.202

It may be useful to the reader to have a historical framework for our consideration of IDEs, so we should discuss their role in the early twentieth century study of biological systems. Vito Volterra was an eminent Italian mathematician (1860–1940) whose career was effectively ended by his refusal to sign the oath of allegiance to the Fascist government in 1931. It is worth noting that only 12 Italian university professors refused to sign; a small number but one that is understandable given the consequences (essentially exclusion from the Italian intellectual and scientific communities). One of Volterra's principal interests was mathematical biology—particularly the dynamic behavior of populations in conflict which is frequently referred to as the predator–prey problem. Let us preface this part of our discussion by considering a simple system consisting of (initially) known populations of foxes (F) and rabbits (R).

Prey species usually reproduce rapidly since interactions with predators will invariably diminish their numbers. A very simple model for the prey population might be written

$$\frac{dR}{dt} = a_1 R - a_2 RF. \tag{9.9}$$

In contrast, predators do not usually prosper by excessive breeding since they will suffer from too much competition for available prey (however, the foxes do benefit from encounters with prey):

$$\frac{dF}{dt} = -b_1 F + b_2 RF. \tag{9.10}$$

You will notice the similarity between the two differential equations, but with a reversal of signs, of course. This elementary model is deterministic; given values for the constants and the initial populations, the future numbers of foxes and rabbits are set for all time, t. We recognize that this cannot be correct. One obvious but trivial objection is that a fractional rabbit is not physically realizable. More importantly, there are aspects of animal behavior that are not reflected by these first-order, ODEs; that is, some behavioral phenomena are *not deterministic*. At the same time, we may be able to learn something useful about the dynamic behavior of such simple systems that will benefit us later. We will use eqs. (9.9) and (9.10) to compute $R(t)$ and $F(t)$, and then we will cross-plot the dependent variables to produce the phase-space trajectory.

The data provided in Figure 9.2 illustrate that an increase in the number of rabbits results in growth of the predator (fox) population. Of course, this increase in predators diminishes the number of prey animals, and so on. As we pointed out previously the model employed is not very realistic. Volterra recognized that it was unreasonable to assume that the dynamic behavior of a biological system should depend *solely upon its present state*. Indeed, many common physical systems exhibit dependence upon past events—what Volterra referred to as "*hereditary influences*." An engineer might be more likely to think of a cylindrical rod repeatedly subjected to torsion, or—as we described at the outset—an airliner cabin that has experienced thousands of pressurization cycles; at some point the materials may fail due to cumulative effects of stress.

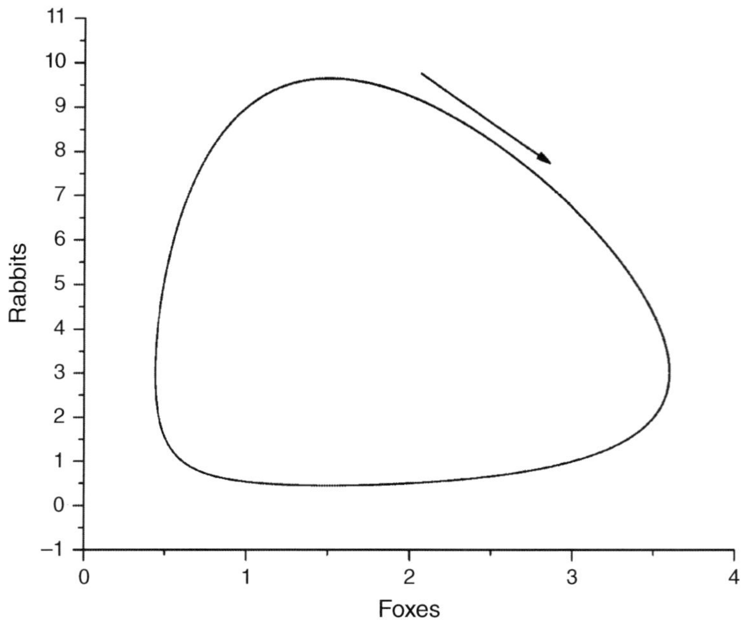

Figure 9.2 Phase-space portrait of the dynamic behavior of the predator–prey problem for foxes and rabbits. Motion on this limit cycle is clockwise, and the parametric values employed were: $a_1 = 1/2$, $b_1 = 1/4$, $a_2 = 1/3$, and $b_2 = 1/12$. The initial populations were nine rabbits and one fox (per acre). The closed cycle means that this system's behavior is *periodic*.

One approach to such problems is to add a time-integral to the model—in essence, a term that will reflect historical influences upon system behavior. We will start with the case of a single population, where an appropriate model might be written:

$$\frac{1}{y}\frac{dy}{dt} = a_1 + a_2 y + \int_{t_0}^{t} K(t,\phi)y(\phi)d\phi. \tag{9.11}$$

The difficulty posed by the integro-differential eq. (9.11) depends mainly upon the nature of the *kernel*, that is, the functional form of $K(t, \phi)$. Before we re-examine our predator–prey (populations in conflict) problem, we can carry out an additional exploration of some of the effects of the addition of a simple history term. We will do this by examining an elementary control strategy that is familiar to many engineers, where part of the control action is determined by the past behavior of the system.

9.2 An Example of Three-Mode Control

We now look at an example from process control that will allow us to better appreciate the significance of the addition of the integral, as illustrated by eq. (9.11) in Section 9.1; we will do so for a problem that is linear with respect to the variable that drives the controller (the error, ε). One familiar and proven strategy for automatic process control is found in the *proportional–integral–derivative* (PID) algorithm. Let ε represent the error detected in an output variable (perhaps temperature, pressure, concentration, flow rate, etc.). P is the controller output (pressure in the case of pneumatic controllers); please note that P is in deviation form—the equilibrium value has been subtracted such that the initial (undisturbed) value for P is zero. In the time domain, our PID algorithm appears as:

$$P = K_C\left(\varepsilon + \frac{1}{\tau_I}\int_0^t \varepsilon dt + \tau_D\frac{d\varepsilon}{dt}\right). \tag{9.12}$$

K_C is the controller gain, τ_I is the integral time, τ_D is the derivative time, and ε is the measured error. We have three parameters that will affect the behavior of this system. The corrective action taken is dependent upon the instantaneous error, the history (time- integral) of the error, and the derivative of the error. It should be clear that the integral action will take into account the past behavior of the system (deviations from the setpoint value), and the derivative action will anticipate what is about to happen at an instant in time. For this reason, $K_C\tau_D\frac{d\varepsilon}{dt}$ is often referred to as *anticipatory control*. It is convenient to formulate a *transfer function* for eq. (9.12) using the Laplace transform—this takes us from the time-domain to the s-plane:

$$\frac{P(s)}{\varepsilon(s)} = K_C\left(1 + \frac{1}{\tau_I s} + \tau_D s\right). \tag{9.13}$$

You will observe that integration with respect to time has been replaced by division by s and that differentiation with respect to time has been replaced by multiplication by s. You will also note that the problem we are considering here is an inverse one, that is, given some form for the error, ε, we try to adjust the controller parameters to obtain the desired response, $P(t)$. Now let us assume a particular functional form for the error; we will start with a sinusoidal error such that $\varepsilon(t) = A\sin(\omega t)$ and $\varepsilon(s) = \frac{A\omega}{s^2+\omega^2}$. Therefore,

$$P(s) = \frac{\frac{A\omega K_C}{\tau_I}(\tau_I\tau_D s^2 + \tau_I s + 1)}{s(s^2 + \omega^2)}. \tag{9.14}$$

Here is where the effect of adding integral control becomes really apparent; the order of the polynomial (in s) in the denominator has been increased by 1. This also means that in the complex plane, a pole has been placed at the origin; you may recall that a complex root with a positive real part implies unstable oscillatory behavior. In the control literature, root locus is a graphical technique used to identify roots of the characteristic equation (a polynomial in s). When complex roots exist, parts of the dynamic response of the system will be governed by terms of the form: $e^{(a+ib)t} = e^{at}(\cos bt + i\sin bt)$. Thus, when a is positive, the system response will be oscillatory with *increasing* amplitude (unstable).

We can invert eq. (9.14) by partial fraction expansion, or choose to work strictly in the time-domain so that our results will be immediately transparent. Our particular interest is to explore the impact of the system's history (manifested in the integral control) upon its behavior. We will achieve this by varying the value of τ_I from 10 to 1 (remember, τ_I has an *inverse* impact upon the integral term).

Our objective with Figure 9.3 is to learn a little bit about how the "history" term affects the solution of eq. (9.12). We see that as the importance of the time integral is increased (τ_I is *decreased* from $10 \to 1$), the oscillatory nature of the response is enhanced and the phase lag (shift of the peaks to the right) is increased. In fact, one of the more interesting features of PID control is that just about any system can be destabilized if enough integral action is added–if τ_I is made small enough! As a practical matter we note that the large response, $P(t)$, seen in Figure 9.2 would almost certainly saturate the final control element (typically a control valve). Saturation can also result from a prolonged history of error (a persistent error condition) which will cause the integral, $\int_0^t \varepsilon dt$, to continue to accumulate resulting in what is referred to in the control literature as integral "windup." This condition is usually avoided in practice by switching the PID algorithm from the *position form* to the *velocity form*.

9.3 Population Problems with Hereditary Influences

It will prove advantageous to return to eq. (9.11) in Section 9.1, with a slight modification, such that we have (initially) a one-variable problem:

$$\frac{dy}{dt} = a_1 y + a_2 y^2 + y\int_0^t K(t-\phi)y(\phi)d\phi. \tag{9.15}$$

You will observe that a specific form has been chosen for the kernel in this case—this is the kernel that Volterra referred to as belonging to the *closed-cycle* class. Following Davis (1962), we approximate the kernel by writing

$$K(z) = K_0 + K_1 z + \frac{1}{2}K_2 z^2 + \dots, \tag{9.16}$$

then we assume that the only K that is non-zero is K_0 (we are truncating our approximation of the kernel). We substitute into eq. (9.15), divide by y, and then differentiate, resulting in:

$$y\frac{d^2y}{dt^2} = \left(\frac{dy}{dt}\right)^2 + a_2 y^2\frac{dy}{dt} + K_0 y^3. \tag{9.17}$$

It proves convenient to let

$$y = -\frac{a_1 u}{a_2}, t = \frac{\tau}{a_2}, \quad \text{and} \quad \lambda = \frac{K_0}{-a_1 a_2},$$

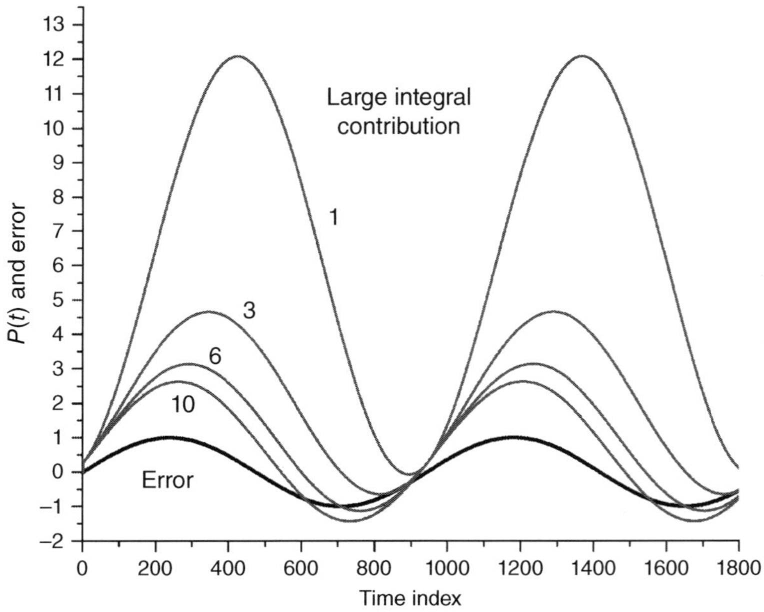

Figure 9.3 Illustration of the response of eq. (9.12) to sinusoidal error (the error is the heavy curve at the bottom that oscillates between −1 and + 1). Results for four different values of τ_I are shown, 1, 3, 6, and 10.

which yields

$$u\frac{d^2u}{d\tau^2} = \left(\frac{du}{dt}\right)^2 - u^2 + \lambda u^3. \tag{9.18}$$

Davis notes that no closed-form solution is known for this ODE, so a numerical solution is required. The intriguing feature of eq. (9.18) is the profound impact that the parameter, λ, has upon the solution. We will explore the solution of this ODE with increasing λ's in Figure 9.4.

The principal task of interest to us now is to assess what happens when hereditary influences are added to the case of *two populations in conflict*. It is likely that some profound changes—relative to the results shown in Figure 9.2 for foxes and

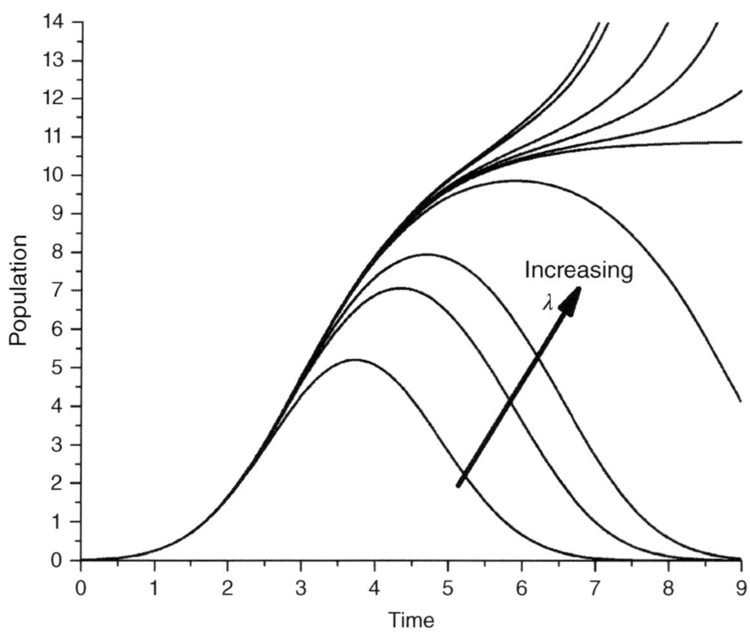

Figure 9.4 Solutions of the ordinary differential eq. (9.18), with increasing values of the parameter, λ. The solution in the "middle" that approaches a horizontal asymptote is often referred to as the growth, or logistic, curve.

rabbits—will occur. The model we will consider takes the form:

$$\frac{dX}{dt} = aX - bXY - K_1X\int_0^t Y(\phi)d\phi, \tag{9.19}$$

and

$$\frac{dY}{dt} = -\alpha Y + \beta XY + K_2Y\int_0^t X(\phi)d\phi. \tag{9.20}$$

Using the approach described previously for the single-population example, we obtain the system of ODEs:

$$\frac{d^2X}{dt^2} = \frac{1}{X}\left(\left(\frac{dX}{dt}\right)^2 - bX^2\frac{dY}{dt} - K_1X^2Y\right) \tag{9.21}$$

and

$$\frac{d^2Y}{dt^2} = \frac{1}{Y}\left(\left(\frac{dY}{dt}\right)^2 + \beta Y^2\frac{dX}{dt} + K_2XY^2\right). \tag{9.22}$$

We will confine our attention to the case for which $a = b = 2$, $K_1 = -0.05$, and $\alpha = \beta = 1$ with $K_2 = -0.05$. The initial values for X and Y will be taken as 1 and 2, respectively. Of course, this approach has generated a fourth-order system—we are now solving a pair of second-order ODEs requiring additional initial conditions; we can obtain values for the first derivatives directly from eqs. (9.19) and (9.20) by noting that for $t = 0$ the integrals disappear. Therefore the initial values for X and Y can be used to obtain $(dX/dt)_{t=0}$ and $(dY/dt)_{t=0}$. We discover that the solution for this system is very different from the case reviewed in the introduction where hereditary influences were neglected. It is evident from Figure 9.5 that *the phase-plane trajectory has an evolving shape and the populations are unstable.*

There are several important lessons to be drawn from this study of populations in conflict. Foremost, the addition of hereditary influences can render the populations aperiodic and unstable. Moreover, by making one of the K's positive, and one negative, we can also obtain extinction of one species and proliferation of the other. The reader may wish to investigate the possibilities by setting $K_1 = -0.05$ and $K_2 = +0.05$, which results in a trajectory that collapses along the y-axis with very small values for X occurring intermittently. The effects obtained with various combinations of K's will form the basis of a student exercise at the end of the chapter.

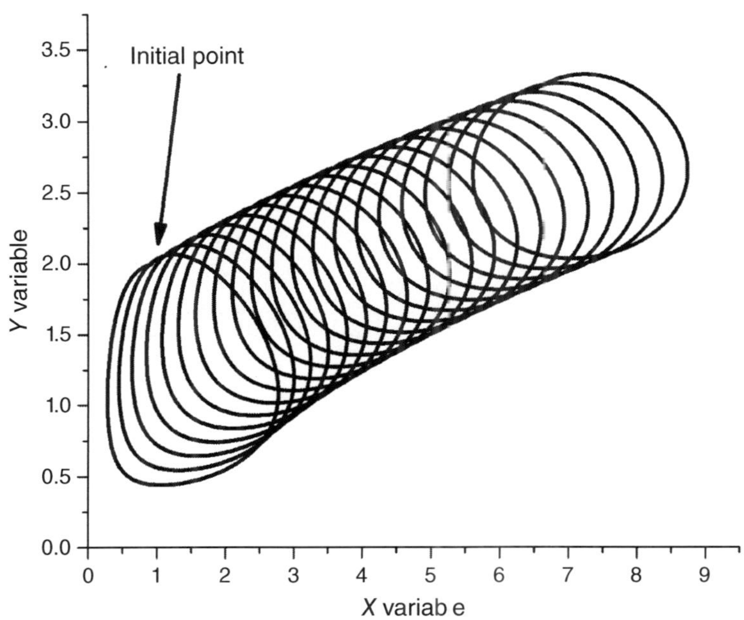

Figure 9.5 Phase-plane portrait for two populations in conflict with hereditary influences. Motion in this case is counter-clockwise from the initial point which is $(X,Y) = (1,2)$. The constants K_1 and K_2 are set to -0.05 for this example.

9.4 An Elementary Solution Strategy

In the previous section we converted our IDEs into second-order ODEs that we solved by methods already familiar to us. However, there is a cost associated with this approach: The differentiation produced a fourth-order system requiring additional initial conditions. We now want to look at an IDE example to illustrate how such problems might be solved when we retain the integral but employ a simple discretization. Consider the nonlinear IDE:

$$\frac{du}{dx} = -1 + \int_0^x u^2(\phi)d\phi, \tag{9.23}$$

for which $u(x = 0) = 0$ and $0 \leq x \leq 1$. This equation has been solved by Batiha et al. (2008) using the *variational iteration method* (VIM), so we will have a convenient comparison. We replace the derivative, du/dx, with the first-order forward difference:

$$\frac{du}{dx} \cong \frac{u(x + \Delta x) - u(x)}{\Delta x}, \tag{9.24}$$

and the integral is approximated by the summation of rectangles each of width, Δx (a Riemann sum). We choose $\Delta x = 0.001$ and proceed from $x = 0$ to $x = 1$. Let us compare our results with those provided by Batiha et al.

x	VIM, (Batiha et al., 2008)	Approximate Discretization
0.0000	0.000000	0.000
0.0938	−0.0937935	−0.094
0.2188	−0.2186091	−0.218
0.3125	−0.3117064	−0.311
0.4062	−0.4039385	−0.404
0.5000	−0.4948226	−0.495
0.6250	−0.6124315	−0.612
0.7188	−0.6969446	−0.697
0.8125	−0.7771007	−0.777
0.9062	−0.8519654	−0.852
1.0000	−0.9205578	−0.921

We can see from the tabulated results that this elementary discretization has produced quite acceptable agreement with the published solution for this IDE. We would do well to wonder how the results of such a simple procedure would compare with a case for which the *analytic solution* can be easily determined, resulting in a more definitive test. We will do this by looking at the linear IDE,

$$\frac{du}{dx} + 3u + 10\int_0^x u(\phi)d\phi = 3x\,\exp(-x), \text{with } x \geq 0 \text{ and } u(0) = 0. \tag{9.25}$$

We observed previously that a linear equation of this type can be solved readily through use of the Laplace transform, and we carry out this process once again. We begin by taking the Laplace transform of each term in the equation (noting that $u(x = 0) = 0$):

$$su(s) + 3u(s) + 10\frac{u(s)}{s} = \frac{3}{(s+1)^2}. \tag{9.26}$$

We isolate $u(s)$:

$$u(s) = \frac{3s}{(s+1)^2(s^2 + 3s + 10)}, \tag{9.27}$$

then by partial fraction expansion:

$$\frac{\mathbf{3s}}{(\mathbf{s}+\mathbf{1})^2\left(\mathbf{s}+\frac{3}{2}+\frac{\sqrt{31}}{2}\mathbf{i}\right)\left(\mathbf{s}+\frac{3}{2}-\frac{\sqrt{31}}{2}\mathbf{i}\right)} = \frac{\boldsymbol{A}}{(\mathbf{s}+\mathbf{1})^2} + \frac{\boldsymbol{B}}{\mathbf{s}+\mathbf{1}} + \frac{\boldsymbol{C}}{\mathbf{s}+\frac{3}{2}+\frac{\sqrt{31}}{2}\mathbf{i}} + \frac{\boldsymbol{D}}{\mathbf{s}+\frac{3}{2}-\frac{\sqrt{31}}{2}\mathbf{i}} \tag{9.28}$$

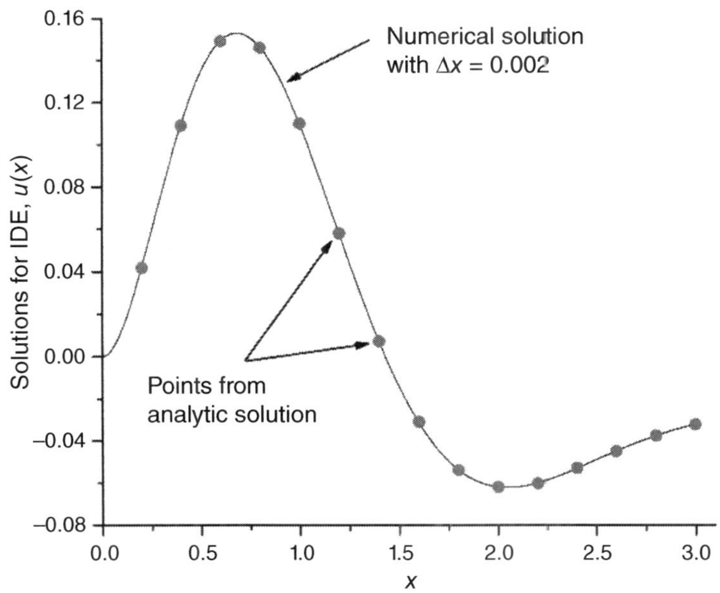

Figure 9.6 Comparison of the analytic solution (filled gray circles) of the integro-differential eq. (9.25) with the approximate solution obtained by discretization.

and we find that $A = -\frac{3}{8}$, $B = 0.4218$, and $C = -0.2109 + 0.0295i$ (D is the complex conjugate of C). *Please note our treatment of the repeated roots in the expansion (9.28).* Thus, we can now invert our transform:

$$u(x) = -\frac{3}{8}xe^{-x} + 0.4218e^{-x} - 0.4218e^{-\frac{3}{2}x}\cos\frac{\sqrt{31}}{2}x + 0.059e^{-\frac{3}{2}x}\sin\frac{\sqrt{31}}{2}x. \tag{9.29}$$

To provide a comparison we will employ the very same discretization scheme we used for the first (nonlinear) example in this section, using a step-size (Δx) of 0.002. Both solutions are provided in Figure 9.6. The analytic case is shown as the filled circles, and the approximate computed values are represented by the solid curve. The typical discrepancy between the two solutions is on the order of about 0.4%. For all but the most exacting applications this would be completely satisfactory.

9.4.1 A Practice Exercise with an Elementary IDE

Let's consider the following linear IDE which is a modified version of an example given by Wolfram™:

$$\frac{dy}{dt} = 1 + \sin(7t) + \int_0^t y(\tau)d\tau, \tag{9.30}$$

with $y(0) = -1$ and $0 \le t \le 3$. We want to solve this IDE two different ways: numerically using the elementary discretization described above and analytically using the Laplace transform. We *must remember to subtract the initial condition in the time domain for the analytic solution since* $L\left\{\frac{dy}{dt}\right\} = sy(s) - y(t=0)$. Note that an approximate numerical solution can be accomplished with a common tool like Excel™ using this sequence of steps repetitively:

```
sum=sum+y*dt
            ynew=dt*(1+SIN(a*tt)+sum)+y
                tt=tt+dt
                    y=ynew
```

Of course, **a** = 7 and the initial value for **y** is −1; the frequency of the *sine* wave is cause for concern with our crude discretization since changes in slope will be frequent on the time interval of interest. Accordingly, you may find that **dt** will need to be very small; alternatively, you may want to employ higher-order approximations recognizing that the revised scheme will not be self-starting. Assume we want to answer the following questions:

1) For what values of t will $y(t) = 1/8$?
2) At what time will $y(t) = 1$?

Answer for part 1): t = 1.2872, 1.6133, and 1.7917.
Answer for part 2): t = 2.8041

9.4.2 Extending the Solution Strategy to Higher-Order Equations

The elementary solution strategy described here can be extended to higher-order equations in some cases. For example, let us consider the IDE:

$$y'' = 3 + 2e^{-t} - \frac{1}{4}t + \int_0^t y(\tau)d\tau \tag{9.31}$$

for $0 \le t \le 2$ with $y(0) = 0$ and $y'(0) = 0$. Equation (9.31) meets the necessary conditions for solution by Laplace transform, so we will have a means for evaluating our discretized procedure:

$$s^2 y(s) = \frac{3}{s} + \frac{2}{s+1} - \frac{\frac{1}{4}}{s^2} + \frac{y(s)}{s}.$$

Therefore,

$$y(s) = \frac{5s^2 + \frac{11}{4}s - \frac{1}{4}}{s(s^4 + s^3 - s - 1)}, \tag{9.32}$$

and the denominator can be factored (we can use *cSolve* with the *TI-nspire*) to yield

$$s(s+1)(s-1)\left(s + \frac{1}{2} - \frac{\sqrt{3}i}{2}\right)\left(s + \frac{1}{2} + \frac{\sqrt{3}i}{2}\right).$$

We equate (9.32) with the partial fraction expansion:

$$= \frac{A}{s} + \frac{B}{s+1} + \frac{C}{s-1} + \frac{D}{s + \frac{1}{2} - \frac{\sqrt{3}i}{2}} + \frac{E}{s + \frac{1}{2} + \frac{\sqrt{3}i}{2}}.$$

You may want to verify that $A = 1/4$, $B = 1$, $C = 5/4$, and $D = -5/4 + 0.866i$; of course, E is merely the complex conjugate of D. We invert this expansion to get back to the time domain, and we will need the Euler relationship, $e^{(\alpha + i\beta)t} = e^{\alpha t}[\cos(\beta t) + i\sin(\beta t)]$, for the D and E terms. The solution for the IDE eq. (9.31) can now be written out:

$$y(t) = \frac{1}{4} + e^{-t} + 1.25e^{t} - 2.5e^{-\frac{1}{2}t}\cos\left(\frac{\sqrt{3}}{2}t\right) - 1.732e^{-\frac{1}{2}t}\sin\left(\frac{\sqrt{3}}{2}t\right) \tag{9.33}$$

and we note $y(0) = 0$ as it must. It is clear that eq. (9.33) is unstable as y will increase exponentially with time due to the third term on the right-hand side.

Our plan is to see if we can verify eq. (9.33) numerically using a very simple discretization; we will use a second-order central difference for the second derivative:

$$y'' \cong \frac{y_{i+1} - 2y_i + y_{i-1}}{(\Delta t)^2} \tag{9.34}$$

and a Riemann sum for the integral. We will isolate the *new* value of y (subscript $i + 1$) from eq. (9.34) and assume that the old value (y_{i-1}) is 0 initially since the system described by eq. (9.31) was at equilibrium up to $t = 0$. The BASIC code required to perform the computation is simply:

```
#COMPILE EXE
#DIM ALL
    REM *** solution of IDE y''=3+2exp(-t)-t/4+INTEGRAL(y(t)dt)
            GLOBAL tt,dt,sum,yold,y,ynew AS DOUBLE
FUNCTION PBMAIN
            yold=0:y=0:dt=0.001:tt=dt:sum=0
            OPEN "c:SECONDordIDEv.dat" FOR OUTPUT AS #1
```

```
100 REM *** continue
            sum=sum+y*dt
                ynew=dt^2*(3+2*EXP(-tt)-1/4*tt+sum)+2*y-yold
                    PRINT tt,ynew
                    WRITE#1,tt,ynew
                        tt=tt+dt
                            yold=y
                            y=ynew
                        IF tt>2.0001 THEN 200 ELSE 100
200 REM *** continue
            SLEEP 6000
            CLOSE:END
```

If we use a small Δt (and you will note above that $dt = 0.001$) with sufficient numerical precision, we find that our procedure will work as demonstrated by this abbreviated table:

Time, t	$y_{analytic}$	$y_{numerical}$
0.1	0.0246	0.0249
0.4	0.3784	0.3792
0.6	0.8317	0.8329
0.8	1.4503	1.4518
1.0	2.2331	2.2349
1.2	3.1865	3.1887
1.4	4.3247	4.3274
1.8	7.2611	7.2651

The agreement is good, with a typical discrepancy of about 0.07%, and if we reduce Δt to 0.0005, the error at $t = 1.8$ is only 0.03%. Since eq. (9.31) can be solved analytically with just a little effort, one might wonder about the motivation for obtaining an approximate numerical solution. Let us suppose that we were interested in exploring the effects of changing the contribution of the integral term in the IDE; in particular, suppose we needed to identify the prefactor that would yield $y = 0$ for $t = 2$. Or, alternatively, suppose we wanted to identify the change required to produce $y'(t = 2) = 0$. Although we *could* obtain the answers sought through use of Laplace transform, that would require a significant time commitment. If we use the discretized approach we can answer such questions much more rapidly and to illustrate we will present some results graphically in Figure 9.7.

The results illustrated by Figure 9.7 *could* have been obtained by Laplace transform just as we demonstrated initially. However, using this simple discretization we can quickly evaluate more complicated versions of eq. (9.31) with some assurance of success; for example, suppose that a term like y^2 or $y^2\exp(-t)$ appeared on the right-hand side. Our elementary discretization would allow us to explore those cases with relative ease.

Recommended exercise: Carry out the procedure described above and identify how the integral term would need to be changed to produce y(t = 2) = 7. You should be able to show the required modification of the integral term is: $-0.8149\int_0^t t(\tau)d\tau$. *Then, add a nonlinear term to the RHS of the equation:*

$$y'' = 3 + 2e^{-t} - 5.75y^2 - \frac{1}{4}t + \int_0^t y(\tau)d\tau$$

with $y(0) = 0$, $y'(0) = 0$, *and* $0 \le y \le 2$. *What is the maximum value of y(t), and when does it occur? You should find: 1.483 at* $t = 1.065$. *Finally, if you would like a bit more practice with the Laplace transform approach, change eq. (9.31) to:*

$$y'' = 3 + 2e^{-t} - \frac{1}{4}t - 2.82\int_0^t y(\tau)d\tau$$

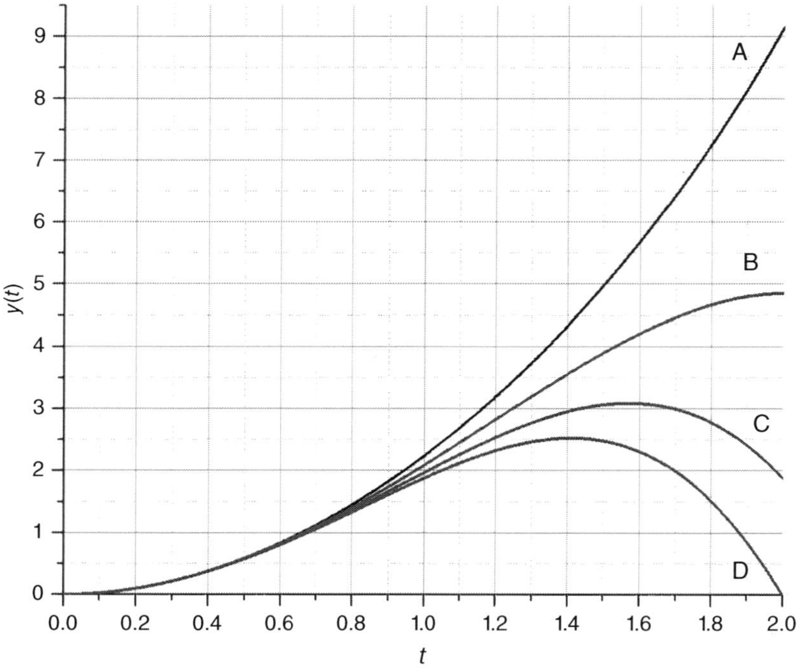

Figure 9.7 Numerical results for IDE eq. (9.31) with different contributions from the integral term obtained by changing the constant (prefactor). The curve labeled "A" is the original equation with $+\int_0^t y(\tau)d\tau$, for "B" the prefactor was −2.82, for "C" −6, and for "D" −8.372. Thus, in order to obtain $y(t = 2) = 0$, the integral term had to be changed to $-8.372\int_0^t y(\tau)d\tau$.

and see if you can confirm that $y'(t = 2) = 0$. You should find that

$$y(t) = -0.0887 - 1.0989e^{-t} + 1.6739e^{-1.4128t} - 0.486e^{0.7064t}\cos(1.2235t) + 1.315e^{0.7064t}\sin(1.2235t).$$

9.5 VIM: The Variational Iteration Method

We saw some limitations of the rough discretization introduced in Section 9.4—particularly in examples like the *Practice Exercise* where the IDE exhibited pronounced changes in curvature over the interval of interest. This suggests that a better solution strategy may be needed for more "difficult" equations, and this is why we cited some results obtained using *VIM*. Pioneering work by He (1999 and 2000) led to rapid exploitation of the technique, and the efforts of He (2007) and He and Wu (2007), Batiha et al. (2008), and *many others* indicated that VIM might be especially well suited to the solution of IDEs. It is, therefore, appropriate that we deviate from our course a little in order to provide a discussion of VIM. So that we better understand the VIM, we must spend a few moments discussing Lagrange's multiplier method, which we saw for the first time in Chapter 2.

Let us suppose that we have two variables that are to be combined such that:

$$x_1^{\,2} + 2x_2^{\,3} = 30. \tag{9.35}$$

We are seeking values for the two variables that cause the product (x_1x_2) to be a maximum, subject to the constraint, eq. (9.35). These values can be identified through use of the Lagrange multiplier, λ. We write

$$F = x_1x_2 + \lambda\left(x_1^{\,2} + 2x_2^{\,3} - 30\right). \tag{9.36}$$

This expression is differentiated with respect to each of the *three* variables, and the partial derivatives are set equal to zero:

$$\frac{\partial F}{\partial x_1} = x_2 + 2\lambda x_1 = 0 \tag{9.37}$$

$$\frac{\partial F}{\partial x_2} = x_1 + 6\lambda x_2^{\,2} = 0 \tag{9.38}$$

$$\frac{\partial F}{\partial \lambda} = x_1^{\,2} + 2x_2^{\,3} - 30 = 0. \tag{9.39}$$

These three equations are solved, and the reader may wish to show that $x_1 = 4.24264$, $x_2 = 1.81712$, and $\lambda = -0.21415$; accordingly, the maximum product subject to the constraint is $x_1 x_2 = 7.70939$. Next, we examine how the Lagrange multiplier technique is modified for application to an *ODE*. Suppose we have the ODE,

$$\frac{du}{dx} + u^2 = 2, \tag{9.40}$$

with $u(0) = 1/4$ and $0 \le x \le 1$. In this case the ODE can be solved yielding:

$$\frac{\sqrt{2} + u}{\sqrt{2} - u} = \exp(2\sqrt{2}(x + 0.126326)). \tag{9.41}$$

We will calculate a few values to have a comparison readily available.

x	u(x)
0	0.25
0.13495	0.50
0.29129	0.75
0.496899	1.00
0.85885	1.25

Now we are set; let our ODE, eq. (9.40), be rewritten: $u_n{}' + u_n{}^2 - 2$. The successive iterations are determined from:

$$u_{n+1} = u_n + \int_0^x \lambda \left(u_n' + u_n{}^2 - 2\right) ds. \tag{9.42}$$

The Lagrange multiplier, λ, can be identified from the stationary condition(s), and for linear ODEs *λ can be determined exactly*. For the nonlinear eq. (9.40), we will *presume* $\lambda = -1$ and proceed. We take $u_0 = 1/4$ (satisfying the initial condition) and find by integration, $u_1 = \frac{1}{4} + \frac{31}{16}x$. Therefore,

$$u_2 = \frac{1}{4} + \frac{31}{16}x - \int_0^x \left(\frac{31}{32}s + \frac{961}{256}s^2\right) ds, \tag{9.43}$$

which yields

$$u_2 = \frac{1}{4} + \frac{31}{16}x - \frac{31}{64}x^2 - \frac{961}{768}x^3. \tag{9.44}$$

If we let $x = 0.29129$, we find $u \approx 0.74235$; setting $x = 0.4969$ produces $u \approx 0.9396$. These two results have corresponding errors of about 1% and 6%. For many applications this level of error would be tolerable, but additional iterations will be necessary if we require improved accuracy.

Let us consider one more example to ensure our familiarity with this very powerful technique. Suppose we have the ODE,

$$\frac{du}{dx} = -\frac{x^3}{(u+1)^2}, \quad \text{with } u(x=0) = 1. \tag{9.45}$$

The analytic solution is known,

$$3x^4 + 4(u+1)^3 = 32, \ \text{ or alternatively }, u = \left[\frac{32 - 3x^4}{4}\right]^{1/3} - 1. \tag{9.46}$$

Our succession of estimates is obtained from

$$u_{n+1} = u_n - \int_0^x \left(u_n' + \frac{s^3}{(u_n+1)^2}\right) ds. \tag{9.47}$$

If we start with $u_0 = 1$, we can very easily find: $u_1 = 1 - \frac{x^4}{16}$. Therefore,

$$u_2 = 1 - \frac{x^4}{16} - \int_0^x \left[-\frac{s^3}{4} + \frac{s^3}{\left(2 - \frac{1}{16}s^4\right)^2}\right] ds, \tag{9.48}$$

and

$$u_2 = 3 + \frac{64}{x^4 - 32}. \tag{9.49}$$

We can compare these estimates with the analytic solution in the table below:

$x =$	0.50	1.00	1.25	1.50	1.75
$u(x)$ analytic	0.996086	0.935438	0.834018	0.613829	−0.011526
$u_1(x)$	0.996094	0.93750	0.847412	0.683594	0.413818
$u_2(x)$	0.996086	0.935484	0.834809	0.624130	0.170782

Except for higher values of x, the second iteration does a remarkable job of representing the solution of this ODE.

We want to facilitate the application of VIM more broadly, and this will require that we make use of some elements from the *calculus of variations*. Because our goal is to use VIM as a tool we will pursue the approach described by Ji-Huan He in his interview in *ScienceWatch*, July 2008: We want to present the technique in such a way that "*....anyone who knows nothing of variational theory in mathematics can apply the method......*" For the reader who is curious about other applications of the calculus of variations, an introduction is provided at the end of this text in Chapter 11. In addition, Robert Weinstock's (1974) book is very useful, and there are very good on-line primers available including James Nearing's (University of Miami) work. There is also a nice appendix in Kenneth Huebner's (1975) book that was specifically written for engineers.

We generalize this part of our discussion by rewriting the ODE we wish to consider as:

$$Lu + Nu = g(x), \tag{9.50}$$

where L and N are linear and nonlinear operators, respectively. Now we construct a *correction functional*:

$$u_{n+1}(x) = u_n(x) + \int_0^x \lambda[Lu_n(\phi) + N\tilde{u}_n(\phi) - g(\phi)]d\phi. \tag{9.51}$$

The nonlinear terms are to be removed by substituting $\tilde{u}$, a restricted variation such that $\delta(\tilde{u}) = 0$, and the Lagrange multiplier, λ, is identified using variational theory. To find the optimal value for λ we write:

$$\delta u_{n+1}(x) = \delta u_n(x) + \delta \int_0^x \lambda[Lu_n(\phi) + N\tilde{u}_n(\phi) - g(\phi)]d\phi, \tag{9.52}$$

and make the correction functional *stationary*. We can illustrate the process with an example (the Lienard equation) provided by Matinfar et al. (2008). Suppose we have the nonlinear ODE,

$$\frac{d^2u}{dt^2} + au + bu^3 + cu^5 = 0, \tag{9.53}$$

where a, b, and c are real constants. We now write the correction functional as before:

$$u_{n+1}(t) = u_n(t) + \int_0^t \lambda\left[u_n''(\phi) + au_n(\phi) + b\tilde{u}_n^{\,3}(\phi) + c\tilde{u}_n^{\,5}(\phi)\right]d\phi. \tag{9.54}$$

The restricted variations are used on the nonlinear terms, so they can be eliminated resulting in:

$$\delta u_{n+1}(t) = \delta u_n(t) + \delta \int_0^t \lambda\left[u_n''(\phi) + au_n(\phi)\right]d\phi. \tag{9.55}$$

The integral is handled by integration by parts which yields three stationary conditions for identification of the optimal λ's:

$$\lambda = \phi - t \tag{9.56}$$

$$\lambda = \frac{1}{\sqrt{-a}} \sin h(\sqrt{-a}(\phi - t)) \tag{9.57}$$

and

$$\lambda = \sin(\phi - t). \tag{9.58}$$

Any one of the three choices will work—they will only differ by the speed of convergence of the correction formula. If we choose the first of the trio of λ's, our iterative process will be based upon:

$$u_{n+1}(t) = u_n(t) + \int_0^t (\phi - t)\left[u_n'' + au_n + bu_n^{\,3} + cu_n^{\,5}\right] d\phi. \tag{9.59}$$

An initial approximation for $u_{n=0}$ is all that is required to get started with the iterative process. Matinfar et al. used MATLAB for their computations with *a,b,* and *c* set equal to −1, 4, and − 3, respectively. For the initial (trial) function they used

$$u_0 = u(t=0) + \left(\frac{du}{dt}\right)_{t=0}(t),$$

and their results were in excellent accord with the closed-form solution. It should be noted that there are many examples of the application of He's method in the recent literature; just a few of the differential equations that have been solved this way include the Kawahara equation by Ganji et al. (2007), the Laplace equation by Jassim (2012), the nonlinear oscillator by He and Wu (2007), and Sturm–Liouville equations by Altintan and Ugur (2012). VIM has even been applied to model contamination (the spread of pollution) in a system of inter-connected lakes—see Merdan (2009).

Of course, our real objective in this chapter is the solution of IDEs. The procedure is wholly analogous to our treatment of ODEs above, and to begin, let us take

$$\frac{du}{dx} = f(x) + \int_0^x g(\phi, u(\phi), u'(\phi))d\phi, \tag{9.60}$$

where $f(x)$ is the source term. We build a *correction functional* just as before:

$$u_{n+1}(x) = u_n(x) + \int_0^x \lambda\left[u_n'(s) - f(s) - \int_0^s g(\phi, \tilde{u}(\phi), \tilde{u}'(\phi))d\phi\right]ds. \tag{9.61}$$

Our task now is to choose the optimal Lagrange multiplier, λ. Remember that the *tilde* in eq. (9.61) denotes a restricted variation; therefore

$$\delta u_{n+1}(x) = \delta u_n(x) + \delta\int_0^x \lambda(s)\left[(u_n)_s(s) - f(s) - \int_0^s g(\phi, u(\phi), u'(\phi))d\phi\right]ds. \tag{9.62}$$

The integration is performed and the stationary conditions are used to identify λ, which in this case is −1. Therefore the iteration formula is

$$u_{n+1}(x) = u_n(x) - \int_0^x \left[u_n'(s) - f(s) - \int_0^s g(\phi, u(\phi), u'(\phi))d\phi\right]ds. \tag{9.63}$$

We will now apply this technique to a nonlinear problem considered by Batiha et al. (2008):

$$\frac{du}{dx} = 1 + \int_0^x u(\phi)\frac{du}{d\phi}d\phi, \tag{9.64}$$

given that $(0 \le x \le 1)$ and $u(x=0) = 0$. Remember that we select a trial function such that the boundary conditions are satisfied, and in this case an obvious function is $u_0(x) = x$. Once again $\lambda = -1$ so the iteration formula is

$$u_{n+1}(x) = u_n(x) - \int_0^x \left[\frac{du_s}{ds} - 1 - \int_0^s u(\phi)\,\frac{du}{d\phi}d\phi\right]ds. \tag{9.65}$$

Batiha et al. reported an incorrect sequence of results in the original paper, and the errors have been corrected here:

$$u_1(x) = x + \frac{1}{6}x^3 \tag{9.66}$$

and

$$u_2 = x + \frac{1}{6}x^3 + \frac{1}{30}x^5 + \frac{1}{504}x^7. \tag{9.67}$$

We will also employ discretization with eq. (9.64) to provide a comparison.

x	$u(x)$ Discretized	$u(x)$ VIM approx
0	0.0000	0.0000
0.0938	0.0940	0.09394
0.2188	0.2207	0.22056
0.3125	0.3178	0.31769
0.4062	0.4178	0.41774
0.5000	0.5220	0.52189
0.6250	0.6693	0.66894
0.7188	0.7879	0.78729
0.8125	0.9156	0.91416
0.9062	1.0548	1.05159
1.0000	1.2086	1.20198

In this example, Batiha et al. (2008) used three VIM iterations, but the values reported in their original table are incorrect. The corrected values for the second iteration are shown in the table above, and you will immediately note that these values are in remarkable agreement with the results computed using discretization.

We now look at an example with a higher-order derivative provided by He and Wu (2007), a fourth-order IDE:

$$\frac{d^4y}{dx^4} = x(1+e^x) + 3e^x + y(x) - \int_0^x y(s)ds. \tag{9.68}$$

We are given that $y(0) = 1$, $y(1) = 1 + e = 3.71828$, $y''(0) = 2$, and $y''(1) = 3e = 8.15485$. Of course we might think about discretizing this equation and trying to solve it just as we did in some of the previous examples. However, the combination of the fourth-derivative and the split boundary conditions renders this approach quite unappealing. We will be better served by applying VIM to this IDE.

For this linear case the exact solution is known, $y = 1 + x\exp(x)$, facilitating comparison. He and Wu suggest a trial function,

$$y_0(x) = 1 + \exp(x)[a + bx + cx^2 + dx^3], \tag{9.69}$$

and clearly $a = 0$ so that $y(0) = 1$. One of the most important contributions of the He and Wu (2007) paper is that appropriate iteration formulae are given for a wide variety of problems—including this one. For example, given the equation,

$$\frac{d^2u}{dx^2} + f(u, u', u'') = 0, \tag{9.70}$$

the first-order approximation is given by

$$u_1 = u_0(x) + \int_0^x (s-x)f(u, u', u'')ds. \tag{9.71}$$

We can simplify matters for our example by, letting

$$f(x) = \int_0^x \int_0^\xi \left[s(1+e^s) + 3e^s + y(s) - \int_0^s y(s)ds \right] d\xi dx. \tag{9.72}$$

The iteration formula can now be written more compactly and applied to obtain y_1:

$$y_1(x) = y_0(x) + \int_0^x (s-x)f(s)ds \tag{9.73}$$

He and Wu found that

$$y_1 = (0.0803x^3 - 0.2113x^2 + 2.302x - 6.9014)e^x + 0.0164x^4 + 0.7162x^3 + 2.3418x^2 + 5.5516x + 7.9014. \tag{9.74}$$

Please be advised that the first-order approximation as given by He and Wu contains an error (their equation number 37); the coefficient provided for x^2 is *not* 10.3418 (it should be 2.3418 as indicated immediately above). In this case we can easily compare the analytic solution with the first-order approximation; some values are provided in the table below and the agreement is excellent—generally within about 0.5%.

x	$1+x\exp(x)$	Approximation, $y_1(x)$
0.0	1.00000	1.0000
0.2	1.24428	1.2390
0.4	1.59673	1.5891
0.6	2.09327	2.0867
0.8	2.78043	2.7770
1.0	3.71828	3.7183

VIM provides us with an extraordinarily powerful tool for the solution of *both* ODEs and IDEs.

9.6 Integro-Differential Equations and the Spread of Infectious Disease

IDEs figure prominently in deterministic models of epidemics. Let us begin by considering a simple model for a population consisting of two types of individuals: those who are susceptible to infection (S) and those already infected (I). We assume the disease is not fatal so there is no death rate, and we exclude birth during the time period of interest for simplicity. An elementary model for this situation might consist of two ODEs:

$$\frac{dS}{dt} = -\beta SI \tag{9.75}$$

and

$$\frac{dI}{dt} = +\beta SI. \tag{9.76}$$

The meaning of this model construct is that the number of susceptibles decreases as a result of interaction with infected individuals, and the number of infectives increases due to interaction. For this example we will take the rate constant, β, to be positive. Since there is neither birth nor death, the total population size is constant, i.e., $S+I=C_1$. Therefore,

$$\frac{dI}{dt} = \beta I(C_1 - I). \tag{9.77}$$

We will take the total population to be 100 individuals, and the initial number of infected to be 3; we will use different values for the rate constant to carry out a numerical exploration of eq. (9.77). The results we obtain are predictable: No matter what positive value is selected for the parameter, β, the infection will spread to the entire population, and the total number of infectives will rapidly approach 100. This behavior is revealed by Figure 9.8.

We saw that the model given by eq. (9.77) did not produce very realistic results; for one thing the infection spread throughout the population because the interaction between susceptibles and infectives was mandated by the differential equation. It completely failed to account for one of the conditions of modern life that has significantly affected the transmission of disease: ease of travel. The very simple model we described above does not take into account the movement of infected individuals—and the intercontinental spread of HIV and COVID-19 revealed how very important this factor is. So, what might we do to incorporate the movement of infected individuals into the modeling?

"Diffusion" approximations have been used to try to model the spread of epidemics through what is often referred to as the Fisher–Kolmogorov–Petrovskii–Piskunov (FKPP) equation:

$$\frac{\partial I}{\partial t} = D\frac{\partial^2 I}{\partial x^2} + KI(1 - I). \tag{9.78}$$

We will think of I in this case as the fraction of infected individuals. D is a kind of diffusion coefficient and K is a rate constant. The FKPP model is tractable, but it displays an unphysical characteristic that we will now demonstrate. Suppose

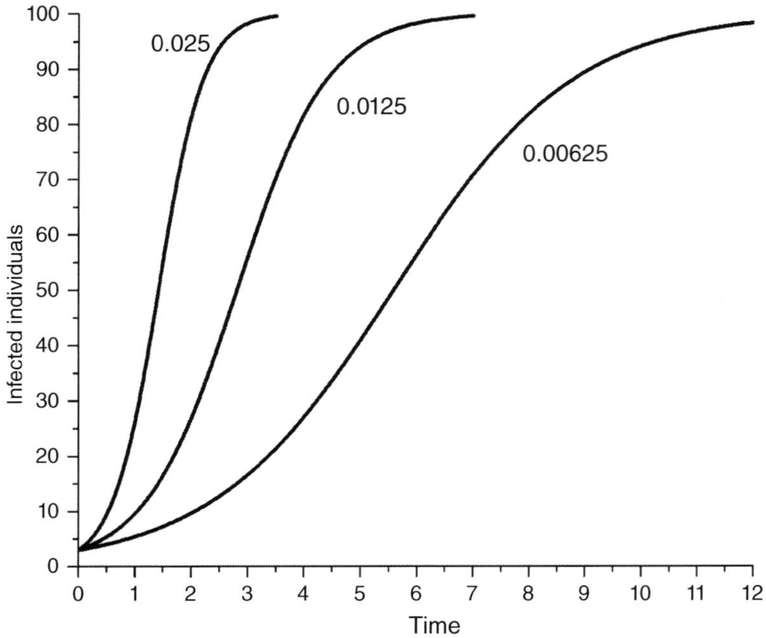

Figure 9.8 Increase in the number of infected individuals for β's of 0.00625, 0.125, and 0.025. As β increases, we see an increasingly rapid approach to a population that is 100% infected.

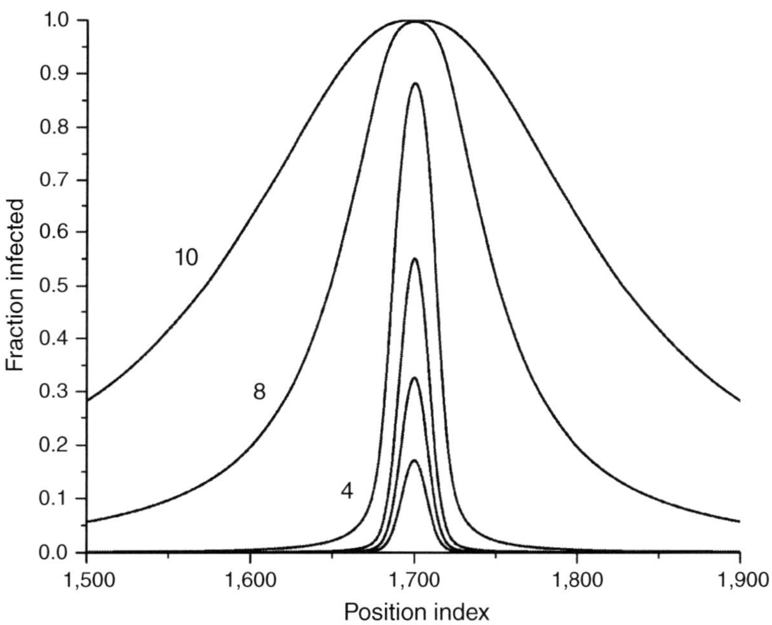

Figure 9.9 Solutions of eq. (9.78) for a fixed time using values for K ranging from 0.1 to 10 (specifically 0.1, 1, 2, 4, 8, and 10).

we have an initial cluster (spike) of infectives at a particular spatial location. Our plan is to solve eq. (9.78) employing different values for K to explore the effect of the rate constant upon the solution. All other parameters are fixed.

The data in Figure 9.9 reveal the failure of the FKPP model as applied to epidemics; the speed of propagation increases as the rate constant, K, increases. In fact, the diffusion model suggests that the rate of spread of disease will become infinite if K is allowed to increase without bound. Fedotov (2001) points out that this behavior is unphysical and that the origin of this problem is due to the parabolic scaling associated with the FKPP equation. He notes that the correct scaling for a propagating front (in our case the interface between infectives and susceptibles) must be *hyperbolic*. Thus, a different modeling approach is required.

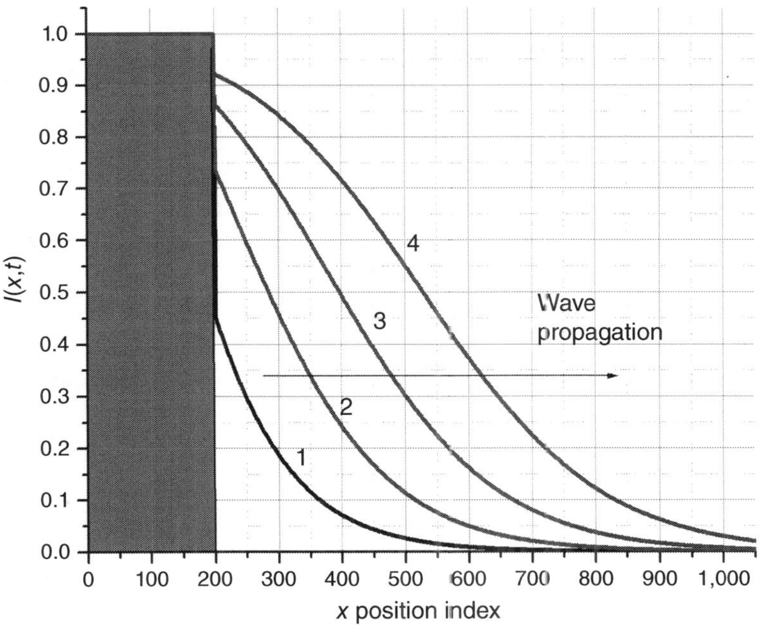

Figure 9.10 Emergence of the traveling wave from the initial block of infectives. The wave continues to move to the right with increasing time. This model is far more realistic than the one posed by the FKPP equation.

We will take $M(x)$ to be the rate at which infected individuals leave position x and travel to a new location in the spatial domain. We will also assume that M is a positive constant. Our model now has the form,

$$\frac{\partial I}{\partial t} = \beta I(C_1 - I) - MI + M \int K(x, y)I(y, t)dy. \tag{9.79}$$

$K(x,y)$ is the kernel; it is a density function for the fraction of infected individuals who *leave position y destined for x*. This density function has the characteristics

$$K(x, y) \geq 0 \text{ and } \int K(x, y)dx = 1. \tag{9.80}$$

Medlock and Kot (2003) assumed that the kernel was of the convolution type,

$$K(x, y) = k(x - y) = k(u) = \frac{1}{2\alpha} \exp\left(-\frac{|u|}{\alpha} \right), \tag{9.81}$$

where u is the separation between the x and y positions. They set $M = 1$, $\beta = 1$, and $\alpha = 0.7698$; I, the fraction of infected individuals was taken to be 1 in the center of the domain. The problem thus posed by eq. (9.79) is formidable. The repeated evaluation of the integral is computationally expensive, so Medlock and Kot used a Runge–Kutta scheme for the time derivative and then the fast Fourier transform (FFT) for the integral. The solution of this problem takes the form of a traveling wave as illustrated in the accompanying Figure 9.10. At small t's the traveling wave emerges from the original block of infectives and then moves to the right (positive x-direction) with increasing time. The results shown in this figure were obtained using an Euler scheme on the time derivative (with a small time step) and the trapezoidal rule for the integral. This numerical procedure is inefficient computationally, but very simple to apply and execute.

The behavior we see in Figure 9.10 is very different from the FKPP model we discussed previously. The transmission of disease with the movement of infected individuals is much more wave-like; the boundary between infectives and susceptibles travels in the x-direction, but the characteristics of the wave are essentially unchanged as it emerges from the initial block of infectives.

9.7 Examples Drawn from Population Balances

We observed at the beginning of the chapter that many important industrial processes are multiphase, often consisting of a continuous fluid phase carrying countable entities. These dispersed objects may be particles, droplets, bubbles, cellular entities, or possibly some combination of several of these. Furthermore, these countable entities are commonly experiencing birth, death, growth (aggregation), breakage (or comminution), interphase transport, etc. These processes may result in changes to extensive variables such as number, mass, or volume, and such changes are often crucial with respect to process performance and control. An illustration of such a process is shown below; here, very small (colloidal) clay particles are aggregating in a concentric-cylinder Couette device to form very large flocs (agglomerates). There is particle growth in this system, but little or no breakage due to the highly ordered nature of the hydrodynamic environment. One of the interesting features of this particular system is that elements of the fluid phase become trapped in the interstices of the aggregate such that the apparent volume of the dispersed phase *increases significantly* during aggregation. Thus neither volume nor mass would be conserved in a system- similar to that shown in Figure 9.11.

One of the formidable challenges offered by problems of this type is the incredible range seen in property values (such as entity size). When colloidal clay particles are flocculated in a more typical environment (inhomogeneous turbulence), we find entity sizes ranging all the way from about 1 μm to several mm (the difference, therefore, is three or even four orders of magnitude). To compound the problem, the number densities for very small particles can be 5–10 orders of magnitude larger than those for very large aggregates formed in the coagulation. This is exactly the situation illustrated by the comparison shown in Figure 9.12.

It will be useful for us to preface our continuing discussion of the application of population balances to multiphase processes by spending a little time considering entity *birth* and *death*. Often when we have a population of individuals (countable entities) the appearance of new individuals (birth) may occur. In biological systems this might be by reproduction, cell division, etc. In crystallization, new entities—though *very small ones*—appear by nucleation. But at a finite specified size, crystals are actually born by the breakage of larger parents. Of course at any instant in time, some existing individuals may disappear (death). Generally, both birth and death may depend in some manner upon the size of the existing population—large numbers may lead to many births and deaths. The reader may find Chapter 8 in Bailey (1964) to be a valuable introduction to this topic, should he/she wish to know more. Let us suppose that the probability that a population of individuals will total n at time, t, is $p_n(t)$. And we will assume that the probability that a given entity will produce a new one in the time interval, Δt, is $\lambda \Delta t$. Now, how could we arrive at a population of n entities at time $t + \Delta t$ taking only birth processes into account? We could come from the $n - 1$ state by a birth, or we could simply remain at the n-state if no births occurred over the time interval. Let us write this down:

$$p_n(t + \Delta t) = p_{n-1}(t)\lambda(n - 1)\Delta t + p_n(t)(1 - \lambda n \Delta t). \tag{9.82}$$

We subtract $p_n(t)$ and divide by Δt, taking the limit as $\Delta t \to 0$; the result is a *differential–difference equation*,

$$\frac{dp_n}{dt} = \lambda(n - 1)p_{n-1} - \lambda n p_n. \tag{9.83}$$

Figure 9.11 Floc formation in Couette device in which the outer cylinder is rotating (inner cylinder at rest). The large aggregates seen in this image consist of clay (kaolin) particles, polymer flocculant, and water trapped in the interstitial spaces. Measurements of the density of these aggregates reveal that the overwhelmingly dominant component is the trapped water.

Figure 9.12 (a and b). Batch flocculation of colloidal kaolin in a stirred 1-L reactor. The disparity in entity size is revealed by this comparison; we see mainly *primary* clay particles in the dispersion shown on the left, and mostly large aggregates on the right. The number density for the small particles on the left is on the order of 1×10^7 per cm^3, while for the case on the right one might find only 1–10 large aggregates per cm^3.

Since we start with a non-zero population, say n_0, at $t = 0$, the initial condition for eq. (9.83) is $p_{n_0}(t = 0) = 1$. This gives us a set of equations that we can solve in succession, beginning with

$$\frac{dp_{n_0}}{dt} = -\lambda n_0 p_{n_0}. \tag{9.84}$$

The model described above is often referred to as the Yule–Furry birth process. We say that such a birth process is homogeneous because the transition probability (λ) is constant—it does not depend upon time. We might ask ourselves how birth processes are changed if the rate of birth depends upon the population to the second power; i.e., when the birth rate per individual depends upon population size (we are requiring interaction between individuals). Under these circumstances the total population grows according to

$$\frac{dn}{dt} = \lambda n^2; \tag{9.85}$$

resulting in

$$n = \frac{1}{\frac{1}{n_0} - \lambda t}. \tag{9.86}$$

The consequence of this model is divergent (explosive) growth where $n \to \infty$ as $\lambda t \to 1/n_0$. Of course, neither the Yule–Furry process nor the explosive growth scenario accounts for the disappearance of individuals (death). Suppose we now add the assumption that individuals die at a rate that is proportional to their number—let the probability that an individual will die during a time interval, Δt, be $\mu\Delta t$. Once again we think about how we can get to the n-state (that is, where we have n-individuals). This could occur due to a birth (coming from $n - 1$), or from a death (coming from $n + 1$), or by having *no* births or deaths over the time interval, Δt (and thus remaining at n). Just as before, this model leads to a differential–difference equation,

$$\frac{dp_n}{dt} = \lambda(n-1)p_{n-1} - (\lambda + \mu)np_n + \mu(n+1)p_{n+1}. \tag{9.87}$$

The initial condition is that $n = n_0$ at $t = 0$ such that $p_{n_0}(t = 0) = 1$. It is worth noting that if the death rate exceeds the birth rate in the combined model, extinction is guaranteed. Among other possibilities in population problems are immigration and emigration. In the case of the latter, the effects of exiting individuals can be handled by simply adjusting the death rate upward. For immigration, the simplest possibility is that the rate of influx is *independent* of the present population, $n(t)$, and therefore the likelihood of a new individual appearing over the time interval, Δt, will be written: $\nu\Delta t$. Thus the probability that the total number of individuals will *increase* by one over Δt is

$$(\lambda n(t) + \nu)\Delta t. \tag{9.88}$$

Once again we consider how a population can arrive at the n-state over a time interval, Δt, now taking into account birth, death, and immigration: Of course we can grow from $n-1$ by birth or immigration, we could remain at n, or we could decline in number from $n+1$. And, just as we saw previously, the result is a (set of) differential–difference equation(s); we should point out that Ji-Huan He has observed that the VIM—as described previously in this chapter—can also be used to advantage for this problem type.

In the preceding discussion of birth and death processes our objective was the determination of probability that a population would total n individuals. But in the process industries our concern is more likely to be centered upon the distribution of entity volume, size, age, etc., and how those distributions evolve with time in different physicochemical environments. Thus we seek a framework that we might employ to model such phenomena in view of their enormous practical importance. The reader with interest in this area is urged to consult D. Ramkrishna's book, *Population Balances: Theory and Applications to Particulate Systems in Engineering*. We will begin with the simplest possible one-dimensional problem type, and we will adopt Ramkrishna's notation.

Suppose we have an initial population of particles that are distributed homogeneously in space. Furthermore, let us assume that we have *only particle growth* occurring and that the growth rate at a given size, x, is $\dot{X}(x,t)$. Our focus is the particular size range, $a \le x \le b$; if the number density function, $f_1(x,t)$, changes in this size interval, then it will either be due to growth into (a,b) from below, or growth out of (a,b) at the upper end. We can write a description for the dynamic behavior of this growth process:

$$\frac{d}{dt}\int_a^b f_1(x,t)dx = \dot{X}(a,t)f_1(a,t) - \dot{X}(b,t)f_1(b,t). \tag{9.89}$$

This dynamic balance is equivalent to

$$\frac{\partial f_1(x,t)}{\partial t} + \frac{\partial}{\partial x}[\dot{X}(x,t)f_1(x,t)] = 0. \tag{9.90}$$

We will take the number density function to be (initially):

$$f_1 = x\exp(-x), \tag{9.91}$$

and we will assume that the growth rate is described by

$$\dot{X} = \frac{C_1}{C_2 + x^2}. \tag{9.92}$$

Of course, the latter eq. (9.92) means that the growth rate is strongly damped by large particle size. We will make a sequence of calculations to see how the number density function will evolve with time under these conditions.

The data provided in Figure 9.13 show how the growth rate limitation at larger entity size is constraining the movement of the number density with respect to larger x. The population is becoming increasingly concentrated at intermediate sizes. If we lessen the impact of the size-limited growth rate (let $\dot{X} = \frac{C_1}{C_2+x}$, for example) we obtain more rapid movement of the number density function toward larger sizes; this is illustrated by Figure 9.14.

Of course, the model we have been considering is unrealistic in that it only describes entity growth. In many systems breakage is a real possibility, and this leads us to re-write eq. (9.90) with the addition of a combined birth and death term to account for the net generation of particles in the size range of interest:

$$\frac{\partial f_1(x,t)}{\partial t} + \frac{\partial}{\partial x}[\dot{X}(x,t)f_1(x,t)] = h(x,t). \tag{9.93}$$

Let us spend a moment thinking about particle breakage. If the breakage rate is dependent upon hydrodynamic conditions, then there may exist a size threshold below which no breakage occurs. For example, in turbulent flows we can use the dissipation rate per unit mass, ε, to assess the strain rate at different eddy scales (sizes). We usually argue that a disruptive eddy must have a scale comparable to the particle or entity size (much larger eddies simply entrain the particle, and very small eddies do not affect it at all). Naturally, larger eddies carry more energy, but they occur at a fixed location less frequently. One way to establish a breakage condition is to propose a balance between disruptive hydrodynamic energy and the restoring force (this could be surface tension, for example, in the case of a suspended droplet). It may now be apparent to you how difficult the characterization of $h(x,t)$ really is: We will need a breakage criterion that takes into account the energetics, we will need a rate at which these energetic events occur, and we will need to know the number and sizes of the

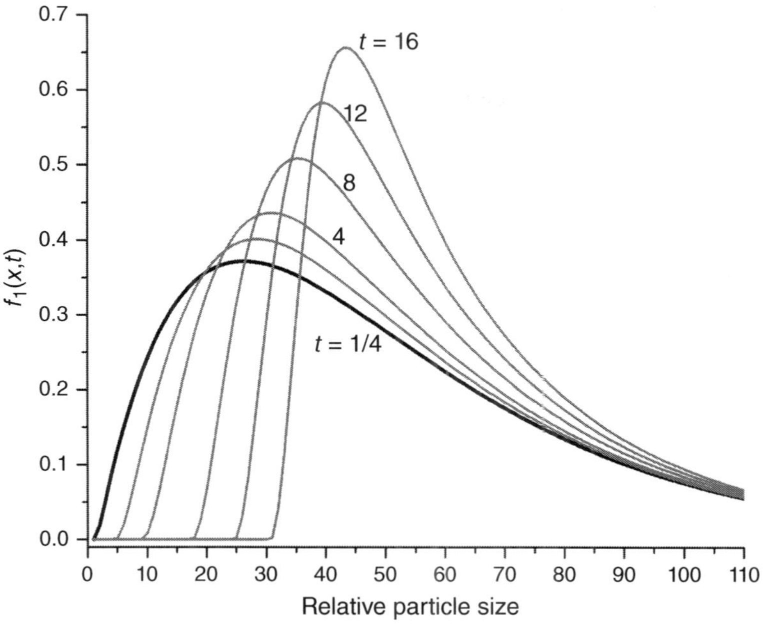

Figure 9.13 Change in the number density function for the growth-only case. The heavy black curve is for $t = 1/4$. Note how the small sizes are disappearing as the density function becomes increasingly narrow.

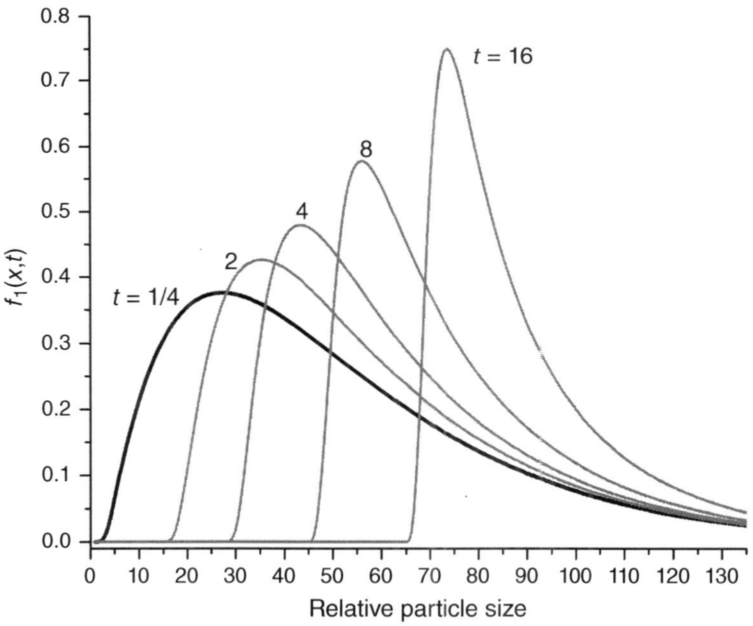

Figure 9.14 Change in the number density function (growth only) using a growth rate expression with less large-size attenuation.

fragments produced by the disintegration. Since we are talking about process elements that are almost always both non-linear and stochastic, these characterizations will not be easy. A major challenge in the construction of population balance models is to produce a result that works *and* is actually grounded in sound physics.

We should explicitly describe some of the difficulties that arise in the characterization of $h(x,t)$. If, at a given size, x, entities disappear by breakage, then $h(x,t)$ is a "sink" term consisting of a breakage rate, $b(x,t)$, multiplied by the density function: $b(x, t)f_1(x, t)$. We will replace the right-hand side of eq. (9.93) with this product and impose a size threshold where the breakage rate becomes large, thereby limiting the movement of the density function to the right (toward larger sizes). Some computed results are provided in Figure 9.15.

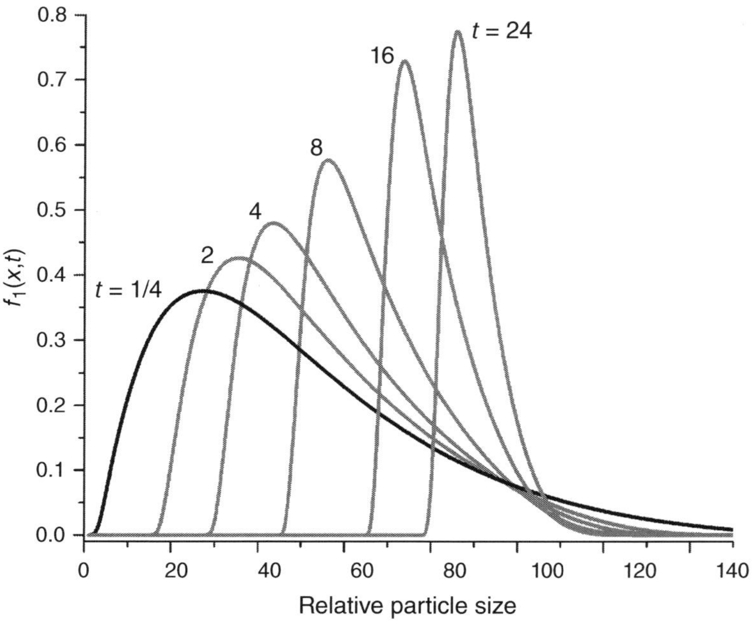

Figure 9.15 Growth process with breakage imposed at a threshold size (corresponding to a relative particle size index of 100). Distributions are shown for $t = 1/4$, 2, 4, 8, 16, and 24. Note how the distributions at $t = 16$ and 24 are being constrained at the upper end.

Normally when breakage is occurring larger entities are increasingly likely to break, and some of those events will generate new entities at size, x. This phenomenon represents a "source" of particles at size x. Now things are a good deal more complicated. When one of these larger entities breaks some number of daughter particles will be formed, and we will represent the average number of fragments with $\nu(x', t)$, where $x' > x$ of course. It is clear that ν must be at least 2 (this is what we call binary breakage), but under sufficiently energetic circumstances it could be much larger. Obviously this also means that we must know where in x-space these fragments will be distributed. We will let $P(x|x')$ be the probability that a larger entity at size x' creates a fragment at size x. The disintegration of *any* larger entity might produce a fragment (or daughter particle) at size x, so we must take *all larger particles* into account:

$$\frac{\partial f_1(x,t)}{\partial t} + \frac{\partial}{\partial x}[\dot{X}(x,t)f_1(x,t)] = \int_x^\infty \nu(x')b(x')P(x|x')f_1(x',t)dx - b(x)f_1(x,t)\,. \tag{9.94}$$

There are some clear restrictions placed upon the conditional probability, $P(x|x')$, including $P(x|x') \geq 0$ and $P(x|x') = 0$ if $x > x'$. In models of this type where both growth and breakage are taken into account, an important concern is that eq. (9.94) is consistent with an appropriate conservation principle. For example, if the total mass in the system is constant, then the first moment of the density function cannot vary, that is,

$$\mu_1 = \int_0^\infty x f_1(x,t)dx \tag{9.95}$$

must not be a function of time. Let us use a very simple illustration to highlight some of the problems encountered with conservation. Suppose we are using mass to characterize entity size and that we have two particles, each of mass M_1, that combine to form a single new particle. Then: $M_1 + M_1 = M_2$ and of course, $M_2 = 2M_1$.

Now suppose we have a liquid–liquid system where the coalescence of two identical droplets occurs; we also assume we are vitally interested in droplet size and volume: $V_1 + V_1 = V_2$. Of course, $R_2^3 = [R_1^3 + R_1^3]$ which implies $d_2 = 1.26d_1$.

Next, assume two small aggregates formed by flocculation in an aqueous system collide and affiliate. These flocs are typically very irregular, and each structure contains a significant amount of interstitial water; moreover, when they combine even more fluid will be trapped in the porous spaces between the contacting aggregates and neither mass nor volume will be conserved. Let us assume that the additional fluid volume amounts to 75% of the volume of the colliding flocs (this percentage is actually on the small side):

$$V_2 = (1.75V_1 + 1.75V_1) = 3.5V_1. \tag{9.96}$$

Under this scenario, we find: $d_2 \cong 1.52d_1$. Naturally when breakage of such an entity occurs, some of that trapped, interstitial fluid is released, so both volume and mass will be lost. It is effectively impossible to enforce a conservation principle in a system of this type, so there is little benefit to working with mass or volume. Therefore in the following discussion we will turn our attention to a straight number balance.

9.7.1 Particle Size in Coagulating Systems

Let us begin by providing a general verbal description of the balance we wish to formulate, and we will consider systems of the types illustrated by Figures 9.11 and 9.12. We will develop a number balance and let (*a*,*b*) be the particular size range of interest.

The rate of change of population in the interval (*a*,*b*) will be determined by:

Production in (a,b) by the growth of smaller entities
Depletion in (a,b) by growth of entities in the interval (beyond b)
Production of fragments in (a,b) by breakage of larger entities
Depletion in (a,b) by breakage of entities from the interval

Remember that these terms are specific to flocculating systems of the type we cited immediately above. In more general situations, particle growth could occur by surface reaction, coating, deposition, etc. We observed previously that birth in crystallization processes might appear to be the result of nucleation; however, we must remember that when saturation is exceeded, the nuclei that are formed initially are extraordinarily small. Thus entity birth at a *finite size* in crystallization may actually be the result of breakage of larger particles.

Continuous crystallizers, flocculators, etc. may operate at steady state such that the number density of entities is virtually constant. A batch process, in contrast, may yield a significant change in number density. For example, in the batch coagulation of colloidal clay particles, the number density may begin at 10^7 or 10^8 per cm^3 (for primary particles), and then decline by several orders of magnitude during the flocculation process. This is exactly what Figure 9.12 reveals.

If particle growth occurs solely by aggregation, then particle–particle contact will be necessary, and a mechanistic description will require a model for the collision rate between entities. For fluid-borne particles collision can result from Brownian motion, laminar or turbulent shear, particle inertia, and differential sedimentation. The functional forms for these collision rates are:

$$\beta_{i,j} = \frac{2kT}{3\mu}\left[\frac{1}{V_i^{1/3}} + \frac{1}{V_j^{1/3}}\right]\left(V_i^{1/3} + V_j^{1/3}\right) \text{ Brownian motion} \tag{9.97}$$

$$\beta_{i,j} = \frac{4}{3}(R_i + R_j)^3\frac{dU}{dy} \text{ Laminar shear} \tag{9.98}$$

$$\beta_{i,j} = 1.3\left(\frac{\varepsilon}{\nu}\right)^{1/2}(R_i + R_j)^3 \text{ Isotropic turbulence} \tag{9.99}$$

$$\beta_{i,j} = 5.7\left(R_i^3 + R_j^3\right)|\tau_i - \tau_j|\frac{\varepsilon^{3/4}}{\nu^{1/4}} \text{ Turbulent inertia} \tag{9.100}$$

$$\beta_{i,j} = \pi\alpha(R_i + R_j)^2(V_{Si} - V_{Sj}) \text{ Differential sedimentation} \tag{9.101}$$

In these rate expressions, R is entity radius, V is volume, and V_S is settling velocity. We observe that the last two collision rate eqs. (9.100) and (9.101), require *disparity in particle size* to be important. For turbulent inertia there must be a difference in characteristic times for the participating particles, where $\tau =$ (mass of particle)/($6\pi\mu R$). For differential sedimentation, there must be a difference in settling velocities, $V_{Si} - V_{Sj}$. In a batch process where we start with all primary particles (monodisperse conditions) these two collision mechanisms will not be important at small times.

Note that in eq. (9.98) dU/dy is the velocity gradient (strain rate) associated with a highly ordered laminar flow; in eq. (9.99), ε is the dissipation rate per unit mass and ν is the kinematic viscosity of the fluid phase. Each of these collision rates has the dimensions, L^3/t. Thus, when the expressions are multiplied by the number densities of the participating particles, n_i and n_j, we obtain the number of collisions per unit volume per unit time. We will look at an example for particle collisions in isotropic turbulence: Suppose we have two classes of spherical particles, $d_i = 5\,\mu\text{m}$ and $d_j = 6\,\mu\text{m}$, and each class has a number density of 10^5 particles per cubic cm. We let the dissipation rate per unit mass be $400\,\text{cm}^2/\text{s}^3$ and the fluid medium be water ($\nu = 0.01\,\text{cm}^2/\text{s}$). We want to calculate the collision rate produced by isotropic turbulence:

$$\beta_{i,j}n_i n_j = (1.3)\left(\frac{400}{0.01}\right)^{1/2}(2.5\times10^{-4} + 3.0\times10^{-4})(10^5)(10^5) = 24.66\ \text{collisions}/(\text{cm}^3\text{s}). \tag{9.102}$$

Now we will compare with the rate produced by laminar shear with $dU/dy = 400\ 1/\text{s}$:

$$\beta_{i,j}n_i n_j = 43(2.5\times10-4+3.0\times10-4)3(400)(10^5)(10^5) = 887.33\ \text{collisions}/(\text{cm}^3\text{s}). \tag{9.103}$$

One of the characteristics of the highly ordered shear field created by the Couette device depicted in Figure 9.10 is that there is virtually no breakage of aggregates. In such cases we only need to model particle growth. Therefore, an appropriate balance can be written:

$$\frac{dn(V)}{dt} = \frac{1}{2}\int_0^V \beta(V-u,u)n(V-u)n(u)du - n(V)\int_0^\infty \beta(V,u)n(u)du, \tag{9.104}$$

where $u < V$. The first term on the right-hand side accounts for production by collision of particles smaller than V; the second term on the right accounts for a loss of entities of size V due to additional growth by contact with other particles.

Solutions for eq. (9.104) are often sought through discretization; for example, Farley and Morel (1986) used the set of equations:

$$\frac{dn_i}{dt} = \frac{1}{2}\sum_{i+j=k}\alpha(i,j)\beta(i,j)n_i n_j - n_k\sum_{i=1}^{m}\alpha(i,k)\beta(i,k)n_i, \tag{9.105}$$

where $\alpha(i,j)=1$ if $i\neq$ j, but 2 if $i=j$. Discretization of a population balance in which growth and breakage are occurring raises an obvious question: How many partitions should be used and how wide should they be? A popular choice is geometric spacing (multiples of 2); if aggregation is occurring and mass is conserved then the affiliation of two particles of class 1 will result in an entity with mass doubled: $M_1 + M_1 = 2M_1 = M_2$. Ramkrishna (2000) points out that a coarse discretization makes it impossible for the model to be internally consistent. In other words a collision event will not necessarily produce an entity in the next larger class or bin. One method that has been used to accommodate this problem is to employ weighting fractions so that only a portion of $i-j$ affiliations produce an entity in the next larger bin. Therefore, we should probably regard a set of discrete balances eq. (9.105) as semi-quantitative; the problem thus posed is more easily solved as it involves merely a set of simultaneous ODEs, but compromises have been made. We will illustrate this by simulating an aerosol system.

Let us assume we have an aerosol for which particle growth occurs solely by Brownian motion, but particles are lost from the control volume due to sedimentation. We will use eight particle classes with bins centered about the diameters, 0.375, 0.75, 1.5, 3, 6, 12, 24, and 48 μm. We will solve eight simultaneous ODEs of the form given by eq. (9.105), but with the addition of a loss term based upon the settling velocity (which is size-dependent). The collision rate will be described by eq. (9.97). For case 1, we will initially populate the first four (smallest) bins with 1×10^7 particles per unit volume each. For case 2, we will heavily populate the smallest (0.375) bin, but the others will have *much smaller* number densities. This will cause a profound change in the dynamic behavior of the numbers of small particles. In Figure 9.16a,b, the number density in the smallest (0.375) bin is shown with a heavy black curve.

9.7.2 Application of the Population Balance to a Continuous Crystallizer

In this section we look at a continuous crystallizer operated at steady state. Liquor containing the solute is fed to a well-mixed vessel and the suspension containing crystals is withdrawn from the apparatus at the same rate. We write the balance for crystals of size, s:

$$\frac{d}{ds}(Ln) = -\frac{n}{\theta} + B - D, \tag{9.106}$$

where B and D are birth and death functions, respectively. n is the number density of crystals, θ is the average holding time, and L is the linear crystal growth rate. Death (disappearance of crystals of a given size) is the result of breakage, and D is assumed to have the functional form,

$$D(s) = kn(s)s^m, \tag{9.107}$$

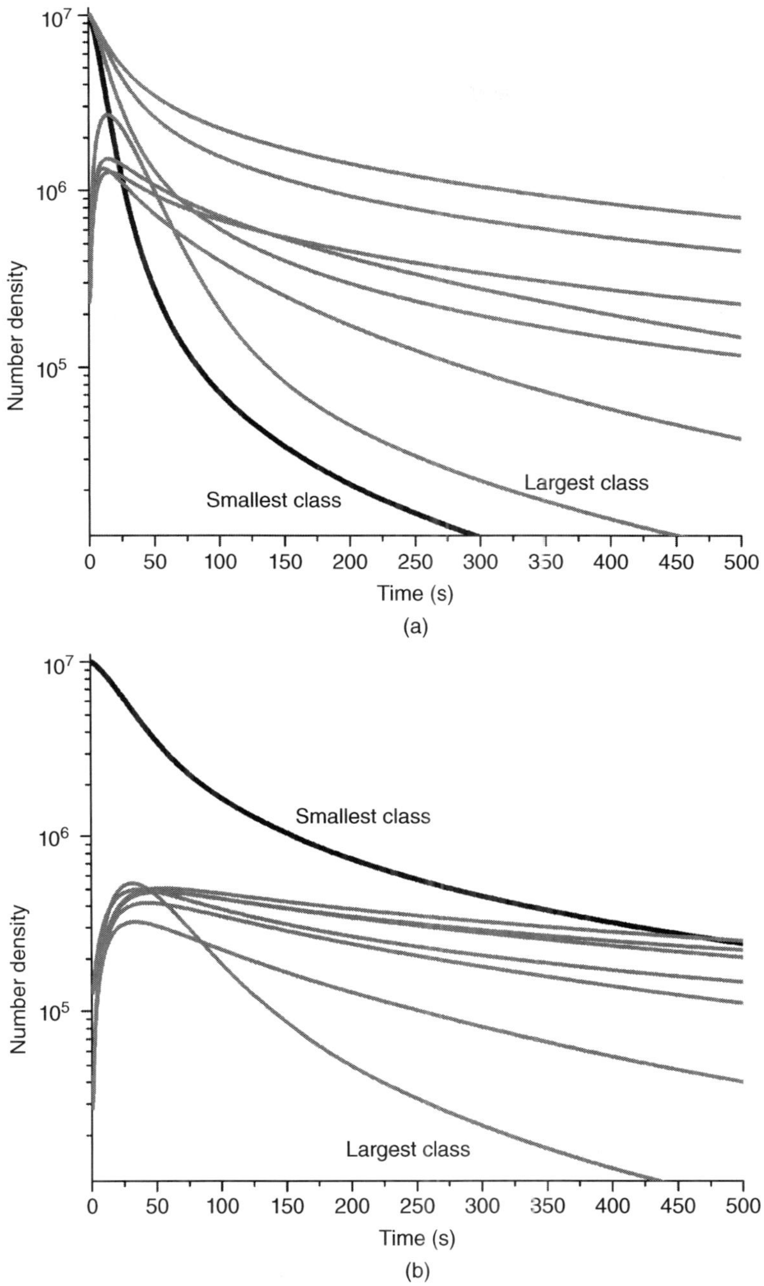

Figure 9.16 (a) Case 1: Dynamic behavior from the discretized model with the first four (smallest) bins initially populated at 1×10^7 particles per cm^3. Because of the increased opportunity for small particle collisions, the 0.375 class decreases rapidly. (b) Case 2: Dynamic behavior from the discretized model with only the 0.375 bin heavily populated. The opportunity for small particle collisions has been severely reduced.

where k and m are empirical constants. The "birth" of crystals of a given size is assumed to occur by breakage of larger entities, and it is *convenient to assume* that the breakage process is binary (the breakage of a parent particle results in the production of two daughter particles). The birth function is written:

$$B(s) = \int_0^1 P(\varepsilon)\left[D\left(\frac{s}{\varepsilon}\right) + D\left(\frac{s}{1-\varepsilon}\right)\right] d\varepsilon. \tag{9.108}$$

$P(\varepsilon)$ is a probability density and $P(\varepsilon)d\varepsilon$ is the likelihood that the breakage of a crystal of size, s, will produce a crystal in the size range, εs to $(\varepsilon + d\varepsilon)s$. Obviously ε is a fractional quantity ranging from 0 to 1. Among the idealizations used in developing this model are the assumption that the breakage process is strictly binary (a bit of a stretch) and that the linear growth rate (L) is a constant. The latter assumption is referred to as McCabe's law, and it is based upon the idea that crystal growth is more likely to be dependent upon supersaturation than upon crystal size. Randolph and Larson (1988) note that this is often observed to be true. The birth and death functions can be inserted into eq. (9.106), and once a functional form for $P(\varepsilon)$ is chosen, a solution may be sought. The simplest case, $P(\varepsilon) = 1$, corresponds to uniform binary breakage, and it is convenient to recast the model in dimensionless form by setting:

$$x = \frac{s}{\theta L} \quad y = \frac{n(s)}{n(0)} \quad \text{and} \quad k_1 = k\theta(L\theta)^m. \tag{9.109}$$

The result is the IDE,

$$\frac{dy}{dx} + (1 + k_1 x^m)y = k_1 \int_0^1 \left[\left(\frac{x}{\varepsilon}\right)^m y\left(\frac{x}{\varepsilon}\right) + \left(\frac{x}{1-\varepsilon}\right)^m y\left(\frac{x}{1-\varepsilon}\right)\right] d\varepsilon. \tag{9.110}$$

Of course, x is dimensionless crystal size and y is dimensionless crystal population density; since y has been normalized with $n(0)$, $y(x = 0) = 1$. This formidable problem has been solved by Singh and Ramkrishna (1977) using the *method of weighted residuals* (MWR). They constructed their trial functions with Laguerre polynomials. We can gain some limited appreciation for the behavior of $y(x)$ by assuming that the birth rate is zero. Then, with the right-hand side of eq. (9.110) gone,

$$\frac{dy}{dx} = -(1 + k_1 x^m)y, \tag{9.111}$$

and a solution is easily found:

$$y = C_1 \exp\left[-\left(x + \frac{k_1}{m+1}x^{m+1}\right)\right].$$

Singh and Ramkrishna used this result (with $C_1 = 1$) as the weight function for their solution by MWR, and they obtained the numerical results shown in the table below. The values selected for the parameters were $m = 4$ and $k_1 = 0.006173$.

x	y
0.0	1.0000
0.2	0.8191
0.5	0.6085
1.0	0.3743
1.5	0.2332
2.0	0.1461
2.5	0.8972×10^{-1}
3.0	0.5141×10^{-1}
3.5	0.2545×10^{-1}
4.0	0.9701×10^{-2}
4.5	0.2529×10^{-2}
5.0	0.3700×10^{-3}
5.4	0.4565×10^{-4}
6.0	0.5746×10^{-6}

Ramkrishna (2000) reviews the use of MWR as a solution technique for population balance equations in Chapter 4 of his book.

9.8 Conclusion

IDEs appear in many important problems in the applied sciences, ranging from biology to separation processes. For linear IDEs the Laplace transform can be extremely useful, and it can produce exact solutions. However, many of the IDEs of interest to us are nonlinear, and rarely can these equations be solved analytically. Our purpose in this chapter was to provide some alternative solution techniques so that students confronted by an IDE have options. In this vein, the VIM developed by He and coworkers is especially useful, and *highly* recommended. The reader interested in applying VIM may find the recent article by Prajapati et al. (2012) very helpful. They solved Abel's integral equation using four different initial guesses for the trial function, $y_0(x)$, and they were able to show that all four produced convergent series solutions with small absolute errors. Their paper contains sufficient detail to be a valuable aid to the analyst who is new to VIM.

However, by no means is VIM the only viable alternative; for example, for nonlinear Fredholm IDEs of the form

$$\frac{dy}{dt} = f(t) + \int_0^1 k(t,s)F(y(t))ds, \tag{9.112}$$

with $0 \le t \le 1$ and $y(t=0) = y_0$; a few of the solution methods that have appeared in the literature are: *homotopy perturbation, rationalized Haar functions, hybrid functions and collocation, and the Tau method with Chebyshev and Legendre bases.* For an example of application of the latter see Pour-Mahmoud et al. (2005).

There are other prominent IDEs that have been intensively studied; two examples where several alternative solution procedures have been devised include the neutron transport equation and the equation for radiative heat transfer. In the case of the latter for a medium that absorbs, emits, and scatters (anisotropically):

$$\frac{dI}{ds} + \beta I = \kappa_a I_b + \frac{\kappa_s}{4\pi}\int_{4\pi} I(s, \Omega')\phi(\Omega', \Omega)d\Omega'. \tag{9.113}$$

I is the intensity of the radiation and Ω is the radiation (vector) direction, and for eq. (9.113), $0 \le s \le L$ and $I = I_0$ for $s = 0$. Zhao and Liu (2007) investigated this IDE and reported that many numerical methods exhibit unphysical oscillations in this particular application. They transformed the equation to obtain a second-order differential equation that was of the diffusion type (with predictable behavior when solved numerically). You will observe that this is an approach that we have illustrated several times in this chapter; where feasible, this practice can often make the solution of such problems a good deal easier.

Problems

9.1 We want to solve the IDE,

$$\frac{du}{dt} - 3u + 6\int_0^t u(\tau)d\tau = t$$

with $u(t=0) = 0$. Find solutions by both the approximate discretization technique and the method of Laplace transform. Verify that your results correspond to the behavior shown in the following table for $(0 \le t \le 4)$.

t	$u(t)$
0.5	0.191
1.0	0.972
1.5	1.989
2.0	0.930
2.5	−6.157
3.0	−18.418
3.5	−16.104
4.0	44.332

9.2 Find the solution for the IDE,

$$\frac{dy}{dt} + 5\int_0^t \cos[2(t-u)]y(u)du = 10,$$

given that $y(t = 0) = 2$. Compare your result with

$$y(t) = \frac{1}{27}(24 + 120t + 30\cos(3t) + 50\sin(3t)).$$

9.3 Find the solution for the IDE,

$$\frac{dy}{dx} = 1 + 2x - y + \int_0^x x(1 + 2x)\exp[s(x - s)]y(s)ds,$$

given that $y(x = 0) = 1$. Verify that $y(1.75) = 21.3809$.

9.4 Use two methods for the solution of the IDE

$$\frac{dy}{dt} = 1 - \int_0^t y(s)ds,$$

given that $y(0) = 0$ and $0 \le t \le 1$. We are particularly interested in obtaining an accurate value for $y(0.5555)$.

9.5 The IDE,

$$\frac{dy}{dx} + y = \int_0^x \exp(s - x)y(s)ds,$$

has been solved for $0 \le x \le 1$ with $y(0) = 1$.

Some of the results obtained are given in the table below:

x	Approximate $y(x)$
0.05	0.95333
0.10	0.91019
0.15	0.87116
0.20	0.83585
0.25	0.80390
0.30	0.77499
0.40	0.72517
0.50	0.68438
0.60	0.65099
0.70	0.62366
0.75	0.61192
0.80	0.60129
0.85	0.59167
0.90	0.58298
0.95	0.57511
1.00	0.56799

Determine how accurate these results are.

9.6 Using the method of your choice, solve the IDE,

$$\frac{du}{dt} + 3u + 10\int_0^t u(\tau)d\tau = 4\exp(-t)$$

with $u(0) = 0$. Find both the maximum and the minimum for $u(t)$, and the times at which they occur.

Answer: maximum at $t = 0.314$

9.7 Find both numerical and analytic solutions for the linear IDE:

$$\frac{dy}{dt} + 3y + 2\int_0^t y(\tau)d\tau = 2\exp(-3t),$$

with $y(0) = 0$. What is the maximum value for $y(t)$ and when does it occur?

9.8 Solve the IDE

$$\frac{dy}{dt} = 1 + \sin(4t) + \int_0^t y(\tau)d\tau,$$

with $y(0) = 0$. When does $y(t)$ exceed 4?

9.9 Consider the nonlinear IDE

$$\frac{du}{dt} = \frac{1}{\varepsilon}\left(u - u^2 - u\int_0^t u(\tau)d\tau\right)$$

with $\varepsilon = 1/10$, $u(0) = 0.1$ and $0 \le t \le 1$. This equation has been solved with an approximate method described by Daliri Birjandi, Saberi-Nadjafi, and Ghorbani (*An Efficient Method for a Class of Nonlinear Volterra Integro-Differential Equations, Hindawi JAM, volume 2018, ID 7461058*). Try a discretization of this equation and see if you can confirm that $u(t = 0.6) = 0.736$ and $u(t = 1) = 0.52$. At what value of t does the maximum u occur?

9.10 One of the examples worked in the text as part of the discussion of VIM was the ODE,

$$\frac{du}{dx} = -\frac{x^3}{(u+1)^2}, \quad \text{with } u(x=0) = 1.$$

The analytic solution is known,

$$3x^4 + 4(u+1)^3 = 32, \text{ or alternatively, } u = \left[\frac{32 - 3x^4}{4}\right]^{1/3} - 1.$$

With VIM, the succession of estimates is obtained from

$$u_{n+1} = u_n - \int_0^x \left(u_n' + \frac{s^3}{(u_n+1)^2}\right) ds.$$

Since $u_0 = 1$, we can very easily find: $u_1 = 1 - \frac{x^4}{16}$. Therefore,

$$u_2 = 1 - \frac{x^4}{16} - \int_0^x \left[-\frac{s^3}{4} + \frac{s^3}{\left(2 - \frac{1}{16}s^4\right)^2}\right] ds,$$

and

$$u_2 = 3 + \frac{64}{x^4 - 32}.$$

Find u_3 for this problem and compare its performance with the table of values that accompanied the example.

9.11 Although vacuum tubes that were once so common in TV sets and radios seem to have disappeared, they are still in demand for certain applications. For example, the triode vacuum tube is preferred for guitar amplifiers, and the 12AX7 and ECC83 are widely available (for preamps) from guitar shops and music stores. The transient performance of a triode vacuum tube can be modeled with an IDE:

$$L\frac{dI}{dt} + RI + \frac{1}{C}\int_0^t I(\tau)d\tau - M\frac{dI_A}{dt} = E\sin(\omega t).$$

L, R, and C are the inductance, resistance, and capacitance, respectively. I_A is the grid current and M is the mutual inductance. E is the driving voltage. The grid current is related to I (approximately) by an algebraic equation; we will *assume* that M is very small so that the fourth term on the left-hand side of the equation can be neglected. We would like to solve the simplified IDE using the following parameters:

$$R = 65{,}000\ E = 240\ L = 1{,}175\ C = 0.00025\ \omega = 30\ \text{Hz}$$

What will the maximum value for I be (in milliamps), and when will it occur given that $I(t = 0) = 0$?
Answer: Just less than 1.5 mA at t = 14.2 ms.

9.12 The difficult problem of two populations in conflict was introduced in Section 9.3. The solution strategy proposed there involved differentiation of the IDEs, resulting in a pair of nonlinear second-order ODEs (that is, a fourth-order system). However, this is not the only approach for problems of this type. Let us return to the original equations:

$$\frac{dX}{dt} = aX - bXY - K_1 X \int_0^t Y(\tau)d\tau$$

and

$$\frac{dY}{dt} = -\alpha X + \beta XY + K_2 Y \int_0^t X(\tau)d\tau.$$

We take $a = b = 2$ and $\alpha = \beta = 1$ with $K_1 = -0.05$ and $K_2 = -0.05$; the initial values will be $X = 1$ and $Y = 2$. We will discretize the equations and solve the IDEs numerically. An elementary solution procedure is presented here:

```
#COMPILE EXE
#DIM ALL
    REM *** Two populations in conflict with heredity
    REM *** Simultaneous integro-differential equations
    GLOBAL x,xold,y,yold,tt,dt,sum1,sum2,K1,K2,a,b,alpha,beta,zz AS DOUBLE
    GLOBAL kcount AS SINGLE
FUNCTION PBMAIN
        xold=1.0:yold=2.0:x=1:y=2:dt=0.000025:a=2:b=2:alpha=1:beta=1:K1=0.05:K2=-0.05
        OPEN "c:TWOintegDIFF.dat" FOR OUTPUT AS #1
    100 REM *** continue
            sum1=sum1+(y+yold)/2*dt
                x=dt*(a*xold-b*xold*yold-K1*xold*sum1)+xold
                    sum2=sum2+(x+xold)/2*dt
                y=dt*(-alpha*yold+beta*xold*yold+K2*yold*sum2)+yold
                            tt=tt+dt
                                xold=x:yold=y
                                    kcount=kcount+1
                                    IF kcount>74 THEN 140 ELSE 160
    140 REM *** continue
                                PRINT tt,x,y
                                    WRITE#1,tt,x,y
                                        kcount=0
    160 REM *** continue
                            IF tt<55 THEN 100 ELSE 200
    200 REM *** continue
                INPUT "end program?";zz
                IF zz>0 THEN CLOSE
                END

END FUNCTION
```

Begin by reproducing the results you see here in Figure P9.12, then repeat the analysis letting $K_1 = -0.05$ and $K_2 = +0.05$. What is the ultimate outcome in this case?

9.13 We saw that the FKPP equation was an unrealistic diffusion model for the influence of travel upon the spread of infectious disease. An alternative is the model proposed by Medlock and Kot (*Mathematical Biosciences*, 184:201, 2003) in which $M(x)$ is the rate at which infected individuals leave position x, and travel to a new location in the

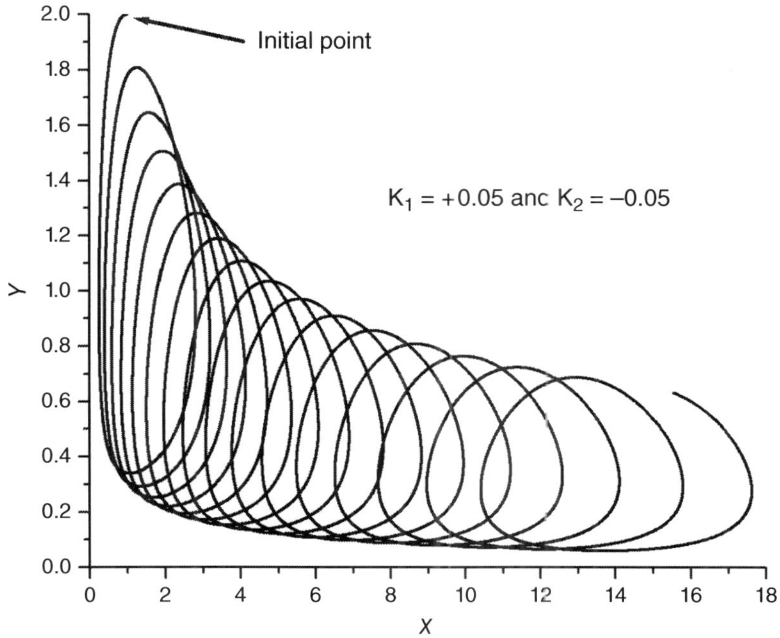

Figure P9.12 Phase-plane portrait of two populations in conflict; notice how the oscillations with respect to X are increasing in amplitude as the cycles in Y are steadily diminishing.

spatial domain. We take M to be a positive constant. The model has the form,

$$\frac{\partial I}{\partial t} = \beta I(C_1 - I) - MI + M\int K(x,y)I(y,t)dy. \tag{9.114}$$

$K(x,y)$ is the kernel; it is a density function for the fraction of infected individuals who leave position y destined for x. This density function has the characteristics

$$K(x,y) \geq 0 \text{ and } \int K(x,y)dx = 1. \tag{9.115}$$

Medlock and Kot (2003) assumed that the kernel was of the convolution type,

$$K(x,y) = k(x-y) = k(u) = \frac{1}{2\alpha}\exp\left(-\frac{|u|}{\alpha}\right), \tag{9.116}$$

where u is the separation between the x and y positions. They set $M = 1$, $\beta = 1$, and $\alpha = 0.7698$; I, the fraction of infected individuals was taken to be 1 in the center of the domain. The problem thus posed by eq. (9.114) is formidable. The repeated evaluation of the integral is computationally expensive, so Medlock and Kot used a Runge–Kutta scheme for the time derivative and the fast Fourier transform (FFT) for the integral. We will let our domain extend from $x = 0$ to some very large x, and we will place a block of infected individuals at the origin. Our goal is to explore the propagation of the "wave front" of infected people. Let the initial block for I occupy 5% of the domain. Use the Medlock–Kot parameters and compute the evolution of the "infected" wave. How long will it take the center portion of the wave to travel five times (5x) the width of the initial block of infectives?

9.14 Ahmed and Elzaki (On the comparative study of IDEs using difference numerical method, *Journal of King Saud University*, 38:84, 2020) applied their approximate solution technique to the nonlinear IDE,

$$\frac{dv}{dt} = -1 + \int_0^t v^2(\tau)d\tau$$

with $v(0) = 0$ and $0 \leq t \leq 1$. The solution they developed was

$$v(t) \cong -t + \frac{t^4}{12} - \frac{t^7}{252} + \frac{t^{10}}{6{,}048} - \frac{t^{13}}{157{,}248} + \ldots$$

Discretize the IDE and solve it numerically and compare your results with the approximate solution above. A more interesting variation of this IDE is

$$\frac{dv}{dt} = -1 + 50 \int_0^t v^2(\tau)d\tau.$$

Solve this equation; what is the minimum value of $v(t)$, and at what t does it occur?

Answer: -0.3107 *at* $t = 0.436$

9.15 In an emulsification process, data show that the droplet (number) density function is approximately described by

$$f(d) = \frac{1}{16}d^2 \exp(-d/2).$$

Therefore, when $d = 4\,\mu\text{m}$, $f(d) = 0.1353$; for $d = 11\,\mu\text{m}$, $f(d) = 0.0309$. The probability density is shown as a function of droplet size immediately below.

Measurements show that the total number density is 1×10^7 droplets per cubic cm. Therefore, for droplets with diameters between 8 and 12 μm, the cumulative probability is about 17%; i.e., there are about 1.7×10^6 droplets *in this size range* per cubic cm. For many applications, the probability distribution for *volume* is more useful than $f(d)$ given in Figure P9.15. Convert the data shown above to $f(V)$. Where (at what volume) is the maximum in the distribution now located?

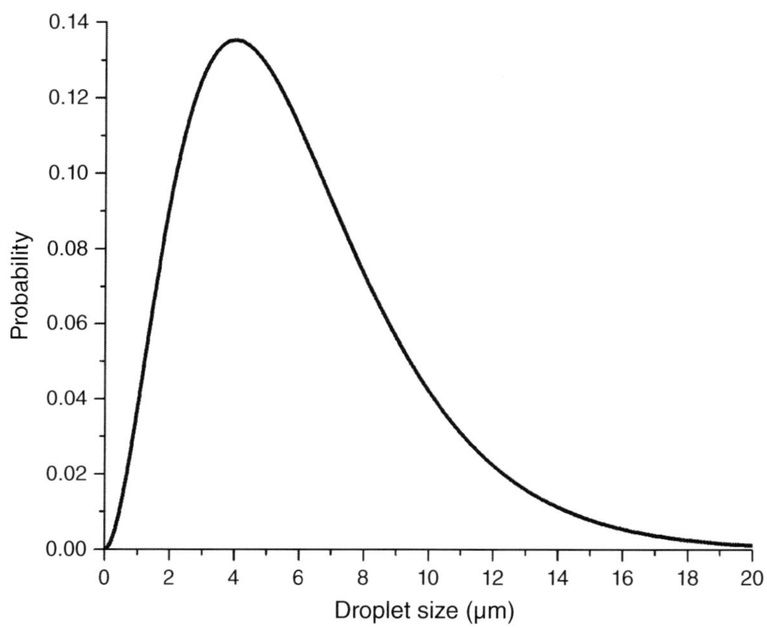

Figure P9.15 Droplet (number) density function for problem 9.15.

9.16 Solve the dynamic balance for a pure growth process,

$$\frac{\partial f_1(x,t)}{\partial t} + \frac{\partial}{\partial x}[\dot{X}(x,t)f_1(x,t)] = 0,$$

assuming that the *initial* density function is $f_1 = x\exp(-x)$, but that the growth rate is given by

$$\dot{X} = \frac{C_1}{C_2 + x^3}.$$

Let $C_1 = 0.75$ and $C_2 = 2.0$. Prepare a figure that illustrates the evolution of the number density function.

9.17 Solve the discretized model, eq. (9.105) in Section 9.7, for an aerosol with eight particle classes using the example in the text as your guide. Center the size bins on 0.375, 0.75, 1.5, 3, 6, 12, 24, and 48 μm. Neglect sedimentation, but compare the results obtained with *three different* collision mechanisms: Brownian motion (as illustrated in Section 9.7.1), laminar shear with $dU/dy = 300\,\text{s}^{-1}$, and isotropic turbulence with $\varepsilon = 1{,}000\,\text{cm}^2/\text{s}^3$. Which mechanism results in the most rapid reduction in number density of small particles? Let the initial concentrations be 1×10^7 particles per cm^3 in each of the three smallest classes (let n_i for the five larger classes be zero initially).

9.18 Yüzbasi and Sezer (*Mathematical Methods in the Applied Sciences*, online, 2012) studied a linear Volterra IDE with a weakly singular kernel:

$$y''(x) - x^2 y'(x) = 6x - \frac{1}{2}x^4 - \frac{16}{5}x^{5/2} +$$

$$\int_0^x \frac{y'(t)}{\sqrt{x-t}}dt - \frac{5}{4}\int_0^x xty''(t)dt,$$

where $y(0) = 1, y'(0) = 0$, with $0 \le x$ and $t \le 1$. They note that the exact solution for this problem is $y(x) = x^3 + 1$. Solve this Volterra IDE using the method of your choice and compare your solution with the exact result.

9.19 Consider the IDE, $\frac{dy}{dt} = \int_0^t y(u)\cos(t-u)du$, with $y(0) = 1$. Your colleague has developed his own method for solving such equations, and he reports that $y(t = 1.5) = 2.135$ and that $y(t = 3) = 4.499$. Use the method of your choice to confirm or refute the results of your colleague.

9.20 In 2013 an on-line discussion occurred in which solution strategies were sought for the *partial* IDE,

$$\frac{\partial u}{\partial t} = \frac{\partial^2 u}{\partial x^2} - u^3 + u + B\left(u_0 - \frac{1}{L}\int_0^L u dx\right),$$

given that $L = 4$, $B = 10$, and $u_0 = ½$. For this problem, $0 \le x \le 4$ and the initial distribution of u consists of block centered at $x = 2$ with a width of ½ and an amplitude of 1 (one); u is zero elsewhere. Try the following approach: Discretize the equation (with respect to x) to get a system of n-ODEs of the form, $\frac{du}{dt} \cong \frac{u_{n+1}-2u_n+u_{n-1}}{(\Delta x)^2} - u_n{}^3 + u + B\left(u_0 - \frac{1}{L}\sum u_n\Delta x\right)$, divide L into 40 pieces, and then solve the set of ODEs to find the time evolution of the nonlinear system. It has been reported that this problem exhibits diffusive behavior. Is that borne out by your results?

9.21 We used an elementary discretization scheme to solve several IDEs in this chapter, and we confirmed our solution strategy by comparison with some simple cases for which the analytic solution could be found. We also compared it to a few nonlinear cases where published numerical or approximate solutions were available. Our objective here is to apply the method to a more difficult nonlinear IDE:

$$\frac{du}{dt} + 5u^2 - 7\int_0^t u(\tau)^2 d\tau = -\frac{1}{2},$$

with $u(0) = 1$ and $0 \le t \le 4$. What is the minimum value of u, and where is it located? Where does $u(t) = 1$ (other than $t = 0$ of course)? An illustration of how one might apply discretization to find an approximate solution is provided here:

```
#COMPILE EXE
#DIM ALL
    REM *** Nonlinear IDE solution by discretization
    REM *** Euler method with summation of rectangles
    REM *** Problem 9.21, du/dt+5*u^2-7*INTEGRALu^2dt=-1/2
            GLOBAL u,u0,tt,dt,sum, zz AS SINGLE
FUNCTION PBMAIN
            tt=0:u0=1:dt=0.001
                OPEN "c:IDEnonlinPB21.dat" FOR OUTPUT AS #1
```

```
100 REM *** continue
            sum=sum+u0^2*dt
                u=dt*(-1/2-5*u0^2+7*sum)+u0
                    tt=tt+dt
                        PRINT tt,u
                        WRITE#1,tt,u
                            u0=u
                                IF tt<4 THEN 100 ELSE 200
200 REM *** continue
            INPUT "end program?";zz
            IF zz>0 THEN CLOSE
            END
END FUNCTION
```

Notice that we are using 4,000 steps for this calculation (since $\Delta t = 0.001$). In any problem of this type where the solution is unknown, it is prudent to reduce the step size and repeat the calculation to see if there are any significant discrepancies. Finally, reverse the sign on the right-hand side such that

$$\frac{du}{dt} + 5u^2 - 7\int_0^t u(\tau)^2 d\tau = +\frac{1}{2},$$

and solve the problem again. At what value of t does $u = 3$?

Answer for part one: $u(t = 0.607) = 0.354$ and $u(t = 2.48) = 1$.

9.22 A model was developed that represents the productivity (f) of a process as a function of throughput rate (t):

$$\frac{df}{dt} = -\frac{af}{t^2}\left(\int_{t_0}^{t} \tau^2 f(\tau)d\tau + M\right),$$

where $f\left(t_0 = \frac{1}{4}\right) = 1$, $a = 1/8$, and $M = -13/2$. Find the maximum value of f and the value of t for which it occurs. What is the value of t where $f < 0.5$?

Answer part one: $f = 10.9833$ for $t = 1.273$.

9.23 Consider continuous, constant-volume culture process in which the substrate (with nutrients for the micro-organisms) is added at the same rate at which the liquid medium is withdrawn. The objective is the production of species "A," but we assume that the bioreactor gets contaminated by a faster growing species, "B." This situation will result in competition between A and B and because B grows more rapidly, the production of A will be curtailed and the concentration of nutrients (S) will be reduced. As B proliferates the concentration of the desirable species will diminish, and may in fact, be reduced to zero. A simplified model for this process consists of three ODEs:

$$\frac{dA}{dt} = k_1 A - \gamma A \tag{9.117}$$

$$\frac{dB}{dt} = k_2 B - \gamma B \tag{9.118}$$

$$\frac{dS}{dt} = \gamma(S_F - S) - k_1 A/y_1 - k_2 B/y_2 \tag{9.119}$$

where k_1 and k_2 are the specific growth rates for A and B, respectively, and they are affected by nutrient concentration:

$$k_1 = 0.32S/(2 + S) \text{ and } k_2 = 0.8S/\left(\frac{3}{2} + S\right).$$

S_F is the feed concentration of nutrients ($S_F = 25$) and y_1 and y_2 are the yield coefficients for the two competing species. Let

$$A(t = 0) = 11, B(t = 0) = 0.02, y_1 = 2, y_2 = 1, \gamma = 0.2, \text{ and } S(t = 0) = 0.5.$$

Use the model equations given above and determine the time at which A attains its maximum concentration.

Answer: $A = 22.33$ at $t = 13.16$ hr.

Interactions of micro-organisms in mixed cultures can be very complicated, and Bungay (*Computer Games and Simulation for Biochemical Engineering,* Wiley-Interscience, 1985) points out that the interaction terms in bioreactor models are not well characterized. Let us suppose that the production of the desirable species, A, is directly affected by the presence of B and, furthermore, that this interaction is cumulative. The term to be added to (9.117) is

$$-b_1 B \int_0^t A(\tau)d\tau$$

with $b_1 = 0.0025$. All other parameters have the same values used previously. Extinction of A is now possible and obviously the concentration of A cannot be less than zero. What is the maximum value for A and when does it occur?

9.24 In Section 9.4 we discussed a simple discretization that would allow us to quickly find approximate solutions for some IDEs. For first-order IDEs we used a first-order forward difference for the derivative and a Riemann sum for the integral. This worked well enough for some IDEs, but we observed that for cases in which $y(t)$ exhibited significant changes in curvature and slope it might be necessary to use higher-order approximations. Of course, such a modified algorithm would not be self-starting. Consider the IDE:

$$\frac{dy}{dt} = 1 + 4\sin(9t) + \int_0^t y(\tau)d\tau \tag{9.120}$$

with $y(0) = -1$ and $0 \le t \le 3$. Now assume we want to accurately identify values of time, t, for which $y = 1/2$ and $y = 3$. Because of the frequency of the *sine* term (9 rad/t) we can expect more than 4 cycles on the time interval 0–3; furthermore, the amplitude of 4 will guarantee that the oscillations will be substantial. Even if we make Δt very small the first-order scheme may not produce useful results. Develop a higher-order discretization for (A) and find a numerical solution that will allow you to identify values of t for which $y(t) = 1/2$ and $y(t) = 3$. Compare your results with those obtained from the analytic solution (which you can find readily using the Laplace transform).

9.25 In Section 9.1 we described the circumstances (cyclic pressurization–depressurization) leading to the failure of the De Havilland Comet 1. Although the idea that damage to a particular structure experiencing repetitive stress was *cumulative* had been put forward by Palmgren in 1924, engineers were slow to incorporate failure prediction (based upon stress history) as part of the design process.

In an eerily similar, but more recent tragedy, the OceanGate submersible, Titan, imploded on June 18, 2023, during a dive to the wreck site of the RMS Titanic. The construction of the Titan—a cylindrical carbon fiber composite hull (12.7 cm thick) with hemispherical titanium end caps—had been questioned several years prior to the accident. Voiced concerns included the use of the carbon fiber material, the thickness of the Plexiglas™ view port, and the adhesive bonding of the carbon fiber cylinder to the titanium end caps. Of course, the pressure differences experienced by a craft diving to the Titanic are far greater than those experienced by the Comet 1 as the depth of the Titanic wreck site is about 12,500 ft which corresponds to an approximate pressure of 5,600 psi (about 386 bar). Carbon fiber composites are anisotropic and not typically used for large compressive loads; each dive (and the Titan had made more than 20) would have produced minute cracks and some water ingress in the cylindrical hull material. The damage experienced by the structure was certainly *cumulative*, but Stockton Rush (OceanGate's CEO) had resisted suggestions than the hull be scanned periodically to detect flaws or delaminations occurring in the composite material. Apparently Rush believed that the acoustic monitoring system inside the craft would give ample warning of any impending failure (it has been stated by sources familiar with OceanGate that on June 18 Rush realized there was a problem and dropped ballast in an effort to stop the descent—this was confirmed in the USCG hearing). Unfortunately, the hull failure was extremely rapid; Dr. Ronald Wagner (of the Technical University of Braunschweig) performed nonlinear simulations of likely failure modes of the craft that are instructive—the implosion once initiated was complete in less than 40 ms making any reaction on the part of the crew utterly impossible.

Review the recent literature regarding structures exposed to cyclic stress (leading to damage and ultimately failure): Describe models that have been put forward in which IDEs are prominent. A possible starting point is the review by Nguyen et al., An energy-based peridynamic model for fatigue cracking, *Engineering Fracture Mechanics*, Volume 241, 2021.

References

Altintan, D. and Ugur, O. (2012). *Variational Iteration Method for Sturm-Liouville Differential Equations*. Ankara, Turkey: Institute of Applied Mathematics, Middle East Technical University.

Antipov, Y.A. and Gao, H. (2000). Exact Solution of Integro-Differential Equations of Diffusion Along a Grain Boundary. *Quarterly Journal of Mechanics and Applied Mathematics*. 53: 645.

Bailey, N.T.J. (1964). *The Elements of Stochastic Processes*. New York: John Wiley & Sons.

Batiha, B., Noorani, M.S.M., and Hashim, I. (2008). Numerical Solutions of Nonlinear Integro-Differential Equations. *International Journal of Open Problems in Computer Science and Mathematics* 1: 34.

Cont, R. and Voltchkova, E. (2005). Integro-Differential Equations for Option Prices in Exponential Levy Models. *Finance and Stochastics* 9: 299.

Davis, H.T. (1962). *Introduction to Nonlinear Differential and Integral Equations*. New York: Dover Publications.

Farley, K.J. and Morel, F.M.M. (1986). Role of Coagulation in the Kinetics of Sedimentation. *Environmental Science and Technology* 20: 187.

Fedotov, S. (2001). Front Propagation into an Unstable State of Reaction-Transport Systems. *Physical Review Letters* 86: 926.

Ganji, D.D., Nourollahi, M., and Rostamian, M. (2007). A comparison of Variational Iteration Method with Adomian's Decomposition Method in Some Highly Nonlinear Equations. *International. Journal of Science and Technology* 2: 179.

He, J.H. (1999). Variational Iteration Method—A Kind of Nonlinear Analytical Technique: Some Examples. *International Journal of Non-Linear Mechanics* 34: 699.

He, J.H. (2000). Variational Iteration method for Autonomous Ordinary Differential Systems. *Applied Mathematics and Computation* 114: 115.

He, J.H. (2007). Variational Iteration Method—Some Recent Results and New Interpretations. *Journal of Computational and Applied Mathematics* 207: 3.

He, J.H. and Wu, X.-H. (2007). Variational Iteration Method: New Development and Applications. *Computers and Mathematics with Applications* 54: 881.

Huebner, K.H. (1975). *The Finite Element Method for Engineers*. New York: John Wiley & Sons.

Jackiewicz, Z., Rahman, M., and Welfert, B.D. (2008). Numerical Solution of a Fredholm Integro-Differential Equation Modeling θ-Neural Networks. *Applied Mathematics and Computation* 195: 523.

Jassim, A.M. (2012). A Modified Variational Iteration Method for Schrodinger and Laplace Problems. *International Journal of Contemporary Mathematical Sciences* 7: 615.

Matinfar, M., Hosseinzadeh, H., and Ghanbari, M. (2008). A Numerical Implementation of the Variational Iteration Method for the Lienhard Equation. *World Journal of Modeling and Simulation* 4: 205.

Medlock, J. and Kot, M. (2003). Spreading Disease: Integro-Differential Equations Old and New. *Mathematical Biosciences* 184: 201.

Merdan, M. (2009). He's Variational Iteration Method Applied to the Solution of Modeling the Pollution of a System of Lakes. *DPU Fen Bilimleri Dergisi* 18: 59.

Pour-Mahmoud, J., Rahimi-Ardabili, M.Y., and Shahmorad, S. (2005). Numerical Solution of the Volterra Integro-Differential Equations by the Tau Method with Chebyshev and Legendre Bases. *Applied Mathematics and Computation* 170: 314.

Prajapati, R.N., Mohan, R., and Kumar, P. (2012). Numerical Solution of Generalized Abel's Integral Equation by Variational Iteration Method. *American. Journal of Computational Mathematics* 2: 312.

Ramkrishna, D. (2000). *Population Balances: Theory and Applications to Particulate Systems in Engineering*. San Diego: Academic Press, Inc.

Randolph, A.D. and Larson, M.A. (1988). *Theory of Particulate Processes*, 2nde. San Diego: Academic Press, Inc.

Singh, P.N. and Ramkrishna, D. (1977). Solution of Population Balance Equations by MWR. *Computers and Chemical Engineering* 1: 23.

Weinstock, R. (1974). *Calculus of Variations with Applications to Physics and Engineering*. New York: Dover Publications.

Zhao, J.M. and Liu, L.H. (2007). Second-Order Radiative Transfer Equation and its Properties of Numerical Solution using the Finite-Element Method. *Numerical Heat Transfer, Part B* 51: 391.

10

Time-Series Data and the Fourier Transform

10.1 Introduction

At the very beginning of this course, we pointed out that some problems are not deterministic. Many real processes include nonlinear, stochastic components that defy conventional modeling efforts. And yet, it is essential that scientists and engineers be able to interpret and analyze such phenomena so that some level of confidence with respect to outcomes can be achieved. The purpose of this chapter is to provide a few tools that can be used in such scenarios.

In the applied sciences it is common to record experimental observations using a time interval Δt (and in the context of this chapter we refer to this as sampling an analog waveform). Such a sequence of data is called a time series. We might collect these data to support a model or a hypothesis, or we might obtain them with the hope that a suitable model could be identified later. It is possible that such data might reveal particular functional behavior that could be periodic. Alternatively, the data might contain multiple periodicities or, in the worst case, exhibit fluctuations without a discernable period—that is, the phenomenon under study might be chaotic. Let us examine some meteorological data from the Bradford Station (UK) in Figure 10.1.

In the form shown above, these data are not very revealing. But, if we change to a line plot and expand the horizontal (time) axis, a very different picture emerges.

Of course we see the expected behavior in Figure 10.2; over a time period of 100 months there are about 8+ cycles in the data set; the annual variation is now completely obvious. The peak (and trough) temperatures occur at 12-month intervals.

Annual (local) climate variations are simple enough that no tools are necessary to see and comprehend periodic behavior of the kind shown in Figure 10.2. We do not have to look far, however, to find greater challenges. Let us consider the Dow Jones Industrial Average (DJIA) from 1896 to 2024 presented in Figure 10.3 (these data were obtained from *S&P Global*). There are several interesting features in this graph; for example, you can clearly see the rampant speculation in the 1920s ending with the market crash initiated on "Black Monday," October 28, 1929. Then, note how the post-war recovery was arrested by a period of stagnation from about 1965 through 1982—the DJIA entered the 1960s over 600 and during 1982 was as low as about 770. If those numbers are indicative of the stock prices of interest to us (assuming we were long in our positions), then we were certain to be disappointed since the annual increase in value was only about 1.3%. In more recent times, the effect of the global financial crisis of 2007–2008 (brought about by deregulation in the banking sector) is quite apparent. This disaster resulted in the collapse of Bear Stearns, Lehman Brothers, Washington Mutual, and many others. Our efforts to interpret these data are very much complicated by the near-random effects of geopolitical tensions, ill-advised government policies, market manipulations, and automated "robot" trading. Please note that the magnitude of market fluctuations in recent years appears attenuated due to the semi-log presentation of the data; this was necessary to cover the entire DJIA range (about 40–42,000).

Fluctuations in the DJIA are apparent in Figure 10.3, but there does not seem to be any obvious regular, periodic behavior. This is important, because if we were able to identify a periodicity (or periodicities) in these data we might be able to anticipate the peaks and valleys, and become very, very rich. What we need is a tool, or a method, by which we could identify periodicities in time-series data when they exist. *Harmonic* (or Fourier) analysis provides us with a means to do this—we can assess fluctuations in a time series by comparison with sinusoids. Even more useful is *spectrum analysis* (see Bingham et al., 1967 for example), which allows us to identify the tendency for oscillations of a given frequency to appear.

Before we discuss spectrum analysis, however, we need to make some observations about the types of data we will be seeing. Suppose we are measuring a phenomenon like point velocity, v, in a turbulent flow, and we collect these data over some time period, T. The length of this period of observation might be dictated by the storage capacity of the recording

Problem Solving in Engineering: Analytical Mathematics and Numerical Analysis, First Edition. Larry A. Glasgow.

Companion website: www.wiley.com/go/glasgow/problemsolvingengineering

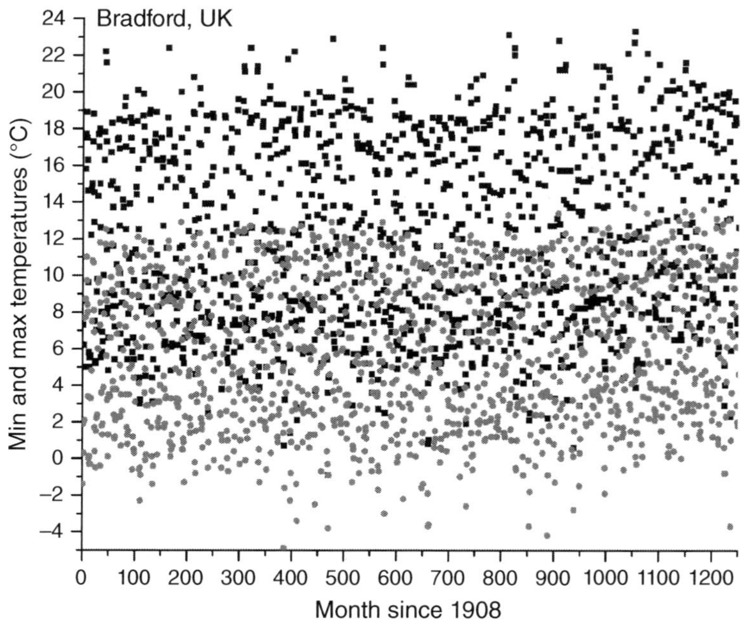

Figure 10.1 A scatter plot of the minimum and maximum (average) temperatures recorded at the Bradford Station (UK) for every month beginning in 1908. The maximum temperatures are the filled black squares, and the minimum temperatures are the filled gray circles. These data were obtained from the metoffice.gov.uk site.

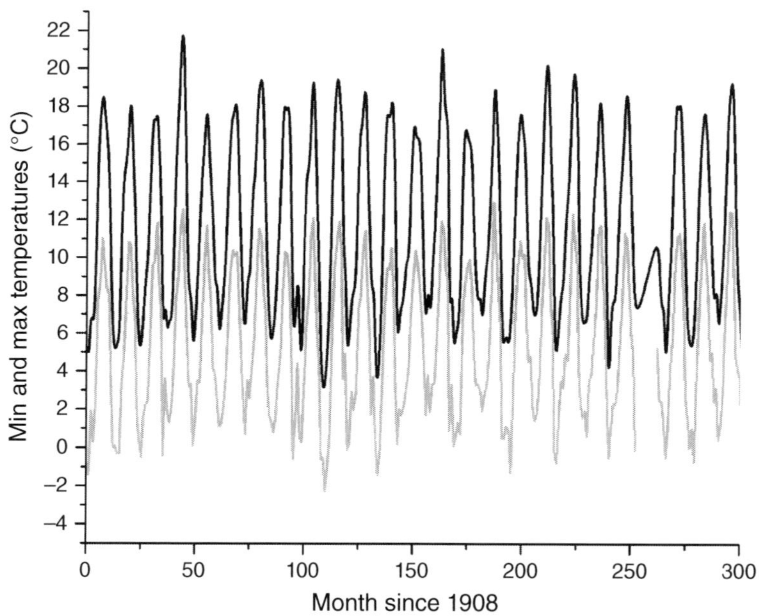

Figure 10.2 Line plot of temperature data from the Bradford Station with an expanded time axis. The black curve represents the average maximum temperatures, and the gray curve the average minimum temperatures.

instrument, but let's assume we can repeat the data collection period N times. This set of N time-series records is called an *ensemble*. If the first two moments of the observed variable do not change as the set size (N) becomes very large, the data are said to be *weakly stationary*. That is, the mean and the variance of the measured velocity, v, will be the same at any instant in time. Unfortunately, many phenomena of interest to engineers and scientists are *nonstationary*; for examples, consider motion of a marine pier subjected to wave action, an earthquake with aftershocks, or the oscillation of an aircraft wing encountering atmospheric turbulence. In such cases, no two data sets will yield the same results, and this is typical of many events involving fluctuating signals. We can contrast such outcomes with experiments in laboratory settings where we may

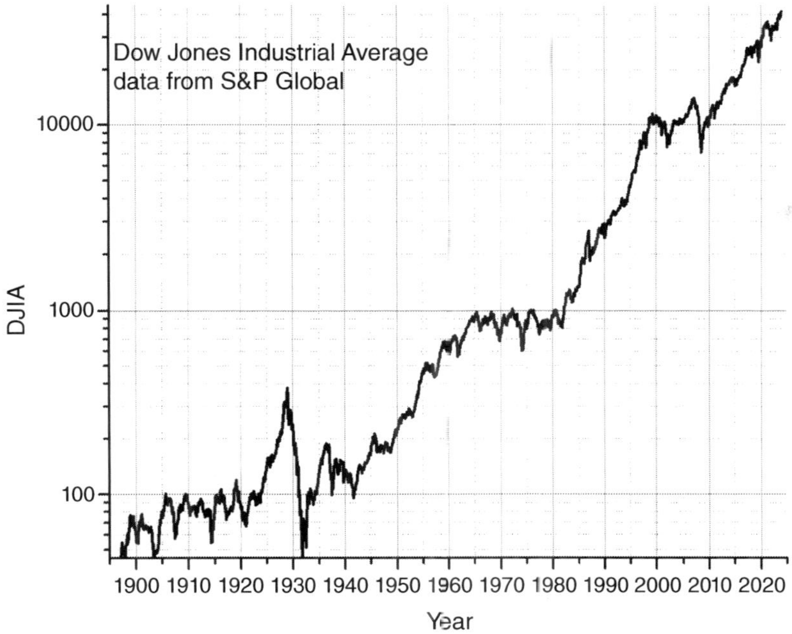

Figure 10.3 The Dow Jones Industrial Average from 1896 to 2024. These data were obtained from *S&P Global*. The market collapse in 1929 and the global financial crisis of 2007–2008 are prominent features. The overall trend of the DJIA is roughly approximated by 50 exp (0.004592 [y-1896]$^{3/2}$).

be able to control the parameters and ensure the data are stationary. This is important because proper analysis of nonstationary data is much more difficult, and for this reason we will treat all of the examples in this chapter as quasi-stationary. Finally, we make two other observations regarding our data collection process: The signal may include *noise* emanating from circumstances beyond our control, and this will yield a data record contaminated with fluctuations *not related* to the phenomenon of interest. Most measurements made by engineers and scientists are analog (A), and they require A-to-D (digital) conversion to permit computer processing. Ramirez (1985) notes that A-to-D conversion carried out by a device like a digital storage oscilloscope (DSO) may result in time jitter (this occurs when recording does not start at the exact same point on a captured waveform) and quantizing noise when an analog value does not get represented exactly. Quantizing errors are non-random, and they occur automatically with any digitizing instrument that employs finite word length (which is all of them). Such noise is small amplitude, and it can be further reduced by using a digitizer with greater word length whenever possible. Note that *any* measurement of a random variable ϕ may include errors that we can characterize in a general way as either *random* or *bias*.

Bias errors are systematic (and usually of the same size) and can result from a data-processing calculation or an improperly calibrated measurement instrument. Clearly, if a measurement operation or process includes significant bias error then an estimate of the mean value, $\overline{\phi}$, will not have much significance, and any subsequently computed statistics will be wrong.

Random errors arise from the limitations of our sampling process—which of course must be finite. For example, suppose we are sampling some stationary random variable, ϕ, and we obtain a sequence of 100 values. This data set will allow us to calculate both a sample mean, $\overline{\phi}$, and a variance,

$$\overline{\phi} = \frac{1}{100}\sum_{i=1}^{100}\phi_i \quad \text{and} \quad \sigma_{100}^2 = \frac{\sum_{i=1}^{100}(\phi_i - \overline{\phi})^2}{100}. \tag{10.1}$$

The random error associated with our observations of the variable, ϕ, is the standard deviation, $\sqrt{\sigma^2}$, and its value will be affected by the accuracy of our estimate for $\overline{\phi}$. Now if we repeat the data collection process a second time we will find that the computed mean and variance from the second set will differ from the first. Continuing this process, suppose we collect some large number, N, of 100 value sequences; we can use this ensemble to obtain a *grand* or *pooled* mean:

$$\overline{\phi}_{\text{pooled}} = \sum_{i=1}^{N}\frac{\overline{\phi}_i}{N}, \tag{10.2}$$

and naturally this will be a better estimate of the population mean than any of the individual $\overline{\phi}_i'$s. Similarly, if we can obtain only one data set, but we collect 1,000 or 10,000 values instead of 100 we would expect to obtain better results. We will explore random errors in time-series data further in an exercise at the end of this chapter.

Finally, we emphasize that waveform detection may be affected by the limitations of the recording device. For example, suppose we are using a digital storage oscilloscope (DSO) that can store 4,000 points at some sampling interval, Δt. Thus, the total period of observation is $4{,}000\Delta t$, and it is certainly possible that a long-period oscillation might not be observed. Both the size of Δt and the length and/or number of data records may impact spectral analysis, and the errors associated with noise and the finite sampling period will affect the interpretation of the results. An extensive treatment of the consequences of sampling errors upon the analysis of the data is beyond the scope of our discussion, but greater detail is available in any number of texts and monographs. The reader interested in this subject may find Chapters 2 and 11 in Bendat and Piersol (1980) particularly useful.

10.2 A Nineteenth-Century Idea

The detection of solar and lunar periodicities by man dates back many thousands of years. But during the eighteenth and nineteenth centuries scientists and mathematicians began to look at many other phenomena with the idea that some of them—such as earthquakes—might occur periodically. Knott (1897), for example, concluded that there was *probably* a connection between lunar tides and earthquakes by means of a Fourier series. Schuster (1897) however argued that one should consider the *relative magnitudes* of the Fourier coefficients compared to those obtained if the events in question were perfectly random.

Schuster proposed the following assessment: Suppose one has a time series, $y(t)$, and computes the integrals,

$$A = \int_{t}^{t+T} y \cos \kappa t dt \quad \text{and} \quad B = \int_{t}^{t+T} y \sin \kappa t dt, \tag{10.3}$$

where T is a given time interval. We now define R as:

$$R = \sqrt{A^2 + B^2}. \tag{10.4}$$

If $y \approx \cos \kappa t$, then R will increase as the time interval, T, increases. However, if $y \approx \cos \lambda t$, where $\lambda \neq \kappa$, then R will fluctuate about some constant value as T increases. We will explore this proposal using a function, $y(t)$, that consists of a limited number of sinusoids Figure 10.4.

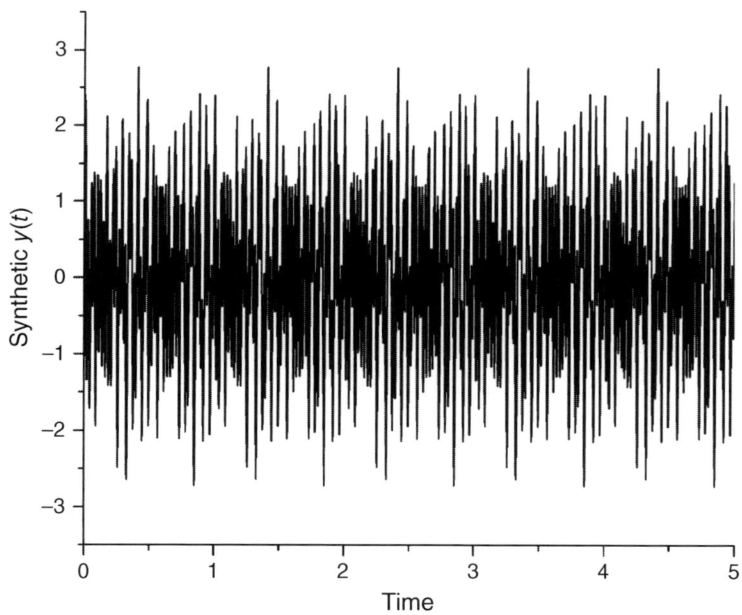

Figure 10.4 A synthetic signal constructed from four sinusoids with radian frequencies, 30π, 34π, 58π, and 84π. The fact that there are only a few characteristic frequencies is obvious.

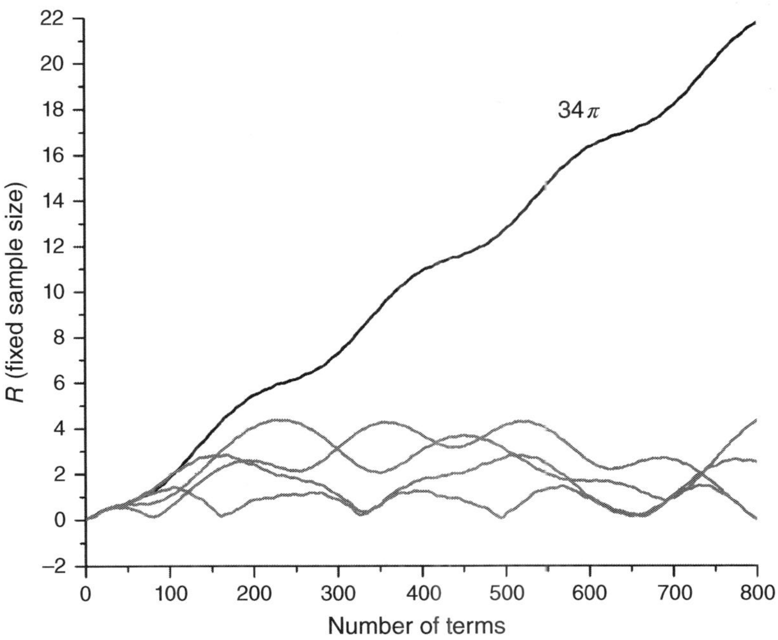

Figure 10.5 Application of Schuster's test to the data shown in Figure 10.4. The calculations were made for a single value of T using five radian frequencies: 29π, 31.5π, 34π, 36.5π, and 39π. One of the characteristic frequencies of the data set shown in Figure 10.4 has been clearly identified; the curve for 34π is growing without limit.

Our plan is to fix the time interval, T, and then compute R for five different frequencies with 34π in the center of the range tested. We can then increase T and repeat the calculation as many times as we wish.

We find through the curves shown in Figure 10.5 that the test recommended by Schuster is capable of identifying periodicities in a given data set. As a practical tool, however, it is severely limited; it would be computationally expensive to test a time series for every possible frequency and in the Victorian age—when the test was devised—it would have been nearly impossible.

In the middle of the twentieth century the importance of *spectrum estimation* in electrical engineering applications (such as radio, radar, and electronic communications) drove an intensified interest in the Fourier transform. This culminated in the Cooley–Tukey algorithm, and variants, which are broadly known as the *fast Fourier transform* (or FFT). James Cooley (1987) provided a wonderful first-hand account of the development of the FFT, and he pointed out that Gauss had grasped the essential ideas in the nineteenth century. Of course the pace at which the use of the discrete Fourier transform (DFT) expanded was driven by the development and wide-spread use of digital computers in the twentieth century. Before we discuss the FFT, however, we need to explore some other aspects of fluctuating signals.

10.3 The Autocorrelation Coefficient

Suppose we observe a signal, $y(t)$, that fluctuates about some constant mean value (as we noted previously such a process is said to be statistically stationary). For the synthetic data shown in Figure 10.4, it is clear that the mean is approximately zero. We let these observations be represented by

$$y(t) = \overline{y} + y', \tag{10.5}$$

where $\overline{y}$ is the mean and y' is the fluctuation. We will define an *autocorrelation coefficient* in the following way:

$$\rho(\tau) = \frac{\overline{y'(t)y'(t+\tau)}}{\overline{y'^2}}. \tag{10.6}$$

The quantity in the denominator is referred to as the *mean-square* fluctuation; naturally the square root of this quantity is the *rms* fluctuation. If the time offset, τ, is zero, then the autocorrelation coefficient is equal to one, $\rho(\tau = 0) = 1$. How $\rho(\tau)$ behaves depends for larger τ of course upon the nature of $y(t)$. If y is periodic (or consists of a set of periodic functions) then

$\rho(\tau)$ will show strong correlation at distinct values of τ; in such a situation the oscillatory correlation is said to be *ringing*. If on the other hand the signal under observation is random or chaotic, then the fluctuations should become uncorrelated as the offset (or time delay) becomes large and $\rho(\tau \to \infty) = 0$.

Historically the correlation coefficient has been used extensively in fluid mechanics, and especially in the study of turbulence. The development of the hot-wire anemometer in the twentieth century produced large sets of time-series data that were observed in real time with oscilloscopes and processed by analog instruments such as spectrum analyzers. It is important though that we recognize that the autocorrelation has a significant limitation: It is *not* reversible. That is, since one *cannot retrieve* the original data from the autocorrelation, the process of obtaining $\rho(\tau)$ automatically entails a loss of information. Consequently, the autocorrelation is of little value for tasks associated with signal processing and image manipulation.

Nevertheless, we will begin exploration of the data set shown in Figure 10.4 by computing the autocorrelation coefficient. Since these data were constructed from a limited set of sinusoids, we should expect an *oscillatory* $\rho(\tau)$.

The data we used to compute the autocorrelation coefficient shown in Figure 10.6 were constructed from four periodic functions added together and previously illustrated in Figure 10.4. The computation of $\rho(\tau)$ reveals that there were *regular fluctuations* in the time-series data; note the strong recurring peaks in the correlation at (number of time intervals) 19, 28, 38, 48, 57, 67, etc. This suggests that something with a period of about 9–10Δt was present in the time series, corresponding to a frequency range of about 250–279 rad/s. Of course, we know that there was a very strong sinusoidal component present at 263.9 (84π) rad/s—right in the middle of the indicated band of frequencies.

In simple cases of this type (where a regular oscillation in the time series is obvious) one might consider developing a model by fitting a periodic function(s) to the data directly. For example, suppose through our examination of the data we concluded that a function of the type,

$$y(t) = A\sin(\omega t) + B\cos(\omega t), \tag{10.7}$$

might adequately represent the behavior. Assuming we were able to estimate the radian frequency directly, we could then minimize the sum of the squares of the deviations by setting:

$$F = \sum (y - A\sin(\omega t) - B\cos(\omega t))^2 \tag{10.8}$$

and setting the derivatives equal to zero:

$$\frac{\partial F}{\partial A} = 0 \quad \text{and} \quad \frac{\partial F}{\partial B} = 0. \tag{10.9}$$

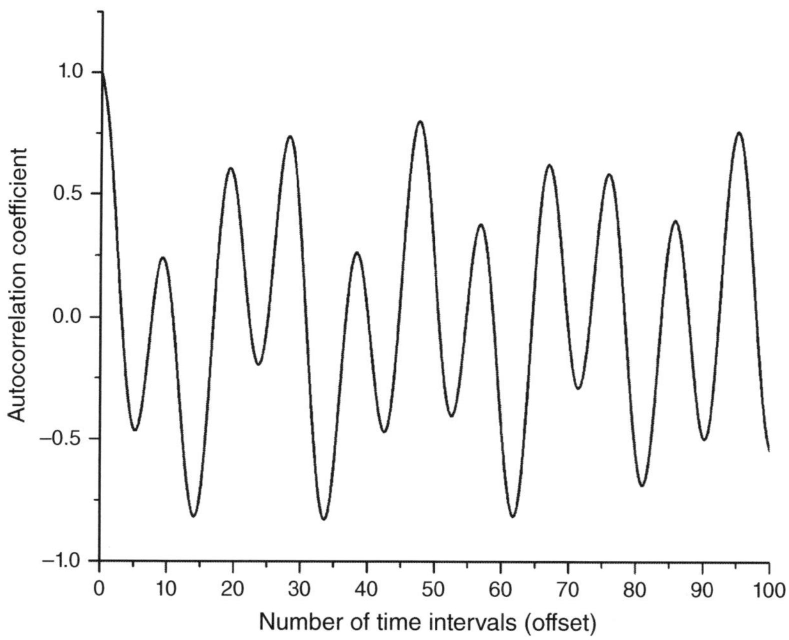

Figure 10.6 Autocorrelation coefficient for synthetic signal comprising four sinusoids. The expected *ringing* behavior is apparent.

Therefore, we could use our data to seek solutions for the equations:

$$-2\sum(y - A\sin(\omega t) - B\cos(\omega t))\sin(\omega t) = 0 \tag{10.10}$$

and

$$-2\sum(y - A\sin(\omega t) - B\cos(\omega t))\cos(\omega t) = 0. \tag{10.11}$$

This least-squares estimation technique is fine, as long as we have a sharply defined periodic behavior where the radian frequency (or a very small number of radian frequencies) is obvious. In such cases we are merely trying to identify the amplitude and the phase. Of course we could add the radian frequency to the determination if necessary. For many problems of interest to us, however, the observed behavior will be altogether too complicated to let direct model development be a realistic alternative.

10.3.1 Example of Periodogram Construction

Assume we have time-series data $x(t)$, and in particular let them be generated by

$$x(t) = \cos(\omega_1 t) + 1.25\sin(\omega_2 t) + Z \tag{10.12}$$

where $\omega_1 = 0.333$ rad/t and $\omega_2 = 0.409$ rad/t. Let Z be a uniform random number between zero and 1, and therefore the expected value of $x(t)$—which we will estimate with $\bar{x}$—will be ½. We will let t assume integer values from 1 to 1,024 (2^{10}). We will set

$$A = \frac{2}{n}\sum(x_i - \bar{x})\cos(\omega t) \quad \text{and} \quad B = \frac{2}{n}\sum(x_i - \bar{x})\sin(\omega t), \tag{10.13}$$

and let $R(\omega)^2 = A(\omega)^2 + B(\omega)^2$. The *periodogram* for the time-series data, $x(t)$, is $I = \frac{n}{8\pi}R(\omega)^2$. In order to determine I we must find A and B by summation for the range of frequencies of interest. If we take $\Delta t = 1$ s, then $f_{Nyq} = 1/2$ Hz, or π rad/s; we cannot obtain any information for frequencies larger than π rad/s, but this is of no matter since the two frequencies of importance in our data set are 0.333 and 0.409 rad/s. Our plan is to initialize ω, perform the indicated summation to find $R(\omega)^2$, increment ω, and repeat. Since we know exactly how the time-series data were generated we will terminate the calculations when $\omega = 0.85$ rad/s. Therefore, if we select $\Delta\omega = 0.001$ rad/s the summations required to determine A and B will be performed each 850 times. This process is computationally laborious, and it would not be feasible if we were trying to identify periodicities in unfamiliar time-series data with the requirement that results be obtained *very rapidly*. For this example, however, the time required is of no consequence and, in any event, a modern PC processor with a compiled BASIC program requires only a fraction of a second for execution of this task. The results are illustrated in Figure 10.7.

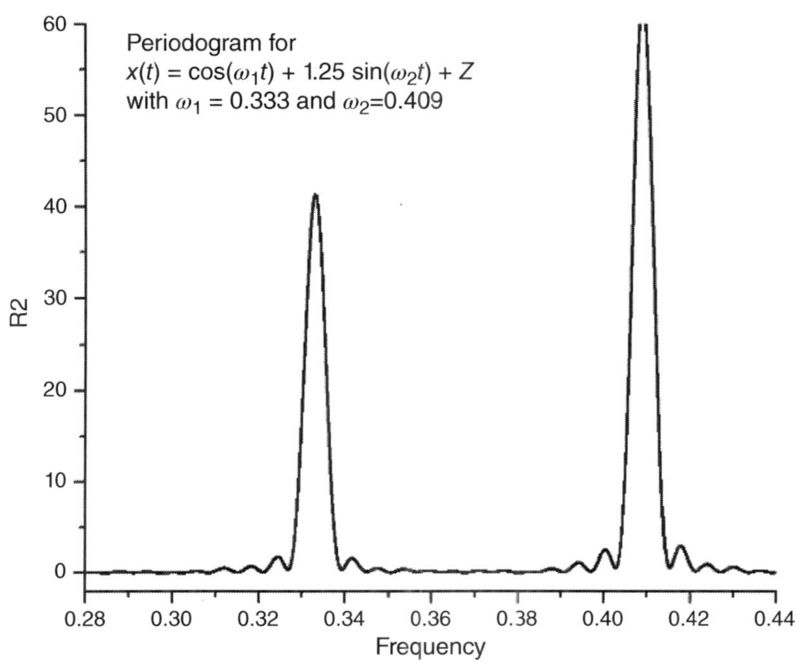

Figure 10.7 Periodogram for $x(t) = \cos(\omega_1 t) + 1.25\sin(\omega_2 t) + Z$.

We find that the periodogram contains two dominant peaks centered at 0.333 and 0.409 rad/s as it must; we will have occasion to make use of the procedure sketched here later in this chapter. H. J. Wearing (who uses spectral analysis in the context of infectious disease ecology) provided a meaningful synopsis of periodogram construction: *The periodogram gives us a quantitative measure of the contribution of individual frequencies to the time-series regression.*

10.4 A Fourier Transform Pair

We saw previously that the autocorrelation coefficient, $\rho(\tau)$, though not reversible, could reveal much information about the time-series data under study. In fact, the autocorrelation and the *power spectral density* (or spectrum), $S(\omega)$, contain the same information, and they are related through the Fourier transform pair:

$$\rho(\tau) = \int_{-\infty}^{+\infty} \exp(i\omega\tau)S(\omega)d\omega \quad \text{and} \quad S(\omega) = \frac{1}{2\pi}\int_{-\infty}^{+\infty} \exp(-i\omega\tau)\rho(\tau)d\tau. \tag{10.14}$$

Since the autocorrelation coefficient is an even function, and since negative frequencies do not hold any physical meaning for us, it is standard practice to write the *one-sided* spectrum:

$$S_1(\omega) = \frac{1}{\pi}\int_{0}^{\infty} \cos(\omega\tau)\rho(\tau)d\tau. \tag{10.15}$$

We have already computed $\rho(\tau)$ for the data shown in Figure 10.4, so why not take this result and use it to evaluate the spectrum using eq. (10.15)? We will do this, but we must keep in mind three important limitations:

1) We will not be able to learn anything about fluctuations occurring more rapidly than the *Nyquist cut-off frequency*. For a sample interval of Δt,

$$f_{Nyq} = \frac{1}{2\Delta t}. \tag{10.16}$$

For the data shown in Figure 4, $\Delta t = 0.0025$ s, so $f_{Nyq} = 200$ Hz or 1256 rad/s.

2) We cannot find out anything about fluctuations that occur so slowly that one cycle does not fit within our total time of observation. For infrequent (very-long-period) oscillations we must acquire a *very lengthy* set of time-series data.

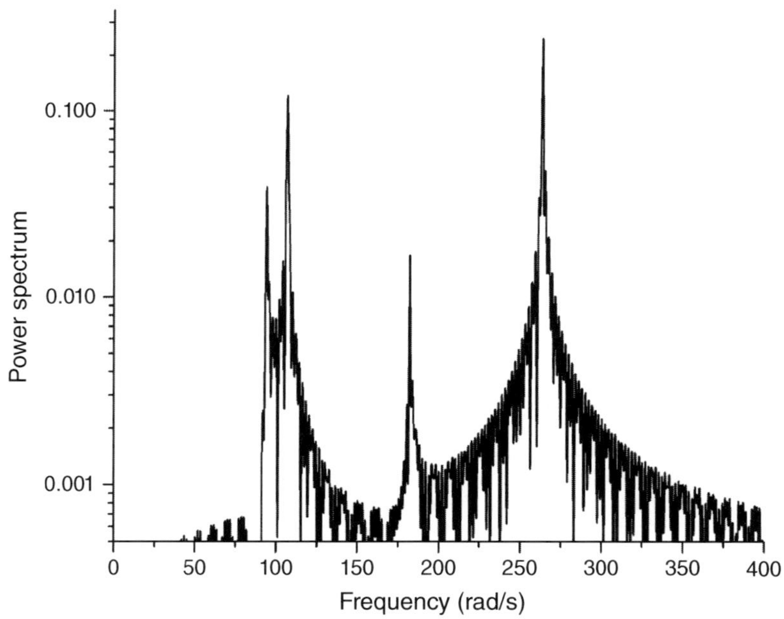

Figure 10.8 Computed spectrum for data of Figure 10.4. The important radian frequencies in the data set were approximately 94.2, 106.8, 182.2, and 263.9 rad/s. All four appear as dominant features in the spectrum shown here. The oscillation at 263.9 rad/s had the largest amplitude of the four sinusoids. Make particular note of the broadening apparent at the sides of the spectral peaks; we will discuss that further a little later in this chapter.

3) Since the time-series data are sampled for a finite time period, our calculation of the autocorrelation coefficient must be regarded as an estimate.

The spectrum shown in Figure 10.8 was computed by repeated integration of the estimated autocorrelation coefficient, after $\rho(\tau)$ had been obtained from the experimental data by calculation of the mean, the fluctuation for each observation, and the mean-square fluctuation. Although this example demonstrates that such a procedure will work, it is computationally expensive and for many applications, simply too time consuming. We need a process that will allow us to obtain the spectrum more rapidly, and the approach we will employ is the fast Fourier transform (FFT).

10.5 The Fast Fourier Transform

Brigham (1988) describes the FFT as *ubiquitous*, and indeed, he provides a list of about 75 applications (which are incredibly diverse). To highlight just a few, consider that the FFT has been used for: studies of aircraft wing flutter, acoustic imaging, diagnosis of airway obstruction, echo or reverberation elimination, speech recognition, chromatography, phased-array antenna analysis, video bandwidth compression, and image restoration. Quite recently the FFT has been used in the study of microseismic events triggered by hydraulic fracturing in gas-bearing shales; see Warpinski et al. (2012) for example spectra.

We mentioned previously that the rapid growth of applications for the FFT was driven by the digital computer and by the 1965 paper published by Cooley and Tukey. Their algorithm is widely used by technical professionals around the world. Since the 1960s many versions of the algorithm have been developed, and a fair number of contributors to the literature have pointed out that Gauss actually originated the idea in the early years of the nineteenth century. However, the reality is that James Cooley, Richard Garwin, and IBM (their employer) were mainly responsible (assisted by the concurrent development of fast analog-to-digital converters) for the explosive growth in the number of applications of the FFT. Rockmore (1999) and Cooley (1987) have also noted that some of the initial urgency was driven by the cold war due to the need for seismic monitoring (the detection of nuclear explosions) and acoustic detection and identification of submarines.

If we have N observations of a fluctuating signal, $y(t)$, then we can think of the Fourier transform as an interpretation of that signal in terms of sinusoids of different frequencies. Specifically, as Bloomfield (1976) points out, we can write

$$A_0 = \frac{1}{N}\sum_{n=0}^{N-1} y_n = \bar{y} \tag{10.17}$$

$$A_j = \frac{2}{N}\sum_{n=0}^{N-1} y_n \cos\omega_j t \tag{10.18}$$

and

$$B_j = \frac{2}{N}\sum_{n=0}^{N-1} y_n \sin\omega_j t. \tag{10.19}$$

If the sequence of observations consists of an *even* number, then

$$A_{N/2} = \frac{1}{N}\sum_{n=0}^{N-1} y_n \cos\omega_{N/2} t = \frac{1}{N}\sum_{n=0}^{N-1} (-1)^n y_n. \tag{10.20}$$

The time-series data can now be represented, if N is even, with

$$y_n = A_0 + \sum_{j<N/2} A_j \cos(\omega_j t) + B_j \sin(\omega_j t) + (-1)^n A_{N/2}. \tag{10.21}$$

A compact, short-hand representation of the DFT for sampled data (in the form of $X[j]$, a complex vector of length, N) can be written:

$$S(k) = \sum_{j=0}^{N-1} X(j) W_N^{jk}, \quad \text{where } W_N = \exp\left(\frac{2\pi i}{N}\right), \tag{10.22}$$

If the sampled waveform is periodic, or contains multiple periodicities, the Fourier transform will exhibit a sharp peak, or peaks. If the time-series data are not periodic, then the Fourier transform will be a continuous function of frequency.

10.5.1 Discrete Fourier Transform (DFT) Example

The logic required to carry out a DFT is quite simple, and to illustrate this we will take the (32 data point) code provided by Ramirez (1985) and modify it to increase its usefulness. The original program was written for the BASIC *interpreter*, and its execution in that mode was slow (an interpreter carries out instructions line by -line *without* converting them to machine language). However, we plan to *compile* our modification so that the executable can rapidly process larger time-series data sets. Let us assume a waveform:

$$x(t) = \cos\left(\frac{2\pi(12.5)t}{100}\right) + \sin\left(\frac{2\pi(8.333)t}{100}\right) + 3Z - \frac{3}{2}. \tag{10.23}$$

Z is a uniform random number between 0 and 1, and therefore we subtract 3/2 so that the expected value for $x(t)$ would be 0. The time parameter, t, will assume integer values (i) from 0 to 511; therefore, the "100" appearing in the denominator was added simply for scaling. The lines in this elementary program have been numbered for easy identification, and the program is set up to treat a time series with 512 discrete values:

```
#COMPILE EXE
#DIM ALL
    REM *** Adaptation of Ramirez 32-point DFT (BASIC interpreter)
    REM *** See "The FFT: Fundamentals and Concepts" by R.W. Ramirez
            GLOBAL pi,ag AS SINGLE
            GLOBAL i,k,n,ndata AS INTEGER
FUNCTION PBMAIN
    DIM x(3000) AS SINGLE:DIM xr(3000) AS SINGLE:DIM xi(3000) AS SINGLE
1            OPEN "c:DFTramirezMOD.dat" FOR OUTPUT AS #1
2                RANDOMIZE TIMER
3                pi=3.14159:ndata=512
4        FOR i=0 TO ndata-1
5        x(i)=COS(2*pi*12.5*i/100)+SIN(2*pi*8.333*i/100)+3*RND-1.5
6                            NEXT i
7            FOR k=0 TO ndata-1
8                FOR n=0 TO ndata-1
9                    ag=2*pi*k*n/ndata
10                        xr(k)=xr(k)+x(n)*COS(ag)/ndata
11                            xi(k)=xi(k)-x(n)*SIN(ag)/ndata
12                                NEXT n:NEXT k
13            FOR i=0 TO ndata-1
14                PRINT i,xr(i),xi(i)
15                WRITE#1,i,SQR(xr(i)^2),xi(i)
16                    NEXT i
17                        SLEEP 8000
18                        CLOSE:END
    END FUNCTION
```

Line 2 provides the random number generator with a new seed for each execution, the number of data points (ndata = 512) is set in line 3, the time-series data are generated in lines 4–6, the determination of the real and imaginary parts of the DFT occurs in lines 7 through 12, and the results are written to a file "*DFTramirezMOD.dat*" in line 15. Note that the real part of the transform has been written as the square root of $xr(i)^2$ so that only positive values will be stored. This program could be easily modified to handle longer time series if desired.

10.5.1.1 A Recommended Exercise

If you have experience with a high-level language, modify the code provided above appropriately for your software and carry out the computation. You should see strong spectral features at index positions $i = 43$ and $i = 64$ (note that $64\pi/256 = 0.785 = [2\pi]12.5/100$). Then, alter line 15 to simply store $xr(i)$ rather than $\sqrt{xr(i)^2}$; graph the results and note the change in the DFT plot. Why does this behavior occur?

The main problem with this DFT program is that the number of arithmetic operations for a time series with N observations scales as N^2. Thus, if one had 512,000 data points (the reason for choosing this number will be made clear in a moment), about 2.6214×10^{11} operations would be required. Modern PC processors are capable of about 5×10^9 to 1×10^{10} Flops (floating-point operations per second), thus very roughly: $\frac{2.6214\times10^{11}}{7\times10^{9}} \approx 37.5$ s. While not a ridiculous amount of time, this is far too long for processes requiring near-instantaneous feedback.

However, John Tukey had shown that if the N observations can be written as a product, $N = ab$, then the Fourier series could be written as an a-term series of b-term (each) subseries. This meant that the scaling for the required number of computations would change from N^2 to $(a + b)N$. Richard Garwin relayed this information to James Cooley and persuaded him to work on algorithm development and programming. The result, now known as the Cooley–Tukey algorithm for the FFT, actually reduces the scope of the computations to $N\log(N)$; if N is 512,000, then the computational effort is reduced to 2.92×10^6 operations. The comparison is striking; for a modern PC we would now have about $\frac{2.92\times10^{6}}{7\times10^{9}} = 0.0004$ s for the required processing time. As Cooley (1987) later described, the development of the FFT algorithm meant that the Fourier transform could now be extended to very large problems. An additional motivation for his work was the interferometer data brought to Cooley by Janine Connes; her husband, the astronomer Pierre Connes, had constructed an interferometer that could produce a time series (a data record) consisting of (the then incredibly large number) 512,000 points. The power of the Cooley–Tukey (1965) algorithm was such that it made the processing of lengthy time series of this type routine. We should note that FFT calculations are *recursive*, which simply means that the next term in a sequence is computed from previous terms. If one were using a computer with limited word length (for example, the original IBM PC used 16 bit logic with an 8 bit bus) the limited numeric precision could be a problem. Contemporary PC's use 64 bit architecture, so this is less of a concern.

For the analyst who must perform FFTs on time-series data, the options are many. There are literally hundreds of FFT codes available in the public domain, written in high-level languages including FORTRAN, BASIC, PASCAL, PYTHON, and C. For convenient working examples, see Chapter 8 in Brigham (1988). There are also numerous websites devoted to the Fourier transform. Moreover, many commercial software packages have FFT capabilities including Mathcad™, Matlab™, MS-Excel™, and Origin™. Thus just about any collection of time-series data can be analyzed, and the signal content in the frequency domain can be determined.

We will look at an illustration to underscore this point. Suppose we have a signal consisting of four sinusoids added together with a random fluctuation (Z):

$$x(t) = \sin\frac{\pi t}{2} + \sin\frac{\pi t}{10} + \sin\frac{\pi t}{5} + \sin\frac{2\pi t}{3} + Z \tag{10.24}$$

Make note of the radian frequencies employed here: $\pi/10$, $\pi/5$, $\pi/2$, and $2\pi/3$. We will allow t to assume integer values from 0 to 1,023 such that we have 2^{10} samples; then we will determine the Fourier transform, first by using Mathcad™. We will use the fast DFT function, *fft*(v). This is a radix-2 transform program; by radix-2 we mean that the number of data points, N, can be written as $N = 2^m$ data points (where m is an integer).

Implementation of Mathcad™ fft(v):

$i := 0..102:$

$$x_i := \sin\left(3.1416 \cdot \frac{i}{2}\right) + \text{rnd}(1) + \sin\left(i \cdot \frac{3.1416}{10}\right) + \sin\left(i \cdot \frac{3.1416}{5}\right) + \sin\left(i \cdot \frac{3.1416}{1.5}\right)$$

$c := \text{fft}(x)$

$N := \text{last}(c)$

$N := 511$

$j := 0..N$

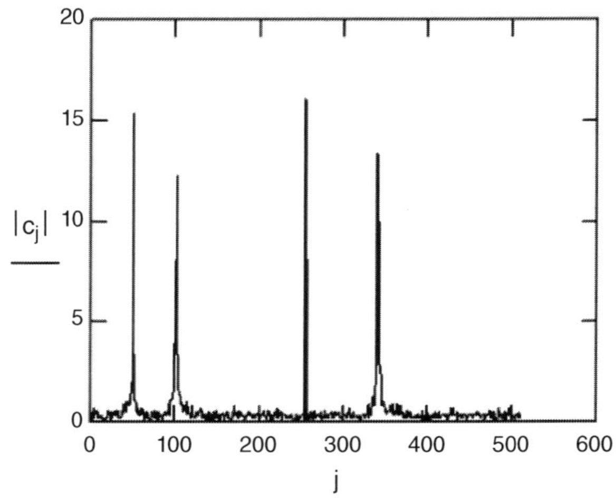

Note where the peaks are located (the j-index values) in this spectrum: approximately 50, 101, 255, and 340. The corresponding radian frequencies are determined from $\frac{j\pi}{512}$, so for example, $\frac{255\pi}{512} = 1.565 \cong \frac{\pi}{2}$.

Now we repeat the analysis of the time series, computing the Fourier transform by integration of the autocorrelation coefficient, using eq. (10.18) from Section 10.4. You will note that the two spectra shown (one from Mathcad and one from the integration of $\rho(\tau)$) are essentially identical, differing mainly due to the normalization employed. Although the results are the same, the processing speed was radically different; the computational result shown in Figure 10.9 required *several seconds,* while the execution of the *fft* function in Mathcad™ appeared to be almost instantaneous. This is the reason the FFT has become so important—it is now possible to compute Fourier transforms for large time-series data sets, and do so almost in real time. Of course, this speed is crucial to signal processing applications.

Finally, we will also test a modified version of the BASIC FFT (Fourier transform) code provided by Brigham (1988) on the very same data sequence and present the result in Figure 10.10.

In the preceding example the spectra were determined three ways: using the built-in capability, *fft*(v), in Mathcad™, by computing the autocorrelation coefficient, $\rho(\tau)$, directly from the time-series data and then integrating the results, and

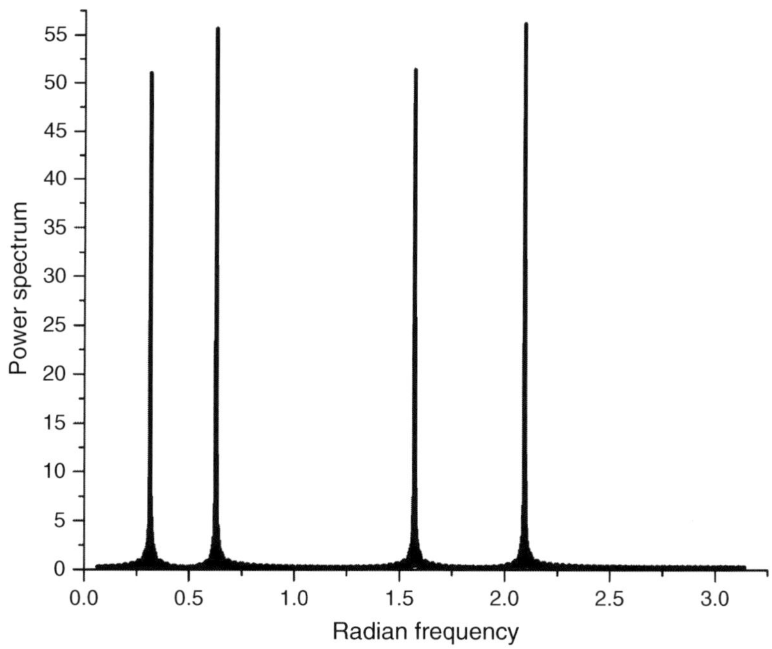

Figure 10.9 Power spectrum for the signal produced by four sinusoids, eq. (10.24). In this case, the autocorrelation coefficient, $\rho(\tau)$, was computed directly from the data and the Fourier transform was found by integration. Note that $\pi/10 = 0.314$, $\pi/5 = 0.628$, $\pi/2 = 1.571$, and $2\pi/3 = 2.094$.

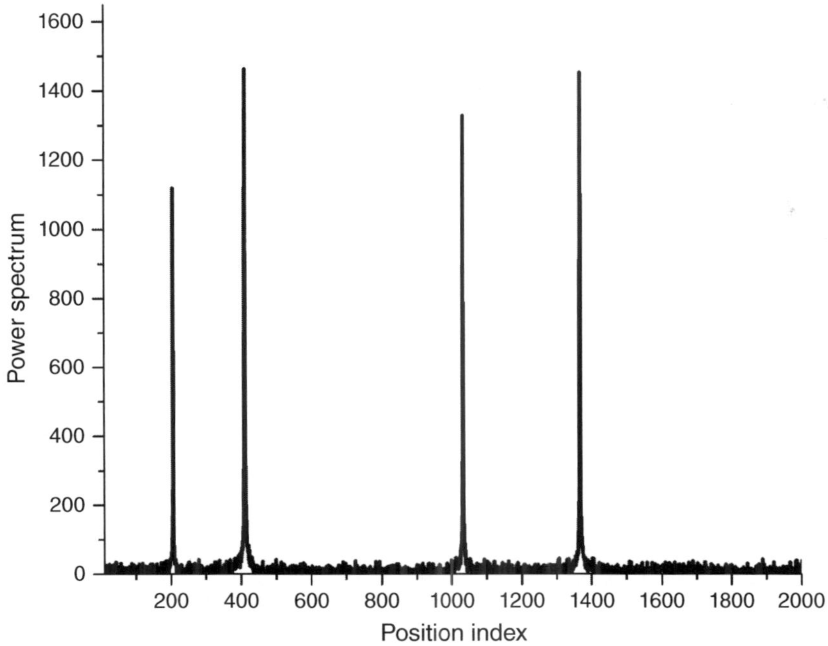

Figure 10.10 Spectrum from a modified version of Brigham's BASIC DFT (radix-2) program. Frequency is determined from the position index by: $\frac{j\pi}{2048}$. Therefore, the second spectral spike in Figure 10.10 corresponds to about $405\pi/2{,}048$ or roughly 0.2π radians/t. Recall that the signal that was transformed consisted of four sinusoids with radian frequencies of 0.1π, 0.2π, 0.5π, and 0.667π.

by using a radix-2 DFT code that was modified from a version of the program supplied by Brigham (1988). The resulting spectra are very nearly identical, and all three procedures have clearly identified the four dominant frequencies present in the time-series data. We should clarify the distinctions between the DFT and the FFT: Generally, DFT codes do not require that the number of samples be an integer power of 2 (although radix-2 DFT's are common), and they save intermediate results during processing, requiring more memory. FFT programs take advantage of the efficiencies obtained by having the number of samples be 2^n, and as a result they tend to be significantly faster in execution.

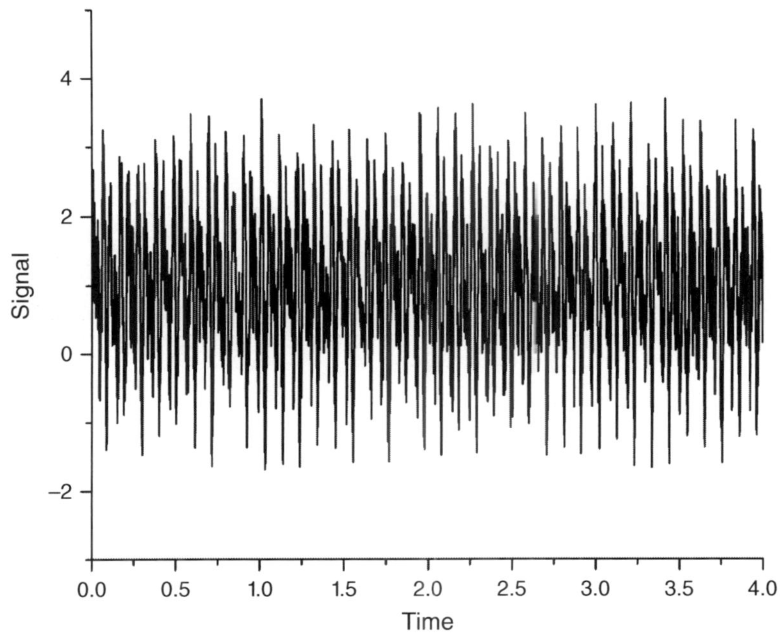

Figure 10.11 Synthetic signal consisting of 11 sinusoids with a random contribution added.

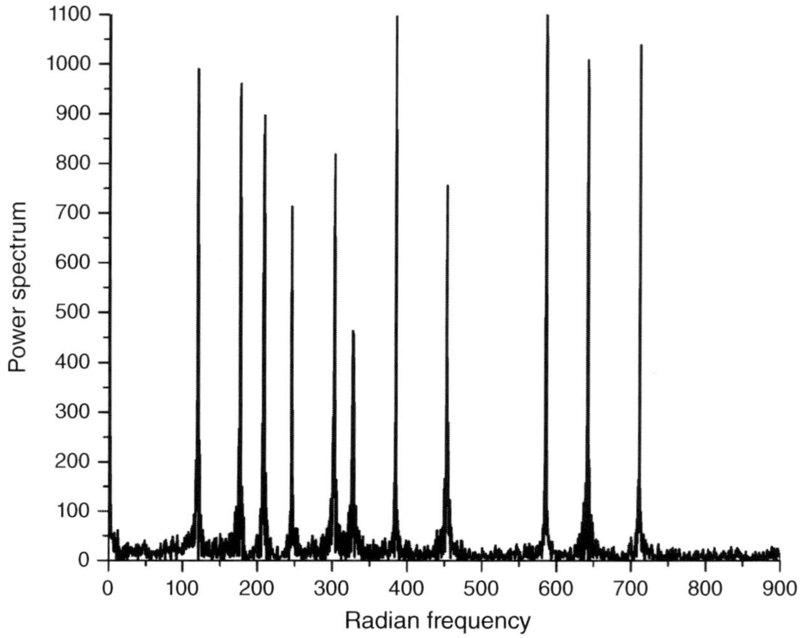

Figure 10.12 The computed spectrum has captured all 11 sinusoidal contributions; the original frequencies were 119.38, 175.93, 207.3, 245, 301.59, 326.73, 383.28, 452.4, 584.34, 640.89, and 710 rad/s.

The example we explored above did not present much challenge for the FFT/DFT procedure. Let us look at a more complicated signal (one with 11 distinct frequencies) that also includes a random contribution. To further complicate things we will truncate the time series at 2,800 values and then append 1,296 zeros to reach 2^{12} samples when we compute the spectrum. The raw data appear as shown in Figure 10.11.

Now we will compute the Fourier transform to see if all of the different elements of this signal have been captured. Keep in mind that we have added a random component *and* filled out the data set (to reach 4,096, or 2^{12}, observations) with zeros. We know that the kinds of abrupt changes in the data that are generated by adding a random contribution and by appending zeros to obtain 2^m total samples will produce additional contributions in the spectrum that may complicate identification of the crucial periodicities.

The computed transform shown in Figure 10.12 reveals how powerful the DFT really is; we constructed a synthetic signal that was a bit more difficult, and the algorithm has successfully identified all 11 of the contributing sinusoids. And this is despite the fact that the data series was truncated (filled out with zeros) and that a strong random contribution was added to the signal.

10.6 Aliasing and Leakage

We described the Nyquist cut-off frequency previously, noting that if the sample interval, Δt, was *too large*, we would not be able to obtain any useful information about a high-frequency (oscillating) waveform. More generally, too large a Δt produces *aliasing*, which yields a distorted spectrum. Suppose we were interested in the *sum* of the two sinusoids shown in Figure 10.13.

If we had the poor judgment to begin sampling at $\pi/2$, and then further to employ an interval of 2π, the sampling of the *sum of the two waveforms* would consist only of the sequence: 1.75, 1.75, 1.75, etc., and absolutely none of the important dynamic behavior would be revealed. It is critical that we pick a sampling Δt that can capture any oscillating behavior that may be occurring. Naturally, if we compute a DFT for time-series data sampled too infrequently, we will not get the correct spectrum. Suppose for example that we sampled the *sum* of the two waveforms illustrated in Figure 10.13 using a time interval of $\pi/4$; we would see the sequence, 0.75, −0.0429, 1.75, −0.0429, 0.75, −1.457, −0.25, −1.457, 0.75, −0.0429, 1.75, −0.0429, 0.75, etc. These equally -spaced samples could be represented by *several* sinusoids; the consequence is that the spectrum folds, or overlaps. If *aliasing* is suspected, the easiest fix is to halve the sample interval, compute the spectrum, and note any changes, and then repeat the process until the spectrum does not change.

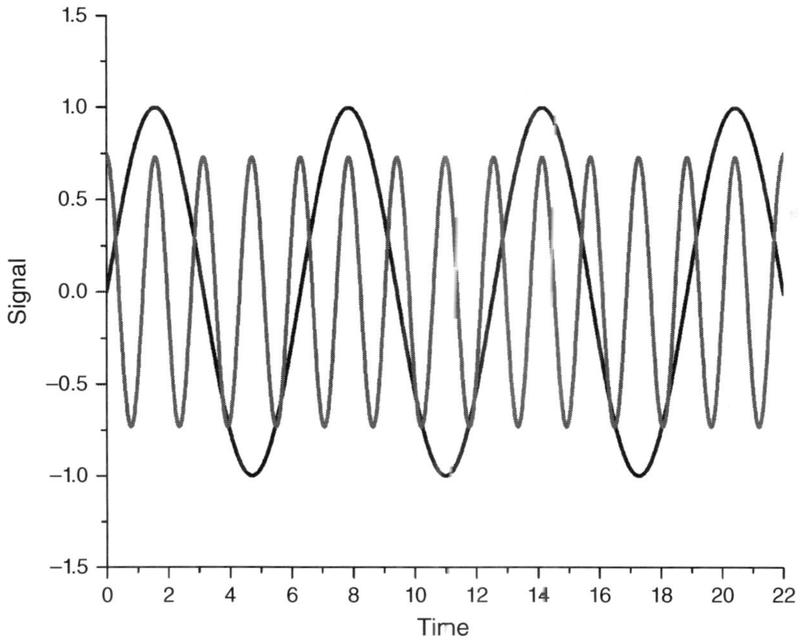

Figure 10.13 A plot illustrating $\sin(\omega t)$ and $0.75\cos(4\omega t)$, with $\omega = 1$.

Leakage appears in a DFT spectrum in the form of what are called "sidelobes." This feature of computed spectra results when a particular harmonic component leads to nonzero transform values at *other* frequencies. It can be exacerbated by the fairly common practice of filling out an incomplete data set with zeros to execute a radix-2 DFT; for example, if we had 1,750 observations, we might add 298 zeros to get to 2,048 (2^{11}). Any phenomenon that produces sharp (abrupt) changes in the time domain will result in additional spectral components in the frequency domain. Let us illustrate this point by generating time-series data with the simple function,

$$y(t) = \sin(120t) + 0.85\cos(180t). \tag{10.25}$$

We apply the DFT with the result shown in Figure 10.14a. Next we truncate the time series and append 1,320 zeros to provide a total sample of 4,096 for Figure 10.14b. Finally, we add a strong contribution to the time series utilizing the random number generator, and include the 1,320 zeros at the end to generate the spectrum shown as Figure 10.14c.

The first thing we should note about Figure 10.14a is that although the time-series data were generated with the sum of sine and cosine terms (only) the resulting spectrum is not exactly what we would expect had we obtained a *continuous* transform. A pure sinusoid in time should yield a δ–function in frequency space. Next, when we truncated the time series by taking 2,776 sampled values with 1,320 zeros appended, we produced sharp oscillations on the shoulders of the two main peaks (Figure 10.14b). And finally, in Figure 10.14c we have used the random number generator to create discontinuities in the time domain which have in turn produced many additional frequency components in the spectrum. The important conclusion that we should take from this example is that discrete sampling in the time domain can yield discontinuities that, upon DFT, yield other contributions in frequency space. These contributions appear as additional peaks or sidelobes in the spectra that are computed. One remedy—which is pretty obvious but usually impractical—is to eliminate all of the abrupt changes that may appear in the time domain.

Another common source of difficulty is *time-domain truncation* which occurs when the number of discrete sample data points is limited. In some cases our ability to obtain time-series data may be constrained (perhaps the hardware has limited storage capacity). This can adversely affect the accuracy of the DFT, and we will illustrate this problem using a signal comprised of two clean sinusoids:

$$y(t) = \cos\left(\frac{2\pi}{5.99}t\right) + \sin\left(\frac{2\pi}{2.88}t\right). \tag{10.26}$$

We will sample this function using integer spacing ($\Delta t = 1$), and we will begin with a data record length of 1,024. We will then compute the DFT for $N = 64$, and repeat using $N = 16$.

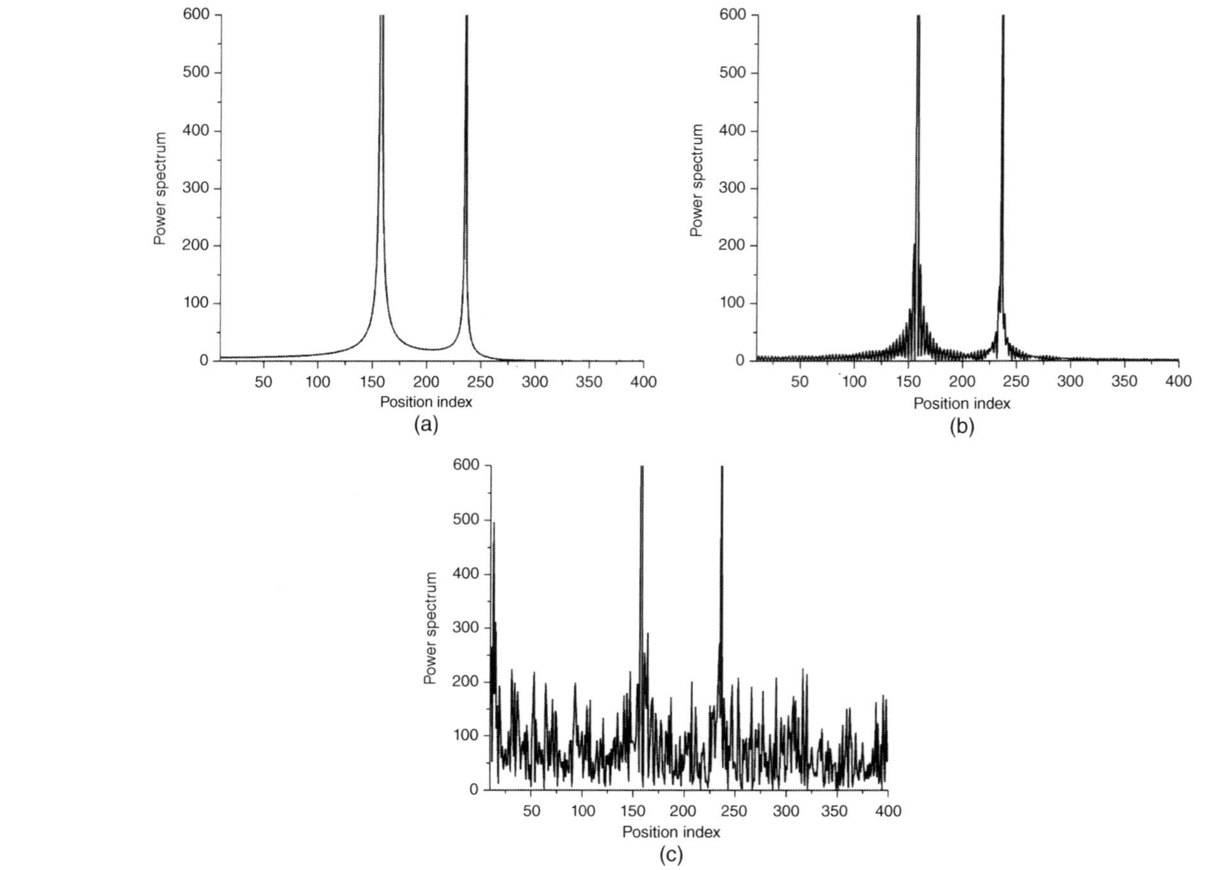

Figure 10.14 (a–c) Comparison of spectra obtained by DFT for time-series data generated by modifications to sin(120 t) + 0.85 cos(180 t). The frequencies are obtained from $\frac{j(500\pi)}{2048}$, such that $j = 155$ corresponds to 120 and $j = 235$ corresponds to 180.

The spectrum shown in Figure 10.15a ($N = 1{,}024$) is a close approximation of the (continuous) Fourier transform. In this case the spectral peaks are located at j-index values of 172 and 357, corresponding to frequencies of $\frac{172}{512}\pi = 0.336\pi$ and $\frac{357}{512}\pi = 0.697\pi$; these values are very close to the actual frequencies employed, 0.334π and 0.694π. But as we decrease the number of samples (time-domain truncation) we get an increasingly poor representation of the two δ-functions we expected to see. Note how the spectra are broadened and the peak amplitudes diminished. In the case of Figure 10.15c ($N = 16$) we have connected the discrete points with spline fitting, but even so our estimate of the spectrum is of no value. Although the best remedy for time-domain truncation is to increase N, there are alternatives for leakage control and we will discuss one of them next.

In the preceding examples we saw that abrupt, or sharp, changes in the time domain, or time-domain truncation, might produce unwanted features in the computed spectra. If we could use a data window (sometimes referred to as a *fader*) to smooth out the shoulders, we might be able to eliminate some undesirable features from computed spectra. One possibility is to use a *taper* on the time-series data and *then* transform the tapered data. An example given by Bloomfield (1976) is the window,

$$W_t = \frac{1}{2}\left[1 - \cos\left(\frac{2\pi\left(t + \frac{1}{2}\right)}{n}\right)\right] \tag{10.27}$$

where t corresponds to the discrete times in the sampled data (consisting of n points), $t = 0, \ldots, n-1$. Please note that in this context, when we speak of n points, we actually mean $n\Delta t$. Suppose for example that our original time-series data form a step with an amplitude of 1. When we apply the *cosine window* to the data we get a smooth transition that is "bell" shaped, as shown in Figure 10.16.

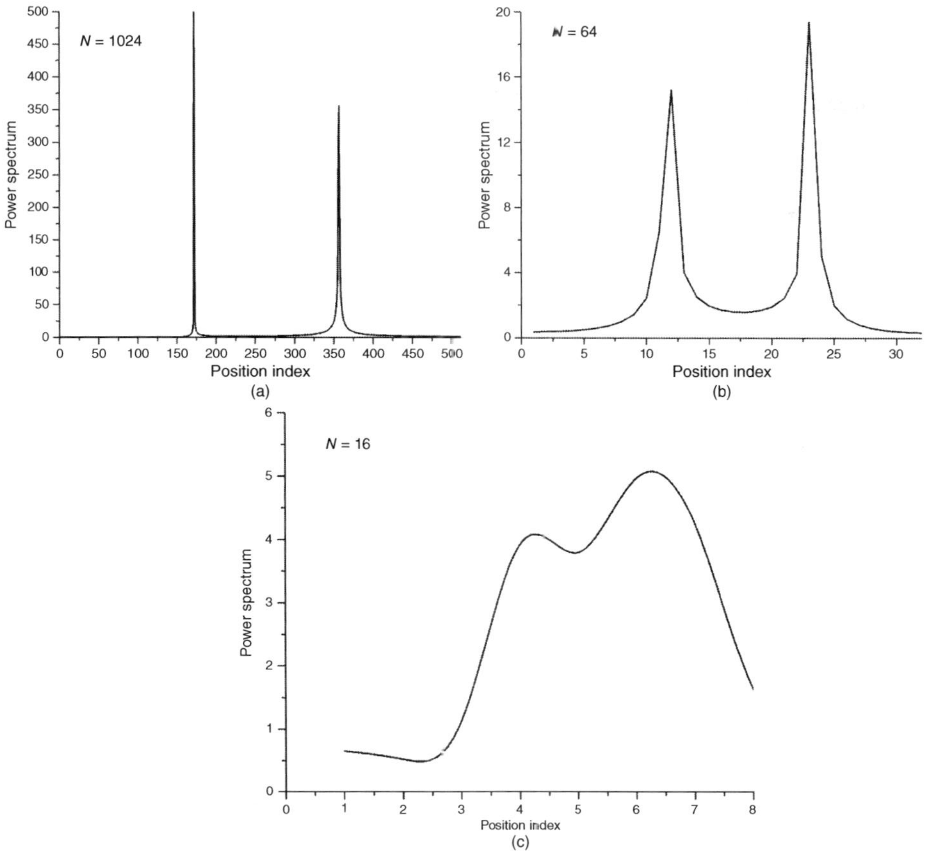

Figure 10.15 (a–c) Comparison of spectra computed by DFT for record lengths of 1,024, 64, and 16 samples. The test signal was $y(t) = \cos\left(\frac{2\pi}{5.99}t\right) + \sin\left(\frac{2\pi}{2.88}t\right)$.

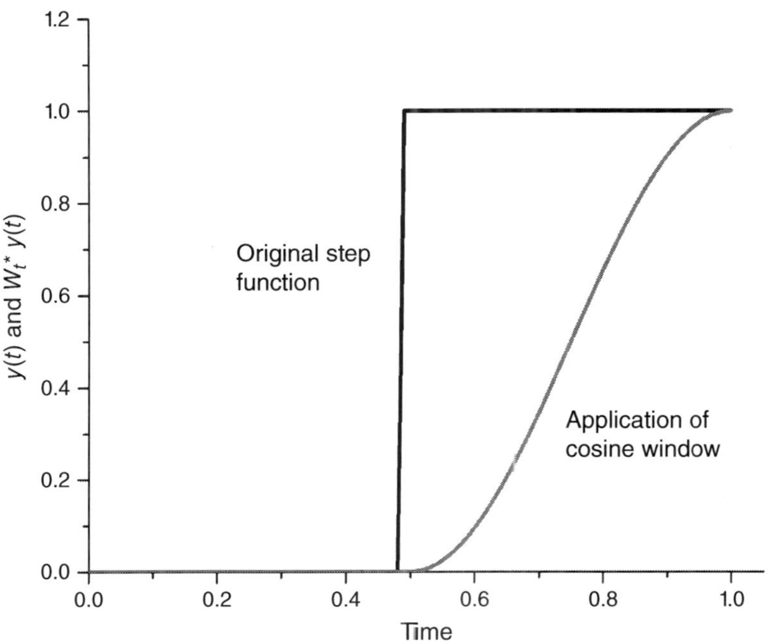

Figure 10.16 Application of the cosine window to a step change in the time-series data.

It is clear from inspection of Figure 10.16 that the windowed (or tapered) data are significantly different from the original step. In many cases we would rather limit the influence of the taper so as to preserve *most* of the character of the original data. Tukey (1967) suggested a modification of the data window to:

$$\frac{1}{2}\left[1-\cos\left(\frac{\pi\left(t-\frac{1}{2}\right)}{m}\right)\right] \quad t=0,\ldots,m-1 \tag{10.28}$$

$$W_t = 1 \quad t=m,\ldots,n-m-1 \tag{10.29}$$

$$\frac{1}{2}\left[1-\cos\left(\frac{\pi\left(n-t+\frac{1}{2}\right)}{m}\right)\right] t=n-m,\ldots,n-1 \tag{10.30}$$

Remember that m and n are to be interpreted as $m\Delta t$ and $n\Delta t$. With this approach, the fraction of the data that will be tapered is $2\,m/n$; a smaller fraction of windowed data will naturally yield *less* improvement in leakage control. Tukey recommended that m be chosen such that $\frac{2m}{n} \approx 20\%$.

We want to look at a detailed example that will first demonstrate the problems created by leakage and then reveal how we might diminish them using tapering. Suppose first that we have time-series data in the form of a square wave (a rectangular window) with a maximum amplitude of one (1). We will then apply a sinusoidal taper to remove both sharp edges, but leave the center intact. Specifically, we will shave off one-third of the rectangular window at both shoulders and leave the center third intact with its original value of one (1). A comparison of the original rectangular window with the modified shape (where the sharp edges have been removed) is shown in Figure 10.17.

Our plan is to apply the DFT to both the original square wave and the tapered wave form so we can see the differences in the spectra; the results appear in Figure 10.18.

The leakage produced by the original square wave is significant; however, we immediately see that trimming one-third of the pulse off (at each shoulder) results in far less leakage while preserving the really important characteristics of the spectrum. It will be left as a student exercise to explore the effects of more aggressive tapers upon the results.

The effectiveness of windowing with respect to the DFT can be assessed by looking at the size of the largest side lobe relative to the central peak height and the rate of decay (sometimes called "roll-off") in the successive side lobes. Window selection is also affected by circumstance: Two waveforms adjacent in frequency space with comparable amplitudes will

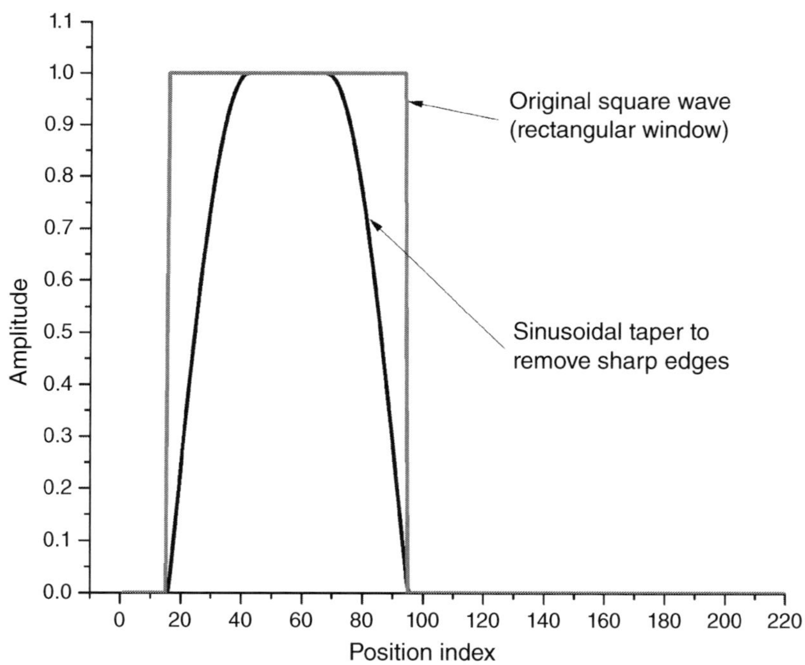

Figure 10.17 The initial square wave (rectangular window) in shown in gray and the tapered waveform in black. The tapered edges correspond to one-third (at each top edge) of the original rectangle. The center third is left alone with an amplitude of one (1).

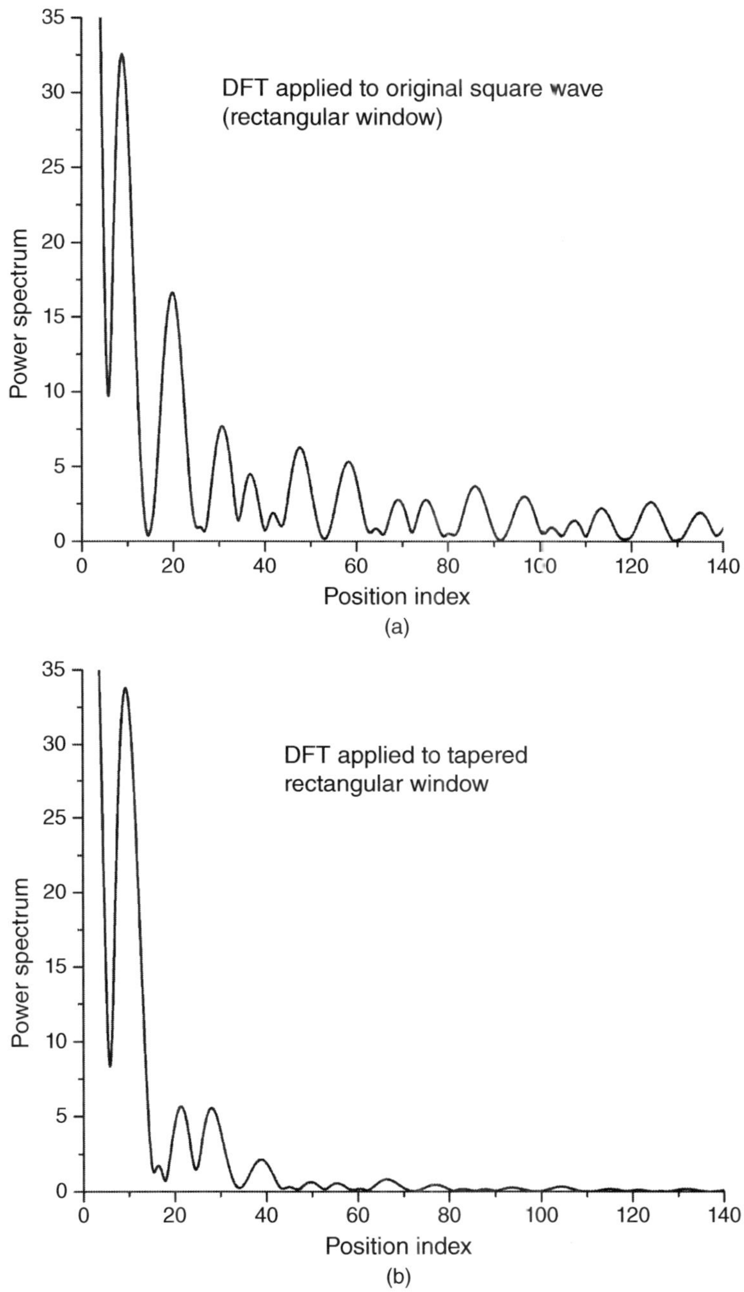

Figure 10.18 Spectra obtained from the original time-series data (a), and from the tapered data set in which the sharp-edged shoulders have been trimmed off (b).

need a different window than two adjacent waveforms with greatly different amplitudes. We explore the latter situation immediately below.

10.6.1 Proximate Signals with Significant Amplitude Disparity

A time series may contain multiple periodicities with similar frequencies, but with dramatically different amplitudes. For illustration, we first consider a modification of an example given by Dodson-Robinson et al. (2022):

$$y(t) = \cos(2\pi f_1 t) + 0.017\sin(2\pi f_2 t), \tag{10.31}$$

where $f_1 = 1.7/50$ and $f_2 = 1.2/50$. *Recommended for self-study*: Let t assume integer values from 1 to 1,024 so that a radix-2 DFT can be applied and then use the Fourier analysis capability of Excel to produce the transform. Note that the amplitude of the *cosine* part of the signal is nearly 59 times larger than the *sine* contribution and the frequencies f_1 and f_2 are relatively close together. Assuming both periodicities are important to our analysis we should be concerned that the low-amplitude signal may be overwhelmed in the resulting spectrum; this is borne out by transform which reveals a very large peak at index position 36 and a very small "peak" at index 23. Indeed, the relative magnitudes are approximately 451/6.72 (the quotient is about 67); had the signal contained even a small amount of noise—which is inevitable—it is certain that the low-amplitude spectral feature would have been overlooked. Solomon (1991) provides an even more severe example in which the amplitudes of two adjacent sinusoids vary by a factor of 1,000 (thus, $20 \log_{10} 1{,}000 = 60$ dB, an enormous difference); he compares the effect of different window types upon Fourier analysis of the signal, $128 \sin(2\pi(20.60)t) + 0.128 \sin(2\pi(23.65)t)$. With no pretreatment of the data (i.e., a rectangular window) the small amplitude peak at 23.65 Hz is completely missing from a conventional DFT computation (you may also want to verify this). *However*, the four-term Blackman-Harris window,

$$W = a_0 - a_1 \cos\frac{2\pi n}{N} + a_2 \cos\frac{4\pi n}{N} - a_3 \cos\frac{6\pi n}{N} (0 \le n \le N-1) \quad \text{with}$$
$$a_0 = 0.35875 \quad a_1 = 0.48829 \quad a_2 = 0.14128 \quad \text{and} \quad a_3 = 0.01168 \tag{10.32}$$

can result in a spectrum in which the small-amplitude oscillation is readily identified. We will illustrate this with an example, but before we do we note that Ramirez (1985) provides a graphical comparison of a number of window types in his Chapter 7. And Harris (1978) has given a more extensive catalog of windows in his paper. These presentations are useful for the student new to Fourier analysis.

Assume we have an interest in the signal,

$$x(t) = \sin(0.985t) + \frac{1}{10}\cos(0.8375t) + bZ \tag{10.33}$$

where Z is a random number $(-1 \le Z \le +1)$ which we include to simulate noise in the waveform. Note that the amplitudes in eq. (10.33) differ by an order of magnitude, and the lower frequency (0.8375 rad/t) is only 15% smaller than 0.985 rad/t of the *sine* term. We look at three cases: moderate noise with no tapering (a rectangular window), high noise level with no tapering, and the four-term Blackman-Harris window applied to $x(t)$ with moderate noise. *Let us re-emphasize*: The amplitude of the *cosine* portion of the waveform is just 10% of the *sine* term (Figure 10.19).

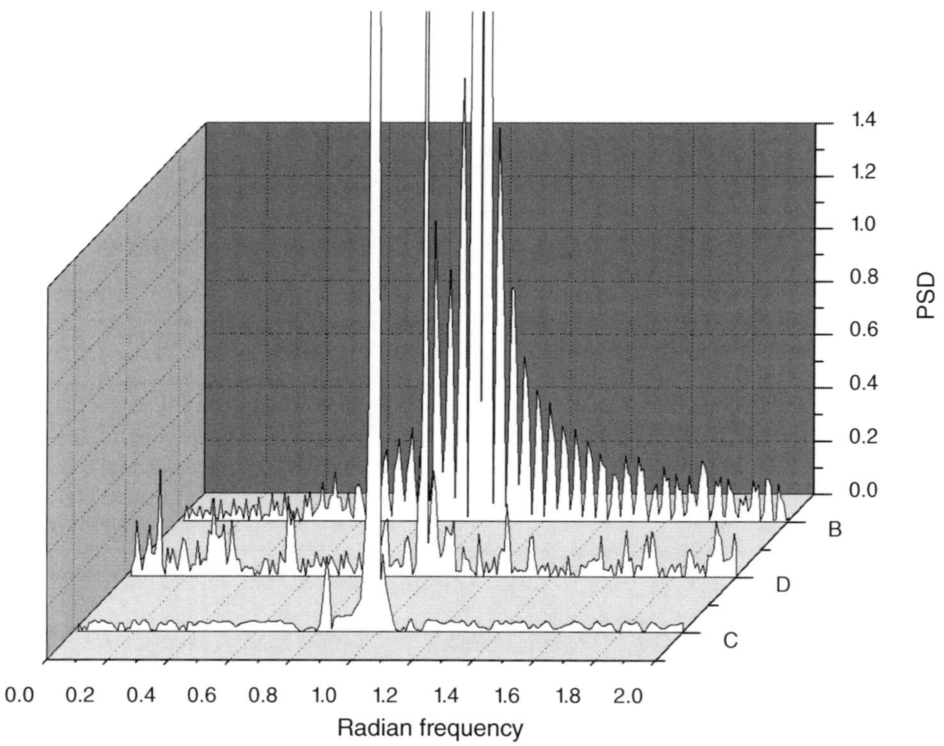

Figure 10.19 Comparison of spectra for three cases: B) moderate noise and no tapering, D) high level noise and no tapering, and C) moderate noise but with application of the four-term Blackman–Harris window.

For spectrum B the sidelobes and the noise make identification of the low-amplitude portion of the signal at 0.8375 rad/s problematic. In the case of D, although there is a peak at 0.8375, it is smaller than many of the spurious components generated by the noise, and its significance might easily be overlooked. The application of the Blackman–Harris window for C resulted in a dominant feature at 0.985 rad/s and a much smaller—yet unmistakable—contribution at 0.8375. Identification of the latter spectral feature is possible because the effects of the included noise have been significantly attenuated.

10.7 Smoothing Data by Filtering

We saw previously how time-series data could be windowed (or tapered) to minimize the impact of sharp irregularities. We can also use digital filtering to smooth very irregular (noisy) data. Let us return to eq. (10.34) in Section 10.3 and let

$$y(t) = \bar{y} + y'; \tag{10.34}$$

Since the mean $\bar{y}$ is nearly constant (or at least varies only very slowly), we could run a moving, three-point average through the data set, replacing $y(t)$ with z:

$$z(t) = \frac{1}{3}[y(t-1) + y(t) + y(t+1)]. \tag{10.35}$$

Naturally we cannot apply this to the end-points of the data. Let us construct a synthetic data set to see how well this works; we will use a random number generator to produce noise which we superimpose upon a single sinusoid. We will also filter the data with the moving, three-point average to gauge its impact upon the noisy data (Figure 10.20).

There are software packages available that include moving average computations, and one example which we will discuss at greater length in Section 10.9 is the *Physics Toolbox Sensor Suite Pro* by Vieyra Software.

Our main interest, of course, is how the filtering affects the computed spectrum. Figure 10.21a shows the spectrum for the unfiltered (noisy) data, and Figure 10.21b shows the effects of two passes of the linear filter (the moving three-point average) upon the computed spectrum.

The filter that was employed to produce the results seen in Figure 10.21b was the simple three-point moving average. Although we used the moving average to lessen the effects of noise upon the computed spectrum, we note that there are other important applications in which the moving average is routinely used on fluctuating signals. Technical analysts on Wall Street use the 200-day moving average (MA) on stock and commodity prices to gauge market trends and momentum.

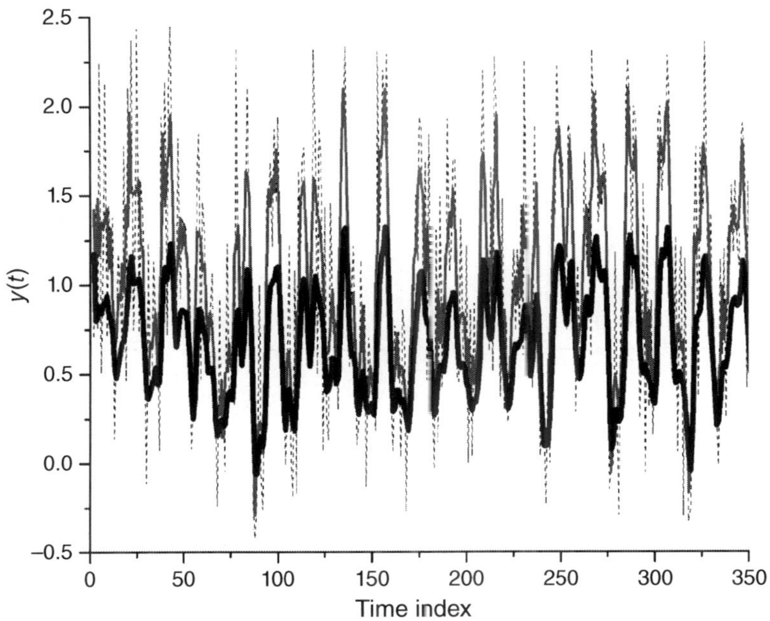

Figure 10.20 Original noisy data (dotted), filtered once (continuous gray), and filtered twice (heavy continuous black). The time index has been expanded to better reveal the behavior of $y(t)$.

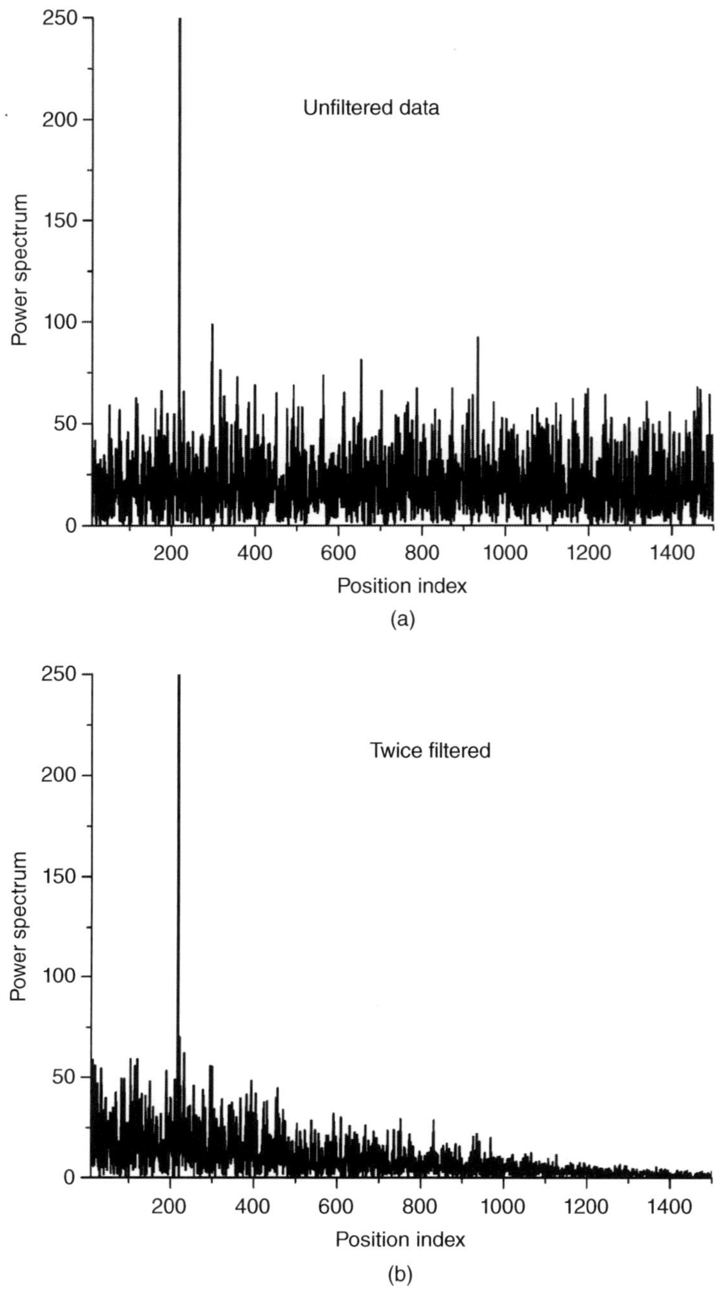

Figure 10.21 Comparison of the spectra for the original, noisy data (top, a), and then computed for the data set after two applications of the linear filter (bottom, b). Many of the additional frequency components that were introduced by the abrupt fluctuations have been attenuated or even eliminated.

Often 200- and 50-day moving averages are used together, and crossings of the two are considered to be particularly significant; when the 50-day moving average *rises above* the 200-day, this is considered as a "buy" signal and is referred to as a "golden" cross. In contrast, when the 50-day *falls below* the 200-day MA, that is taken as an indicator of a bear market and is called a "death" cross. A common objection to the 200-day MA in this context is the heavy reliance upon old data that may not accurately reflect the most recent influences upon stock price. The exponential moving average (EMA) is sometimes used to minimize this problem; the EMA uses a weighting multiplier, k, defined as $\frac{2}{(n+1)}$, where n is the number of daily observations, and thus if a closing stock price is monitored for 15 days, $k = 2/16$. The EMA is then calculated as $\text{EMA} = kP(t) + \text{EMA}(t-1)(1-k)$, where $P(t)$ is the present price and $\text{EMA}(t-1)$ is the EMA from the previous day. Many traders rely upon these moving averages to provide guidance for investment decisions.

We discussed the EMA in the context of fluctuating stock prices above, but let's expand on that with a detailed treatment of a model waveform:

$$x(t) = \frac{1}{3}\sin(0.2t) + \frac{4}{7}\cos(0.39t) + bZ, \tag{10.36}$$

where Z is a uniform random number with $-1 \leq Z \leq +1$. Consequently, if this waveform is sampled for a suitably long time period we should find $\bar{x} \cong 0$. We can apply the exponential filter to $x(t)$ to obtain $y(t)$:

$$y(t) = wx(t) + (1 - w)y(t - \Delta t). \tag{10.37}$$

As noted previously, larger values for w place greater importance upon the current value of the variable x, and small values emphasize earlier x's. For example, if we let $b = 3$ and use a new value of Z with each time step, a significant amount of noise will appear in the waveform eq. (10.36). If we then compute the periodogram for $x(t)$ the expected peaks at $\omega = 0.2$ and 0.39 will appear, but we will see many other significant features between about 0.4 and 0.9 rad/t as well. Now, if we process this time series using eq. (10.37) with $w \gtrsim 0.5$ the amplitude of the oscillations will be reduced, but most of the fine-scale structure will be preserved. On the other hand, if we employ eq. (10.37) with a small value for w, say 0.25 or less, the amplitude of oscillations will be *severely* reduced, and much of the fine-scale structure in the original waveform will be obliterated. It is an instructive exercise for the reader to experiment with the waveform eq. (10.36), apply the exponential filter eq. (10.37), and then compute the periodogram for both small and large values of w. One could also use the *Fourier analysis* capability of Excel to rapidly explore a range of values for w. For example, if you let t assume values from 0 to 1,023 and set $w = 0.5$, the spectrum of the EMA-processed data will contain a large peak at index 34 ($34\pi/512 = 0.208$) and a much smaller peak at about index 65 ($65\pi/512 = 0.398$), with the latter only slightly more significant than peaks arising from the included noise.

There are many other options for lessening the impact of noise appearing in time-series data. One that is quite effective at attenuating noise but preserving the essential character of the oscillations is the digital ***triangular*** filter. The idea is to place greatest weight upon the central point, and decreasing weight upon more distant data points (thus, the triangular filter addresses one of the criticisms of the 200-day moving average described immediately above); one example of a triangular filter is:

$$z(t) = \frac{1}{9}[y(t-2) + 2y(t-1) + 3y(t) + 2y(t+1) + y(t+2)]. \tag{10.38}$$

Note that the area of the triangle is $\frac{1}{2}bh = \left(\frac{1}{2}\right)(6)\left(\frac{3}{9}\right) = 1$. Of course, other weighting schemes can be employed as well. We can get a better sense of how effectively eq. (10.38) works by applying it to a test set of data in Figure 10.22.

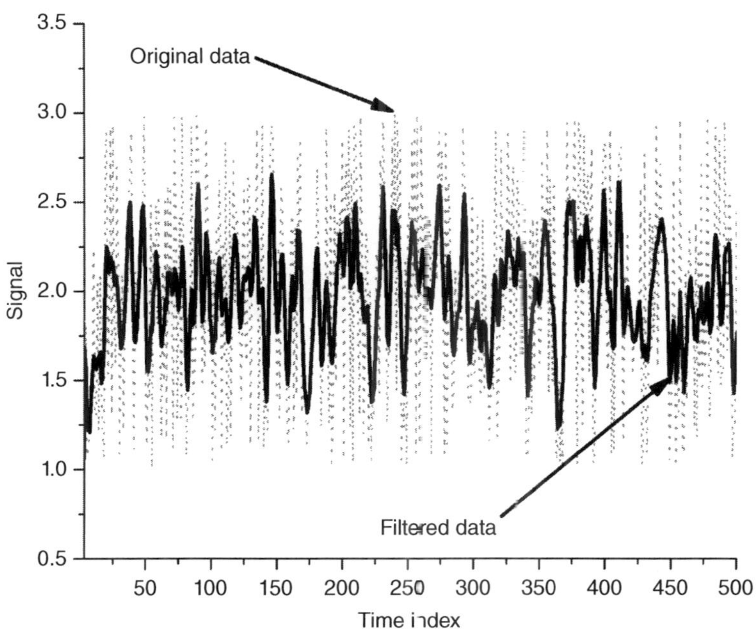

Figure 10.22 Effects of a *single* application of the digital triangular filter to a noisy data set. For severely noisy signals, the triangular filter (10.38) can be run through the data set repeatedly.

We looked at the effects of data filtering above in a practical way, by examining spectra computed from the original noisy data and then from the filtered data. There is another, very convenient approach to assessing the effects of filter application. Suppose we have a simple signal in the form of a sinusoid which we write as:

$$y(t) = A\cos(\omega t + \phi), \tag{10.39}$$

which is the real part of $A\exp[i(\omega t + \phi)]$. For the linear filter (the three-point moving average) we have

$$\begin{aligned} z(t) &= \frac{1}{3}[y(t-1) + y(t) + y(t+1)] = \frac{1}{3}A[\cos(\omega t - \omega + \phi) + \cos(\omega t + \phi) + \cos(\omega t + \omega + \phi)] \\ &= \frac{1}{3}A[\exp(i(\omega t - \omega + \phi)) + \exp(i(\omega t + \phi)) + \exp(i(\omega t + \omega + \phi))] \end{aligned} \tag{10.40}$$

This expression can be written more compactly as

$$= \frac{1}{3}A\exp[i(\omega t + \phi)][\exp(-i\omega) + 1 + \exp(i\omega)]. \tag{10.41}$$

Of course, *cos* is an even function, so we can write the real part of eq. (10.41) as

$$= \frac{1}{3}A\cos(\omega t + \phi)(1 + 2\cos\omega). \tag{10.42}$$

When we compare this output with the original signal, eq. (10.39), it is clear that we are modifying it with the factor, $\frac{1}{3}(1 + 2\cos\omega)$. We see immediately that a signal component with $\omega = 2\pi/3 = 2.0944$ will be *completely removed* by the filter (since $\cos\left(\frac{2\pi}{3}\right) = -\frac{1}{2}$).

10.8 Modulation (Beats)

We have now seen on numerous occasions that the FFT/DFT is capable of revealing periodicities in time-series data. Let us begin this part of our discussion by considering a synthetic signal consisting of:

$$y(t) = Z + \cos(82t) + \cos(89t) + \sin(196t) + \sin(207.3t). \tag{10.43}$$

Notice that the *two pairs* of radian frequencies are close together. If we compute the DFT for these data we see four concentrated peaks in the spectrum, just as expected.

Now we will carry out the DFT computation again, but this time we use the *product* of the two cosine contributions added to the *product* of the sine contributions. Notice from Figure 10.23 that the pair of cosine contributions is separated by 7 rad/s, and the sine contributions are separated by 11.3 rad/s. This will prove to be very significant.

When the DFT is applied to the sum of the cosine–cosine product and the sine–sine product, new spectral components appear at about 7 and 12 rad/s. These are *beat* frequencies that arise from amplitude modulation. If we multiply $\cos(\omega_1 t)\cos(\omega_2 t)$, where the two radian frequencies ω_1 and ω_2 are fairly close together, then we produce a modulated wave with fluctuating amplitude—notice the peak in Figure 10.24 located at 7 rad/s. You will also notice that the spectral content at 82–89 rad/s has disappeared entirely. Clearly, harmonic analysis (the DFT) has failed to reveal all of the periodicities of the original data. What we need here is called *complex demodulation*, and we will illustrate it with an elementary example. Let us assume we have a function described by

$$y(t) = A(t)\cos(\omega t + \phi(t)). \tag{10.44}$$

We will allow both the amplitude function, $A(t)$, and the phase, $\phi(t)$, to vary but at frequencies significantly lower than ω, in particular:

$$A(t) = 3\sin\left(\frac{\omega}{5}t\right) \quad \text{and} \quad \phi(t) = 4\sin\left(\frac{\omega}{6}t\right) \tag{10.45}$$

The resulting signal is illustrated in Figure 10.25. If we perform a harmonic analysis upon these data, what would we see? If you have a few moments, try setting $\omega = 0.777$ rad/t in eq. (10.44) and compute the spectrum using the time-varying amplitude and phase given by eq. (10.45). *Very strong* spectral components will appear at 0.104, 0.362, 0.543, 0.623, 0.675, 1.009, 1.190, 1.322 and 1.451 rad/t, but *not* at 0.777.

We will take eq. (10.44) and rewrite it as

$$y(t) = \frac{1}{2}A(t)[\exp(i(\omega t + \phi)) + \exp(-i(\omega t + \phi))]. \tag{10.46}$$

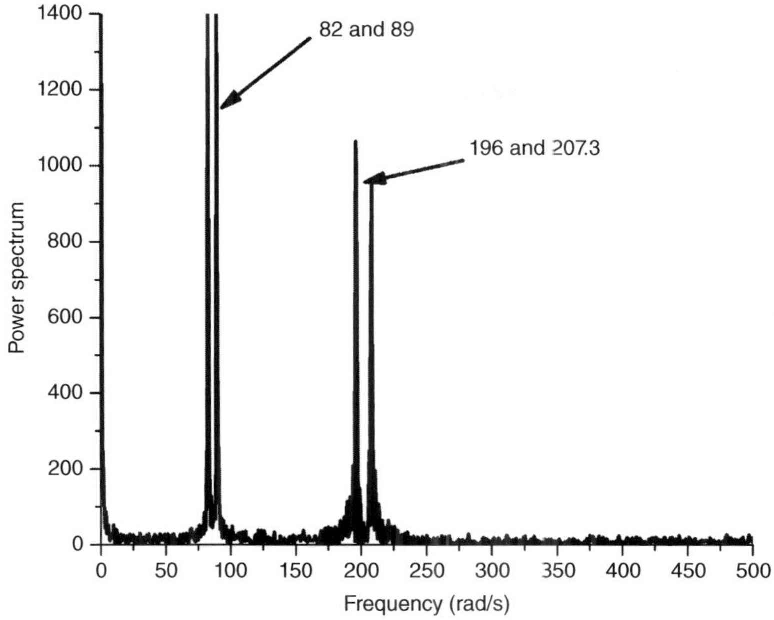

Figure 10.23 DFT computed for the simple sum of four periodic functions with frequencies of 82, 89, 196, and 207.3 rad/s.

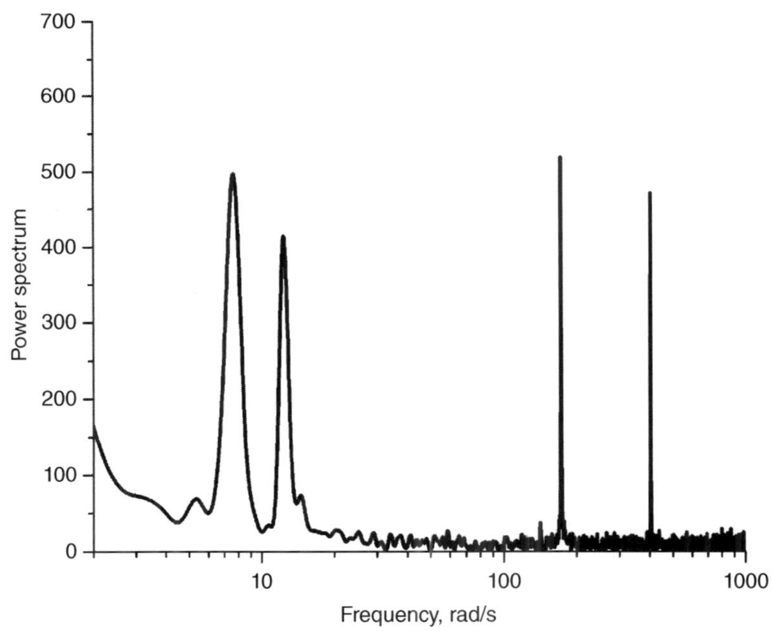

Figure 10.24 DFT computed for two *products* added together, the cosine pair and the sine pair. The frequency axis has been changed to a log scale to better reveal the two low-frequency components.

Now suppose we form a new signal from $y(t)$ by writing

$$z(t) = y(t)\exp(-i\omega t),$$

the result is

$$z(t) = \frac{1}{2}A(t)[\exp(i\phi) + \exp(-i(2\omega t + \phi))]. \tag{10.47}$$

Note that the first term on the right-hand side varies slowly—remember that $\phi(t) = 4\sin\left(\frac{1}{6}\omega t\right)$. The frequency for the second term in eq. (10.47) is 2ω, so it varies rapidly. If we apply a low-pass filter to the signal to remove all content with

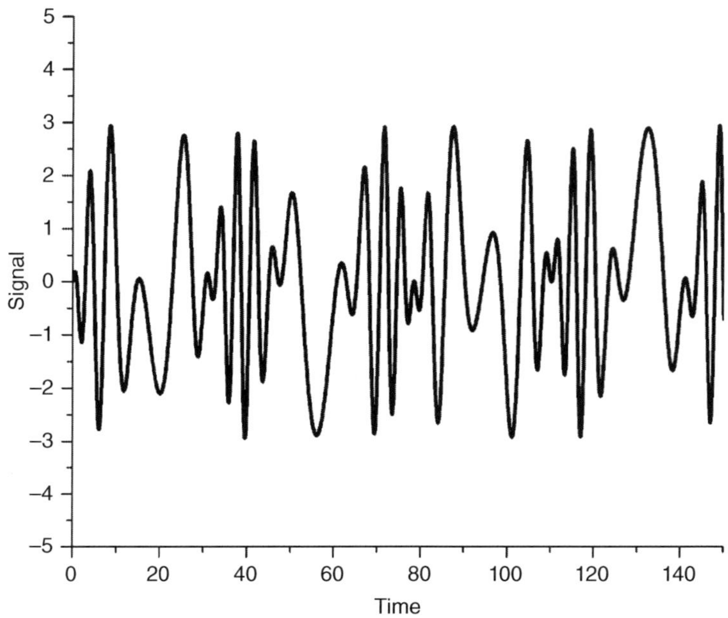

Figure 10.25 An oscillating function with frequency, ω, for which both the amplitude and the phase vary sinusoidally, but at much lower frequencies.

frequencies larger than ω, we are left with:

$$z_f(t) = \left[\frac{1}{2}A(t)\exp(i\phi(t))\right]_f. \tag{10.48}$$

The subscript "f" indicates that the signal has been smoothed by application of the filter.

Now the amplitude is just the magnitude of z_f, and we merely divide eq. (10.48) by $A(t)$ to isolate $\phi(t)$.

10.9 Some Important Examples

10.9.1 Using a Smartphone for Collection of Time-series Data

Whatever one may think of the impact of smartphones upon social interactions, the sensors built in to the devices allow nearly anyone to make scientific data collection part of their normal routine. The *Physics Toolbox* by Vieyra Software is an *app* available for both *Android* and *iOS* devices; it includes g-Force Meter, Linear Accelerometer, Gyroscope, and Barometer. Data can be collected and saved as comma delimited files (.csv) and subsequently shared with others. For example, Figure 10.26 contains measurements of the lateral acceleration experienced by the operator of an agricultural tractor performing fieldwork at an average speed of 4 mph.

The time-series data can then be processed to facilitate interpretation of the measurements. The autocorrelation coefficient and spectrum were computed for these data. It was found that the autocorrelation coefficient, $\rho(\tau)$, was still greater than 0.7 after a 10τ offset, and the spectrum shown in Figure 10.27 indicated that most of the signal energy occurred at frequencies less than about 10 Hz. This is fully expected given the slow speed of the working tractor.

The total signal energy can be determined by integration,

$$\int_{\omega_1}^{\omega_{Nyq}} S(\omega)d\omega, \tag{10.49}$$

where the lower limit (ω_1) is determined by the length of the time series and the upper limit is the Nyquist frequency set by the sample interval, Δt, $f_{Nyq} = 1/(2\Delta t)$. Therefore the fraction of energy in any given frequency range, ω to ω_2, can be determined by the quotient:

$$\int_{\omega}^{\omega_2} S(\omega)d\omega \Big/ \int_{\omega_1}^{\omega_{Nyq}} S(\omega)d\omega. \tag{10.50}$$

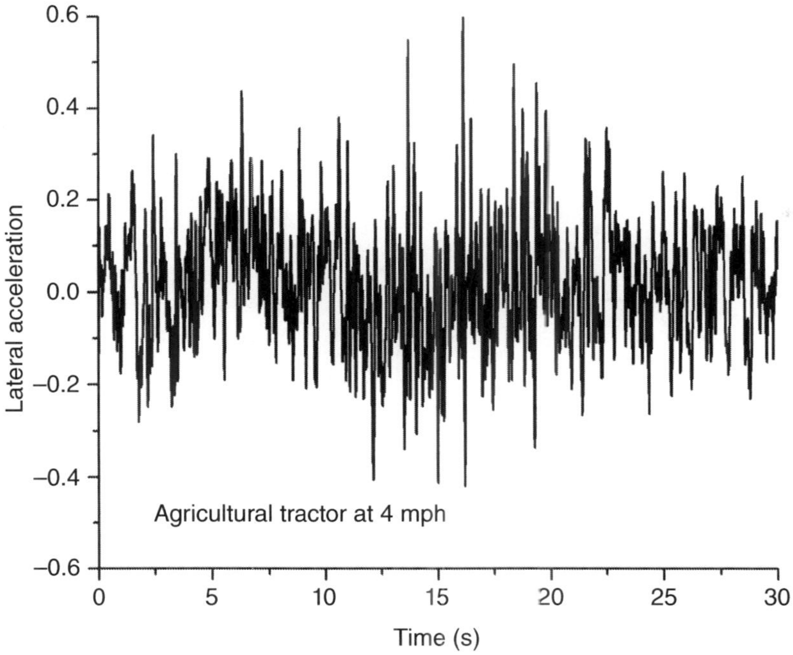

Figure 10.26 Time-series data for lateral acceleration measured with *Physics Toolbox* by Vieyra Software using an *Android* device.

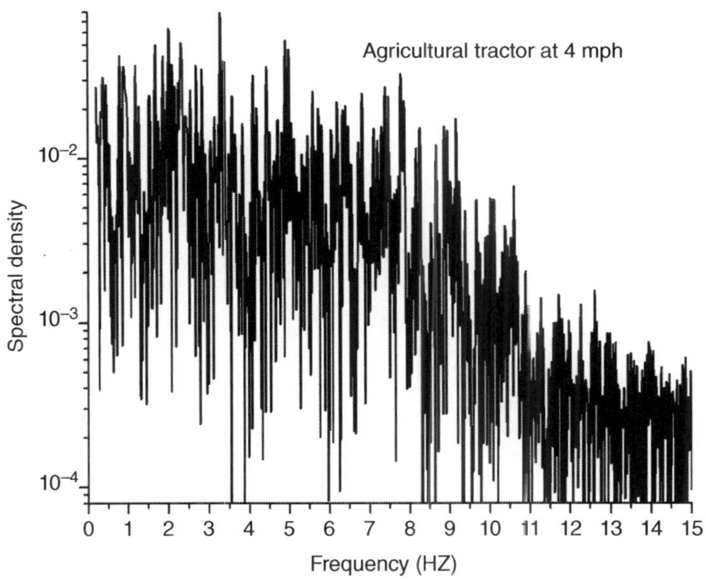

Figure 10.27 Spectrum, $S(\omega)$, computed for lateral acceleration experienced by operator of an agricultural tractor operated at 4 mph.

The fractions of total signal energy for several values of ω_2 are provided here:

Upper Limit Frequency (Hz)	Fraction of Total
3	0.36
5	0.58
10	0.85
30	0.92
40	0.95

We can of course determine the fraction of total signal energy contained in any given frequency range; for example, the spectral content between 2 and 30 Hz is 71% of the total.

10.9.2 Bridges and Structural Integrity

Bridges are indispensable to modern commerce, yet few of us think much about their stability and structural integrity. But perhaps we should given the history of catastrophic bridge failures; a few examples follow:

Ashtabula River Bridge (Ohio)	12/29/1876	92 deaths
Tay Rail Bridge (Scotland)	12/28/1879	75 deaths
Tacoma Narrows Bridge (Washington)	11/7/1940	Tubby (the dog)
Whangaehu River Rail Bridge (New Zealand)	2/24/1953	151 deaths
I-35 W Bridge (Minnesota)	8/1/2007	13 deaths
Morandi Bridge (Italy)	8/14/2018	43 deaths

Some of the bridges in this abbreviated list were not in service long prior to failure—the Tacoma Narrows Bridge had only been open for 6 months when a sustained 42 mph wind brought it down. The Tay River bridge was opened to train traffic in 1878 and failed 18 months later. In contrast, the I-35 W Bridge in Minneapolis had been in service for 40 years when the combination of rush hour traffic, construction equipment, and an important design flaw led to its failure. Of course, the question that technical professionals responsible for bridge safety seek to answer is: How do we determine—in an efficient manner—whether or not a particular bridge is safe?

Unquestionably, the best-known example of excessive bridge oscillation is the original Tacoma Narrows Bridge which was completed in 1940 and failed spectacularly the same year. We are fortunate because F. B. Farquharson (Professor of Civil Engineering at the University of Washington) had the presence of mind to borrow a 16 mm motion picture camera and record the events leading to failure of the Tacoma Narrows Bridge. Prior to failure the dominant mode of oscillation had been a longitudinal sine wave with a frequency of 0.63 Hz in which the bridge deck itself—though moving vertically—remained horizontal. However, on November 7, 1940, this rate suddenly slowed to about 0.2 Hz and the longitudinal motion was subsequently overwhelmed by an abrupt transition to severe torsional oscillation in which a section of the bridge would pitch up to nearly 45° and then twist downwards to −45°. Damage to one of the vertical supporting cables resulted, leading to a lateral imbalance of the restoring force and destructive twisting and complete failure. Although this has been attributed to the riser cables (on the high side) going slack and then that portion of the bridge deck falling under the influence of gravity, Professor Farquharson ventured onto the bridge during the torsional oscillation and observed that the cables *did not* go slack during the severe torsional motion (for a modern review of the failure, see Green and Unruh, *American Journal of Physics*, 74:706, 2006). There has been frequent speculation in the non-technical literature that vortex shedding was responsible for the disaster but this is incorrect; the Strouhal number (dimensionless vortex shedding frequency) for Tacoma Narrows Bridge indicates that vortices shed from the bridge deck were not responsible for oscillations leading to failure. That is, the popular explanation that vortex shedding led to resonance by matching the fundamental frequency of the structure is wrong.

Like all structures, bridges have fundamental frequencies. However, as Nagayama et al. (*Bridge Natural Frequency Estimation by Extracting the Common Vibration Component From the Responses of Two Vehicles*, 6th International Conference on Advances in Experimental Structural Engineering, 2015) point out, the natural frequencies of most bridges are unknown. What is well documented though is the band of vibration frequencies associated with heavy traffic on bridges, which is generally 1.5–4.5 Hz. Consequently engineers try to design bridges that will not experience resonance in this range.

Although single-point measurement of vibrations in bridges under traffic conditions is relatively easy, determining the fundamental frequency of the structure from such data is not. Figure 10.28 is an excerpt from accelerometer data obtained on a conventional beam bridge during the passage of a heavy truck at approximately 45 mph (or 73 kph):

Spectral analysis is not really necessary for these data as it is abundantly clear that the main vibration is occurring at about 12 Hz, although there is some finer-scale, low-amplitude structure in the vertical component that may be very important.

Gaunt and Sutton (*Highway Bridge Vibration Studies*, 1981) reported fundamental frequencies for two-span, steel beam bridges that ranged from 2.19 to 3.73 Hz, and Mohseni et al. (*Latin American Journal of Solids and Structures*, 2014) cited

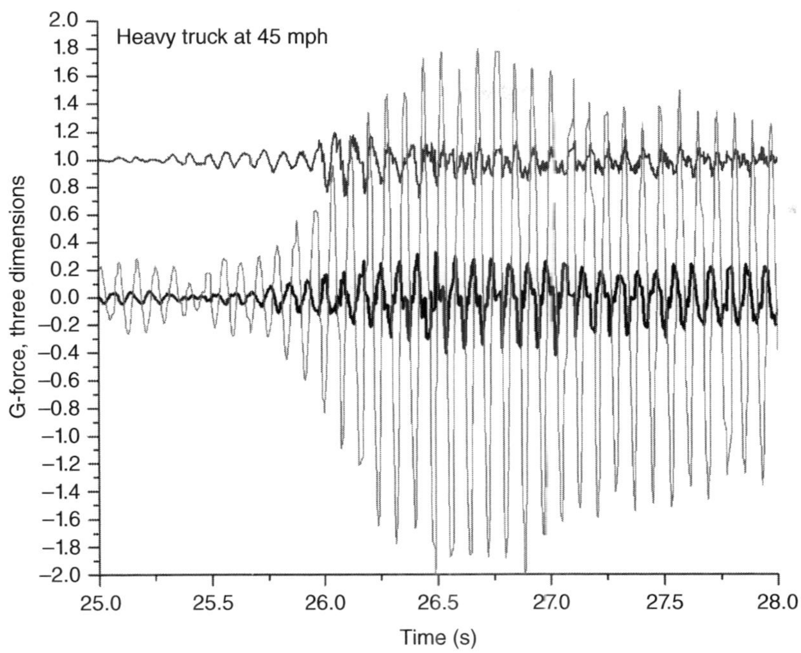

Figure 10.28 G-forces recorded on the middle span of a beam bridge over the Kansas River during passage of a heavy truck (flatbed carrying construction equipment).

an empirical correlation for straight box-girder bridges:

$$f_B \cong \frac{234.2}{L}, \tag{10.51}$$

where L is the length of one span in feet. Although these data provide an *indication* of the magnitude of f_B, they are not really useful in the present context because structural engineers would like to use a change in a bridge's fundamental frequency as a measure of bridge health; such changes can be caused by beam damage, scour (erosion) at the base of piers, damage to the deck, or corrosion of steel components (as in the case of the Morandi Bridge in Italy). However, *these changes may be subtle* so that detecting, say, a 15% change in f_B would require that the fundamental frequency be known with precision and monitored regularly. It is impractical to have accelerometers permanently installed on large numbers of bridges, so structural engineers have sought alternatives that would not impede traffic. Typically this means processing an accelerometer signal obtained from inside a moving vehicle. Of course this raises a number of complicating issues including irregularities in the roadway and the excitation and damping of the vehicle's suspension.

Nagayama et al. devised a clever method based upon driving two vehicles, each equipped with a triaxial accelerometer, across a bridge at speeds between 10 and 30 km/h; the *cross-spectra* computed from the two data records produced a dominant spectral peak at 2.17 Hz (the fundamental frequency of the structure). The underlying idea here is that different vehicles have different fundamental frequencies, and the vehicle–bridge interactions will be different for each vehicle and their response to road bed roughness would not be identical. Therefore, the cross-spectrum would tend to emphasize vibrations that were common to both, that would of course include f_B. The tested bridge was a 59 m, simply supported steel (box-girder) span in Chiba Prefecture, Japan. We should also note that many researchers, including Nagayama et al., have observed the effects of vehicle speed upon accelerometer data; often higher speed tends to increase the spectral densities at higher frequencies, making it difficult or impossible to identify f_B reliably.

10.9.3 Two Signals Received Simultaneously

Circumstances arise where we need to investigate two simultaneous time-varying signals; we let the *fluctuations about the mean values* be represented by $x(t)$ and $y(t)$. Assume we would like to know if the two time series are correlated, and if so, how the signal energy is distributed with respect to frequency. We define a cross-correlation coefficient as:

$$\gamma_{xy}(\tau) = \frac{\overline{x(t)y(t+\tau)}}{(\overline{x^2})^{1/2}(\overline{y^2})^{1/2}}, \tag{10.52}$$

with the root-mean-square (rms) fluctuations appearing in the denominator; of course, $\gamma_{xy}(\tau = 0) = 1$. This is precisely what was done in the Nagayama et al. (2015) study cited in the preceding discussion. They mounted accelerometers in two separate vehicles to produce vibration histories for a steel (box-girder) bridge in Japan. The cross-correlation and cross spectra yielded the bridge's fundamental frequency. There are many other applications as well, for example, in studies of turbulent flows one might assess the Reynolds stress by measuring the correlation between different velocity vector components (v_x and v_y typically).

Very recently astrophysicists have been intrigued by the possibility of obtaining incredibly weak simultaneous signals emanating from cosmic events occurring at vast distances from our solar system. Gravitational waves—which Einstein identified as ripples in spacetime—were proposed by Poincaré as being analogous to electromagnetic waves early in the twentieth century. But because gravity is so very weak, and because the mediating particle has eluded detection, their existence could not be empirically demonstrated *until* the two LIGO interferometers (Hanford, WA and Livingston, LA) were constructed. These *very large* instruments provide us with some terrific data to work with in this discussion: two signals obtained simultaneously at widely separated locations. We will be using model representations (from Einstein's theory) of

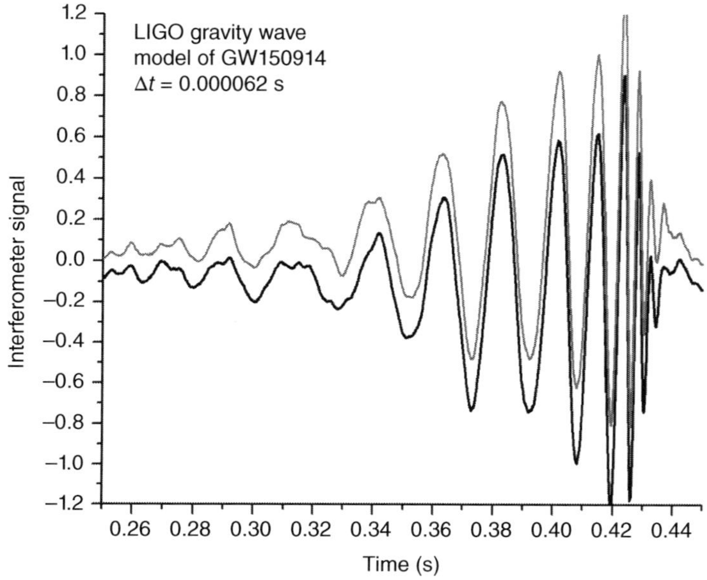

Figure 10.29 Models of interferometer strain for event GW150914 detected and recorded at Hanford and Livingston.

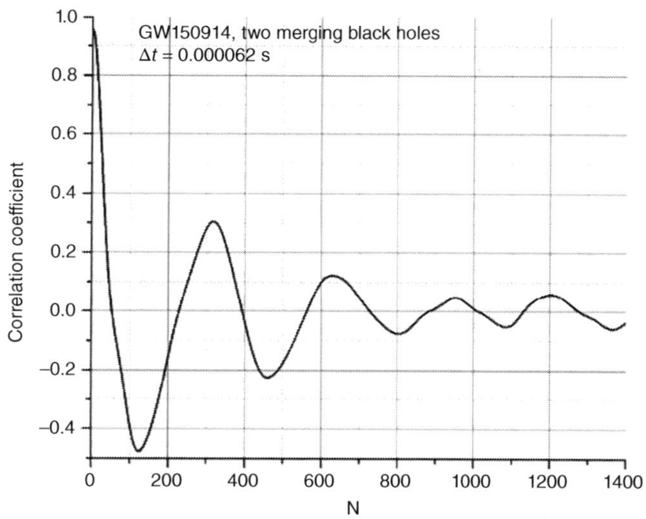

Figure 10.30 Computed correlation coefficient, $\gamma(\tau)$, for modeled event GW150914. Note that peak occurring at about 310 (Δt) corresponds roughly to 52 Hz. $\gamma(\tau)$ exhibits the oscillation expected of signals with strong periodicities.

the recorded event, GW150914 (the merging of two black holes). Please note that the agreement between the raw data and two model curves is spectacular—the models are just smoother. The model data we will be using are shown in Figure 10.29; the time ranges from about 0.25 to 0.45 s and the interval between "samples" is 0.000062 s. Therefore, the cut-off frequency is $1/(2\Delta t) = 8{,}064$ Hz, though the signals of interest occur at much lower frequencies. The modeled interferometer signals from Hanford and Livingston are highly correlated—as they should be when a gravity wave is detected.

The observed signal frequency for time ranging from 0.38 to 0.4 s is about 56 Hz. The computational structure for the correlation coefficient calculations is provided in Figure 10.30.

```
#COMPILE EXE
#DIM ALL
    REM *** cross-correlation computation for LIGO signals
    REM *** two signals, u(t) and v(t)
FUNCTION PBMAIN
        GLOBAL sumu,sumv,ubar,vbar,sumupsq,sumvpsq,upmnsq,vpmnsq,uprms,vprms
             AS SINGLE
        GLOBAL sumcor,dt,tt,zz AS SINGLE
        GLOBAL n,ilag,nsamples AS INTEGER
            DIM u(5000) AS SINGLE
            DIM v(5000) AS SINGLE
                DIM uprm(5000) AS SINGLE
                DIM vprm(5000) AS SINGLE
                    DIM corrco(5000) AS SINGLE
                        dt=0.01:nsamples=3350
                        OPEN "d:LIGOdata comma.dat" FOR INPUT AS #1
    REM *** input two sets time series data, comma delimited
            FOR n=1 TO nsamples
                INPUT#1,tt,u(n),v(n)
                    NEXT n
                    CLOSE
    REM *** obtain means for each
                    FOR n=1 TO nsamples
                        sumu=sumu+u(n)
                        sumv=sumv+v(n)
                            NEXT n
                                ubar=sumu/nsamples
                                vbar=sumv/nsamples
                                    PRINT ubar,vbar
    REM *** obtain fluctuation values, mean-square and rms
                    FOR n=1 TO nsamples
                        uprm(n)=u(n)-ubar
                        vprm(n)=v(n)-vbar
                            NEXT n
                                FOR n=1 TO nsamples
                                    sumupsq=sumupsq+uprm(n)^2
                                    sumvpsq=sumvpsq+vprm(n)^2
                                        NEXT n
                                            upmnsq=sumupsq/nsamples
                                            vpmnsq=sumvpsq/nsamples
                                                uprms=SQR(upmnsq)
                                                vprms=SQR(vpmnsq)
                                                    PRINT uprms,vprms
    REM *** correlation coefficient calculation
```

```
                sumcor=0
                ilag=1
    100 REM *** continue
                    FOR n=1 TO nsamples-ilag
                        sumcor=sumcor+uprm(n)*vprm(n+ilag)
                            NEXT n
                                corrco(ilag)=sumcor/(nsamples-ilag)/(uprms*vprms)
                                    ilag=ilag+1
                                    sumcor=0
                                        IF ilag<1750 THEN 100 ELSE 200
    200 REM *** continue
                OPEN "c:corrcoef.dat" FOR OUTPUT AS #1
                        FOR n=1 TO 1750
                            PRINT n,corrco(n)
                            WRITE#1,n,corrco(n)
                                NEXT n
                        INPUT "end program now?";zz
                        IF zz>0 THEN CLOSE
                        END
END FUNCTION
```

There are other important applications for which we might want to record and utilize two simultaneous signals, and one is the determination of *coherency*. The coherence provides a statistical measure of the relationship between a pair of signals, and it has been used in applications as diverse as geophysical phenomena and neurobiology. Consider a single-input, single-output system (SISO) where the input (or forcing function) is $x(t)$ and the output (system response) is $y(t)$. Let $S_{xx}(f)$ be the power spectral density (or autospectrum) for x and $S_{yy}(f)$ the power spectral density for y. Assume we also compute the cross-spectrum using x and y, let it be denoted by $S_{xy}(f)$. We will define a coherence function, $C^2_{xy}(f)$ as

$$C^2_{xy}(f) = \frac{|S_{xy}(f)|^2}{S_{xx}(f)S_{yy}(f)}, \tag{10.53}$$

where $0 \leq C^2_{xy}(f) \leq 1$ for $0 \leq f \leq \infty$. Some authors, including Bloomfield (1976), refer to eq. (10.53) as the *squared coherency*. If $C^2_{xy}(f) = 1$, then the input and output functions are linearly related for all t and all f, but if $C^2_{xy}(f)$ is less than 1, this is an indicator of the presence of nonlinearities, noise in the system, or possibly other inputs affecting system response. Additionally, if the coherence is zero then the variables $x(t)$ and $y(t)$ are unrelated.

Equation (10.53) requires determination of the cross spectrum, $S_{xy}(f)$:

$$S_{xy}(f) = \int_{-\infty}^{+\infty} \gamma_{xy}(\tau)e^{-2\pi i f\tau}d\tau, \quad \text{estimated as} \tag{10.54}$$

$$S_{xy}(f) \cong \frac{1}{N}\sum_{t=0}^{N-1}(x_t - \bar{x})e^{-2\pi i f t}\sum_{s=0}^{N-1}(y_s - \bar{y})e^{-2\pi i f s}. \tag{10.55}$$

A problem in the estimation of spectra is an uncontrolled (unknown) variance that may be too large to allow us to properly interpret the result. Obviously one way to minimize this would be to collect many sets of time-series data, process each one, and then average the results. This is impractical from a time-invested standpoint, and it may even be impossible for some types of investigations. An alternative is to divide a time-series data set into a number of overlapping segments and then process each segment separately. This scheme was proposed by Peter Welch (1967) and is widely known as Welch's method. Typically, both the number of segments (k) and the extent of the overlap must be chosen. For example, if we had a series of 4,096 data points and we chose a segment length of 256 with *no* overlap, then the number of segments would of course be 16. If we specified a 25% overlap, then $k = \frac{4{,}096-256/4}{256-256/4} = 21$; with a 50% overlap, $k = 31$. We emphasize that the number of segments employed has a significant impact upon spectral resolution for noisy signals; this was demonstrated clearly by Solomon (see Sandia Report SAND91-1533, December 1991, pages 17–19) and Dodson-Robinson (2022).

To determine the cross-spectrum for two time series, $x(t)$ and $y(t)$, we first divide the data into segments as indicated above, and then Fourier transform each segment to produce a set of S_{xx}'s and S_{yy}'s; the segments must be windowed to minimize undesirable leakage. For each segment pair we compute the product of S_{xx} and the *complex conjugate* of S_{yy}: S_{xx} S_{yy}*, and sum this product over all of the segments. The real part of the result is called the *cospectrum*, *C*, and the imaginary part is the *quadrature* spectrum, *Q*. The resulting amplitude for each segment is $\sqrt{C^2 + Q^2}$ and the coherence is

$$C_{xy}^2(f_k) = \frac{C^2(f_k) + Q^2(f_k)}{S_{xx}(f_k)S_{yy}(f_k)}. \tag{10.56}$$

You will recall that the periodogram for a time series was introduced in Section 10.3. In the case of two time series, $x(t)$ and $y(t)$, the DFTs are represented by S_{xx} and S_{yy}, and the periodograms of the two signals are proportional to $[S_{xx}(\omega)]^2$ and $[S_{yy}(\omega)]^2$. The *cross periodogram* is given by

$$I_{xy}(\omega) = \frac{n}{2\pi} S_{xx}(\omega)S_{yy}^*(\omega) \tag{10.57}$$

(the asterisk indicates the complex conjugate). Note that the cross periodogram is generally complex, it may not be positive, and it is not symmetric.

10.9.3.1 Suggested Exercise: Computing the Cross Spectrum for Two Signals, *x*(*t*) and *y*(*t*)

There are cases where the cross spectrum proves invaluable, and this is particularly true when we need to learn something about the connection between two signals that—at first—may appear quite unrelated. Let's set about demonstrating in detail how the calculation can be carried out (and I urge the reader to complete this exercise). We assume we have two sets of time-series data, $x(t)$ and $y(t)$, of equal length and recorded with the same time interval, Δt. Specifically, let's take

$$x(t) = \cos\frac{2\pi(12.5)j}{100} + \sin\frac{2\pi(8.333)j}{100} + 4Z - 2 \tag{10.58}$$

and

$$y(t) = \frac{1}{2}\cos\frac{2\pi(12.47)j}{100} + \frac{2}{3}\sin\frac{2\pi(8.11)j}{100} + \frac{3}{4}\cos\frac{2\pi(10)j}{100} + 4Z - 2. \tag{10.59}$$

Again, Z is a uniform random number between 0 and 1. We let j (our time index) assume integer values from 0 to 511 so that we have 512 discrete values for each set. Notice the similarity of the pairs of frequency values, 12.5 and 12.47, and 8.333 and 8.11; this will have implications for our cross spectrum. You will also see that $y(t)$ contains an extra cosine term with a frequency of 10—which has no counterpart in $x(t)$. Thus, this term should not be prominent in the cross *spectrum*.

1) Using an available DFT code (such as Excel, or the example program from Section 10.5), process $x(t)$ and find S_{xx}.
2) Next, process $y(t)$ and compute S_{yy}.
3) Multiply the imaginary part(s) of S_{yy} by -1 to obtain the complex conjugate, S_{yy}^*.
4) Multiply S_{xx} and S_{yy}^*; note that $(a + ib)(c - id) = ac + bd + i(bc - ad)$.
5) The cospectrum, C, is $ac + bd$, and the quadrature spectrum, Q, is $bc - ad$.
6) Find the cross spectrum amplitude, $A = \sqrt{C^2 + Q^2}$.
7) Plot A as a function of the time index, j.

Note: $\frac{2\pi(12.5)}{100} = 0.7854$, $\frac{2\pi(12.47)}{100} = 0.7835$, $\frac{2\pi(8.333)}{100} = 0.5236$, $\frac{2\pi(8.11)}{100} = 0.5096$.

What will your cross spectrum plot, $A(j)$, reveal? A very prominent peak should appear for index values of about 62–64 (remember, $64\pi/256 = 0.785$), and a broader, lower-amplitude peak should appear for $j = 40$–44 ($42\pi/256 = 0.515$). The extra cosine term in $y(t)$ with $2\pi(10)/100 = 0.6283$ (corresponding to an index position of 51) is completely absent, as expected. If you have followed through with this recommended exercise, you may be tempted to also calculate coherence since the required pieces seem to be in place. *However*, the computation will only show $C_{xy} = 1$. Since we had just one data segment, the magnitude and phase at each frequency are set over the time period (0–511). Therefore we cannot effect a meaningful comparison—the time-series data must be divided into segments with each segment pair treated separately and the average computed over the entire set. This is where Welch's (1967) method becomes really important. If you are familiar with Matlab™, Gille (UCSD-SIOC 221A, Lecture 15) shows exactly how to carry the process out.

Computing the cross spectrum and the coherence would be laborious if one were starting from scratch, but as we saw in the recommended exercise, it is manageable if DFT software is available. Furthermore, there are software tools, including Matlab, Origin, XLSTAT by Lumivero (for Excel), SIGVIEW by SignalLab, and the NCAR Command Language (the function is *cohsq*) that can be used to compute $C_{xy}^2(f)$. Note that the NCAR Command Language is available for free download as it was developed under the auspices of the National Science Foundation; it does require knowledge of *Python* however.

10.9.4 Applications of Coherence

The features of coherence described above suggest that $C^2_{xy}(f)$ might be used as a diagnostic tool for the analysis of machines and structures, and one application that has been discussed in the literature is the investigation of vibrations in aerospace hardware. For example, R. L. Barnoski (1969) studied the response of a beam (10 cm long and 4.6 mm thick) consisting of a viscoelastic material bonded between two steel strips, forcibly excited at the secured end; accelerometers were used to measure both the input (forcing function) the response at the free end. Both power spectral densities S_{xx} and S_{yy} exhibited peaks at 60 Hz reflecting powerline contamination (a common occurrence whenever 110 V AC powered instruments are used for laboratory measurements); both also contained peaks at about 165 and 820 Hz corresponding to resonant frequencies. The computed coherence function was very close to 1 from about 20 Hz to around 400 Hz (suggesting a linear relationship between excitation and system response), although it did reveal a dip at 500–600 Hz and a precipitous decline between 800 and 900 Hz.

More recently Bejger and Gawdzinska (2015) explored the use of coherence in the rapid and nondestructive testing of composite castings (an aluminum alloy poured into a felt and carbon fiber preform). They used acoustic excitation of the castings and computed the coherence between the response of flawed samples containing voids and a flawless specimen (as determined by computed tomography, or CT scan). The coherence function exhibited oscillatory behavior with very low values at particular frequencies including 3, 15, and 19 kHz. Additionally, the *amplitude ratio* between input and output signals revealed large magnitude peaks at 22, 23, and 25 kHz. These researchers concluded that coherence might provide an effective tool for revealing flaws in composite castings, and do so in near real time; of course, the motivation for this type of testing is that a defective component could be removed from the assembly line in a production process.

Coherence is also being used in astronomy in efforts to identify habitable planets by comparison of observed time series for a stellar activity indicator, $x(t)$, and a radial velocity signal, $y(t)$. Dodson-Robinson et al. (2022) described efforts to sort out detected signals by identifying frequencies common to both time series, and they noted that the main difficulty is obtaining high-quality estimates for S_{xy}. A serious problem in this application is that the sample intervals for stellar activity indicators and radial velocity signals are often different, and therefore conventional FFT methods require interpolation of irregularly spaced data points onto an evenly spaced grid. Dodson-Robinson and her coworkers have developed a software package (*NWelch*) to handle such data without the approximations introduced by interpolation. This software is freely available at https://github.com/sdastro/NWelch, and it comes with installation instructions and a description of software functionality.

10.9.5 Seismometry and Time-series Data

The Loma Prieta earthquake occurred on October 17, 1989, and the epicenter was about 10 mi northeast of Santa Cruz, CA (very near Mt. Loma Prieta). Its magnitude on the Richter scale was 6.9 and the main shaking lasted for about 20 s. For comparison, the devastating San Francisco earthquake of 1906 had an estimated magnitude of about 7.9 with rupture occurring on the San Andreas fault over a distance of 296 mi. Of course by the 1980s there were monitoring stations all over the United States, so recorded data from the Loma Prieta event abound. Figure 10.31 shows a seismometer data set from USGS (corrected CRA-1 data, 16,000 points); the first 5,000 points are included in the graph.

An obvious challenge in areas prone to earthquakes is the design and construction of buildings that are less susceptible to catastrophic damage. In the Loma Prieta quake 63 people were killed; 3,757 injured; and 18,306 houses were damaged with 963 destroyed. Part of Interstate 880 collapsed, and a section of roadbed fell from the Bay Bridge (data obtained from the *California Department of Conservation*). One positive consequence (an opinion shared by many Bay area residents) was the demolition of the elevated Embarcadero Freeway that scarred the waterfront in San Francisco for decades.

The nature of the soil underneath a building is very important to the determination of how it responds to a seismic event; bedrock is better than saturated soils where liquefaction can cause large-scale movement. In fact, there have been cases where earthquake stresses upon saturated soils caused entire buildings to topple. From a structural standpoint, larger buildings in earthquake-prone areas are often constructed with base isolators that either absorb part of the energy or prevent it from being transferred to the building (using low-friction plates that can slide back and forth under the structure). Serious problems can occur in cases where there is resonance between the seismic waves and the building—of course flexibility is crucial, and wood and steel structures generally react better to shaking than masonry

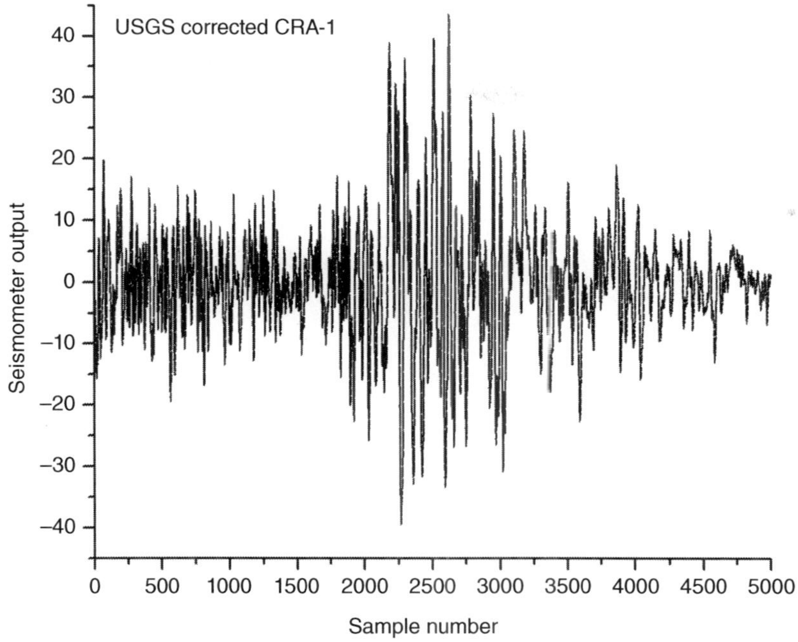

Figure 10.31 USGS data from the Loma Prieta earthquake in 1989. The large-amplitude oscillations in the center of the graph occurred at a frequency of about 5–6 Hz. The autocorrelation coefficient for these data exhibits the oscillation ("ringing") expected of this type of signal.

and unreinforced concrete construction. In the large-magnitude quake in Mexico City on September 19, 1985, many 6–15-story buildings were damaged, whereas both very low and very tall structures suffered less. Higher-frequency, low-amplitude waves affect short buildings and low-frequency, high-amplitude waves affect tall buildings (*IRIS,* an NSF collaboration, has a building oscillation seismic simulation, *BOSS*, that is illustrative). The *Architectural Institute of Japan* (AIJ) has published approximate data on the fundamental frequencies of buildings with steel structure, and some are reproduced here:

Height (m)	**Frequency (Hz)**
30	1.7
50	1.0
80	0.63
100	0.50
200	0.25

Since many buildings in California are instrumented, some specific data on fundamental frequencies of large structures in earthquake zones are available. For example, the Transamerica Pyramid in San Francisco is 850 ft tall with $f_{\text{fund}} = 0.3$ Hz and the Pacific Park Plaza in Emeryville is 308 ft tall with $f_{\text{fund}} = 0.38$ Hz. Warren Hall on the CSU Hayward campus was 200 ft tall with $f_{\text{fund}} = 0.76$ Hz; it was demolished in 2013 because it was determined to be seismically vulnerable (and located near the Hayward fault). The fundamental frequency of a building can be estimated if one knows both the stiffness, K, and the mass, M: $f = \frac{1}{2\pi}\sqrt{\frac{K}{M}}$, and empirically, f is found to be inversely proportional to height as indicated by the AIJ data above.

Ground-motion frequencies in earthquakes are usually in the 0.1 to perhaps 20 Hz range with 1–10 Hz being very common. In Figure 10.32, a power spectrum computed by Fourier transform using the data of Figure 10.31 is presented. The dominant frequencies are 1, 4, 5, and 6 Hz, and these are primarily above the range that would lead to resonance in large, modern buildings in the Bay area.

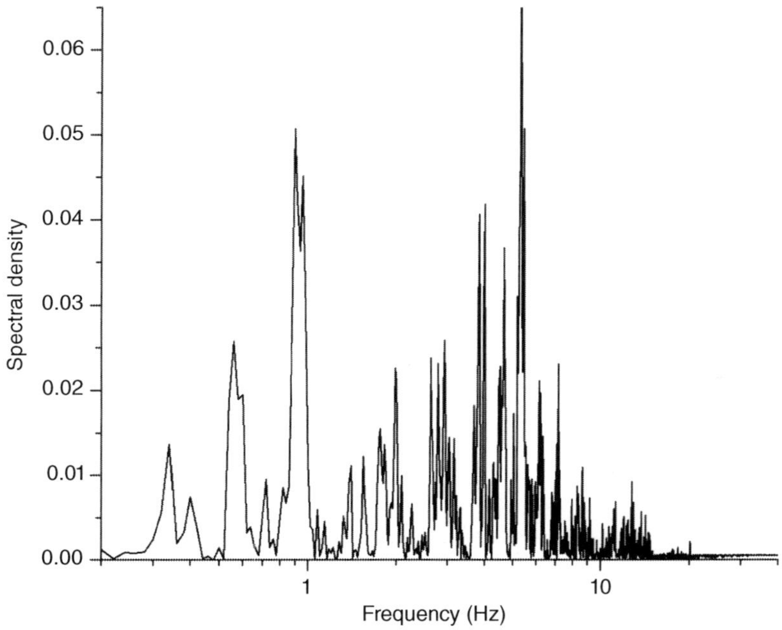

Figure 10.32 Computed spectrum for USGS data from Loma Prieta earthquake of 1989. The bulk of the signal energy is located between 0.5 and 10 Hz. The cutoff frequency for these data was about 140 Hz, well beyond the range typical of earthquakes.

The logic employed for these calculations is provided here:

```
#COMPILE EXE
#DIM ALL
    REM *** Autocorrelation and Fourier transform
    REM *** Input USGS Loma Prieta earthquake data
            GLOBAL dt,f,sum,vbar,vmnsq,pi,S,zz AS SINGLE
           GLOBAL i,j,n,ndel,nlimit AS INTEGER

FUNCTION PBMAIN
            DIM v(8500) AS SINGLE
            DIM corr(7000) AS SINGLE
                        nlimit=8000:pi=3.1416
    OPEN "c:USGS Loma Prieta 1989 corr CRA1 FORMATTED.txt" FOR INPUT AS #1

                        FOR n=1 TO nlimit
                            INPUT#1,v(n)
                                PRINT v(n)
                            NEXT n
                                CLOSE
    REM *** determine dt and vbar
                        dt=0.0034
                            FOR n=1 TO nlimit
                                sum=sum+v(n)
                                NEXT n
                                    vbar=sum/nlimit
                                        PRINT dt,vbar
    REM *** find mean-square fluctuation
                sum=0
                        FOR n=1 TO nlimit
```

```
                        sum=sum+(v(n)-vbar)^2
                        NEXT n
                            vmnsq=sum/nlimit
                                PRINT vmnsq
                    OPEN "c:USGSlomaPriCORREL.dat" FOR OUTPUT AS #1
    REM *** find correlation coefficient
                            sum=0:ndel=1
    200 REM *** continue
                                FOR n=1 TO nlimit-ndel
                                sum=sum+(v(n)-vbar)*(v(n+ndel)-vbar)
                                    NEXT n
                                        corr(ndel)=sum/(nlimit-ndel)/vmnsq
                                        PRINT ndel,corr(ndel)
                                        WRITE#1,ndel,corr(ndel)
                                            ndel=ndel+1
                                            sum=0
                                        IF ndel<6500 THEN 200 ELSE 300
    300 REM *** continue
                CLOSE
                OPEN "c:LOMApriSPECTR.dat" FOR OUTPUT AS #1
        REM *** begin spectrum computations, f in HZ
                        sum=0:f=0.03
    400 REM *** continue
                            FOR n=1 TO 6500
    sum=sum+COS(f*2*pi*n*dt)*corr(n)*dt
                                    NEXT n
    S=sum/pi
                                        PRINT f,SQR(S^2)
                                        WRITE#1,f,SQR(S^2)
                                        f=f+0.02
                                        sum=0
                                            IF f<50 THEN 400 ELSE 500

    500 REM *** continue
            INPUT "Shall we continue?";zz
            IF zz>0 THEN CLOSE
            END
END FUNCTION
```

10.9.6 Decaying Turbulence in a Box

Previously we alluded to the fact that time-series data are generated frequently in engineering and scientific investigations. One area where this is particularly important is the study of turbulent flows, where a number of instruments have been developed that can monitor fluid velocities at a point in space. Examples include laser-Doppler velocimetry (LDV) and hot-wire anemometry (HWA). These devices produce signals as functions of time and in the case of HWA, the signal is usually voltage. The response time for a hot-wire anemometer is short, so high-frequency fluctuations can be observed without difficulty.

In a study of decaying turbulence in a rectangular enclosure, a hot-wire anemometer was used to measure point velocity; motion of the air was induced with a centrifugal blower (producing a large-scale recirculation). The blower was turned off at about 2.3 s, and decay of the turbulence was recorded (Figure 10.33).

The mean velocity for the initial period of operation (the stationary data from 0 to 2.3 s) was 6.4 m/s, and the root-mean-square (rms) fluctuation was 2.9 m/s; therefore, $\frac{v_{rms}}{\overline{V}} = 45\%$. The spectrum for this initial period of operation

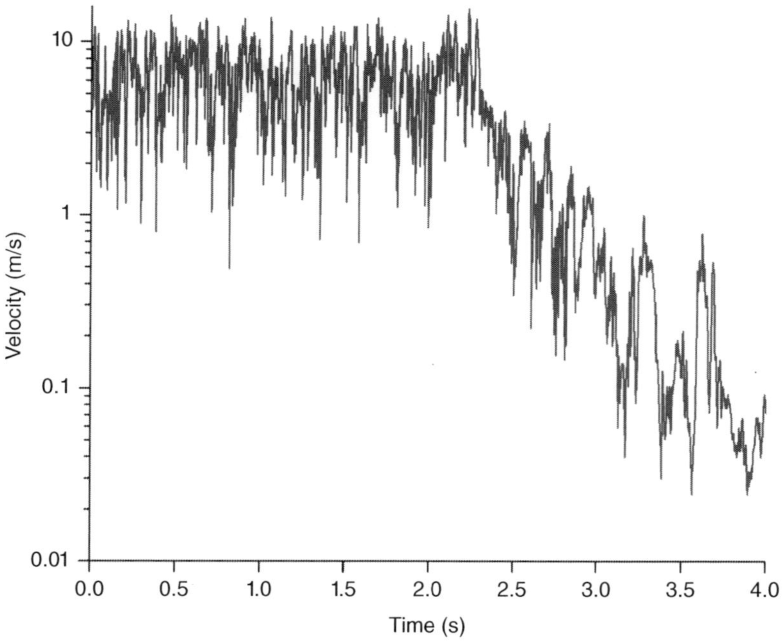

Figure 10.33 Recorded velocity for a turbulent, recirculating flow in a box. The blower was shut off at about $t = 2.3$ s leading to a protracted period of decay. Note that just 1.7 s after shut down the mean velocity has fallen to about 0.06 m/s. The sample interval was $\Delta t = 0.002$ s, so that the Nyquist frequency was 250 Hz.

reveals that most of the signal energy is located between 5 and 80 Hz with dominant peaks at about 7 and 8 Hz. Because there were only 1,119 data points in the initial phase of operation the detection of very-low-frequency motions is constrained (1,119 data points corresponds to only 0.24 s). The spectral density at frequencies between 80 and 250 Hz is quite low.

10.9.7 Bubbles and the Gas–Liquid Interface

When a gas is introduced into a liquid, whether it be due to local pressure changes, injection through a simple orifice, by jet aeration, or by liquid drops impacting a gas–liquid surface, a number of processes are set in motion that are capable of producing acoustic noise. An example of such phenomena is cavitation occurring on propellers and pump impellers due to large localized pressure reductions (recall problem 6.14). Not only will cavitation produce noise, but it may also result in severe damage due to bubble collapse—should you have a chance to walk through Aquatic Park in San Francisco, look for the propellers on display from the ferry, *Klamath*, which exhibit the pitting resulting from years of cavitation. Cavitation bubbles are normally undesirable and occur inadvertently, but in the process industries and in wastewater treatment, bubbles may be introduced into a liquid to promote mass transfer and mixing. In aerated reactors, for example, air bubbles may form at and detach from simple orifices, or they may be introduced through surface agitation or jet aeration. Such processes may be accompanied by bubble detachment from orifices, followed by shape oscillations, bubble breakage and coalescence, and ultimately bubble disengagement at the free surface. Small air bubbles in aqueous media are generally spherical, but other shapes are also common depending upon size (shape affects buoyancy and drag forces, of course). An image of air bubbles in water produced by jet aeration illustrates the variety of shapes seen in two-phase (gas–liquid) processes.

Notice how the small bubbles shown in Figure 10.34 are spherical, while larger bubbles tend to be ellipsoidal. The spherical bubbles oscillate with a characteristic frequency determined by surface tension (τ), the density of the surrounding liquid (ρ_1), and bubble size (R):

$$s^2 = \frac{12\tau}{\rho_1 R^3}. \tag{10.60}$$

Thus, for a 0.5 mm air bubble in water, $s^2 = \frac{(12)(72)}{(1)(0.025)^3} = 5.53 \times 10^7$ rad^2/s^2 and therefore $s = 1{,}183$ Hz. Of course this frequency is in the acoustic range, so these oscillations can produce detectable sound; we should also observe that much research has been devoted to augmenting gas–liquid mass transfer rates by using acoustic noise to induce bubble oscillation.

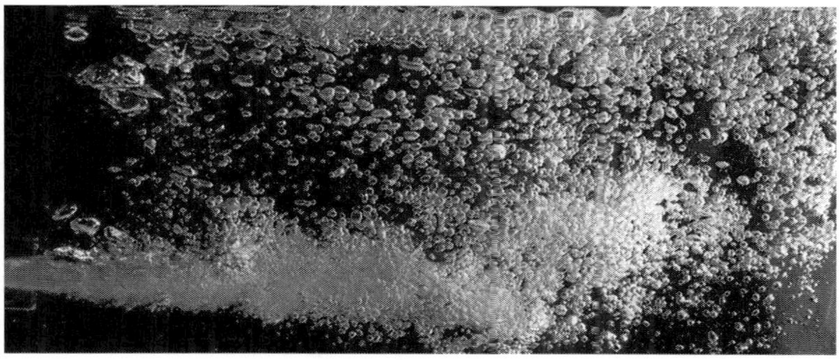

Figure 10.34 Characteristic shapes of air bubbles in water produced by jet aeration. The very large population of bubbles at the top of the image at the gas–liquid interface affords many opportunities for coalescence to occur prior to bubble rupture.

A number of experiments were carried out in which air was introduced through a sieve plate into column of a liquid medium consisting of water and glycerol. The objective was to learn more about energetic phenomena associated with bubbles in aqueous media of higher viscosities, and the possible impact of those energetic motions upon entrained entities (perhaps cells, or cells being grown upon microcarriers). A microphone was used to monitor sounds produced both at the top of the column (above the interface) and at the bottom at the same level as the sieve plate sparger. Glycerol slightly decreases the surface tension when added to aqueous solutions, but the effect is small. At the top of the column the sounds produced are mainly associated with disengagement processes (coalescence, bubble breakage, cavity collapse, and droplet ejection and impact). At the bottom of the column the noises mainly come from bubble formation, detachment, and shape oscillations. Figure 10.35 shows how the signal energy was distributed with respect to frequency at modest gas (air) rates as detected at the top of the column. The distribution of signal energy at the *bottom* of the column adjacent to the sieve plate sparger was significantly different both at low and high frequencies.

The spectrum for disengagement processes at the top of the column, Figure 10.35, shows significant signal energy at 60, 70, 115, 220, 300–500, 1,100, 1,750, and 2,020 Hz. In contrast, the spectrum obtained at the bottom of the column indicates a broad band of important frequencies between 300 and 600 Hz, with a very sharp, narrow, contribution centered at 1,605 Hz. It is clear that these detected acoustic signals do not merely represent spherical bubble oscillations; after all, we know that 400 Hz would correspond to the fundamental oscillation of a spherical bubble with $R = 0.0515$ cm. Many of the bubbles

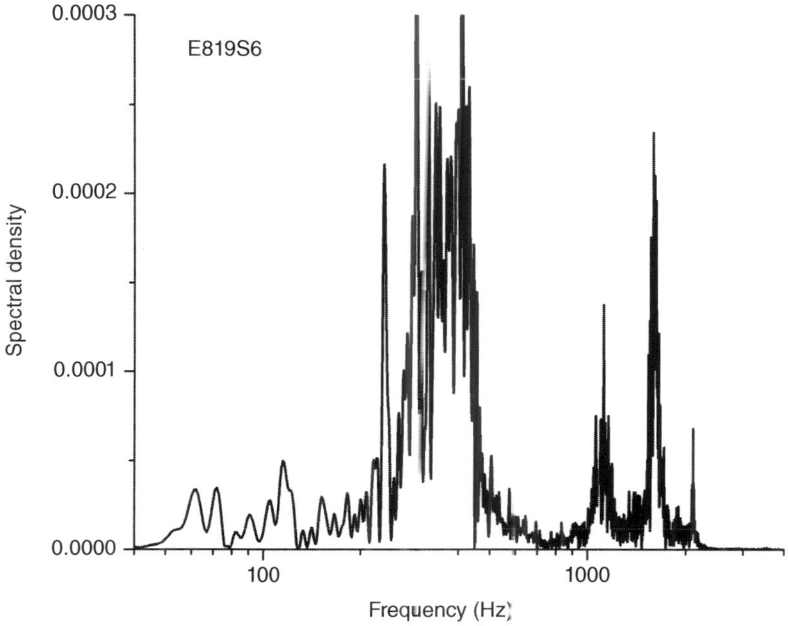

Figure 10.35 Spectrum for the recorded acoustic noise at the top of the column where bubble disengagement processes are the dominant sound producers.

formed at the sieve plate in this aqueous glycerol solution were larger than that. Furthermore, when bubbles arrive at the free surface some may coalesce before film thinning produces an unstable interface followed by rupture. When large bubbles break they typically throw small liquid “film droplets” laterally. When small bubbles break, the cavity created at the liquid surface due to the higher pressure in the bubble’s interior will collapse; this collapse can eject larger “jet” droplets usually with a nearly vertical trajectory. And often these breakage processes at the free surface subsequently produce much smaller bubbles capable of generating higher-frequency sounds (the next time you pour a soft drink onto ice, listen carefully at the top of the glass).

10.9.8 Shock and Vibration Events in Transportation

When cargo is moved by ship, rail, or truck, it is subjected to shock and vibration. If that cargo is delicate, such phenomena may lead to damage and loss; consequently, much effort has been expended in measuring strains and accelerations in transport environments. For example, in the case of railroad transport low-frequency motions arise from irregular track surfaces; a low spot opposite a high spot will lead to rail car rocking. Shust (Objective Engineers, Inc., 2007) found that rocking and vertical bounce from spring deflection occurred at frequencies mainly under 4 Hz. On the other hand, vibrations measured in a locomotive bearing box revealed important energy content at 440, 500–540, 607, and 978 Hz. A source of violently abrupt motions in railroad transport is car coupling. Magnuson and Wilson (1977) measured acceleration spectra for coupling events occurring at an impact velocity of 5.25 mph (about 99% of car-coupling impacts occur at speeds less than 11 mph). One of the more interesting features of the coupling shock event is that accelerations of significant energy are produced at *higher* frequencies. Figure 10.36 illustrates a spectrum redrawn from data presented by Magnuson and Wilson.

Shust notes that a common railroad car design specification (the ability to withstand 4 g accelerations) is frequently violated by the accelerations produced at higher frequencies generated by coupling events. This is a critical phenomenon with bearing upon events completely unrelated to railroad car coupling: If energy from impacts or collisions is transferred to higher frequencies (as indicated by Figure 10.36), this may result in unusually large accelerations at frequency ranges not expected from the original event.

Because tank cars are used extensively to transport hazardous and flammable liquids their response to impact is especially important. In the first half of the twentieth century most tank cars had a full-frame design, but in 1954 the Union Tank Car Company switched to frameless construction where the ends of the tank shell rest upon sill pads mounted on top of the trucks—in essence, the tank itself became the car frame. It is known that overspeed impacts involving modern tank cars can result in crack formation and propagation in the sill pad. The Federal Railroad Administration’s

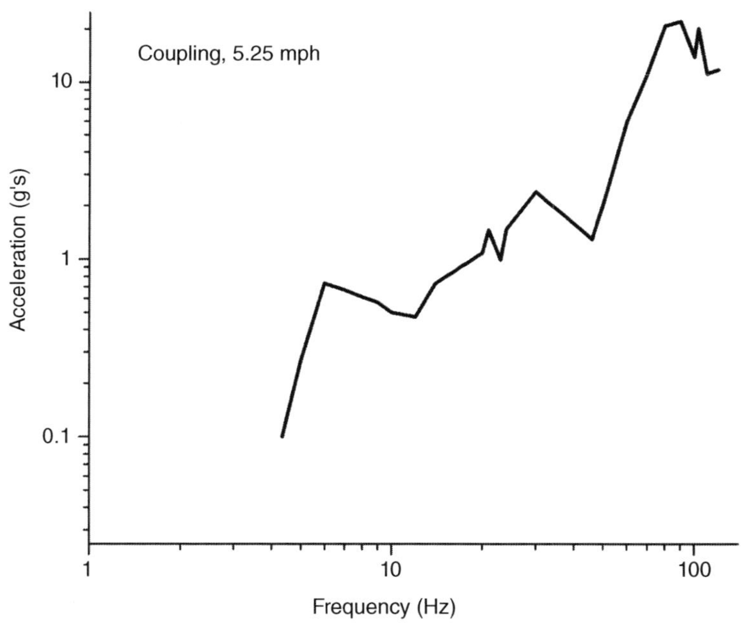

Figure 10.36 Spectrum of measured accelerations resulting from car coupling at 5.25 mph adapted from Magnuson and Wilson. Note the large measured accelerations occurring at about 100 Hz.

(*Research Results*) RR 19-17 (2019) contains an image of an extreme case in which complete lateral failure of the sill occurred.

Lyons et al. (DOT VNTSC-FRA-07-03, 2007) used 37 accelerometers mounted on a tank car to measure both longitudinal and vertical accelerations occurring during coupling events. Tests were conducted with the tank car empty, and filled with 22,577 gal of detergent alkylate (the filled tank car weighed 263,000 lbs). The coupling events were created by running the tank car into a stationary consist of hopper cars, and the observed waveforms were sampled at a time interval of 0.0002 s, producing a Nyquist frequency of 2,500 Hz. Coupling experiments for a full car at speeds ranging from 7.2 to 7.5 mph revealed vertical accelerations (at the top center of the car) exceeding ± 50 g's. The longitudinal accelerations were smaller, mostly within the range of ± 15 g's. The authors note that measurements made with accelerometers mounted at the impact end of the tank car included significant noise, so the middle of the car provided the best signal-to-noise compromise. Spectra were computed from the time-series data, and the results obtained for the empty tank car with a peak coupler force of 1,004,000 lbs were particularly interesting, revealing significant spectral peaks at about 10, 240, 280, 360, 400, 460, 500, 550, and 600 Hz. Lyons et al. also estimated the natural frequencies for bending, radial, and axial modes for a simple thin-walled cylinder; the first five modes for bending were 36, 144, 325, 577, and 902 Hz, and for radial vibration the estimated natural frequency was 616.5 Hz. It should be noted that the response of the couple (a Cardwell Westinghouse Mark 50 draft gear) to impact varied with severity of the collision. This type of draft gear has friction wedges that can slip, leading to spring compression, and after 3.25 in. of travel, direct transmission of force.

Lamoreaux et al. (1981) (DOE contract DE-AC04-76DP000789) investigated shock and vibration events associated with shipping spent (nuclear) fuel casks on flat-bed rail cars. A coupling occurring at 10.7 mph for a flat car loaded with a 40 ton cask produced accelerations as large as ± 300 g's. The duration of the event was a little over 100 ms, and the main frequency of the recorded oscillations was about 480 Hz. Fortunately, this example could be characterized as extreme since—as noted previously—virtually all railroad car couplings occur at speeds less than 11 mph. Data presented in the FRA's RR (Research Results) 19-17 (2019) show that the impact force due to tank car coupling rises dramatically as the speed exceeds 7.5 mph. In fact, the data presented show a typical coupling force of 500,000–600,000 lb_f at 7.5 mph, but as much as 1.5×10^6 lb_f at 9 mph.

Time-series data collected in rail transport scenarios reveal that large strains (and perhaps very large accelerations) are especially likely to be generated by shock events like coupling or even worse, derailment. In some cases, the accelerations may exceed design specifications for the vehicle or car. This can pose a serious problem when hazardous materials are being transported; a ruptured tank car, for example, might disrupt revenue service and even lead to evacuations (and liability) in populated areas. Recent tragic accidents in Casselton, ND, where 400,000 gal of crude oil were spilled, and the devastating explosion and fire in Lac-Megantic, Quebec, resulting from tank car brake failure have highlighted the need for re-examination of tank car design. The FFT/DFT is absolutely indispensable in investigations of this type, where improved car design criteria can be developed to enhance transportation safety.

Truck transport also results in shock events that occur through travel over pavement irregularities, potholes, railroad crossings, etc. Magnuson and Wilson (1977) reported spectra from over-the-road tests performed with seven different tractor–trailer combinations. They included measurements for all three axes (vertical, transverse, and longitudinal) in their study, and as one might expect vertical accelerations were the largest by a significant margin. Large vertical accelerations are evident in the data, but unlike shocks experienced by railcars during coupling events, these data do not show the same degree of transfer of energy to higher frequencies. In fact, the peak vertical accelerations occur at about 18 Hz where the maximum is nearly 10 g's.

10.10 Conclusion and Some Final Thoughts

Harmonic analysis has been a standard tool in engineering practice for decades, but it may be difficult for the student new to this area to appreciate the significance. For additional emphasis, let us consider an engineering study in aviation that gives us one more example of the role of the Fourier transform in the context of public safety. In 1968 G. R. Slusher authored FAA Report NA-68-27, "*Reciprocating Engine and Exhaust Vibration and Temperature Levels in General Aviation Aircraft*." Safety, of course, is a critical concern in general aviation and one problem area—known to have caused fatal accidents—is compromised exhaust system integrity. The environment for exhaust systems on reciprocating aircraft engines is incredibly hostile due to the combination of thermal stresses, engine vibration, and chemical attack by combustion products. A mechanical failure of the exhaust system could result in engine fires or carbon monoxide infiltration into the passenger cabin (both have occurred in general aviation). Slusher found, and not surprisingly, that vibration levels were highest during takeoff when the engines were operating at maximum power. Furthermore, since the study focused upon horizontally opposed reciprocating engines, the largest amplitude vibrations were lateral (the direction of both piston travel and valve motion). The data revealed that the largest lateral accelerations (at high speed, 2,331 rpm) were about ± 20 g's, and many of the spectra showed important signal contributions at 1,500–2,000 Hz. Slusher (1968) noted that the time-series data (engine vibration waveforms) were periodic, but consisted of a complex and continuous frequency spectrum with some very energetic content in certain frequency ranges. The study demonstrated that the materials that were being used for exhaust systems on general aviation aircraft in the 1960s (often types 321 and 347 stainless steels) were not ideal for the application (Inconel 625 is often used on modern aircraft engine exhaust systems). Though prosaic, this is just one more example of how the Fourier transform has been used in engineering practice to enhance public safety.

Finally, let us observe that the Fourier transform (and specifically the FFT) has become enormously important to modern life through signal processing, communications, analytic chemistry, seismic event detection and interpretation, collision warning systems for automobiles, process control, and so on. The list of applications of the FFT/DFT is incredibly long, and many of the technological marvels we take for granted depend upon very fast determination of the Fourier transform of time-series data. Let us think a little bit about the future with a final example we had previously mentioned in passing, speech recognition. Try recording yourself speaking the words "yes" and "no" using either a microphone with a DSO, or the *Physics Toolbox Sensor Suite Pro* software on your smartphone.

You should find that the "y" in "yes" is nearly a distinct frequency, but the "e" in the middle and the "s" at the end are significantly different. Also, the "s" portion of "yes" has much smaller amplitude but much higher frequency. In fact, the "s" is one unambiguous characteristic that distinguishes the two words. In the case of the spoken "no", the "n" sound (about the first 8 cycles) will be sharply defined at around 127 Hz. In contrast, it will be apparent that the "o" sound involves several frequencies. Now imagine that we wished to issue instructions to a computer verbally and have those instructions—in near real time—be translated into executable code. It will be necessary for the machine to recognize the individual phones (a phone is a single speech sound) and assemble them into an intelligible instruction. If you think about how people actually speak, often running words together, failing to enunciate certain letter combinations clearly, etc., you can understand what an incredibly difficult problem speech recognition really is; and to make it worse, in critical applications the system would have to be 100% reliable. The data you have collected suggest that distinguishing between a spoken "yes" and a spoken "no" would be easy, but interpreting an entire sentence flawlessly would be far more difficult. This underscores what a remarkable achievement a device like Amazon's Alexa™ really is—through a combination of cloud computing and Amazon's voice recognition service Alexa responds to instructions and inquiries in milliseconds and often does so without error. Of course, some machine learning is required where Alexa must make "educated guesses" based upon customer preferences and past interactions. Artificial intelligence has assumed a prominent role in speech recognition with the incorporation of neural networks, perceptual linear predictive (PLP) features, and Viterbi searching (used to determine the most likely word in a speech signal). We will have an opportunity to take a direct approach to the detection of speech patterns in a student exercise where we will use the FFT to examine four words, all with at least some similar sounds.

Finally, let us observe that engineering investigations often involve the measurement of time-series data, and in many cases these data may superficially appear to be random. This is where the power of Fourier analysis can be employed to reveal important periodicities disguised in the clutter and noise. The FFT/DFT can identify signal energies at critical frequencies, allowing us to expose hazards, diagnose system "health" and perhaps even anticipate system failures.

Problems

10.1 Apply Schuster's test to the oscillating waveform:

$$y(t) = \sin(39\pi t) + \sin(51\pi t) + \sin(59\pi t) + \sin(79\pi t).$$

Examine the discrete frequencies 57, 58, 59, 60, and 61π and prepare a plot similar to Figure 10.5 in Section 10.2.

10.2 Consider a set of sampled values consisting of 1,024 observations. These data are uniformly zero until $n = 85$, where a ramp occurs with amplitude increasing at a rate corresponding to 0.05 per sample. This "waveform" then has an amplitude of 1 until $n = 175$; for $n > 175$, $y(t) = 0$, i.e., the amplitude reverts to zero. Compute the DFT for this waveform without tapering. How severe is the leakage?

10.3 Apply the DFT to a square wave (a rectangular window) of amplitude one (1) then apply a split cosine (bell-shaped) window to the data leaving the center 80% of the data unaffected. Finally, allow the window to shrink so that the center amplitude falls below 1–0.75. What impact does this have upon the DFT? Is leakage completely eliminated?

10.4 By an inviscid analysis (see Lamb, *Hydrodynamics, 6th Edition, 1932*, Chapter IX) it can be shown that bubbles and (immiscible) droplets immersed in liquids have a characteristic frequency of oscillation which is given by

$$s^2 = n(n+1)(n-1)(n+2)\frac{\tau}{[(n+1)\rho + n\rho_1]R^3}.$$

For a 0.2 cm diameter water droplet surrounded by air, τ (the surface tension) is about 72 dynes per cm, $\rho = 1$ g/cm, and $\rho_1 \approx 0$. If the most important mode of vibration is that for which $n = 2$, then:

$$s^2 = \frac{(24)(72)}{(3)(0.1)^3} = 576,000\ \text{rad}^2/s^2 \quad \text{and therefore,} f \cong 120\ \text{Hz}.$$

If, on the other hand, an air bubble of the same size is surrounded by water, the frequency of vibration is approximately:

$$s^2 = \frac{(12)(72)}{(1)(0.1)^3} = 864,000\ \text{rad}^2/s^2, \quad \text{and thus} f \cong 148\ \text{Hz}.$$

If bubbles were formed at a single orifice at a rate of 79 per minute and if their sizes ranged from 0.9 mm (diameter) to 2.39 mm, what dominant frequencies would you expect to see in a computed spectrum (data obtained from a hydrophone or a sensitive pressure transducer)?

10.5 Dodge (NASA Contractor Report, CR-1880, September 1971) carried out a study of slosh suppression in LOX tanks; of course, sloshing is critical to stability and control of rockets using cryogenic liquids (oxygen and hydrogen) for propulsion. The work focused upon the use of lightweight, flexible plastic baffles for dampening of the periodic forces created by the moving liquid. In the experimental work, a shake table was used to oscillate a 76 cm diameter tank containing liquid nitrogen. The natural slosh frequency in this tank was cited as 1.05 Hz, and the amplitude obtained was a little less than 4 cm. A load cell was used to record the decay of the liquid motion within the tank. *Suppose* the data record obtained from such an experiment appeared as shown in Figure P10.5.

Use the DFT to see if you can identify all of the frequencies appearing in this data set. The data will be provided to you separately in electronic format. Does the rapid attenuation cause any significant problems? The sample interval, Δt, is 0.02 s.

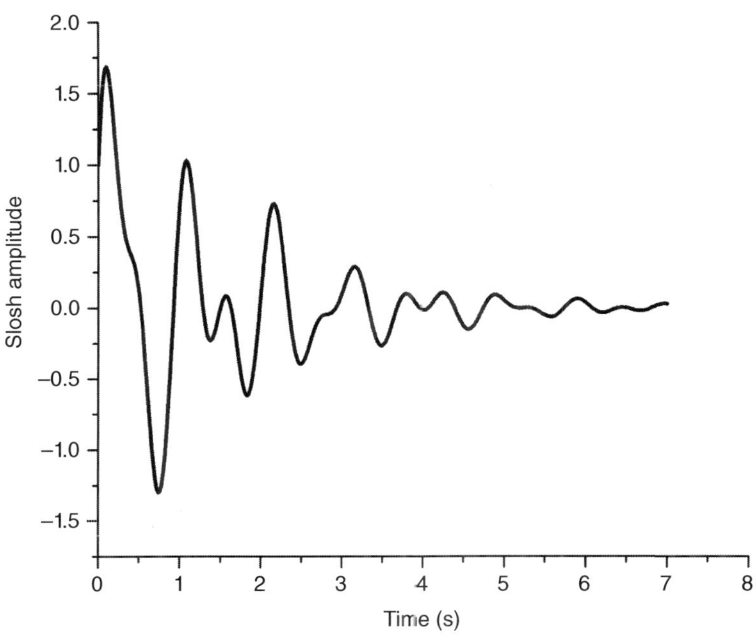

Figure P10.5 Slosh amplitude for cryogenic liquids in tanks.

10.6 Baseball bats, golf clubs, tennis racquets, and the like vibrate upon impact with a ball. In some cases the resulting stinging sensation (transmitted to the hands) can be severe, and in baseball the amplitude of the vibration can be such that the batter's hand(s) may even lose contact with the bat handle. This is a phenomenon that has seen much study by physicists including Howard Brody at the University of Pennsylvania (*American Journal of Physics*, 58:756, 1990) and Rod Cross at the University of Sydney (*American Journal of Physics*, 66:772, 1998). It turns out that wooden baseball bats typically have a couple of fundamental frequencies of about 500–600 Hz and about 150–200 Hz. Naturally, the precise vibration frequencies depend upon where the ball is contacted on the bat, how the bat is secured, and where the piezoelectric vibration sensors are mounted. *Suppose* a ball impact test on a wooden Louisville Slugger™ produced the data set illustrated in Figure P10.6.

Perform a harmonic analysis of these data and determine the fundamental frequencies produced by the impact of a ball upon the wooden bat. The data will be supplied to you separately in electronic format. The sample interval in this case was 0.0005 s (0.5 ms).

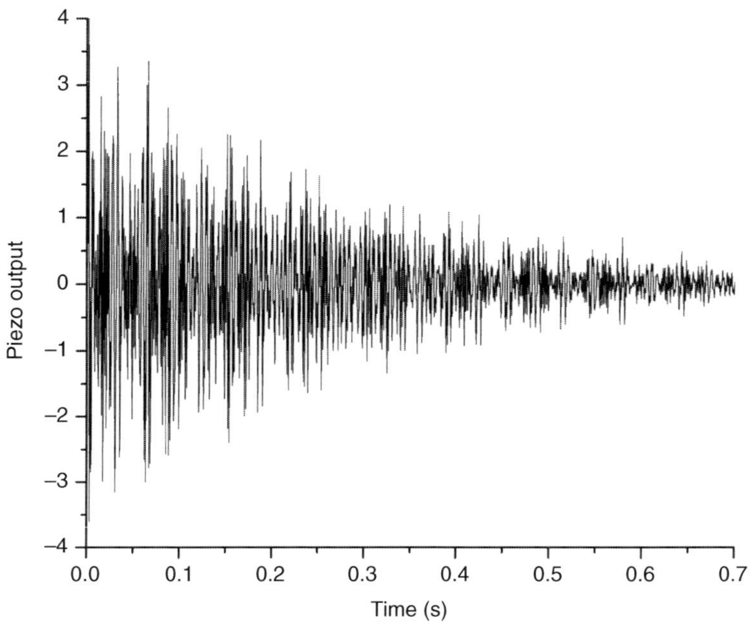

Figure P10.6 Data of the type obtained from a piezoelectric sensor mounted on the barrel of a wooden baseball bat.

10.7 Miles et al. (Torsional and Bending Vibration Measurement on Rotors using Laser Technology, *Journal of Sound and Vibration*, 226:441, 1999) studied both bending and torsional vibrations occurring in the crankshaft of a 2 L diesel engine (with four cylinders) running under load at both 600 and 750 rpm. They used laser light scattered off of the face end of the crankshaft pulley to detect the motion at the end of the crankshaft. Because of the proximity of the measurement point to the number one cylinder, one could expect firing in that cylinder to have a major impact upon the motion. Each segment of piston travel corresponds to 180° of rotation on the crankshaft, so the power pulses on the number one cylinder are separated by (4)(180) = 720°. Therefore, if the engine is running at 600 rpm (10 revolutions per second) power pulses on the number one cylinder should occur about every 0.2 s. This is the pattern revealed by Figure 17 of the Miles et al. paper. Their computed spectrum is shown as (their) Figure 18, and it exhibits important signal content at *many discrete frequencies including* 9, 32, 42, 55, 87, 211, 300, 352, 367, 413, 432, 454, 471, and 644 Hz, among others. *Construct a set of sinusoids with appropriate amplitudes and frequencies* such that the DFT produces a spectrum comparable to Figure 18 of Miles et al. Which components of the oscillating signal must have the largest amplitudes in order for the two spectra to be comparable?

10.8 W. C. Shust (Objective Engineers, Inc., 2007) studied shock and vibration in railroad operations. In one part of the study vibrations in an operating locomotive bearing box were measured. Some simulated data with similar periodicities are reported in the Figure P10.8 immediately below. Perform a DFT on these data (supplied separately), and identify the important contributions to the spectrum. What are the most significant frequencies in the data set?

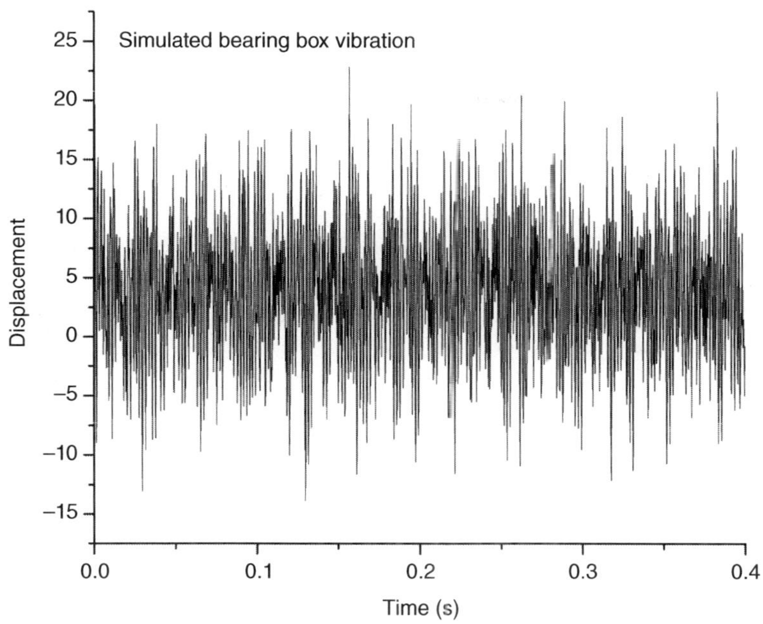

Figure P10.8 Simulated locomotive bearing box vibrations, with $\Delta t = 0.00025$ s. Therefore, $f_{Nyq} = 1/(2\Delta t) = 2{,}000$ Hz.

10.9 The failure of the Tacoma Narrows (TN) bridge in November 1940 led to multidisciplinary efforts to better understand oscillations occurring in suspended structures, and how those vibrations might lead to structural failures. Although *vortex shedding* has frequently been identified as the cause of the TN bridge-deck's oscillations, this explanation is incorrect. At the time of the failure, Professor Farquharson (of the University of Washington) observed that the destructive *torsional* oscillation was occurring at 0.12 Hz—a frequency much lower than that arising from vortex shedding from a suitably sized bluff body (with a wind speed of 42 mph). The aeroelastic torsion that doomed the bridge was nothing at all like the benign vertical "galloping" that was observed even while the bridge was still under construction. The failure of the TN bridge had an important positive result, however; it forced bridge designers and engineers to better understand (and design for) aeroelastic oscillations.

Abdel-Ghaffar and Scanlan (Ambient Vibration Studies of Golden Gate Bridge: I. Suspended Structure, *Journal of Engineering Mechanics, ASCE*. 11:463, 1985) studied vertical, torsional, lateral, and longitudinal motions of the center span of the Golden Gate bridge in response to normal excitations (wind, vehicular traffic, and wave action). In Figure P10.9, a spectrum obtained from span station 3 has been reconstructed from the published paper (the quality of the original figure was poor).

Construct a set of sinusoids, that when Fourier-transformed, will reproduce the essential characteristics of this spectrum. Use a random number generator to produce some fine-scale structure in the spectrum and compare your results with Figure P10.9.

10.10 Recordings of a human voice speaking the words "example," "execution," "expansion," and "exponent" are illustrated in Figure P10.10. Obtain comparable data for these four words using the oscilloscope function available with Physics Toolbox Sensor Suite Pro (this enhanced version must be purchased) or similar sound recording capability. Apply the FFT to all four data sets. Are the spectral characteristics different enough to allow you to distinguish between the words? What are the essential differences, and are you confident that based upon your spectra you could identify all four words without error?

10.11 In recent years the frequency and cumulative effects of concussions resulting from collisions in dynamic sports like football have been revealed to be a much more serious problem than most observers thought. It has become all too evident that even mild traumatic brain injuries (TBI) may have cumulative effects leading to chronic traumatic encephalopathy (CTE) which may manifest itself years later in memory loss, behavioral changes, and even suicide. For many years the standard mechanism for assessment of head trauma has been the Gadd severity index (GSI)

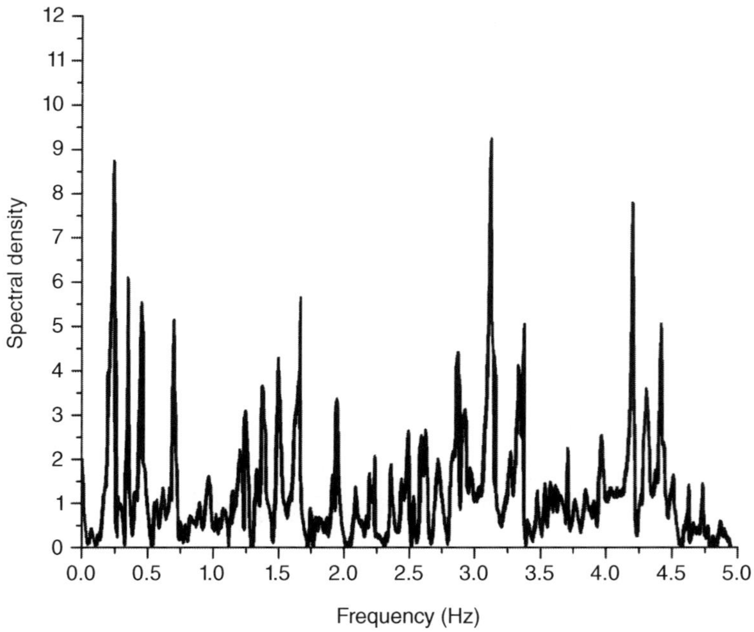

Figure P10.9 Spectrum for torsional motions of the center span. The frequency scale is arithmetic (Hz) and the two highest peaks in the right-half of the spectrum occur at 3.1 and 4.18 Hz. The major contribution at the left edge is centered at 0.225 Hz, approximately. This spectrum is a reconstruction of the data presented by Abdel-Ghaffar and Scanlan (1985).

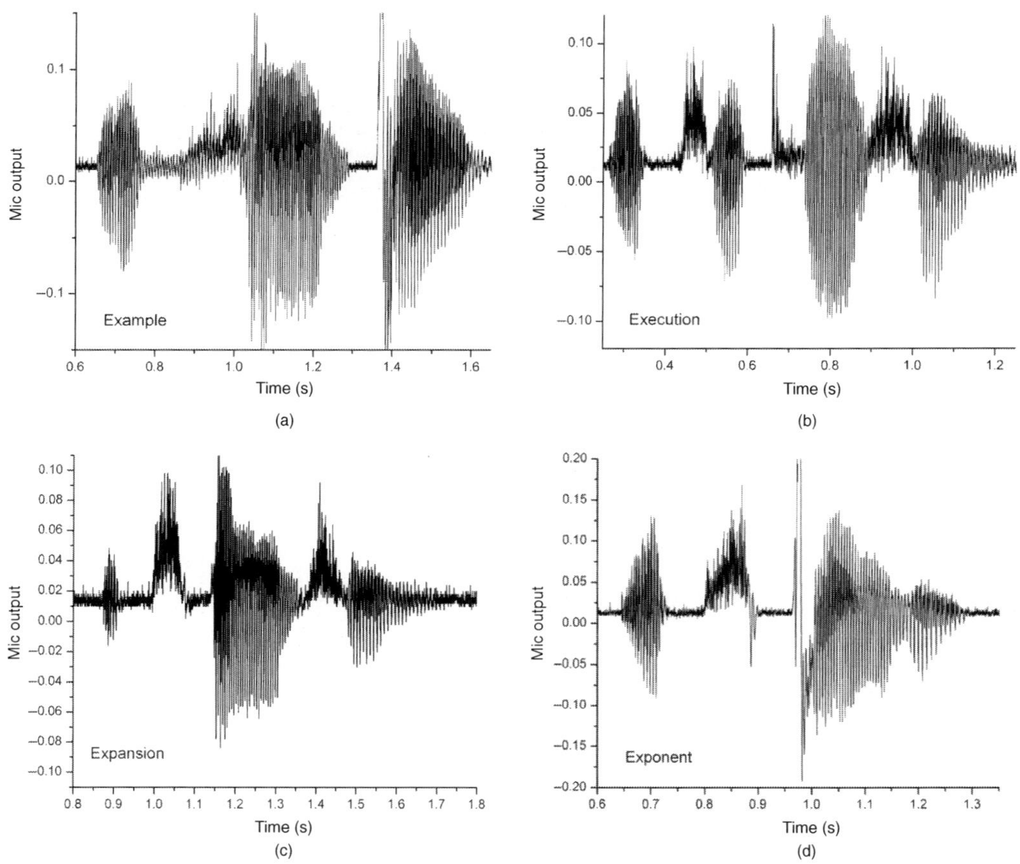

Figure P10.10 (a–d) Time-series data for four spoken words, "example" and "execution," followed by "expansion" and "exponent" at the bottom.

which is determined from the linear acceleration (at the headform center of gravity) and the duration of the event:

$$\mathrm{GSI} = \int_0^T a^{5/2}dt.$$

Throughout the literature of head trauma the threshold value for the severity index has been cited as 1,000 with a typical event duration of about 15 ms (0.015 s). An alternative, but closely related, measure is the head injury criterion, or HIC:

$$\mathrm{HIC} = (t_2 - t_1)\left[\frac{1}{t_2 - t_1}\int_{t_1}^{t_2} a(t)dt\right]^{5/2}.$$

Studies have shown that concussions in professional football (the NFL) typically result from impact velocities on the order of 9.3 ± 1.9 m/s (note that this corresponds to about 20.8 mph). This raises an interesting question: Is it possible that significant energy is being produced at higher frequencies as a result of football-related collisions? You may recall that near the end of this chapter (see Figure 10.36) we saw exactly that phenomenon occurring in railroad car-coupling events, where measured spectra revealed large accelerations at much higher frequencies (close to 100 Hz, for example). This transfer of energy was obtained at much lower velocities (about 5.25 mph) and Shust (Shock and Vibration in Rail and other Transport Modes, 2007) further noted that energy transfer in car-coupling events routinely resulted in the 4 g design specification being exceeded. There have been many published studies in which linear accelerations were measured for head/helmet collisions in football. Accelerometers have been placed in helmets, mouthpieces, and earbuds; however, spectral data for gauging energy transfer in football collisions are rare. If the dangers of CTE are to be minimized or even eliminated, it seems certain that such data must be collected. Search the latest literature to see what spectral data have appeared for helmet impacts. Do the spectra show the type of energy transfer seen in railroad car-coupling events? If they do not, what do the spectra reveal about the dynamics recorded in the accelerometer data?

10.12 Reciprocating machinery can generate very strong vibrations that may—in exceptional cases—result in damage or even a catastrophic incident. For one example, see the G. R. Slusher FAA Report cited in Section 10.10 regarding horizontally opposed engines (and their impact upon exhaust system integrity) in light aircraft. The question we want to consider here is whether it is possible to use harmonic analysis to detect mechanical problems (lubrication failure, bearing eccentricities, etc.) before they reach the crisis point. To explore the feasibility further, we will obtain some accelerometer data using the *Physics Toolbox* app on a smartphone and record vibrations on the top surface of a reciprocating air compressor:

This particular compressor is belt-driven by a 240 V AC motor turning at 3,450 rpm; there is a speed reduction such that the twin-cylinder compressor shaft rotates at 2,000 rpm (about 33 Hz). We process these data by Fourier transform to obtain the spectrum:

Use your smartphone with the *Physics Toolbox* app and obtain data similar to Figure P10.12a for some operating machine that has reciprocating motions. Compute the spectrum and see if expected frequencies are being identified. **SAFETY FIRST!** *Do not attempt to collect data from any operating machine without first obtaining permission and studying the critical safety issues.*

If the data collection required in this problem is not feasible for you, make use of engine vibration data given at enDAQ.com/pages/measurement-analysis-vibration-from-car-engine-during-morning-commute. These data were obtained from the top of a well-worn Saab 9-3 four-cylinder engine, and they were recorded for a period of about 1,100 s. Use the z-component of the waveform; the spectrum should reveal strong contributions at 10–14 Hz, 30 Hz, and 50–70 Hz, with broad distribution between about 400–1,600 Hz. The engine vibrations were sampled at 10,000 Hz, so the data files are large; much of the interesting behavior occurs in the first 300 s, so you may want to limit your work to those data.

10.13 ***Confer with your instructor—if a safe bridge cannot be identified do not attempt this exercise!***

Investigate your locality and identify bridges that have pedestrian walkways. *Do not carry out the instructions for this problem if you cannot isolate yourself completely from the hazards of vehicular traffic. ONLY attempt this problem if you can access a bridge with a protected pedestrian walkway!*

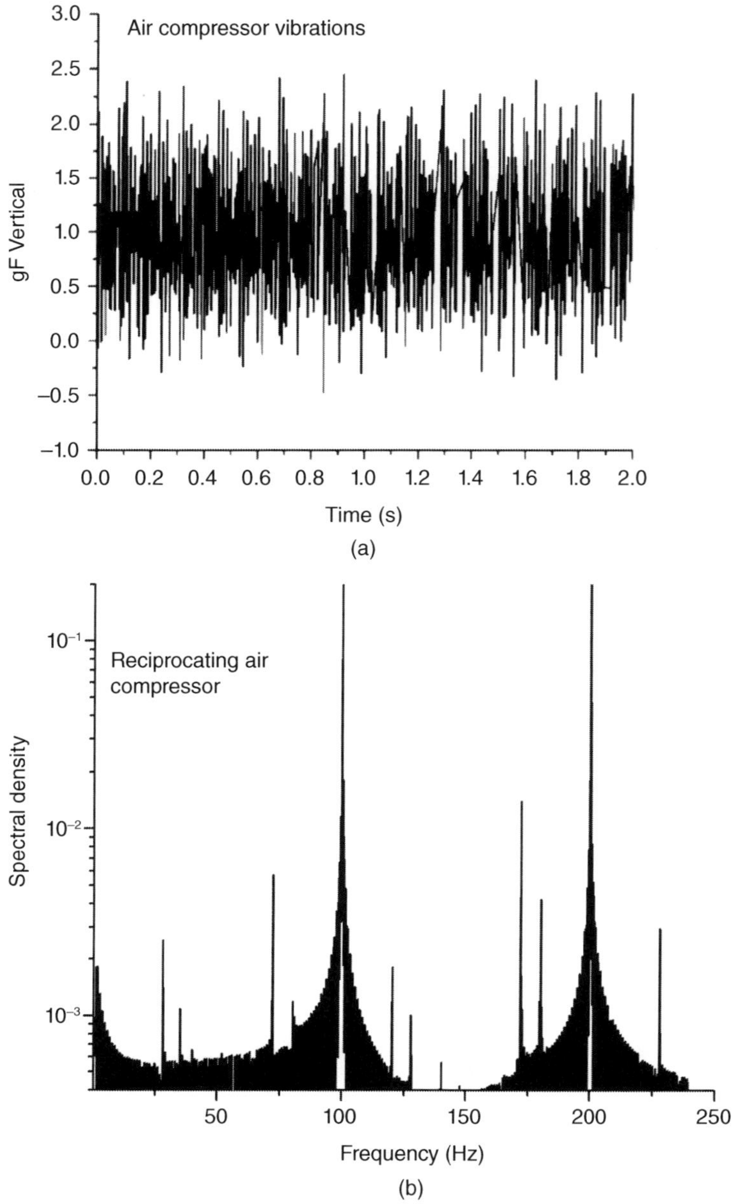

Figure P10.12 (a) Time-series data recorded with an accelerometer positioned on the top of a reciprocating air compressor (11.5 SCFM at 90 psia). (b) Computed spectrum for time-series data obtained with an accelerometer from the top of an air compressor.

Position yourself at or near the center of the longest span of the bridge, and find a suitable horizontal surface for the measurement (perhaps a post, pier, or railing). Using the accelerometer in your smartphone with the *Physics Toolbox* app, record the response of the bridge to vehicular traffic. Try to capture data when there is a mix of vehicle types and record the data for a period of about 45 s. "Share" the data by sending them to your email account as a comma delimited (.csv) file; then use appropriate software to compute the spectrum and identify the important frequencies. Does it appear that the bridge's fundamental frequency can be identified from your results?

If you cannot identify a *suitably safe* bridge for study, you may make use of the data provided in Figure P10.13 or refer to problem 10.14.

10.14 As an alternative to the previous exercise, you may prefer to use accessible data from the *enDAQ* website. Tina Fabiano used their S3-E100D40 shock and vibration sensor to measure to measure the structural response of a bridge on the MBTA Lowell line when a commuter train passed over. The data she and her colleague collected are labeled "*VIBRAT data train-accel.*" Use the *y*-component data as they are the most interesting.

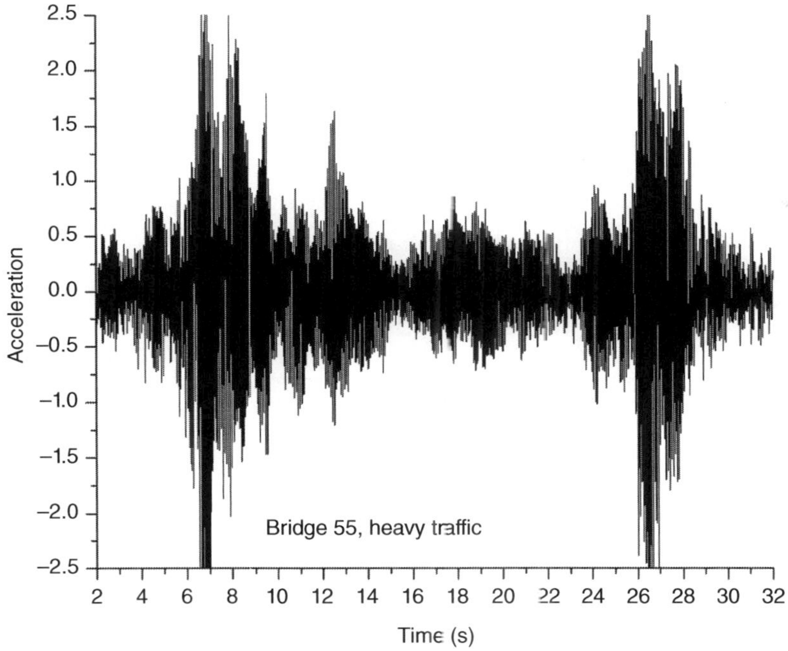

Figure P10.13 Accelerometer data for Bridge 55, a beam bridge with curvature.

10.15 Parkinson's disease (PD) is a neurodegenerative disorder that affects more than 6.2 million people worldwide. It is expected to become even more prevalent with the aging populations found in nearly all of the developed world. Very recently researchers concerned with onset PD have begun to investigate the use of subtle changes in speech as a diagnostic tool. Yifei Men (2013) reported a study, "The Parkinson's Voice Initiative: Early Diagnosis for Parkinson's Disease through Speech Recognition" in which a single sustained phonation ("ahhh.....") was used for this purpose; this work was described in the *Stanford Journal of Public Health*. N. D. Pah et al. (*Scientific Reports,* 2022) in "Phonemes Based Detection of Parkinson's Disease for Telehealth Applications" reported the use of recorded spoken vowels "a-e-i-o-u" to detect onset PD. And F. Amato et al. published "An Algorithm for Parkinson's Disease Speech Classification Based on Isolated Words Analysis" in *Health Information Science and Systems,* 9:32 (2021).

Review the most recent work in this area including applications of AI (artificial intelligence). Can you identify a speech recognition approach that appears to offer the most potential for detection of early onset PD? Will the Fourier transform play a significant role in clinical tools developed for PD diagnosis?

10.16 *You must use a second person as the automobile driver in this exercise—there is no justification for inattentive driving.* Use your smartphone with the *Physics Toolbox app* and place the phone so that the bottom is in direct contact with the dash surface. Measure and record accelerations in all three directions as the vehicle travels over a rough road at modest velocities (certainly less than 50 km/hr). Studies of vibrations detected in traveling vehicles make it clear that high speeds will needlessly complicate interpretation of the results. Compute spectra from the recorded time series. Are you able to distinguish between vibrations due directly to the road surface and the characteristic oscillations of the vehicle itself?

10.17 We briefly discussed errors associated with the collection of time-series data in Section 10.1. In this exercise we will explore some issues arising with random errors in greater detail. First, construct a synthetic signal consisting of two periodic waveforms accompanied by noise as illustrated here:

$$\phi = 1.2 + \sin(f_1 t) + \cos(f_2 t) + Z,$$

where $f_1 = 60\ \text{Hz} = 377\ \text{rad/s}$ and $f_2 = 39.5\ \text{Hz} = 248.2\ \text{rad/s}$. Z is a uniform random number selected such that $-1 \leq Z \leq +1$. We will collect "data" for $0 \leq t \leq 10$ s. It is obvious that over an extensive period of time the average value for this signal *must* be 1.2. Approach this exercise as follows: "Collect" data for 10 s and start by obtaining

100 samples with $\Delta t = 0.1$ s. We acknowledge that this is inadequate since *we must sample at an interval of 0.0083 s in order to capture a 60 Hz waveform*. In fact, if we collect 500 data points over the 10 s (so that $\Delta t = 0.02$ s) and plot the data, it will *appear* that there is a significant oscillation with a frequency of about 10 or 11 Hz. This is an illusion that underscores a possible pitfall when we are trying to analyze time-series data from a poorly understood system; we may not know if our sampling interval, Δt, is small enough. Our plan is to repeatedly double the number of samples collected over 10 s and complete the following table:

n	Mean	σ^2	σ
100	1.2407	6.967	2.639
200			
400			
800			
1,600			
3,200			
6,400			
12,800			

The first row of data for $n = 100$ is for illustration purposes only—you will *not* get the same numbers of course. Note that for $n = 12{,}800$, $\Delta t = 0.00078$ s such that the Nyquist frequency is 640 Hz. There are two additional parts to this problem: First, repeat the $n = 3{,}200$ case five to ten times to obtain an ensemble and determine the pooled mean. How close is it to the expected value of 1.2? Second, find the Fourier transform for one of your data sets that contains at least 1,600 points. Can you clearly identify the periodic contributions to the spectrum at 60 and 39.5 Hz?

10.18 The V-22 (or MV-22) Osprey—an aircraft that can hover like a helicopter and then by tilting its rotors fly like a fixed-wing aircraft—has had a problematic service life with numerous fleet groundings to sort out mechanical issues. One unsolvable problem is the high heat and velocity downwash that can scour a landing area and entrain dirt and debris, damaging the engines and obscuring pilot visibility. Recently incidents have also occurred due to hard-clutch engagement problems; in these failures clutch slippage on one side of the aircraft leads to a sequence of events in which uneven side-to-side power distribution can cause roll and even loss of control. Note that the Osprey cannot glide in the event of an in-flight failure nor can it auto-rotate down to a survivable hard landing. Consequently, such events have too frequently ended with loss of life.

The V-22 was equipped with a "health" monitoring system called VSLED (vibration, structural life, and engine diagnostics) during the infancy of the program; the developers also adopted a philosophy of "on condition maintenance." This meant that instead of regularly scheduled maintenance, service on the aircraft would *only* be performed when a fault was detected by the data collection system. The architecture of the vibration monitoring system was described in detail by D. A. Dousis at the 2000 IEEE Aerospace Conference (Proceedings, vol. 6, pages 245–261). Interestingly, and probably unfortunately, the VSLED system provided for—*but did not include*—gearbox vibration monitoring. Some early problems were detected successfully, including a loose right-hand side hanger bearing which was revealed by a logged trend of increasing vibration amplitude.

Recently it has been discovered that the input quill assembly (the component that connects the engine to the gearbox) wears out much more rapidly than previously thought and, indeed, the military is now replacing it in all V-22s after just 800 hr of flight time. Military spokespersons have emphasized that this new maintenance requirement is "a mitigation and not a fix." As each V-22 costs approximately $100 million, there is immense pressure to find and implement a permanent solution that will improve fleet availability. In March 2023 Bell-Boeing was awarded a contract for $53.6 million to develop a gearbox vibration monitoring system to provide earlier detection of degrading gearbox components. Of course, this is the aspect of the V-22's operational life that is of interest to us.

i) Is it possible that either the FFT or the coherence function can be used to detect gearbox degradation?
ii) The changes that are likely to occur in either the power spectrum or $C^2_{xy}(f)$ may be subtle, so how difficult will it be to distinguish between normal wear-and-tear and incipient failure? You may find online videos produced

by *Reliability Maintenance Solutions Ltd.* (RMS) to be useful as they illustrate some changes seen over time in power spectra measured on rotating equipment.[1]

iii) Search the recent literature and identify other efforts to monitor the physical status of machines with rotating components through collection and processing of time-series data. Some contemporary commercial products for this purpose include *VibroSens* by VibroSystM, the *Vibration Monitoring System* by SKF Inc., and Emerson's *AMS Machinery Manager*.

10.19 Three spectra were computed for a sinusoidal waveform with increasing noise (Z) levels:

$$y(t) = \sin(0.985t) + \cos(0.375t) + aZ,$$

where $-1 \leq Z \leq +1$. Of course, we can expect important spectral contributions centered around the two radian frequencies, 0.375 and 0.985 rad/s. The PSDs are illustrated in this figure (Figure P10.19):

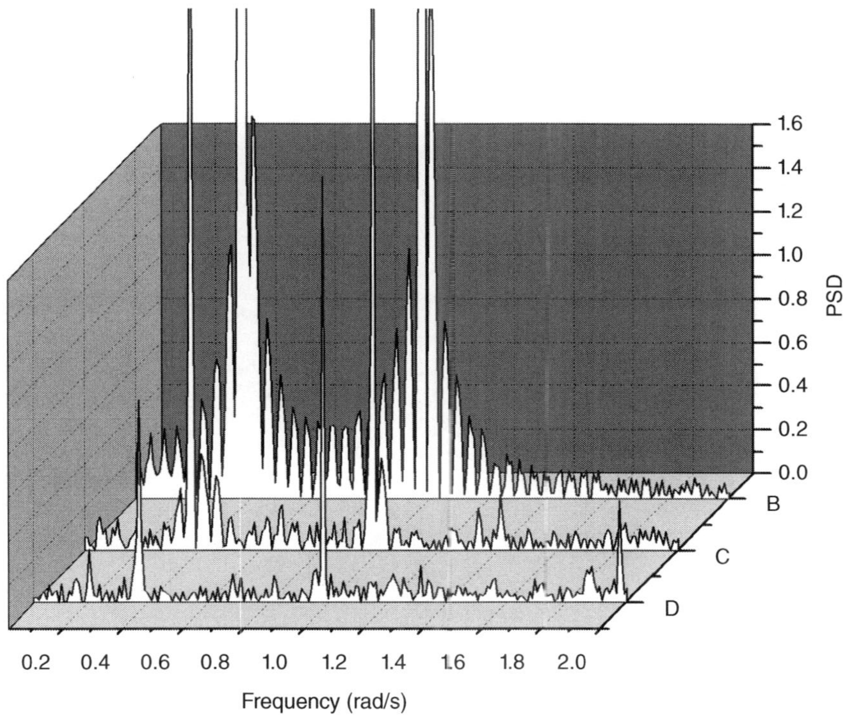

Figure P10.19 Three spectra with two distinct frequencies, 0.985 and 0.375 rad/s, and significant noise.

The noise levels for these spectra increase from back to front, and they can be characterized as mild, moderate, and severe (B, C, D). Leakage is very apparent for B, and this is an expected outcome for a predominantly sinusoidal waveform. But there are two other important points to note here: First, as the noise level is increased the peaks associated with the two dominant frequencies are attenuated, and second, the increasing levels of random noise generate other spectral components that might be misidentified as phenomena of interest. This is particularly apparent in D where spectral peaks can be seen at 0.2 and about 1.98 rad/s. If we recognize that a recorded signal is contaminated with significant noise, then we need to take steps to mitigate the problem. Develop a sinusoidal signal with a limited number of frequencies that includes increasing levels of noise; as the random contribution to the signal is increased you will see behavior very similar to that shown in the figure above. Select the worst case and apply a digital filter (of your choosing) to the data set, then recompute the spectrum to reveal the improvement.

1 A video produced by RMS Ltd. shows the evolution of the spectrum for a combustion fan on a thermal oxidizing unit; in this example, the changes detected over time by vibration analysis of a bearing are very clear. The drive-end bearing history revealed degradation culminating in a peak-to-peak signal of 59 g's. When the bearing was removed from service and inspected, the severity of the spalling on the bearing race was obvious.

10.20 In 1967 Peter Welch (*IEEE Transactions on Audio and Electroacoustics*, AU-15, 70–73) described a technique for obtaining the spectrum for time-series data in which the data were divided into segments, and the periodograms for each were averaged to produce an improved estimate of the spectrum. There are numerous examples in the literature that illustrate the power of this approach, particularly in cases where the data are accompanied by significant levels of noise. Suppose we have time-series data generated by

$$x(t) = \sin(1.5t) + 0.9\cos(0.55t) + 16Z,$$

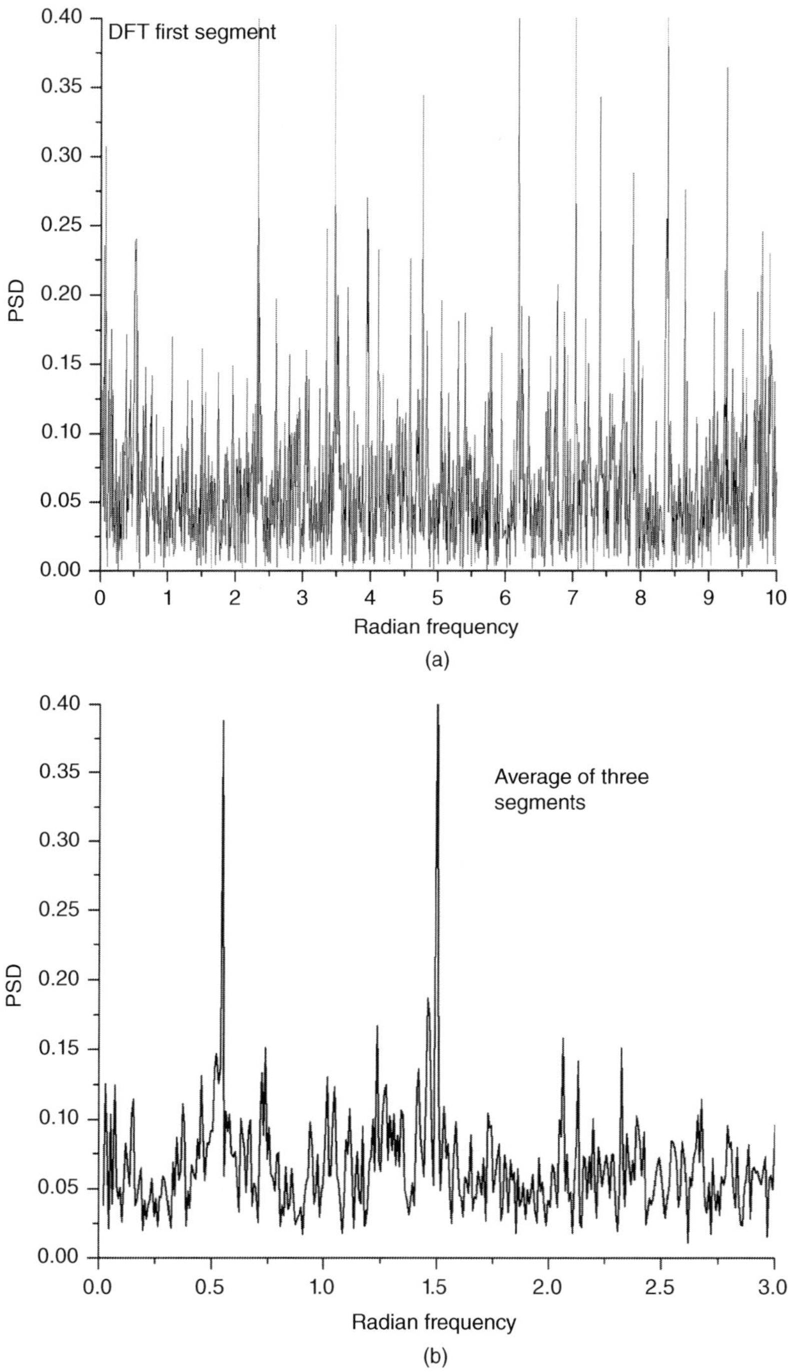

Figure P10.20 Comparison of spectra for a single data segment versus the average of three data segments (with expanded scaling).

where Z is a uniform random number between -1 and $+1$. Of course, the expected value for $x(t)$ would be zero for a suitably lengthy data record. Let $\Delta t = 0.25$ s such that the Nyquist frequency is 2 Hz or 12.6 rad/s, and compute at least 4,096 data points with this expression. Note that the values computed for $x(t)$ will be different in every realization assuming the random number generator is given a different seed each time. Thus, no two trials will produce the same results.

Divide the data record into an integer number (n) of non-overlapping segments and compute the DFT for each. Then, compare the spectrum from a single segment to the *average* obtained from the n-spectra. You should see something like the results illustrated in Figure P10.20:

Only three segments were used in this example, but even so the results are striking. For the single segment DFT on the left, the spectral feature at 1.5 rad/s is completely overshadowed by noise-generated features, and the peak at 0.55 rad/s is surrounded by spurious contributions. On the right, the average of the three DFTs produces dominant peaks at 0.55 and 1.5 rad/s as we expect from $x(t)$. We would anticipate even better results if a larger number of segments had been used for the averaging.

10.21 For climate change deniers, one cold winter season "disproves" the scientific consensus that the planet is warming and concomitantly the severity of storms is increasing. One fact that is irrefutable is that the CO_2 concentration in the atmosphere has increased from about 280 ppm at the onset of the industrial revolution to approximately 419 ppm at the end of 2023—an increase of about 50%. Although many people have observed that the global climate has experienced large oscillations over the millennia and therefore short-term trends are unimportant, others—often with ulterior motives—are all too willing to use limited data sets when it suits their purpose. The data provided below (114 samples) represent the total annual snowfall (in inches) in Buffalo NY for each winter since 1910 (each entry represents the winter ending in that year). Notice that the total for the winter ending in 2023 was the seventh largest on record. Perform a Fourier analysis of this very limited data set. Can your results be used to support *any* conclusion regarding climate? Identify the practical difficulties affecting your interpretation of the results, and include a discussion of statistical significance.

1910	126.4	1950	88.7	1990	93.7
1911	82.4	1951	71.4	1991	57.5
1912	78.1	1952	83	1992	92.8
1913	51.1	1953	53.9	1993	93.2
1914	90.9	1954	89.9	1994	112.7
1915	76.2	1955	84.6	1995	74.6
1916	104.5	1956	105.2	1996	141.4
1917	87.4	1957	113.7	1997	97.6
1918	110.5	1958	124.7	1998	75.6
1919	25	1959	114.5	1999	100.5
1920	69.3	1960	115.6	2000	63.6
1921	53.5	1961	102.4	2001	158.7
1922	39.8	1962	101.4	2002	132.4
1923	63.6	1963	89.8	2003	111.3
1924	46.7	1964	71.5	2004	100.9
1925	72.9	1965	70.9	2005	109.1
1926	74.6	1966	98.3	2006	78.2
1927	83.6	1967	55.5	2007	88.9
1928	80.7	1968	66.1	2008	103.8
1929	60.3	1969	78.4	2009	100.2
1930	79	1970	120.5	2010	74.1
1931	74.4	1971	97	2011	111.8
1932	49.6	1972	109.9	2012	36.7

1933	54.7	1973	78.8	2013	58.8
1934	71.8	1974	88.7	2014	129.9
1935	49.1	1975	95.6	2015	112.9
1936	103.9	1976	82.5	2016	55.1
1937	51.6	1977	199.4	2017	76.1
1938	82.4	1978	154.3	2018	112.3
1939	83.6	1979	97.3	2019	118.8
1940	77.8	1980	68.4	2020	69.2
1941	79.3	1981	60.9	2021	77.2
1942	89.6	1982	112.4	2022	97.4
1943	85.5	1983	52.4	2023	131.2
1944	58	1984	132.5	2024	
1945	120.7	1985	107.2	2025	
1946	110.5	1986	114.7	2026	
1947	65.4	1987	67.5	2027	
1948	39.9	1988	56.4	2028	
1949	40.1	1989	67.4	2029	

10.22 In Section 10.9 we introduced the notion that anyone with a smartphone could use the *Physics Toolbox* or *Physics Toolbox Sensor Suite Pro* to collect time-series data for a wide variety of phenomena. This could take the form of monitoring a machine or a structure for important periodicities, or it could be as mundane as measuring cadence during a physical activity or workout. It is a variation of the latter that is of interest in this exercise. Athletes of all abilities have experienced stresses created by repetitive motions, and it can be particularly problematic in cases involving foot strikes upon surfaces. Use your smartphone to measure *g*-forces experienced while descending stairs or (if you are an avid runner or jogger) measure and compare *g*-forces experienced running across grass surfaces versus those created by paved surfaces like concrete or asphalt. Note that A. Ferro-Sanchez et al. (An Analysis of Running Impact on Different Surfaces for Injury Prevention, *Int. J. Environ. Res. Public Health*, 14:6405, 2023) conducted a study using 30 amateur runners and three different surfaces (grass, synthetic track, and concrete); they found small differences in both mean and peak accelerations with the latter approaching 4–5 *g*'s on concrete surfaces. Report both the time-series data and the Fourier transform.

10.23 In Section 10.9 we considered some instances where two sets of time-series data might be acquired simultaneously, and then some conclusion could be formed based on an appropriate analysis and comparison.

On April 17, 2018, a Boeing 737-700 operated by Southwest Airlines (Flight 1380) experienced a contained engine failure when fan blade 13 in the left CFM56-7B engine failed due to a low-cycle fatigue crack (the crack initiated in the dovetail which is part of the blade root). The fan diameter in this engine is 61 in. and its normal rotation rate is 3,200 rpm; therefore, the blade-tip velocity would be about 850 ft/s. The incident occurred at an altitude of 32,000 ft where the atmospheric pressure is only 4.2 psi (or 217 mmHg), and debris from the fan cowl struck the fuselage near a cabin window causing rapid depressurization and one fatality when a passenger was partially ejected from the cabin. Up to that point a routine maintenance procedure involved inspection of fan blades visually using a fluorescent penetrant (which would be absorbed into a crack subsequently revealed by illumination). But this fan blade failure, and a very similar one in 2016, underscored the need for improved crack detection. Several new technologies have been developed for this purpose, including the measurement of radial blade-tip clearance using capacitive displacement sensors. Woike and Abdul-Aziz (NASA TM-2010-216239, March 2010) described exploration of this technique using NASA Glenn Research Center's spin rig which is capable of rotating a simulated turbine disk at speeds up to 12,000 rpm. The idea here is that minute—but detectable—speed-dependent deformations would occur during rotation. They compared signals obtained from a pristine disk with that from an artificially notched disk (to simulate presence of a crack). The *x*–*y* plots of the vibration vector were significantly different for the two tested disks.

A consequence of the CFM56 fan blade failure was an NTSB recommendation that Boeing redesign the fan cowl structure and latching system as the existing design had a significant weak point at the six o'clock (bottom) position. They also recommended adding eddy current inspection (ECI) to the fan blade overhaul procedure for enhanced detection of newly initiated cracks that would not be revealed by surface inspection. The ECI technique utilizes two coils (one for excitation and one as receiver); an alternating current flows through the excitation coil generating a magnetic field which induces an eddy current in the conductive part being inspected. Typically the test instrument compares the transmitted and received signals, and if a crack is present in the component the propagation of the eddy current changes. The penetration depth of the testing depends upon the conductivity of the part, its magnetic permeability, and the frequency employed (lower frequencies provide greater penetration). Although the frequency is the only test parameter that can be adjusted, ECI can detect some fatigue cracks that are not apparent through visual inspection.

Review recent developments in fan blade crack detection and prepare a brief report of cases where a comparison of two signals provides the basis for corrective maintenance.

10.24 Electric utilities around the world have been erecting wind farms in efforts to diminish their carbon dioxide emissions. In Kansas many of the installations use the GE 1.6–82.5 (1,600 KW) unit with a swept area of about 5,345.6 m^2 and a rated wind speed of 10 m/s. The cut-in speed is 3.5 m/s and the cut-out speed is 25 m/s (about 55 mph) with a normal rotation rate of 16.8 rpm corresponding to a blade-tip velocity of about 72.5 m/s. The newer units use light-weight carbon fiber blades, and General Electric anticipates a service life of 30 years. Recently GE introduced the 1.6–100 turbine with a 100 m rotor and an improved capacity factor; the blades on this unit have trailing-edge serrations to produce a smaller sound footprint.

On the Great Plains the environment for these wind turbines can be incredibly hostile with ambient temperatures ranging from −20 °F to around 120 °F coupled with extreme precipitation events. The harsh conditions can lead to a number of problems with blade integrity including carbon fiber fracture, delamination, and both fore- and aft- blade pitting. A challenge in the operation of wind farms is detection of blade damage *before* the problem becomes catastrophic. Acoustic monitoring of wind turbines is the subject of ongoing study, and it is appealing to utilities because it is a passive technique where the non-contact acoustic sensors can be permanently placed and the data analyzed remotely. As one might expect, there are problems with this approach and among the more vexing are: 1) aerodynamic noise generated by the blades contains a large number of components (distributed over a broad range of frequencies) unrelated to blade damage, 2) higher wind velocities produce increased background noise complicating signal interpretation, and 3) optimal placement of the acoustic sensors is not trivial. Recently, Ding et al. (Acoustic-Signal-Based Damage Detection of Wind Turbine Blades—A Review, *Sensors*, 23:4987, 2023) described efforts to use acoustic signals for blade diagnostics, and they briefly discuss some work focused upon use of the FFT for this purpose. Prepare a brief review of the most recent research (using the Fourier transform) in this area and identify acoustic frequencies that have been cited as being related to blade damage.

10.25 The Earth's magnetic field is incredibly dynamic, and it is affected by a host of phenomena including solar storms, atmospheric disturbances (including lightning), the molten core of the planet, and diurnal variations in sunlight. Furthermore, localized disturbances are created by power grids and even by electrical appliances in your kitchen. The British Geological Survey (BGS) collects high-resolution geomagnetic data at several observatories in the UK, including Lerwick, Hartland, and Eskdalemuir. In a recent UK educational project, fairly sensitive magnetometers capable of measuring magnetic field strength in the nano-Tesla (nT) range were placed in secondary schools to allow students to monitor geomagnetic fluctuations (*search Raspberry Pi magnetometer at Edinburgh*). A comparison of the measurements made by the school instrument with the BGS equipment at Eskdalemuir for April 3, 2015, shows remarkably good correlation. Most smartphones are equipped with magnetometers and the *Physics Toolbox Sensor Suite Pro* can capture and display sufficiently large geomagnetic fluctuations (usually in the μT range). This sensitivity is not sufficient for nT measurements, but will allow one to observe fluctuations caused by electric appliances such as fans or microwave ovens. Use your phone to measure magnetic fluctuations in your home environment; which appliances create the largest variations?

I found that a small electric fan produced fluctuations with maximum amplitudes of about 5–10 μT at frequencies mainly in the 1–30 Hz range.

References

Barnoski, R.L. (1969). Ordinary Coherence Functions and Mechanical Systems. *AIAA Journal of Aircraft* 6: 372.

Bejger, A. and Gawdzinska, K. (2015). An Attempt to Use the Coherence Function for Testing the Structure of Saturated Composite Castings. *Metalurgija* 54 (2): 361.

Bendat, J.S. and Piersol, A.G. (1980). *Engineering Applications of Correlation and Spectral Analysis*. New York: Wiley Interscience.

Bingham, C., Godfrey, M.D., and Tukey, J.W. (1967). Modern Techniques of Power Spectrum Estimation. *IEEE Transactions on Audio and Electroacoustics* AU15: 56.

Bloomfield, P. (1976). *Fourier Analysis of Time Series: An Introduction*. New York: John Wiley and Sons.

Brigham, E.O. (1988). *The Fast Fourier Transform and Its Applications*. Englewood Cliffs, NJ: Prentice Hall.

Cooley, J.W. (1987). The Re-Discovery of the Fast Fourier Transform Algorithm. *Mikrochimica Acta (Wien)* III (33).

Cooley, J.W. and Tukey, J.W. (1965). An Algorithm for the Machine Calculation of Complex Fourier Series. *Mathematics of Computation* 19: 297.

Dodson-Robinson, S.E., Delgado, V.R., Harrell, J., and Haley, C.L. (2022). Magnitude-squared Coherence: A Powerful Tool for Disentangling Doppler Planet Discoveries from Stellar Activity. *The Astronomical Journal* 163: 169.

Gaunt, J.T. and Sutton, C.D. (1981). *Highway Bridge Vibration Studies*. Joint Highway Research Project, Purdue University.

Green, J.T. and Unruh, W.G. (2006). The Failure of the Tacoma Bridge: A Physical Model. *American Journal of Physics* 74: 706.

Harris, F.J. (1978). On the Use of Windows for Harmonic Analysis with the Discrete Fourier Transform. *Proceedings of the IEEE* 66: 51.

Knott, C.G. (1897). On Lunar Periodicities in Earthquake Frequency. *Proceedings of the Royal Society* 60: 457.

Lamoreaux, G. H., Trujillo, A. A., and C. F. Magnuson. Truck and Rail Shock Environments During Normal Transport. *CONF-811074-2, DOE Contract DE-AC04-76DP000789 and NRC Contract A-1049-6* (1981).

Lyons, M. L., Riddell, W. T., and K. W. Koch. Analysis of Accelerations Measured During Full-Scale Tank Car Impact Tests. *DOT/FRA/ORD 07/08 Final Report* (2007).

Magnuson, C. F. and L. T. Wilson. Shock and Vibration Environments for Large Shipping Containers on Rail Cars and Trucks. *Sandia Laboratories, SAND76-0427* (1977).

Mosheni, I., Rashid, A.K., and Kang, J. (2014). A Simplified Method to Estimate the Fundamental Frequency of Skew Continuous Multicell Box-Girder Bridges. *Latin American Journal of Solids and Structures* 11: 649.

Nagayama, T., Reksowardojo, A. P., Su, D., Mizutani, T., and C. Zhang. Bridge Natural Frequency Estimation by Extracting the Common Vibration Component from the Responses of Two Vehicles. *6th International Conference Advances in Experimental Structural Engineering*, University of Illinois (2015).

Ramirez, R.W. (1985). *The FFT: Fundamentals and Concepts*. Englewood Cliffs, NJ: Prentice-Hall.

Rockmore, D.N. (1999). *The FFT—An Algorithm the Whole Family Can Use*. Dartmouth College: Department of Mathematics and Computer Science.

Schuster, A. (1897). On Lunar and Solar Periodicities of Earthquakes. *Proceedings of the Royal Society* 61: 455.

Shust, W.C. (2007). *Shock and Vibration in Rail and other Transport Modes*. PPT presentation: Objective Engineers, Inc.

Slusher, G. R. Reciprocating Engine and Exhaust Vibration and Temperature Levels in General Aviation Aircraft. *FAA Report NA-68-27* (1968).

Solomon Jr. O. M. PSD Computations Using Welch's Method. *Sandia Report SAND91-1533* (1991).

Tukey, J.W. (1967). An Introduction to the Calculations of Numerical Spectrum Analysis. In: *Spectral Analysis of Time Series* (ed. B. Harris). New York: Wiley & Sons.

Warpinski, N. R., Du, J., and U. Zimmer. Measurements of Hydraulic-Fracture-Induced Seismicity in Gas Shales. *SPE 15197, SPE Hydraulic Fracturing Technology Conference* (2012).

Welch, P.D. (1967). The Use of Fast Fourier Transform for the Estimation of Power Spectra: A Method Based on Time Averaging over Short, Modified Periodograms. *IEEE Transactions on Audio and Electroacoustics* AU-15: 70.

11

An Introduction to the Calculus of Variations and the Finite Element Method

11.1 Some Preliminaries

About 100 years ago Andrew R. Forsyth wrote a very appealing introduction to the subject of this chapter in *Calculus of Variations (COV)* published by the University Press (1927):

> "…*the calculus of variations deals with one of the earliest problems of ordinary experience. The requirement was, and is, to obtain the most profitable result from imperfectly postulated data; and the data may possibly be subjected to conditions, which are likewise imperfectly postulated. When data and conditions are expressed in analytical form, the necessary mathematical calculations cannot be effected directly, because of some essential deficiency in the information. The gap has to be filled before resolution of the problem is attained*…"

With this introduction in mind, we can summarize our present objective with the following question: How does one obtain an optimal result for problems in which the key relationship between the pertinent variables may be unknown, and where some or all of the variables may be subject to constraints? This is precisely where the COV becomes an invaluable tool.

11.1.1 Elementary Principles of Extrema

We assume a function $y(x)$ is continuous on the interval $a \leq x \leq b$ and that the derivative, $y'(x)$, exists everywhere with the possible exception of point, p. If $y'(p) = 0$, or if $y'(p)$ does not exist, then $x = p$ is a *critical* point. You will remember from your study of calculus that a *necessary* condition for $y(x)$ to be a *local extremum* is that $y'(x) = 0$. However, the first derivative test does not reveal whether that point corresponds to a maximum or a minimum, nor does it reveal if the point x corresponds to a global extremum on the interval of interest. This is where the second derivative test is crucial to our investigation of $y(x)$. Suppose, for example, that $y'(x)$ and $y''(x)$ exist everywhere on the interval and that at some point we find $y'(p) = 0$. If $y''(p) < 0$ then $x = p$ is a *relative* maximum; if $y''(p) > 0$, then $x = p$ is a *relative* minimum.

Consider the polynomial,

$$y(x) = \frac{1}{2} + 5x - 2x^2 + \frac{1}{4}x^3, \tag{11.1}$$

with the first and second derivatives,

$$y' = 5 - 4x + \frac{3}{4}x^2 \text{ and } y'' = -4 + \frac{3}{2}x. \tag{11.2}$$

We are interested in the behavior of this cubic equation (illustrated in Figure 11.1) for $0 \leq x \leq 4.5$:

Applying the first derivative test, we find that $y' = 0$ at $x = 6/3 = 2$ and at $x = 10/3$ and, consequently, we know these two points are *relative* extrema. At $x = 2$, $y'' = -1$ and at $x = 10/3$, $y'' = +1$; this means that $x = 2$ corresponds to a relative maximum and $x = 10/3$ is a relative minimum. Furthermore, we also find that y'' is zero at $x = 8/3$, indicating a point of inflection. Note that the *global* minimum occurs at $x = 0$ where $y = 1/2$; the *global* maximum occurs at $x = 4.5$ where $y = 5.28125$. The significance of this brief review is that solving $f'(x) = 0$ to identify stationary points is closely related to a key part of COV, namely, finding stationary values for an integral, I.

Problem Solving in Engineering: Analytical Mathematics and Numerical Analysis, First Edition. Larry A. Glasgow.

Companion website: www.wiley.com/go/glasgow/problemsolvingengineering

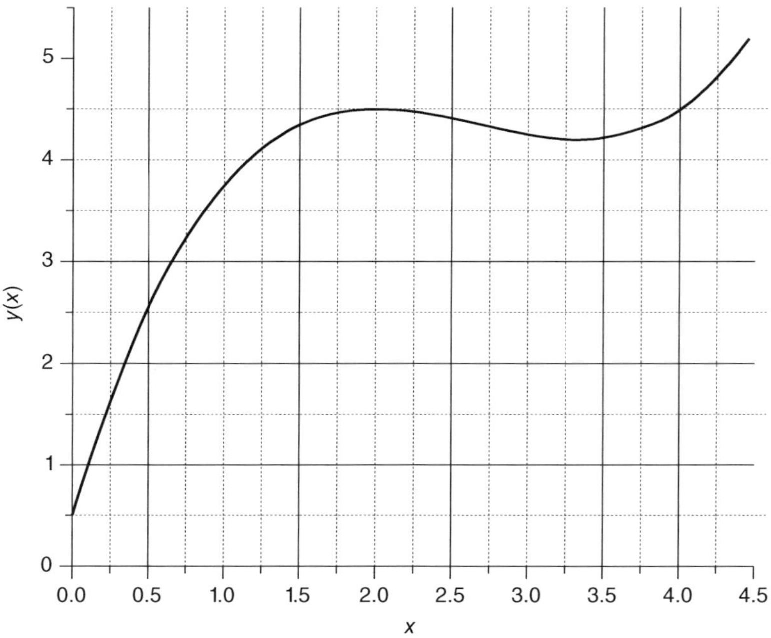

Figure 11.1 Graph of $y(x) = \frac{1}{2} + 5x - 2x^2 + \frac{1}{4}x^3$ for $0 \leq x \leq 4.5$.

11.1.2 Principle of Least Action

Many of the elementary problems we study in engineering physics can be posed in terms of variational principles. For example, suppose we launch a particle with some initial velocity at an angle (with horizontal) of θ. This entity is subjected only to the influence of gravity; there are no dissipative forces. We know this particle will have a parabolic trajectory where in mid-flight it will reach a maximum elevation corresponding to its largest potential energy. At that point it will also have its smallest kinetic energy as the vertical component of the velocity vector will be zero. This is a *conservative system*, so the sum of kinetic and potential energies will be constant, but what will the *difference* between the kinetic and potential energies look like?

Specifically, suppose we launch the particle from the origin with an initial velocity of 18 m/s at an angle of 34°. The horizontal component of velocity will be constant at 14.92 m/s (remember, there is no drag in this conservative system), and the initial value of the vertical component of the velocity will be 10.07 m/s. For simplicity we take the mass of this entity to be 1 kg; consequently, the initial value of the kinetic energy (KE) will be 162 kg m^2/s^2, and of course the initial potential energy (PE) will be zero. We will terminate the flight at $t = 1.9$ s, when $x = 28.35$ m and $y = 1.44$ m as indicated on the right-hand side of Figure 11.2.

Now, assume we cause the particle to travel a more complicated path to the terminal point while requiring the total flight time to *remain at* 1.9 s. Obviously it will be necessary for the initial kinetic energy to be larger if we are to conclude the

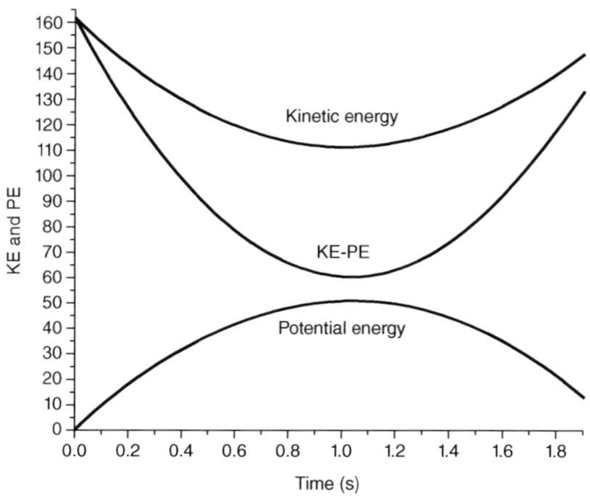

Figure 11.2 Potential and kinetic energies of an object launched with an initial velocity of 18 m/s at an angle (with horizontal) of 34°. The flight of the object ends at $t = 1.9$ s at the point ($x = 28.35$, $y = 1.44$).

flight at the specified point at the required time. Consider the integral of the difference between KE and PE:

$$\int_0^{t_f} \left[\frac{1}{2}mV^2 - mgy\right] dt. \tag{11.3}$$

For the initial scenario the value of this integral is 173 kg m^2/s, which is the *minimum* value for this problem in mechanics. A more complicated pathway—meeting our requirements for final position and time—will result in a larger value. This is an example of the *principle of least action*, also referred to as the principle of stationary action. More generally, for problems in mechanics the time integral of the Lagrangian function, L, must be a minimum:

$$\int_0^{t_f} L(q, \dot{q}, t)dt, \tag{11.4}$$

where the q's in (4) are functions related to KE and PE (velocity and vertical position). So, the nature of the problem we are describing is one in which the system's motion produces a trajectory given by $x(t)$ and $y(t)$ that will make this integral a minimum. Determining this optimal pathway is a very common application of the COV.

You may recall that we previously made use of some concepts from the *COV* in our discussion of *VIM* (the variational iteration method). We will now expand on that very brief treatment to illustrate some other applications of *COV*. We will mainly follow the conventions employed by Weinstock (1974) and Fox (1987). It is important to note that our approach will focus upon *conservative* fields for which the force, F, is related to a potential energy function, ϕ, such that $F = -\nabla\phi$. This is a significant limitation since real processes always involve friction (dissipative losses) and, therefore, cannot be conservative.

Consider a function, $y = f(x)$, in the x–y plane that joins the two points, (x_1,y_1) and (x_2,y_2) together. We recall from elementary calculus that the length of the pathway connecting these two points is:

$$L = \int_{x_1}^{x_2} \sqrt{1 + \left(\frac{dy}{dx}\right)^2}\, dx. \tag{11.5}$$

To reveal the nature of the problem we wish to contemplate more clearly; let us select two specific points:

$(x_1,y_1) = (1,4)$ and $(x_2,y_2) = (4,9)$.

Now suppose the function, $f(x)$, connecting these two points is a straight line:

$$y = \frac{5}{3}x + \frac{7}{3}. \tag{11.6}$$

We want to evaluate L from eq. (11.5), recognizing of course that the value produced by eq. (11.6) will be the minimum possible length. But we also want to have a couple of comparisons available, so in addition we will let $y(x)$ be given by the two equations:

$$y = \frac{1}{3}x^2 + \frac{11}{3} \text{ and } y = \frac{7}{x} - \frac{797}{240}x + \frac{77}{240}x^3. \tag{11.7a and b}$$

Now we will compute the arc length from point 1 to point 2 using eq. (11.5) for all three cases; the results are 5.83, 5.91, and 8.98, respectively. The straight line provided the minimum length as we expected. The second-degree equation exhibits a little curvature, so its value for L is a bit longer, and the third-degree equation—which is a good deal more complicated—provides a significantly longer L as we would expect. If you would like to review the use of eq. (11.5), fit the two given points with the equation, $y = a(1 - e^{-bx})$. Show that $a = 10.586$ and $b = 0.475$ resulting in $L = 5.923$.

Imagine now a situation in which we wanted to identify the minimum value of the integral eq. (11.5) but had no idea of the form of the function, $y(x)$. This certainly sounds like a much more difficult task, and we have an additional complication: in technical problems our objective will almost certainly be far more complicated than the mere distance between two points in the x–y plane as given by eq. (11.5).

As we suggested above our real interest is the more general case where we need to identify some function, $y(x)$ where $y_1 = y(x_1)$ and $y_2 = y(x_2)$, but with the stipulation that the integral,

$$I = \int_{x_1}^{x_2} F(x, y, y')dx \tag{11.8}$$

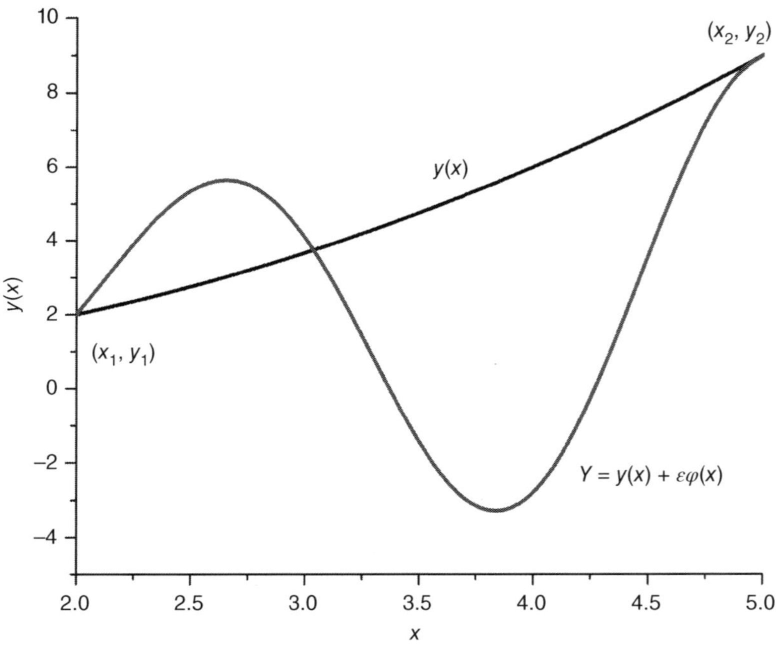

Figure 11.3 Illustration of the "family" of comparison functions. If $\varepsilon = 0$, then we get the behavior shown by $y(x)$.

be a minimum or maximum. The function $y(x)$ that meets this requirement is called an *extremal*, and the integral eq. (11.8) is referred to as a *functional* with one independent variable, x. We will proceed in the following way: We form a family of comparison functions using a single parameter, ε,

$$Y(x) = y(x) + \varepsilon\phi(x). \tag{11.9}$$

We require that $\phi(x)$ be a differentiable function that disappears at the end-points x_1 and x_2 so that $\phi(x_1) = 0$ and $\phi(x_2) = 0$. Now we rewrite eq. (11.8) as

$$I(\varepsilon) = \int_{x_1}^{x_2} F(x, Y, Y')dx \tag{11.10}$$

and note that by $Y'(x)$ we mean $Y'(x) = y'(x) + \varepsilon\phi'(x)$. Of course, if $\varepsilon = 0$ everywhere, then eq. (11.10) is exactly the same as eq. (11.8). The essential point here is that no matter what form $\phi(x)$ has, the extremizing function, $y(x)$, will be a member of the comparison function family. Let us illustrate this scenario with a graph, Figure 11.3.

If we differentiate eq. (11.10) with respect to ε and set $\varepsilon = 0$, then $\frac{dI}{d\varepsilon} = 0$. Thus,

$$I'(0) = \int_{x_1}^{x_2} \left(\frac{\partial F}{\partial y}\phi + \frac{\partial F}{\partial y'}\phi' \right) dx. \tag{11.11}$$

By integrating by parts and noting that the result must be valid for all ϕ, Weinstock (1974) shows that the function we are seeking must satisfy the *Euler–Lagrange differential equation*:

$$\frac{d}{dx}\left(\frac{\partial F}{\partial y'}\right) - \frac{\partial F}{\partial y} = 0, \tag{11.12}$$

which was derived by Euler in the eighteenth century. We can perform the indicated differentiation to see the equation in its entirety:

$$\frac{d^2y}{dx^2}\frac{\partial^2 F}{\partial y'^2} + \frac{dy}{dx}\frac{\partial^2 F}{\partial y'\partial y} + \frac{\partial^2 F}{\partial y'\partial x} - \frac{\partial F}{\partial y} = 0.$$

The Euler–Lagrange equation is a necessary—but not sufficient—condition for $y(x)$ to be an extremal; at this stage we do not know if we have found a minimum, a maximum, or a stationary point (at a stationary point the derivatives of the function vanish). For a simple example of a stationary point, consider the behavior of $f(x) = x^3$ with $f'(x) = 3x^2$; clearly at $x = 0$ $f' = 0$, but this point does not correspond to an extremum. The differential eq. (11.12) may be written in an equivalent form that is often useful:

$$\frac{d}{dx}\left[F - y'\frac{\partial F}{\partial y'}\right] = \frac{\partial F}{\partial x}. \tag{11.13}$$

Should x *not appear* in F explicitly, then

$$F - y'\frac{\partial F}{\partial y'} = C. \tag{11.14}$$

Equation (11.14) is known as the *Beltrami identity* (named for Eugenio Beltrami, an Italian mathematician of the nineteenth century). And should y *not appear* explicitly, then $\frac{\partial F}{\partial y} = 0$, and eq. (11.13) is simply: $\frac{d}{dx}\left(\frac{\partial F}{\partial y'}\right) = 0$, such that

$$\frac{\partial F}{\partial y'} = C. \tag{11.15}$$

11.2 Notation for the Calculus of Variations

Let us return to eq. (11.8) in Section 11.1 to retrieve the function $F(x, y, y')$, but now we take the independent variable, x, to be constant. We define the *variation of F* as:

$$\Delta F = F(x, y + \varepsilon\phi, y' + \varepsilon\phi') - F(x, y, y'). \tag{11.16}$$

We can employ a truncated Taylor series expansion for $F(x, y + \varepsilon\phi, y' + \varepsilon\phi')$:

$$\cong F(x, y, y') + \frac{\partial F}{\partial y}\varepsilon\phi + \frac{\partial F}{\partial y'}\varepsilon\phi' + \ldots \tag{11.17}$$

This means that the variation of F is

$$\Delta F \cong \frac{\partial F}{\partial y}\varepsilon\phi + \frac{\partial F}{\partial y'}\varepsilon\phi' + \text{neglected terms} \tag{11.18}$$

In the literature of the COV, the variation of F is written for the first two terms on the right-hand side of (11.18):

$$\delta F = \frac{\partial F}{\partial y}\varepsilon\phi + \frac{\partial F}{\partial y'}\varepsilon\phi'. \tag{11.19}$$

The reason this is important is because a *necessary* condition for the *integral of the function*, $F(x, y, y')$, to be an extremum is:

$$\delta\int_{x_1}^{x_2} F(x, y, y')dx = 0. \tag{11.20}$$

That is, the first variation of the integral must vanish. Though conceptually attractive, we bear in mind an observation made by Courant and Hilbert (1989):

> "....*in the calculus of variations the existence of an extremum for a particular problem cannot be taken for granted. A special existence proof is needed for the solution of each problem or class of problems.*"

Courant and Hilbert provide examples of *COV* problems that while appropriately developed, *do not* have solutions.

11.3 Brachistochrone Problem

The utility of the COV is often demonstrated with the "brachistochrone" problem that was posed by Johann Bernoulli as a challenge to European mathematicians in 1696. This example serves another purpose as well since it conveys a clearer picture of Sir Isaac Newton's formidable powers; it is extremely difficult three centuries later to grasp how far Newton had surpassed his contemporaries in mathematics and physics. It was reported by Newton's biographers (much of what we know about Newton has come down to us from John Conduitt and William Stukeley) that Newton received the brachistochrone problem in the evening after returning from work at the Royal Mint (Newton served as Master of the Mint for 10 years); Newton developed concepts for the COV *and* solved the brachistichrone problem, completing his work at 4 a.m. the next morning. He insisted that the solution be published anonymously (and it was, by the Royal Society in January 1697), but it has been reported that when Bernoulli saw it, he remarked that *we recognize the lion by his claw.*

The physical picture for the brachistochrone problem is as follows: We consider two points in a vertical plane connected by a ribbon or wire that can assume any shape desired. An object of mass, m, slides down the wire under the influence of gravity. The process is frictionless so that at any point in time the sum of potential and kinetic energies is constant; there are *no dissipative forces in play*, only gravity. The question to be addressed is what shape should the wire have such that the sliding mass reaches the bottom position in the least possible time?

Let the initial point be the origin and the lower point be P corresponding to position (x_2, y_2). Since there are no dissipative forces in this process we note by energy balance, $PE_0 + KE_0 = PE_2 + KE_2$. We assume that the mass was initially at rest, so the kinetic energy term on the left-hand side is zero:

$$mgy_0 = mg(y_0 - y) + \frac{1}{2}m\left(\frac{ds}{dt}\right)^2, \tag{11.21}$$

and therefore, $\frac{ds}{dt} = \sqrt{2gy}$. Of course, s is the length of the arc connecting the initial and final points. The total time required for the mass to slide from the origin to its final position can be obtained from the quotient of the path length and the object's velocity (keep in mind that we are starting at the origin, so initially $x = 0$):

$$T_{\text{req}} = \int_{x=0}^{x=x_2} \frac{ds}{\sqrt{2gy}}. \tag{11.22}$$

We already know that the length of an arc is given by eq. (11.5) in Section 11.1; therefore,

$$T_{\text{req}} = \frac{1}{\sqrt{2g}} \int_0^{x_2} \frac{\sqrt{1 + y'^2}}{\sqrt{y}} dx. \tag{11.23}$$

The objective is to find the function $y(x)$ that minimizes this integral, recognizing that this is exactly the type of COV problem we described previously. One could propose different functions and calculate the total time for each case (as we did in the example at the beginning of this chapter), but we would not know whether the $y(x)$ so identified was really the extremal we are seeking. By comparison with the form for eq. (11.8) in Section 11.1, we see

$$F = \frac{\sqrt{1 + y'^2}}{\sqrt{y}},$$

and thus F *does not* explicitly depend upon x. Consequently we can make use of eq. (11.14) in Section 11.1:

$$\frac{\sqrt{1 + y'^2}}{\sqrt{y}} - y'\left[\frac{y'}{\sqrt{1 + y'^2}\sqrt{y}}\right] = C, \tag{11.24}$$

and accordingly,

$$y' = \frac{dy}{dx} = \sqrt{\frac{1/C - y}{y}}. \tag{11.25}$$

We set $A = 1/C$ and change the y-variable: $y = A \sin^2 \theta$. The reader may wish to show that

$$x = A\left[\theta - \frac{1}{2}\sin 2\theta\right]. \tag{11.26}$$

The constant of integration that should appear here is equal to zero since the curve must go through the origin (where $x = y = 0$). The expressions for x and y give us the parametric equations for a cycloid; for example, when $\theta = \pi/6$, $x = 0.0906A$ and $y = 0.25A$, and when $\theta = \pi/4$, $x = 0.2854A$ and $y = 0.5A$. The value for A is selected to ensure that the curve passes through the desired point, (x_2, y_2).

We will now work through a detailed Brachistochrone example to better illustrate the process described above. A particle initially located at the origin is allowed to fall under the influence of gravity to the new position, $(x,y) = (10,10)$; the process is frictionless. The positive y-direction is *down*, the x-axis is horizontal, the distances given are in meters, and $g = 9.8\ \text{m/s}^2$. If the particle fell along a path given by a *straight line*, then $y = x$ and $y' = 1$, and the time required would be 2.02 s. If $y = 10x - 0.09x^3$, the time required would be 4.381 s. Of course, the problem with a straight-line trajectory is that at the instant of release the particle is only subjected to 0.707 mg; if the initial frictionless descent was more nearly vertical, the entity would experience greater acceleration which would result in a shorter transit time for the frictionless process.

The simplest type of variational problem is one in which we want to find *the* $y(x)$ that makes the integral $\int_{x_1}^{x_2} F(x,y,y')dx$ a minimum. We will examine a group of functions given by

$$y(x) = y_0(x) + \varepsilon\eta(x) = y_0 + \delta y, \tag{11.27}$$

where the final term on the right-hand side represents a small variation of the function. Equation (11.27) is substituted into the integral of F:

$$\phi(\varepsilon) = \int_{x_1}^{x_2} F(x, y + \varepsilon\eta,\ y' + \varepsilon\eta')dx \tag{11.28}$$

This integral will be a minimum when $\varepsilon = 0$ so

$$\phi'(0) = 0 = \int_{x_1}^{x_2} \left(\frac{\partial F}{\partial y}\eta + \frac{\partial F}{\partial y'}\eta'\right) dx \tag{11.29}$$

The second term is integrated by parts resulting in

$$\phi'(0) = \int_{x_1}^{x_2} \left(\frac{\partial F}{\partial y} + \frac{d}{dx}\frac{\partial F}{\partial y'}\right) \eta dx \tag{11.30}$$

The first variation of the integral must vanish, and it will disappear for *all* η (or δy) only if

$$\left(\frac{\partial F}{\partial y} + \frac{d}{dx}\frac{\partial F}{\partial y'}\right) = 0. \tag{11.31}$$

The total derivative with respect to x is taken resulting in Euler's differential equation:

$$y''\frac{\partial^2 F}{\partial y'^2} + y'\frac{\partial^2 F}{\partial y \partial y'} + \frac{\partial^2 F}{\partial x \partial y'} - \frac{\partial F}{\partial y} = 0. \tag{11.32}$$

In our case, $F(y,y') = \sqrt{\frac{1+y'^2}{2gy}}$, which is *independent of* x and consequently $F - y'\frac{\partial F}{\partial y'}$ = constant. Since $\frac{\partial F}{\partial y'} = \frac{1}{2}(1+y'^2)^{-\frac{1}{2}}(2y')/\sqrt{2gy}$, we can isolate y' resulting in

$$y' = \frac{dy}{dx} = \sqrt{\frac{1 - 2c^2gy}{2c^2gy}} = \pm\sqrt{\frac{k-y}{y}} \tag{11.33}$$

where $k = \frac{1}{2c^2g}$. Therefore the extremal, $y(x)$, for the minimum time of descent is given by

$$x = \int \frac{ydy}{\sqrt{ky - y^2}} = -\sqrt{ky - y^2} + \frac{k}{2}\cos^{-1}\frac{k-2y}{k}. \tag{11.34}$$

The parameter, k, must have a value that allows the curve to go through ($x = 10, y = 10$), so we can either solve eq. (11.34) for those coordinate values or we can solve the two parametric equations: $2x = 20 = k(\theta - \sin\theta)$ and $2y = 20 = k(1 - \cos\theta)$. You may want to confirm that k = 11.458 and θ = 2.412, producing the following points for the extremal curve (which is also depicted in Figure 11.4):

y	1	2	3	4	5	6	7	8	9	10
x	0.202	0.589	1.117	1.781	2.586	3.548	4.694	6.074	7.777	9.999

11.4 Other Examples

11.4.1 Minimum Surface Area

Again we assume that we have some curve in the x–y plane that connects the two pairs of points, (x_1, y_1) and (x_2, y_2). If we rotate this curve about the x-axis, then a surface is generated that has some particular surface area, as well as some enclosed

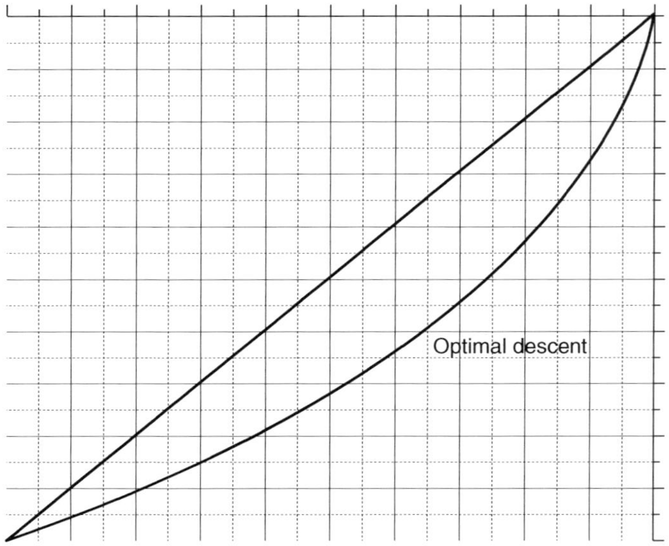

Figure 11.4 Solution for the brachistochrone problem for a particle falling from the origin to the point (10, 10) under the influence of gravity. A straight line is provided for comparison. Interestingly, the exponential curve, $y = 11(1 - \exp(-0.2398x))$, is fairly close to the extremal—it produces a transit time of 1.862 s. The reader may want to confirm that the optimal transit time is about 8% less than this t.

volume. We know from elementary calculus that this generated surface has an area given by

$$A = \int_{x_1}^{x_2} 2\pi y(x)\sqrt{1 + \left(\frac{dy}{dx}\right)^2}\, dx. \tag{11.35}$$

Therefore, if a straight line extended from the origin to the point (4,5), then $y(x) = \frac{5}{4}x$, and if we rotate this line around the x-axis we find

$$A = 2\pi\left(\frac{5}{4}\right)\sqrt{\frac{41}{16}}\left.\frac{x^2}{2}\right|_0^4 = 100.58.$$

Of course we could confirm this number by use of the mensuration formula for the lateral surface area of a cone ($A_{lat} = \pi RL = (3.1416)(5)\sqrt{41} = 100.58$). But our goal in this section goes beyond a determination of surface area; we want to determine the *minimum surface area* formed by the revolution of $y(x)$ about the x-axis. Therefore, we need to identify the *minimum* value of the integral,

$$I = 2\pi \int_{x_1}^{x_2} y\sqrt{1 + y'^2}dx. \tag{11.36}$$

Observe once again that the integrand does *not* depend explicitly upon x; thus, we can turn immediately to Section 11.1, $F - y'\frac{\partial F}{\partial y'} = C$:

$$y\sqrt{1 + y'^2} - \frac{yy'^2}{\sqrt{1 + y'^2}} = C. \tag{11.37}$$

Now we isolate dy/dx to show:

$$\int \frac{dy}{\sqrt{\frac{y^2}{C^2} - 1}} = x + B. \tag{11.38}$$

We substitute $z = y/C$ to find

$$C\int \frac{dz}{\sqrt{z^2 - 1}} = x + B,$$

and we notice that the denominator of the integral is a quadratic irrationality which would normally call for trigonometric substitution. However, in this case we might recognize the appropriate antiderivative since the integral is one that appears in every elementary calculus book, and therefore,

$$\cosh^{-1}(z) = \frac{x}{C} + B_1 \text{ resulting in } y = C\cosh\left[\frac{x}{C} + B_1\right]. \tag{11.39}$$

The intriguing feature of this *catenary* is that if we force the curve through the initial point (x_1, y_1) we *might not be able* to pass through the end-point (x_2, y_2); that is, the end-point might lie outside the envelope of this catenary family. Weinstock discusses this particular case which reveals a limitation of the theory we have employed above, and he notes that the actual minimum surface area for this situation is given by the *Goldschmidt discontinuous solution*—the catenary family will provide relative or absolute minima for surface area *only if* the end-point is sited appropriately (in general, the end-point must be above or to the left of the catenary envelope). We will consider this situation further at the end of Section 11.7 as part of our treatment of systems with surface tension.

Before we leave this discussion of the catenary, we should make note of the fact that the COV can be used to show that an inverted catenary (the catenary arch) has the lowest possible internal stress. Of course, catenary arches have been used in buildings for centuries, and a modern example is the Sheffield Winter Garden in South Yorkshire. The very well-known Gateway Arch in St. Louis, Missouri, is an example of a flattened catenary.

11.4.2 Systems of Particles

We are concerned here with a collection of n-particles in a conservative system for which the potential energy function, V, can be used to obtain the three force components; for the n-th particle (located at position x, y, and z) we would have:

$$F_x{}^n = -\frac{\partial V}{\partial x_n}, F_y{}^n = -\frac{\partial V}{\partial y_n}, \text{ and } F_z{}^n = -\frac{\partial V}{\partial z_n}. \tag{11.40}$$

The total kinetic energy, T, for this system can be obtained from:

$$T = \frac{1}{2}\sum_{i=1}^{n} m_i \left(\left(\frac{dx}{dt}\right)^2 + \left(\frac{dy}{dt}\right)^2 + \left(\frac{dz}{dt}\right)^2\right). \tag{11.41}$$

In classical mechanics we would describe the behavior of a particle(s) with Newton's law, $F = ma = m\frac{d^2s}{dt^2}$. We imagine that at two different times, say, t_1 and t_2, we have two different configurations—the particles have moved around subject to $F = ma$. As we noted at the beginning of this chapter the *principle of least action*[1] states that integral of the Lagrangian function over this time interval must have a minimum value. Since the Lagrangian function for this system is merely the difference between kinetic and potential energies, we focus on the *x*-*y* plane and require that

$$I = \int_{t_1}^{t_2} (KE - PE)dt = \int_{t_1}^{t_2} \left(\frac{1}{2}m(\dot{x}^2 + \dot{y}^2 - mgy\right) dt \tag{11.42}$$

be an extremum (minimum). The problem thus posed is one of determining the paths traveled by the particles that make this integral a minimum. We let the particle velocities be represented by $\dot{q}_1, \dot{q}_2, \dots \dot{q}_n$ and the coordinate locations be given by $q_1, q_2, \dots q_n$. Naturally the kinetic energy of the system is a function of the velocities, the positions, and time, while the potential energy is a function of the coordinates and time. The requirement that the integral eq. (11.42) be a minimum is often referred to as *Hamilton's principle* which is applicable to the motion of any *conservative* system. This principle leads to Lagrange's general equations of motion which we write as:

$$\frac{d}{dt}\frac{\partial T}{\partial \dot{q}_i} - \frac{\partial}{\partial q_i}(T - V) = 0 \text{ or } \frac{\partial L}{\partial q_i} - \frac{d}{dt}\left(\frac{\partial L}{\partial \dot{q}_i}\right) = 0 \text{ (for } i = 1, 2, \dots n) \tag{11.43}$$

where the q_i's are generalized coordinates. These eq. (11.43) constitute a set of simultaneous second-order differential equations whose solution(s) reveal the generalized coordinates as functions of time. Solution, of course, requires that the initial values for both q and $\dot{q}$ be known.

---

1 The philosophical underpinning of the principle of least action was given by Euler:

> "As the construction of the universe is the most perfect possible, being the handiwork of an all-wise Maker, nothing can be met with in the world in which some maximal or minimal property is not displayed. There is, consequently, no doubt but that all of the effects of the world can be derived by the method of maxima and minima from their final causes as well as from their efficient ones."

11.4.3 Vibrating String

In the previous section we considered a situation in which points of mass (particles) were distributed in space. However, Hamilton's principle can also be applied to cases in which mass is continuously distributed. To illustrate, let us consider a string anchored at positions $x = 0$ and $x = L$. The string is flexible, and if it is physically perturbed, it can oscillate normal to its equilibrium (or undisturbed) axis. We will let the string's displacement be ϕ and its mass per unit length be m. The total kinetic energy of this string of length L is then:

$$T = \frac{1}{2}\int_0^L m\left(\frac{\partial \phi}{\partial t}\right)^2 dx. \tag{11.44}$$

The string is perfectly elastic, so any disturbance will result in an increase in length. We will assume that the potential energy of the string is directly related to its elongation, and therefore by eq. (11.5) in Section 11.1, this change in length is given by:

$$\int_0^L \sqrt{1+\left(\frac{\partial \phi}{\partial x}\right)^2}\, dx - L. \tag{11.45}$$

If the increase in string length caused by the displacement is sufficiently small, then

$$\sqrt{1+\left(\frac{\partial \phi}{\partial x}\right)^2} \approx 1 + \frac{1}{2}\left(\frac{\partial \phi}{\partial x}\right)^2$$

such that

$$V \cong \frac{k}{2}\int_0^L \left(\frac{\partial \phi}{\partial x}\right)^2 dx,$$

and therefore,

$$T - V = \frac{1}{2}\int_0^L \left[m\left(\frac{\partial \phi}{\partial t}\right)^2 - k\left(\frac{\partial \phi}{\partial x}\right)^2\right] dx. \tag{11.46}$$

By the principle of least action, the first variation of I should be zero:

$$I(\phi) = \frac{1}{2}\int_{t_1}^{t_2}\int_0^L \left[m\left(\frac{\partial \phi}{\partial t}\right)^2 - k\left(\frac{\partial \phi}{\partial x}\right)^2\right] dxdt. \tag{11.47}$$

Note that the functional depends *only* upon the two derivatives (with respect to t and x); it does not involve x or t or even ϕ explicitly. The Euler–Lagrange equation for this case (we will use subscripts to indicate derivatives with respect to x and t) is

$$\frac{\partial F}{\partial \phi} - \frac{\partial}{\partial x}\left(\frac{\partial F}{\partial \phi_x}\right) - \frac{\partial}{\partial t}\left(\frac{\partial F}{\partial \phi_t}\right), \tag{11.48}$$

which means that the function ϕ which makes the first variation of I disappear, must be governed by:

$$\frac{\partial^2 \phi}{\partial t^2} = c^2\frac{\partial^2 \phi}{\partial x^2}, \tag{11.49}$$

where the characteristic velocity, c, is obtained from the quotient of k/m. Equation (11.49) tells us that the string's displacement must follow the one-dimensional wave equation, exactly as we expected.

11.4.4 Laplace's Equation

We have seen previously that problems concerned with conservative fields (no dissipative effects) can often be described with Laplace's equation, for example, in two dimensions,

$$\nabla^2\phi = 0 \text{ or } \frac{\partial^2 \phi}{\partial x^2} + \frac{\partial^2 \phi}{\partial y^2} = 0. \tag{11.50}$$

We may also remember that eq. (11.50) has a broad range of applications in electrostatics, heat transfer, gravitation, and hydrodynamics. For cases in which the field variable is specified around the boundary, we have what is known as a Dirichlet problem. Such problems can be expressed alternatively in variational form,

$$I(\phi) = \iint F(x, y, \phi, \phi_x, \phi_y)dxdy, \tag{11.51}$$

and for the 2D Laplace equation the variational principle is known:

$$I(\phi) = \iint \frac{1}{2}\left(\left(\frac{\partial \phi}{\partial x}\right)^2 + \left(\frac{\partial \phi}{\partial y}\right)^2\right)dxdy. \tag{11.52}$$

Our goal in this case is to identify the potential function that causes the first variation of I to disappear. The solution procedure is analogous to the Ritz method which is discussed extensively in Sections 11.4.5 and 11.5. The main problem with this approach is that the required work load is significant; thus, we would normally use this technique for solution of the Laplace equation only when a closed-form solution cannot be found, but we absolutely must obtain an analytic approximation.

11.4.5 Boundary Value Problems

In a typical variational problem we try to identify a function, $y(x)$, that leads to a minimum value of the integral just as we saw at the beginning of this chapter,

$$I(y) = \int_{x_1}^{x_2} F(x, y, y')dx.$$

A *direct approach* to this problem involves the construction of a sequence of functions, $y_1, y_2, y_3, \ldots y_n$ such that as $n \to \infty$ we obtain an extremum for I. The set of functions that achieves this objective is called a *minimal* (or *minimizing*) *sequence.* There is a powerful technique that can be used for this purpose, and we will look at a detailed example below; however, as Smith (1953) indicates, we need to keep two very important points in mind: First, the function sequence must be selected with care to obtain a suitably rapid rate of convergence, and second, the minimal sequence—even as n becomes very large—may *not necessarily* converge to the actual solution of the variational problem.

The *COV* can be used to solve certain boundary value problems through use of a technique we discussed previously in this course, the *Rayleigh–Ritz* (or simply Ritz) *method*. We can best illustrate this with a simple example for which the analytic solution is easily determined. Suppose we have a differential equation,

$$\frac{d^2\phi}{dx^2} = -f(x) = a + bx, \tag{11.53}$$

with $f(x = 0) = 0$ and $f(x = 1) = 7$; further, we presume a form for the solution, $\phi = 11x - 17x^2 + 13x^3$. This means that $a + bx = -34 + 78x$. The behavior of the function $\varphi(x)$ is illustrated in Figure 11.5.

In the application of the Ritz method to this boundary value problem, we set out to find a function that minimizes the integral:

$$I(\phi) = \int_0^1 \left[\frac{1}{2}\left(\frac{d\phi}{dx}\right)^2 - f(x)\phi(x)\right] dx. \tag{11.54}$$

We proceed with the understanding that our objective is to identify an analytic approximation for $\phi(x)$ as quickly and easily as possible. We let the trial function be

$$\phi \cong x[C_1 + C_2(1 - x) + C_3(1 - x)^2 + C_4(1 - x)^3 + \ldots]. \tag{11.55}$$

The boundary conditions are automatically satisfied if we take $C_1 = 7$. The trial function is substituted into eq. (11.54), and the integration is carried out to obtain I, which depends upon the undetermined constants, C_2 through C_n. Values are obtained for these C's by differentiating I and setting the derivatives equal to zero:

$$\frac{\partial I}{\partial C_2} = 0, \quad \frac{\partial I}{\partial C_3} = 0, \ldots, \frac{\partial I}{\partial C_n} = 0. \tag{11.56}$$

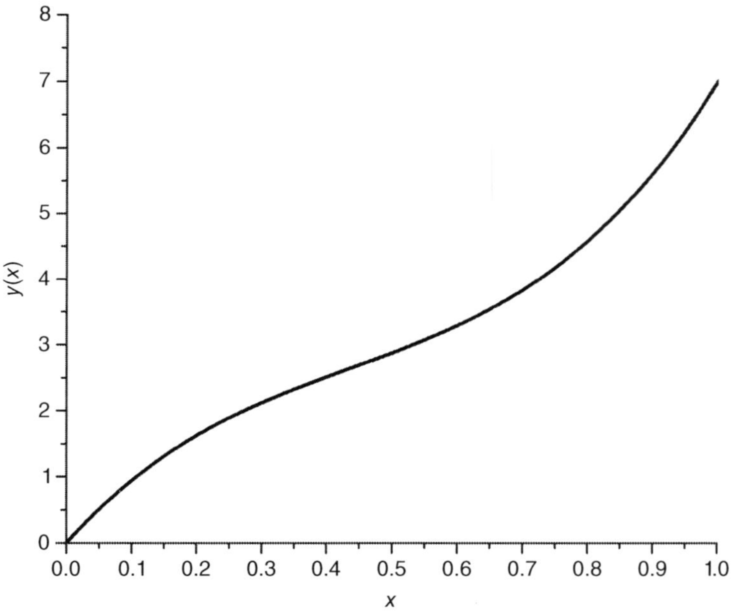

Figure 11.5 $\phi(x)$ for illustration of the Ritz method. The reader should integrate eq. (11.53) and verify the solution.

Naturally, the more terms that are retained in the trial function, the better the result is likely to be. This example provides a terrific opportunity for the reader to explore the Ritz method without expending too much time or effort, as we shall now see. We will begin by truncating our approximation (we know the result will not meet critical inspection, but it will be simple enough to be executed rapidly): $\phi \approx x[7 + C_2(1-x)]$. Of course this quadratic form cannot duplicate the point of inflection we see in Figure 11.5, but it will reveal the essence of the technique. Now we compute the value of the integral (11.54) using plausible values for the parameter, C_2. The results are illustrated by Figure 11.6.

Using $\phi = x[7 - 2.25(1-x)]$, we obtain:

x	0.0	0.2	0.4	0.6	0.8	1.0
$\phi(x)$	0	1.04	2.26	3.66	5.24	7

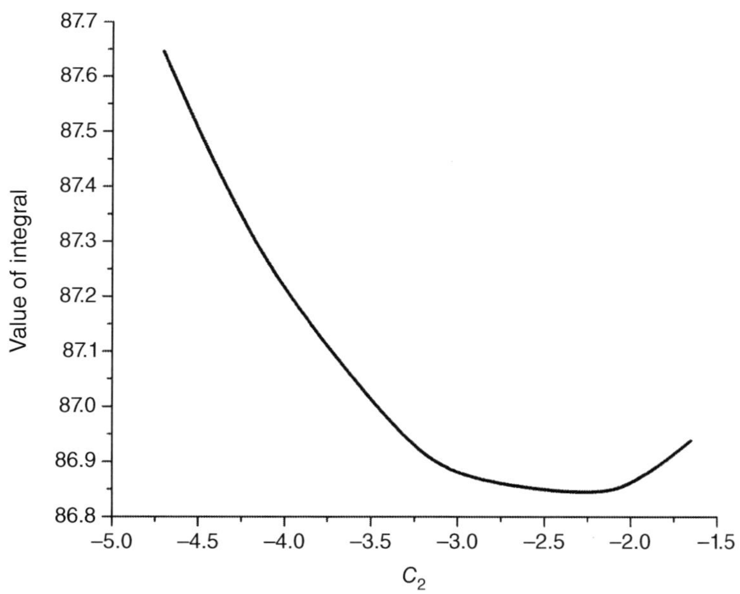

Figure 11.6 Calculated values for the integral eq. (11.54); the minimum occurs when C_2 is approximately equal to −2.25.

We can compare the results in this abbreviated table with the behavior of the function illustrated in Figure 11.5. For example, at $x = 0.4$, the correct value is about 2.5, while at $x = 0.8$, we should have gotten about 4.5 or 4.6. The reader is urged to verify this result and then add the C_3 term and repeat! How much better is the approximation when the polynomial is third degree?

This direct approach to the solution of variational problems can be applied to more difficult cases if we use a little ingenuity, and we will illustrate this using an example developed by Ritz and discussed in detail by Smith (1953). Suppose we are interested in a problem of the type,

$$I = \int_{x_1}^{x_2} (p_1 y'^2 + 2p_2 yy' + p_3 y^2)dx. \tag{11.57}$$

Our objective, of course, is to find $y(x)$ that makes $I(y)$ an extremum (a minimum); we stipulate that p_1, p_2, and p_3 be either constants or functions of x (but they *do not* depend upon y). We also require that y' be finite and continuous for the interval $x_1 \leq x \leq x_2$, and we are given that $y(x_1) = \beta_1$ and $y(x_2) = \beta_2$. It is quite useful in a case like this to employ a transformation that simplifies the boundary conditions. Consider the function:

$$y(x) = \frac{x - x_2}{x_1 - x_2}\beta_1 + \frac{x - x_1}{x_2 - x_1}\beta_2 + \phi(x), \tag{11.58}$$

and observe that $\phi(x)$ must disappear (be equal to zero) at both end-points. This is extremely convenient, and the required behavior suggests a function type that is suitable for the minimizing sequence; for example, let the approximations for $\phi(x)$ have the form:

$$\phi_1 = C_1 \sin\frac{\pi(x - x_1)}{x_2 - x_1}. \tag{11.59}$$

Therefore,

$$\phi_n = C_1 \sin\frac{\pi(x - x_1)}{x_2 - x_1} + C_2 \sin\frac{2\pi(x - x_1)}{x_2 - x_1} + \dots C_n \sin\frac{n\pi(x - x_1)}{x_2 - x_1}. \tag{11.60}$$

Of course the final step is to determine the constants C_1, C_2, etc. that make $I(y)$ a minimum. The precise forms for p_1, p_2, and p_3 would have to be known before this process could be carried out.

There is another direct method for determining the minimizing sequence that is described in Chapter IV of Courant and Hilbert (1989). It involves discretizing the integral as given by eq. (11.8) in Section 11.1 using finite difference approximations and dividing the interval (x_1, x_2) into m pieces such that

$$\int_{x_1}^{x_2} F(x, y, y')dx \approx \sum_{i=0}^{m} F\left(x_i, y_i, \frac{y_{i+1} - y_i}{\Delta x}\right)\Delta x.$$

This is now simply an ordinary minimum problem handled in piecewise fashion. Should higher derivatives appear in the integral, they would be replaced by forward difference approximations.

11.5 The Rayleigh–Ritz Method and Sturm–Liouville Problems

Assume we have identified an integral of the form

$$I = \int_a^b F(x, y, y', y'', \dots)dx, \tag{11.61}$$

and we want to determine $y(x)$ such that this integral is stationary. We will assume that $y(x)$ can be represented as

$$y \cong \sum_{i=0}^{n} a_i f_i(x). \tag{11.62}$$

The functions, $f_i(x)$, are appropriately chosen for the problem, and the coefficients (a_i's) are undetermined. We substitute eq. (11.62) into eq. (11.61) and integrate, obtaining an expression for I which you will recall will be stationary if the coefficients are chosen such that

$$\frac{\partial I}{\partial a_0} = 0,\ \frac{\partial I}{\partial a_1} = 0,\ \frac{\partial I}{\partial a_2} = 0, \text{etc.} \tag{11.63}$$

These derivatives provide a set of simultaneous equations that can be solved to produce the assumed solution eq. (11.62). Unless we are working a very simple problem, or we are incredibly fortunate, we should not trust the approximate solution in critical applications.

Now consider extremizing the integral,

$$I = \int_{x_1}^{x_2} [p(x)y'^2 - q(x)y^2]dx + c_1(y(x_1))^2 + c_2(y(x_2))^2 \tag{11.64}$$

subject to the condition, $\int_{x_1}^{x_2} g(x)y^2dx = 1$. In the case of a *fixed end-point* problem where $y(x_1) = y(x_2) = 0$, this integral (11.64) is

$$I = \int_{x_1}^{x_2} [p(x)y'^2 - q(x)y^2]dx. \tag{11.65}$$

We use the Lagrange multiplier $(-\lambda)$ to make the constraint part of the extremal resulting in

$$I - \int_{x_1}^{x_2} \lambda g(x)y^2dx = \int_{x_1}^{x_2} [p(x)y'^2 - q(x)y^2 - \lambda g(x)y^2]dx \tag{11.66}$$

The ordinary differential equation (ODE) that must be satisfied by the extremizing function is

$$\frac{d}{dx}\left[p(x)\frac{dy}{dx}\right] + [q(x) + \lambda g(x)]y = 0. \tag{11.67}$$

You can easily verify this by taking the integrand from eq. (11.66) as F, and applying the Euler equation: $\frac{d}{dx}\left(\frac{\partial F}{\partial y'}\right) - \frac{\partial F}{\partial y} = 0$; this will result in eq. (11.67). This ordinary differential eq. (11.67), with $p(x)$, $q(x)$, and $g(x)$ specified (either continuous functions of x or constants), is an example of a Sturm–Liouville (SL) problem. You may remember that we find solutions for SL problems for specific values—*eigenvalues*—of λ. The corresponding $y(x)$'s are *eigenfunctions*. Familiar Sturm–Liouville differential equations include:

$$y'' + \lambda y = 0$$

$$y'' + \frac{1}{x}y' + \lambda y = 0$$

$$(1 - x^2)y'' - 2xy' + \lambda(\lambda + 1)y = 0 \tag{11.68a,b,c}$$

We recognize that eq. (11.68b) is a form of Bessel's equation and eq. (11.68c) is the Legendre equation. Let's examine the elementary eq. (11.68a) in detail for $0 \le x \le 1$; we begin by solving the ODE analytically with $y(0) = y(x = 1) = 0$ (a fixed end-point problem):

$$(D^2 + \lambda)y = 0, \quad \text{so}$$

$$\left(D + i\sqrt{\lambda}\right)\left(D - i\sqrt{\lambda}\right) = 0, \quad \text{and therefore}$$

$$y = A\sin\sqrt{\lambda}x + B\cos\sqrt{\lambda}x. \tag{11.69}$$

Of course $B = 0$, but since the solution is periodic, there are an infinite number of λ's that will result in $y(1) = 0$. The first of those—the smallest eigenvalue—will be $\sqrt{\lambda} = \pi$ so $\lambda = 9.8696$.

Now we write down a truncated approximation from eq. (11.62):

$$y \approx a_0 + a_1x + a_2x^2 + a_3x^3. \tag{11.70}$$

Generally, we cannot know in advance if this limited series will be adequate to represent the solution over the interval of interest. In this case the analytic solution is known, so we will be able make an assessment at the end. We satisfy the boundary conditions by requiring

$$a_0 = 0 \text{ and } a_1 + a_2 + a_3 = 0 \text{ so that } a_3 = -(a_1 + a_2).$$

As an aside, we note that the boundary conditions would have been automatically satisfied by rewriting eq. (11.70) as $y \approx x(1-x)(a_0 + a_1x + a_2x^2 + \ldots)$. We differentiate (11.70) and replace a_3:

$$y' = a_1 + 2a_2x - 3(a_1 + a_2)x^2. \tag{11.71}$$

The smallest eigenvalue is obtained by extremizing the quotient:

$$\lambda = I = \int_0^1 y'^2 dx \Big/ \int_0^1 y^2 dx. \tag{11.72}$$

We introduce eqs. (11.70) and (11.71)—both squared—into eq. (11.72) and perform the indicated integrations. Since we have two unknown coefficients (remember a_3 has been written in terms of a_1 and a_2) we need two equations:

$$\frac{\partial I}{\partial a_1} = 0 \text{ and } \frac{\partial I}{\partial a_2} = 0. \tag{11.73}$$

Forming the quotient given by eq. (11.72) and determining the derivatives indicated by eq. (11.73) could charitably be called tedious, with errors nearly inevitable (in fact, an erroneous solution is given by Spiegel, 1971). This process is not recommended as a student exercise for this reason, but should you wish to try it you should find for eq. (11.72):

$$\lambda = I = \frac{168a_1^2 + 126a_1a_2 + 28a_2^2}{16a_1^2 + 11a_1a_2 + 2a_2^2}. \tag{11.74}$$

At this point we would still need to obtain the derivatives eq. (11.73), *but* if we divide through by a_2^2 and set $\beta = a_1/a_2$ we obtain $\frac{168\beta^2+126\beta+28}{16\beta^2+11\beta+2}$. This quotient has a single parameter, β, so we can readily determine the distinct minimum value of I, which is $70/7 = 10$ when $\beta \cong -1$. Therefore, since the analytic value is 9.8696, our estimate of the first eigenvalue has an error of just 1.3%.

It is also possible to modify the example so that the analysis can be performed in just a few minutes if we impose a further truncation: Let $y = a_1x + a_2x^2$ so that $y' = a_1 + 2a_2x$; then square these expressions and integrate as indicated by eq. (11.72). You should confirm

$$\lambda = \frac{a_1^2 + 2a_1a_2 + \frac{4}{3}a_2^2}{\frac{1}{3}a_1^2 + \frac{1}{2}a_1a_2 + \frac{1}{5}a_2^2}. \tag{11.75}$$

Since $a_2 = -a_1$, one can show this quotient is $(1/3)/(1/30) = 10$—a triumph of luck over analytical prowess. Finally, let's select an amplitude for eq. (11.69), say, $A = 4$; we let our approximation match this sinusoid at $x = 1/2$, resulting in the following comparison:

x	$y = 16(x - x^2)$	$y = 4\sin(\pi x)$
0	0	0
¼	3	2.828
½	4	4
¾	3	2.828
1	0	0

One other observation about this example: The ODE is simple enough that we know much about $y(x)$, and since *sine* is an odd function, it might make more sense to let $y = a_1x + a_3x^3$. We will leave this for an exercise at the end of the chapter, and you will be able to compare your result with the estimates tabulated above.

Next, we turn our attention to eq. (11.68b), which we noted previously is a form of Bessel's equation: $y'' + \frac{1}{x}y' + \lambda y = 0$. We will require $y(0)$ to be finite and $y(1) = 0$. This is another convenient example for us since we know the solution is:

$$y = C_1 J_0\left(\sqrt{\lambda} x\right) + C_2 Y_0\left(\sqrt{\lambda} x\right). \tag{11.76}$$

Since y must be finite for $x = 0$, we require $C_2 = 0$. Of course J_0 is oscillatory and because the first zero occurs for $\sqrt{\lambda}x = 2.4048$ we also know the eigenvalue we are seeking is $\lambda_1 = 5.7831$.

We will let $y \cong a_0 + a_2x^2 + a_4x^4$ *because* we realize that J_0 is an *even function*. Usually when we employ the Rayleigh–Ritz method we would not have this information in advance and we might contemplate what our outcome would be should we make a more common approximation. Proceeding, we wish to minimize either

$$I = \frac{\int_0^1 xy'^2 dx}{\int_0^1 xy^2 dx}, \tag{11.77}$$

or, alternatively, letting $F = \int_0^1 xy'^2 dx$ and $G = \int_0^1 xy^2 dx$, we minimize F subject to G for the unknown parameters. Let's explore this approach first:

$$0 = \frac{\partial}{\partial a_0}(F - \lambda G), \quad 0 = \frac{\partial}{\partial a_2}(F - \lambda G), \quad \text{and } 0 = \frac{\partial}{\partial a_4}(F - \lambda G). \tag{11.78}$$

If we employ these eq. (11.78), we can perform the differentiations, eliminate the unknown coefficients a_0, a_2, and a_4, and obtain a quadratic equation for λ. Since $y(1) = 0$, we can use $a_0 + a_2 + a_4 = 0$ to remove one of the unknowns (a_0); therefore,

$$F - \lambda G = \left\{a_2^2 + \frac{8}{3}a_2a_4 + 2a_4^2 - \lambda\left(\frac{1}{6}a_2^2 + \frac{5}{12}a_2a_4 + \frac{4}{15}a_4^2\right)\right\}. \tag{11.79}$$

We differentiate the right-hand side with respect to a_2:

$$2a_2 + \frac{8}{3}a_4 - \lambda\left(\frac{1}{3}a_2 + \frac{5}{12}a_4\right) = 0, \tag{11.80}$$

and separately with respect to a_4:

$$\frac{8}{3}a_2 + 4a_4 - \lambda\left(\frac{5}{12}a_2 + \frac{8}{15}a_4\right) = 0. \tag{11.81}$$

Each of these expressions is rearranged to provide the quotient $\frac{a_2}{a_4}$, and the two ratios are equated:

$$\frac{-4 + \frac{8\lambda}{15}}{\frac{8}{3} - \frac{5\lambda}{12}} = \frac{-\frac{8}{3} + \frac{5\lambda}{12}}{2 - \frac{\lambda}{3}}.$$

This yields the quadratic equation:

$$\lambda^2 - 42.663\lambda + 213.316 = 0, \tag{11.82}$$

with the two solutions, 5.784 and 36.879. Of course, the smaller of this pair is the eigenvalue we are seeking.

Now we will return to the numerator of the quotient eq. (11.77) where we have $\int_0^1 x(2a_2x + 4a_4x^3)^2 dx$, and for the denominator, $\int_0^1 x(a_0 + a_2x^2 + a_4x^4)^2$. We perform the integrations and eliminate a_0 since $a_0 = -(a_2 + a_4)$; the result is

$$I = \frac{a_2^2 + \frac{8}{3}a_2a_4 + 2a_4^2}{\frac{1}{6}a_2^2 + \frac{5}{12}a_2a_4 + \frac{4}{15}a_4^2}. \tag{11.83}$$

Normally we would obtain the two needed equations by setting $\frac{\partial I}{\partial a_2} = \frac{\partial I}{\partial a_4} = 0$; however, if our main interest is the identification of eigenvalues we can avoid the messy differentiation of this quotient. Let $\beta = \frac{a_2}{a_4}$ and divide (11.83) by a_4^2, resulting in an equation with a single parameter:

$$I = \frac{\beta^2 + \frac{8}{3}\beta + 2}{\frac{1}{6}\beta^2 + \frac{5}{12}\beta + \frac{4}{15}}. \tag{11.84}$$

We can perform a simple one-dimensional search which reveals a distinct minimum at $\beta = -3.57$, corresponding to $I_{min} = \lambda_1 = 5.7841$; our estimate for the first eigenvalue is in error by only 0.02%. We would also like to compare our approximate $y(x)$ with the analytic solution. We will require $y(0) = 4$, and since $\frac{a_2}{a_4} = -3.57$ and $a_0 = 4$, we find:

$$y \cong 4 - 5.556x^2 + 1.556x^4. \tag{11.85}$$

In this case the Rayleigh–Ritz method provides an excellent result:

x	$4 - 5.556x^2 + 1.556x^4$	$4J_0(2.4048x)$
0	4	4
0.2079	3.763	3.754
0.4158	3.086	3.061
0.6238	2.074	2.047
0.8317	0.901	0.896
1	0	0

Our approximate solution for Bessel's equation is extraordinarily good (the worst case is an error of about 1.3%) partly because we *knew in advance* that the solution of the ODE was an even function. As we pointed out previously, this is information we might not have in typical cases where we were employing the Rayleigh–Ritz method. You might briefly revisit this scenario assuming we had initiated the example problem with the trial function (11.70),

$$y = a_0 + a_1x + a_2x^2 + a_3x^3.$$

Here are some preliminaries to help you get started: We choose $y(0) = 1$ and require $y(1) = 0$ so that $a_0 = 1$ and $-1 = a_1 + a_2 + a_3$. Differentiating, $y' = a_1 + 2a_2x + 3a_3x^2$, squaring and multiplying by x, we can determine the numerator for I by integration. The result is

$$\frac{1}{2}a_1^2 + \frac{4}{3}a_1a_2 + \frac{3}{2}a_1a_3 + a_2^2 + \frac{12}{5}a_2a_3 + \frac{3}{2}a_3^2, \tag{11.86}$$

and we can eliminate a_3 since $a_3 = -(1 + a_1 + a_2)$, leaving two undetermined coefficients.

11.6 Contemporary COV Analyses of Old Structural Problems

11.6.1 Flexing of a Rod of Small Cross Section

We will begin this part of our discussion by considering a thin rod with a length that extends from $x = 0$ to $x = L$; this rod will be subject to flexing (transverse vibrations). It is easy enough to visualize this situation—imagine a strand of uncooked spaghetti for example that is oriented vertically between your finger tip and a table top. By pressing down on the end of the pasta you can observe the lateral deformation, and ultimately, failure as the load is increased. It is this transverse flexing that we wish to consider. We will let $u(x,t)$ represent transverse displacement such that the transverse velocity is $\frac{\partial u}{\partial t}$, and we will take the mass of the rod per unit length to be m. Therefore, the total kinetic energy of the transverse motions can be written as

$$T = \frac{1}{2}m\int_0^L \left(\frac{\partial u}{\partial t}\right)^2 dx. \tag{11.87}$$

The total potential energy due to strain is determined from the second derivative (with respect to x) of the displacement:

$$V = \frac{1}{2}EJ\int_0^L \left(\frac{\partial^2 u}{\partial x^2}\right) dx, \tag{11.88}$$

where J is the moment of inertia. Thus by Hamilton's principle, we have

$$I = \frac{1}{2}\int_{t_1}^{t_2}\int_0^L \left(m\left(\frac{\partial u}{\partial t}\right)^2 - EJ\left(\frac{\partial^2 u}{\partial x^2}\right)^2\right) dxdt. \tag{11.89}$$

The task confronting us is to choose the function $u(x, t)$ that yields the extremum of I and, at the same time, satisfies the conditions imposed at the ends of the rod. For a structural member (like our strand of spaghetti) the ends can be *free*, *hinged*, or *clamped*. By free, we mean that neither u nor the slope, $\left(\frac{\partial u}{\partial x}\right)$, is constrained to a particular value at $x = 0$ (or $x = L$). For the hinged condition, u is fixed but the slope is arbitrary (this is the likely situation when we load the pasta strand by pressing down), and for the clamped condition both the displacement, u, and the slope have particular values. The integral is extremized by creating a set of comparison functions such as: $U = u(x, t) + \varepsilon\phi(x, t)$, and it is necessary that the function $\phi(x, t)$ be zero at both times, t_1 and t_2 (you should recognize that this is exactly the process that is illustrated by Figure 11.3). If the rod is hinged at the ends, it is necessary that $\phi(x = 0, x = L)$ be zero, but the slope, $\frac{\partial \phi}{\partial x}$, is arbitrary. Our interest is the case where $\varepsilon = 0$ and thus, $I'(0) = 0$. For the sake of compactness we will let the integrand in eq. (11.89) be represented by F (which includes the prefactor of ½), and therefore

$$I'(0) = \int_{t_1}^{t_2}\int_{0}^{L}\left(\frac{\partial F}{\partial \dot{u}}\dot{\phi} + \frac{\partial F}{\partial u_{xx}}\phi_{xx}\right) dxdt = 0. \tag{11.90}$$

Weinstock (1974) shows that through repeated integration by parts one can obtain the differential equation that governs the displacement of the rod (you may recall the analogous development of the Euler–Lagrange differential equation at the beginning of this chapter):

$$m\frac{\partial^2 u}{\partial t^2} + EJ\frac{\partial^4 u}{\partial x^4} = 0 \text{ for } 0 \le x \le L. \tag{11.91}$$

This is the equation of motion for transverse motions of the thin rod. If the ends of the rod are hinged, then it is necessary that $u = 0$ and $\frac{\partial^2 u}{\partial x^2} = 0$. We can use the product method to seek a solution for eq. (11.91) as we will now demonstrate by proposing that $u = f(x)g(t)$; this results in two ODEs:

$$g'' = -\lambda^2 g \text{ and } f'''' - \frac{\lambda^2}{\beta}f = 0, \text{ where } \beta = \frac{EJ}{m}. \tag{11.92}$$

The solution for the first of this pair is just: $g = A\sin\lambda t + B\cos\lambda t$, and we should recognize that what we have here is an eigenvalue–eigenfunction problem. Furthermore, to obtain this result we have assumed that there can be no negative values for λ. To solve this problem we would need to find the eigenfunctions (the f's) for the sequence of identified λ's that satisfy the necessary conditions at the ends of the rod. We will leave this problem for the end of the chapter as a student exercise. What we want to do now, though, is to transition from the transverse displacement of a loaded strand of spaghetti, to a much more fundamental question. If we need to use a column as a building support, what kind of column will give us the maximum strength while minimizing the amount of material required? In other words, is there an optimal column *shape*?

11.6.2 The Optimal Column Shape

Let us consider the design of a supporting column subjected to a vertical (axial) load; such columns were a staple of classical architecture (think of the Parthenon in Athens). If such a column is subjected to an extreme compressive load it will fail or *buckle*. Structural engineers use Euler's formula to estimate the critical buckling load,

$$F_c = \frac{EJ\pi^2}{L^2}, \tag{11.93}$$

where E is Young's modulus of elasticity, J is the moment of inertia of the cross-section, and L is the column length. Typically, the modulus of elasticity for carbon steel is about 30×10^6 psi, for concrete it is about 3×10^6 psi, and for Teflon™ about 75,000 psi. The second moment of an area A (moment of inertia) about the x-axis is $J = \int_A y^2 dA$. The load that a column can carry also depends upon how it is secured at the ends (at $x = 0$ and $x = L$); as we noted above, columns are commonly *free*, *hinged*, or *clamped* at the ends. These boundary conditions play a critical role in determining the form of the first buckling mode. Letting y represent the transverse displacement, the simple linear model for the buckling of a long, straight column is

$$\frac{d^2y}{dx^2} + \frac{F}{EJ}y = 0, \tag{11.94}$$

and the solution for eq. (11.94) is just: $y = A\sin\sqrt{\beta}x + B\cos\sqrt{\beta}x$, where $\beta = \frac{F}{EJ}$. If $y = 0$ at both ends, then $B = 0$ and $\sqrt{\beta} = \frac{n\pi}{L}$; i.e., the shape of the buckling mode is revealed and the eigenvalues yield $F = EJ\frac{n^2\pi^2}{L^2}$. Such an analysis works well enough for *slender* columns (for steel columns, *slender* means $L/R \ge 140$), but it fails to address the more complete design question, namely: *If the total volume of material and the column length are fixed, what shape or profile can support the greatest*

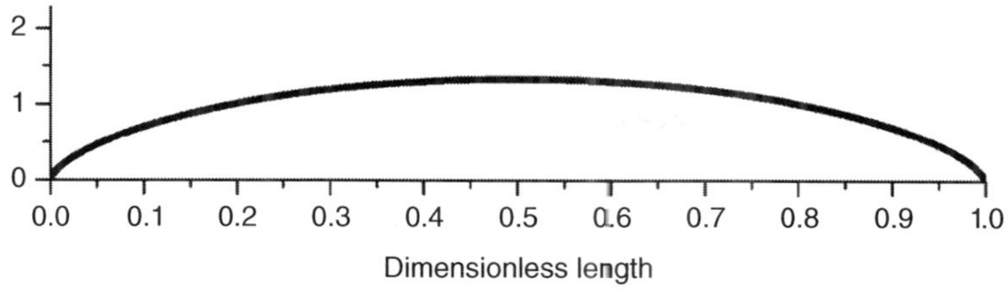

Figure 11.7 Optimal column shape if both ends are hinged (profile shown in horizontal orientation) as determined by Keller's (1960) solution.

load? In other words, we want to evaluate columns where the cross-sectional area, A, can vary with length but with the requirement that $V = \int_0^L A(x)dx$. We should think of this enterprise as finding the $A(x)$ that meets this volume requirement and *maximizes* the smallest eigenvalue of F.

In the eighteenth century Euler worked on the problem of finding the form of $y(x)$ that would lead to minimum stored energy. With the assumption that E and J were constant, Euler was able to minimize

$$\int_0^L \frac{EJ|y''|^2}{(1+|y'|^2)^{9/4}}dx - \lambda \int_0^L (1+|y'|^2)^{1/2}dx. \tag{11.95}$$

The second part of this expression is the length constraint with the Lagrange multiplier, λ. Cox (1992) notes that Euler was also able to determine the critical load for certain non-uniform columns for which $J(x) = (a + bx/L)^q$. The exponent, $q = 2$, corresponds to cylindrical columns for which A varies linearly over the length, L (a tapered cylinder).

It turns out that the question of the strongest column shape (for fixed volume and length) has been revisited repeatedly over the 200 years that have passed since Lagrange "proved" that the answer was a right-circular cylinder. Many have suspected that Lagrange's cylinder was incorrect; Cox makes the very practical observation that an optimal column shape should have large A where bending might be expected, and reduced A where little bending would occur. Keller (1960) revisited the problem and found that if both ends of the column were hinged, the optimal shape is actually a "stunted" cycloid (a bit like a sausage that is overly plump in the middle). Because the column is hinged at the ends, A at both $x = 0$ and $x = L$ is reduced—that is, the column does not have to resist bending at the ends. Cox notes that this shape, illustrated in Figure 11.7, is stronger than a cylinder of the same length and volume by more than 30%. But what about a column that is clamped at both ends? Tadjbakhsh and Keller (1962) examined that case and found a solution for which $A(x) \to 0$ at two interior locations, $x = 1/4$ and $x = 3/4$. Of course, this means that $y''(x)$ has two singularities, and these two points can be thought of as interior hinges. For a time it was generally accepted that the strongest clamped–clamped column would have two interior locations where the cross-sectional area would vanish; however, Olhoff and Rasmussen (1977) discovered that the Tadjbakhsh–Keller solution was incorrect. In the earlier work $y'(x)$ was assumed to be continuous (it does not need to be). Moreover, when Olhoff and Rasmussen used the $A(x)$ profile from the 1962 paper in the conventional eigenvalue formulation they discovered that the critical compressive load was actually *less* than that for a uniform column.

Olhoff and Rasmussen (1977) developed the appropriate equations, recognizing that a bimodal formulation was necessary, and solved the problem numerically. They point out that the equation set provides a coupled, nonlinear, integro-differential eigenvalue problem. Their results showed that a column with clamped-end conditions would have two interior regions where the cross-section would diminish significantly but *A(x) would not approach zero*; these contractions occur at fractional length positions of 1/4 and 3/4 (an excellent illustration is provided by Cox (1992) as his Figure 5). Olhoff and Rasmussen found that a column thus designed is almost 33% stronger than a uniform (cylindrical) column of the same height and volume. Cox and Overton (1992) used non-smooth analysis to *prove* that A need not be continuous, but that it would have a lower (and non-zero) bound. Their results, which are also summarized by Cox, confirmed the numerical calculations of Olhoff and Rasmussen: The strongest clamped–clamped columns have *two* contractions, where A is reduced significantly relative to a uniform cylindrical column.

11.7 Systems with Surface Tension

Surface tension, σ, is a property of interfaces; it acts as a negative pressure in such a way as to minimize interfacial area. Work must be performed against surface tension in order to expand the interface and if the area is increased by an amount, dA, then the required work is $dW = \sigma dA$. Values for σ are usually reported as force per unit length, typically dynes/cm and for the air–water interface, $\sigma = 72$ dynes/cm. As we noted above, surface tension creates a pressure difference across a curved interface, and we can think of this in the following way: The tendency for a bubble to contract, that is, to decrease its surface area, must be countered at equilibrium by an increase in pressure on the concave (interior) side. For a spherical bubble the pressure difference is given by $\Delta p = \frac{2\sigma}{r}$ such that for a 0.5 mm diameter bubble in water, $\Delta p = \frac{(2)(72)}{(0.025)} = 5,760$ dynes/cm^2 (576 Pa), or 5.76×10^{-3} bar. Since the action of surface tension is to minimize surface area, it is clear that the situation is very similar to that posed in Section 11.4. Let us illustrate this point.

Suppose we take two circular loops of wire, separated by a distance $2h$, but fixed so that the two circles are parallel. We let the circles be located at $x = \pm h$ with $y^2 + z^2 = R^2$. Now, we dip this wire framework into a basin of soapy water and then remove it. What shape will the resulting film have? The surface we are describing is a surface of revolution created by rotating a curve, $f(x)$ about the x-axis such that $f(-h) = R$ and $f(+h) = R$. For this case,

$$I = 2\pi \int_{-h}^{+h} f\sqrt{1 + f'^2} dx, \tag{11.96}$$

and our task is to minimize I. This is equivalent to solving the Beltrami identity:

$$\frac{f'^2 f}{\sqrt{1 + f'^2}} - f\sqrt{1 + f'^2} = C, \tag{11.97}$$

or consolidating, $\frac{-f}{\sqrt{1+f'^2}} = C$. We can do a little rearranging and then integrate, yielding $\frac{1}{C}x + a_1 = \cosh^{-1}\left|\frac{f}{\alpha}\right|$, such that

$$f = C\cosh\left(\frac{x}{C} + a_1\right). \tag{11.98}$$

You will see immediately that this is the catenary discussed previously in this chapter. The validity of this result is demonstrated beautifully in the little book of interfacial experiments by C. V. Boys (*Soap Bubbles: Their Colours and the Forces which Mold Them*); see Figure 27 on page 55 of the Dover reprint of the "new and enlarged" edition of 1911. Boys used two large glass rings to draw out the soap film and he notes that "....the film is so far curved as to have a most elegant waist." He also points out that the surface thus formed is a *catenoid*—just as we demonstrated above. One might suspect that a limit exists as the soap film catenoid is drawn out by increased separation of the two circular rings—as the waist of catenoid pinches inward, an unsustainable condition is reached and the film ruptures leaving just the two circular films on the rings. Deison Préve has a very nice demonstration of this *minimum surface catenoid collapse* on YouTube; the video is a physical illustration of the *Goldschmidt discontinuous solution* referred to previously in this chapter in Section 11.4.

MacNeish (1906) points out that two types of solutions exist for the problem of minimizing a surface of revolution, catenaries and straight lines. The latter was first reported by the German mathematician C.W.B. Goldschmidt in 1831, and this discontinuity raises an intriguing question: Can a maximum separation of the two circular ends be identified? That is, at what distance may the two rings be separated before the catenary solution fails?

Before we attack the stability question, let us return to eq. (11.98) and note that the two constants, C and a_1, are determined by passing the catenary through the point pairs (x_1, y_1) and (x_2, y_2):

$$y_1 = C\cosh\left(\frac{x_1}{C} + a_1\right) \text{ and } y_2 = C\cosh\left(\frac{x_2}{C} + a_1\right).$$

So for example, if we use (1,3) and (3,1), we find (using the *TI-nspire*) $C = 0.9638$ and $a_1 = -2.8394$. The catenary thus defined exhibits a minimum for the range $1 \le x \le 3$, and you might take a moment to verify that it occurs at $x = 2.7366$. The really important observation here is that *we cannot find a catenary for every pair of points*; MacNeish identified the boundary in the *x-y* plane that separates catenary from discontinuous solutions.

We can effect a simplification by placing the origin equidistant from the two circular rings so that one circle is located at $-x_0$ and the other at $+x_0$. Because *cosh* is an even function, the equation for the minimum surface catenary becomes $y = C\cosh\left(\frac{x}{C}\right)$. The question we now consider concerns identifying the maximum value of the ratio, x/y, for which catenary solutions are possible. We follow the approach laid out by Eric Weisstein (*archiv.lib.msu.edu* 1996). Let

$$p = 1/C \text{ so that } yp = \cosh(px). \tag{11.99}$$

At the maximum permissible value for x, $\frac{dx}{dp} = 0$, and we let this maximum value of x be represented by x_m.

$$\frac{d}{dp}[yp = \cosh(px)] \text{ yielding } y = \sinh(px)\left[x + p\frac{dx}{dp}\right]. \tag{11.100}$$

At x_m, $dx/dp = 0$ so $y_m = x_m \sinh(px_m)$, and since $py_m = \cosh(px_m)$, we can divide the former by the latter obtaining

$$px_m = \coth(px_m). \tag{11.101}$$

A solution for eq. (11.101) can be found using the *TI-nspire*: $px_m = 1.19968$. Since $\frac{y_m}{x_m} = \sinh(px_m) = 1.5089$, we find $\frac{x_m}{y_m} = 0.6627$. Therefore, the *maximum distance between the two circular rings* for which catenoid solutions exist is 1.3254 times the radius (R) of the circular ends.

11.8 Less Familiar COV Applications

The COV has been used for some problems that a student new to COV might not expect. For example, Ewing (1969) described a simplified model of a rocket's trajectory where the question addressed concerned how the fuel load should be burned in order to achieve the maximum height. Because the amount of fuel required for rocket flight is large, the mass of the vehicle changes dramatically; for example, at liftoff a Saturn V (a three-stage rocket used for the Apollo lunar missions) weighed about 6.2 million pounds, but the combined empty weight of the three stages was only 425,000 lbs. The Space X Starship stack (fueled) weighs about 11 million pounds with the mass of the propellant and oxidizer exceeding 10 million pounds. If we write a highly simplified model incorporating thrust and gravity:

$$m(t)\frac{d^2z}{dt^2} + c\frac{dm}{dt} + m(t)g = 0, \tag{11.102}$$

and specify the initial and final masses, m_0 and m_f, what function $m(t)$ yields the maximum value of z? We assume that the flight profile is vertical and that the surface at $z = 0$ is flat. Therefore, our model is the Lagrange problem

$$J = \int_{t_0}^{t_f} \frac{dz}{dt}dt, \tag{11.103}$$

where our objective is to make the integral a global maximum. Ewing recommends setting $\phi = cln(m)$ so that

$$\frac{d\phi}{dt} = c\frac{dm/dt}{m}; \tag{11.104}$$

thus, if we let $v = \frac{dz}{dt}$, the force balance becomes

$$\frac{dv}{dt} + \frac{d\phi}{dt} + g = 0. \tag{11.105}$$

One of the main difficulties here concerns $d\phi/dt$; this derivative is proportional to the thrust per unit mass, and it must be allowed to have discontinuities. Furthermore, since m is decreasing throughout the fuel burn, $d\phi/dt$ must range between some large negative value and zero:

$$-a \le \frac{d\phi}{dt} \le 0. \tag{11.106}$$

Solution is facilitated by the introduction of an auxiliary variable, ψ, and the condition immediately above is recast as:

$$a\left(\sin\frac{d\psi}{dt} - 1\right) - 2\frac{d\phi}{dt} = 0. \tag{11.107}$$

Ewing observes that if the thrust per unit mass is continuous, then this COV problem has no solution. Proceeding to the solution is formidable, and the reader interested in pursuing it should consult Section 5.12 in Ewing's book. The maximum height achieved by the rocket is given by

$$z_{\max} = (2a - g)\frac{at^2}{g}. \tag{11.108}$$

Therefore, if $a = 50$ and burnout is attained after 25 s, then

$$z_{\max} = (100 - 32.17)\frac{50(25^2)}{32.17} = 65{,}890 \text{ ft}. \tag{11.109}$$

Lefkowitz and Eckman (Application and Analysis of a Computer Control System, *Trans. ASME, series D,* 81 :569, 1959) studied a batch chemical process in which a sequence of chemical reactions occurred:

$$\mathrm{A + H \rightarrow B} \text{ and } \mathrm{B + H \rightarrow C}. \tag{11.110}$$

The kinetic descriptions, in terms of mole fractions, are:

$$\frac{dx_A}{dt} = -k_1 x_A \text{ and } \frac{dx_B}{dt} = k_1 x_A - k_2 x_B. \tag{11.111}$$

Of course, the sum of the mole fractions is 1 (one): $x_A + x_B + x_C = 1$. Normally we would expect the rate constants to vary according to the Arrhenius law; however in this case, although the temperature is constant, *the pressure can be manipulated* to effect changes in k_1 and k_2:

$$k_1 = a_1 p^{n_1} \text{ and } k_2 = a_2 p^{n_2}. \tag{11.112}$$

In a process like this we may have several objectives; for example, we might seek to maximize the production of the desired product "B" and minimize formation of the less desirable "C." Alternatively, we might want to minimize processing time to facilitate turnaround and maximize the number of batches prepared per day. One might even presume that both—maximize "B" and minimize processing time—would be in the best interest of the operation. Let us begin by exploring this problem for the case in which *pressure is constant*. We will take the *initial* mole fraction for "A" to be 1 so that $x_B(0) = x_C(0) = 0$. We expect the concentration of "B" to rise to some maximum value and then gradually decline with the additional formation of "C." If we let the process continue indefinitely we will of course end up with all "C." It is apparent from the kinetic descriptions of the process that ideally we would like k_1 to be very large and k_2 to be small—this would accelerate the formation of "B" and diminish the rate of formation of "C." Characteristic behavior for this batch process (employing constant values for all parameters) is illustrated below.

Now, let us *assume* that $x_B = 0.767$ is the desired terminal composition for this batch process. Since the rate constants are pressure dependent, it is likely that we could adjust the pressure dynamically and achieve this terminal point in less time. Since there are an infinite number of possible pressure "pathways" for this chemical process, the problem is to choose the optimal $p(t)$ that yields the desired mole fraction x_B in the minimum time. As we pointed out previously, any operational change that increases the discrepancy between k_1 and k_2 could be advantageous.

Lefkowitz and Eckman approached the general problem by setting

$$v = \ln\left(\frac{x_{A0}}{x_A}\right), u = x_B/x_A, \text{ and } m = 1 - k_2/k_1. \tag{11.113}$$

These substitutions can be introduced into the rate equations and the time variable can be removed from the problem yielding:

$$\frac{du}{dv} = mu + 1. \tag{11.114}$$

If m is constant (the pressure does not vary), then this ODE can be integrated,

$$\frac{1}{m}\ln(mu+1) = v + c_1 \text{ so that } u(v) = \frac{c_1\exp(mv) - 1}{m}, \tag{11.115}$$

and we can think of $u(v)$ as the pathway for the batch process. For the example illustrated by Figure 11.8, the initial values for x_A and x_B were 1 and 0, respectively, and the terminal values of the mole fractions were $x_A = 0.081$, $x_B = 0.767$, and $x_C = 0.152$. The rate constants were chosen such that $m = 1 - 0.1 = 0.9$, and the final values for u and v were 9.457 and 2.512; the table below summarizes the operating pathway for this isobaric process:

v	0	0.5	1.0	1.5	2.0	2.5	2.512
u	0	0.631	1.621	3.175	5.611	9.431	9.457

Of course, our objective is to minimize the time required for the batch operation by changing p dynamically. The total time for the batch operation is given by the integral:

$$t_{\text{batch}} = \alpha \int_0^{\ln\left(\frac{x_{Ao}}{x_{Af}}\right)} (1-m)^{1-\beta} dv \tag{11.116}$$

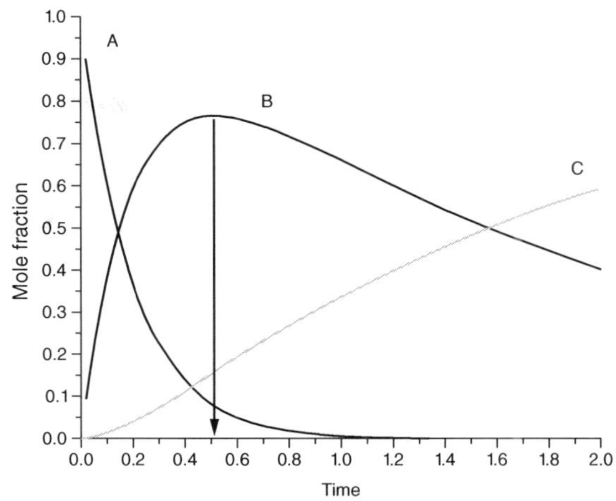

Figure 11.8 Typical behavior of the batch chemical process in which "B" is the desired product. The maximum concentration of "B" occurs at about $t = 0.5$ with $x_B \approx 0.767$ for the chosen parameter values.

where

$$\alpha = a_2^{\beta n_1/n_2}/a_1^{\beta} \text{ and } \beta = \frac{n_2}{n_2 - n_1}. \tag{11.117}$$

This Lagrange problem has a side condition, eq. (11.114), and the terminal conditions, $v_0 = 0$, $v_f = \ln(x_{A0}/x_{Af})$, $u(v_0)$ and $u(v_f)$. The Euler–Lagrange equation provides the necessary condition for minimizing the integral (11.116), and the optimal relationship between u and v that results is

$$\frac{du}{dv} = \frac{1-m}{\beta u}. \tag{11.118}$$

With specified initial and final conditions (see the table for constant p operation), we have the two eqs. (11.114) and (11.118) required for determination of $u(v)$ and $m(v)$. Unfortunately, Ewing (1969) notes that no elementary solution is known for this two-equation problem, so a solution must be sought numerically for each specific case.

For the example problem presented by Lefkowitz and Eckman, $0 \le v \le 1.6$, $u(0) = 0.6$, and the initial value for m was 0.5. We can verify their results numerically by obtaining the simultaneous solution of eqs. (11.114) and (11.118):

v	u	m	k_2/k_1
0	0.6	0.5	0.5
0.1	0.735	0.553	0.447
0.2	0.882	0.593	0.407
0.4	1.212	0.648	0.352
0.6	1.599	0.684	0.316
0.8	2.052	0.709	0.291
1.0	2.585	0.728	0.272
1.2	3.209	0.742	0.258
1.4	3.942	0.752	0.248
1.6	4.801	0.761	0.239

Therefore, the minimum processing time for the specified parameters will require adjusting the pressure in the batch vessel to provide the rate constant ratio shown in the far-right column since

$$\frac{k_2}{k_1} = \frac{a_2}{a_1}p^{n_2-n_1} \text{ and therefore } p = \left[\frac{a_1}{a_2}\frac{k_2}{k_1}\right]^{\frac{1}{n_2-n_1}}. \tag{11.119}$$

Processing time can be determined from these data by integration:

$$t = \alpha \int_0^v (1-m)^{1-\beta} dv. \tag{11.120}$$

11.9 The Connection Between COV and the Finite Element Method

The finite element method (FEM) as it is used in engineering applications had its origin in the analysis of structures. In particular, FEM developed in response to structural problems that involved a continuum (as opposed to those, for example, that concerned a truss with a finite number of beams and connecting points). A. Hrenikoff (Solution of Problems in Elasticity by the Framework Method, *Journal of Applied Mechanics*, Volume 8, 1941) proposed discretizing the continuum by dividing it into a finite number of elements. Initially many applications of FEM were focused upon structures, but now the technique is employed by virtually all branches of science working with continuum mechanics. The COMSOL™ website is an ideal place for the reader to get a sense of the breadth of these uses.

So what are the essential characteristics of the FEM? In short, FEM is a discretization process in which a governing equation(s) with infinite degrees of freedom is converted to an approximation with finite degrees of freedom. The region of interest is subdivided into cells (or elements) and the dependent variable(s)—say ψ_i—in that element is represented by an approximation with unknown parameters; the solution for the set of unknown parameters usually entails simultaneous linear algebraic equations. Of course, we must have continuity of the dependent variable across element boundaries.

Three FEM categories have been identified by Norrie and de Vries (1973): *nodal*, *coefficient*, and *cell*. In the nodal approach the unknown parameters are the values of ψ_i at the cell nodes. For the coefficient method the unknowns are the values of coefficients in the functional approximation, and in the cell method the parameters are average values of ψ_i along the faces. For a problem with n-independent variables, the number of required nodes per cell is $n+1$; for example, for an equilibrium problem in two dimensions (in the x-y plane), $n = 2+1 = 3$, and the elements will be triangular. For steady-state problems in three dimensions $n+1 = 4$, so the cells will be tetrahedrons.

Nodal FEMs are probably the most commonly used, and there are at least three different approaches as to how the simultaneous equations are developed:

i) A variational principle is used to obtain a set of equations written in terms of the nodal values. Of course, this means that the variational principle must either be known for the particular problem or derived; that is, the desired solution must yield an extremum value for some functional, $I(\psi)$.
ii) The residual (R)—the difference between the actual solution and the approximation—is required to be small in some sense, producing the necessary equations. You may recall that we discussed using the residual previously in efforts to find approximate solutions for differential equations.
iii) The approximate solution of the problem, written in terms of the trial functions and the nodal values for each cell, is substituted directly into the governing equation.

Generally speaking, we are trying to approximate the solution for our dependent variable, ψ, with an equation of the form:

$$\psi_n \cong \sum_{i=1}^{n} C_i \phi_i,$$

where ϕ_i are the trial functions and C_i are the unknown parameters.

Previously in this chapter we used the Ritz method to find an approximate solution for a boundary value problem, and you will recall that the trial function was valid over the *entire* domain (from $x = 0$ to $x = 1$). This is a distinguishing characteristic with regard to the FEM where the trial functions (or interpolating functions) are valid only in a piecewise manner. Our intent with FEM is to break the domain into a collection of pieces and then use the trial function(s) to represent the solution over each piece separately. So an important difference between the Ritz method and FEM is now clear—in FEM the trial function does *not* have to satisfy the boundary conditions.

Suppose, for example, that we have a continuous function in the x-y plane: $\psi = f(x, y)$. Our goal is to represent this function in an approximate fashion over the entire domain, but do so with a collection of small pieces. We will use triangular elements and, for the moment, assume that the field variable is known at the three vertices, which we will label 1, 2, and 3. The simplest feasible representation (simplest in the sense that it allows for the linear variation of ψ with changes in the two independent variables) for ψ will be taken as:

$$\psi \cong C_1 + C_2 x + C_3 y. \tag{11.121}$$

You can see immediately that the problem confronting us is one of choosing the "best" values for C_1 through C_3 for our small triangular piece. We will place the vertices at (1,1), (2,1), and (2,2) and we observe that the area of this triangle is ½. We will also arbitrarily take $\psi = x^2y^2$ such that the node values for the dependent variable will be 1, 4, and 16, respectively.

Since we know that value of the field variable at each of the vertices (or nodes), we write:

$$\psi_1 = 1 = C_1 + C_2(1) + C_3(1), \tag{11.122}$$

$$\psi_2 = 4 = C_1 + C_2(2) + C_3(1), \tag{11.123}$$

and

$$\psi_3 = 16 = C_1 + C_2(2) + C_3(2). \tag{11.124}$$

We use these three equations, solving them simultaneo.6usly, to determine the unknown parameters. The result is: $C_1 = -14$, $C_2 = 3$, and $C_3 = 12$; these choices provide the correct values at the vertices (or nodes) as we required. But what about the value of ψ at other points *within* the triangular element? Our approximation is $\psi \cong -14 + 3x + 12y$, so obtaining a comparison should be easy.

x	y	$\Psi_{\text{estimated}}$	Ψ_{actual}
1.5	1.0	2.5	2.25
1.5	1.25	5.5	3.5156
1.5	1.5	8.5	5.0625
1.75	1.0	3.25	3.0625
1.75	1.25	6.25	4.7852
1.75	1.5	9.25	6.8906
2.0	1.5	10.0	9.0
2.0	1.75	13.0	12.25

Although the values for ψ are correct at the vertices, some of the estimated values in the interior of the triangular element are in error by as much as 60% (the average error for the points tested in the table above is 27.6%). This performance is not acceptable, and we have two options: We could increase the degree of the trial function polynomial (we could make it quadratic with respect to x and y, for example), or we could refine (subdivide) the triangular element. In the case of the latter and using our example immediately above, we simply draw a line from (1,2) to (1.5,1.5), producing *two* triangles of equal size. Naturally we have doubled our workload as we must now determine values for two sets of parameters, but this is of little consequence for an automated calculation. If necessary, we can continue subdividing the element until the linear trial (or interpolation) function gives satisfactory agreement over each element. When we are solving elliptic partial differential equations (PDEs) it is usually obvious where the steepest gradients will be located, so the analyst often has a very good idea where the element size might need to be reduced.

We determined the three needed parameters (the C_n's) from eqs. (11.122)–(11.124) by Gaussian elimination, but equivalently,

$$C_1 = \frac{\psi_1(x_2y_3 - y_2x_3) + \psi_2(x_3y_1 - x_1y_3) + \psi_3(x_1y_2 - x_2y_1)}{2A}, \tag{11.125}$$

$$C_2 = \frac{\psi_1(y_2 - y_3) + \psi_2(y_3 - y_1) + \psi_3(y_1 - y_2)}{2A}, \tag{11.126}$$

and

$$C_3 = \frac{\psi_1(x_3 - x_2) + \psi_2(x_1 - x_3) + \psi_3(x_2 - x_1)}{2A}. \tag{11.127}$$

The subscripts 1, 2, and 3 refer to the three vertices, and we are proceeding from the (1,1) vertex in the counter-clockwise sense; therefore, vertex 2 is located at (2,1) and vertex 3 is at point (2,2). A is the area of the triangular element, $\frac{1}{2}bh = \frac{1}{2}(1)(1) = \frac{1}{2}$, so the denominator in each (11.125), (11.126), and (11.127) is 1 (one). We will introduce (11.125), (11.126), and (11.127) into eq. (11.121) and write the result in general form:

$$\psi(x,y) = \frac{a_i + b_ix + c_iy}{2A}\psi_i + \frac{a_j + b_jx + c_jy}{2A}\psi_j + \frac{a_k + b_kx + c_ky}{2A}\psi_k \tag{11.128}$$

where

$$a_i = x_j y_k - x_k y_j, b_i = y_j - y_k, \text{ and } c_i = x_k - x_j. \tag{11.129}$$

Following Huebner (1975) we define the linear *shape functions* as:

$$N_n = \frac{a_n + b_n x + c_n y}{2A}. \tag{11.130}$$

Therefore for *each* triangular element we have three vertex values and three shape functions:

$$\psi = \begin{matrix} \psi_i \\ \psi_j \\ \psi_k \end{matrix} \text{ and } N = [N_i, N_j, N_k], \tag{11.131}$$

and thus: $\psi(x, y) = N_i\psi_i + N_j\psi_j + N_k\psi_k$. Our approximation for the variation of the field variable over the complete domain is then

$$\psi(x, y) = \sum_{s=1}^{M} [N^{(s)}]\{\psi^{(s)}\}, \tag{11.132}$$

where M is the total number of triangular elements. The remaining task is to identify the nodal (vertex) values of the dependent (or field) variable, ψ, so that the functional $I(\psi)$ is an extremum, where

$$I(\psi) = \sum_{s=1}^{M} I^{(e)}\psi^{(e)}. \tag{11.133}$$

The first variation of the functional, I, must be zero, and therefore for each element, e, we have

$$\frac{\partial I^{(e)}}{\partial \psi_j} = 0, \tag{11.134}$$

and the j-index refers to the nodes (or vertices) of each element. Equation (11.134) provides a set of simultaneous equations that, when solved, yield the problem's solution in terms of the nodal values of ψ.

Now let's apply the *nodal* FEM to the ODE,

$$\frac{d^2y}{dx^2} - y = 0 \text{ for } 0 \le x \le 2, \tag{11.135}$$

with $y(0) = 1$ and $y(2) = 7.389$. Since this ODE is easy to solve, we will have a convenient comparison:

$$(D^2 - 1)y = (D + 1)(D - 1)y = 0 \text{ and therefore}$$

$$y = C_1 \exp(-x) + C_2 \exp(+x). \tag{11.136}$$

From the boundary conditions we find $C_1 = 0$ and $C_2 = 1$, so $y = e^x$. We will divide the region into four finite elements with five nodes; node 1 corresponds to $x = 0$ where $y = 1$. Assume the dependent variable, y, is to be approximated over each cell by the linear equation:

$$\hat{y} = a_n + a_m x. \tag{11.137}$$

Let the first two *nodal* values for y be represented by y_1 and y_2 at the corresponding positions, x_1 and x_2:

$$y_1 = a_{11} + a_{12}x_1 \text{ and } y_2 = a_{11} + a_{12}x_2. \tag{11.138}$$

We subtract the second of the pair from the first to find the slope of the approximating line,

$$a_{12} = \frac{y_1 - y_2}{x_1 - x_2}, \text{ and then the intercept: } a_{11} = y_2 - a_{12}x_2. \tag{11.139}$$

Taking these back to our linear approximation and writing the result for any element, e,

$$\hat{y}_e = y_{e+1}\left[1 - \frac{(x - x_{e+1})}{(x_e - x_{e+1})}\right] + y_e\frac{(x - x_{e+1})}{(x_e - x_{e+1})}. \tag{11.140}$$

Note that the right-hand side of this equation involves the independent variable, x, as well as the *locations* and *values* for the nodes. The values of the dependent variable, y, at the nodes, e and e+1, are multiplied by the *shape* functions. We are employing four elements or cells, so 2/4 = ½ and $x_e - x_{e+1} = -$ ½; thus,

$$\hat{y}_e = -2(x - x_{e+1})y_e + (1 + 2x - 2x_{e+1})y_{e+1}. \tag{11.141}$$

Of course, this is a *nodal* FEM method, so the unknowns in this equation are the values of y at the interior nodes. We can write down the trial function for each of the four elements:

$$y_{e1} = -2\left(x - \frac{1}{2}\right)y_1 + (1 + 2x - 1)y_2$$

$$y_{e2} = -2(x-1)y_2 + (1 + 2x - 2)y_3$$

$$y_{e3} = -2\left(x - \frac{3}{2}\right)y_3 + (1 + 2x - 3)y_4$$

$$y_{e4} = -2(x-2)y_4 + (1 + 2x - 4)y_5. \qquad (11.142)$$

As noted above, the interior nodal values, y_2, y_3, and y_4 are unknown. We should check the boundary values for y_{e1} and y_{e4} for verification; if $x = 0$ we get $y_{e1} = y_1$, and if $x = 2$ we find $y_{e4} = y_5$, as we should. The sum of these four element equations represents the trial solution over the domain. Now we must find a functional that satisfies Euler's equation—the necessary condition for the existence of an extremum. You will recall that Euler's equation appeared in Section 11.2:

$$y''\frac{\partial^2 F}{\partial y'^2} + y'\frac{\partial^2 F}{\partial y \partial y'} + \frac{\partial^2 F}{\partial x \partial y'} - \frac{\partial F}{\partial y} = 0.$$

Next, we note that *if* $F = y^2 + (y')^2$, then Euler's equation yields $2y'' - 2y = 0$, which—after dividing by 2—is the ODE we are trying to solve. Therefore, the solution $y(x)$ we are seeking must be an extremum of

$$I = \int_0^2 [y^2 + y'^2]dx. \qquad (11.143)$$

This is the variational principle that is at the heart of our solution procedure and herein lies the problem with this approach to *nodal* FEM: *we must be able to either find or deduce the appropriate variational principle to proceed.* For many problems of interest to engineers the correct form for eq. (11.143) may be unknown or undiscoverable. In such cases the analyst might choose either the *residual* or *direct* methods for *nodal* FEM. In the residual approach we require the difference between our approximation and the exact solution to be small in some sense. For the direct method, the trial function is substituted directly into the differential equation that is to be solved.

Since we were able to identify a variational principle for our example, we can continue: We introduce the trial function into eq. (11.143) and perform the integration; this can be done for each individual element since the limits of integration have been specified, e.g., 0 to ½, ½ to 1, and so on. For the first element, the integration from $x = 0$ to $x = 1/2$ yields

$$I_{e1} = 2.1667y_1^2 - 3.8333y_1y_2 + 2.1667y_2^2, \qquad (11.144)$$

and then the process is repeated for elements 2, 3, and 4, resulting in

$$I = I_{e1} + I_{e2} + I_{e3} + I_{e4}. \qquad (11.145)$$

The three unknowns are the *interior* nodal values, so in order for the functional to be stationary it is necessary that

$$\frac{\partial I}{\partial y_2} = \frac{\partial I}{\partial y_3} = \frac{\partial I}{\partial y_4} = 0, \qquad (11.146)$$

and this produces the three simultaneous equations that we must solve to complete the example. This process is left to the reader as an exercise; see if you can verify that $y_2 = 1.635$ and $y_4 = 4.461$. A comparison of the nodal values for our FEM solution with the analytic values is provided here:

x	y_{FEM}	y_{analyt}
0	1	1
0.5	1.635	1.649
1.0	2.696	2.718
1.5	4.461	4.482
2.0	7.389	7.389

At the mid-point the error is about 0.81%, so our coarse discretization has worked surprisingly well. Some observations are in order here at the conclusion of this example:

i) Four elements are sufficient to produce a reasonable solution for this ODE—of course the ODE was not at all66 problematic.
ii) Attacking a problem of this type from scratch using the nodal FEM is not a trivial undertaking.
iii) The use of a variational principle was feasible in this case only because we were able to identify the appropriate F from the Euler equation. For many—perhaps most—important engineering problems this will not be possible. For these reasons, *residual* or *direct* FEM approaches are more likely to be useful to the analyst tasked with solving either a new, an unfamiliar, or an atypical problem.

Next, we explore use of this method for an elementary problem in steady-state heat conduction in two dimensions, where the governing equation is

$$\frac{\partial^2 T}{\partial x^2} + \frac{\partial^2 T}{\partial y^2} = 0. \tag{11.147}$$

We will use a single triangular element so that the process is completely transparent with a linear (with respect to x and y) trial function. Let the vertices of the triangle be located at (x,y) positions, (0,0), (2,1), and (1,2). We let the top correspond to vertex 1 where the temperature is 100°, so $T(x = 1, y = 2) = 100$. Both bottom vertices, and the bottom edge, are maintained at 50°. Now we proceed in clockwise fashion from vertex 1, writing the three equations:

$$a_1 + b_1(1) + c_1(2) = 1 \quad a_1 + b_1(2) = 0 \quad a_1 = 0. \tag{11.148}$$

For vertex 2:

$$a_2 + b_2(2) + 0 = 1 \quad a_2 = 0 \quad b_2 + 2c_2 = 0. \tag{11.149}$$

And for vertex 3:

$$a_3 = 1 \quad a_3 + b_3 + c_3(2) = 0 \quad a_3 + b_3(2) = 0. \tag{11.150}$$

In each case you will note that the equation for the "home" vertex is set equal to one (1), and it is zero at the other two vertices. Thus for the first set of equations where $(x = 1, y = 2)$ we find $c_1 = \frac{1}{2}$, for the second set $(x = 2, y = 0)$ we find $b_2 = \frac{1}{2}$ and $c_2 = -\frac{1}{4}$. The reader should verify that for vertex three (3): $a_3 = 1$, $b_3 = -\frac{1}{2}$, and $c_3 = -\frac{1}{4}$. We can now write out our approximation:

$$T \approx 100\left(\frac{1}{2}y\right) + 50\left(\frac{1}{2}x - \frac{1}{4}y\right) + 50\left(1 - \frac{1}{2}x - \frac{1}{4}y\right). \tag{11.151}$$

Looking at a few selected points, we obtain: $T(x = 1, y = 1) = 75°$, $T(x = ½, y = ½) = 62.5°$, $T(x = 2/3, y = 4/3) = 83.33°$, $T(x = 5/3, y = 2/3) = 66.67°$, and $T(0 \le x \le 2, y = 0) = 50°$. Should we wish to improve the quality of our estimated solution, we could place an inverted triangle inside the original domain with vertices at (1,0), (1.5,1), and (0.5,1); the result is four equal-sized triangles inside the original figure. If we subsequently divided each of those four triangles into four equal pieces using the same procedure, we would end up with 16 elements—all identical triangles. By refining the mesh we can improve the quality of the solution at the cost of additional simultaneous equations (to be solved).

There are *many* examples in the literature in which the FEM is applied to elliptic PDEs; de Vries and Norrie (1971) is especially useful as they illustrate how the variational problem is posed for the Laplace equation (with emphasis upon potential flow) for different types of boundary conditions. There are also many examples of simple FEM codes for readers wishing to start from the ground up; examples are included in Huebner (1975) and Reddy and Gartling (1994). For complex problems requiring FEM—particularly if the analyst has time constraints—commercial codes are probably worth the cost of acquisition. One such program that is very well known is COMSOL™ which was originally marketed as FEMLAB™. This software was designed to find solutions for multiphysics problems such as buoyancy-driven flows arising in free convection. The user interface allows one to rapidly explore the effects of changing the boundary conditions, refining the mesh, etc. Such programs can be used to solve quite difficult problems with complicated boundaries; an example (viscous flow around a rectangular obstruction) computed with COMSOL is shown in Figure 11.9 (a and b). Please keep in mind that one should fully test some benchmark cases related to any problem of interest before relying upon a solution produced by such software.

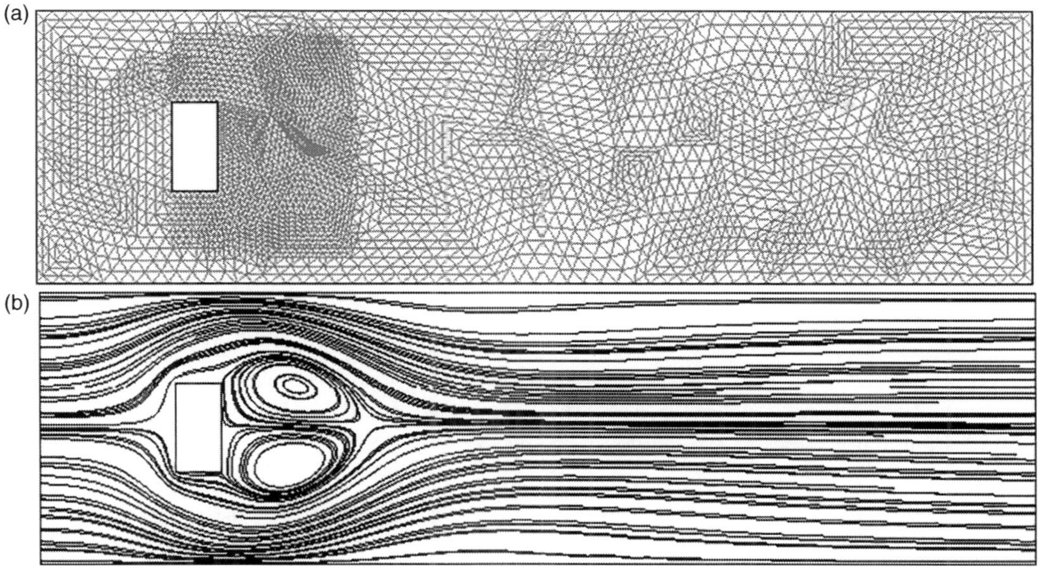

Figure 11.9 (a and b) Flow around a rectangular obstruction at low Reynolds number. The discretization (triangular elements) are shown at the top (11.9a) and the streamlines are shown in 11.9b. The inflow boundary condition (left-edge) was set to constant velocity and the horizontal boundaries had the slip/symmetry condition applied. Notice how the element size was reduced near the obstruction. These results were obtained with the FEM using COMSOL™.

11.10 Conclusion

The COV is extremely useful for problems in which an *extremal* is being sought and the *functional* is simple enough to be dealt with analytically. Unfortunately, in practice this often turns out to be quite restrictive because for a problem of interest: 1) the variational principle may not be known, 2) the existence of an extremum may not be guaranteed, 3) solution for a properly posed problem may not even be possible, and 4) the problem may be highly nonlinear such that identifying the minimizing sequence requires solution of large sets of nonlinear algebraic equations. To illustrate these points, consider the integral,

$$I = \int_{-1}^{+1} x^4 \left(\frac{dy}{dx}\right)^2 dx \tag{11.152}$$

Now suppose we wanted to identify a continuous function, $y(x)$, with a piecewise continuous derivative, that would minimize (11.152); we require that $y(-1) = -1$ and that $y(+1) = +1$. Courant and Hilbert (1989) indicate that although the integral can be made small, no appropriate function $y(x)$ can be identified that would cause the variation of (11.152) to vanish.

Nevertheless, *COV* is an important tool of historical significance that can be quite valuable in the right context; as we have seen, for example, *COV* ideas are used routinely in the application of the FEM to the solution of PDEs. And FEM has emerged as an invaluable asset for the analysis of problems in acoustics, electromagnetics, fluid dynamics, geophysics, heat transfer, mass transfer, optics, quantum mechanics, structural mechanics, wave propagation, etc. The interested reader can get a sense of the breadth of possibilities by reviewing the proceedings from any of the recent COMSOL™ conferences.

Finally, we want to emphasize that for an individual analyst developing a broadly capable FEM program is a formidable undertaking—particularly if starting from scratch. However, there are numerous options available for multiphysics problems including both commercial and open-source software. For the latter, options include *Elmer FEM*, *FEniCS Project*, and *FreeFEM*; both the *FEniCS Project* and *FreeFEM* have examples and tutorials available online to allow a novice user to quickly grasp how important practical problems can be solved. There is no doubt that commercial software packages like COMSOL™ are easier to use, but some of the open-source alternatives can provide comparable solutions for difficult problems. The caveat is that the learning curve is certain to be both steeper and longer for the free software.

Problems

11.1 In *Variational Methods for the Solution of Problems of Equilibrium and Vibration* (*Bull. Am. Math. Soc.*, 49 :2165, 1943) R. Courant points out that the equivalence between boundary value problems involving PDEs and the COV has been studied for a very long time. He notes that both Lord Rayleigh and Walther Ritz suggested the variational approach to such problems might be reduced to a much simpler extremum problem in which the number of undetermined parameters would be manageably finite. In part II of his paper Courant describes how one goes about solving a variational problem numerically by constructing a minimizing sequence (this is what is often referred to as the Rayleigh–Ritz method, as we described previously). He points out that the convergence of this process can be improved by adding a "sensitizing" functional to I. Read Courant's paper and determine how the addition of the "sensitizing" term will impact the nature of the simultaneous equations that must be solved. Courant suggests that an important objection to the Rayleigh–Ritz method is the difficulty the analyst has in assessing the accuracy of the approximation. Explore this topic and determine if substantive progress has been made in the years that have passed since Courant's original presentation in Washington in the spring of 1941.

11.2 In Section 11.5 we discussed the Rayleigh–Ritz method applied to Sturm–Liouville problems. One of the examples presented was the elementary second-order ODE, $y'' + \lambda y = 0$ with $y(0) = 0$ and $y(1) = 0$. Here we consider a variation of that problem but with adjustments to make it a little easier to obtain the correct result. Once again the ODE is:

$$\phi'' + \lambda\phi = 0$$

but with $\phi(x = 0) = 1$ and $\phi\left(x = \frac{\pi}{3}\right) = 0$. We *speculate* that an *even* function might be appropriate here, so we let

$$\phi \approx a_0 + a_2x^2 + a_4x^4.$$

Of course, we know immediately that $a_0 = 1$. Furthermore, since $\phi\left(x = \frac{\pi}{3}\right) = 0$, we have a simple relationship between a_2 and a_4. Using

$$\lambda = \frac{\int_0^{\pi/3} \phi'^2 dx}{\int_0^{\pi/3} \phi^2 dx},$$

find the smallest λ and the values for a_2 and a_4 that produce it.

11.3 The variational principle for the Laplace equation was given earlier in this chapter in Section 11.4. Suppose we wanted to explore the finite element solution of a two-dimensional, steady conduction problem in an *anisotropic* medium that includes a thermal energy source term, that is,

$$\frac{\partial}{\partial x}\left(k_x\frac{\partial T}{\partial x}\right) + \frac{\partial}{\partial y}\left(k_y\frac{\partial T}{\partial y}\right) + S = 0,$$

where S might be a function of position, $S = f(x,y)$. We approximate the dependence of T upon position with

$$T(x,y) \approx \sum_{j=1}^{n} T_j^e \phi_j^e(x,y),$$

where the T_j^e's are the node values of T and the ϕ_j^e's are the approximating functions for each finite element. For a weighted-residual approach to this problem, we could write

$$0 = \iint W\left[\frac{\partial}{\partial x}\left(k_x\frac{\partial T}{\partial x}\right) + \frac{\partial}{\partial y}\left(k_y\frac{\partial T}{\partial y}\right) + f(x,y)\right] dxdy.$$

The T's that appear here are the *approximations* of course. Our intent is to use this weighted-residual form to generate a set of simultaneous algebraic equations that, when solved, will yield the approximate nodal values for T. You will notice, however, that we have said nothing about the boundary conditions that must accompany the original PDE. Investigate how *both* Dirichlet- and Robin's-type boundary conditions are implemented in this weighted-residuals treatment of this conduction problem. You will find Chapter 2 in Reddy and Gartling (1994) to be helpful in this exercise.

11.4 Courant and Hilbert (1989) describe other direct means by which a minimizing sequence might be determined. Suppose we have

$$I = \int_0^b \int_0^a \left[\left(\frac{\partial \phi}{\partial x} \right)^2 + \left(\frac{\partial \phi}{\partial y} \right)^2 \right] dxdy,$$

which is to be minimized subject to the condition,

$$1 = \int_0^b \int_0^a \phi^2 dxdy.$$

The allowable comparison functions must disappear on the rectangular boundary, of course. Suppose we represent ϕ using a Fourier series:

$$\phi \cong \sum_{m,n=1}^{\infty} B_{mn} \sin \frac{m\pi x}{a} \sin \frac{n\pi y}{b}.$$

The task now is to identify the coefficients, B_{mn}'s, that produce the minimum of I. Show that

$$I = \frac{ab\pi^2}{4} \sum_{m,n=1}^{\infty} B_{mn}^2 \left(\frac{m^2}{a^2} + \frac{n^2}{b^2} \right),$$

then use the condition that $1 = \int_0^b \int_0^a \phi^2 dxdy$ to show that the only non-zero B_{mn} is B_{11}, which is equal to $\frac{2}{\sqrt{ab}}$. Finally, show that the minimum value of the integral is

$$\pi^2 \left(\frac{1}{a^2} + \frac{1}{b^2} \right).$$

11.5 In the previous exercise we explored a direct solution technique for a problem that included an auxiliary condition. We will now examine a very specific case with an appended condition. Suppose we need an extremum for the integral:

$$I = \int_{x_1}^{x_2} F(x, y, y')dx,$$

but subject to the condition that:

$$\int_{x_1}^{x_2} G(x, y, y')dx = C.$$

This variational problem described above is equivalent to finding the extremum for

$$I^* = \int_{x_1}^{x_2} [F(x, y, y') + \lambda G(x, y, y')]dx,$$

where λ must be determined from the auxiliary condition. Now let us focus on the *x*-*y* plane, where we seek the *minimum* length curve between points (x_1, y_1) and (x_2, y_2) that yields a *specific area* under the curve, represented by A. We already know that the length of the curve was given in Section 11.1 by eq. (11.5), $L = \int_{x_1}^{x_2} \sqrt{1 + y'^2}dx$. The area under the curve (that we seek) is of course: $\int_{x_1}^{x_2} ydx = A$. Suppose that the specific points of interest to us are (0,0) and (5,3); if we connect these points with a straight line, we form a triangle with an area of 15/2. But for our problem we will require that $A = 15$, and this makes it apparent that the curve we seek must be convex-up. It is also clear from the preceding description that we need to identify the extremum for

$$I^* = \int_0^5 \left[\sqrt{1+y'^2} + \lambda y\right] dx.$$

Smith (1953) shows that in order to make the first variation of I^* disappear, it is necessary that

$$\frac{d}{dx}\left[\frac{y'}{\sqrt{1+y'^2}}\right] = \lambda.$$

Determine the exact nature of the curve connecting (0,0) with (5,3) that yields upon integration an area of 15. What is the numerical value of λ?

11.6 *Isoperimetric* problems appear frequently in the literature of COV. The origin is found in the tale of *Dido* (who may or may not have been an actual historical figure). By legend, Dido fled to north Africa to escape her brother, Pygmalion. Upon arrival she asked the Berber *Iarbas* for a spot of land for her party. They agreed that she could claim whatever land might be encircled by a single oxhide—which she cut into very thin strips as to make a very long cord; hence the phrases "cut a dido" or "cutting didoes." In this manner she was able to claim a small hill which according to legend became Carthage. For the Dido problem we wish to maximize the area given by $I = \int_a^b y dx$, while the total length, $L = \int_a^b \sqrt{1+y'^2}dx$, must equal some particular value. Of course the function $y(x)$ must produce a closed path that contains the desired area. Let $H = y + \lambda\sqrt{1+y'^2}$ and use the Euler equation, $\frac{\partial H}{\partial y} - \frac{d}{dx}\left(\frac{\partial H}{\partial y'}\right) = 0$ to find the solution for this problem (we should get the equation for a circle, of course).

11.7 In Section 11.5 we examined the elementary eigenvalue problem, $y'' + \lambda y = 0$, with $0 \le x \le 1$. We specified fixed end-points with both values for y set to zero. We noted that the solution was *sine*, an odd function, so perhaps a different choice for the trial function would make sense. Let $y = a_1x + a_3x^3 + a_5x^5$ and repeat the Rayleigh–Ritz calculation, comparing your results with the text example.

Answer: the first eigenvalue is 9.8896

11.8 Consider the ODE, $\frac{d}{dx}\left[(1-x^2)\frac{dy}{dx}\right] + \lambda y = 0$. You may recognize that this is eq. (11.68c) of Section 11.5, the Legendre equation. Posed as a variational problem we have

$$I = \int_{-1}^{+1} (1-x^2)\left(\frac{dy}{dx}\right)^2 dx,$$

with the condition

$$1 = \int_{-1}^{+1} y^2 dx.$$

Let us assume that an appropriate trial function is

$$y = a_0 + a_1x + a_2x^2 \text{ such that } y' = a_1 + 2a_2x.$$

Therefore,

$$1 = \int_{-1}^{+1} y^2 dx \text{ becomes } \phi = -1 + 2a_0^2 + \frac{4}{3}a_0a_2 + \frac{2}{3}a_1^2 + \frac{2}{5}a_2^2.$$

Three solutions are possible in this problem; you can find the three characteristic values from the quotients:

$$(\partial I/\partial a_0)/(\partial\phi/a_0),$$

$$(\partial I/\partial a_1)/(\partial\phi/a_1), \text{ and}$$

$$(\partial I/\partial a_2)/(\partial\phi/a_2).$$

The values for a_0, a_1, and a_2 that make I stationary can then be determined; verify that:

$$\frac{0}{4a_0 + \frac{4}{3}a_2} = \frac{\frac{8}{3}a_1}{\frac{4}{3}a_1} = \frac{\frac{32}{15}c_2}{\frac{4}{3}a_0 + \frac{4}{5}a_2} = \lambda.$$

Find the *three* solutions and finish this problem; one of them is obvious from the intermediate quotient, clearly $\lambda = 2$. *Hint: For example, consider the third quotient shown immediately above. Let* $a_2 = -3a_0$; *we find* $\frac{-\frac{96}{15}a_0}{\frac{4}{3}a_0 - \frac{12}{5}a_0} = \frac{-\frac{96}{15}}{-\frac{16}{15}} = 6.$

11.9 Consider the second-order Sturm–Liouville problem:

$$f'' + \frac{1}{x}f' + \lambda^2(1 - x^2)f = 0$$

for $0 \leq x \leq 1$ with $f(0) = 1$, $f(1) = 0$, and $f'(0) = 0$. This differential equation arises in the study of the thermal entrance length for laminar flow in a tube. Find accurate values for the first four eigenvalues using whatever method you deem appropriate, and determine and plot the corresponding eigenfunctions. A numerical method will need to make a provision for the term $\frac{1}{x}f'$ for $x = 0$. *If one were interested only in the first eigenvalue, and if an approximate value was sufficient then the Rayleigh–Ritz approach might be considered. But in this case accuracy is required, and some alternative method is probably necessary.*

11.10 We have seen how the Rayleigh–Ritz method could be used to obtain eigenvalue estimates—and this can work well if the ODE is simple enough. That was the case with problem 11.2; with a judicious choice of trial function and a bit of luck we were able to get a surprisingly accurate value for λ. This approach is much less likely to work well for more complicated ODEs of higher order. Consider the fourth-order equation,

$$f'''' + 16f'' + \lambda^2 f = 0 \text{ for } 0 \leq x \leq \pi.$$

We know

$$f(0) = 0 \text{ and } f'(0) = 1 \text{ and } f''(0) = 0.$$

We also know $f(\pi) = 0$. We would like to determine two eigenvalues for this problem, and we begin by noting that the ODE is linear. This will allow us to find out something about the function's behavior *before* we attempt anything more complex: We use linear differential operator notation:

$$(D^4 + 16D^2 + \lambda^2)f = 0,$$

and let λ^2 assume an arbitrary value, say 12.5, for example. Now we employ *cSolve* on the TI-nspire to find roots: $\pm 3.89568i$ and $\pm 0.907552i$. What does this tell us about this particular case? The function f is oscillatory *and* we could obtain analytic solutions for various values of λ^2. This line of attack is not recommended for obvious reasons.

References

Boys, C.V. (1959). *Soap Bubbles: Their Colours and the Forces which Mold Them*. New York: Dover Publications.

Courant, R. and Hilbert, D. (1989). *Methods of Mathematical Physics*, vol. 1. New York: John Wiley & Sons.

Cox, S.J. (1992). The Shape of the Ideal Column. *The Mathematical Intelligencer* 14: 16.

Cox, S.J. and Overton, M.L. (1992). The Optimum Design of Columns Against Buckling. *SIAM Journal on Mathematical Analysis* 23: 287.

de Vries, G. and Norrie, D.H. (1971). The Application of the Finite Element Technique to Potential Flow Problems. *Journal of Applied Mechanics, Transactions of the ASME* 38: 798.

Ewing, G.M. (1969). *Calculus of Variations with Applications*. New York: Dover Publications.

Fox, C. (1987). *An Introduction to the Calculus of Variations*. New York: Dover Publications.

Huebner, K.H. (1975). *The Finite Element Method for Engineers*. New York: John Wiley & Sons.

Keller, J.B. (1960). The Shape of the Strongest Column. *Archive for Rational Mechanics and Analysis* 5: 275.

Lefkowitz, I. and Eckman, D.P. (1959). Application and Analysis of a Computer Control System. *Transactions ASME, Series D* 81: 569.

MacNeish, H.F. (1906). Concerning the Discontinuous Solution in the Problem of the Minimum Surface of Revolution. *Annals of Mathematics* 7: 72.

Norrie, D.H. and de Vries, G. (1973). *The Finite Element Method*. New York: Academic Press.

Olhoff, N. and Rasmussen, S.H. (1977). On Single and Bimodal Optimum Buckling Loads of Clamped Columns. *International Journal of Solids and Structures* 13: 605.

Reddy, J.N. and Gartling, D.K. (1994). *The Finite Element Method in Heat Transfer and Fluid Dynamics*. Boca Raton: CRC Press.

Smith, L.P. (1953). *Mathematical Methods for Scientists and Engineers*. New York: Dover Publications.

Spiegel, M.R. (1971). *Advanced Mathematics for Engineers and Scientists*. New York: McGraw-Hill Book Company.

Tadjbakhsh, I. and Keller, J.B. (1962). Strongest Columns and Isoperimetric Inequalities for Eigenvalues. *Journal of Applied Mechanics* 29: 159.

Weinstock, R. (1974). *Calculus of Variations with Applications to Physics and Engineering*. New York: Dover Publications.

Index

Problem Solving in Engineering: Analytical Mathematics and Numerical Analysis, First Edition. Larry A. Glasgow.

Companion website: www.wiley.com/go/glasgow/problemsolvingengineering